VIIIB	IB	IIB	IIIA	IVA	VA	VIA	VIIA	Noble gases
								2 He 4.003
			5 B 10.81	6 C 12.011	7 N 14.007	8 O 15.999	9 F 18.998	10 Ne 20.179
			13 Al 26.982	14 Si 28.086	15 P 30.974	16 S 32.06	17 Cl 35.453	18 Ar 39.948
28 Ni 58.71	29 Cu 63.546	30 Zn 65.37	31 Ga 69.72	32 Ge 72.59	33 As 74.922	34 Se 78.96	35 Br 79.904	36 Kr 83.80
46 Pd 106.4	47 Ag 107.868	48 Cd 112.40	49 In 1.14.82	50 Sn 118.69	51 Sb 121.75	52 Te 127.60	53 I 126.905	54 Xe 131.30
78 Pt 195.09	79 Au 196.966	80 Hg 200.59	81 Tl 204.37	82 Pb 207.19	83 Bi 208.2	84 Po (~210)	85 At ~210	86 Rn (~222)

67 Ho 164.930	68 Er 167.26	69 Tm 168.934	70 Yb 173.04	71 Lu 174.97
99 Es (254)	100 Fm (257)	101 Md (256)	102 No (254)	103 Lr (257)

CHEMICAL PRINCIPLES
SECOND EDITION

RICHARD E. DICKERSON

California Institute of Technology

HARRY B. GRAY

California Institute of Technology

GILBERT P. HAIGHT, JR.

University of Illinois

CHEMICAL PRINCIPLES

SECOND EDITION

W. A. Benjamin, Inc. · Menlo Park, California
London · Amsterdam · Don Mills, Ontario · Sidney

CHEMICAL PRINCIPLES
SECOND EDITION

ISBN 0-8053-2364-3
ABCDEFGHIJ-DO-79876543

W. A. Benjamin, Inc.
Menlo Park, California 94025

PREFACE

This second edition differs from the first in a complete reorganization and rewriting of the chapters on chemical equilibrium. The one clear message obtained from conversations and correspondence with a great number of chemistry teachers is that most people prefer an elementary, nonthermodynamic introduction to equilibrium early in the textbook where it can support the laboratory work. Therefore, we have presented the part of equilibrium theory and acid–base equilibria that does not depend on thermodynamics in a new Chapter 5. The chapter following the one on thermodynamics now represents, not an introduction to equilibria, but a tying together of the concept of free energy with the ideas of Chapter 5. The treatment of solubilities and solubility products has been shortened and simplified in Chapter 5, and the chapter on oxidation–reduction equilibria and electrochemistry (Chapter 17) is totally new. We have taken advantage of the resetting of type to make minor changes and to update at many places, but the remainder of the text is basically that of the first edition. We considered going over completely to the International System of Units (SI) and replacing calories by joules throughout. But we decided that this would be premature for the present edition; too much of the world's scientific business is still conducted in calories, and will be for some years. The advantage of joules is a negative one—there is much in the way of conversion factors that no longer need be remembered. But during a period of transition a well-prepared student must know both systems; thus we have included a brief discussion of SI units in Appendix 1.

As we stated in the first edition, this text has been written to be read by the student. It is not a reference work or a compendium of information. Neither is it a source book for students or lecturers. Our goal in writing it has been to be clear, rather than to be exhaustive. We feel that a good text should be a second lecturer to the student; it should discuss ideas and suggest analogies rather than merely tabulate information. Because we want the student to read this text and to feel that someone is telling him about chemistry, we have been somewhat more discursive in places than a reference text generally is. Chemistry is interesting, and the people who do it are interesting. We have tried to present chemistry as a living and growing thing instead of a dry collection of knowledge. The text is not intended to be a history of chemistry, but we have not hesitated to bring in history when it helped to show how chemists think (and occasions when they did not think).

Because this text is a teaching device, the order of topics is dictated by pedagogy rather than by any systematic chemical organization. Few people can appreciate or even remember a new concept upon first exposure. Only by constant repetition, enlargement, and integration into what we know and are learning do new ideas become real.

As an example, the central concept of atoms and molecules is presented three times, at three levels of complexity (and of difficulty). The first presentation is of the basic concept, as it arose in Dalton's time. We show how powerful such a simple idea can be in explaining chemical phenomena. The second exposure is to the prequantum mechanics atom—the atom of Cannizzaro, van't Hoff, and G. N. Lewis. This treatment focuses on the periodic table, which is regarded not as a consequence of electronic structure but as an observed fact that must be explained. Finally, the quantum mechanical picture of atoms and molecules is introduced. By this time, the student is prepared to accept quantum theory. Schrödinger's equation is seen, not as a complicated and abstruse mystery, but as the logical answer to a dilemma that the student himself can appreciate.

The topics of early chapters are interwoven into later chapters. This is common practice and is inevitable if chemistry has a logical structure. But more difficult topics that must necessarily be postponed to later chapters are anticipated at every possible point in the earlier material. Both the terminology and the need for the material are introduced before the topic is discussed in detail.

We think that there are justifications for this approach in good teaching practice. When a child learns to read, he is not kept from books until his classes begin. He sees and learns from letters and words; by the time his formal reading instruction begins, he has an idea of the process he will be learning and the advantages that it offers. Similarly, a beginner profits from having an intuitive grasp of the *concepts* of thermodynamics (energy, entropy and disorder, free energy) before he undertakes a systematic treatment of the relationships between them. It is unfair to saddle a student with a strange set of terms and concepts and, simultaneously, to present him with an unfamiliar mode of manipulating them. Hence, the basic ideas of thermodynamics are woven into the early part of the book, and the chapter on thermodynamics is less of an exposition of the subject than an orderly resumé and outline of chemical applications. In the same way, the properties of metals, the chemistry of the transition metals, the nature of covalent bonding, the structures of coordination compounds, and bonding in nonmetals all are discussed in several places with increasing thoroughness.

Nuclear, organic, and biochemistry are given more careful attention than in most introductory texts. Beginning chemistry texts usually lean too heavily on the side of inorganic and physical chemistry and assume that these other topics will be studied later. If the introductory course is the last chemistry course that a nonmajor will ever take, however, he cannot be left in the dark about such important matters as the chemistry of nuclear reactions, or the compounds and processes of living organisms. The science major, too, will benefit from his initial exposure to these fields; he will be better motivated for later studies by knowing something of the goals of these areas. Chapters 12 and 13, which cover organic and biochemistry, and nuclear chemistry, extend the cyclic or iterative teaching method of this book to the courses that the science major will take later. The amount of new material in these chapters is necessarily large, and may be intimidating to the unprepared student. It should be emphasized to him that these chapters are written to give him an overview of a new field and are not to be memorized. It is better to understand one generalization and thereby predict three out of five facts correctly, at this point, than it is to remember all five facts but not understand them.

The return to one topic several times and the weaving together of topics are not only realistic in terms of the unitary nature of chemistry, but constitute good teaching practice. This plan is more sensible than organizing a book with a chapter of this, a chapter of that, followed by a chapter of something else. Our approach makes referencing to a specific topic more involved; however, as we have stated, this text is not a reference book, but a teaching device.

Chemical Principles is arranged into four parts:

I.   The Beginnings of Chemistry: The Idea of Atoms

II. Classical Ideas of Structure and Bonding

III. The Quantum Revolution

IV. Chemical Dynamics

Because of this arrangement, the text can be used by students with a broader variety of background preparation than is usually the case. The material in Parts I and II includes the topics of atoms, molecules, moles, stoichiometry, equilibria, the periodic table, structure, and bonding that often are the core of a good high school course. Although the level of treatment here is higher, this organization means that an honors section of unusually well-prepared freshmen could begin at Part III (Chapter 8) and use the first seven chapters as a review. Conversely, beginning students who have had no high school chemistry whatever will find that they can quickly orient themselves to the material and catch up with their classmates with the help of the book by Jean D. Lassila *et al.*, *Programed Reviews of Chemical Principles*.

Just as *Chemical Principles* has been designed as a second lecturer, so Wilbert Hutton's *A Study Guide to Chemical Principles* has been designed as a second recitation section. The study guide expands upon points in the first seven chapters that student reviewers of the text manuscript found to be difficult. It discusses techniques of problem-solving and mathematical methods. It provides annotated solutions, with explanations, for problems of the type given in the text. Additional problems (with solutions) for each chapter are provided in *Relevant Problems for Chemical Principles*, by Ian S. Butler and Arthur E. Grosser.

The *Chemical Principles Teaching System* for freshman chemistry, designed for the first edition of this text, was an experiment: What will be the outcome if you bring together the authors of a textbook, a programed supplement for strengthening weak background preparation, a study guide, an advanced problems book, and a teachers' manual and have them write their respective components in daily contact and consultation? We think that the experiment was successful in that the components of the teaching system are genuinely integrated. None of them is an add-on item written later by someone uninvolved in the preparation of the original text. Because many teachers and students seemed to like the results of our experiment, all supplements of the teaching system have been revised to follow the second edition of *Chemical Principles*. We appreciate the contributions of the authors of the other components of the teaching system—Drs. Lassila, Hutton, Butler, and Grosser. We also thank Dr. Fred Anson, who advised us on the new equilibrium chapters, and Dr. Richard L. Keiter, who wrote many new problems for every chapter of the text and is coauthor (with Dr. Hutton and Ellen A. Keiter) of the revised teachers' manual.

The viewpoint expressed in this book is one of a healthy empiricism. In the words of J. J. Thomson, ". . . a theory is more a matter of policy than a creed." The data are presented first, then the theory is introduced as a means of accounting for the data, and not the reverse. When observations and theory are compared, it is the theory that is tested and not the observations. This is one reason why topics such as atomic theory are treated in several cycles; the inadequacies of one cycle demand a theory that heralds the next cycle.

Chemistry thus develops an inevitability that is often lacking in a more intensively structured treatment.

This text is addressed not only to the person who will make chemistry a profession, but to the many people who will have to make decisions about chemical matters that will affect the quality and sometimes the length of their lives. To paraphrase an old epigram about war and politics, chemistry is too important a matter to be left to the chemists. People who will never operate a reactor, or synthesize an insecticide, or fluoridate a water supply, or design an internal combustion engine, must know the consequences— good and bad—that will result when other people choose to do or not to do such things. In the mansion of the universe, we are living in a very small room and the more we see that the neighboring rooms are uninhabitable, the more important it is that we know enough to keep this one tolerable.

RICHARD E. DICKERSON
HARRY B. GRAY
GILBERT P. HAIGHT, JR.

CONTENTS

WHY ARE YOU IN THIS COURSE?

Why study chemistry? What does a chemist do that inspires you to be one or prompts you to learn something about the field even if you do not plan to pursue it as a career? In the past, the answer often has been given in terms of the many important products that have come from the laboratory: dyes, petrochemicals, plastics, fertilizers, drugs, synthetic fibers. Older texts are filled with photographs of blast furnaces and rayon spinning mills, and eulogies to the Haber process for the production of ammonia and the Solvay process for sodium carbonate.

But times change, and so do people's sense of values. Material comfort and a colored plastic telephone do not seem as centrally important as they once did. Synthetic rubber now hardly seems like one of the higher manifestations of the human spirit. Indeed, many of our once-heralded achievements have backfired on us. We can travel from one place to another rapidly at the cost of polluting the air and filling it with noise. We can manufacture cheap paper to support widespread literacy at the cost of killing off the water life downstream. Our hopes for abundant nuclear power are clouded by the problems of thermal pollution. We keep the wheels of transportation turning, but blacken our coastlines with escaped oil to do so. We eradicate insects to aid our crops and then find that we also have killed the robins and contaminated the salmon in Lake Michigan. The genie of chemistry seems to be a malevolent spirit who accompanies each gift with a trap that leaves us with a new problem for every one we solve.

Most of these traps have evolved because we looked at each technical advance in isolation and paid too little attention to the ultimate effects of each new development. The enthusiasm of past generations for the "wonders of chemistry" was sincere but naive. We should not turn away from science now, but should learn to use it more intelligently. We desperately need a generation of scientists who are committed to the wise use of their discoveries. Moreover, we need a generation of nonscientists who know enough about chemistry and physics to anticipate the outcome of technical decisions and to compute long-range costs and benefits as well as short-term gains. There never has been a time when it was more important for the nonscientist to understand chemistry and physics, for there never has been a time when political and economic judgments were as likely to get us into scientific trouble. The past few years have witnessed a wholesale involvement in public affairs by a new generation of lawyers, of which "Nader's Raiders" are only one example. There has been no such mass interest yet on the part of scientifically trained people, yet scientific competence is becoming more necessary to decide matters of public policy. Perhaps in another generation the proper entry into government and politics should not be a degree in law but in general science.

One of the painful realities that faces all of us is leaving the security of our childhood. As children, we never worried about where our home came from or who would provide our food and clothing. These things were just there, in the natural order of life.

If we left our room in a mess, somehow, in some way, it all would be put right.

The planet that we all have been living on is a very small room. Like children, we have accepted its gifts as inexhaustible and free. We have littered our room with garbage—solid, liquid, and gaseous—and have trusted that it all will disappear somewhere. Yet we are entering a troubled intellectual adolescence, in which we are realizing that these assumptions are not true. If the planet is to remain livable, someone must keep it so. There are no such things as either endless resources or infinite capacity for waste disposal. One man's garbage inevitably becomes someone else's raw materials. One of the tasks of the chemist in the coming years is to create workable plans by which we can live together on this planet, and the job of the scientifically literate citizen is to make it possible for such plans to be put into action. The Greeks were ingenious in imagining torments for their fallen heroes, but even they did not imagine Prometheus finally drowning in garbage.

So far we have been talking about what we should do with chemistry. But what can chemistry do? Just as we are beginning to look at life on this planet as a whole, so we are beginning to look at the chemistry of an entire living organism. At last chemists are beginning to have something concrete to say about that most intricate of chemical systems, a living creature. Francis Crick, who together with James Watson discovered the molecular structure of the hereditary material of life, DNA, was a physical chemist. The deciphering of the nucleic acid code, or the system by which the information for building a living organism is stored in DNA, was a triumph of biochemists. When Arthur Kornberg and his colleagues succeeded in copying the complete DNA of a virus and in demonstrating that this synthetic genetic material would build a new virus as well as natural DNA would, they did so with the intimate cooperation of enzyme chemists and molecular biologists. Organic and biochemists now are able to synthesize vitamins, hormones, and enzymes in a way that would have seemed incredible ten years ago. Penicillin, insulin, and even the enzyme ribonuclease have been made synthetically. Physical chemists and biochemists can solve the three-dimensional structures of enzymes and can construct atomic models of them. With these models as a starting point, enzyme chemists can make more progress than ever before in understanding how the catalytic action of enzymes takes place.

Why would we want to know about chemistry at all? The goal of knowledge in chemistry, as in any other area, is control. If we know how hormones act, perhaps we can control their action. If we understand enzymatic catalysis, perhaps we can correct metabolic failures such as phenylketonuria, in which the inability to metabolize one key substance can lead to feeblemindedness in an infant. If we learn enough about the chemistry of DNA and genetic information transmission, perhaps we can detect and cure mongolism, which is produced by an extra chromosome early in the life of the embryo. Even more dramatic hereditary engineering has been proposed, but we need to distinguish between the verbs "can" and "should." As R. S. Morison has said,

> In a short time we will be able to design the genetic structure of a good man. There is some uncertainty about the exact date, but no doubt that it will come before we have defined what a good man is.[1]

New examples of chemical influences—both natural and artificial—on behavior continually are coming to light. Two rare

---

[1] R. S. Morison, "Science and Social Attitudes," *Science* **165,** 150 (1969).

chemicals in the bloodstream have a suggestive but unproven connection with schizophrenia. Large doses of the common lactic acid can produce anxiety neuroses in humans, and the behavior-changing effects of mescaline and LSD are a matter of concern. Before LSD became a cult, it was a research tool in the study of artificial schizophrenia.

Rats are most like humans in their clannishness and their reaction to overcrowding. Their sense of community identity and their enmity to strangers is strong (one is tempted to add "and human"). Experimenters actually have brought about a reduction in the rat population of a city block area by introducing new rats. The reason, interestingly enough, is chemical. The presence of new and alien rats on the territory of an existing group leads to fighting, stress, and anxiety. But it is not the fighting that lowers the numbers of rats. When a rat is made neurotic by conflict and overcrowding, its body secretes a hormone that reduces sexual aggressiveness in males and interferes with pregnancy in females. Therefore, the birth rate falls and the pressures that led to the anxiety are thereby eased. Such chemical control of behavior clearly is adaptive and advantageous, at least for wild populations of rats. Is part of our behavior similarly subject to chemical control? The answer is certainly yes. What we do about it is a tougher question. Giving everyone tranquilizers is no answer to the problem; they do not even permanently relieve the symptoms. In a sense things are more difficult when there are quick chemical responses to psychological and social problems, for they can weaken the pressures to find solutions to the real ills. Alcohol as an escape from real problems is less dangerous than LSD because the euphoria induced by alcohol is so clearly second-rate and temporary.

In the past our control of our environment has been as haphazard and uncertain as our control of the chemistry of our bodies. The very permanence of the products of chemical technology has brought trouble. So long as we built with materials that were collected rather than synthesized, our debris stood a good chance of blending back into the environment without leaving permanent scars. Wood and cloth will rot, organic matter will be eaten by microorganisms, iron will rust, and glass will shatter and mix with the natural silicates that make up the soil. But aluminum remains intact long after iron has disappeared. Polyethylene and most other plastics will neither decompose nor be eaten by microorganisms. Synthetic detergents have created foaming rivers downstream from sewage disposal plants because they cannot be degraded by bacteria in the way that soaps can. Biodegradable detergents are more expensive. At what point do we decide that the expense of these biodegradable compounds is less than the damage to the environment in terms of fish killed and streams polluted? And who pays the cost? Do we similarly regulate the use and discarding of inert materials such as aluminum and polyethylene, or do we find microorganisms to eat plastic? (This is a tough assignment. What would polyethylene-eating microorganisms have done to sustain themselves in the millions of years before Lavoisier?)

Insecticides such as DDT have proven embarrassingly effective. Their resistance to chemical breakdown is an advantage to the farmer who wants one spraying to last a long time, but a disadvantage to the higher organisms in which the DDT concentration builds up with time to near-lethal or lethal doses. In one marsh on the Long Island shore, where spraying with DDT for mosquito control has been carried out for twenty years, the plankton have accumulated 0.04 parts per million (ppm) of DDT by wet weight. But the clams that eat the plankton have 0.42 ppm of DDT, the minnows have 1.0 ppm, and the seagulls that eat both clams and minnows have as much as 75.0 ppm. Another tenfold increase in this concentration of insecticide in the food chain would lead to death, as it has done for smaller birds in parts of the midwest. The hopes of Great Lakes fishermen that the introduction of Coho salmon from the Pacific Northwest would bring on a renaissance in sport fishing in the area were dimmed when the flesh of the fish was found to have a high DDT concentration because of drainoff from agricultural land around the lakes. No one intended for the seagulls on Long Island or the Coho salmon in Lake Michigan to accumulate DDT, but the unintended happened. Ironically enough, many pests tend to flourish under such circumstances because they are lower down the food chain and have a shorter lifetime; hence, they do not necessarily accumulate so much insecticide. Meanwhile, the higher animals that formerly kept them in check die off.

What do we do about DDT? How can we balance the increase in insect-free crop production and the decrease in insect-borne diseases like malaria against the contamination and death of higher animals that keep other pests in check? If we decide to forbid a course of action that offers immediate financial return to a farmer because the ultimate damage to society is greater, do we owe him compensation for our action? If so, who pays? Or do we convince him that no compensation is called for because he had no right to pollute the environment to his own gain in the first place? Our hypothetical "farmer" himself is a deceptive abstraction. The subsistence-level farmer in Bangladesh and the conservation-minded farmer in Michigan balance off dead wildlife versus increased crop yield in a somewhat different way.

These questions are not going to be resolved by a panel of scientists, no matter how well informed. But neither can they be solved well by government policymakers, congressmen, or corporation advisory boards whose members are not literate in the field of chemistry. In the past, ignorance, if not bliss, was at least moderately harmless. Now, it can be disastrous. If the choice had to be made for the next generation between teaching chemists chemistry or teaching nonchemists chemistry, we could almost say that the latter course of action would be preferable.

From the preceding statements, chemistry may seem only like a scientific way of managing the planet. But man does not live by carbon dioxide-foamed wheat starch product alone. There is also the satisfaction of knowing what we are, and where we are, and where we came from. How did life evolve from nonliving chemical matter on this planet? How did this chemical matter itself arise? We cannot turn back the clock and watch the process, but we can set up what we believe to be primitive earth conditions and study the reactions that are likely to have taken place. We can see how the raw materials of living systems could have arisen naturally, and why more complex chemical assemblages would have been stable and long-lived. We can understand, in principle, how assemblages so complex that they must be called "living" would have developed. To a limited extent we can check our experimental paleochemistry with the evidence of mineral deposits laid down at various stages of the history of the earth. The apparently inhospitable conditions for life on the moon and Venus, and possibly on Mars, are disappointing, but they do not eliminate the fundamental question: "Given the proper conditions, is the evolution of life natural and virtually inevitable, or is its appearance on earth a fortuitous accident?" We can design and carry out experiments that help to answer this question even if only one planet in our solar system were proven to have the proper conditions.

Whether or not the moon reveals much about life, the chemistry of its rocks will allow us to reconstruct the history of the solar system. Early reports on lunar rock samples showed a far higher concentration of high-melting-point metals than in any terrestrial ores. Does this mean that the moon solidified at high temperatures, at which much of the lighter material boiled away and was lost? Does the contrast between earth and moon mean that the moon was a wanderer captured by the earth rather than a daughter that was formed as the earth was? The answers to such queries will come, in part, from detailed chemical comparisons of the materials of earth and moon. Such efforts will not keep Lake Michigan from being polluted or make it possible for earth to feed 10,000,000 more people, but they will provide a stretching of the human spirit that our species sorely needs.

Knowing where we came from and how we developed has an effect on how we think about ourselves. The revolution in thought that sometimes is symbolized by Copernicus and Galileo, which removed man from the center of the universe to one of several planets around a rather obscure star, shaped the patterns of thought of the citizens of Europe for generations. Man lives by ideas more than his pragmatist representatives in mid-twentieth century America like to admit. We now are slowly piecing together a new picture of man and his universe that is based on what we are learning in cosmology, astronomy, physics, geochemistry, molecular biochemistry, and behavioral biology. This new picture of man will be as influential to future generations as the Renaissance picture of man was in its time. Chemistry has much to contribute to this picture of the nature of man and of his origins.

To the question, "Why study chemistry?," there is a practical answer and an intangible answer. The practical answer is not the same as that of a generation ago; in part, today's answer is a need to make up for the blunders of the past. But precisely because the problem is more complicated, it is more interesting. We can begin to see wholes rather than parts, and the organization of a whole is almost always more interesting than the collection of parts. The intangible answer arises from the things we can know from chemistry that we had no hope of knowing a generation ago: what life is, where it came from, how it operates, what our solar system is like, and how it arose. A man can be overwhelmed by a surfeit of knowledge, but understanding can be a source of strength. For the first time, in chemistry, we are on the verge of understanding.

Ernest Rutherford, in one of his less charitable moments, remarked that there are two kinds of science: physics and stamp collecting. Lavoisier and Dalton's atomic theory brought chemistry one step above stamp collecting. The quantum revolution of the 1920's and 30's set chemistry on the road to becoming a science, and the current studies of the chemistry of life promise to bring the field to the level where Rutherford's partisan figure of speech will have to be revised.

Chemistry is never an end in itself. Whenever we have regarded it in this light, we usually have ended by misusing it. We must define our goals on other grounds. But the techniques of chemists, if used wisely and with enough foresight as to the second-

and third-order side effects of chemical applications, can help us to reach goals that are not otherwise obtainable. Better living is not achieved by an accumulation of better things. But better living for the entire human family, if we are wise enough, can be achieved through chemistry.

PART ONE

THE BEGINNINGS OF CHEMISTRY: THE IDEA OF ATOMS

The first step forward in modern chemistry was the recognition that atoms existed. Today this seems like no exceptional revelation; yet the great physical chemist Wilhelm Ostwald (1853–1932) did not believe in atoms, and as late as 1910 it was possible for Alexander Smith, professor of chemistry at the University of Chicago, to write in his book *Inorganic Chemistry* (Century Co., New York, 1910):

> The language of chemists has become so saturated with the phraseology of the atomic and molecular hypotheses, that we speak in terms of atoms and molecules as if they were objects of immediate observation. It must be reiterated, therefore, that this language is figurative, and must not be taken literally. The atomic hypothesis provides a convenient form of speech, which successfully describes many of the facts in a metaphorical manner. But the handy way in which the atomic hypothesis lends itself to the representation of the characteristic features of a chemical change falls far short of constituting a proof that atoms have any real existence.

This is the viewpoint that has led some well-known scientists and philosophers of science to maintain that an electron, a component of atoms, is not a real object in itself, but is a concept created by experimenters to account for the movement of pointers on dials, and that only these raw observations have true reality.

An extension of this frame of mind would be the assertion that I, myself, am the only real thing in the material universe, of whose existence I am sure; and this paper, this typewriter, and all other objects and people around me are, at best, only my mental constructs or theoretical entities by which I can correlate and account for the sensory data that I receive. If I choose to account for the touch impressions at my fingertips, the clicking sounds I hear, the smell of ink and rubber, and the visual impressions of brown enameled metal and white paper with black symbols as evidence for a "typewriter," then "typewriter" is clearly a convenient organizing concept for these sensory inputs. Nevertheless, the absolute existence of a typewriter is not thereby proven.

This attitude has been found in philosophical circles since the middle ages and earlier, and is called "solipsism." It has had an extensive but largely sterile history; for by denying reality it tends to discourage efforts to manipulate reality. Even though such an extreme form of thought now is generally rejected, atomic theory has had to fight against its own form of chemical solipsism. However, after all of the scientific advances in atomic and nuclear physics and chemistry in the last thirty years, and the revolutionary impact that they have had on our lives, the evidence for the reality of the "atom" is surely as firm as that for the "typewriter." The means of detection are just more sophisticated.

The first two chapters deal with the way in which the idea of atoms evolved and became compelling in terms of available evidence. We discuss how chemists learned to measure the weights of atoms and to decide on the atomic makeup of simple molecules, and how a very simple theory of moving molecules can explain many of the properties of gases.

1 ATOMS, MOLECULES, AND MOLES

In the trial scene in *Alice in Wonderland*, the White Rabbit, called to the witness stand, asks, "Where shall I begin, please?" The answer is straightforward: "Begin at the beginning, and go on till you come to the end, then stop." But we shall begin in the middle, with a description of what atoms are like, *before* saying anything about how we know that atoms exist. When we examine the evidence for atomic structure in later sections, you will have at least an idea of the goal of the effort. The result hopefully will be to make this text more comprehensible than most of Lewis Carroll's books. (The White Rabbit's evidence did not fare very well: "If any one of them can explain it," said Alice, "I'll give him sixpence. *I* don't believe there's an atom of meaning in it.")

1–1 THE STRUCTURE OF ATOMS AND THE CONCEPT OF MOLES

An atom (ours, not Carroll's) consists of a *nucleus*, surrounded by one or more negatively charged particles called *electrons*. Most of the mass of the atom is in the nucleus; an electron is only 1/1836 the mass of the lightest nucleus, that of hydrogen. The nucleus is quite small compared with the overall size of the atom. Interatomic

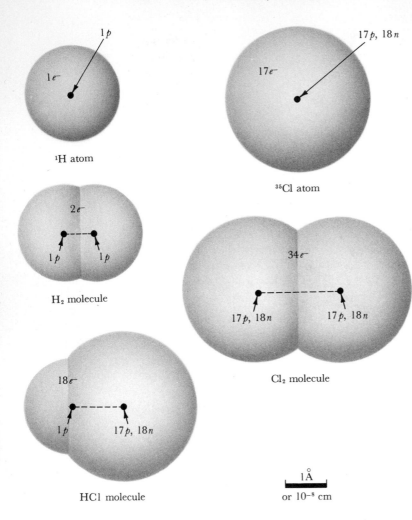

¹H atom

H₂ molecule

HCl molecule

³⁵Cl atom

Cl₂ molecule

1Å
or 10⁻⁸ cm

Figure 1–1. Relative sizes of some atoms and simple molecules. The effective radii of H and Cl have been obtained from measurements of how close unbonded atoms come when packed together in a crystal. We can find the distances between nuclei in H_2, Cl_2, and HCl by a variety of methods, including infrared spectroscopy, microwave spectroscopy, and x-ray crystal structure analysis. The nuclei here are not to scale; they are far too large. (p stands for proton, n, neutron, and e^-, electron.)

distances in crystals and bonding distances in molecules indicate that the radius of an atom typically is about 1 to 2.5 Å [one angstrom unit, abbreviated Å, is 10^{-8} centimeters (cm)], whereas particle scattering experiments place the nuclear radii in the neighborhood of 10^{-5} Å (Figure 1–1). Thus, if an atom were the size of the earth, its nucleus would be approximately 200 feet in diameter.

The nucleus of an atom contains *protons* and *neutrons*. Protons and neutrons have similar mass, 1.67×10^{-24} grams (g) or 1.01 *atomic mass units* (amu), but they differ in charge. One proton has a positive charge that exactly balances the charge on an electron, whereas a neutron has no charge. Since an atom is electrically neutral, the number of protons in the nucleus must equal the number of electrons around it. This number is known as the *atomic number*. And since the number and arrangement of electrons around the nucleus are responsible for the chemical properties of the atom, all atoms with the same atomic number are classified as the same chemical *element*. Each element is identified by a one- or two-letter symbol that usually is derived from its English or Latin name. For example, hydrogen has the symbol H. Carbon, oxygen, and nitrogen are designated by C, O, and N. Calcium, cerium, and chromium are given two letters to distinguish them from carbon: Ca, Ce, Cr.

Isotopes

Although they have the same number of electrons and protons, different atoms of one element may differ in the number of neutrons in their nuclei (Table 1–1). These varieties of atoms of the same element are called *isotopes*. Atoms of chlorine have 15 to 23 neutrons, although the only isotopes of chlorine that exist in appreciable amounts in nature are the ones with 18 (75.5%) and 20 neutrons (24.5%). The total mass of the atom, or its *atomic weight*, is very nearly the sum of the masses of protons and neutrons. Since protons and neutrons have approximate masses of 1 on the atomic mass unit scale, the atomic weight is almost equal to the sum of the number of protons and neutrons in the nucleus.

In Table 1–1, the experimental atomic weights of isotopes are not integers, as the preceding arguments would lead you to expect. This is because some of the mass of the particles in a nucleus is used up as the energy to bind the particles together. Another minor factor is the neglect of the weights of the electrons in our approximate calculations. Neither factor is important here, so long as you realize why the atomic weights do not have their ideal integral values. The atomic mass unit scale used in Table 1–1 is based on a value of exactly 12 amu for the isotope of carbon that has six protons and six neutrons.

A specific isotope is represented by the symbol of the element with the atomic number as a subscript just before the symbol and the atomic weight as a superscript just before (or sometimes just after) the symbol. Therefore, the standard carbon isotope, for instance, is $^{12}_{6}C$, and the two most common isotopes of chlorine are $^{35}_{17}Cl$ and $^{37}_{17}Cl$. Since the atomic symbol and atomic number are redundant, the latter often is omitted: ^{35}Cl, ^{37}Cl.

The atomic weight of an element is the weighted average, in terms of natural abundance, of the atomic weights of the isotopes. For chlorine, the average atomic weight is

$$0.755 \times 34.97 \text{ amu} + 0.245 \times 36.97 \text{ amu} = 35.45 \text{ amu}$$

A table of elements, symbols, atomic numbers, and atomic weights is given in the inside back cover. As you can see, over 40% of the elements have atomic weights that differ from integers by more than ±0.1 amu; hence, these elements have significant amounts of at least two naturally occurring isotopes. The chemical properties of isotopes of an element are virtually identical. An isotopic composition seldom is altered in a chemical reaction.

Table 1–1.   Elementary Particles and Typical Atoms

Particle	Number Electrons	Protons	Neutrons	Atomic number	Mass Atomic mass units	Grams	Total charge (electron units)
Electron,[a] e^-	1	—	—	—	$\frac{1}{1823}$	9.109×10^{-28}	−1
Proton, p	—	1	—	—	1.007	1.673×10^{-24}	+1
Neutron, n	—	—	1	—	1.009	1.675×10^{-24}	0
Hydrogen atom, $^{1}_{1}H$ or H	1	1	0	1	1.008		0
Deuterium atom, $^{2}_{1}H$ or D	1	1	1	1	2.014		0
Tritium atom, $^{3}_{1}H$ or T	1	1	2	1	3.016		0
Hydrogen ion, H^+	0	1	0	1	1.007		+1
Helium atom, $^{4}_{2}He$	2	2	2	2	4.003		0
Helium nucleus or alpha particle, He^{2+} or α	0	2	2	2	4.002		+2
Lithium atom, $^{7}_{3}Li$	3	3	4	3	7.016		0
Carbon atom, $^{12}_{6}C$	6	6	6	6	12.000		0
Oxygen atom, $^{16}_{8}O$	8	8	8	8	15.995		0
Chlorine atom, $^{35}_{17}Cl$	17	17	18	17	34.969		0
Chlorine atom, $^{37}_{17}Cl$	17	17	20	17	36.966		0
Naturally occurring mixture of Cl	17	17	18 or 20	17	35.453		0
Uranium atom, $^{234}_{92}U$	92	92	142	92	234.04		0
Uranium atom, $^{235}_{92}U$	92	92	143	92	235.04		0
Uranium atom, $^{238}_{92}U$	92	92	146	92	238.05		0
Naturally occurring mixture of U	92	92	varied	92	238.03		0

[a] The value $\frac{1}{1823}$ is atomic mass in amu; $\frac{1}{1836}$ is the proton/electron mass ratio.

Gram-atom

The number of grams of an element that is numerically equal to its atomic weight in amu is called one *gram atomic weight*, or one *gram-atom* (g-atom). Thus 1.008 g of hydrogen, 12.01 g of carbon, 35.45 g of chlorine, and 207.19 g of lead each are 1 g-atom of the respective elements. Since atomic weights are *relative* weights of the atoms from which elements are made, the *same number of atoms* will be in 1 g-atom of any substance. This is the reason why gram-atoms are such useful measuring units for elements, even in the absence of a knowledge of what that number of atoms per gram-atom might be. We just as well could use pound-atoms or ton-atoms. Thus 12.01 tons of carbon and 207.19 tons of lead would be one ton-atom of each, and both would contain the same number of atoms (although quite a few more than in a gram-atom).

We do know the number of atoms in 1 g-atom of an element with some precision. There are 6.022169×10^{23} atoms in 1 g-atom. This number is called *Avogadro's number*, N. The justification for this number, as for every other statement in this section, will have to wait until later. Nevertheless, there are several straightforward ways of calculating Avogadro's number from physical data. (If we accept the value given for N, how many atoms will there be in a ton-atom of an element?)

When atoms combine to form a *molecule*, the collection of all such molecules, or the *compound*, will show distinctive chemical properties, which may not be related in an obvious manner to the properties of the elements from which it is built. In other words, chemical properties are not additive properties. In contrast, the *molecular weight* of a compound is simply the sum of the atomic weights of all atoms in one molecule. The *molecular formula* of a substance shows how many of each type of atom are present in a molecule. The molecular formula for water is H_2O, and the one for benzene is C_6H_6. The approximate molecular weight of water, H_2O, is $1 + 1 + 16 = 18$; the one for benzene, C_6H_6, is $6 \times 12 + 6 \times 1 = 78$. Using more significant figures, we calculate the more precise molecular weight of vitamin B_1, $C_{12}H_{18}Cl_2N_4OS$, to be

$$
\begin{array}{lrl}
\text{C:} & 12 \times 12.01 = & 144.12 \text{ amu} \\
\text{H:} & 18 \times 1.01 = & 18.18 \\
\text{Cl:} & 2 \times 35.45 = & 70.90 \\
\text{N:} & 4 \times 14.01 = & 56.04 \\
\text{O:} & 1 \times 16.00 = & 16.00 \\
\text{S:} & 1 \times 32.06 = & \underline{32.06} \\
& \text{Total:} & 337.30 \text{ amu}
\end{array}
$$

Mole

The number of grams of a compound that is numerically equal to its molecular weight in amu is called one *gram molecular weight*, one *gram-mole*, or, more commonly, one *mole*. (Ton-moles, pound-moles, and ounce-moles are rare enough to be disregarded.) Again, the advantage of such units is that 1 mole of any molecular compound will contain the same number of molecules, namely 6.022×10^{23}. [To be precise, 1 mole now is *defined* as the amount of substance of a system containing as many elementary entities as there are atoms in 0.012 kg (exactly 12 g) of carbon-12.] A reaction in which the amounts of reactants and products are expressed in moles can be thought of as a scale-up by 6.022×10^{23} times the molecular reaction. The reaction of 2 moles of hydrogen gas with 1 mole of oxygen gas has a significance at the molecular level that the reaction of 2 g of hydrogen with 1 g of oxygen does not.

Example. If 2.00 g of hydrogen gas (H_2) react with 1.00 g of oxygen gas (O_2), which component is in excess, and how much of it is unused?

Solution. Both of the gases exist as diatomic molecules, H_2 and O_2. The *chemical equation* for the reaction is

$$2H_2 + O_2 \rightarrow 2H_2O$$

Two molecules of hydrogen will react with one molecule of oxygen to give two molecules of water; similarly, 2 *moles* of hydrogen (4.00 g) will react with 1 *mole* of oxygen (32.0 g) to produce 2 *moles* of water (36.0 g). Yet we are given only[1]

$$\text{Hydrogen:} \quad \frac{2.00 \text{ g}}{2.00 \text{ g mole}^{-1}} = 1 \text{ mole of } H_2$$

$$\text{Oxygen:} \quad \frac{1.00 \text{ g}}{32.0 \text{ g mole}^{-1}} = 0.0313 \text{ mole of } O_2$$

Since 0.0313 mole of O_2 will react with only $2 \times 0.0313 = 0.0626$ mole of H_2, there will be an excess of 0.9374 mole of H_2, or 0.9374 mole $\times$ 2 g mole^{-1} = 1.875 g of H_2. The chemical equation requires two parts (molecules or moles) of hydrogen to one of oxygen. In this example, there is an excess of hydrogen because the measuring unit, gram, is not related directly to the number of molecules. A gram of hydrogen contains 16 times as many molecules as a gram of oxygen.

Example. How many grams of hydrogen will react with 100 g of carbon to form benzene, C_6H_6? How much benzene will be produced?

Solution. The molecular weight of benzene is $6 \times 1.01 + 6 \times 12.01 = 78.12$ amu. (The value of 78 used previously, although useful when thinking about molecular structure, is not accurate enough for quantitative calculations.) The chemical reaction is

$$6C + 3H_2 \rightarrow C_6H_6$$

The amount of carbon reacting is

$$\frac{100 \text{ g}}{12.01 \text{ g mole}^{-1}} = 8.32 \text{ moles of C atoms} \\ (8.32 \text{ g-atoms of C})$$

By the chemical equation, *half* as many moles of H_2 gas will be needed, or 4.16 moles of H_2. This equation is

$$4.16 \text{ moles} \times 2.02 \text{ g mole}^{-1} = 8.40 \text{ g of hydrogen gas}$$

---

[1] A negative exponent will be used throughout this book to indicate a unit in the denominator. Thus, "feet per second" will be written "ft sec^{-1}," "grams per square centimeter" will be written "g cm^{-2}," and "calories per degree per mole" will be written "cal deg^{-1} mole^{-1}."

Also by the chemical equation, one sixth as many moles of benzene are produced as moles of carbon are consumed. So the amount of benzene produced will be $8.32/6 = 1.386$ moles, or

$$1.386 \text{ moles} \times 78.12 \text{ g mole}^{-1} = 108.4 \text{ g of benzene}$$

As a check on arithmetic, note that 100 g of carbon and 8.40 g of hydrogen produce 108.4 g of benzene. Furthermore, within the limits of error of this slide rule calculation, matter is neither created nor destroyed. This is what is meant when a chemical equation is said to be *balanced*. The same number of each type of atom appears on each side of the equation.

Exercise. Starting with 10.0 g of silver (Ag) and 1.00 g of sulfur, how many grams of silver sulfide can be formed by the reaction

$$2Ag + S \rightarrow Ag_2S$$

Which starting material will be left over, and how many grams?

(*Answer:* 7.73 g Ag_2S; 3.27 g Ag left.)

Challenge. There are additional exercises on mole and weight relationships in chemical equations in Butler and Grosser, *Relevant Problems for Chemical Principles*.[2] You may want to try their Problems 1–15 to 1–20 to see if you can apply these relationships to the use of baking powder, wine bottle manufacture, blast furnace operation, and lunar exploration.

How Do We Know?

The preceding statements about atomic structure and weight, although correct, all have been made in a very dogmatic way, with no justification of any kind. How do we know that one atom of carbon is 12 times as heavy as one atom of hydrogen? It is not easy to think of a way to weigh single atoms. How can we decide when two quantities of two elements contain the *same* number of atoms, so we can find their relative atomic weights by weighing these quantities? There is no obvious, straightforward way to count atoms into piles. How do we obtain atomic numbers for the elements? Why should atoms with the same atomic number but different atomic weights (isotopes) have so nearly identical chemical properties that we give them the same symbol and group them all together as one element? Expressed differently, why is atomic number more fundamental than atomic weight in controlling chemical properties? How do we know that the negative charges in an atom are on the outside, and the positive charges are grouped in a central nucleus?

---

[2] Throughout this text, we shall refer you to the supplementary book by Butler and Grosser for challenging problems that are related to chemical principles. The complete reference is Ian S. Butler and Arthur E. Grosser, *Relevant Problems for Chemical Principles*, W. A. Benjamin, Menlo Park, Calif., 1974, 2nd ed.

Why can we say that this nucleus is so small relative to the size of the atom? And what do we mean by an atomic radius: Is not the size of an atom as difficult to measure as its weight? What laboratory measurements can be related to such microscopic dimensions, and how can we be sure that the relationship is correct?

So long as we are being difficult, how do we know that atoms exist at all, and that everything said so far is not the product of the chemist's hyperactive imagination? Alchemists explained chemical reactions in terms of mythological figures or planets (the distinction was not clear in their own minds) that they associated with the reagents: gold with the sun, copper with Venus, iron with Mars, tin with Jupiter, and lead with Saturn. In what way are atoms more successful models than Greek gods? And how are hydrogen, helium, lithium, beryllium, and so forth really more satisfactory than earth, air, fire, and water?

The rest of this chapter will be a reversion to a stage far earlier than the one summarized in this first section. We shall go back to the time of two men who revolutionized chemistry: Antoine Lavoisier (1743–1794), who demonstrated that the fundamental quantity to follow in a chemical reaction is *mass*, and John Dalton (1766–1844), who proposed that the fundamental units in chemical reactions are *atoms*. Dalton was not the first man to propose the idea of atoms, but he was the first to show in a convincing way that atoms do exist, and that they are a useful basis for understanding chemical reactions.

Before you go on. The concepts of moles and gram-atoms are important and often difficult to understand. You can find additional study material in Lassila, *Programed Reviews of Chemical Principles*, Sections 1–1 and 1–2.[3]

1–2   THE CONCEPT OF AN ELEMENT

One of the oldest ideas in science is that of fundamental materials out of which everything else is made. Empedocles (500 B.C.), in Greece, performed what may be the first recorded chemical analysis. He noted that when wood burns, smoke or air rises first and is followed by flame or fire. A cool surface held near a flame will have moisture or water condense on it, and, after combustion, the remains are ash or earth. Empedocles interpreted combustion as a breakdown of wood into its four elements: earth, air, fire, and water. He and later writers generalized these into the four elements from which all substances were composed in varying proportions (Figure 1–2). Originally, at least, these ideas were not meant to be flights of metaphysical invention

---

[3] This supplementary book is designed especially to help you to understand basic chemical principles. When appropriate, we will refer to pertinent sections in Jean D. Lassila, Gordon M. Barrow, Malcolm E. Kenney, Robert L. Litle, and Warren E. Thompson, *Programed Reviews of Chemical Principles*, W. A. Benjamin, Menlo Park, Calif., 1974, 2nd ed.

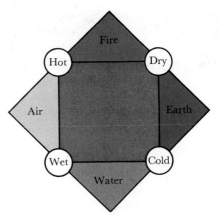

Figure 1–2. The Greeks of the fifth century B.C. pictured all material substances as composed of different proportions of the four basic elements: earth, air, fire, and water. These elements shared, in pairs, the properties of heat or cold and wetness or dryness: Earth was cold and dry, water was cold and wet, air was hot and wet, and fire was hot and dry.

but were attempts to explain observations. Later, among the Greek, Arabic, and Medieval alchemists, the ideas became imbued with mysticism in a process which, although interesting, does not concern us here. Earth, air, fire, and water were abandoned as fundamental elements, but varying sets of what we now would call elements or simple compounds were chosen by different alchemists as the fundamental materials of nature.

Aristotle (384–322 B.C.) gave a theoretical definition of an element that, even now, hardly can be improved:

> Everything is either an element or composed of elements.... An element is that into which other bodies can be resolved, and which exists in them either potentially or actually, but which cannot itself be resolved into anything simpler, or different in kind.

However, this definition leaves unanswered the question of how you recognize an element when you encounter one. Robert Boyle (1627–1691) gave a more practical definition, which states, in effect, that *an element is a substance that always will gain weight when undergoing chemical change.* This statement must be understood in the sense in which it was intended. For example, when iron rusts, the iron oxide produced weighs more than the original iron. Yet the weight of the iron *and* the oxygen that combines with it is exactly the same as the weight of the final iron oxide. Conversely, when the red powder of mercuric oxide is heated, oxygen gas is emitted, and the silvery liquid mercury that remains weighs less than the original red powder. But if the decomposition takes place in an enclosed flask, one sees that there is no *overall* loss of weight during the reaction. (It was a century after Boyle when Lavoisier made careful weighing experiments that demonstrated the conservation of mass in such reactions.)

By Boyle's definition, mercuric oxide could not be an element, for it can be decomposed into parts, each of which is lighter than the original substance. Mercury provisionally could be called an element, at least until the day when someone else succeeded in separating it into components. Until the present century of spectroscopy and other laboratory techniques, it was easy to prove that a substance was *not* an element, but impossible to prove that one was. As the famous German chemist Justus von Liebig wrote, in 1857, "The elements count as simple substances not because we know that they are so, but because we do not know that they are not."

The rare earths provide an example of the difficulties of proving by purely chemical means that a substance is an element. In 1839, Mosander extracted a new element from cerium nitrate and named it *lanthanum* (from the Greek "*lanthanein*," to lie hidden). Two years later he showed that his lanthanum-containing preparation contained a second substance, which he christened the element *didymium* (from the Greek "*didymos*," or twin). In 1879, Lecoq de Boisbaudran isolated another substance from the didymium preparation, *samarium*, and all of these were accepted as chemical elements. But didymium vanished from the rolls of chemistry in 1885, when von Welsbach separated it into two new elements, *neodymium* ("new twin") and *praseodymium* ("green twin"). It is only because we now have the periodic table, and understand the principles behind its construction, that we can say that there can be *no* other new elements between hydrogen, $_1H$, and element 105.

What kinds of substances are elements? The first to be recognized correctly as such were the metals. Gold, silver, copper, tin, iron, platinum, lead, zinc, mercury, nickel, tungsten, and cobalt all are metals. In fact, all but 22 of the 105 known elements have metallic properties. Five of the nonmetals (helium, neon, argon, krypton, and xenon) were discovered in a minor residual mixture of gases when all the nitrogen and oxygen in air were removed. Chemists thought that these "noble" gases were inert to chemical combination until 1962, when it was shown that xenon combines with fluorine, the most chemically active nonmetal. The other chemically active nonmetals are either gases (such as hydrogen, nitrogen, oxygen, and chlorine) or brittle, crystalline solids (such as carbon, sulfur, phosphorus, arsenic, and iodine). Only one nonmetallic element, bromine, is liquid under ordinary conditions.

1–3 COMPOUNDS, COMBUSTION, AND THE CONSERVATION OF MASS

Three very familiar classes of chemical compounds are *salts*, typically formed by the combination of metals with nonmetals ($NaCl$, $MgSO_4$, $CaCl_2$, KNO_3), *acids*, often combinations of hydrogen with nonmetals or of oxides of nonmetals with water (HCl, HNO_3, H_2SO_4, H_3PO_4), and *bases*, typically the combinations of metal oxides with water ($NaOH$, $Ca(OH)_2$, KOH). We shall expand on this oversimplified classification of compounds later.

Nonmetals form compounds with one another as well as with metals. Carbon, the most versatile of the nonmetals, is a constituent of well over a million compounds that have been isolated and characterized. It is the key element in the compounds that make up organic and living matter. Compounds composed only of nonmetals are often, but certainly not always, volatile liquids, easily melted solids, or gases at ordinary temperature and pressure. In contrast, most salts are hard, crystalline compounds melting at high temperatures.

Compounds

Most eighteenth-century chemists were devoted to preparing and describing pure compounds, and to decomposing them to the elements from which they are formed. The great advances were in the chemistry of gases, the last substances to be understood well. Solids and liquids were easy to identify and differentiate, but the idea of different kinds of "airs" came slowly. In 1756, Joseph Black completely changed chemists' ideas about "airs" when he showed, in his M.D. thesis at Edinburgh, that limestone could be decomposed to quicklime and a gas (which we now know as carbon dioxide, CO_2), and that *the process could be reversed*. This demonstration proved that there were different kinds of gases, and that they could take part in chemical reactions just as well as liquids and solids. One of his contemporaries, John Robinson, wrote the following:

> He had discovered that a cubic inch of marble consisted of about half its weight of pure lime and as much air as would fill a vessel holding six wine gallons. . . . What could be more singular than to find so subtle a substance as air existing in the form of a hard stone, and its presence accompanied by such a change in the properties of the stone?

In the following years, Henry Cavendish discovered hydrogen (1766), Daniel Rutherford found nitrogen (1772), and Joseph Priestley invented

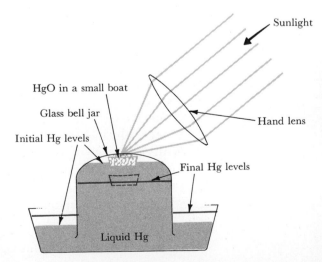

Figure 1–3. Priestley's apparatus for preparing oxygen gas. Mercuric oxide in a small pan floating on the surface of the mercury bath is decomposed to liquid mercury and oxygen by solar heat. The arrangement of the mercury bath and bell jar prevents loss of the gas evolved.

carbonated water and identified nitrous oxide ("laughing gas"), nitric oxide, carbon monoxide, sulfur dioxide, hydrogen chloride, ammonia, and oxygen. In 1781, Cavendish proved that water was a combination of only hydrogen and oxygen after he had witnessed Priestley explode the two gases in what Priestley later recalled as "a random experiment to entertain a few philosophical friends." The discovery of oxygen (Figure 1–3) led Antoine Lavoisier to the overthrow of the predominant idea of eighteenth-century chemistry, the phlogiston theory. The process by which this theory was shattered illustrates the great importance of *quantitative measurements* in chemistry.

Phlogiston

When Empedocles watched wood burn, he was impressed with the idea that something *left* the wood so only a light, fluffy ash remained. It generally was accepted that combustion was the decomposition of a substance accompanied by a loss of weight. Metal oxides are usually less dense and less compact than the metals from which they come. Even when it was known that the oxide was heavier than the original metal, a confusion between density (weight per unit volume) and weight itself compounded the error. The Germans Johann Becher and Georg Stahl proposed, in 1702, that all combustible material contains a common element called *phlogiston*, which escapes when the material burns. According to their theory:

1) Metals, when heated, lose phlogiston and become *calces*.
2) Calces, when heated with charcoal, reabsorb phlogiston and become metals again. The charcoal is necessary because the original phlogiston has become scattered through the surrounding atmosphere and lost.
3) Charcoal, therefore, must be very rich in phlogiston.

By this theory, a match is extinguished when placed in a closed bottle because the air in the bottle becomes saturated with phlogiston; respiration in living organisms is a purification process in which phlogiston is removed; a mouse under a bell jar eventually dies when the air around him has absorbed all the phlogiston it can.

Think about these ideas for a while. So long as you make no weighing experiments, this theory explains combustion as well as our present ideas, and seems to agree with common-sense observations about the appearance of metals and calces. Jean Rey, in France, had demonstrated that tin gains weight when it burns, but chemists were not accustomed to attaching as much importance to weight as we are now; hence, the significance of Rey's work was overlooked. Stahl gave an even cleverer answer in 1723:

> The fact that metals when transformed into their calces increase in weight, does not disprove the phlogiston theory, but, on the contrary, confirms it, because phlogiston is lighter than air, and, in combining with substances, strives to lift them, and so decreases their weight; consequently, a substance which has lost phlogiston must be heavier than before.

It is no wonder that hydrogen, when it was discovered, was hailed as the first preparation of pure phlogiston! Again there was a confusion between the two ideas of weight and of density (in terms of buoyancy).

Mass Conservation

Lavoisier discovered that, when heated, mercuric calx lost weight and produced free mercury and a gas. He measured the volume of gas released. Then he showed that when mercury was reconverted to calx, the same volume of this gas was reabsorbed and there was a weight increase equal to the earlier loss. On the basis of these careful weighing experiments, Lavoisier proposed that combustible materials burn by *adding* oxygen, thus increasing in weight. ("Oxygen" was his name for the gas. Priestley called it "dephlogisticated air" since it apparently could absorb even more phlogiston than atmospheric air.) Lavoisier demonstrated that the products of burning wood, sulfur, phosphorus, charcoal, and other substances were gases whose weight always exceeded that of the solids that burned. His rebuttal to the metallurgical explanations of Becher and Stahl was as follows:

1) Metals combine with oxygen from the air to form calces, which are oxides.
2) Hot charcoal removes oxygen from calces to form a metal and a gas that then was called "fixed air" (CO_2).
3) Charcoal, therefore, does not combine with the *metal;* rather, it removes the oxygen that previously had been combined with the metal in the calx.

The key to this theory was the chemical balance. Lavoisier was the first chemist to realize the importance of the principle of the conservation of mass. In his *Traité Elémentaire de Chimie,* he wrote:

> We must lay it down as an incontestable axiom, that in all the operations of art and nature, nothing is created; an equal quantity of matter exists both before and after the experiment. . . . Upon this principle, the whole art of performing chemical experiments depends.

Lavoisier was a businessman first and a chemist second. His full-time occupation was as a member of the *Ferme générale,* an agency that collected taxes on a commission basis for the French government before the revolution. One of his biographers has called his conservation of mass dictum the "principle of the balance sheet," and has claimed to see the origin of this in his role as tax collector. Be that as it may, in 1794, his connection with the *Ferme générale* cost him his life.[4]

---

[4] Haled before a revolutionary tribunal because of his past aristocratic associations, Lavoisier heard Coffinhal, president of the tribunal, reject a plea for clemency: "The Republic has no need of chemists and savants. The course of justice shall not be interrupted." This was surely one of the most serious governmental cutbacks in the history of science.

Lavoisier published his textbook, *Traité Elémentaire de Chimie*, in 1789, and it would be difficult to overemphasize the impact that it had on chemistry. In addition to laying down the principle of mass conservation in chemical reactions and overthrowing the phlogiston theory, the book contained in an appendix what is essentially our present system of nomenclature. For a generation, therefore, chemistry became "the French science" (the phrase lingered longer in France than elsewhere).

1–4 DOES A COMPOUND HAVE A FIXED COMPOSITION?

After Lavoisier, chemists began an intensive study of quantities in chemical reaction, that is, masses. The distinction between compounds and mixtures or solutions gradually became clear. A feud developed between those who claimed that the ratios of elements in compounds were fixed, and those who believed that a continuous range of proportions was possible. The French chemist Berthollet cited alloys of metals in support of the idea of a variable composition. But J. L. Proust, in Madrid, insisted on compounds with fixed composition, and correctly rejected alloys as solid solutions, not compounds:

> Let us recognize, therefore, that the properties of true compounds are invariable as is the ratio of their constituents. Between pole and pole, they are found identical in these two respects; their appearance may vary owing to the manner of aggregation, but their [chemical] properties never. No differences have yet been observed between the oxides of iron from the South and those from the North. The cinnabar of Japan is constituted according to the same ratio as that of Spain. Silver is not differently oxidized or muriated in the muriate of Peru than in that of Siberia.

The dispute between Berthollet and Proust had the good effect of sending chemists to the laboratory to prove[5] the ideas of one or the other camps, and incidentally to compile rapidly a body of knowledge about chemical composition. Of course, Proust was right; yet there are solid crystalline materials in which, because of defects in the crystal structure, there may not be quite the same ratio of atoms as predicted by the ideal chemical formula. For example, iron sulfide has actual compositions that vary from $Fe_{1.1}S$ to $FeS_{1.1}$, depending on how the sample is prepared. Such substances are called *nonstoichiometric solids*, although it has been suggested that they should be called "berthollides" after the loser in the debate just discussed.

Equivalent Proportions

Between 1792 and 1802, an obscure German chemist named Richter made an important discovery that was ignored almost completely by his contem-

---

[5] The orthodox viewpoint is that they went to the laboratory to decide between two conflicting theories. Let us be honest: Scientists are people, and science seldom is conducted in such a nonpartisan vacuum.

poraries. His idea was the one of *equivalent proportions:* The same relative amounts of two elements that combine with one another also will combine with a third element (assuming that the reactions are possible at all). This concept is easy to understand from a few examples.

1 g of hydrogen combines with 8 g of oxygen to form water.

1 g of hydrogen combines with 3 g of carbon to form methane.

1 g of hydrogen combines with 35.5 g of chlorine to form hydrogen chloride.

1 g of hydrogen combines with 25 g of arsenic to form arsine.

The chemical reactions and formulas (which were not known at the time) are, in fact,

$$2H_2 + O_2 \rightarrow 2H_2O$$
$$2H_2 + C \rightarrow CH_4$$
$$H_2 + Cl_2 \rightarrow 2HCl$$
$$3H_2 + 2As \rightarrow 2AsH_3$$

Using modern atomic weights, you should verify that the preceding statements about weights involved in the reactions are true.

Richter's law of equivalent proportions states that, if carbon and oxygen combine, they should do so in the ratio of 3 to 8 by weight. This is true for what we now know to be CO_2. If they react, carbon and chlorine should do so in the ratio of 3 to 35.5, and this is true for the liquid that we now know as carbon tetrachloride, CCl_4. In a similar way, arsenic forms $AsCl_3$ and As_2O_3, and chlorine and oxygen form Cl_2O.

Combining Weights

A *combining weight* can be defined for each element as the weight of the element that combines with 1 g of hydrogen. Or if no hydrogen compound exists, it is the weight that combines with 8 g of oxygen or with one combining weight of some other element that does form a hydrogen compound. In this way a branching network of reactions can lead to a table of combining weights for all the elements. Richter's principle, if true, assures us that there will be no contradictions within the table. Such a set of combining weights is shown in Figure 1–4 and Table 1–2.

There is one disastrous flaw to this scheme, which is why no one took Richter very seriously. The flaw is that many elements appear to have *more than one* combining weight. Carbon forms a second oxide (we know it now as carbon monoxide, CO), in which the ratio of carbon to oxygen is only 3/4. Either the combining weight of carbon has risen to 6, or that of oxygen has fallen to 4. In ethane the combining weight of carbon is 4, in ethylene it is 6, and in acetylene it is 12. The expected oxide of sulfur, SO, does not appear, and in the two most common oxides (SO_2 and SO_3) sulfur has combining weights of 8 and $5\frac{1}{3}$ (Figure 1–5).

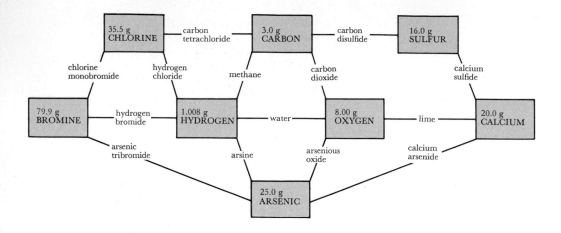

Figure 1–4. Weights of elements that combine with one another in forming the compounds indicated. We can predict from the diagram that 25.0 g of arsenic, for example, will combine with 35.5 g of chlorine or 16.0 g of sulfur. This, in fact, does occur. Arsenic and chlorine react in the predicted weight ratio to give arsenic trichloride $(AsCl_3)$, whereas arsenic and sulfur react in the predicted ratio to give the compound arsenous sulfide (As_2S_3).

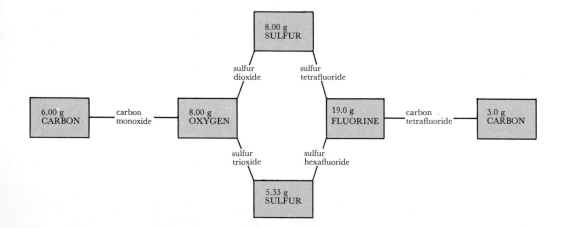

Figure 1–5. Variable combining weights for sulfur and carbon. Note that in this figure and in Figure 1–4 there are three combining weights for sulfur: 5.33 g, 8.00 g, and 16.0 g. These weights are in ratios 2/3/6. The two combining weights for carbon are in 1/2 ratio.

Table 1–2.   Experimental Combining Weights from Simple Compounds

Element	Combining weights (and sources)[a]			
H	1 (by definition)			
O	8 (H_2O)			
C	3 (CH_4, CO_2)	4 (C_2H_6)	6 (CO, C_2H_4)	12 (C_2H_2)
Cl	$35\frac{1}{2}$ (HCl, CCl_4)			
Br	80 (HBr, CBr_4, ClBr)			
As	25 (AsH_3, As_2O_3, $AsBr_3$)			
S	16 (H_2S, CS_2)	8 (SO_2)	$5\frac{1}{3}$ (SO_3)	
Ca	20 (CaO, CaS, Ca_3As_2)			
N	$4\frac{2}{3}$ (NH_3)	$3\frac{1}{2}$ (NO_2)	7 (NO)	14 (N_2O)

[a] Although the compounds were known in Dalton's time, the chemical formulas were not. Apparent anomalies in combining weights are in color.

Nitrogen is particularly troublesome. In ammonia it has a combining weight of $4\frac{2}{3}$, and in the three oxides known since Priestley's time its combining weights are $3\frac{1}{2}$, 7, and 14. If you know chemical formulas, the combining weights are easy to calculate, and you should be able to check them. But if you know *only* the combining weights, could you deduce the formulas? The significance of the ratios of elements in compounds was obscured even more by the habit of reporting composition in percent by weight; it was John Dalton who developed the trick of writing them as ratios to one common element and setting up combining weight tables, as we still do. With the three oxides of nitrogen reported by Sir Humphry Davy as containing 29.50%, 44.05%, and 63.30% nitrogen by weight, no one noticed that the nitrogen was combining in the relative ratios of 1 to 2 to 4. (These percentages are Davy's experimental values. What are the correct percentages?) By 1802, it was established that compounds had fixed compositions, and that there could be several such definite compositions between the same two elements. Yet no one knew why, or where to go from there.

1–5   JOHN DALTON AND THE THEORY OF ATOMS

John Dalton, a science (or "natural philosophy") teacher in the Manchester schools, was compelled by data such as in Section 1–4 to propose a theory of atoms, which, in 1802, he presented to the Literary and Philosophical Society of Manchester and published three years later. His theory was as follows:

1) All matter is made up of atoms. These are the ultimate particles, and are indivisible and indestructible.

2) All atoms of a given element are identical, both in weight and in chemical properties.

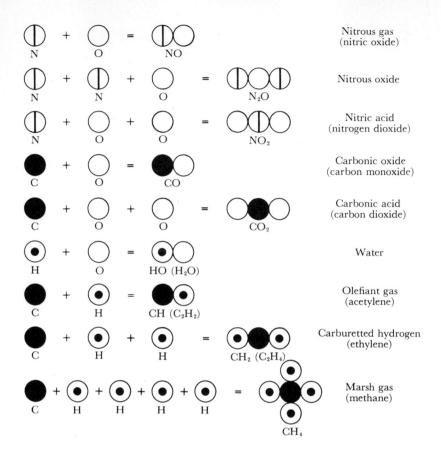

Nitrous gas
(nitric oxide)

Nitrous oxide

Nitric acid
(nitrogen dioxide)

Carbonic oxide
(carbon monoxide)

Carbonic acid
(carbon dioxide)

Water

Olefiant gas
(acetylene)

Carburetted hydrogen
(ethylene)

Marsh gas
(methane)

Figure 1–6. Dalton's original symbolism for reactions that form simple compounds. Modern symbols appear beneath them. The names to the right are Dalton's. When either the formula or name of the product is different today, the modern formula or name is in parentheses.

3) Atoms of different elements have different weights and different chemical properties.

4) Atoms of different elements can combine in simple whole numbers to form compounds.

5) When a compound is decomposed, the recovered atoms are unchanged and can form the same or new compounds.

Dalton also emphasized weights, as did Lavoisier; further, Dalton invented a convenient symbolism for atoms, as shown in Figure 1–6 and discussed in Section 2–10. In Dalton's usage, the symbol for hydrogen represents more than merely an unspecified amount of hydrogen. It represents either one *atom* of hydrogen or some standard weight of hydrogen containing a standard number of atoms (such as the gram atomic weight containing Avogadro's number of atoms). Chemical formulas and equations therefore are not merely symbolic, but quantitative.

An Ancient Idea

The idea of atoms was far from new. Democritus and the Epicurians in Greece had proposed an atomic theory, about 400 B.C., that contains virtually all of Dalton's ideas. The original writings are lost, but we know of this theory from the attacks of its opponents and from a long poem written, in 55 B.C., by a Roman Epicurian, Lucretius. (The poem is entitled *De rerum natura*, or "On the Nature of Things.") After that time, the ideas of atomism drifted in and out of alchemy for 2200 years without making a significant impact on it. Isaac Newton and Lavoisier both believed in atoms, but more as philosophical concepts or figures of speech that helped in thinking about reactions than as a theory requiring experiment.

There is an important point here that cannot be overstressed. A theory in science is important *if, and only if*, it makes the understanding of the behavior of the real world clearer. Describing bronze as a substitutional alloy of tin and copper is superior to describing it as the confluence of Jupiter and Venus, in alchemical terminology, because the tin–copper theory suggests experiments by which the properties of bronze might be explained, predicted, and even improved, whereas the "celestial marriage" theory leaves you nowhere. But perhaps it is less apparent that Democritus' atomic theory, and even Newton's, was not much of an improvement on this celestial marriage idea; it was Dalton's measurements, explanations, and predictions that made atomic theory valuable.

Fixed Ratios

Dalton took the table of combining weights as his point of departure and asked why the ratios of elements in compounds should be fixed. His answer was that *a compound consists of a large number of identical molecules, each of which is built up from the same small number of atoms, arranged in the same way*. Yet Dalton still needed to know how many atoms of carbon and oxygen combined in each molecule of an oxide of carbon, and how many hydrogen and oxygen atoms combined in a water molecule. Lacking any other guide, he proposed a "rule of simplicity" that started him off well but eventually landed him in serious trouble. The most stable two-element molecule, he reasoned, would be the simple diatomic one, AB. If only one compound of two elements were known, it would be an AB compound. Next most stable would be the tri-atomic molecules, AB_2 and A_2B. If only two or three compounds of two elements were known, they would be of these three types. This rule was one of those principles of economy, like the minimization of energy in mechanics or the principle of least action in physics, which are sometimes right, and sometimes wrong. Dalton was wrong.

Table 1–3.   *The Possible Choices of Chemical Formulas Open to Dalton for the Oxides of Carbon and Nitrogen*

	C/O Mass ratio	Possibility 1	Possibility 2	
Oxide A	3/4	CO	C_2O	
Oxide B	3/8	CO_2	CO	
Atomic weight of C (assuming O = 8)		6	3	

	N/O Mass ratio	Possibility 1	Possibility 2	Possibility 3
Oxide A	$3\frac{1}{2}/8$	NO	NO_2	NO_4
Oxide B	7/8	N_2O	NO	NO_2
Oxide C	14/8	N_4O	N_2O	NO
Atomic weight of N (assuming O = 8)		$3\frac{1}{2}$	7	14

Dalton began by mistakenly assuming from his rule of simplicity that water had a diatomic formula, HO. This made the *atomic weight* of oxygen equal to its combining weight of eight (all relative to one for hydrogen). He then turned to the oxides of carbon and nitrogen; the possible choices are shown in Table 1–3. (All atomic weights in this discussion are based on the true numerical values, and not on Dalton's values. He was a notoriously poor experimentalist: The atomic weight of oxygen, even on his own terms, began at 6.5 and slowly worked up to 8.) One oxide of carbon had a C/O ratio of 0.75, and the other had a ratio of 0.375. If the first oxide were CO—he assumed that one of them had to be—then as Table 1–3 shows, the other would be CO_2. Thus, the atomic weight of carbon would be 6. If the second oxide were CO, the first would have to be C_2O. (Can you prove this?) Then carbon would have an atomic weight of 3. Since oxide A was more stable to decomposition, he argued that this one must be CO, and correctly chose possibility 1. For the oxides of nitrogen, he similarly ruled out possibilities 1 and 3 because the five-atom molecules clashed with his rule of simplicity; and he again made the correct assignment of an atomic weight of 7 for nitrogen.

Dalton should have sensed trouble as soon as he came to ammonia. He assumed by the rule of simplicity that the molecular formula was NH. However, since $4\frac{2}{3}$ g of nitrogen combine with 1 g of hydrogen, this assumption would have meant an atomic weight of $4\frac{2}{3}$ for nitrogen, a value in conflict with the number 7 calculated from the oxides. As an alternative, he could have kept the atomic weight of 7 and worked out the formula for ammonia:

Hydrogen: $\dfrac{1 \text{ g of hydrogen}}{1 \text{ g mole}^{-1}} = 1$ mole of hydrogen atoms

Nitrogen: $\dfrac{4\frac{2}{3} \text{ g of nitrogen}}{7 \text{ g mole}^{-1}} = 0.667$ mole of nitrogen atoms

With the molar ratio of hydrogen to nitrogen (and therefore the ratio of atoms as well) being 1/0.667, or 3/2, the chemical formula would have to be N_2H_3, N_4H_6, or some higher multiple. Such a result would have shaken Dalton's faith in the rule of simplicity, and might have forced him to go back and find the right track. Yet he was undone by the poor quality of his experimental data. His initial value for the combining weight of oxygen was 6.5, which he raised to 7 in 1808. Davy increased it to 7.5, and Proust finally arrived at the correct figure (given Dalton's assumptions) of 8. Dalton refused to believe their values (a stubborn attitude for such a poor experimentalist), and all of the nitrogen calculations described here were carried out by Dalton with a nitrogen atomic weight of 5 rather than 7.

Law of Multiple Proportions

It is easy to be critical of a man who has gone astray because of bad data. But the real achievement of the atomic theory, which made people accept it almost at once, was not the calculation of atomic weights. It was that the atomic theory explained perfectly a relationship between elements forming more than one compound, an explanation that had lain unnoticed in the published literature for over fifteen years. This was Dalton's *law of multiple proportions*.

The law of multiple proportions states that if two elements combine to form more than one compound, then the amounts of one element that combine with a fixed amount of the other will differ by factors that are the ratios of small whole numbers. (Or you can multiply the amounts by a suitable constant and produce a set of integers). Since we have been using combining weights, perhaps a more meaningful statement is that if an element shows several combining weights, these weights will differ among themselves by ratios of small whole numbers. For example, the combining weights of carbon in Table 1–2 differ in the ratios of 3/4/6/12 or, more revealingly, in the ratios of 1/4 to 1/3 to 1/2 to 1. The combining weights for sulfur are in the ratios of 1 to 1/2 to 1/3, and the ones for nitrogen are 1/3 to 1/4 to 1/2 to 1 in NH_3, NO_2, NO, and N_2O. Dalton's explanation of these simple fractions was that one, or two, or any small number of atoms could combine with another kind, but that a molecule with 1.369 . . . atoms combined with 1 atom of another was physically impossible according to atomic theory.

The combining weights differ by small whole number fractions because the atoms combine in small whole numbers.

A search through the chemical literature showed that this law was the universal rule. It is one thing to prove your theory with new data that you have collected, but it is much more impressive to prove it with everyone else's; this is what Dalton did. The acceptance of the atomic theory was rapid and almost unanimous.

1–6 EQUAL NUMBERS IN EQUAL VOLUMES: GAY-LUSSAC AND AVOGADRO

As chemists tried to deduce formulas for more and more compounds, the flaws in Dalton's atomic weights and in his rule of simplicity became more and more obvious. No one could come up with a dependable method of deciding on chemical formulas. Of the three pieces of molecular information —combining weights of the elements, atomic weights of the elements, and molecular formulas—any one could be calculated if the other two were known. Yet only one, the combining weight, was directly measurable. Dalton's wrong assumptions about formulas led to wrong atomic weights, which led back again to wrong formulas for new compounds. Between 1850 and 1860 more than 13 different formulas were assigned to acetic acid, the common acid of vinegar. The confusion was so great that some chemists despaired for the atomic theory. Jean Dumas wrote:

> If I were in charge, I would efface the word atom from science, for I am persuaded that it goes far beyond experience, and chemistry must never go beyond experience.

The great German chemist Friedrich Wöhler complained, even as early as 1835, that

> . . . organic chemistry just now is enough to drive one mad. It gives me the impression of a primeval tropical forest, full of the most remarkable things, a monstrous and boundless thicket, with no way of escape, into which one may well dread to enter.

Gay-Lussac

However, the key to the dilemma was already in the chemical literature, and had been since 1811. In 1808, Joseph Gay-Lussac (1778–1850) began a series of experiments with the *volumes* of gases that react. He found that equal volumes of HCl gas and ammonia form neutral, solid ammonium chloride. An initial excess of either gas is left at the end of the reaction. Two volumes of hydrogen react with one of oxygen to form two volumes of steam; three volumes of hydrogen with one of nitrogen yield two volumes of ammonia; and one volume of hydrogen with one of chlorine produce two volumes of

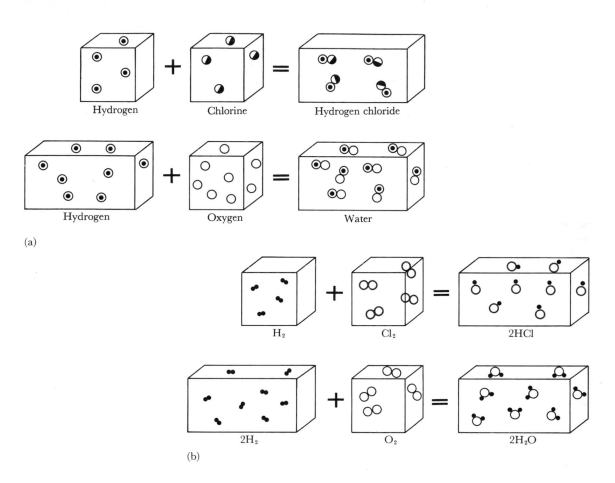

Figure 1–7. *Gay-Lussac's results on the combining volumes of gases and the explanations by (a) Dalton and (b) Avogadro. Gay-Lussac found that one volume of hydrogen and one of chlorine produce two volumes of HCl gas, and that two volumes of hydrogen react with one of oxygen to produce two volumes of steam. (a) Dalton argued that, if the volume of HCl is twice the volume of either hydrogen or chlorine, then there must be half as many molecules per volume unit in the HCl. Similarly, if there are N molecules of hydrogen per unit of volume, and if each of these produces a water molecule in the same total volume, there also will be N molecules per volume unit in water. But only half the volume of oxygen is required, so the density of oxygen must be 2N molecules per volume unit. Thus, in hydrogen chloride, hydrogen, chlorine, water, and oxygen, the numbers of molecules per volume unit are N/2, N, N, N, and 2N. (b) Avogadro proposed that each molecule of hydrogen, chlorine, or oxygen contained two atoms. With this assumption, all of the participants in the HCl reaction have the same number of molecules per volume unit of gas. Applying this same assumption to the water reaction leads to a new formula for water, H_2O, and ultimately to a complete revision of Dalton's atomic weight scale.*

HCl gas. In these and other experiments, in which the gas reactions were usually explosions triggered by a spark in an enclosed container, he always found that gases react in simple whole number units of volume. Moreover, the products, if gases, also have simple whole numbers of volume units, provided that the products are brought back to the same temperature and pressure after the explosion as the initial gases. Gay-Lussac was a cautious man and a protégé of Berthellet, who did not believe in compounds with fixed compositions. Gay-Lussac drew no conclusions in his *Memoire*, but the possibility of a connection with Dalton's atomic theory was apparent.

Avogadro

Dalton used Gay-Lussac's data to "prove" that equal volumes of gas do not have equal numbers of molecules, another wrong turn like his rule of simplicity. Dalton's argument is illustrated in Figure 1–7. The Italian physicist Amedeo Avogadro (1776–1856) saw another path. He began by assuming that equal volumes of gas (at the same temperature and pressure) contain *equal numbers of molecules*. As Figure 1–7 shows, this assumption requires that gases of the reactive elements such as hydrogen, oxygen, chlorine, and nitrogen be composed of two-atom molecules instead of single, isolated atoms. If Avogadro had been believed when he published his ideas in 1811, a half century of confusion in chemistry would have been avoided. To many, though, his ideas seemed like one shaky assumption (equal numbers in equal volumes) buttressed by an even shakier one (diatomic molecules). At that time, most ideas of chemical bonding were based on electricity, and it was difficult to understand how two *identical* atoms could do anything but repel each other. And if they do attract, why do they not form H_3, H_4, and higher aggregates? (Remember this point; we shall have to explain it, also.) Jöns Jakob Berzelius used data on the vapors of sulfur and phosphorus to undercut Avogadro. Yet Berzelius did not realize that these were examples of just such higher aggregates (S_8 and P_4). Avogadro himself did not help matters by mixing terminology so much that it sometimes appeared as if he were splitting hydrogen atoms ("elementary molecules") rather than separating atoms in a diatomic molecule ("integral molecules").

1–7 CANNIZZARO AND A RATIONAL METHOD OF CALCULATING ATOMIC WEIGHTS

By 1860, the confusion about atomic weights was so widespread that nearly every chemist of any repute had his own private method of writing chemical formulas. August Kekulé (the inventor of the Kekulé structure for benzene) called a conference in Karlsruhe, Germany, to try to reach some kind of an agreement. The man who settled the entire issue was the Italian Stanislao Cannizzaro (1826–1910), who based a rigorous method for finding atomic weights on the long-ignored work of his countryman Avogadro.

Table 1–4.   Cannizzaro's Method for Determining Molecular Weights, Atomic Weights, and Formulas (Assuming that Molecular Weight Is Proportional to Gas Density: $M = kD$)

(1) Assume that the atomic weight of H is 1.0, and that of the diatomic H_2 molecule is 2.0.

(2) Assume that Avogadro's deductions that oxygen is diatomic and water is H_2O are correct. Since the combining weight of oxygen in water is 8.0, the atomic weight of oxygen must be 16.0, and the molecular weight of O_2 is 32.0.

(3) Evaluate the constant, k, from the gas densities of H_2 and O_2:

Gas	Density, D (g liter^{-1})	Molecular weight, M (g mole^{-1})	$k = \dfrac{M}{D}$ (liter mole^{-1})
H_2	0.0899	2.0	22.25
O_2	1.429	32.0	22.40
		Average value:	22.33

(4) Evaluate the molecular weights of a series of compounds containing the elements whose atomic weights are to be determined. Use analytical percent composition data to calculate weights of elements *per molecular unit*. Look for largest common divisor, in weights per molecular units, for each element.

	Density, D (g liter^{-1})	Molecular weight: $M = kD$	Elemental composition percent by weight			Weight per molecular unit			Probable formula
			C	H	Cl	C	H	Cl	
Methane	0.715	16.0	74.8	25.0	—	12.0	4.03	—	CH_4
Ethane	1.340	29.9	79.8	20.2	—	23.9	6.04	—	C_2H_6
Benzene	3.48	77.8	92.3	7.7	—	71.8	6.00	—	C_6H_6
Chloroform	5.34	119.1	10.05	0.844	89.10	12.0	1.01	106.2	$CHCl_3$
Ethyl chloride	2.88	64.3	37.2	7.8	55.0	23.9	5.02	35.4	C_2H_5Cl
Carbon tetra- chloride	6.83	152.6	7.8	—	92.2	11.9	—	141.0	CCl_4
			Greatest common factor =			12.0	1.0	35.3	

Carbon occurs only in multiples of 12.0 (slide rule accuracy), so 12 is an acceptable atomic weight. But so are 6, 4, or any other common factor. Cannizzaro's atomic weights will be either the correct values or, at worst, integral multiples of them.

Cannizzaro's method of determining atomic weights is illustrated for carbon and chlorine in Table 1–4. The first step is to determine the molecular weights of as many compounds as possible containing the elements under study. Avogadro had shown how this could be done; for if equal volumes of all gases have the same number of molecules (at the same temperature and pressure), then *the gas density will be proportional to the molecular weight,*

$$M = kD \tag{1-1}$$

in which M = molecular weight in grams mole^{-1}, and D = density in grams liter^{-1}. The proportionality constant, k, can be determined from gases of known molecular weight. Then this expression is used to find molecular weights for new compounds whose gas density or vapor density can be measured.

The second step in the analysis is to use analytical data on the composition by weight to calculate the *weight per molecular unit* of each element for each compound. The third step is to scan the weights per molecule for the elements, one element at a time, and to select the largest common factor for each. For the example in Table 1–4, all of the carbon weights are multiples of 12, all of the hydrogens are multiples of 1, and all of the chlorines are multiples of 35.3. These factors then are either the true atomic weights of C, H, and Cl, or integral multiples of them. If, after analysis of more carbon compounds in a similar fashion, just *one* compound has a weight of carbon of 6 units, then the atomic weight of carbon must be reduced to 6; hence, the formulas at the right of the table become C_2H_4, C_4H_6, $C_{12}H_6$, C_2HCl_3, and so forth. However, if many carbon compounds have been examined and the common factor has never fallen below 12, it is reasonably safe to adopt 12 as the atomic weight of carbon.

Cannizzaro's achievement was the last link in the chain of logic that began with Proust and the law of constant composition. The battle was over, save for the computing. Accurate atomic weights now could be found for any element that appeared in compounds having measurable vapor densities. With these atomic weights, the percentage composition of a new compound would lead unambiguously to the chemical formula. The *mole* was defined as we have stated it in Section 1–1, that is, the number of grams of a compound equal to its molecular weight on Cannizzaro's scale (which is the one we use today, with improvements in accuracy). It was realized that a mole of any compound would have the same number of molecules. Although the value of that number was not known, it was named *Avogadro's number*, N, in belated recognition of his contribution.

Example. A white salt contains 22.5% sulfur, 32.4% sodium, 45.1% oxygen, and no other elements. Calculate the simplest possible formula for the salt.

Solution. The easiest starting point is to assume any convenient quantity of the material for purposes of calculation. In 100 g of the material there will be (using atomic weights from the table on the inside back cover)

$$\text{Sodium:}\ \frac{32.4\ \text{g Na}}{23.0\ \text{g mole}^{-1}} = 1.41\ \text{moles Na atoms}$$

Oxygen: $\dfrac{45.1 \text{ g O}}{16.0 \text{ g mole}^{-1}}$ = 2.82 moles O (NOT moles of O_2!)

Sulfur: $\dfrac{22.5 \text{ g S}}{32.06 \text{ g mole}^{-1}}$ = 0.702 mole S atoms

The ratios of these three compounds are 2/4/1 (0.70 is a common factor). Thus, the empirical composition is Na_2O_4S or, as it is more commonly written, Na_2SO_4, and the material is sodium sulfate.

Exercise. A reddish-brown solid contains 18.81% potassium, 47.00% platinum, and 34.20% chlorine by weight. What is its simplest formula?

(*Answer:* K_2PtCl_4.)

1–8 ATOMIC WEIGHTS FOR THE HEAVY ELEMENTS: DULONG AND PETIT

One problem remained: What does one do about the heavy elements, especially metals, that cannot be prepared readily in gaseous compounds? The problem can be illustrated by considering lead.

Example. The combining weight of lead in lead oxide is 51.8 g. What is the atomic weight of lead?

Solution. 103.6 g of lead combine with 16 g, or 1 g-atom ($\frac{1}{2}$ mole), of oxygen, but we can go no further without knowing the chemical formula. Hence, we are caught in the same vicious circle that Cannizzaro escaped for the light atoms. *If* the formula is PbO, then the atomic weight of lead is 103.6. But if the formula is Pb_2O, the atomic weight is 51.8, and if PbO_2, 207.2. Can you show that, in general, if the formula for lead oxide is Pb_xO_y, the atomic weight of lead will be 103.6 (y/x)? The problem has several solutions.

Example. Silver oxide is 93.05% silver by weight. What is the atomic weight of silver?

Solution. If we assume, for simplicity, a specimen sample of 100 g, there will be 93.05 g of silver for every 6.95 g of oxygen. The combining weight of silver (amount per 8.00 g of oxygen) is then 93.05 g × (8.00 g/6.95 g) = 108.2 g. One gram-atom of oxygen combines with twice this amount, or 216.4 g of silver. The choice of atomic weight now is limited to a set of multiples or fractions of 108.2 g, depending on what we assume the formula to be.

Formula:	Ag_2O	Ag_3O_2	AgO	Ag_2O_3	AgO_2
Atomic weight:	108.2	144.3	216.4	324.6	432.8

We need some means of deciding among these values.

Table 1–5. Dulong and Petit's Data on the Molar Heat Capacities of Solid Elements[a,b]

Element	Specific heat (cal deg^{-1} g^{-1})	Atomic weight	Molar heat capacity (cal deg^{-1} mole^{-1})
Bi	0.0288	212.8	6.128
Au	0.0298	198.9	5.926
Pt	0.0317	188.6	5.984
Sn	0.0514	117.6	6.046
Zn	0.0927	64.5	5.978
Ga	0.0912	64.5	5.880
Cu	0.0949	63.31	6.008
Ni	0.1035	59.0	6.110
Fe	0.1100	54.27	5.970
Ca	0.1498	39.36	5.896
S	0.1880	32.19	6.048

[a] All data are the original values (with atomic weights changed to a scale in which O = 16.0). Do modern atomic weights make the molar heat capacities more similar?

[b] Reproduced by permission from J. B. Conant, *Harvard Case Histories in Experimental Science,* Harvard University Press, Cambridge, 1957, Vol. 1, p. 305.

Pierre Dulong (1789–1838) and Alexis Petit (1791–1820) discovered such a means in 1819, but it largely had been overlooked in the general confusion that attended chemistry at that time. They made a systematic study of all physical properties that possibly could have a correlation with atomic weight, and found a good one in the specific heats of solids.

The specific heat of a substance is the number of calories needed to raise the temperature of 1 g of the substance by 1 degree centigrade (°C). (One calorie is defined as the heat required to raise 1 g of water from 14.5°C to 15.5°C. Thus, the specific heat of water is 1 cal deg^{-1} g^{-1}.) The product of atomic weight and specific heat of an element would be the heat required to raise 1 g-atom by 1°C, or the *molar heat capacity*. Dulong and Petit noted that, for many solid elements, the molar heat capacity was very close to 6 cal deg^{-1} mole^{-1} (Table 1–5). Because a mole of any substance has the same number of molecules (or a gram-atom of metal the same number of atoms), this experimental finding is evidence that the process of heat absorption is related more strongly to the *number of atoms* of matter present than to the *mass* of matter. Later work on the theory of heat capacities of solids has shown that there should be such a constant molar heat capacity for simple solids. Dulong and Petit gave no explanation, however.

Since no reason was advanced for this phenomenon, at the time it was regarded by most chemists as being as dubious as the rule of simplicity (which was wrong) or Avogadro's principle of "equal volumes/equal number" (which was right). It was not until Cannizzaro prepared the way with light atoms that the method of Dulong and Petit was appreciated for heavy atoms.

We now can choose among the possible precise values of atomic weight derived from analytical data, by using an approximate value obtained from the method of Dulong and Petit.

Example. The specific heats of lead and silver, as tabulated by Dulong and Petit, are 0.0293 and 0.0557 cal deg^{-1} g^{-1}, respectively. Choose the proper atomic weights in the previous examples.

Solution

$$\text{Approximate atomic weight of lead} = \frac{6}{0.0293} = 200$$

$$\text{Approximate atomic weight of silver} = \frac{6}{0.0557} = 100$$

The correct choices from the previous examples must be 207.2 for lead and 108.2 for silver; the chemical formulas then are PbO_2 and Ag_2O.

Exercise. A common cobalt-containing mineral, linnaeite, contains 58.0% cobalt and 42.0% sulfur by weight. The specific heat of cobalt metal is 0.1037 cal deg^{-1} g^{-1}. Assuming that you know the atomic weight of sulfur to be 32.06, compute the atomic weight of cobalt and the correct empirical formula for linnaeite.

(*Answer:* 59, Co_3S_4.)

1-9   COMBINING CAPACITIES AND EMPIRICAL FORMULAS

With Dalton's atomic theory, and with the contributions of Avogadro, Dulong and Petit, and Cannizzaro, it became possible to deduce, in a straightforward manner, atomic weights for elements from chemical analyses and physical data such as vapor densities and specific heats. Calculations such as these have given us the table of atomic weights shown on the inside back cover. The next great task of chemistry was to explain the formulas that could be derived.

The most primitive concept in chemical bonding probably is the idea of *combining capacity*, sometimes called "valence." The combining capacity of an element in a given compound is defined as the ratio of its true atomic weight to its combining weight in that compound:

$$\text{Combining capacity} = \frac{\text{atomic weight}}{\text{combining weight}}$$

Hydrogen has a combining capacity of one, by definition. Oxygen has a combining capacity of two in H_2O and most other compounds, but a combining capacity of one in hydrogen peroxide, H_2O_2. In Table 1–2, Cl and Br have combining capacities of one, Ca of two, and As of three; carbon shows several combining capacities: four, three, two, and one. Sulfur has a combining capacity of two in H_2S, four in SO_2, and six in SO_3. Nitrogen has a combining capacity of three in ammonia, four in NO_2, two in NO, and one in N_2O. Notice in these examples how the total combining capacity of one element exactly balances the total combining capacity of the other. In SO_3, sulfur with a combining capacity of six balances three oxygen atoms having a capacity of two each.

This concept was the first step toward a theory of chemical bonding. The second step, to which we shall return in Chapter 3, was to assign plus and minus signs to these combining capacities so that the *sum* of such "signed" capacities for a molecule is zero. Signed combining capacities, for reasons that will become clearer in Chapter 3, are called *oxidation numbers*.

It is now a relatively uncomplicated process to determine what is called the *empirical formula* or *simplest formula* for a compound from analytical data and the table of atomic weights. This is the formula for the relative numbers of atoms in a compound, expressed with no common factor (i.e., smallest whole-number ratio) in the numbers of atoms. H_2O is a *molecular formula*, for it states that two atoms of hydrogen combine in a molecule with one atom of oxygen. It is also an empirical formula, for the numbers of atoms, two and one, have no common factor. The molecular formula for benzene is C_6H_6; six carbon atoms form a hexagon, with a hydrogen atom attached to each carbon atom. But this could not be known from analytical data alone. Benzene's *empirical formula* is CH.

Example.   1.000 g of tin metal burns in air to give 1.270 g of tin oxide. What is the empirical formula of the oxide?

Solution.   The gram of tin combines with 0.270 g of oxygen. The number of moles of each element is given by

$$\frac{\text{weight Sn}}{\text{at. wt Sn}} = \frac{1.000 \text{ g}}{118.7 \text{ g mole}^{-1}} = 0.00843 \text{ mole Sn atoms}$$

$$\frac{\text{weight O}}{\text{at. wt O}} = \frac{0.270 \text{ g}}{16.00 \text{ g mole}^{-1}} = 0.01686 \text{ mole O (atoms, not } O_2 \text{ molecules)}$$

There are 2 moles of oxygen atoms for every mole of tin atoms, or two atoms of oxygen for every one of tin; so the *empirical formula* is SnO_2. If we want to know whether this is also the molecular formula, or whether the molecule is really SnO_2, Sn_2O_4, Sn_3O_6, or some higher multiple, we must have recourse to some physical measurements other than simple analytical data.

Example.   One gram of butane gas (which contains only carbon and hydrogen) burns in air to give 3.03 g of carbon dioxide (CO_2) and 1.55 g of water (H_2O). What is the molecular formula for butane?

Solution. The products are

$$\frac{\text{wt } CO_2}{\text{mol wt } CO_2} = \frac{3.03 \text{ g}}{44.01 \text{ g mole}^{-1}} = 0.0686 \text{ mole of } CO_2$$

$$\frac{\text{wt } H_2O}{\text{mol wt } H_2O} = \frac{1.55 \text{ g}}{18.02 \text{ g mole}^{-1}} = 0.0860 \text{ mole of } H_2O$$

Because each mole of CO_2 has 1 g-atom of carbon, and each mole of H_2O has 2 g-atoms of hydrogen, the original ratio of hydrogen to carbon in butane must have been

$$\frac{n_H}{n_C} = \frac{2 \times 0.0860}{0.0686} = 2.50 \qquad (n = \text{number of atoms})$$

and the formula could be C_2H_5, C_4H_{10}, or some multiple of these. But the additional information, from Cannizzaro's gas-density method, that the molecular weight must be approximately 58 g decides the issue in favor of C_4H_{10}.

A recent series of studies involving removal of oxygen from the fairly ordinary compound potassium perrhenate ($KReO_4$) led to the report of the synthesis of KRe. The preparation could not be separated easily from the compounds KOH and H_2O, but investigators thought that they had made 95% pure $KRe \cdot 4H_2O$. (Compounds with water associated in definite ratios are common.) The compound KRe excited much theoretical interest among scientists, who made many elaborate physical measurements to determine its properties. One of these measurements showed that the compound contained some chemical bonds involving only rhenium and hydrogen. The analyses (still on impure material) were reevaluated with the help of analogies with other known compounds of metals with hydrogen, and formulas such as $KReH_4 \cdot 2H_2O$ and $K_6Re_2H_{14} \cdot 6H_2O$ were assigned. The compound still appeared to be unusual, and chemists renewed their efforts to prepare pure material. An improved preparation method yielded material that had the formula K_2ReH_8. The magnetic properties of the material were not as expected for a compound with this formula. Finally, research chemists found evidence that clearly showed that there are nine hydrogen atoms for every rhenium atom in the compound, and the formula K_2ReH_9 was established in 1963, more than 10 years after the initial preparation of the so-called KRe.[6] It was simply too difficult to prepare and analyze the compound accurately with techniques available at the time of its discovery. Accurate determination of the number of hydrogen atoms in a formula when the number is large is quite difficult. For K_2ReH_9, an error of 0.4% in an analysis would change the conclusion concerning the number of hydrogen atoms from eight to nine or nine to ten.

---

[6] You can read more about this topic in A. P. Ginsburg, "Hydride Complexes of the Transition Metals," Chapter 3 in *Transition Metal Chemistry*, R. L. Carlin, Ed., Marcel Dekker, New York, 1965, Vol. 1.

Consider the compound heptane, one of the many found in oil and natural gas. Analysis shows that, by weight, it is 83.9% carbon and 16.1% hydrogen. For convenience, we assume a 100-g sample of heptane to calculate the carbon–hydrogen mole ratio:

$$\frac{83.9 \text{ g}}{12.0 \text{ g mole}^{-1}} = 6.99 \text{ moles of carbon atoms}$$

$$\frac{16.1 \text{ g}}{1.008 \text{ g mole}^{-1}} = 15.97 \text{ moles of hydrogen atoms}$$

Mole ratio C/H = 0.438

Looking at this ratio of integers, we can see that it corresponds to C_7H_{16}, but it is also very close to C_3H_7 and C_4H_9. Given normal uncertainties concerning the purity of the compound and the accuracy of the analysis, it is evident that analysis alone is insufficient to determine the formula. Similarly, the two compounds acetylene and benzene each are 92.3% C and 7.7% H. This percentage gives a 1/1 mole ratio of C to H and an empirical formula of CH.

Table 1–6.   *Formulas from Vapor Densities and Elemental Analyses*[a]

Compound	Vapor density (g liter^{-1})	Molecular weight	Weight per mole (grams)		Moles per mole of compound	
			C	H	C	H
Heptane	4.48	100	84	16.1	7	16
Acetylene	1.16	26	24	2.02	2	2
Benzene	3.48	78	72	6.05	6	6

[a] The weight per mole is the product of the molecular weight and the percent composition by weight. The numbers in the last two columns are obtained by dividing the weight of element per mole of compound by the weight per mole of the element (atomic weight). The numbers in the last two columns represent the subscripts in the formulas C_7H_{16}, C_2H_2, and C_6H_6.

Cannizzaro's (actually Avogadro's) method of vapor densities can be used to help us to decide the formulas of all three compounds. Measurements on many gases have produced a value of 22.414 liters per mole for the molar volume of an ideal gas at 0°C and 1 atmosphere (atm) pressure. This quantity is k in Table 1–4. The product of k and the vapor density yields the molecular weight (Table 1–6), and this product together with the analytical data always give the correct formula.

Before you go on. If you think that you could profit from more drill in writing empirical formulas from weight data, you can find a self-study sequence of questions and answers on the topic in Lassila's book, Section 1–3.

1-10 SUMMARY

We now are able to give satisfactory answers to many of the questions raised at the end of Section 1–1. We know now that atoms exist. (Or at least we know that someone who chooses to maintain that matter only behaves "as if atoms existed" has a long and difficult list of "as if" statements to explain some other way.) We know, thanks to Avogadro, how to count atoms of different elements into piles with equal numbers. And we know, thanks to Cannizzaro, how to determine atomic weights, at least on a relative scale.

An element is a collection of identical atoms, and a compound is a collection of identical molecules, each formed from small whole numbers of atoms. The law of constant composition, the law of equivalent proportions, and the law of multiple proportions were attempts to summarize behavior that now is obvious from atomic theory. With analytical data and the table of atomic weights, we can calculate the empirical formula of a compound; and with the aid of supplementary information, such as vapor density, we can obtain the true molecular formula.

Many of our original questions are still unanswered. Thus far we have ignored atomic numbers and all questions about the structure of atoms, the mechanisms of chemical bonding, and the reasons for the individual behavior of the elements. These are some of the tasks that lie ahead.

1-11 POSTSCRIPT: JOSEPH PRIESTLEY AND BENJAMIN FRANKLIN

Joseph Priestley (1733–1804) is, after Lavoisier, one of the most interesting personalities of this period in chemistry. A Unitarian minister in Leeds and Birmingham in the north of England, and later a suspected sympathizer with the French Revolution, he was constantly under attack for his heretical religious and political views. He published widely, both in chemistry (which brought him fame) and in politics and religion (which brought him notoriety). Although an innovator and a careful experimentalist, he did not have the theoretical grasp of chemical principles that his younger French colleague did. Priestley's oxygen experiments inspired Lavoisier to initiate the research that demolished the phlogiston theory, but Priestley himself continued to hold stubbornly to the phlogiston theory until his death.

In 1791, on the second anniversary of the storming of the Bastille, rioting against suspected "republicans" broke out in Birmingham. The city was under mob rule for three days. Priestley's church, home, laboratory, apparatus, and manuscripts were burned, and he had to flee in disguise to Worcester. After that, he spent three unhappy years in London, and finally emigrated to the United States. He was offered both a professorship and a church, but declined both, choosing to live in relative seclusion for his last decade.

Priestley and Benjamin Franklin were in continual correspondence, the former about his gas chemistry, the latter about his voltaic piles and Leyden jars. In a letter to Benjamin Vaughan, in 1788, Franklin wrote:

> Remember me affectionately to good Dr. Price and to the honest heretic Dr. Priestley. I do not call him *honest* by way of distinction, for I think all the heretics I have known have been virtuous men. They have the virtue of fortitude, or they would not venture to their own heresy; and they cannot afford to be deficient in any of the other virtues, as that would give advantage to their enemies; and they have not, like orthodox sinners, such a number of friends to excuse or justify them. Do not, however, mistake me. It is not to my good friend's heresy that I impute his honesty. On the contrary, it is his honesty that has brought him the character of heretic.

William Cobbett, a royalist who published his antirepublican sentiments during and after the American Revolution in a broadside called *Peter Porcupine's Gazette*, wrote this account of the riots and their aftermath:

> Some time after the riots, the Doctor [Joseph Priestley] and the other revolutionists who had had property destroyed, brought their actions for damages, against the town of Birmingham. The Doctor laid his damages as £4122 11s 9d sterling, of which sum £420 15s was for works in manuscript, which, he said, had been consumed in the flames. The trial of this cause took up nine hours: the jury gave verdict in his favour, but curtailed the damages to £2502 18s. It was rightly considered that the imaginary value of the manuscript works ought not to have been included in the damages; because the Doctor being the author of them, he in fact possessed them still, and the loss could be little more than a few sheets of dirty paper. Besides, if they were to be estimated by those he had published for some years before, their destruction was a benefit instead of a loss, both to himself and his country. The sum then of £420 15s being deducted, the damages stood at £3701 16s 9d; and it should not be forgotten, that even a great part of this sum was charged for an apparatus of philosophical instruments, which in spite of the most unpardonable gasconade of the philosopher ("You have destroyed the most truly valuable and useful apparatus of philosophical instruments that perhaps any individual, in this or any other country, was ever possessed of, in my use of which, I annually spent large sums, with no pecuniary view whatever, but only in the advancement of science, for the benefit of my country, and of mankind."—Letter to the Inhabitants of Birmingham), can be looked upon as a thing of imaginary value only, and ought not to be estimated at its cost any more than a collection of shells or insects, or any other of the frivola of a virtuoso.

Eleven years before the tragedy at Birmingham, Franklin wrote a letter to Priestley that is both perceptive in its forecasts and pessimistic in its time scale, and fully as relevant now as it was in 1780.

> I always rejoice to hear of your still being employed in experimental researches into nature, and of the success you meet with. The rapid progress *true* science now makes, occasions my regretting sometimes that I was born too soon. It is impossible to imagine the height to which may be carried, in a thousand years, the power of man over matter. We may perhaps learn to deprive large masses of their gravity, and give them absolute levity, for the sake of easy transport. Agriculture may diminish its labor and double its produce; all diseases may by

sure means be prevented or cured, not excepting that of old age, and our lives lengthened at pleasure even beyond the antediluvian standard. O that moral science were in as fair a way of improvement, that men would cease to be wolves to one another, and that human beings would at length learn what they now improperly call humanity!

SUGGESTED READING

O. T. Benfey, Ed., *Classics in the Theory of Chemical Combination*, Dover, New York, 1963.

J. B. Conant and L. K. Nash, *Harvard Case Histories in Experimental Science*, Harvard University Press, Cambridge, 1957. Eight critical developments in the experimental sciences, presented as case studies for nonscientists. Extensive reproductions of the original papers. Excellent for understanding the human side of scientific progress. Includes Boyle, the phlogiston theory, the nature of heat, the atomic theory, Pasteur and fermentation, and the nature of electricity.

W. F. Kieffer, *The Mole Concept in Chemistry*, Reinhold, New York, 1962.

H. M. Leicester, *The Historical Background of Chemistry*, Dover, New York, 1971. A very readable introduction to the subject, including the Greeks, the Arabic and Medieval alchemists, the rise of the new chemistry after Lavoisier and Dalton, and the progression of new chemical ideas to the era that we shall call the "quantum revolution."

Lucretius, *The Nature of the Universe* (De rerum natura), Penguin Books, London, 1967. A good prose translation by R. E. Latham. The best record that we have of the atomic theory of Democritus and the Epicureans.

D. McKie, *Antoine Lavoisier: Scientist, Economist, Social Reformer*, Macmillan, New York, 1962. A good account of a scientist and man of public affairs in the midst of a revolution.

J. W. Mellor, *A Comprehensive Treatise on Inorganic and Theoretical Chemistry*, Wiley, New York, 1922. A multivolume treatise, of which the first seven chapters of Volume 1 form a good history of chemistry. More details, more quotations from original sources than the book by Leicester.

T. Thomson, *The History of Chemistry*, Colburn and Bentley, London, 1830. Particularly interesting for its account of alchemy and its immediate successors, and because Dalton and the atomic theory constitute the *last* chapter of the book. Written by a friend and scientific mentor of Dalton.

QUESTIONS

1 What elements do the following symbols represent: C, H, O, S, N, Cl, As, Pb, Na, K, Re, Ag? Which are metals and which are nonmetals?

2 What is an isotope, and how do isotopes differ in chemical properties?

3 What is a mole, and why are moles useful units for measuring chemical substances?

4 What is the difference between an element, a compound, and a solution? Classify the following substances into one of the three categories: ammonia, iron, brass, copper, air, bronze.

5   A. D. Risteen is quoted by Mellor as saying, as late as 1895, that "I cannot see what warrant there is for assuming that when a weight A of one substance combines with another whose weight is B, the weight of the resulting compound is universally and necessarily A + B." Ignoring the subtlety of nuclear mass–energy conversions, can you prove that Risteen is mistaken? Why should mass be conserved, and not some other of the many physical properties, such as volume, density, or temperature?

6   How was the idea of the conservation of mass fatal to the phlogiston theory?

7   What is a nonstoichiometric solid?

8   Use the simple atomic theory to explain the following observed behavior to a skeptic:

a) the law of constant composition;

b) the law of equivalent proportions;

c) the law of multiple proportions.

9   What is Dalton's "rule of simplicity"? How did it lead to trouble?

10   Suppose that, in 1812, you had been offered two conflicting hypotheses:

a) The gaseous elements exist as single atoms. Not all gases necessarily have the same number of molecules per liter at a given pressure and temperature.

b) All gases at the same pressure and temperature have the same number of molecules per liter. The common gaseous elements have atoms that associate in pairs.

Which alternative would you have chosen? Can you really justify your choice, other than by twentieth-century hindsight?

11   What did Cannizzaro do with Avogadro's hypothesis that made it acceptable when Cannizzaro presented it fifty years later?

12   Why is the atomic weight of oxygen sixteen on Cannizzaro's scale but eight on Dalton's?

13   How can the combining weight of an element be calculated from atomic weights and chemical formulas?

14   How can a chemical formula be deduced from a knowledge only of atomic weights and the combining weights of the elements in the compound?

15   How can the atomic weight be deduced from the combining weight of an element and the formula of the compound in which it appears?

16   Cannizzaro's method can establish an upper limit for an atomic weight but never a lower limit. Why?

17   Show that, if the combining weight of silver in silver oxide is 108.2 g, and if the formula is Ag_xO_y, then the atomic weight of silver is $216.4(y/x)$.

18   Why does the constancy of the product of specific heat and atomic weight for metals suggest that heat absorption is more a property of the number of molecules of matter rather than the number of grams?

19   What is the combining capacity of the S atom in the following compounds: H_2S, SO, SO_2, SO_3, S_8?

20   Which of the following are empirical formulas: H_2S, N_2O_4, CH_4, C_2H_6, C_3H_8, C_6H_6?

21   How can a chemical formula be obtained from only analytical data and percent composition and atomic weights? What auxiliary data might help to derive a molecular formula from an empirical formula?

PROBLEMS

1 How many atoms are there in 0.00745 g of tungsten wire?

2 What is the weight of 7.63×10^{20} atoms of arsenic?

3 The atomic weight of thulium is 169. What is the weight of one average atom of thulium?

4 A box contains 8.50 g of carbon tetrabromide (CBr_4) vapor. How many molecules of CBr_4 are in the box?

5 How many nitrogen atoms are in 0.0150 mole of N_2O_4?

6 The following isotopes of magnesium occur in nature:

Isotope	Atomic weight (amu)	Percent abundance
$^{24}_{12}Mg$	23.9850	78.70
$^{25}_{12}Mg$	24.9858	10.13
$^{26}_{12}Mg$	25.9826	11.17

What is the atomic weight of naturally occurring magnesium?

7 If 0.87 mole of a substance weighs 5.30 g, what is the molecular weight of the substance?

8 What is the weight in grams of 3.20 moles of propane gas (C_3H_8)? How many molecules are contained in this weight?

9 What is the weight in tons of 1 ton-mole of sulfur? How many atoms of sulfur are in this amount?

10 What is the molecular weight of the amino acid alanine, $C_3H_7O_2N$?

11 What is the molecular weight of H_2CO_3?

12 Until 1961 the chemists' atomic weight scale was based on the assignment of an atomic weight of exactly 16 to the isotopic mixture of oxygen occurring in the earth's atmosphere. The atomic weight scale presently is based

on $^{12}_6C$. Is Avogadro's number the same now as it was before 1961? If it is not the same, by how much is it different?

13 In 1962, the first of a series of compounds of xenon, Xe, with fluorine and oxygen was discovered. What is the molecular weight of potassium xenate, K_6XeO_6? What is its xenon composition in percent by weight?

14 How many grams and gram-atoms of oxygen, O, are in 0.0100 mole of ascorbic acid (vitamin C), $C_6H_8O_6$?

15 What is the weight percent of C in ethane, C_2H_6?

16 What is the weight percent of each element in $K_2Cr_2O_7$?

17 The formula for trinitrotoluene (TNT) is $C_7H_5N_3O_6$. Calculate the weight percent of each element in TNT.

18 One of the components of Portland cement contains 52.7% calcium, 12.3% silicon, and 35.0% oxygen. What is the empirical formula of this substance?

19 When a 5.00-g sample of element A reacts completely with a 15.0-g sample of element B, compound AB is formed. When a 3.00-g sample of element A reacts with an 18.0-g sample of element C, compound AC_2 is formed. The atomic weight of element B is 60.0. Calculate the atomic weights of elements A and C.

20 A metal, M, forms an oxide, M_2O_3, containing 68.4% of the metal by weight. Calculate the atomic weight of M.

21 A compound is 22.9% sodium, 21.5% boron, and 55.7% oxygen. What is its empirical formula?

22 The molecular weight of an oxide of phosphorus is 284. Elemental analysis shows that the compound contains 43.6% phosphorus. Determine the molecular formula of the compound.

23 How many moles of XeF_6 can be made from 0.0320 g of xenon and 0.0304 g of fluorine?

24 When silver metal and powdered sulfur are heated together, solid, black Ag_2S forms. A mixture contains 1.73 g of Ag and 0.540 g of S. Determine whether one element is partly unused when the other is completely converted to Ag_2S. How many grams of an element, if any, remain after the reaction?

25 Iron reacts with oxygen to produce oxides of different compositions, depending on the experimental conditions. The percentages of iron in three of these oxides are 77.73, 72.36, and 69.94. With these data, illustrate the law of multiple proportions.

26 Copper reacts with bromine to form black, solid $CuBr_2$. If 2.13 g of Cu react with 10.3 g of Br_2, what is the weight of product? If any reactant remains, how much?

27 How many moles of ethanol, C_2H_5OH, can be made from 3.00 g of carbon?

28 The degree to which various pollutants are present in air, water, food, and so on, often is expressed in parts per million (ppm) by weight or volume. How many moles of lead are present in a human circulatory system that contains 6.00 liters of blood with a lead concentration of 0.200 ppm by weight and a density of 1.05 g cm^{-3}? (The blood of the average American has a lead concentration of 0.2 ppm; a lead concentration of 0.8 ppm is considered to indicate lead poisoning.)

29 If 1.00 g of a metal reacts with 3.98 g of oxygen, what is the combining weight of the metal?

30 If 10.0 liters of hydrogen (H_2) and 10.0 liters of oxygen (O_2) are allowed to react, what will be the volume of water vapor produced if the reaction is carried out at constant temperature and pressure? What volume of which reactant will be left over?

31 Under the same pressure and temperature, ethylene gas (C_2H_4) is how many times as dense as helium (He) gas?

32 A compound contains 40.0% carbon, 53.3% oxygen, and 6.67% hydrogen. At the same pressure and temperature as in Table 1–4, the vapor density of this compound is 2.68 g $liter^{-1}$. Calculate the empirical formula, the molecular weight, and the molecular formula.

33 Toluene (a compound of hydrogen and carbon) can be burned to produce carbon dioxide and water vapor. When this combustion occurs, one volume of toluene vapor reacts with nine volumes of O_2 to give seven volumes of CO_2 and four volumes of water vapor. What is the simplest formula for toluene?

34 A compound contains 65.45% carbon, 29.06% oxygen, and 5.49% hydrogen by weight. A 3.3-g sample of the compound yields 672 ml of vapor, measured at the same temperature and pressure as in Table 1–4. Determine the empirical formula, the molecular weight, the molecular formula, and the number of molecules in 3.3 g.

35 Three compounds are known of carbon and another element, X. The weight percent of X and the vapor density at the same pressure and temperature as in Table 1–4 are as follow:

Compound	Weight percent X	Vapor density (g $liter^{-1}$)
A	86.4	3.92
B	82.6	6.16
C	61.4	2.77

What is the largest possible atomic weight for X? If this value is correct, what are the molecular formulas for compounds A, B, and C? What other values for molecular weights of X are possible? Look at the periodic table inside the front cover and try to identify X. What are the most likely values for the atomic weight of X and for the molecular formulas?

36 An amount of 3.70 g of metal combines with 1.94 g of O_2 gas. The specific heat of the metal is 0.14 cal g^{-1} deg^{-1}. What is the combining weight of the metal? What is the atomic weight?

37 When a 1.00-g sample of a copper oxide reacts with hydrogen gas, the products are water and 0.799 g of copper metal. The specific heat of copper is 0.0921 cal deg^{-1} g^{-1}. What is the atomic weight of copper?

38 A 1.00-g sample of uranium reacts with 0.0126 g H_2 gas. If the specific heat of U is 0.027 cal g^{-1} deg^{-1}, what is the atomic weight of U? What is the combining capacity of U?

39 The specific heat of an unknown metal, M, is 0.0295 cal g^{-1} deg^{-1}. The metal forms an oxide in which 8.50 g of metal combine with each gram of O_2. Find the combining weight, atomic weight, and combining capacity of the metal. What is the formula of the metal oxide? Can you identify the metal from the periodic table?

Scientific research consists in seeing what everyone else has seen, but thinking what no one else has thought.
A. Szent-Gyorgyi

2 THE GAS LAWS AND THE ATOMIC THEORY

Although gases were the last substances to be understood chemically, they were the first substances whose physical properties could be explained in terms of simple laws. It is fortunate that when this most tenuous state of matter is subjected to changes in temperature, it behaves according to much simpler rules than do solids and liquids. The importance of our ability to capture and weigh gases already has been demonstrated in our discussion of the overthrow of the phlogiston hypothesis. Moreover, one of the best tests of the atomic theory is its ability to account for the behavior of gases: This is the story of the present chapter.

Given any trapped sample of gas, we can measure its mass, its volume, its pressure against the walls of a container, its viscosity, its temperature, and its rate of conducting heat and sound. We also can measure the rate at which it effuses through an orifice into another chamber, and the rate at which it diffuses through another gas. In this chapter we shall show that these properties are not independent of one another; further, we shall discuss the laws that correlate the results of several of these measurements. Then we shall show how a theory which assumes that matter consists of molecules in motion can explain the behavior of a gas.

2-1 AVOGADRO'S LAW

In Chapter 1 we encountered a disguised form of the first law describing gas behavior. This law is Avogadro's observation that, at a given temperature and pressure, the number of molecules of *any gas* in a specified volume will be the same. Consequently, the number of molecules, and also the number of moles, n, is proportional to the volume of the gas:

$$n = kV \qquad \text{(at constant } P \text{ and } T) \tag{2-1}$$

in which k is a proportionality constant. We shall be looking for other such relationships for gases that relate the pressure, P, the volume, V, the temperature, T, and the number of moles in a sample, n.

2-2 THE PRESSURE OF A GAS

If a glass tube, closed at one end, is filled with mercury, and the open end inverted in a pool of mercury as in Figure 2-1(a), the level of mercury in the tube will fall until the mercury column stands about 760 millimeters (mm) above the surface of the pool. The pressure produced at the pool surface by the weight of the mercury column is balanced exactly by the pressure of the surrounding atmosphere. Because there is a balance of opposing pressures, more mercury will not flow into or out of the tube. A device such as this can measure atmospheric pressure, as Evangelista Torricelli (1608–1674) first realized. He showed that it was the *pressure* at the bottom of the mercury column that mattered, and not the total weight of mercury; thus, the height of mercury in a barometer tube is independent of the size or shape of the tube. Atmospheric pressure at sea level supports a column of mercury 760 mm high. This pressure is 1 standard atmosphere (atm), or standard pressure, and the unit of 1 mm of mercury has been named a *torr* in honor of Torricelli. (Given the density of mercury of 13.59 g cm^{-3}, you should be able to show that 1 atm pressure, or 760 torr, corresponds to a pressure of 1033 g cm^{-2}. As a help, imagine that the mercury cylinder has a cross-sectional area of 1.00 cm^2. What is standard atmospheric pressure in pounds per square inch?) With accurate reading devices, we can determine the height of a mercury column to a fraction of a millimeter. But modern experimental techniques often involve high-vacuum equipment that maintains pressures in the range 10^{-3} to 10^{-9} torr. At such low pressures, mercury manometers are useless, and one must find other means of measuring pressure.

2-3 BOYLE'S LAW RELATING PRESSURE AND VOLUME

Robert Boyle, who gave us an operational definition of an element, also was interested in phenomena occurring in evacuated spaces. When devising vacuum pumps for removing air from vessels, he noticed a property familiar to anyone who has used a hand pump for inflating a tire or football, or who

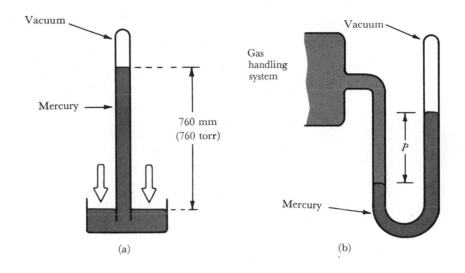

Figure 2–1. Measuring gas pressure. (a) Torricellian barometer. When a mercury-filled tube is inverted in a dish of mercury, the level in the tube falls, thereby leaving a vacuum at the top of the tube. Only a trace of mercury vapor is present in the space at the top. The height of the column is determined by the pressure of the atmosphere on the mercury in the reservoir. (b) In a gas-handling system, pressure P (in millimeters of Hg or torr) is determined by measuring the difference in heights of the two mercury columns of a manometer. If the system is evacuated completely, the levels are equal.

has squeezed a balloon without breaking it: As air is compressed, it pushes back with increased vigor. Boyle called this the "spring of the air," and measured it with the simple device shown in Figure 2–2(a) and (b).

Boyle trapped a quantity of air in the closed end of the J-tube as in Figure 2–2(a), and then compressed it by pouring increasing amounts of mercury into the open end (b). At any point, the total pressure on the enclosed gas is the atmospheric pressure *plus* that produced by the excess mercury, which has height h in the open tube. Boyle's original pressure–volume data on air are given in Table 2–1. Although he did not take special pains to keep the temperature of the gas constant, it probably varied only slightly. Boyle did note that the heat from a candle flame produced a drastic alteration in the behavior of air.

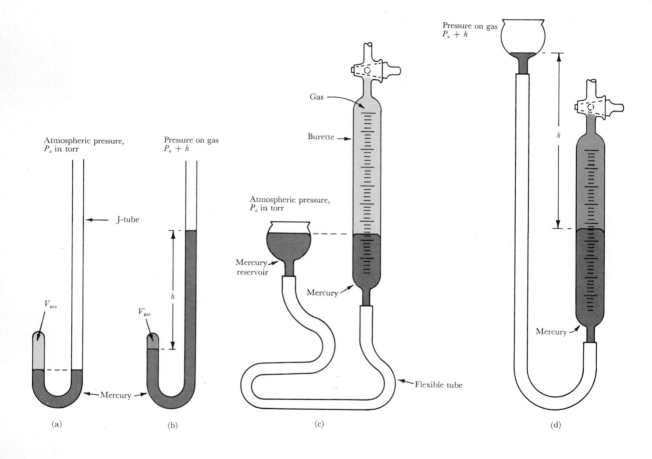

Figure 2–2. Dependence of volume of gas sample on pressure. (a) The simple J-tube apparatus used by Boyle to measure pressure and volume. When the height of the column is equal in the open and closed tubes, the pressure exerted on the gas sample is equal to atmospheric pressure. (b) The pressure on the gas is increased by adding mercury to the tube. (c) The gas burette, a device employing the same principle as the J-tube apparatus. The gas is at atmospheric pressure. (d) The pressure on the gas is increased by raising the mercury reservoir. In (a) and (b) the cross section of the J-tube is assumed constant, so the height of the gas sample is a measure of volume. In (c) and (d) the volume of the gas is measured by the calibrated burette.

Table 2–1.   Boyle's Original Data Relating Pressure and Volume for Atmospheric Air[a]

Volume (index marks along uniform bore tubing)[b]		Pressure (inches of mercury)[c]	$P \times V$
A	48	$29\frac{2}{16}$	1400
	46	$30\frac{9}{16}$	1406
	44	$31\frac{15}{16}$	1408
	42	$33\frac{8}{16}$	1410
	40	$35\frac{5}{16}$	1412
	38	37	1408
	36	$39\frac{5}{16}$	1416
	34	$41\frac{10}{16}$	1420
	32	$44\frac{3}{16}$	1416
	30	$47\frac{1}{16}$	1414
	28	$50\frac{5}{16}$	1410
	26	$54\frac{5}{16}$	1412
	24	$58\frac{13}{16}$	1414
	23	$61\frac{5}{16}$	1411
	22	$64\frac{1}{16}$	1411
	21	$67\frac{1}{16}$	1410
	20	$70\frac{11}{16}$	1415
	19	$74\frac{2}{16}$	1410
	18	$77\frac{14}{16}$	1403
	17	$82\frac{12}{16}$	1410
	16	$87\frac{14}{16}$	1407
	15	$93\frac{1}{16}$	1398
	14	$100\frac{7}{16}$	1408
	13	$107\frac{13}{16}$	1395
B	12	$111\frac{9}{16}$	1342

[a] Reprinted by permission from J. B. Conant, *Harvard Case Histories in Experimental Science,* Harvard University Press, Cambridge, 1957, Vol. 1, p. 53.

[b] End data points A and B correspond to those labels on Figure 2–3.

[c] The height, *h*, in Figure 2–2(b), plus $29\frac{1}{8}$ inches for atmospheric pressure.

Analysis of Data

After a scientist obtains data such as those in Table 2–1, he then attempts to infer a mathematical equation relating the two mutually dependent quantities that he has measured. One technique is to plot various powers of each quantity against one another until a straight line is obtained. The general equation for a straight line is

$$y = ax + b \tag{2-2}$$

in which x and y are variables and a and b are constants. If b is zero, the line passes through the origin.

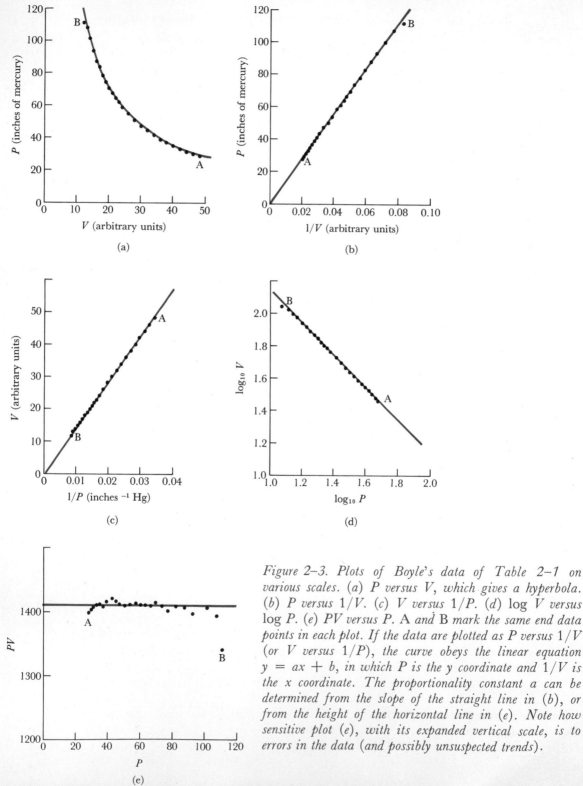

Figure 2–3. *Plots of Boyle's data of Table 2–1 on various scales.* (a) *P versus V, which gives a hyperbola.* (b) *P versus* $1/V$. (c) *V versus* $1/P$. (d) *log V versus log P.* (e) *PV versus P. A and B mark the same end data points in each plot. If the data are plotted as P versus* $1/V$ *(or V versus* $1/P$*), the curve obeys the linear equation* $y = ax + b$, *in which P is the y coordinate and* $1/V$ *is the x coordinate. The proportionality constant a can be determined from the slope of the straight line in* (b)*, or from the height of the horizontal line in* (e)*. Note how sensitive plot* (e)*, with its expanded vertical scale, is to errors in the data (and possibly unsuspected trends).*

Figure 2–3 shows several possible plots of the data for pressure, P, and volume, V, given in Table 2–1. The plots of P versus $1/V$ and V versus $1/P$ are straight lines through the origin. A plot of the logarithm of P versus the logarithm of V is also a straight line with negative slope of -1. From these plots the equivalent equations are deduced:

$$P = \frac{a}{V} \tag{2–3a}$$

$$V = \frac{a}{P} \tag{2–3b}$$

and

$$\log V = \log a - \log P \tag{2–3c}$$

These equations represent variants of the usual formulation of Boyle's law: *For a given mass of gas, the pressure is inversely proportional to the volume if the temperature is held constant.*

When the relationship between two measured quantities is as simple as this one, it can be deduced numerically as well. If each value of P is multiplied by the corresponding value of V, the products all are nearly the same for a single sample of gas at constant temperature (Table 2–1). Thus,

$$PV = a \simeq 1410 \tag{2–3d}$$

Equation 2–3d represents the hyperbola obtained by plotting P versus V [Figure 2–3(a)]. This experimental function relating P and V now can be checked by plotting PV against P to see if a horizontal straight line is obtained [Figure 2–3(e)].

Boyle found that for a given quantity of gas at constant temperature, the relationship between P and V is given reasonably precisely by

$$PV = \text{constant} \qquad \text{(at constant } T \text{ and } n) \tag{2–4}$$

for all gases.

Before you go on.   To be sure that you understand Boyle's law, you may want to try Items 1–14 in Section 2–1 of Lassila, *Programed Reviews of Chemical Principles.*

2–4 CHARLES' LAW RELATING VOLUME AND TEMPERATURE

We know that air expands on heating, thereby decreasing its density. For this reason, balloons rise when inflated with warm air. About 100 years after Boyle derived his law, Jacques Charles (1746–1823), in France, measured the effect of changing temperature on the volume of an air sample. This measurement can be done quite easily with the device in Figure 2–4. Some sample data are plotted in Figure 2–5; these show that a graph of V versus T is a straight line with an extrapolated intercept of $-273°$ on the centigrade scale

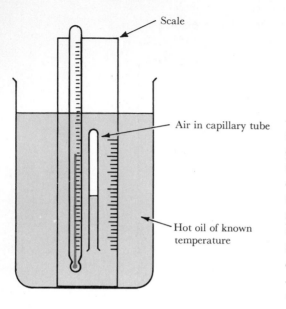

Scale

Air in capillary tube

Hot oil of known
temperature

*Figure 2–4. Experimental determination of the rela-
tionship between volume and temperature of a gas. The
apparatus consists of a small capillary tube and a
thermometer mounted on a ruled scale and immersed in
a hot oil bath. As the system cools, the oil rises in the
tube, and the length of air space and temperature are
measured at intervals. For a tube of constant bore,
the length of the air space is a measure of the gas
volume. So long as the bottom of the air space in the
capillary is maintained at the same depth below the
surface of the oil bath, the pressure in the capillary will
be constant.*

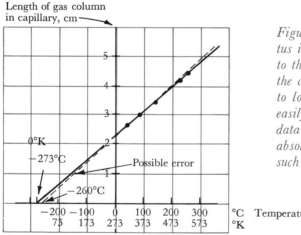

Length of gas column
in capillary, cm

0°K
−273°C

Possible error

−260°C

*Figure 2–5. A plot of data, obtained with the appara-
tus in Figure 2–4, showing that volume is proportional
to the absolute temperature. Just such a plot employing
the centigrade scale of temperature originally was used
to locate the absolute zero of temperature. Notice how
easily a small error in the slope of the line through the
data points could produce a large error in the value of
absolute zero. It should be clear that, if at all possible,
such long extrapolations should be avoided.*

of temperature, or −460° on the Fahrenheit scale. Charles expressed his
law as

$$V = c(t + 273)$$

in which V is the volume of a gas sample, t is the temperature on the centigrade scale, and c is a proportionality constant.

Later, Lord Kelvin (1824–1907) suggested that the intercept of −273°
represented an absolute minimum of temperature below which it is not pos-

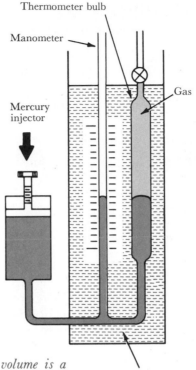

Figure 2–6. *A simple gas thermometer. The gas volume is a measure of the absolute temperature. The scale can be calibrated with the freezing point (0°C) and boiling point (100°C) of water. The mercury is injected into or removed from the apparatus to maintain constant atmospheric pressure.*

sible to go. Scientists now use Kelvin's absolute scale of temperature with $0°K = -273.16°C$ and $0°C = 273.16°K$. Charles' law is expressed as

$$V = cT \qquad \text{(at constant } P \text{ and } n) \tag{2–5}$$

in which T is the absolute temperature in degrees Kelvin (i.e., $t + 273$). Equation 2–5 indicates that *at constant pressure the volume of a given mass of gas is directly proportional to the absolute temperature.* For light gases such as hydrogen and helium, Charles' law is so accurate that gas thermometers often replace mercury thermometers for precise temperature measurement (Figure 2–6). A mercury thermometer calibrated to read 0°C in a water–ice mixture and 100°C in boiling water is inaccurate by as much as 0.1 degree (deg) at intermediate points, whereas a hydrogen thermometer is much more accurate throughout this region.

2–5 THE COMBINED GAS LAW

The three gas equations that we have encountered so far all may be written in terms of the proportionality of volume to another quantity:

$$V \propto n \qquad \text{(at constant } P \text{ and } T\text{)} \qquad\qquad \text{(Avogadro's law)}$$

$$V \propto \frac{1}{P} \qquad \text{(at constant } T \text{ and } n\text{)} \qquad\qquad \text{(Boyle's law)}$$

$$V \propto T \qquad \text{(at constant } P \text{ and } n\text{)} \qquad\qquad \text{(Charles' law)}$$

Therefore, the volume must be proportional to the product of these three terms, or

$$V \propto \frac{nT}{P}$$

or

$$PV = nRT \qquad\qquad\qquad (2\text{–}6)$$

in which R is the proportionality constant. This last equation is known as the *ideal gas law*. It contains all of our earlier laws as special cases and, in addition, predicts more relationships that can be tested. For example, Gay-Lussac verified the prediction that, at constant volume, the pressure of a fixed amount of gas is proportional to its absolute temperature. (In effect, Equation 2–6 is a definition of the ideal gas; the differences between real gases and the hypothetical ideal gas are discussed in Section 2–8.)

The gas law is often useful when expressed in the form of ratios of *before* and *after* variables. For example, suppose that a fixed amount of gas at constant temperature is compressed from P_1 to P_2, with volumes V_1 and V_2. Then $P_1 = nRT/V_1$, $P_2 = nRT/V_2$, and the pressure ratio and volume ratio are related by

$$\frac{P_2}{P_1} = \frac{V_1}{V_2} \qquad (T, n \text{ constant}) \qquad\qquad (2\text{–}7)$$

This is Boyle's law in ratio form. In a similar way, Charles' law states that the ratio of volume before and after matches the temperature ratio at constant pressure:

$$\frac{V_2}{V_1} = \frac{T_2}{T_1} \qquad (P, n \text{ constant}) \qquad\qquad (2\text{–}8)$$

Increasing by a factor the number of moles of gas at constant temperature and pressure increases the volume by the same factor:

$$\frac{V_2}{V_1} = \frac{n_2}{n_1} \qquad (P, T \text{ constant}) \qquad\qquad (2\text{–}9)$$

And increasing by a factor the number of moles of gas at constant temperature in a tank of fixed volume increases the pressure inside the tank by the same factor:

$$\frac{P_2}{P_1} = \frac{n_2}{n_1} \qquad (T, V \text{ constant}) \tag{2-10}$$

You should be able to derive these equations easily from the ideal gas law, as well as the analogous equation that expresses Gay-Lussac's observations about pressure and temperature at constant volume.

The numerical value of the gas constant, R, depends on the units in which pressure and volume are measured (assuming that only the absolute, or Kelvin, temperature scale is used). If pressure is in atmospheres and volume is in liters, then $R = 0.082054$ liter atm deg^{-1} mole^{-1}. If pressure is in torr and volume is in cm^3, then $R = 62{,}361$ torr cm^3 deg^{-1} mole^{-1}. But as you can see from Appendix 1, R also can be expressed in the units of erg deg^{-1} mole^{-1} or cal deg^{-1} mole^{-1}. We shall show in Chapter 15 that the product PV has the units of work or energy.

Standard Temperature and Pressure

Physical scientists have discovered that it is convenient to compare volumes of gases involved in physical and chemical processes. Such comparisons are interpreted most easily if they are made at the same temperature and pressure, although it generally is inconvenient to make all *measurements* under such carefully controlled conditions; 0°C (273°K) and 1.0 atm (760 torr) have been designated arbitrarily as *standard temperature and pressure* (STP). Thus, if we measure the volume of a sample of gas at any condition, by employing the combined gas law we easily can calculate the volume it would have as an ideal gas at STP. This calculated volume is useful even if the substance itself becomes a liquid or solid at STP.

Example. In an experiment, 300 cm^3 of steam are at 1.0 atm pressure and 150°C. What is the ideal volume at STP?

Solution. From Equation 2–8,

$$V_{STP} = V_1 \frac{T_{STP}}{T_1}$$

$$= 300 \text{ cm}^3 \times \frac{273°\text{K}}{423°\text{K}} = 194 \text{ cm}^3$$

This is the volume that the steam would occupy at STP, *if* it behaved like an ideal gas instead of condensing.

At STP, 1 mole of an ideal gas occupies 22.414 liters, as can be seen from the ideal gas law:

$$\text{Volume per mole} = \frac{V}{n} = \frac{RT}{P}$$

$$= \frac{0.082054 \text{ liter atm deg}^{-1} \text{ mole}^{-1} \times 273.16 \text{ deg}}{1.0000 \text{ atm}}$$

$$= 22.414 \text{ liters mole}^{-1}$$

This volume often is called a *standard molar volume* and is the constant, *k*, in the Cannizzaro calculations of Chapter 1.

Ideality and Nonideality

The equations describing the various gas laws are exact mathematical expressions. Measurements of volume, pressure, and temperature more accurate than those of Boyle and Charles show that gases only *approach* the behavior that the equations express. Gases depart radically from so-called ideal behavior when under high pressure or at temperatures near the boiling point of the corresponding liquids. Thus, the gas laws, or more precisely the ideal gas laws, accurately describe the actual behavior of a real gas only at low pressures and at temperatures far above the boiling point of the substance in question. In Section 2–8 we shall return to the problem of how to correct the simple ideal gas law for the behavior of real gases.

Before you go on. A thorough understanding of the laws governing ideal gas behavior is necessary both as an introduction to the kinetic molecular theory of gases and as a tool in practical chemistry. If you have any doubts about your ability to handle gas-law problems at the end of this chapter, work through the entire Review 2 of Lassila, *Programed Reviews of Chemical Principles*.

2–6   KINETIC MOLECULAR THEORY OF GASES

At STP, 1 mole of carbon dioxide gas occupies 22.2 liters, whereas the same amount of Dry Ice (solid CO_2) has a volume of only 28 cm^3 (assuming a density of Dry Ice of 1.56 g cm^{-3}). This greater volume of a gas, plus the fact that a gas is compressed or expanded so easily, suggests strongly that much of a gas is empty space. Then how does a system that is mostly empty space exert pressure on its surroundings? Experiments such as the one in Figure 2–7 indicate that molecules are moving, and moving in straight lines. They also collide with the walls of the container, with one another, and with any other objects that may be in the container with the gas (Figure 2–8). As we shall see, the collision with the container walls produces pressure. It is unnecessary to assume any forces between molecules and container to account for pressure.

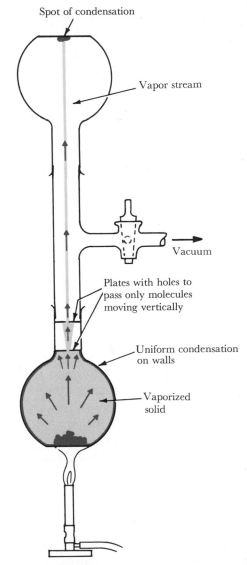

Figure 2–7. An experiment to test whether gas molecules move in straight lines. Two flasks are joined by a straight tube with a side arm and stopcock. The bottom flask contains material, such as iodine, that can be vaporized by heating. The flasks are evacuated and the material heated, thereby producing vapor. Molecules leave the solid in random directions, and condensation occurs uniformly over the entire surface of the bottom flask. However, only molecules moving vertically can pass through the collimating holes in the connecting tube and into the top flask. These molecules pass straight through and form a single spot directly opposite the source material. A high vacuum (low pressure) is required to prevent molecular collisions from randomizing molecular motion in the connection tube and upper flask.

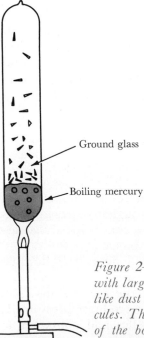

Figure 2–8. An experiment demonstrating collisions of gas molecules with large solid particles. Particles of ground glass are kept suspended like dust particles in air by bombardment with moving mercury molecules. The heavy molecules (mainly monatomic Hg) leaving the surface of the boiling mercury have high kinetic energy, some of which is transferred to the glass particles on collision.

We can explain many observed properties of gases by a simple theory of molecular behavior that was developed in the latter half of the nineteenth century by Ludwig Boltzmann (1844–1906), James Clerk Maxwell (1831–1879), and others. This *kinetic molecular theory* has three assumptions:

1) A gas is composed of molecules that are extremely far apart from one another in comparison with their own dimensions. They can be considered as essentially point objects or small, hard spheres. (The *shapes* of molecules are neglected.)

2) These gas molecules are in a state of constant random motion, which is interrupted only by collisions of the molecules with each other and with the walls of the container.

3) The molecules exert no forces on one another or on the container other than through the impact of collision. Furthermore, these collisions are *elastic;* that is, no energy is lost as friction during collision.

Our experience with colliding bodies such as a tennis ball bouncing on pavement is that some kinetic energy is lost on collision; the energy is transformed into heat as a result of what we call *friction*. A bouncing tennis ball gradually "dies down" and comes to rest because its collisions with the pavement are subject to friction and are therefore *inelastic*. If molecular collisions involved friction, the molecules gradually would slow down and lose kinetic energy, thereby hitting the walls with decreasing change of momentum so the pressure would drop slowly to zero. This process does not happen; therefore, we must postulate that *molecular collisions are frictionless, that is, perfectly elastic*. In other words, the total kinetic energy of colliding molecules remains constant.

The Phenomenon of Pressure and Boyle's Law

This simple model is adequate to explain pressure and to provide a molecular explanation of Boyle's law. Consider a container, which we will make cubical for simplicity, with a side of length l. Suppose that the container is evacuated completely except for one molecule of mass m that moves with a velocity v having components v_x, v_y, and v_z parallel to the x, y, and z edges of the box (see Figure 2–9).[1]

---

[1] If the idea of the breakdown of a vector such as velocity into its three components, v_x, v_y, and v_z, is unfamiliar, there is an explanation that, although less exact, leads to the same answer. This is to assume that since the motions of a molecule in the x, y, and z directions are unrelated, we can think of the molecules as being divided into three groups: one third moving in the x direction, one third in the y direction, and one third in the z direction. The pressure from one molecule on the YZ wall is then $P_x = mv^2/V$ (analogous to Equation 2–11). The pressure from all of the molecules moving in a direction perpendicular to that wall is $N/3$ times this value, or

$$P_x = \frac{N}{3} \frac{\overline{mv^2}}{V}$$

as in Equation 2–19. The rest of the proof is the same.

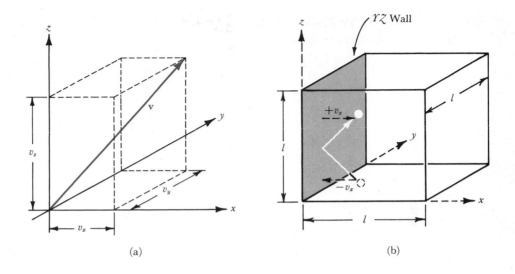

(a)

(b)

*Figure 2–9. (a) The velocity of a molecule of gas resolved into components. We determine the components of the velocity vector **v** by dropping perpendiculars from the head and tail of the vector to the coordinate axes. (b) Collision of molecule with wall showing change in direction of x component of velocity.*

Let us look first at what happens when the molecule rebounds from a collision with one of the YZ walls, which are perpendicular to the x axis.

Pressure is force per unit area, and force is the rate of change of momentum with time. When a molecule bounces off the shaded wall in Figure 2–9(b), it exchanges momentum of $2mv_x$ with the wall; for the particle begins with momentum in the x direction of $-mv_x$ and ends with momentum $+mv_x$. The velocity components in the y and z directions are not changed during a collision with the YZ wall and do not enter into the calculation. No matter how many collisions the molecule has with an XY or an XZ wall along the way, if the x component of velocity is v_x, the molecule will return to collide with the original YZ wall in a time $2l/v_x$. If the molecule transfers momentum of $2mv_x$ every $2l/v_x$ seconds, then the rate of change of momentum with time, or the force, is

$$f_x = \frac{2mv_x}{2l/v_x} = \frac{mv_x^2}{l}$$

The force per unit area, or the pressure, is

$$P_x = \frac{mv_x^2}{l \cdot l^2} = \frac{mv_x^2}{l^3} = \frac{mv_x^2}{V} \tag{2–11}$$

since the area of the wall is l^2, and the total volume of the box is $V = l^3$.

Similarly, for the other walls,

$$P_y = \frac{mv_y^2}{V} \tag{2-12}$$

$$P_z = \frac{mv_z^2}{V} \tag{2-13}$$

If the box now contains N molecules rather than just one,

$$P_x = N \frac{\overline{mv_x^2}}{V} \tag{2-14}$$

$$P_y = N \frac{\overline{mv_y^2}}{V} \tag{2-15}$$

$$P_z = N \frac{\overline{mv_z^2}}{V} \tag{2-16}$$

in which the quantities $\overline{v^2}$ are the *averages* over all molecules of the squares of the velocity components, since we cannot assume that all molecules have the same velocity.

The total velocity of a molecule is related to its velocity components by

$$v^2 = v_x^2 + v_y^2 + v_z^2 \tag{2-17}$$

If the motions of the individual molecules are truly random and unrelated, the average of the square of the velocity component in each direction will be the same. There will be no preferred direction of motion in the gas:

$$\overline{v_x^2} = \overline{v_y^2} = \overline{v_z^2} = \tfrac{1}{3}\overline{v^2} \tag{2-18}$$

(As before, the bars over v^2 indicate averages over all molecules.) An immediate consequence of this randomness of motion is that the pressure will be the same on all walls, a fact that certainly agrees with our observations of real gases. Rewriting Equations 2–14, 2–15, and 2–16 in terms of $\overline{v^2}$ gives

$$P_x = \frac{N}{3} \frac{\overline{mv^2}}{V} \qquad P_y = \frac{N}{3} \frac{\overline{mv^2}}{V} \qquad P_z = \frac{N}{3} \frac{\overline{mv^2}}{V}$$

and

$$P_x = P_y = P_z = P = \frac{N}{3} \frac{\overline{mv^2}}{V} \tag{2-19}$$

or

$$PV = \frac{N}{3} \overline{mv^2} \tag{2-20}$$

This last expression looks very much like Boyle's law. Boyle's law maintains that the product of pressure and volume for a gas is constant *at constant temperature;* our derivation from the simple kinetic molecular theory states that the PV product is constant for a given *mean velocity* of gas molecules. If the theory is correct, the mean velocity of the molecules of a gas cannot depend on either pressure or volume, but only on temperature. The mean molecular kinetic energy is $\bar{\epsilon} = \frac{1}{2}m\overline{v^2}$; further, if N is Avogadro's number, the kinetic energy of 1 mole of molecules is $E_k = N\bar{\epsilon}$. For a mole of gas, the PV product of Boyle's law is proportional to the kinetic energy per mole:

$$E_k = N\bar{\epsilon} = \frac{1}{2} Nm\overline{v^2} \tag{2-21}$$

Multiplying and dividing the right term by three and rearranging gives

$$E_k = \left(\frac{3}{2}\right)\left(\frac{1}{3}\right) Nm\overline{v^2} = \left(\frac{3}{2}\right)\left(\frac{N}{3}\right) m\overline{v^2} \tag{2-22}$$

Comparison with Equation 2–20 shows that

$$PV = \frac{2}{3} E_k \tag{2-23}$$

The combination of this derivation from the kinetic theory and the observed ideal gas law (Equation 2–6) tells us that the kinetic energy per mole is directly proportional to the temperature. Or, reversing the statement, absolute temperature, T, is an indication of the kinetic energy of gas molecules and ultimately of their mean square velocity. For 1 mole of an ideal gas, $PV = RT$. Substitute the value for PV given in Equation 2–23:

$$E_k = \frac{3}{2} RT \tag{2-24}$$

But $E_k = N\bar{\epsilon}$, in which $\bar{\epsilon} = \frac{1}{2}m\overline{v^2}$; therefore,

$$T = \frac{2}{3}\frac{N}{R}\frac{1}{2} m\overline{v^2} = \frac{M\overline{v^2}}{3R} \tag{2-25}$$

in which the molecular weight is $M = Nm$. In short, *temperature is a measure of motion of molecules.* If we heat a gas and raise its temperature, we do so by increasing the mean square velocity of its molecules. When a gas (or any other substance) cools, its molecular motion diminishes. This molecular motion need not be confined to movement of whole molecules from one place to another, which is the picture that we have drawn for an ideal gas. It also can include *rotations* of entire molecules or of groups on a molecule, and *vibrations* of molecules.

We now can see more clearly what happens when kinetic energy of macroscopic objects is dissipated as heat. When a speeding car skids to a halt, its braking is achieved by converting its energy of motion into frictional heat. But this conversion means changing the motion of the large object— the automobile—into increased relative motion of the molecules of the brake shoes and drum, the tires, and the pavement. Instead of having rubber molecules in the tires vibrating relatively slowly but moving rapidly as a unit, we have a heated tire with molecules moving more rapidly relative to one another but without a net direction of motion. The motions of the molecules have become less directional and more randomized.

This behavior is typical of all real processes. It is easy to go from coherent motion (the rolling tire) to incoherent motion (the hot but stationary tire); it is not possible to go the other way without paying a price. As we shall see in Chapter 15, in any real process the disorder of the object under examination, plus all of the surroundings with which it interacts, always will increase. Or in other words, in this world things always get messier. This notion is simply the second law of thermodynamics. The quantity that measures this disorder, and which we shall learn to use later in chemical situations, is called *entropy*, S.

2–7 PREDICTIONS OF THE KINETIC MOLECULAR THEORY

The test of any theory is not its beauty or its internal consistency, but its usefulness in predicting the behavior of real systems correctly. By this criterion, the kinetic molecular theory is a good one, as we shall see.

Molecular Size

The density of solid CO_2 (Dry Ice) is 1.56 g cm^{-3}, so 1 mole of solid CO_2 occupies 44.01 g mole^{-1}/1.56 g cm^{-3} = 28.3 cm^3 mole^{-1}. The volume *per molecule* then is 28.3/6.022 $\times$ 10^{23} cm^3 or 47.0 Å^3. Let us assume for the moment that the CO_2 molecules in Dry Ice can be approximated by closely packed spheres. We will show in the next chapter that such closely packed spheres fill 74% of the space available, with 26% as empty space between the spheres. The radius of such spheres with an overall volume of 47.0 Å^3 per sphere is given by

$$0.74 \times 47.0 = \tfrac{4}{3}\pi r^3 \tag{2–26}$$

or

$$r = 2.02 \text{ Å}$$

In short, the observed density of solid CO_2 is that which would be expected from spheres with a molecular weight of 44 and a radius of 2 Å.

The situation is quite different for gaseous CO_2. The measured density of CO_2 gas at STP is 1.977 g liter^{-1}. The molar volume is 44.01 g mole^{-1}/ 1.977 g liter^{-1} = 22.2 liter mole^{-1}. (Note the deviation from ideal gas behavior of CO_2 at STP.) The volume per molecule is

$$\frac{22{,}200 \text{ cm}^3 \text{ mole}^{-1}}{0.6022 \times 10^{24} \text{ molecules mole}^{-1}} \times \frac{10^{24} \text{ Å}^3}{1 \text{ cm}^3} = 36{,}800 \text{ Å}^3 \text{ molecule}^{-1}$$

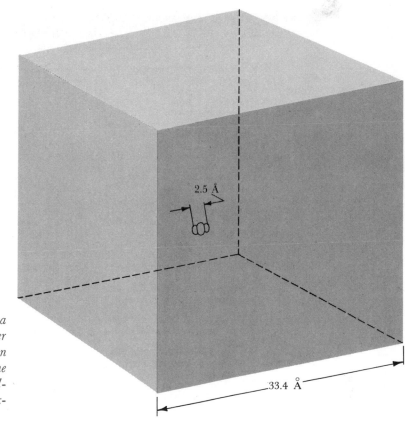

Figure 2–10. The relative size of a CO_2 molecule and the volume per molecule available to it in carbon dioxide gas at STP. Of course, one molecule is not confined to this volume, nor are other molecules excluded from it.

The gas has a molar volume that is 785 times the volume of the solid. The *volume per molecule* in the gas phase corresponds to a cube that is 33.4 Å on a side (Figure 2–10); only one part in 800 of the gas volume actually is filled by molecules.

Molecular Speeds

With nothing more than the elementary kinetic theory presented here, we can calculate the *root-mean-square* (rms) *speed*, v_{rms}, which is the square root of the average of the squares of the speeds of individual molecules. From Equation 2–25, v_{rms} is

$$\sqrt{\overline{v^2}} = v_{rms} = \sqrt{\frac{3RT}{M}} \qquad (2-27)$$

in which R is the gas constant, T is the absolute temperature, and M is the molecular weight. This equation is a good example of the absolute necessity of keeping careful track of units. The gas constant, R, must be used in the form of $R = 8.314 \times 10^7$ ergs $\deg^{-1}$ mole^{-1}, and *not* $R = 0.08205$ liter atm $\deg^{-1}$ mole^{-1}, if the speed is to be expressed in cm sec^{-1}. Since 1 erg = 1 g cm^2 sec^{-2}, the units of $3RT/M$ are

$$\frac{(\cancel{g}\, cm^2\, sec^{-2}\, \cancel{deg^{-1}}\, \cancel{mole^{-1}})(\cancel{deg})}{(\cancel{g}\, \cancel{mole^{-1}})} = cm^2\, sec^{-2}$$

and v_{rms} is in the desired units. At STP the expression is

$$v_{rms} = \frac{26.1 \times 10^4}{M^{1/2}}\ cm\ sec^{-1} \tag{2-28}$$

Oxygen molecules at STP travel at a root-mean-square speed of 46,000 cm sec^{-1} or 1030 miles per hour.

The root-mean-square speed of nitrogen molecules at STP is 49,300 cm sec^{-1} or 493 m sec^{-1}. However, this does not mean that all nitrogen molecules travel at this speed. There is a *distribution* of speeds, from zero to values considerably above 493 m sec^{-1}. As gas molecules collide and exchange energy, their speeds will vary. The actual distribution of speeds in nitrogen gas at 1 atm pressure and three temperatures is shown in Figure 2–11. These curves portray a *Maxwell-Boltzmann distribution* of speeds. The equations for these curves can be derived from the kinetic theory by using probability function arguments. At higher temperatures, the root-mean-square speed increases, as expected from Equation 2–27. But Figure 2–11 shows that the distribution of speeds also becomes more diffuse: There is a greater spread of speeds, and fewer molecules have a speed close to the average value.

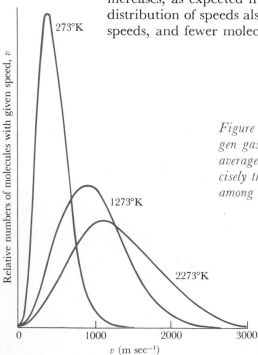

Figure 2–11. The distribution of speeds among molecules in nitrogen gas at three different temperatures. At higher temperatures the average speed is greater, there are fewer molecules that have precisely this average speed, and there is a broader distribution of speed among molecules.

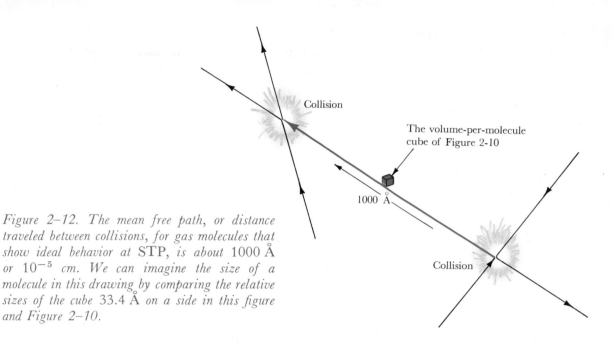

Figure 2–12. The mean free path, or distance traveled between collisions, for gas molecules that show ideal behavior at STP, is about 1000 Å or 10^{-5} cm. We can imagine the size of a molecule in this drawing by comparing the relative sizes of the cube 33.4 Å on a side in this figure and Figure 2–10.

From the size of a molecule, the speed with which it travels, and the density of other molecules around it, we can calculate the *mean free path* (or distance a molecule travels between two successive collisions) and the *collision frequency*. Molecules such as O_2 or N_2 travel an average of 1000 Å between collisions, and they experience approximately five billion collisions per second at STP (Figure 2–12).

Challenge. To see how the principles in this section can be applied to planetary atmospheres, try Problem 2–23 in the Butler and Grosser problems book.

Dalton's Law of Partial Pressures

If each molecule in a gas travels independently of every other except at moments of collisions, and if collisions are elastic, then in a mixture of different gases the total kinetic energy of all of the different gases will be the sum of the kinetic energies of the individual gases:

$$E = E_1 + E_2 + E_3 + E_4 + \cdots$$

Since each gas molecule moves independently, the pressure that each gas exerts on the walls of the container can be derived separately (Equation 2–23):

$$p_1 = \frac{2E_1}{3V} \qquad p_2 = \frac{2E_2}{3V} \qquad p_3 = \frac{2E_3}{3V} \qquad \text{etc.} \tag{2–29}$$

This pressure exerted by one component of a gas mixture is called its *partial pressure*, p. Each of these equations can be rewritten to give kinetic energy in terms of pressure:

$$E_1 = \tfrac{3}{2}p_1V \qquad E_2 = \tfrac{3}{2}p_2V \qquad E_3 = \tfrac{3}{2}p_3V \qquad \text{etc.}$$

Substituting in the energy expression and canceling the $\tfrac{3}{2}V$ terms from both sides of the equation produces

$$P = p_1 + p_2 + p_3 + p_4 + \cdots = \sum_j p_j \tag{2-30}$$

The special sign at the right is a *summation sign*, which is a shorthand way of writing the instructions: Sum all of the terms of the type p_j for all the different values of j. It will be used frequently.

The *total pressure*, then, is the *sum of the partial pressures* of the individual components of the gas mixture, each considered as if it were the only gas present in the given volume. Dalton's *law of partial pressures* was proposed during the gas investigations that eventually led him to an atomic theory.

An important measure of concentrations in a mixture of gases (and in solutions and solids as well) is the *mole fraction*, X. The mole fraction of the jth component in a mixture of substances is defined as the number of moles of the given substance divided by the total number of moles of all substances:

$$X_j = \frac{n_j}{n_1 + n_2 + n_3 + n_4 + \cdots} = \frac{n_j}{\sum_i n_i} \tag{2-31}$$

Another version of Dalton's law is the statement that the partial pressure of one component in a mixture of gases is its concentration in mole fraction times the total pressure. If there are n_j moles of gas j present in a mixture, the partial pressure of that gas is calculable from the ideal gas law:

$$p_j = n_j \frac{RT}{V} = \frac{n_j}{n} \, n \times \frac{RT}{V} \qquad \left(n = n_1 + n_2 + n_3 + \cdots = \sum_i n_i \right)$$

Since $n_j/n = X_j$ is the mole fraction, and $nRT/V = P$ is the *total* pressure, Dalton's law becomes

$$p_j = X_j P \tag{2-32}$$

Example. A gas mixture at 100°C and 600 torr contains 50% helium and 50% xenon by weight. What are the partial pressures of the individual gases?

Solution. First find the number of moles of helium and xenon in any given sample. A convenient sample choice is 100 g. Then the number of moles of each gas is

$$n_{\text{He}} = \frac{50.0\text{ g}}{4.00\text{ g mole}^{-1}} = 12.5 \text{ moles He}$$

$$n_{\text{Xe}} = \frac{50.0\text{ g}}{131.3\text{ g mole}^{-1}} = 0.381 \text{ mole Xe}$$

The next step is to calculate the mole fraction, X_j, of each component:

$$X_{He} = \frac{12.5}{12.5 + 0.381} = 0.970$$

$$X_{Xe} = \frac{0.381}{12.5 + 0.381} = 0.030$$

According to Dalton's law, the partial pressure of each component is expressed as $p_j = X_j P$. Thus, we have

$$p_{He} = 0.970\ P = 0.970(600) = 582 \text{ torr}$$
$$p_{Xe} = 0.030\ P = 0.030(600) = 18 \text{ torr}$$

Note that no total volume was specified, and a convenient but arbitrary sample size was used for calculation purposes. Why is the answer independent of volume? Will the answer change if the temperature is changed?

Often gases are collected over liquids such as water or mercury, as in Figure 2–13. Dalton's law must be applied in such cases to account for partial evaporation of the liquid into the space occupied by the gas.

Example. Oxygen gas generated in an experiment is collected at 25°C in a bottle inverted in a trough of water (Figure 2–13). When the water level in the originally full bottle has fallen to the level in the trough, the volume of collected gas is 1750 milliliters (ml). How many moles of oxygen gas have been collected?

Solution. If the water levels inside and outside the bottle are the same, then the total pressure inside the bottle equals atmospheric pressure, which we will assume for this problem to be 760.0 torr. But at 25°C the vapor pressure of water is 23.8 torr, so the partial pressure of oxygen gas is only 760.0 − 23.8 = 736.2 torr. The mole fraction of oxygen gas in the bottle is 736.2/

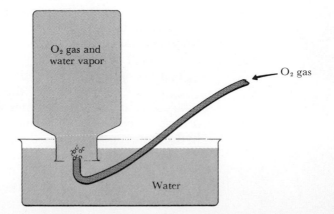

Figure 2–13. When oxygen gas is collected by displacing water from an inverted bottle, the presence of water vapor in the collecting bottle must be recognized when calculating the amount of oxygen collected. The correction is made easily by using Dalton's law of partial pressures.

O₂ gas and water vapor

O₂ gas

Water

760.0 = 0.969 and not 1.000, and the partial pressure of oxygen also is 0.969 atm (i.e., p_{O_2} = 0.969 × 760 torr/760 torr atm^{-1}). The number of moles is

$$n = \frac{PV}{RT} = \frac{0.969 \text{ atm} \times 1750 \text{ cm}^3}{82.054 \text{ cm}^3 \text{ atm deg}^{-1} \text{ mole}^{-1} \times 298 \text{ deg}}$$
$$= 0.0694 \text{ mole}$$

What would the answer have been had the pressure of water vapor been neglected?

Exercise.   On a humid day at 110°F in Galveston, Texas, the vapor pressure of water is 66 torr. What is the water content of the atmosphere, expressed as a mole fraction? Assuming that dry air is 20 mole % O_2 and 80 mole % N_2, what is the water content under the above conditions in percent by weight?

(*Answer:* 0.087; 5.62%.)

Challenge.   To test your ability to relate some of the preceding concepts to air pollution and the design of life-support systems, try Problems 2–25 and 2–27 through 2–31 in the Butler–Grosser book.

Other Predictions of the Kinetic Molecular Theory

Derivations from the kinetic molecular theory that are not much more complicated in principle than the ones we have seen for the gas pressure furnish us with a host of other predictions about the behavior of gases. These predictions have been tested and have encouraged confidence in the theory. A derivation of the probability of a molecule hitting a hole in the wall of a container leads to Graham's law of effusion, which predicts that the rate of leakage of a gas from a small hole in a tank will be inversely proportional to the square root of the molecular weight.

Thomas Graham (1805–1869) observed, in 1846, that the rates of effusion of gases are inversely proportional to the square roots of their densities. Since, by Avogadro's hypothesis, the density of a gas is proportional to its molecular weight, Graham's observation agrees with the kinetic theory, from which we predict that the rate of escape is proportional to molecular velocity or inversely proportional to the square root of the molecular weight (Equation 2–28). However, the law begins to fail at high densities, in which molecules collide several times with one another as they escape through the orifice. The law also fails when there are holes large enough so the gas has a hydrodynamic flow toward the hole, thereby leading to the formation of a jet of escaping gas. But so long as isolated molecules escape by hitting the orifice during their random motions through a stationary gas, the kinetic molecular theory prediction is exact.

From the theory we also can predict correctly the phenomena of gaseous diffusion, viscosity, and thermal conductivity, the three so-called transport properties. In molecular diffusion, mass diffuses from regions of high to low concentration, or down a concentration gradient. Viscosity of a fluid arises because slowly moving molecules diffuse into (and retard) rapidly moving fluid layers, and faster molecules diffuse into (and accelerate) the slow regions. This is a transport of momentum down a velocity gradient. Thermal diffusion is the scattering of rapidly moving molecules into regions of slower ones. It can be described as a transport of kinetic energy down a temperature gradient. In all three cases, the kinetic molecular theory predicts the diffusion coefficient correctly, with best accuracy at low gas pressures and high temperatures. Yet these cases are just the conditions for which the simple ideal gas law is most applicable.

We can summarize the situation by asserting that the elementary kinetic molecular theory, as outlined here, provides a correct explanation of the behavior of ideal gases. It gives us confidence in the reality of molecules, and encourages us to look for molecular modifications of the simple theory that will account for deviations from ideal gas behavior.

2–8 REAL GASES DEVIATE FROM THE IDEAL GAS LAW

If gases were ideal, the quotient PV/RT always would equal one for 1 mole of gas. Actually all real gases deviate, to some extent, from ideal behavior; the quantity $Z = PV/RT$, called the *compressibility coefficient*, is one measure of this deviation. (Z is plotted against pressure for several gases at 273°K in Figure 2–14, and for one gas at several temperatures in Figure 2–15.) We can

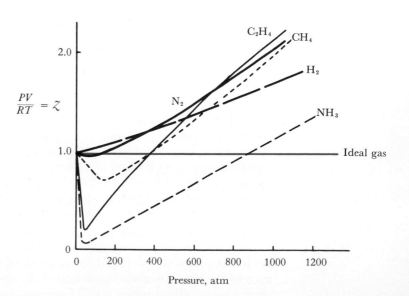

Figure 2–14. Deviations from the ideal gas law for several gases at 273°K, in terms of the compressibility factor $Z = PV/RT$. The dip of Z below 1.0 at low pressures is caused by intermolecular attractions; the rise above 1.0 at high pressures is produced by the shorter range intermolecular repulsions as the molecules, of finite bulk, are crowded closely together.

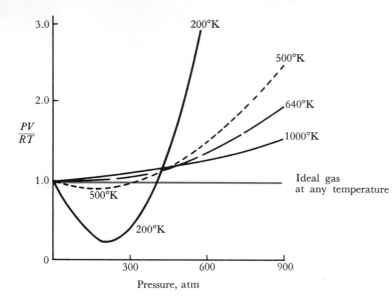

Figure 2–15. PV/RT for 1 mole of methane gas at several temperatures. Note that PV/RT is less than 1.0 at low pressures and greater than 1.0 at high pressures. Ideal gas behavior is approached at high temperatures. PV/RT = Z, the compressibility coefficient.

interpret the behavior of real gases as a combination of intermolecular attractions (which are effective over comparatively long distances) and repulsions caused by the finite sizes of molecules (which become significant only when molecules are crowded together at high pressures). At low pressures—but still too high for ideal behavior—intermolecular attractions make the molar volume unexpectedly low, and the compressibility coefficient is less than one. However, at sufficiently high pressures the crowding of molecules begins to predominate, and the molar volume is greater than it would have been if the molecules were point masses. The higher the temperature (Figure 2–15), the less significant the intermolecular attraction will be in comparison with the kinetic energy of the moving molecules, and the lower will be the pressure at which the bulk factor dominates and Z rises above one.

An equation such as the ideal gas law, $PV = nRT$, is known as an *equation of state* because it describes the state of a system in terms of the measurable variables P, V, and T (Figure 2–16). Other equations of state that have been proposed describe the behavior of real gases better than the ideal gas law. The best known of these equations is the one introduced, in 1873, by van der Waals. Van der Waals assumed that, even for a real gas, there is an ideal pressure, P^*, and ideal volume, V^*, that would apply to the ideal expression $P^*V^* = nRT$; but because of the imperfections of the gas, these were not the measured pressure, P, and measured volume, V. The ideal volume, he reasoned, should be less than the measured volume because the molecules have a finite volume instead of being point masses, and the portion of the container's volume that is occupied by other molecules is unavailable to any given molecule. Therefore, the "ideal" volume should be less

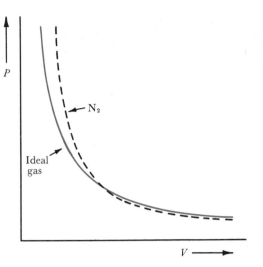

Figure 2–16. Pressure–volume curves for nitrogen and an ideal gas at constant temperature. At low pressures, the molar volume of N_2 is less than for an ideal gas because of intermolecular attraction. At high pressures, the nonzero molecular volume of N_2 makes the volume greater than ideal.

than the measured volume by a constant, b, that is related to molecular size by $V^* = V - b$.

Moreover, a gas molecule subject to attractions from other gas molecules strikes the walls with less force than if these attractions were absent. For as the molecule approaches the wall, there are more gas molecules behind it in the bulk of the gas than there are between it and the wall (Figure 2–17). The number of collisions with the wall in a given time is proportional to the density of the gas, and each collision is softened by a back-attraction factor, which itself is proportional to the density of molecules doing the attracting. Therefore, the correction factor to P is proportional to the square of the gas density, or inversely proportional to the square of the volume: $P^* = P + a/V^2$.

Figure 2–17. Reduction of pressure of a real gas as a result of intermolecular attractions. (a) Gas at low density. (b) Gas at high density. A molecule M in a high-density gas hits the wall with a smaller impact than in a low-density gas because the attractions of its nearest neighbors reduce the force of the impact.

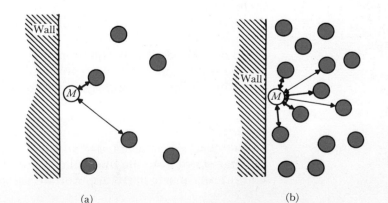

(a)   (b)

Table 2–2.   Measures of Molecular Size Obtained from the Kinetic Theory

	Van der Waals constants		Spherical molecular diameters, d, in Å		
Gas	a (liter² atm mole⁻²)	b (cm³ mole⁻¹)	From van der Waals[a]	From gas viscosity	From density of liquid or solid[b]
Hg	8.09	17.0	2.38	3.60	3.26
He	0.0341	23.70	2.48	2.00	—
H₂	0.2444	26.61	2.76	2.18	—
H₂O	5.464	30.49	2.88	2.72	3.48
O₂	1.360	31.83	2.90	2.96	3.75
N₂	1.390	39.12	3.14	3.16	4.00
CO₂	3.592	42.67	3.24	4.60	4.54

[a] This is a bad approximation for all except Hg and He.

[b] Assuming that the molecules are spheres, which, when most closely packed, fill 74% of the space available. If M is the molecular weight and D is the density, the molecular volume is

$$V_m = \frac{\pi}{6} d^3 = 0.74 \frac{M}{ND}$$

The complete van der Waals equation *for 1 mole* of gas is

$$\left(P + \frac{a}{V^2}\right)(V - b) = RT \tag{2–33}$$

The constants a and b are chosen empirically to provide the best relationship of the equation to the actual PVT behavior of a gas. Even so, the molecular size calculated from this purely experimental b agrees well with the ones obtained by other means (Table 2–2), and gives us confidence that we have the right explanation for deviations from ideality.

Experimentally obtained values of a and b are given in Table 2–2 for several gases, along with several calculations of molecular diameters. We might suppose that the constant b is simply the excluded volume per mole (as in Figure 2–18): $b = 8NV_m = \frac{4}{3}\pi Nd^3$. However, collision is a two-molecule process and this calculation overcounts the excluded volume by a factor of two. The molecular diameters in Table 2–2 were obtained from the b values by having b equal to $4NV_m$, in which $V_m = (\pi d^3/6)$ is the volume of one molecule.

The van der Waals equation is applicable over a much wider range of temperatures and pressures than is the ideal gas law; it is even compatible with the condensation of a gas to a liquid.

2–9 SUMMARY

In this chapter we have tested the idea of atoms and molecules by examining to what extent a simple molecular theory can explain the properties of gases. Gases are a particularly appropriate subject at this stage of our discussion,

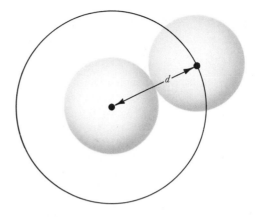

Figure 2–18. The center of no other molecule can come closer than a molecular diameter to the center of a given molecule. The volume around each molecule from which other molecules are excluded then is $\frac{4}{3}\pi d^3$, or eight times the molecular volume of $\frac{4}{3}\pi (d/2)^3$.

because their molecules are so far apart that the forces between molecules are minimal, and to a first approximation can be neglected altogether.

We have learned how the observed behavior of gases leads to a simple *equation of state*, the ideal gas law $PV = RT$ for 1 mole of gas. We also have discovered that an extremely simple model of gases (which states that gases are composed of molecules moving rapidly in straight lines and interacting with one another only when they collide) is sufficient to account for many measurable properties of gases. We have observed that temperature is a measure of molecular motion, in liquids and solids as well as in gases. The irreversibility of the conversion of macroscopic motion to molecular motion has been mentioned as a prelude to what we shall encounter later as the second law of thermodynamics. Finally, we have made reasonable improvements on the ideal gas law to extend the range of pressure and temperature over which it applies to real gas behavior.

2–10   POSTSCRIPT TO GAS LAWS AND ATOMIC THEORY

When the dust has settled after a new discovery, it is all too easy to forget how much controversy and effort went into its development. Thomas Thomson (1773–1852) was Regius Professor of Chemistry at the University of Glasgow, and was the man to whom Dalton turned for help in publicizing his new theory of atoms. In 1830, Thomson published his *History of Chemistry*, which is particularly interesting because many of the participants in the atomic revolution in chemistry were alive, active, and friends of Thomson. In the last chapter of his *History*, Thomson describes the circumstances of the birth of the atomic theory (comments in square brackets are modern):

> In the year 1804, on the 26th of August, I spent a day or two at Manchester, and was much with Mr. Dalton. At that time he explained to me his notions respecting the composition of bodies. I wrote down at the time the opinions which he offered . . . [A brief account of the atomic theory followed.]

Mr. Dalton informed me that the atomic theory first occurred to him during his investigations of olefiant gas [acetylene, C_2H_2] and carburetted hydrogen gas [ethylene, C_2H_4], at that time imperfectly understood, and the constitution of which was first fully developed by Mr. Dalton himself. It was obvious from the experiments which he made upon them, that the constituents of both were carbon and hydrogen, and nothing else. He found further, that if we reckon the carbon in each the same, then carburetted hydrogen gas contains exactly twice as much hydrogen as olefiant gas does. This determined him to state the ratios of these constituents in numbers, and to consider the olefiant gas as a compound of one atom of carbon and one atom of hydrogen; and carburetted hydrogen of one atom of carbon and two atoms of hydrogen. The idea thus conceived was applied to carbonic oxide, water, ammonia, etc.; and numbers representing the atomic weights of oxygen, azote, etc., deduced from the best analytical experiments which chemistry then possessed. Let not the reader suppose that this was an easy task. Chemistry at that time did not possess a single analysis which could be considered as even approaching to accuracy. . . .

In the third edition of my *System of Chemistry*, published in 1807, I introduced a short sketch of Mr. Dalton's theory, and thus made it known to the chemical world. . . . These facts gradually drew the attention of chemists to Mr. Dalton's views. There were, however, some of our most eminent chemists who were very hostile to the atomic theory. The most conspicuous of these was Sir Humphry Davy. In the autumn of 1807 I had a long conversation with him at the Royal Institution, but could not convince him that there was any truth in the hypothesis. A few days after, I dined with him at the Royal Society Club, at the Crown and Anchor, in the Strand. Dr. Wollaston was present at the dinner. After dinner every member of the club left the tavern, except Dr. Wollaston, Mr. Davy, and myself, who stayed behind and had tea. We sat about an hour and a half together, and our whole conversation was about the atomic theory. Dr. Wollaston was a convert as well as myself; and we tried to convince Davy of the inaccuracy of his opinions; but, so far from being convinced, he went away, if possible, more prejudiced against it than ever. Soon after, Davy met Mr. David Gilbert, the late distinguished president of the Royal Society; and he amused him with a caricature description of the atomic theory, which he exhibited in so ridiculous a light, that Mr. Gilbert was astonished how any man of sense or science could be taken in with such a tissue of absurdities. . . . [Wollaston finally convinced Gilbert after a long recital of the chemical evidence.]

Mr. Gilbert went away a convert to the truth of the atomic theory; and he had the merit of convincing Davy that his former opinions on the subject were wrong. What arguments he employed I do not know; but they must have been convincing ones, for Davy ever after became a strenuous supporter of the atomic theory. The only alteration which he made was to substitute *proportion* for Dalton's word, *atom*. Dr. Wollaston substituted for it the term *equivalent*. The object of these substitutions was to avoid all theoretical annunciations. But, in fact, these terms, *proportion, equivalent,* are neither of them so convenient as the term *atom;* and unless we adopt the hypothesis with which Dalton set out, namely, that the ultimate particles of bodies are *atoms* incapable of further division, and that chemical combination consists in the union of these atoms with each other, we lose all the new light which the atomic theory throws upon chemistry, and bring our notions back to the obscurity of the days of Bergman and of Berthollet.

SUGGESTED READING

J. Hildebrand, *An Introduction to Molecular Kinetic Theory*, Reinhold, New York, 1963.

T. L. Hill, *Lectures on Matter and Equilibrium*, W. A. Benjamin, Menlo Park, Calif., 1966. Written at the honors freshman level. The first four chapters, on states of matter, gases, and intermolecular forces, are particularly useful.

W. Kauzmann, *Kinetic Theory of Gases*, W. A. Benjamin, Menlo Park, Calif., 1966. Thorough, clear. Chapters 1 and 2, on equations of state of gases, are especially relevant. Chapter 4 continues the discussion of the distribution of molecular velocities, but requires calculus.

QUESTIONS

1 Why should gases obey simpler laws than liquids or solids?

2 Early hydraulic engineers found that no suction pump could lift water more than approximately 34 feet. Can you explain this phenomenon from the information in this chapter?

3 How did Boyle design his experiment to test the "spring of the air" theory?

4 Why is a plot of experimental data that produces a straight line useful or desirable?

5 How is an absolute scale of temperature defined in terms of gas behavior?

6 Under what conditions does Boyle's law apply? When is Charles' law applicable? How are these laws derived from the complete ideal gas law?

7 What does STP signify, and why is it useful?

8 What molecular explanation can you give for the deviation of real gases from ideal gas behavior? Under what conditions will real gases most resemble ideal behavior?

9 What experimental evidence is there that each of the three assumptions of the kinetic molecular theory of gases is valid?

10 Why can we say that the product PV for an ideal gas is proportional to the kinetic energy, E_k?

11 Why can we say that the temperature is proportional to the square of the speed of the molecules (actually to the mean-square speed)?

12 If the molecules in a liter of hydrogen gas and a liter of oxygen gas are moving with the same mean-square speed, which gas is hotter?

13 What fraction of a typical gas is occupied by the volume of the molecules of which it is composed? What direct physical measurements can tell you this?

14 Why is the gas volume of 22.414 liters significant?

15 How does the speed of sound in air at sea level compare with the root-mean-square speed of the molecules in the air?

16 Which would you expect to be greater, the average speed or the root-mean-square speed? Can you rationalize this from the definitions of the two speeds?

17 What does Dalton's law of partial pressures indicate about the behavior of gases in a mixture?

18 Why does the compressibility coefficient of a real gas deviate above and below 1.00 as it does?

19 How is a measure of molecular size obtained from the van der Waals equation?

PROBLEMS

1 A sample of gas at 25°C occupies 2.34 liters. What will be its volume at 300°C?

2 A gas at initial pressure of 700 torr is allowed to expand until the pressure is 150 torr. What is the ratio of the final to the initial volume?

3 An ideal gas occupies 76 liters at 1.0 atm. What pressure will reduce the volume to 10 liters?

4 At STP, 10.3 g of a gas occupy 453 cubic inches. What is the volume of this sample at 1 atm and 100°C?

5 The temperature of a 0.0100-g sample of chlorine gas (Cl_2) in a 10-ml sealed glass container is raised in an oven from 20°C to 250°C. What is the initial pressure at 20°C? What is the pressure at 250°C?

6 How many molecules of an ideal gas are in 1.000 ml if the temperature is $-80°C$ and the pressure is 10^{-3} torr?

7 What pressure will be exerted by 5.0×10^{13} molecules of an ideal gas in 1.000 ml at 0°C?

8 An amount of 4.4 g of CO_2 is confined in a 2-liter flask at 27°C. What is the pressure inside the flask?

9 One liter each of O_2, N_2, and H_2, all at 1 atm, are forced into a single 2-liter container. What is the resulting pressure?

10 What is the density of XeF_6 gas at STP in g liter^{-1}?

11 One atmosphere of pressure will push a column of mercury to a height of 760 mm when the cross-sectional area of the column is 1.00 cm². What would be the height of the column if its cross-sectional area were 0.500 cm²? (The density of mercury is 13.59 g cm^{-3} and the acceleration due to gravity is 980.7 cm sec^{-2}.)

12 The density of a gas at STP is 1.62 g liter^{-1}. What will be its density at 302°K and 740 torr?

13 A total of 0.750 g of a gas occupies 4.87 liters at 742 torr and 20°C. What is the molecular weight of the gas? What might the gas be?

14 An amount of 1.12 liters of a gas weighs 0.40 g when measured at 0°C and 0.50 atm. The gas is 25% hydrogen and 75% carbon by weight. What is the molecular weight of the gas? What is its empirical formula?

15 A 250-ml sample of a compound with the empirical formula CH_2 weighs 0.395 g at 700 torr and 27°C. What are the molecular weight and molecular formula of the compound?

16 A sample of 0.524 g of a compound fills a volume of 129 ml at 25°C and 753 torr. Chemical analysis shows that it is 23.5% carbon, 2.0% hydrogen, and 74.5% fluorine by weight. What is its molecular formula?

17 A 0.490-g sample of a compound is heated through the successive evolution of the following: 280 ml of water vapor at 182°C and 1 atm, 112 ml of ammonia vapor at 273°C and 1 atm, 0.0225 g of water at 400°C, and 0.200 g of SO_3 at 700°C. At the end of the heating, 0.090 g of FeO remains. Deduce the empirical formula for the compound.

18 A gas mixture contains half argon and half helium, by weight, with a total pressure of 840 torr. What is the partial pressure of each gas in the mixture?

19 A mixture of gases contains 0.5 mole of oxygen, 0.1 mole of hydrogen, and 0.8 mole of nitrogen. The total pressure is 0.8 atm. What is the partial pressure of each gas?

20 The concentration of carbon monoxide (CO) in cigarette smoke is 20,000 parts per million (ppm) by volume. Calculate the partial pressure of CO in 1 liter of cigarette smoke, which exerts a total pressure of 760 torr.

21 A mixture of 3.86 g of CCl_4 and 1.92 g of C_2H_4 at 450°C exerts how many atmospheres of pressure inside a 30-ml metal bomb? How much pressure is contributed by the C_2H_4?

22 One liter of dry air at 760 torr and 86°C is placed in contact with 1.00 ml of liquid water at the same temperature. The volume of the gas phase remains constant throughout the experiment. The vapor pressure of water at this temperature is 451 torr, and its density is 0.97 g ml^{-1}. When equilibrium has been established: (a) What is the partial pressure of air in the vessel? (b) What is the partial pressure of water vapor in the vessel? (c) What is the total pressure in the vessel? (d) How many moles of water will have evaporated? (e) What volume of liquid water, if any, will remain?

23 Consider a molecule moving at the rate of 4000 cm sec^{-1} in a cubical box of length 12.0 cm (see Figure 2–10). How many collisions will the molecule undergo with one wall in one second?

24 One gram of methane, CH_4, was burned to produce CO_2 (gas) and H_2O (liquid). At 25°C, the pressure exerted by the products was 750 torr. The vapor pressure of water at 25°C is 23.8 torr. Calculate the volume of dry CO_2 produced in the reaction.

25 At 25°C the root-mean-square speed of a collection of oxygen molecules is 4.82×10^4 cm sec^{-1}. To what tempera-ture must the sample be raised to increase the root-mean-square speed of this sample to 4.82×10^5 cm sec^{-1} while maintaining a constant volume? By what factor would the pressure be increased as a result of this temperature change?

26 Calculate the root-mean-square speed of a collection of oxygen molecules at 1°K. At what temperature would the root-mean-square speed drop to 1 cm sec^{-1}?

27 A sample of an unknown gas is shown by analysis to contain only sulfur and oxygen. The same gas requires 28.3 sec to effuse through an orifice into a vacuum, whereas an identical number of O_2 molecules pass through the same orifice in 20.0 sec. Determine the molecular weight and formula of the gas.

28 The average speed of O_2 molecules at STP is 1030 miles per hour. What is the average speed of H_2 molecules under the same conditions? Which gas would be more likely to escape from the gravitational attraction of the earth?

29 Argon gas is 10 times as dense as helium gas at the same temperature and pressure. Which gas diffuses faster? How much faster?

30 What is the average kinetic energy of a methane (CH_4) molecule at STP? What is its average kinetic energy at 100°C? What is its root-mean-square speed at 100°C?

31 Demonstrate that the van der Waals equation of state approaches the ideal gas law as the pressure decreases.

32 Two containers of equal volume are filled with hydrogen gas, one at 0°C and 1.00 atm, and the other at 300°C and 5 atm. Compare quantitatively the following properties in the two boxes: (a) number of molecules in each box, (b) average speed of the molecules, (c) molecular collisions per unit area of wall per second, (d) momentum exchange per wall impact, and (e) average kinetic energy per molecule.

PART TWO

CLASSICAL IDEAS OF STRUCTURE AND BONDING

More than one of the histories of science and philosophies of science recount the development of chemistry until the era of Lavoisier and Dalton and then leave it, ostensibly for more alluring subjects. This is a grave blunder; for real chemistry begins, not ends, with these men. The "Daltonian revolution" lasted more than a century, and during this time chemistry evolved from a respectable hobby for schoolmasters, clergymen, and aristocrats to the basis for industries that would change the character of nations. W. H. Perkin's successful synthesis, in 1857, of the first artificial aniline dye led ultimately to a half-century of virtual monopoly, in Germany, of the extremely valuable organic chemical industry based on coal-tar products. This industry produced not only dyes, but resins, polyester plastics, solvents, saccharine, and TNT. Ostwald's process for preparing oxides of nitrogen from ammonia, as a substitute for Chilean nitrates in explosives, prevented the Allied blockade from bringing World War I to a rapid close. (There may be mixed feelings about this, as there are about defoliants and other more recent chemical successes.)

In the late 1800's, the advent of petroleum and petroleum products brought large-scale whaling to an end and changed the economy of the New England seacoast. In the United States after World War I, petrochemicals had the same effect as did coal-tar chemicals in Germany a generation earlier: They revolutionized many aspects of life patterns by the innovations of synthetic rubbers, plastics, textiles, detergents, and agricultural chemicals. The Solvay process for making sodium carbonate, the Frasch process for obtaining sulfur for sulfuric acid ("the pig iron of the chemical industry"), the Haber process for nitrogen fixation, Leo Baekland and his "Bakelite," W. H. Carothers and Nylon: The list of achievements of chemistry in the century and a quarter after Dalton is limitless.

All of these achievements were within the framework of a chemistry that was still primarily a science of observation and classification. The chemist was the naturalist *par excellence*. He observed, measured, and classified the elements in that magnificent summary known as the periodic table. But he had no idea why it existed. He accumulated a great body of knowledge about organic chemical reactions. But *prediction* implied less an appeal to fundamental principles (there were very few) than a guess about the outcome of one reaction based on what happened in several hundred known reactions. The other great monument of this period, along with the periodic table, is Beilstein's *Handbuch der Organischen Chemie*, a reference work that eventually swelled to 64 volumes.

The second revolution in chemistry was the quantum revolution, which is still going on. The application of quantum mechanics to chemistry has shown us why the periodic table is arranged as it is. It also has revealed why atoms bond as they do, and with the geometries that they do, and it is beginning to explain why substances react at the rates with which they do. Part 3 is devoted to the quantum revolution.

However, the earlier ideas were not only easier to arrive at, but they were, and still are, easier to understand upon first encounter. Before we can appreciate what quantum mechanics did for chemistry, therefore, we have to know something about what chemistry is, and what could and could not be explained with the simpler theories. Nothing from the classical era is wrong; it is just incomplete. In Part 2, then, we shall see how the properties of elements were systematized, how elementary ideas of structure, bonding, reactions, and equilibrium arose, and how chemistry became quantitative instead of descriptive.

I came to my professor, Clive, whom I admired very much, and I said, "I have a new theory of electrical conductivity as a cause of chemical reactions." He said, "This is very interesting," and then he said "Good-bye." He explained to me later that he knew very well that there are so many different theories formed, and that they are almost all certain to be wrong, for after a short time they disappeared; and therefore by using the statistical manner of forming his ideas, he concluded that my theory would not exist long.
Svante Arrhenius

3 MATTER WITH A CHARGE

The discovery that atoms of elements have rather specific combining capacities (Chapter 1) for uniting with other atoms to form compounds led immediately to speculation on the nature of the forces holding atoms together. Of the forces of attraction known in the nineteenth century—gravity, electricity, and magnetism—only electrical interaction seemed suitable for interatomic forces. Gravity is far too weak, and the north and south poles of a magnet cannot be separated in the way that positively and negatively charged particles can. It was logical to assume that combining capacity was somehow electrical in nature. This is why Avogadro's proposal that two *like* atoms could attract one another in a diatomic molecule was rejected for so long. In support of the electrical nature of chemical bonding, chemists could point out that many substances (salts) conduct electricity when molten, that these and other substances conduct in an aqueous solution, and that when electric current is passed through these melts or solutions, *chemical changes occur.*

It was apparent even to Dalton and the early atomists that it would be difficult to explain a complex molecular structure in purely electrostatic terms. Even so, chemists have found repeatedly that an electrostatic description of the parts of a chemical system gives an excellent first approximation to the true picture. We must recognize that such a description is an oversimplification; but if we keep this in mind, we can gain considerable insight at the cost of very little labor. In this chapter we shall describe the first concrete evidence that some atoms and molecules can acquire electric charge, and that species with opposite charges often interact to form compounds.

In this chapter we shall be looking at substances whose individual particles are dispersed throughout liquid water. Such preparations are called *solutions*. The dispersed substance is the *solute*, and water is the *solvent*. If some other liquid is used instead of water, we have a *nonaqueous* solution. We normally expect that the individual particles of solute are its molecules, but the ideas in this chapter developed because some solutions show more solute particles than there are solute molecules.

Two common concentration units are used for solutions: *molarity* and *molality*. The molarity of a solution is the number of moles of solute *per liter of solution*. The molality is the number of moles of solute *per kilogram of solvent* (not solution). We will compare these units in more detail in Chapter 4.

3–1 ELECTROLYSIS

In the middle of the nineteenth century, Michael Faraday (1791–1867) performed some classic experiments on a process that was named *electrolysis* (from "electro" and "lysis," or "breaking up with electricity"). If two chemically inert conductors of electricity such as rods of platinum or graphite are immersed in certain types of solutions or melts, and if a voltage or electrical potential is applied across the rods, current will flow and chemical reactions will occur where the rods meet the solution (Figure 3–1). Gases such as O_2 or Cl_2 may bubble from the surface of one rod; H_2 may bubble off or metals may plate on the other. Salts, acids, and bases exhibit this behavior in aqueous solution, as do molten salts. Such compounds were named *electrolytes*. Conductivity of electricity and chemical reactions at the *electrodes* (the conducting rods) always occur together; if there is no chemical change, the solution or melt will not conduct.

Faraday noted that elementary metals and hydrogen usually were produced at the electrode connected to the negative pole of a battery or other potential source; he called this pole the *cathode*. Oxygen, chlorine, and other nonmetals or simple compounds of nonmetals were produced at the other terminal, named the *anode*. (The reasons for the names will become clearer when we see what is happening within the electrolyte.) Table 3–1 lists the observed products of electrolysis in a few dilute solutions.

Table 3–1.   *Products of Electrolysis*

Electrolyte	Cathode product	Anode product
Sulfuric acid (H_2SO_4) in H_2O	H_2	O_2
Sodium sulfate (Na_2SO_4) in H_2O	H_2	O_2
Sodium chloride (NaCl) in H_2O	H_2	Cl_2
Potassium iodide (KI) in H_2O	H_2	I_2
Copper sulfate ($CuSO_4$) in H_2O	Cu	O_2
Silver nitrate ($AgNO_3$) in H_2O	Ag	O_2
Mercuric nitrate [$Hg(NO_3)_2$] in H_2O	Hg	O_2
Lead nitrate [$Pb(NO_3)_2$] in H_2O	Pb	O_2 and some PbO_2
Molten lye (NaOH); not in H_2O	Na	O_2

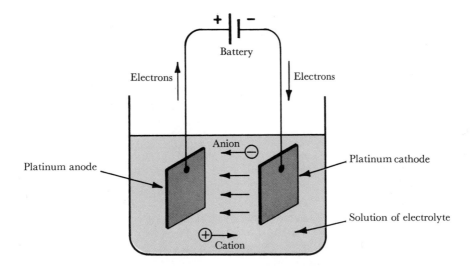

Figure 3–1. Electrolysis. For current to be carried, the fluid must contain charged atoms or molecules (ions). A substance that gives ions in solution, and thus is capable of carrying current, is called an electrolyte. The solution also must contain species capable of being reduced by electrons at the cathode and species capable of being oxidized by losing electrons at the anode. Suppose that the electrolyte solution contains $CuCl_2$, which dissociates to give Cu^{2+} and Cl^- ions. On electrolysis, Cu^{2+} ions move toward the cathode, where they accept electrons ($Cu^{2+} + 2e^- \rightarrow Cu$). Copper metal plates on the platinum cathode. The Cl^- ions move toward the anode, where they lose electrons ($2Cl^- \rightarrow Cl_2 + 2e^-$) and produce chlorine gas.

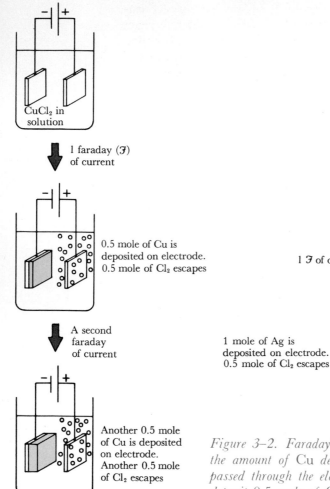

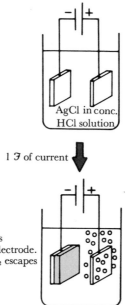

Figure 3–2. Faraday's laws of electrolysis. In the CuCl₂ *experiment, the amount of* Cu *deposited is proportional to the amount of current passed through the electrolysis cell. One faraday ($\mathscr{F}$) of current will deposit 0.5 mole of* Cu *or 1 mole of* Ag *since the combining weight of* Ag *is equal to its atomic weight, whereas the combining weight of* Cu *is half its atomic weight.*

Faraday's Laws of Electrolysis

For electrode processes in which elements are the only products, Faraday observed that:

1) The weight of an element deposited is proportional to the amount of electricity passed through the solution.

2) The weights of various elements deposited by a given quantity of electricity are *proportional to their combining weights.*

Both of these observations are illustrated in Figure 3–2. The second of Faraday's laws implies that there is a given quantity of electricity that will react
with and liberate *one combining weight* of any element. The quantity is called
1 faraday ($\mathscr{F}$) and is equal to 96,500 coulombs. (Physicists define units of
charge from first principles in terms of the forces exerted between bodies with
equal but opposite charges. But the practical international definition of a
coulomb is that amount of electricity which will deposit 0.0011180 g of
silver from a solution of silver nitrate. Can you verify the value of a faraday,
in coulombs, from this information?)

Since the days of Benjamin Franklin, electric currents have been known
to consist of moving charges. Faraday explained his results by suggesting
that molecules of the dissolved or molten electrolytes separate into charged
particles (positive and negative). Furthermore, under the influence of the
applied voltage, they flow toward the electrodes and undergo chemical
change.

3–2 ARRHENIUS' THEORY OF IONIZATION

In the early 1880's, a young Swedish graduate student of physics, Svante
Arrhenius (1859–1927), began a study of electrolytic solutions. He postulated
that *electrolytes on dissolving in water are dispersed by the water, not merely into
separate molecules, but into separate ions of positive and negative charges.* Since the
electrolytes have no *net* charge before they are dissolved, the total number
of positive and negative charges must be equal. This explains the overall
electrical neutrality of the solution. Yet the separation of positive and negative ions gives them a degree of independence. Thus, when an electrical
potential is imposed on such a solution, the ions are free to move—positive
ions to the cathode and negative ions to the anode. At the electrodes, reactions that maintain the charge balance in the solution take place: Charged
ions are changed into neutral atoms or molecules, or ions are produced from
atoms or molecules of the electrode or solvent.

Now we know that electric current is carried by *electrons*, negatively
charged particles whose properties we shall be studying in Section 3–6. The
negative terminal of a battery is the terminal from which electrons flow out;
the *cathode* of an electrolytic cell is the terminal from which these electrons
flow into the solution by means of reactions such as

$$Na^+ + e^- \rightarrow Na$$
$$Cu^{2+} + 2e^- \rightarrow Cu$$
$$2H^+ + 2e^- \rightarrow H_2$$

Such positive ions flow toward the cathode under the influence of the voltage,
and are *reduced* by combining with electrons. At the other terminal, the *anode*,
negatively charged ions give up electrons and are *oxidized*:

$$2Cl^- \rightarrow Cl_2 + 2e^-$$

The material of which the anode is composed also may be oxidized. For example, if the anode were copper metal, the anode oxidation equation would be

$$Cu \rightarrow Cu^{2+} + 2e^-$$

Anodes made of platinum or graphite are not oxidized in this way. Positive ions are called *cations* and negative ions are called *anions*, after the terminals to which they migrate.[1]

Opposition to the Theory of Ionization

Arrhenius had to overcome objections to his theory that were severe enough to jeopardize his chances of obtaining his doctorate. The nature of these objections can be illustrated by considering what the theory implies about dissolved salt (NaCl) in light of the chemical properties of sodium (Na), chlorine (Cl_2), and salt itself. Arrhenius' theory suggests that salt ionizes in water to form separate and nominally independent Na^+ and Cl^- ions. His critics refused to distinguish free Na atoms, which react explosively with water, from Na^+ ions. Similarly, free chlorine (Cl_2) is a deadly poison, whereas dissolved salt is ingested freely with food and is known to aid digestion. Arrhenius asserted vainly that the properties of atoms and ions were quite different. However, his protest was ineffective until the existence of ions could be demonstrated in a manner convincing to all. (In the early 1960's a chemically confused politician in California protested against fluoridation of municipal water supplies by warning against the dangers to health of adding "fluorides, which as any chemist can tell you, is a poisonous gas." His blunder, although belated, was in an illustrious tradition.)

The Charge of an Ion

Arrhenius extended Faraday's experiments to many electrolytes and demonstrated that 1 $\mathscr{F}$ of charge always accompanies one combining weight of element reacted. Therefore, the simplest assumption is that 1 $\mathscr{F}$ of charge contains as many charged units (electrons) as there are atoms in a gram-atom: Avogadro's number, N. If we knew the charge on an electron, we finally could calculate Avogadro's number. Conversely, if we knew Avogadro's number, we would know the charge on the electron. By this assumption, a calcium ion has an atomic weight of twice its combining weight because *two* electrons are used per ion reduced, or 2 $\mathscr{F}$ of charge per mole of calcium:

$$Ca^{2+} + 2e^- \rightarrow Ca$$

---

[1] These terms are from the Greek prefixes *cata*-, meaning "away," and *ana*-, meaning "back." Xenophon's *Anabasis* is translated as "The Long March Back," and an unsuccessful catapult could be called an anapult. At the cathode, electrons flow into the solution, and at the anode they flow out of the solution into the electrical circuit.

A chloride ion loses only one electron in forming chlorine gas, so only 1 $\mathscr{F}$ of charge results when a mole of chloride ion is oxidized. Thus, the combining weight is the same as the atomic weight:

$$Cl^- \rightarrow \tfrac{1}{2}Cl_2 + e^-$$

Table 3–2 summarizes electrolysis data for several elements. The number of faradays of charge required to liberate 1 mole of atoms of the element gives the charge on the ion. The sign depends on whether the ion releases electrons at the anode or accepts them at the cathode.

Table 3–2. *Electrolysis Data*

Product of electrolysis	Electrode	Faradays per mole of atoms deposited	Ion in solution
Silver (Ag)	Cathode	1[a]	Ag^+
Chlorine (Cl_2)	Anode	1	Cl^-
Copper (Cu)	Cathode	2	Cu^{2+}
Hydrogen (H_2)	Cathode	1	H^+
Iodine (I_2)	Anode	1	I^-
Oxygen (O_2)[b]	Anode	2	O^{2-}
Zinc (Zn)	Cathode	2	Zn^{2+}

[a] For example, electrolysis of silver nitrate solution for 1 hour by using a current of 0.5 ampere deposits 2.015 g of silver; $2.015/107.9 = 0.0187$ mole (or g-atom) of silver.

$$\frac{(0.5 \text{ coulomb sec}^{-1}) \times 3600 \text{ sec}}{96,500 \text{ coulombs } \mathscr{F}^{-1}} = 0.0187 \ \mathscr{F}$$

[b] Actually, oxygen (O_2) is produced by a complicated electrode reaction. The species O^{2-} can exist in molten oxides, but in water O^{2-} becomes $2OH^-$ by reaction with a water molecule.

Once the charges on a few ions have been determined, it is simple to obtain, from the formulas of electrolytes, the charges on other ions that do not undergo simple electrolysis reactions. Hence, if we know that Ag^+ is a silver ion, the formula AgF means that the fluoride ion is F^-. If Cl^- is a chloride ion, the formula NaCl requires the sodium ion to be Na^+. The formula $AlCl_3$ similarly requires the aluminum ion to be Al^{3+}, and so on.

In many cases, charged groups of atoms behave as single ionic units. For example, in copper sulfate, zinc sulfate, and many other similar compounds, the sulfate group, SO_4^{2-}, behaves like a simple anion. The electrolyte does not dissociate beyond the state of cation and SO_4^{2-} anion; hence, the sulfate group migrates as a unit. Similarly, the ammonium ion, NH_4^+, behaves in many ways like a simple cation such as Na^+.

Tables 3–3 and 3–4 list several ions. The charge on a simple ion such as Na^+, Al^{3+}, or O^{2-} is its *oxidation state* or *oxidation number*, which is the number of electrons that must be added or removed to reduce or oxidize the ion to a

Table 3-3. Partial List of Ions of Elements

Li^+	Be^{2+}	Al^{3+}	Sn^{4+}	N^{3-}	O^{2-}	F^-
Na^+	Mg^{2+}	Sc^{3+}	Mn^{4+}	P^{3-}	S^{2-}	Cl^-
K^+	Ca^{2+}	Y^{3+}	U^{4+}		Se^{2-}	Br^-
Rb^+	Sr^{2+}	Lanthanide ions	Th^{4+}			I^-
Cs^+	Ba^{2+}	Ga^{3+}	Ce^{4+}			
Cu^+	Mn^{2+}	In^{3+}				
Ag^+	Fe^{2+}	Tl^{3+}				
Tl^+	Co^{2+}	Sb^{3+}				
	Ni^{2+}	Bi^{3+}				
	Cu^{2+}	V^{3+}				
	Zn^{2+}	Cr^{3+}				
	Cd^{2+}	Fe^{3+}				
	Hg^{2+}	Co^{3+}				
	Sn^{2+}					
	Pb^{2+}					

neutral atom

$$Co^{3+} + 3e^- \rightarrow Co \qquad \text{(Oxidation number } +3)$$
$$S^{2-} \rightarrow S + 2e^- \qquad \text{(Oxidation number } -2)$$

This oxidation number for such simple ions is the combining capacity, discussed in Chapter 1, with the addition of a sign such that in a compound with no net charge, the sum of all the oxidation numbers of the ions is zero. (The concept of oxidation number is applicable more generally than to such simple ionic compounds; we shall return to it in Chapter 7.) Positive ions are given the name of the element with the oxidation number indicated by Roman numerals in parentheses [e.g., Fe^{3+} is iron(III) and Fe^{2+} is iron(II)]. An older nomenclature system is still used to name cations, in which the Latin name of an element is modified with a suffix to indicate the oxidation number. If an element forms two positive ions, the name of the one of higher charge ends in "ic," and the one of lower charge, in "ous." Thus Fe^{3+} is called ferric ion; Fe^{2+} is the ferrous ion. The name of a simple anion is derived from the element's name with the suffix "ide." Examples are nitride (N^{3-}), oxide (O^{2-}), and fluoride (F^-). The suffixes "ate" and "ite" are used to distinguish higher and lower oxidation numbers of elements combined with oxygen in anions. For example, nitrate (NO_3^-) and nitrite (NO_2^-); sulfate (SO_4^{2-}) and sulfite (SO_3^{2-}); arsenate (AsO_4^{3-}) and arsenite (AsO_3^{3-}).

Before you go on. You may want to review the nomenclature and concepts related to the electrolysis of molten salts in Review 3 of Lassila's *Programed Reviews of Chemical Principles.*

Table 3–4. Some Common Ions

NH_4^+, Ammonium	$Cu(NH_3)_4^{2+}$, Tetraammine-copper(II)	$Co(NH_3)_6^{3+}$, Hexaammine-cobalt(III)	$Fe(CN)_6^{4-}$, Hexacyano-ferrate(II) (ferrocyanide)	PO_4^{3-}, Phosphate	CO_3^{2-}, Carbonate	OH^-, Hydroxide
$Ag(NH_3)_2^+$, Diammine-silver(I)	UO_2^{2+}, Uranyl			AsO_4^{3-}, Arsenate	SO_4^{2-}, Sulfate	NO_3^-, Nitrate
$(CH_3)_4N^+$, Tetramethyl-ammonium	$Ni(NH_3)_6^{2+}$ Hexaammine-nickel(II)			AsO_3^{3-}, Arsenite	SO_3^{2-}, Sulfite	NO_2^-, Nitrite
NO^+, Nitrosyl				BO_3^{3-}, Borate	CrO_4^{2-}, Chromate	BF_4^-, Fluoroborate
NO_2^+, Nitryl				$Fe(CN)_6^{3-}$, Hexacyano-ferrate(III) (ferricyanide)	$Cr_2O_7^{2-}$, Dichromate	CN^-, Cyanide
					$C_2O_4^{2-}$, Oxalate	ClO^-, Hypochlorite
					$S_2O_3^{2-}$, Thiosulfate	ClO_2^-, Chlorite
					$PtCl_4^{2-}$, Tetrachloro-platinate(II)	ClO_3^-, Chlorate
					$PtCl_6^{2-}$, Chloroplatinate or hexachloro-platinate(IV)	ClO_4^-, Perchlorate
						BrO_3^-, Bromate
						IO_3^-, Iodate
						MnO_4^-, Permanganate
						SCN^-, Thiocyanate
						$C_2H_3O_2^-$, Acetate
						I_3^-, Triiodide

3–3   ELECTRICAL EVIDENCE FOR IONS

The migration of ions can be observed by an experiment such as the one diagramed in Figure 3–3. The Cu^{2+} ion in water has a characteristic blue color that is easy to follow visually. If we make a copper sulfate solution in gelatin (which prevents turbulence, convection, and such irrelevant mixing) with a conducting solution of some other electrolyte connecting it to two electrodes as shown, and if we apply current for several weeks, we will see that the blue color of the Cu^{2+} ion (actually the hydrated copper ion) moves slowly out of the gelatin and into the solution near the cathode. At the same time, the sulfate ion migrates into the solution around the anode, where it can be detected by any of the simple tests for sulfate ion (such as

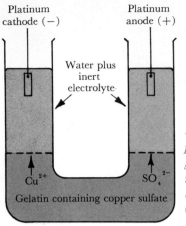

Platinum
cathode (−)

Platinum
anode (+)

Water plus
inert
electrolyte

Cu^{2+} SO_4^{2-}

Gelatin containing copper sulfate

Figure 3–3. Electrolytic migration and deposition of ions. As current is passed through the solution, copper ions (blue in water) move out of the gelatin toward the cathode. Sulfate ions move toward the anode. Upon reaching the cathode, copper ions deposit to produce copper metal while oxygen is formed at the anode; 965 coulombs (0.010 $\mathscr{F}$) produce 0.318 g of copper and 0.080 g of oxygen.

precipitation of $BaSO_4$). The cathode becomes plated with metallic copper. The anode reaction is complex, but yields oxygen gas and liberates enough H^+ ions to balance the charge on the sulfate ions migrating toward it. Thus we have demonstrated that the species to which Arrhenius assigned opposite electric charges do move in opposite directions in an electric field.

3–4   CHEMICAL EVIDENCE FOR IONIZATION

Arrhenius' electrical evidence for ionization failed to convince his professors of the worth of his theory. He therefore turned to the *chemical* behavior of electrolytes. We shall describe several examples of phenomena that he was able to explain with his model of electrolyte solutions.

Colors of Ions in Solution

Many electrolytes that are compounds of metals are colored. For example, $CuSO_4$ is white, $CuCl_2$ is green, and $CuBr_2$ is brown. When dissolved in water, each of these colored salts gives a pale blue solution. Since many sulfates (SO_4^{2-}), chlorides (Cl^-), and bromides (Br^-) produce colorless solutions, Arrhenius suggested that the common blue color observed is due to a single substance, Cu^{2+}. The three copper salts dissociate in solution to yield a different negative ion but the *same* positive ion. Similarly, in water, Co^{2+} is pink, Ni^{2+} is green, Fe^{2+} is colorless, and Mn^{2+} is very pale pink. The Fe^{3+} ion is pale violet in water, although association with Cl^- or OH^- [as in $FeCl^{2+}$, $FeCl_2^+$, and $Fe(OH)(H_2O)_5^{2+}$] can result in a yellow or

Table 3–5.   *Colors of Indicators in Acid and Base Solutions in Water*

Indicator		Base color
Phenolphthalein	Colorless	Pink
Methyl orange	Red	Yellow
Bromthymol blue	Yellow	Blue
Methyl violet	Yellow	Purple

orange solution. Many complex ions have characteristic colors. For example, $PtCl_6{}^{2-}$ is yellow, $Cu(NH_3)_4{}^{2+}$ is dark blue, $I_3{}^-$ is brown, $MnO_4{}^-$ is purple, $Cr_2O_7{}^{2-}$ is orange, and $CrO_4{}^{2-}$ is yellow. In every case, the color of the ion differs from the color of the free element.

Heat of Neutralization of Acids and Bases

Most compounds of hydrogen with nonmetals or with negative ions form electrolytes with the familiar properties of acids. Metal hydroxides (such as NaOH) often act as electrolytes and exhibit the familiar caustic properties of strong alkalis, or bases. Acids, such as HCl, and bases react to neutralize each other and make salts and water. For example,

a) $NaOH + HCl \quad \rightarrow NaCl \ + H_2O$

b) $KOH + HNO_3 \rightarrow KNO_3 + H_2O$

Both of these reactions in aqueous solution yield exactly the same amount of heat per mole of water formed. Furthermore, all acids turn certain organic dyestuffs (indicators) one color, whereas all bases turn them another color (see Table 3–5). Arrhenius stated that all acid solutions contain the hydrogen ion, H^+, and all basic solutions contain the hydroxide ion, OH^-. Thus, indicators exhibit common reactions with H^+ when added to solutions of hydrochloric acid (HCl), sulfuric acid (H_2SO_4), nitric acid (HNO_3), acetic acid (CH_3COOH), phosphoric acid (H_3PO_4), and so forth. In basic solution, indicators react with OH^-, whether added to solutions of NaOH, KOH, NH_4OH, or to $Ca(OH)_2$. Reactions (a) and (b) produce the same heat of reaction per mole of water formed because the reaction is the same in each case, namely,

$H^+ + OH^- \rightarrow H_2O$ (13,600 cal heat liberated)

The other ions do not participate in the reaction.

Precipitation Reactions

If a water-soluble silver salt such as silver nitrate is added to a solution of any chloride that dissolves in water to make a conducting solution, a white precipitate, AgCl, settles:

$$AgNO_3 + NaCl \rightarrow AgCl\downarrow + NaNO_3$$
$$2AgNO_3 + CuCl_2 \rightarrow 2AgCl\downarrow + Cu(NO_3)_2 \quad \text{(blue solution)}$$
$$AgNO_3 + CCl_4 \rightarrow \text{no reaction}$$
$$\underset{\substack{\text{insoluble}\\\text{nonconductor}}}{}$$

(A downward arrow is the conventional sign for a precipitation; an upward arrow indicates the evolution of a gas.)

Such an experiment would suggest that the reaction involves only the chloride anion acting independently of the cation. Arrhenius proposed that the same reaction describes the formation of AgCl independently of the source of *ionic* chloride:

$$Ag^+ + Cl^- \rightarrow AgCl\downarrow$$

According to his theory, the blue color of Cu^{2+} is unaffected because it does not participate in the reaction. The CCl_4 solvent does not dissociate to provide chloride ions, so it does not react with the $AgNO_3$ solution.

In water, HCl reacts with magnesium metal to liberate hydrogen gas, and with silver salts to form silver chloride:

$$Mg + 2H^+ \rightarrow H_2\uparrow + Mg^{2+}$$
$$Ag^+ + Cl^- \rightarrow AgCl\downarrow$$

But in benzene, HCl does not react with either. Because HCl in benzene does not conduct electricity (as it does in water), we deduce that it does not form ions in benzene. Therefore, ionization explains the reactivity of HCl in water and its inertness in benzene.

3–5 PHYSICAL EVIDENCE FOR IONS

Both the electrical and chemical properties of solutions of electrolytes strongly suggested that these substances dissociate into ions in aqueous solutions. But there was additional evidence from the physical properties of solutions of electrolytes.

Abnormal Freezing Point Depression

The addition of a nonvolatile substance to a liquid such as water lowers its freezing point; this is the reason why salt (NaCl or $CaCl_2$) is spread on sidewalks and roads to melt the ice in winter. The *freezing point* of a liquid is the temperature at which a solid and liquid are in equilibrium. In ice, molecules of water continually are striking the ice and being captured, while others are vibrating loose from the ice and entering the solution. The *equilibrium point* is the temperature at which the rates of the two processes are the same; so there is no net increase or decrease in the amount of ice. A foreign substance such as sugar or salt has no effect on the rate at which water molecules

Table 3–6.   Freezing Point Data on Electrolytes[a]

Solute	Molality	$\Delta T_f(°C)$	Mole number, i	Number and kind of ions formed
HCl	0.0100	−0.0360	1.93	2 (H^+, Cl^-)
HNO$_3$	0.0100	−0.0364	1.95	2 (H^+, NO_3^-)
NaOH	0.010	−0.0355	1.90	2 (Na^+, OH^-)
K$_2$SO$_4$	0.010	−0.0501	2.70	3 ($2K^+$, SO_4^{2-})
CaCl$_2$	0.010	−0.0511	2.75	3 (Ca^{2+}, $2Cl^-$)
K$_4$Fe(CN)$_6$	0.0075	−0.0690	4.93	5 [$4K^+$, $Fe(CN)_6^{4-}$]
Co(NH$_3$)$_5$Cl$_3$	0.020	−0.1088	2.93	3 [$Co(NH_3)_5Cl^{2+}$, $2Cl^-$]
Co(NH$_3$)$_6$Cl$_3$	0.010	−0.0643	3.46	4 [$Co(NH_3)_6^{3+}$, $3Cl^-$]
MgSO$_4$	0.010	−0.0308	1.62	2 (Mg^{2+}, SO_4^{2-})
NH$_4$Cl	0.010	−0.0358	1.92	2 (NH_4^+, Cl^-)
CH$_3$COOH	0.010	−0.0193	1.04	2 (H^+, CH_3COO^-)[b]

[a]Some complex electrolytes, such as Co(NH$_3$)$_5$Cl$_3$, are included here to show how the concept of mole numbers can be used to obtain information about the structure of complicated materials.

[b]Only 4% ions, the rest being CH$_3$COOH molecules of acetic acid.

vibrate loose from ice and enter the liquid, but it does impede the reverse process. If only 90% of the molecules in a sugar solution are water, then one collision out of ten will be by a sugar molecule. Thus, the probability of any given collision leading to the capture of a water molecule by the ice is less, and the net effect is to initiate the melting of ice.

To restore equilibrium, the temperature must be lowered until the increased likelihood of capturing the slower moving water molecules compensates for the fact that a smaller proportion of the molecules are water. As we shall see in Chapter 15, the lowering of the freezing point in dilute solutions, ΔT_f, is proportional to the molal concentration of the added substance, m:

$$\Delta T_f = -k_f m \tag{3–1}$$

The molal freezing point depression constant, k_f, varies from one solvent to another but is independent of the nature of the added solute.

Freezing point depression measurements offer one convenient way of determining molecular weights. If we know the number of grams of a solute, and if, from Equation 3–1, we can find the number of moles, then the solute's molecular weight is determined by dividing the number of grams by the number of moles (see Chapter 15 for an example). But what would happen if each molecule were to break into two or more particles when the solute is added to the solvent? Suppose, Arrhenius suggested, that HCl dissolves in water to give free H^+ and free Cl^- ions, independent of one another although equal in number. Then the lowering of the freezing point of water should be *twice* as great as predicted from Equation 3–1.

In Table 3–6 we see that the molal freezing point lowering (ΔT_f) for 0.01-molal HCl is −0.036°C, compared with −0.0186°C for molecular solutes like sugar, alcohol, and glycerin, all of which do not dissociate into ions and hence are nonelectrolytes. The effect is about twice as much as one might expect. This is consistent with Arrhenius' proposal that a molecule of HCl separates into two ions in water.

Let us define the *mole number*, i, as the number of moles of solute particles (ions) that are produced when 1 mole of a substance dissociates. This number also is called the van't Hoff i factor, after Jacobus H. van't Hoff (1852–1911), who studied the effect. The freezing point lowering then will be

$$\Delta T_f = -i \Delta T_0 = -i k_f m \qquad (3\text{--}2)$$

in which ΔT_0 is the expected value for a nondissociating solute. In Table 3–6 are experimental values of i for several electrolytes and an explanation of these values in terms of the ions formed.

The mole number, i, generally increases with decreasing concentration, thereby approaching an "ideal" whole number at infinite dilution. This is the number of ions produced when a molecule dissociates completely. Values less than this ideal value actually are found in most electrolytes. This is not because they do not dissociate into ions, but because these oppositely charged ions still exert attractive forces on one another in solution. Only in the extrapolated limit of infinite dilution are such attractive forces negligible.

We often can predict from the molecular formula what this limiting value of i should be, or how many ions will be formed. For example, HCl obviously can produce no more than two ions. Once we learn from a study of chemical reactions that the groups OH^-, SO_4^{2-}, and NH_4^+ usually act as units, then the limiting values for NaOH, K_2SO_4, $MgSO_4$, and NH_4Cl become understandable. Because we know that CH_3COO^- and $Co(NH_3)_5Cl^{2+}$ each behave as units during ionization, we have valuable information about how these complex ions are put together. The Arrhenius theory of dissociation not only explains why i differs from 1 for electrolytes, but it often permits us to predict the maximum value that i should have.

Degree of Ionization

Electrolytes that appear to be dissociated completely in aqueous solution are known as *strong electrolytes*. These include most salts and bases. Solutions of most inorganic acids also are strong electrolytes. But some acids, especially organic acids such as acetic, appear to be only partially dissociated in aqueous solution and are called *weak electrolytes*. The higher the concentration, the less these weak electrolytes are dissociated (Figure 3–4). For such compounds, the value of i is related to the *degree of dissociation* of the molecules.

Consider the dissociation of an acid HA:

$$HA \rightarrow H^+ + A^- \qquad (3\text{--}3)$$

Let C_0 be the original concentration of nonionized HA, and C_1 be the con-

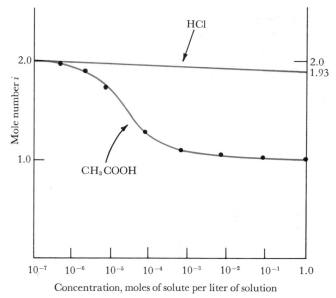

Figure 3–4. Mole number versus concentration for a weak electrolyte, acetic acid (CH_3COOH), *and a strong electrolyte, hydrochloric acid* (HCl). *Acetic acid is ionized completely only in very dilute solutions, whereas hydrochloric acid is ionized completely at all concentrations. The deviation of mole numbers of electrolytes like HCl from whole numbers corresponding to complete ionization are accounted for by the Debye–Hückel theory of interionic attractions. This theory interprets the restrictions on the "independence" of the ions as caused by interionic attractions and repulsions in the solution.*

centration of H^+ formed. Then we have

$$C_0 - C_1 = \text{the concentration of HA remaining nonionized}$$
$$+\ C_1 = \text{the concentration of } H^+$$
$$\underline{+\ C_1 = \text{the concentration of } A^-}$$
$$C_0 + C_1 = \text{the total concentration of solute particles}$$

and

$$\frac{C_0 + C_1}{C_0} = i, \quad \text{the mole number}$$

For the one-step ionization of a simple acid, as in Equation 3–3, the range of C_1 is $0 \leq C_1 \leq C_0$. Thus, i will vary from two for 100% dissociation to one for no dissociation. In this case,

$$\text{Percent dissociation} = 100(i - 1)$$

Arrhenius' simple theory was expressed mainly in terms of water as the solvent. We now appreciate, even more than he, how great a part the solvent plays in the dissociation of a material into ions. In water, HCl dissociates into H^+ and Cl^- ions; in benzene it does not dissociate at all (Figure 3–5). Most solvents that are readily miscible (will mix) with water will produce ionization like water does, although not necessarily as easily. Solvents that we regard as "oily" or organic, and that are immiscible with water, will be poor ionizing solvents. Hydrochloric acid and acetic acid will dissolve in gasoline or kerosene but will not ionize; they will act like simple nondissociating molecules. Salts will not dissolve at all in gasoline or kerosene.

HCl in benzene

HCl in water

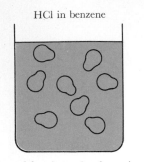

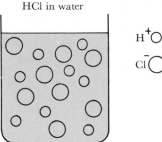

H^+O

Cl^-O

Figure 3–5. The nature of the solvent, benzene or water, has a great effect on the chemical behavior of the solute, HCl.

Normal freezing point depression ($i = 1$).
No reaction with $AgNO_3$, Mg metal,
or at electrodes

Twice normal freezing point depression
($i = 2$)

Solvents aiding dissociation are those whose molecules are *polar* (positively and negatively charged ends). The water molecule is bent; that is, the angle from H to O to H is 105° instead of the 180° of a linear molecule. The O atom has a slight negative charge, and the two H atoms carry a slight positive charge. The molecule can be thought of as a tiny dipole. (A *dipole* is an object with a positive and negative charge separated by a finite distance.) In a polar solvent, these solvent dipoles cluster around the ions. Their negative ends point toward the cations and their positive ends point toward the anions. Therefore, each ion becomes covered with a layer of solvent dipoles like quills on a porcupine: These solvent molecules help to keep the ions separated and stabilized. Ions are *hydrated* if the solvent is water, or *solvated* in other solvents.

Liquid ammonia is the polar solvent with properties most like those of water. Nonaqueous chemistry in liquid ammonia is a flourishing field. It has been suggested that life based on liquid ammonia as a solvent rather than water is a real possibility on colder planets (Jupiter?).

When we write an ion such as Ca^{2+} or Na^+, we should understand that the ion is hydrated in water; it is not a bare positive ion. This is especially true of small ions such as H^+ (which is only a proton) and Li^+, to which the water molecules can come extremely close so the attraction between cation and dipole is strong. The hydration of the proton should be written as something like $H(H_2O)_4{}^+$. But it is conventional to write it as H_3O^+ and to refer to the hydrated hydrogen ion as a *hydronium ion*.

Organic acids such as acetic, picric, or butyric acid will be dissociated only partially, even in water, and will provide a smaller concentration of hydrogen ions than will strong acids such as HCl. The hydrogen ion serves as a catalyst, which accelerates the rates of certain chemical reactions. An important proof of the validity of Arrhenius' theory emerged when it was shown that the amount of dissociation, as measured by the conductivity of a solution, correlated closely with the catalytic effect of the solution on reactions. Examples of these data are given in Table 3–7. Hydrochloric acid accelerates the esterification of formic acid (it is irrelevant at the moment

Table 3–7. Extent of Dissociation and Catalytic Activity of Various Acids in Anhydrous Ethanol, Relative to HCl = 100

Acid	Relative conductivity	Catalytic effect on the esterification of formic acid
Hydrochloric	100	100
Picric	10.4	10.3
Trichloroacetic	1.00	1.04
Trichlorobutyric	0.35	0.30
Dichloroacetic	0.22	0.18

what this reaction is, but you can find out in Chapter 12) more than does dichloroacetic acid because HCl dissociates more, even in ethanol, and produces more H^+ ions.

This catalytic comparison convinced Wilhelm Ostwald, a famous physical chemist, of the merit of Arrhenius' theory of ionization. His enthusiasm influenced Arrhenius' professors, who finally awarded him his doctorate. The incident demonstrates that science is not immune to the effects of internal politics; for Ostwald was a scientist whose opinions carried great weight with his contemporaries.

Challenge. You now should be able to solve problems concerning electrolysis reactions involved in applications such as the lifetimes of flashlight and automobile batteries and the plating of silverware. Try Problems 3–4 to 3–7 in *Relevant Problems for Chemical Principles* by Butler and Grosser.

3–6 GASEOUS IONS

Arrhenius firmly established the importance of ions in solution in chemical reactions. His work suggested the existence of fundamental units of charge and the exchange of these units in chemical reactions. His theories were aided greatly, and the way for their acceptance prepared, by other research being carried out by physicists on electric discharge in gases, a research topic seemingly far removed from solution chemistry.

Gas Discharge Tubes

In the early 1850's, Sir William Crookes (1832–1919) began experimenting with electric discharges through gases in tubes such as shown in Figure 3–6. When the gas is at atmospheric pressure, nothing happens, even with a potential of 10,000 volts (V) between cathode and anode. But as the pressure is lowered by pumping, current begins to flow and the gas begins to glow. At low pressures, the glowing region moves toward the anode, and a dark

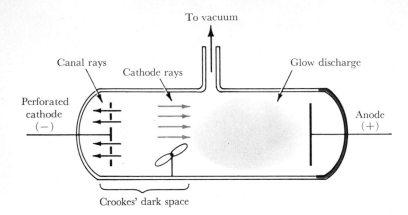

Figure 3–6. A Crookes tube. Cathode rays moving from cathode to anode cause pin-wheels to turn in a Crookes dark space and cause residual gas to glow. At low pressures, glass at the anode end begins to glow as a Crookes dark space appears to extend the length of the tube. Cathode rays are negative electrons and are independent of electrode material or residual gas. Canal rays are positive ions corresponding to the residual gas in the tube. The color of the discharge is characteristic of the residual gas in the tube and moves away from the cathode at high vacuum (low pressure). The glow is produced by collisions of electrons with gas molecules. A Crookes dark space results when these collisions become infrequent. At high vacuum the electrons reach the anode and the glass at the anode end of the tube, without colliding with gas molecules. Under these circumstances the glass itself glows and the anode emits rays.

region appears. A screen coated with zinc sulfide or certain other minerals will glow in this dark space. The light is produced in bursts as if the screen were being bombarded by particles. Metal shields with slits set up on one side or the other of the ZnS screen indicate that the particles are coming from the cathode. A solid object in these cathode rays will cast a shadow on the screen, and a light-weight pinwheel will spin from the bombardment of particles. If a metal plate with a slit is placed in front of the cathode, the cathode rays are emitted in a thin beam. The deflection of this beam by both external magnetic poles and external electrostatically charged plates is evidence that the particles carry a *negative charge*. These particles are called *electrons*. It was observed that the same electrons are obtained regardless of the material from which the cathode is made (so long as it is a conductor of electricity).

The glow in a Crookes tube occurs when electrons strike atoms in the gas and ionize them. At low pressures, electrons travel a greater distance from the cathode before encountering a gas molecule; hence, the glow appears closer to the anode. Electrons that do not strike gas molecules and the electrons that are emitted when gas atoms are ionized move to the anode. In contrast, ionized gas molecules migrate to the cathode. If holes are drilled

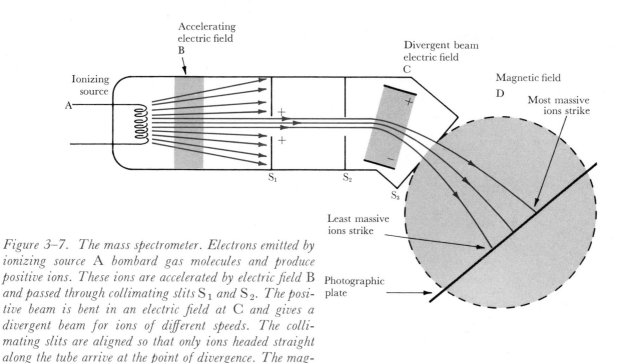

Figure 3–7. The mass spectrometer. Electrons emitted by ionizing source A *bombard gas molecules and produce positive ions. These ions are accelerated by electric field* B *and passed through collimating slits* S_1 *and* S_2. *The positive beam is bent in an electric field at* C *and gives a divergent beam for ions of different speeds. The collimating slits are aligned so that only ions headed straight along the tube arrive at the point of divergence. The magnetic field at* D *refocuses the beam in such a way that all ions of the same* e/m *strike the same spot on the photographic plate.*

in the cathode, some of these ions pass through and strike the back of the tube. These beams of positive ions (as shown by their deflection in magnetic or electrostatic fields) are called *canal rays*. We can measure the relative masses of these ions from their deflection by magnetic fields of known strength; the masses differ according to the gas in the tube. These and similar experiments proved that atoms have negatively charged, very light electrons that are held loosely to the atom. Atoms become positive ions by the *removal of electrons*, and become negative ions by the addition of more electrons.

Mass Spectrometry

The deflection of a beam of particles in a magnetic or electrostatic field is greater for particles with larger charges and smaller masses. This deflection can be used to find the charge-to-mass ratio (e/m) of a particle. In 1897, J. J. Thomson (1856–1940) measured the charge-to-mass ratio of the electron to be -1.76×10^8 coulombs g^{-1}. A descendant of his apparatus, the *mass spectrometer*, is now a valuable research tool for measuring the mass per unit charge of any substance that can be given a positive charge (Figure 3–7). Mass spectrometry offers the most direct measurement of atomic weights of the elements (Cannizzaro was vindicated) and is the method by which the existence of isotopes was proved.

The mass spectrometer can be used to identify extremely minute amounts of material and to separate one isotope from another. By looking at the molecules' masses and also the masses of the fragments into which they break during electron bombardment in the spectrometer, organic chemists often obtain useful information about the probable molecular structure of a substance. During the development of the atomic bomb in the midst of World War II, mass spectrometry separated ^{235}U from ^{238}U, although the extremely low pressures required (10^{-6} torr) were not feasible for large-scale production.

Challenge. The application of mass spectrometry to the determination of the relative abundance of isotopes of an element and the identification of chemical compounds is illustrated by Problems 3–9 to 3–14 in the Butler and Grosser book.

The Charge on the Electron

The preceding comments about mass spectrometers implied that the charge on the electron had been found and that the mass per unit charge measurements could be converted directly into mass readings. But the charge on the electron was not measured until 1911, 14 years after the charge-to-mass ratio was known. In 1911, Robert A. Millikan (1863–1953) published the results of an ingenious experiment, illustrated in Figure 3–8. He irradiated with x rays a spray of tiny oil droplets between two charged plates. Ions that were formed from the air molecules adhered to the oil drops and gave them a charge. Then he varied the electric field until the electric force on a charged oil drop counteracted the force of gravity on the drop, so the drop was suspended. Millikan calculated the charge on the drop from the known mass of the drop and known strength of the electric field. He discovered that the charge on a drop was always equal to some integral multiple of 1.602×10^{-19} coulomb, and proposed that this value was the charge on a single electron.

Calculation of Avogadro's Number

The first value of Avogadro's number was obtained in 1865 by J. Loschmidt, a Viennese schoolteacher. He calculated from the kinetic theory of gases that there are 2.71×10^{19} atoms of a monatomic gas per cubic centimeter, corresponding to 6.07×10^{23} for Avogadro's number. A much better value became available when Millikan's oil-drop experiment revealed the charge on an electron. If a faraday (96,500 coulombs) represents a mole of electric charges, then Avogadro's number is simply the ratio of the faraday to the charge per electron:

$$\frac{96{,}500 \text{ coulombs } \mathscr{F}^{-1}}{1.602 \times 10^{-19} \text{ coulomb unit}^{-1}} = 6.022 \times 10^{23} \text{ units } \mathscr{F}^{-1}$$

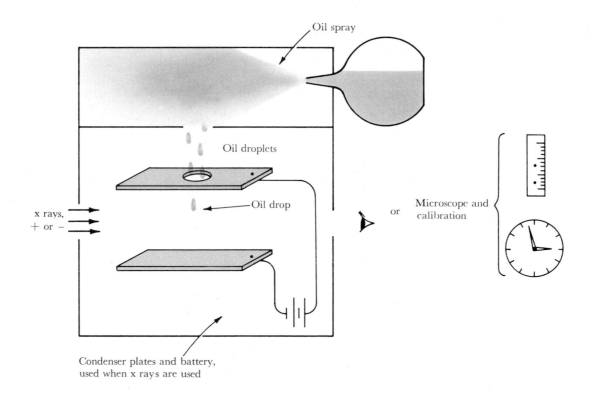

Figure 3–8. Millikan's oil-drop experiment. Tiny droplets of oil are suspended between two plates that may be charged. The drop is allowed first to fall freely through air, and its path as it falls between the condenser plates is monitored. From the data on air resistance, density of the oil, and terminal velocity of the drop, we can calculate its dimensions. The rays ionize the air, and negatively charged particles stick to the drop. The charge on the drop can be determined from the upward velocity of the charged oil drop when the battery is turned on. All charges measured are integral multiples of 1.602×10^{-19} coulomb, which is the charge on an electron.

3–7   IONS IN SOLIDS

Ions occur in solids of two types: *metals* and *salts*. In metals, the positively charged ions are packed tightly together in a regular manner. These ions are surrounded by a sea of mobile electrons that neutralize the charge on the cations but are not bound to specific ions. The ions are not held together by specific connections or *bonds*. Instead, the entire metal is held by the attraction between positive ions and mobile electrons. In salts, cations and anions (which are in the proper numbers to balance their charges) are packed in an orderly array in a crystal. Neither type of ion is free to move about; so salt crystals are not good conductors of electricity in the way that metals, with their free electrons, are. In both metals and salts, there is no preferred direction of interaction between one ion and its neighbors. The construction of a crystalline solid is dictated by the energy involved in the packing of charged spheres. To understand the structures of ionic solids, we must look first at the ways in which spheres can be packed in a regular manner.

Close Packing of Spheres and the Structures of Metals

In 1912, Peter Ewald, a graduate student at the University of Munich, presented a seminar on the interactions of light waves with crystals. A professor at the seminar, Max von Laue, pointed out that *if* x rays were electromagnetic waves like light, and *if* crystals were composed of atoms packed in a regular and orderly fashion in three dimensions (neither fact was known with certainty at the time), then a crystal might diffract x rays in the same way that a ruled diffraction grating diffracts light, which has a longer wavelength than x rays. W. Friedrich tried the experiment, and soon obtained a good diffraction pattern of x rays with a copper sulfate crystal. In England, the father and son team of William and Lawrence Bragg developed x-ray diffraction into a generally applicable tool for determining the atomic structures of crystalline solids.

At the present time, the method has progressed to the point at which the structures of crystallizable proteins such as hemoglobin, with a molecular weight of 68,000, have been determined successfully. The first crystals to be studied were the simpler metals and minerals—ionic compounds of the type that we are concerned with here. The close packing of spheres provides model structures for all metals and most ionic salts and other minerals.

Figure 3–9. Uniform spheres pack into a sheet most efficiently as shown by gray circles; each sphere is surrounded by six nearest neighbors. A second sheet of spheres (color) can be fitted most closely to the original sheet by nesting its spheres over the holes between triads of spheres in the first sheet. A third sheet can be packed closely against the second in two ways: by placing its atoms either directly over those of the first sheet (gray circles) or over the holes marked "3." The first packing scheme, in which the arrangement of the layers can be represented by -1-2-1-2-1-2-, is called hexagonal close packing, abbreviated hcp. The second scheme, with the order of layers as -1-2-3-1-2-3-1-2-3-, is called cubic close packing, ccp, or face-centered cubic, fcc.

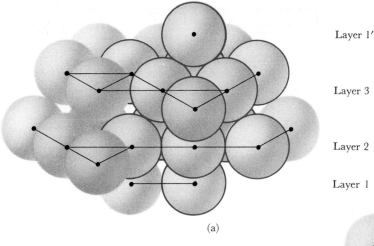

Layer 1′

Layer 3

Layer 2

Layer 1

(a)

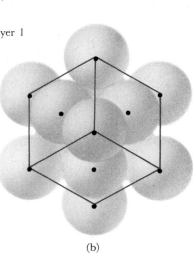

(b)

Figure 3–10. (a) Portions of four layers of a ccp structure in which we see the nesting of spheres. Atoms extracted in (b) are outlined in color. (b) Although it is not easy to see at first, atoms in a ccp array are packed in a cubic unit cell. The atoms are at the corners of a cube and at the center of each cube face. These atoms are in exactly the same orientation as the ones outlined in color in (a), but the cube is easier to see if you turn the page upside down.

If a large number of uniform marbles are packed closely together on a tabletop, they will pack in the manner shown by the gray circles in Figure 3–9. Two such close-packed layers can be stacked against one another most efficiently by nesting them as the colored circles in Figure 3–9. A third layer can pack on the other side of the second layer in exactly the same way as the first; this forms a symmetrical "sandwich." Or the layer can be shifted to a third position unlike either of the first two. This two-layer or three-layer regular repetition constitutes the two close-packed structures in three dimensions: *hexagonal close packed* (hcp) and *cubic close packed* (ccp). In both of these packing schemes, the marbles take up $(\sqrt{2}/6)\pi = 0.74$ of the total volume, with 26% empty space between marbles. Each marble is surrounded by twelve nearest neighbors, and has a *coordination number* of twelve.

It is easy to see why, with six immediate neighbors to a given marble in a plane, and similarly oriented triangles of marbles above and below it, the hcp structure should be called hexagonal. But where in Figure 3–10(a) are the right angles and equal sides of a cube? The answer is in Figure 3–10(b); ccp is identical to an arrangement of marbles or atoms at the corners and centers of the faces of a cube.

A slightly less efficient packing scheme is one in which atoms occupy the corners and the *center* of a cube. This scheme is the *body-centered cubic*, or bcc, structure. In this form the atoms take up $(\sqrt{3}/8)\pi = 0.68$ of the total volume. Each center atom has eight nearest neighbors at the corners of a cube. However, each corner atom also has eight neighbors at the corners of a cube. These neighbors are the center atoms from the eight neighboring cubes. The two positions, centers and corners, are entirely equivalent, and an atom's coordination number in the structure is eight.

These three structures, ccp, hcp, and bcc, account for virtually all metals and alloys. All three have comparable stability, and there is no simple way of predicting which structure a given metal will adopt. Many metals can exist in more than one structure: Ni can be either ccp or hcp; Na can be hcp or bcc; and Fe can be bcc below 912°C, ccp above it, or one of two other structures. The stacking of close-packed layers need not be perfect or confined to the simple hcp and ccp. The metals Am, Ce, La, Pr, and Nd crystallize in a modified close-packing scheme in which the order of layers, in the nomenclature of Figure 3–9, is -1-2-1-3-1-2-1-3-. Sm uses a nine-layer repeat: -1-2-1-2-3-2-3-1-3- and so forth. Li and Na can be hammered into a close-packed structure of disordered layering.

An *alloy* is a solid solution of one metal in another. Alloys can be of two types: *substitutional*, in which atoms of a different metal substitute for atoms of the original structure, and *interstitial*, in which smaller atoms are inserted into gaps between the metal atoms. For example, α-brass is an alloy in which atoms of Zn replace Cu atoms in the ccp lattice of pure Cu. However, the Zn atoms are larger than the Cu and only a little over 30 mole % of Zn can be tolerated in the Cu lattice before it breaks. β-Brass, the stable form in a roughly equimolar alloy, is a random mixture of Cu and Zn atoms in a bcc structure. The strain of the larger Zn atoms in Cu can be relieved in a more open structure, with a coordination number of eight instead of twelve.

There are two types of holes in a close-packed structure, into which alien atoms can fit in interstitial alloys: *octahedral*, between two triangles of atoms on adjacent layers (Figure 3–11), and *tetrahedral*, between a triangle of atoms in one layer and a single atom in the next (Figure 3–12). An atom inserted into an octahedral hole has six nearest neighbors, or a coordination number of six, and an atom in a tetrahedral hole has a coordination number of four. Both types of holes in a ccp structure are shown in Figure 3–13. Octahedral holes also are at the centers of the cube faces in a bcc structure.

Steels are interstitial alloys of iron and small amounts of carbon and other atoms. High-carbon steel, 0.75% to 1.5% carbon, has carbon atoms scattered in the octahedral holes on one set of faces of the bcc iron structure. Fe_4N is an interstitial alloy in which the nitrogen atoms occupy octahedral holes in a ccp iron lattice. In such cases the interstitial atoms make the metal harder and more resistant to bending and deformation by acting like sand in the bearings of an engine; they get in the way and make it more difficult for

Figure 3–11. An octahedral hole (indicated by colored asterisks) in a close-packed structure lies between two triangles of atoms in adjacent layers. The choice of "layers" in a ccp structure is arbitrary. There are four possible choices of pairs of triangles to define the layers. These choices correspond to the four pairs of parallel faces in a regular octahedron. All pairs are completely equivalent. (It is we who draw the layers, and not the crystal.) Drawing (a) emphasizes the layer structure; (b) emphasizes that the six nearest neighbors to the octahedral hole position lie equally far away along three mutually perpendicular axes. (Note: *These spheres should be large enough to touch, but have been reduced in size for clarity.*)

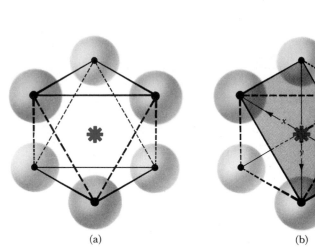

(a)   (b)

Figure 3–12. A tetrahedral hole in a close-packed structure lies between a triangle of atoms in one layer and a closely nested atom in the next. Can you prove that there are exactly as many octahedral holes as there are spheres in a ccp structure, and that there are twice as many tetrahedral holes? Are the numbers the same in a hcp structure?

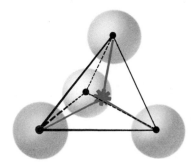

one layer of metal atoms to slip over another when the metal is stretched, hammered, or bent.

Metals are good conductors of electricity because their electrons are mobile and can flow under the influence of an electrical potential. Metals can be soft, malleable, and ductile since, in the absence of bonds in specific directions, deformation is as easy as slipping one layer of marbles over another. The mobility of their electrons also is responsible for the characteristic silvery or reflective sheen that we describe as "metallic." As we shall discover in Chapter 8, electrons confined to the vicinity of one atom or ion can absorb only certain wavelengths of energy; their possible energy states are quantized.

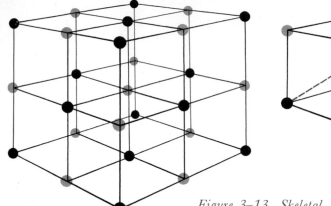

Figure 3–13. Skeletal view of a ccp structure (black atoms) with different atoms (colored) in all of the octahedral holes. Note that the black and colored positions are equivalent; this structure could just as well be described as one in which the black atoms occupy the octahedral holes of a colored-atom ccp framework. You can see this most easily if you realize that the real structure is endless in all three directions, and that only a small portion is shown here. A tetrahedral hole position (colored asterisk) is at the center of each of the smaller cubes, such as the one extracted to the right. The true unit cell, which by exact repetition in three dimensions will build the entire crystal structure, is the large cube to the left and not the small one to the right. Can you see why? If the black atoms represent sodium ions and the colored ones represent chloride ions, then this is the structure of crystalline NaCl.

Yet delocalized electrons, such as the ones in a metal, can absorb and reemit energy over a broad band of wavelengths; hence, they can reflect the light that impinges on the metal surface.

Crystalline Salts

The structure of salts is dictated by a balance between two frequently conflicting tendencies: (1) the tendency to surround an ion having a given charge with as many ions of the opposite charge as possible, and (2) the tendency to avoid close contact between these ions of like charge. In addition, the proportion of ions must be such that there is no net charge on the crystal. In general, cations formed when atoms lose electrons tend to be smaller than anions formed when atoms gain them; and most salt structures can be thought of as variations on close-packed anions with interstitial cations. One important structure-determining factor is the *radius ratio*, the ratio of the ionic radius of cation to anion. Ionic radii for some common ions are given in Table 3–8 and illustrated in Figure 3–14.

Table 3–8.   Ionic Radii for Some Ions[a]

				Positive ions				
Ag^+	1.26	Ba^{2+}	1.35	Al^{3+}	0.50	Ce^{4+}	1.01
Cu^+	0.96	Be^{2+}	0.31	B^{3+}	0.20	U^{4+}	0.97
K^+	1.33	Ca^{2+}	0.99	Bi^{3+}	0.74	Ti^{4+}	0.68
Li^+	0.60	Cd^{2+}	0.97	Cr^{3+}	0.65	Zr^{4+}	0.80
Na^+	0.95	Co^{2+}	0.82	Fe^{3+}	0.67		
NH_4^+	0.42	Cu^{2+}	0.70	Ga^{3+}	0.62		
Rb^+	1.48	Fe^{2+}	0.78	In^{3+}	0.81		
Tl^+	1.44	Hg^{2+}	1.10	La^{3+}	1.15		
Cs^+	1.69	Mg^{2+}	0.65	Tl^{3+}	0.95		
		Mn^{2+}	0.80	Y^{3+}	0.93		
		Ni^{2+}	0.69				
		Pb^{2+}	1.16				
		Sr^{2+}	1.13				
		Zn^{2+}	0.74				

				Negative ions		
Br^-	1.95	O^{2-}	1.40	N^{3-}	1.71
Cl^-	1.81	S^{2-}	1.84	P^{3-}	2.12
F^-	1.36	Se^{2-}	1.98		
H^-	1.54	Te^{2-}	2.21		
I^-	2.16				

[a] Measured in angstroms.

The most common ionic structure is the NaCl structure (Figure 3–13). In this structure, Na^+ ions occupy the corners and centers of the faces of a cube, and Cl^- ions occupy the cube center and the midpoints of the edges. However, these are entirely equivalent sets of positions in the virtually infinite structure of the crystal. Either ion can occupy all of the octahedral holes of a ccp lattice of the other, and the coordination number of each ion will be six. Many alkali halides (Li^+, Na^+, K^+, Rb^+, and Cs^+ with F^-, Cl^-, Br^- or I^-) can occur in the NaCl structure (Table 3–9), as can most of the alkaline earth metals with O^{2-}, S^{2-}, Se^{2-}, and Te^{2-} (Table 3–10), and many other 1/1 ionic substances. Nevertheless, as Figure 3–15(c) illustrates, if the cation/anion radius ratio falls below 0.414 for a cation in an octahedral hole, the anions will touch. This is true for LiCl, LiBr, and LiI, and the result is that their crystal structure is weakened (Figure 3–16).

The cesium cation is so large that more than six anions can group around it without touching. Cesium halides normally have the body-centered cubic structure illustrated in Figure 3–17; for each ion the coordination number is eight. The rubidium halides (except RbF) can be induced to shift to this

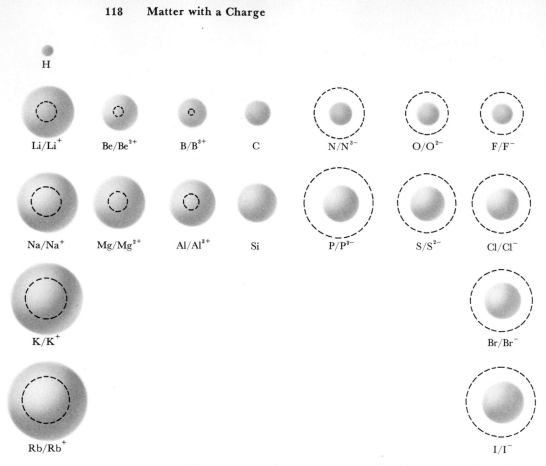

H

Li/Li⁺ Be/Be²⁺ B/B³⁺ C N/N³⁻ O/O²⁻ F/F⁻

Na/Na⁺ Mg/Mg²⁺ Al/Al³⁺ Si P/P³⁻ S/S²⁻ Cl/Cl⁻

K/K⁺ Br/Br⁻

Rb/Rb⁺ I/I⁻

Figure 3–14. Relative atomic radii of elements. Solid spheres represent atoms and dashed circles, ions. Notice that positive ions are smaller than their neutral atoms, and negative ions are larger. Why should this be so?

Table 3–9.   Radius Ratios, r_+/r_-, in Ionic Crystals of the C^+A^- Type[a]

	Cation				
Anion	Li⁺ (0.60)	Na⁺ (0.95)	K⁺ (1.33)	Rb⁺ (1.48)	Cs⁺ (1.69)
F⁻ (1.36)	0.45	0.70	0.98	1.09	1.24
Cl⁻ (1.81)	0.33 [b]	0.52	0.74 [c]	0.82 [d]	0.93 [e]
Br⁻ (1.95)	0.31	0.49	0.68	0.76	0.87
I⁻ (2.16)	0.28	0.44	0.62	0.69	0.78

[a] Quantities in parentheses are ionic radii in angstroms. All structures are of the NaCl type except as noted.

[b] NaCl structure, but anions touching and crystal weakened.

[c] CsCl structure above 20,000 atm pressure.

[d] CsCl structure above 5000 atm pressure.

[e] CsCl structure.

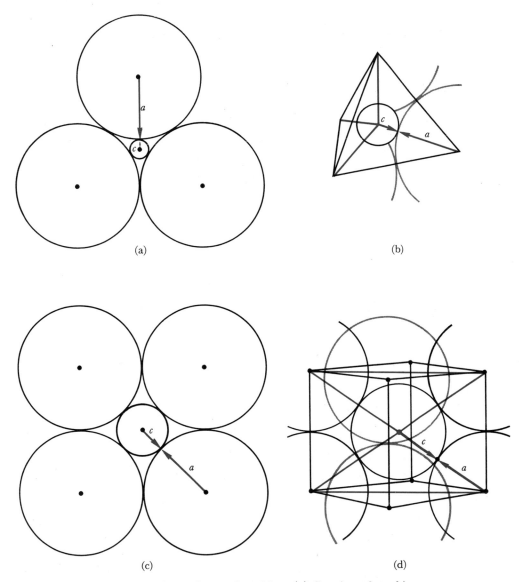

(a)   (b)

(c)   (d)

Figure 3–15. Radius ratios and crystal packing. (a) In trigonal packing as shown here, if the ratio of the radius of the cation, c, to that of the anion, a, is less than $(2\sqrt{3}/3) - 1 = 0.155$, then the similarly charged anions will touch, which is an inherently unstable situation. (b) An ion in a tetrahedral hole cannot keep its neighboring ions apart unless its radius ratio, c/a, is equal to or greater than $\sqrt{3/2} - 1 = 0.225$. (c) The minimum radius ratio for an ion in an octahedral hole is $\sqrt{2} - 1 = 0.414$. (d) The ions at the corners of a simple cube will touch unless the unlike ion at the cube center has a radius ratio of $\sqrt{3} - 1 = 0.732$ or greater.

Table 3–10.   Radius Ratios in Ionic Crystals of the $C^{2+}A^{2-}$ Type[a]

	Cation				
Anion	Be²⁺ (0.31)	Mg²⁺ (0.65)	Ca²⁺ (0.99)	Sr²⁺ (1.13)	Ba²⁺ (1.35)
O²⁻ (1.40)	0.22[b]	0.46	0.71	0.81	0.97
S²⁻ (1.84)	0.17[c]	0.35	0.54	0.61	0.73
Se²⁻ (1.98)	0.15	0.33	0.50	0.57	0.68
Te²⁻ (2.21)	0.14		0.45	0.51	0.61

[a]Quantities in parentheses are ionic radii in angstroms. All structures are of the NaCl type except as noted.
[b] β-ZnS or wurtzite structure.
[c] α-ZnS or sphalerite structure.

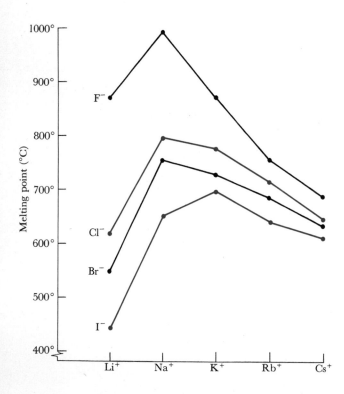

Figure 3–16. The crystal structures of lithium halides are weakened by the close approach of anions around the small cation, and this weakening is manifested in abnormally low melting points. Note that this weakening also begins to appear in NaI, *because the iodide ion is so big relative to the sodium ion.*

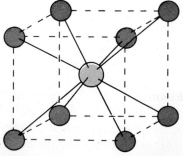

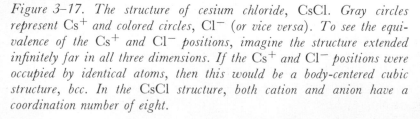

Figure 3–17. The structure of cesium chloride, CsCl. *Gray circles represent* Cs⁺ *and colored circles,* Cl⁻ *(or vice versa). To see the equivalence of the* Cs⁺ *and* Cl⁻ *positions, imagine the structure extended infinitely far in all three dimensions. If the* Cs⁺ *and* Cl⁻ *positions were occupied by identical atoms, then this would be a body-centered cubic structure, bcc. In the* CsCl *structure, both cation and anion have a coordination number of eight.*

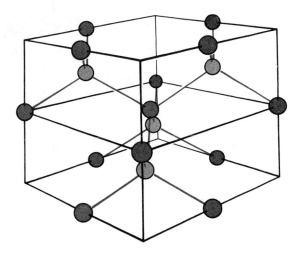

Figure 3–18. The structure of one of the two crystal forms of zinc sulfide, β-ZnS or sphalerite. The Zn occupies the positions of a ccp structure (gray circles), and the S occupies half of the tetrahedral holes. Again, the sites are equivalent; the Zn can be described as occupying half of the tetrahedral holes of a ccp lattice of S atoms. The coordination number of both types of atom is four.

CsCl structure at pressures above 5000 atm, and above 20,000 atm even the potassium halides adopt this form. This eightfold coordination without touching anions is possible for compounds with a radius ratio greater than 0.732 [Figure 3–15(d)].

Most divalent alkaline earth cations combine with divalent anions in the NaCl structure (Table 3–10). However, for a very small cation such as Be^{2+}, the six-coordinated NaCl structure is impossible, and the coordination number drops to four. For BeS, BeSe, and BeTe, the structure in Figure 3–18 results. In this case, anions are in a ccp array, and cations are arranged symmetrically in *half* of the tetrahedral holes. This is known as the β-ZnS structure. BeO crystallizes in a closely related α-ZnS structure, which also has coordination number four but with differently arranged tetrahedra.

We usually can understand more complex crystalline salts and minerals in terms of these simple structures and the packing of spheres (oxygen atoms in many minerals) with small cations inserted in the interstitial holes. Many physical properties of minerals, such as talc, asbestos, mica, and quartz, depend in a particularly simple way on their crystal structure. These minerals furnish clear examples of how structure determines properties. We shall return to the structures of solids briefly in Chapter 14. However, one other aspect of crystal structure is immediately relevant to us: A new way of calculating Avogadro's number.

In one of the most reliable methods of determining Avogadro's number, N, we use only a salt's molecular weight, measured density, and interatomic spacing in the crystal as determined by its x-ray diffraction pattern.

Example. The density of crystalline NaCl is 2.169 g cm^{-3}. From x-ray diffraction analysis we know that NaCl has the structure shown in Figure 3–13, and that the distance from one Na$^+$ to the next along an edge of the cube is 5.64 Å. Calculate Avogadro's number.

Solution. The volume of the cubic cell is

$$V = (5.64 \text{ Å})^3 = 179.1 \text{ Å}^3$$

As it appears in the figure, the cell has four sodium ions and four chloride ions. This may be difficult to comprehend, since you clearly can see 14 sodium and 13 chloride ions. Nevertheless, each of the sodium ions in the centers of the faces is shared between two adjacent cubic units; thus, only one half of each ion is counted with the unit pictured. The six faces of the cell then contain six half-sodiums, or three ions apportioned to this cell. Similarly, the eight sodium ions at the cube corners are each shared with eight cells that meet at a corner. These amount to eight eighths, or one, for a total of four sodium ions per cell. The twelve quarter-chlorides on the cell edges and the one in the interior account for the four chloride ions. Each cell as depicted in Figure 3–13 therefore represents four units of NaCl. The volume per NaCl unit is then 44.78 Å^3, and the volume per mole is $N \times 44.78 \text{ Å}^3 = N \times 44.78 \times 10^{-24} \text{ cm}^3$, in which N is the quantity to be calculated. The weight per mole is the molecular weight of 58.45. (Recall that molecular weights and atomic weights are obtained from combining ratios with hydrogen, and do not depend on a knowledge of Avogadro's number.) The density (ρ) of a NaCl crystal is

$$\rho = \frac{\text{weight}}{\text{volume}} = 2.169 \text{ g cm}^{-3}$$

Thus

$$\frac{58.45 \text{ g}}{N \times 44.78 \times 10^{-24} \text{ cm}^3} = 2.169 \text{ g cm}^3$$

and Avogadro's number is calculated as

$$N = \frac{58.45 \text{ g}}{44.78 \times 10^{-24} \text{ cm}^3 \times 2.169 \text{ g cm}^{-3}} = 6.022 \times 10^{23}$$

3–8 SUMMARY

Primarily because of the inspiration of one man, Svante Arrhenius, the presence and importance of charged ions in many solutions was demonstrated. Sometimes these ions appear to form after the pure substance has been dissolved in a suitable solvent such as water. When the substances are either pure or dissolved in a nonionizing solvent such as benzene, these substances do not appear to be ions. They often have low melting and boiling points (acetic acid is a liquid and HCl is a gas at 25°C), and conduct electricity very poorly when pure or dissolved in a solvent such as benzene. They may be weak electrolytes in aqueous solution and ionize only partially (as acetic acid does), or they may ionize completely as strong electrolytes (HCl). However, in either situation, the ionization is a result of interaction with the solvent.

In contrast, many strong electrolytes appear to be in the form of ions in the solid. The solvent molecules merely separate the already formed ions. Such ionic solids are called salts. They are usually difficult to melt (common table salt, sodium chloride, has a melting point of 800°C and a boiling point of 1413°C) and are only slightly soluble, if at all, in nonionizing solvents like benzene. The presence of ions is indicated by the high electrical conductivity of molten salts. Unlike metals, though, salts do not conduct electricity in the solid, because ions and not free electrons carry current, and the ions in salts are locked into a rigid structure. Most simple compounds of metals with nonmetals are ionic salts.

We have seen that atoms contain small, relatively loosely held units of charge called electrons. Both the mass and the charge of an electron have been measured, and two independent calculations of Avogadro's number have been made.

Metals derive their characteristic properties from the looseness of inter-actions between electrons and positive ions. A positive ion is formed from a neutral atom by removing one or more electrons, and a negative ion is made by adding one or more electrons to a neutral atom. In solid salts, these ions pack in a regular array, in which the packing pattern is influenced heavily by the relative sizes of cation and anion, or the radius ratio.

In short, chemical changes appear to be the large-scale manifestations of the transfer of electrons between atoms. One of our next tasks is to look for regularities and patterns in chemical behavior.

SUGGESTED READING

E. S. Gould, *Inorganic Reactions and Structure*, Holt, Rinehart & Winston, New York, 1962, 2nd ed. Good general reference. Very readable.

J. P. Hunt, *Metal Ions in Aqueous Solution*, W. A. Benjamin, Menlo Park, Calif., 1963.

L. Pauling, *The Nature of the Chemical Bond*, Cornell University Press, Ithaca, N.Y., 1960, 3rd ed. Chapter 11, "The Metallic Bond," and Chapter 13, "The Sizes of Ions and the Structure of Ionic Crystals," are especially relevant.

QUESTIONS

1 What is electrolysis? What are the cathode and anode in an electrolytic cell?

2 What are Faraday's laws of electrolysis? How do they imply a connection between electricity and chemical change?

3 In an electrolytic cell, is the anode the site of reduction, or oxidation? What is meant by the oxidation or reduction of a chemical substance?

4 What evidence is there for the assertion that 1 $\mathscr{F}$ of charge contains 6.022×10^{23} electronic charges?

5 How many faradays of electricity are required to reduce a mole of thallic ion,

Tl^{3+}, to the metal? How many are required to reduce the thallic ion to the thallous, Tl^+?

6 How many ions are produced when the following compounds dissociate in aqueous solution: $AlCl_3$, $NaBr$, $NaOH$, H_2SO_4, HNO_3, NH_4Cl?

7 What is the oxidation number of each of the following substances: Ca, Ca^{2+}, Na^+, I^-, S, S^{2-}?

8 Which of the following ions is the chromous ion and which is the chromic: Cr^{2+}, Cr^{3+}?

9 From the nomenclature of the compounds, in which of the compounds does arsenic (As) have the higher oxidation state: sodium arsenate (Na_3AsO_4) or sodium arsenite (Na_3AsO_3)? Can you justify your choice from what you know of oxidation numbers so far?

10 How do the heats of neutralization of acids and bases suggest the existence of ions in solution?

11 How do the freezing points of salt solutions suggest the existence of ions?

12 How do the colors of solutions of salts suggest the existence of ions?

13 How does Arrhenius' theory account for the different catalytic activities of inorganic and organic acid solutions?

14 How is it established that the cathode rays in a Crookes tube come from the cathode and are negatively charged?

15 How does a mass spectrometer separate ions?

16 What physical properties distinguish salts from metals?

17 What is the difference between ccp and hcp structures?

18 How many octahedral holes are there per close-packed atom in a ccp structure? How many tetrahedral holes are there?

19 What is an alloy? What is the difference between interstitial and substitutional alloys? Can you give examples of each?

20 Why are salts good conductors when molten but not when solid, whereas metals conduct electricity in both states?

21 Suppose that Millikan had observed only oil drops with an *even* number of electronic charges: 2, 4, 6, 8, ... Then he would have thought that the drop with two electrons carried the basic unit of charge, and would have computed a value of 3.204×10^{-19} coulomb for the electron. If you were Millikan, how might you have shown that this was not the case? [*Hint:* Would values of Avogadro's number from other experiments help, or are these values really independent of Millikan's work?]

PROBLEMS

1 A flow of electrons at the rate of 1 coulomb per second is a current of 1 *ampere* (A). How many moles of metallic aluminum can be obtained by passing a current of 1.00 A through molten $AlCl_3$ for 5 h 30 min?

2 How many faradays and coulombs of electricity are required to reduce 0.782 g of Cu^{2+} to metallic copper?

3 How many coulombs of electricity are required to reduce 0.300 mole of Fe^{3+} to Fe?

4 Often the amount of charge passed through a circuit is determined by measuring the mass of solid silver deposited by electrolysis of Ag^+ solution. If a cathode increases in mass by 0.197 g, how many coulombs have passed through the electrolysis cell?

5 A current of 5 A flows for 10 h through a solution containing KI. How many liters of hydrogen gas (STP) are released at the cathode?

6 When aqueous HCl solution is electrolyzed, how long must a current of 0.020 A flow to liberate 0.015 mole of hydrogen gas (STP) at the cathode?

7 When molten NaCl is electrolyzed, the products are Na and Cl_2. However, when aqueous NaCl is electrolyzed, the products are H_2 and Cl_2. The reason for this difference is that H_2O is easier to reduce than is Na^+. Write equations for the two electrolysis reactions. If a current of 4.0 A flows through each cell, how long will it take to produce 1.0 mole of each of the gaseous products in each reaction?

8 A quantity of electricity is passed through a water solution of $AgNO_3$, thereby causing 2.00 g of silver to be deposited at the cathode. How many grams of lead will be deposited if the same quantity of electricity is passed through a solution of $PbCl_2$?

9 Molten $ZnCl_2$ is electrolyzed by passing a current of 3.0 A through an electrolysis cell for a certain length of time. In this process, 24.5 g of Zn are deposited on the cathode. What is the chemical equation for the reaction at the cathode? At the anode? How long does this process take? What volume of Cl_2 gas (STP) is liberated at the anode?

10 One of the major purposes for building the large dams on the Columbia River is to provide cheap hydroelectric power for the electrolytic production of aluminum. The power house at each dam produces approximately 2×10^8 A of electricity at a voltage high enough to decompose molten salts of aluminum(III). What is the daily production of metallic aluminum in kilograms when all of the electricity from one dam is used to produce aluminum? How many dams would be needed for a daily production of 3000 metric tons of Al? (1000 kg = 1 metric ton.)

11 Calculate the number of electrons required to produce a charge of one coulomb.

12 The molal freezing point depression constant, k_f, for water is $1.86°C \, mole^{-1}$. If 0.100 mole of sodium sulfate, Na_2SO_4, is dissolved in 1000 g of water, what will be the freezing point of the solution?

13 When 45 g of glucose are dissolved in 500 g of water, the solution has a freezing point of $-0.93°C$. What is the molecular weight of glucose? ($k_f = 1.86°C \, mole^{-1}$ for water.) If the empirical formula for glucose is CH_2O, what is the molecular formula?

14 What are the freezing points of the following aqueous solutions:

a) 0.1-molal $Al_2(SO_4)_3$
b) 0.05-molal $BaCl_2$
c) 0.20-molal HCl
d) 0.06-molal sodium hexacyanoferrate(II)

15 Write chemical formulas for the following compounds:

a) lithium sulfide
b) calcium phosphate
c) zinc hexacyanoferrate(III)
d) mercuric acetate
e) nitrosyl sulfate
f) ammonium hexacyanoferrate(II)
g) uranyl oxalate
h) magnesium nitrite
i) ferric sulfate
j) stannous chloride
k) chromic fluoride

l) potassium permanganate

m) barium carbonate

n) silver arsenite

16 When 25 g of a compound with the empirical formula $PtCl_4 \cdot 4NH_3$ are dissolved in 1000 g of water, the freezing point of the solution is $-0.34°C$. What does this indicate about the number of ions per molecule? Can you explain why you obtain the answer you do? If you cannot explain it, look at Table 11–1 for a clue.

17 Prove that the radius of an ion in an octahedral hole [Figure 3–15(c)] must be greater than 0.414 times the radius of the surrounding ions if these ions are not to touch one another.

18 In the CsCl structure, the radius ratio between the two types of ions must be at least 0.732 if the larger ions are not to touch. In terms of the data in Table 3–10, why does BaO have the NaCl structure rather than the CsCl structure?

19 The crystalline structure of chromium metal is body-centered cubic. The distance between centers of one chromium atom and the next along an edge of the cube is 2.89 Å. Calculate the density and atomic radius of chromium.

20 Uranium crystallizes in a body-centered cubic lattice. The density of uranium is 18.7 g cm^{-3}. Calculate the atomic radius of uranium.

21 For CsCl calculate the distance between the centers of

a) two chloride ions on a body diagonal.

b) two chloride ions on an edge.

c) two chloride ions on a face diagonal.

And so, nothing that to our world appears,
Perishes completely, for nature ever
Upbuilds one thing from another's ruin;
Suffering nothing yet to come to birth
But by another's death.
Lucretius

4 QUANTITIES IN CHEMICAL CHANGE: STOICHIOMETRY

Stoichiometry is the study of quantities in chemical reactions. The phlogiston theory was completely respectable until Lavoisier made quantitative measurements and proved that metals *gained* rather than *lost* weight on combustion. Chemists after Dalton could be satisfied with water that was HO, and oxygen with a molecular weight of eight, until Gay-Lussac showed that a given volume of oxygen gas combined with twice that volume of hydrogen. Arrhenius' thesis supervisors could ignore his ideas about ions until he measured the freezing point lowering in electrolyte solutions and discovered that it was too large to be accounted for without ionization. Chemistry is a *quantitative* subject, and science without numbers is not science at all. Democritus speculated about atoms, and Lucretius wrote long poems about them for Roman noblemen. But until someone began measuring them, in the nineteenth century, no one was compelled to take them seriously. In this chapter we shall learn some of the techniques for dealing with *quantities* in chemical reactions.

4–1 FORMULAS AND MOLES

We have discussed earlier that each element can be given an atomic weight, which is based on an element's combining weight with hydrogen or with other elements that combine with hydrogen. The scale for a list of atomic weights is arbitrary. Dalton based atomic weights on a value of one for hydrogen. Berzelius and other European chemists for a time used a scale in which the atomic weight of oxygen was 1.0 and hydrogen was 0.125, but they returned to Dalton's conventions later. Current tables are based on an atomic weight of exactly 12 for the most common isotope of carbon, ^{12}C.

The important concepts of moles and gram-atoms were introduced in Chapter 1 and will be used extensively in this chapter. A gram-atom of an element is an amount in grams numerically equal to its atomic weight. There just as easily could be pound-atoms or ton-atoms, but these units are seldom used. One gram-atom of carbon is 12 g, and 12 tons would be a ton-atom. The number of atoms in 1 g-atom is known as *Avogadro's number*, N. We know now that this number is 6.022×10^{23} atoms, but a reasonable value for this number was not obtained until 54 years after Avogadro's hypothesis. The concept of gram-atoms and molecular weights is useful even when we do not know how many molecules there are in a given unit of matter; for it enables us at least to measure the *same* number of molecules of different substances.

The chemical formula that indicates the *relative* numbers of atoms in a compound, with no common divisor among the numbers of atoms, is called the *empirical formula*, or sometimes the "simplest formula." The empirical formula for table salt is NaCl, the one for water is H_2O, and the one for benzene is CH. The first two formulas are acceptable, but the last one is wrong because we know that a benzene molecule is not comprised of one carbon atom and one hydrogen atom. Benzene's *molecular formula*, which reveals not only the ratio of atoms, but also the number of atoms in a molecule, is C_6H_6. If the compound forms discrete molecules, the molecular formula will be either the empirical formula or an integral multiple of it.

We always can find the empirical formula by chemical analyses, but we need more information to derive the molecular formula. Note that for table salt the molecular formula has no meaning. No one Na^+ ion is associated with any particular Cl^- ion in the crystal. If we insist on using a molecular formula to describe the composition of the next most highly organized unit above the atomic level, then the molecular formula of table salt would be Na_xCl_x, in which x is the number of atoms of Na or Cl in the particular salt crystal chosen. In contrast, in the gas phase Na^+ and Cl^- ions associate in pairs, and NaCl is a legitimate molecular formula. It is conventional to use the empirical formula of a salt when writing equations in the same manner that the molecular formula is used for a compound that forms molecules. A chemical formula usually will represent a molecular formula if the compound exists as a discrete molecule, or an empirical formula if it does not.

The sum of the atomic weights of all of the atoms in a molecule (or in an empirical formula for a nonmolecular substance such as NaCl) is called the *molecular weight*. In Chapter 1 we defined a *mole* of a compound as the amount of grams of a substance that is numerically equal to its molecular weight. Since we know that 1 mole of any substance has the same number of molecules, we also can define a mole of a compound as that amount which contains as many molecules as there are atoms in exactly 12 g of ^{12}C. This new definition is an "ideal" definition because it expresses a useful feature of moles: the equal numbers of atoms. Conversely, our former definition is an "operational" definition; it describes the operation by which we can find this desired number. The current value of the number of molecules in a mole is 6.022169×10^{23}. However, it is not as important to know exactly how many atoms are taking part in a reaction as it is to know how to measure equivalent amounts of different reactants.

Example. Three gases are analyzed and have the following elemental compositions:

Gas A: 12 g of carbon for each 1 g of hydrogen

Gas B: 6 g of carbon for each 1 g of hydrogen

Gas C: 4 g of carbon for each 1 g of hydrogen

What are their empirical and molecular formulas?

Solution

Gas A has $(12 \text{ g}/12 \text{ g mole}^{-1}) = 1.0$ mole of C
for every $(1 \text{ g}/1 \text{ g mole}^{-1}) = 1.0$ mole of H,
and an empirical formula of CH.

Gas B has $(6 \text{ g}/12 \text{ g mole}^{-1}) = 0.5$ mole of C
for every $(1 \text{ g}/1 \text{ g mole}^{-1}) = 1.0$ mole of H,
and an empirical formula of CH_2.

Gas C has 0.33 mole of C for every mole of H,
and an empirical formula of CH_3.

From the information given we *cannot* find the molecular formulas.

Example. Add to the previous problem the information that, at STP, the measured densities of the three gases all are within the range 1.2 ± 0.2 g liter^{-1}. Now what are the molecular formulas?

Solution. Since, at STP, 1 mole of gas occupies 22.414 liters, the weight of 1 mole must be 1.2 g liter^{-1} $\times$ 22.414 liters = 27 g, with a possible range of error from $1.0 \times 22.414 = 22.4$ g to $1.4 \times 22.414 = 31.2$ g. Even such approximate density measurements are close enough to allow a choice among the possible molecular weights. The possibilities for gas A are $12 + 1 = 13$

for CH, $2 \times 12 + 2 \times 1 = 26$ for C_2H_2, $3 \times 12 + 3 \times 1 = 39$ for C_3H_3, and so forth. The second choice is obviously right. In a similar way, you can show that the other two gases must be C_2H_4 and C_2H_6.

It is fairly common in science to find, from experimentation, several possible precise values for a quantity such as molecular weight, differing by an unknown integer, and then to choose the correct value on the basis of a much cruder physical measurement that gives little more than the magnitude.

Before you go on. If you still have difficulty with problems involving gram-atoms, moles, and empirical formulas, you will find a review of these basic concepts in Review 1 of Lassila's *Programed Reviews of Chemical Principles.*

4–2 EQUATIONS

A chemical reaction such as the one illustrated by Equation 4–1 can have more than one interpretation

$$C_3H_8 + 5O_2 \rightarrow 3CO_2 + 4H_2O \tag{4–1}$$

The simplest interpretation is that 1 mole of propane gas (C_3H_8) burns with 5 moles of oxygen and yields 3 moles of carbon dioxide and 4 moles of water. Knowing the molecular weights, we can say that 44 g of propane react with 160 g of oxygen to produce 132 g of carbon dioxide and 72 g of water, with no net gain or loss of weight during the reaction.

We also can interpret the equation to mean that one *molecule* of propane gas reacts with five molecules of oxygen to yield three molecules of carbon dioxide and four molecules of water. This description is true in the overall sense, but we would be reading too much into the equation if we assume that the mechanism of the reaction is for one propane molecule and five O_2 molecules to collide simultaneously and react.

We get into more trouble with an equation such as

$$CaCO_3 + 2HCl \rightarrow CaCl_2 + CO_2 + H_2O \tag{4–2}$$

Again, the molar interpretation is valid. Furthermore, we can say that 100.1 g of calcium carbonate, if allowed to react with 72.9 g of hydrogen chloride, will yield 44.0 g of carbon dioxide, 18 g of water, and 111 g of calcium chloride upon drying the resulting solution. Nevertheless, we would be demanding too much from the equation if we claimed that one molecule of calcium carbonate (if such a molecule existed) will react with two molecules of pure HCl in the manner shown. As written, the equation is a device for keeping track of the amounts of material involved; it may or may not also represent the actual mechanism of reaction. A better approximation to what really happens is

$$CaCO_3(s) + 2H^+(aq) \rightarrow Ca^{2+}(aq) + CO_2(g) + H_2O(l)$$

This is a more accurate description of the actual reaction, but is less useful as a record of how much of each compound is involved. [The symbol (g) after a chemical substance in an equation indicates that it is in the gas phase. Similarly, (l) represents a liquid, (s) a solid, and (aq) an ion hydrated in solution. When considering the energy of a reaction, it is essential to specify the physical state in which each of the reactants and products is found.]

We must deduce chemical equations describing reactions from experiments that (a) identify reactants and products, and (b) measure the relative quantities of each reactant and product involved. Simple identification is quite often all that is necessary to deduce an equation. For instance, hydrogen reacts with chlorine to produce hydrogen chloride. Knowing the chemical formulas of reactants and product, we can write $H_2 + Cl_2 \rightarrow HCl$. Then we deduce the correct equation by simply insisting that no hydrogen or chlorine atoms can be created or destroyed:

$$H_2 + Cl_2 \rightarrow 2HCl$$

We now have a *balanced* equation that must describe the overall result if hydrogen and chlorine react and hydrogen chloride is the only product.

For assurance, we should measure the relative amounts of reactants and products before affirming that a reaction occurs according to a particular equation. For instance, the following equations all are balanced:

$$2MnO_4^- + 3H_2O_2 + 6H^+ \rightarrow 2Mn^{2+} + 4O_2 + 6H_2O$$
$$2MnO_4^- + 5H_2O_2 + 6H^+ \rightarrow 2Mn^{2+} + 5O_2 + 8H_2O$$
$$2MnO_4^- + 7H_2O_2 + 6H^+ \rightarrow 2Mn^{2+} + 6O_2 + 10H_2O$$

More equations could be written using the same reactants and products. Careful measurement shows that 2 moles of MnO_4^- and 5 moles of H_2O_2 disappear for every 5 moles of O_2 formed. Therefore, the middle equation describes what actually happens. Thus, it is necessary to identify reactants and products and to measure mole ratios before an equation can be written that adequately describes an actual process. Chemical equations must account for the relative number of moles of each reactant and product, and each side of the equations must contain the same total number of atoms for each kind of atom and the same net charge. *Mass, atoms, and charge all must be conserved.* Then an equation is said to be balanced.

Comparing Quantities by Using Equations

Chemists have observed that hydrogen peroxide (H_2O_2) decomposes into oxygen (O_2) and water (H_2O). We can depict the decomposition qualitatively by stating that H_2O_2 gives H_2O and O_2. To account for all of the quantities of reactants and products involved, we determine by experiment that

2 moles of H_2O_2 give 2 moles of H_2O + 1 mole of O_2

or

$$2H_2O_2 \rightarrow 2H_2O + O_2 \qquad\qquad (4\text{--}3)$$

From the equation and a table of atomic masses we can deduce that

2 moles, or 2 × 34 g, of H_2O_2 give

2 moles, or 2 × 18 g, of H_2O plus

1 mole, or 32 g, of O_2 (22.4 liters at STP)

An equation such as Equation 4–3 indicates the relative numbers of moles of reactants and products involved in a chemical reaction. For 1 mole of any substance we can determine these quantities: (a) the weight of 1 mole, from the chemical formula, and (b) the volume of 1 mole for any gaseous substance, from $PV = nRT$. It therefore follows that if, for any substance appearing in the equation, we are given (a) its weight or (b) its gas volume at stated conditions of pressure and temperature, we can find either of these quantities for any other substance involved in the equation.

The most complicated problems involving quantities of matter often can be dealt with easily by dividing them into sections and then treating them as a series of conversions of units.

Example.   If 2.50 g of H_2O_2 decompose, (a) what weight of H_2O is formed? (b) What volume of O_2 is produced at 750 torr and 300°K?

Solution.   (a) The relationship between H_2O_2 and H_2O is the chemical equation, which is written in terms of moles, not grams. The first step, then, is to convert the quantity of H_2O_2 to moles. Since 1 mole of H_2O_2 is 34 g, multiplication by the quotient 34 g H_2O_2/1 mole H_2O_2 is like multiplying by 1/1. A conversion from moles to grams of H_2O_2 can be set up in the following way:

$$5 \text{ moles } H_2O_2 \times \frac{34 \text{ g } H_2O_2}{1 \text{ mole } H_2O_2} = 170 \text{ g } H_2O_2$$

In this problem we convert grams to moles by multiplying by the inverse of the conversion factor used above:

$$2.50 \text{ g } H_2O_2 \times \left(\frac{1 \text{ mole } H_2O_2}{34.0 \text{ g } H_2O_2} \right) = 0.0735 \text{ mole } H_2O_2$$

The second conversion is between moles of H_2O_2 and moles of H_2O. Since 2 moles of one produce 2 moles of the other, the following conversion is correct:

$$0.0735 \text{ mole } H_2O_2 \times \left(\frac{2 \text{ moles } H_2O}{2 \text{ moles } H_2O_2} \right) = 0.0735 \text{ mole } H_2O$$

Finally, the conversion from moles of H_2O to grams is

$$0.0735 \text{ mole } H_2O \times \left(\frac{18.0 \text{ g } H_2O}{1 \text{ mole } H_2O}\right) = 1.32 \text{ g } H_2O$$

Overall,

$$\frac{2.50 \text{ g } H_2O_2}{34.0 \text{ g } H_2O_2/1 \text{ mole } H_2O_2} \times \frac{2 \text{ moles } H_2O}{2 \text{ moles } H_2O_2} \times \frac{18.0 \text{ g } H_2O}{1 \text{ mole } H_2O} = 1.32 \text{ g } H_2O$$

(b) The second part also can be handled as a series of conversions:

$$2.50 \text{ g } H_2O_2 \times \left(\frac{1 \text{ mole } H_2O_2}{34 \text{ g } H_2O_2}\right) = 0.0735 \text{ mole } H_2O_2$$

$$0.0735 \text{ mole } H_2O_2 \times \left(\frac{1 \text{ mole } O_2}{2 \text{ moles } H_2O_2}\right) = 0.0368 \text{ mole } O_2$$

$$0.0368 \text{ mole } O_2 \times \left(\frac{22.4 \text{ liters } O_2 \text{ at STP}}{1 \text{ mole } O_2}\right) = 0.825 \text{ liter } O_2 \text{ at STP}$$

$$0.825 \text{ liter } \times \left(\frac{300°K}{273°K}\right) = 0.905 \text{ liter} \qquad\qquad \text{(Eq. 2–8)}$$

$$0.905 \text{ liter } \times \left(\frac{760 \text{ torr}}{750 \text{ torr}}\right) = 0.916 \text{ liter} \qquad\qquad \text{(Eq. 2–7)}$$

$$0.916 \text{ liter } \times \left(\frac{1000 \text{ ml}}{1 \text{ liter}}\right) = 916 \text{ ml of } O_2 \text{ at } 300°K \text{ and } 750 \text{ torr}$$

The fourth and fifth steps are not conversions of units as are the others, but are applications of Charles' and Boyle's laws. The first conversion is from grams of H_2O_2 to moles of H_2O_2, then to moles of O_2, and then to liters of O_2 at STP. The fourth step yields the volume of O_2 at 760 torr and $300°K$, and the fifth, the volume of O_2 at 750 torr and $300°K$. Verify for yourself that the units cancel properly at each step, and that each step makes sense. This is a rather intuitive and rapid scheme of going from STP to any other conditions. Prove for yourself that the calculation of the *number* of moles of O_2 and the use of the ideal gas law, $PV = nRT$, lead to the same answer.

Example. What volume of a solution containing 0.30 mole of H_2O_2 per liter is needed to produce 5.00 g of oxygen gas?

Solution

$$5.00 \text{ g } O_2 \times \frac{1 \text{ mole } O_2}{32.0 \text{ g } O_2} \times \frac{2 \text{ moles } H_2O_2}{1 \text{ mole } O_2} \times \frac{1 \text{ liter } H_2O_2 \text{ solution}}{0.30 \text{ mole } H_2O_2}$$

$$= 1.04 \text{ liters of } H_2O_2 \text{ solution}$$

Break the preceding summary of the conversions into individual conversion steps, and satisfy yourself that each step makes sense.

Example.   The following reaction is known to occur:

$$K_2Cr_2O_7 + 8HClO_4 + 6HI \rightarrow 2KClO_4 + 2Cr(ClO_4)_3 + 3I_2 + 7H_2O$$
$$(4\text{--}4)$$

Calculate the quantities of each substance remaining if 2.00 liters of HI gas at STP are added to 5.73 g of $K_2Cr_2O_7$ in 0.250 liter of 1.00-molar $HClO_4$.

Solution.   First we must determine which reactant is consumed completely, by finding the number of moles of each:

$$\frac{5.73 \text{ g } K_2Cr_2O_7}{294 \text{ g } K_2Cr_2O_7/\text{mole } K_2Cr_2O_7} = 0.0195 \text{ mole } K_2Cr_2O_7$$

$$\frac{0.250 \text{ liter } HClO_4 \times 1.00 \text{ mole } HClO_4}{1.00 \text{ liter } HClO_4} = 0.250 \text{ mole } HClO_4$$

$$\frac{2.00 \text{ liters HI}}{22.4 \text{ liters/mole HI}} = 0.0893 \text{ mole HI}$$

The compounds $K_2Cr_2O_7$, $HClO_4$, and HI react in the relative amounts of 1 mole to 8 moles to 6 moles, as given by the chemical equation. In this reaction, 1 mole of $K_2Cr_2O_7$, 8 moles of $HClO_4$, and 6 moles of HI are each one stoichiometric unit of the compound involved. A *stoichiometric unit* of a compound, in a specified reaction, is the number of moles of the compound equal to the numerical coefficient accompanying the compound in the equation. This terminology is convenient because equal numbers of stoichiometric units of all the reactants will react with no leftover material.

In this problem, there are 0.0195 mole of $K_2Cr_2O_7$, 0.250 mole of $HClO_4$, and 0.0893 mole of HI. This is equivalent to

$$0.0195 \text{ mole } K_2Cr_2O_7 \times \frac{1 \text{ stoich. unit}}{1 \text{ mole } K_2Cr_2O_7} = 0.020 \text{ stoich. unit } K_2Cr_2O_7$$

$$0.250 \text{ mole } HClO_4 \times \frac{1 \text{ stoich. unit}}{8 \text{ moles } HClO_4} = 0.031 \text{ stoich. unit } HClO_4$$

$$0.0893 \text{ mole HI} \times \frac{1 \text{ stoich. unit}}{6 \text{ moles HI}} = 0.015 \text{ stoich. unit HI}$$

In terms of stoichiometric units, or units of the reaction as written, there is less HI than either of the other reactants, thus some $HClO_4$ and $K_2Cr_2O_7$ will remain unreacted. When we calculate how much $KClO_4$ is produced when all of the HI reacts, we find that

$$0.0893 \; \cancel{\text{mole HI}} \times \frac{2 \; \cancel{\text{moles KClO}_4}}{6 \; \cancel{\text{moles HI}}} \times \frac{138.6 \; \text{g KClO}_4}{1 \; \cancel{\text{mole KClO}_4}}$$

$$= 4.12 \; \text{g KClO}_4 \; \text{is produced,}$$

$$0.0893 \; \cancel{\text{mole HI}} \times \frac{1 \; \cancel{\text{mole K}_2\text{Cr}_2\text{O}_7}}{6 \; \cancel{\text{moles HI}}} \times \frac{294 \; \text{g K}_2\text{Cr}_2\text{O}_7}{1 \; \cancel{\text{mole K}_2\text{Cr}_2\text{O}_7}}$$

$$= 4.38 \; \text{g K}_2\text{Cr}_2\text{O}_7 \; \text{is used up, and}$$

$$5.73 - 4.38 = 1.35 \; \text{g K}_2\text{Cr}_2\text{O}_7 \; \text{is left over.}$$

Work out the rest of the problem and check your answers by applying the law of conservation of mass. Each setup should be checked carefully for cancellation of units to give the proper units for the quantity sought.

Before you go on. The ability to calculate the relative amounts of reactants and products, by using a balanced equation and a given mass of reagent, is vitally important in chemistry. You will find a review of this topic in Section 4–1 of Lassila's book.

4–3   BALANCING CHEMICAL EQUATIONS

Of the two equations just discussed, the hydrogen peroxide equation could be balanced by inspection. However, given only the reactants and products, it would be a formidable task to balance the second equation (with so many components) by trial and error. If we can recognize, in a system undergoing chemical change, some basic process that is quantified easily, then we can employ this as a starting point for balancing equations.

Acid–Base Reactions

We shall consider acid–base equilibria in the next chapter. For the moment, we want only to see how matter is conserved in acid–base reactions. Arrhenius classified acids as substances that give H^+ (or H_3O^+) in water and bases as substances that give OH^- in water. We shall use a slightly different classification of acid–base reactions, as suggested by the Danish chemist Johannes Brønsted (1879–1947). These reactions involve transfer of hydrogen ions (H^+) from acids to bases.

According to the Brønsted definition, a base is *any* substance that can accept a hydrogen ion in a reaction. An acid is any substance that can donate a hydrogen ion. Once an acid has donated a hydrogen ion, the remaining part of the original acid is a base, since it can *accept* a hydrogen ion. Similarly, once a base accepts a proton, the combination of base plus proton becomes an acid because it is a potential H^+ donor. Such acid–base pairs are called *conjugate pairs*. By this definition, HCl is an acid and Cl^- is

Table 4–1.   Common Acids and Bases

Acids			
HF	Hydrofluoric	H_3BO_3	Boric
HCl	Hydrochloric		
HClO	Hypochlorous		
$HClO_2$	Chlorous	HCOOH	Formic,
$HClO_3$	Chloric		
$HClO_4$	Perchloric		
HBr	Hydrobromic		
$HBrO_3$	Bromic	CH_3COOH	Acetic,
HI	Hydriodic		
HIO_4	*ortho*-Periodic		
H_5IO_6	*para*-Periodic		
H_2SO_3	Sulfurous	Bases	
H_2SO_4	Sulfuric	LiOH	Lithium hydroxide
HNO_2	Nitrous	NaOH	Sodium hydroxide
HNO_3	Nitric	KOH	Potassium hydroxide
H_3PO_2	Hypophosphorous	$Mg(OH)_2$	Magnesium hydroxide
H_3PO_3	Phosphorous	$Ca(OH)_2$	Calcium hydroxide
H_3PO_4	Phosphoric	$Ba(OH)_2$	Barium hydroxide
H_2CO_3	Carbonic	NH_4OH	Ammonium hydroxide

its conjugate base; NH_3 is a base, and $NH_4{}^+$ is its conjugate acid; OH^- is a base and H_2O is its conjugate acid. But H_2O plays a double role; it is also the conjugate base for the acid H_3O^+. Because H^+ is a proton, all of these reactions are called *proton-transfer reactions*. Therefore, if a reaction is of the proton-transfer type, to balance the equation for the reaction we need only ascertain the acid (proton donor) and base (proton acceptor) among the reactants and then make their ratio such that the number of protons donated equals the number received. Hence, a mole of $Ca(OH)_2$ requires 2 moles of HCl to be neutralized:

$$Ca(OH)_2 + 2HCl \rightarrow CaCl_2 + 2H_2O \tag{4-5}$$

Now consider the more complicated reaction

$$Al_2S_3 + H_2O \rightarrow Al(OH)_3 + H_2S \qquad \text{(unbalanced)}$$

Here, the sulfide ion (S^{2-}) is a base that accepts protons from water, which is an acid in this reaction. We see that each water molecule donates only one proton and forms hydroxide ion, but each sulfide ion requires two protons to make H_2S. Therefore, we need $2H_2O$ for each S^{2-}. The equation is

$$Al_2S_3 + 6H_2O \rightarrow 2Al(OH)_3 + 3H_2S \tag{4-6}$$

This and most other equations for acid–base reactions are simple enough to balance by inspection, but knowing the type of change involved also can be of assistance in writing a balanced equation. Moreover, it is helpful if our experience with chemicals enables us to know the common properties of species involved in reactions. In this case, most chemists would know that S^{2-} is a strong proton acceptor capable of reacting with water as above. (Common acids and bases are listed in Table 4–1.)

4–4 SOLUTIONS AS CHEMICAL REAGENTS

Liquid solutions often are convenient media for chemical reactions. The volumes of reactants and products change only slightly in solutions, and the easy flow of liquids simplifies measurement and transfer of quantities. For quantitative chemical work, solution concentrations can be defined readily in terms that enable us to measure the number of moles of a substance simply by determining a volume of solution.

In solutions involving a liquid and a gas or solid, the liquid component commonly is called the *solvent;* the other component is the *solute.* In solutions of two liquids the terminology is less clear. However, the liquid present in the greater amount usually is called the solvent. The *molarity* of a solute is defined as the number of moles of solute in *1 liter of solution.* This amount is not the same as 1 liter of solvent, because the volume of the liquid usually will increase when the solute is added. If 1 mole of NaCl is dissolved in 1 liter of water, the solution will *not* be 1 molar. The mole of NaCl must be dissolved in less than a liter of water, and the total volume then brought up to the 1-liter mark in a volumetric flask.

There is a practical method of estimating final volumes in salt solutions, which is useful even though the answer is only approximately correct. The method involves the assumption that volumes of solution components are *additive* and that a solution's volume is equal to the sum of the volumes of water and of crystalline salt. We can illustrate this assumption by an example.

Example. An amount of 264 g of ammonium sulfate [$(NH_4)_2SO_4$] is dissolved in 1.00 liter of water. What are the approximate final volume and approximate molarity?

Solution. The density of solid ammonium sulfate is 1.77 g cm^{-3}. Therefore, the volume of crystalline ammonium sulfate is

$$264 \text{ g} \times \frac{1 \text{ cm}^3}{1.77 \text{ g}} = 150 \text{ cm}^3 \quad \text{or} \quad 0.150 \text{ liter}$$

If we assume additivity of volumes, then the final volume of the solution would be 1.000 liter + 0.150 liter = 1.150 liters. The number of moles of solute is

$$264 \text{ g} \times \frac{1 \text{ mole}}{132 \text{ g}} = 2.00 \text{ moles ammonium sulfate}$$

The approximate molarity of such a solution is

$$\frac{2.00 \text{ moles}}{1.150 \text{ liters of solution}} = 1.74 \text{ molar}$$

Because ammonium sulfate is a common medium in protein chemistry, tables of molarity based on actual measurements are available. The true molarity of such a solution is 1.80 molar.

Exercise. Again assuming the additivity of volumes, approximately what volume of water will be required, along with 264 g of ammonium sulfate, to make 1.00 liter of a 2.00-molar solution?

(*Answer:* 0.85 liter of water.)

This sort of calculation *should be considered only as an approximate guide to true volumes.* Usually the true volume is less than or greater than the result computed by this method. Yet the method is handy when rough concentrations are needed; furthermore, it illustrates the fundamental difference between molarity and our next concentration measure, molality.

We define the *molality* of a solute as the number of moles of solute in *1000 g of solvent.* The 1.74-molar ammonium sulfate solution from our previous example is 2.00 *molal*, because the density of water is nearly 1 g cm^{-3}, and 1 liter of water is about 1000 g of water at room temperature. For other solvents, we would need to make a correction for the density of the solvent. *Molality* is useful because it is easy for calculating percent composition by weight or mole fraction. In contrast, the conversion from *molarity* to mole fraction requires a knowledge of how much solvent it took to make a *liter of solution.*

In dilute solutions, the volume contributed by the solute can be neglected, and the volume of solvent and of final solution can be assumed to be the same. In such dilute solutions, molality and molarity become the same in water. In other solvents, the correction for solvent density must be made.

Example. A solution of acetic acid in ethanol (C_2H_5OH) is 0.0100 molar. What is the *molality* of the solution?

Solution. The density of liquid ethanol is 0.789 g ml^{-1} = 0.789 kg liter^{-1}. Then the molality is

$$0.0100 \text{ mole liter}^{-1} \times \frac{1 \text{ liter}}{0.789 \text{ kg}} = 0.0127 \text{ mole kg}^{-1}$$
$$= 0.0127 \text{ molal}$$

Dilution Problems

If we dilute a solution by adding more solvent, the number of moles of solute does not change. This number is given by MV, in which M is the molarity (not molality) and V is the volume in liters. If the subscript 1 represents the solution before dilution, and 2 represents the diluted solution,

$$\text{Moles of solute} = M_1V_1 = M_2V_2 \qquad (4\text{--}7)$$

Example. To what volume must 5.00 ml of 6-molar HCl be diluted to make the concentration 0.100 molar?

Solution

$$\frac{6 \text{ moles}}{1000 \text{ ml}} \times 5.00 \text{ ml} = \frac{0.100 \text{ mole}}{1000 \text{ ml}} \times x$$

$x = 300$ ml after dilution (*not* the amount of solvent to be added during dilution).

Exercise. What will be the molarity after dilution, if 175 ml of a 2.00-molar solution are diluted to 1 liter?

(*Answer:* 0.350 molar.)

Before you go on. A practical knowledge of solutions and solution stoichiometry is essential to a chemistry student. Section 4–2 of Lassila's book supplements Section 4–4 of this text by introducing solution reactions and stoichiometry. If you study Section 4–2, you can improve your knowledge of: molarity and dilution problems; definitions of solution, solute, solvent; how to solve problems involving percent-by-weight concentration and gram-formula weight; and how to write equations for reactions of ionic salts in solution.

Chemical Equivalents in Acid–Base Reactions

One gram equivalent weight of an acid is the quantity of acid that can donate 1 mole of protons (H^+) to a base; 1 gram equivalent weight of base is that amount which can accept 1 mole of protons. Chemical equivalents of any acid and any base will exactly neutralize each other. Since one molecule of HCl releases one proton, the molecular weight and equivalent weight are the same. But the equivalent weight of sulfuric acid, H_2SO_4, is half the molecular weight; the equivalent weight of phosphoric acid, H_3PO_4, is one third its molecular weight. Similarly, a mole and a gram equivalent weight of NaOH or KOH are identical, but the equivalent weight of $Ca(OH)_2$ is half its molecular weight. We can appreciate the advantage of equivalent nomenclature in the following neutralization reaction:

$$2H_3PO_4 + 3Ca(OH)_2 \rightarrow 6H_2O + Ca_3(PO_4)_2 \qquad (4\text{--}8)$$

Molecular weights:	98.0	74.1	18.0	310.3
Equivalent weights:	32.7	37.0		

It is not true that 98.0 g of phosphoric acid will react with 74.1 g of calcium hydroxide, for the compounds do not react in 1/1 mole ratio. But 1 equivalent of phosphoric acid reacts completely with 1 equivalent of calcium hydroxide. Because 1 mole of phosphoric acid furnishes 3 moles of protons, the gram equivalent weight is $98.0/3 = 32.7$ g. And since 1 mole of calcium hydroxide furnishes enough hydroxide ion to react with 2 moles of protons, the equivalent weight is half the molecular weight. Equivalent weights in acid–base reactions are similar to combining weights in the synthesis of compounds from elements.

The *normality* of a solution, represented by N, is the number of gram equivalent weights of solute per liter of solution. A 1-molar solution of phosphoric acid is therefore 3 normal; a 0.01-molar solution of calcium hydroxide is 0.02 normal.

Example. An amount of 4.00 g of sodium hydroxide is dissolved in water. The volume of the solution is brought up to 500 ml. What is the normality of the solution?

Solution. The molecular weight of sodium hydroxide is 40.0 g mole^{-1}, so the sample is 0.100 mole. The molarity is

$$\frac{0.100 \text{ mole}}{0.500 \text{ liter}} = 0.200 \text{ molar}$$

Because 1 mole of sodium hydroxide (NaOH) accepts 1 mole of protons, the solution also is 0.200 normal.

Exercise. An amount of 10.0 g of sulfuric acid (H_2SO_4) is mixed slowly with enough water to make a final volume of 750 ml. What is the normality of the solution?

(*Answer:* 0.272 normal.)

Before you go on. The theory of Brønsted acids and bases and the stoichiometry of acid–base reactions is discussed in detail in Sections 4–3 and 4–4 of Lassila's book.

Titration

Chemists frequently compare relative concentrations of chemical equivalents in solutions of reagents by means of titrations (Figure 4–1). When enough acid solution in a calibrated burette has been added to neutralize the base in a sample being analyzed, the number of equivalents of acid and base is the same. A sensitive indicator determines the end point of the neutralization. From the volume of acid solution used and its normality, we can calculate the number of equivalents of base in the sample.

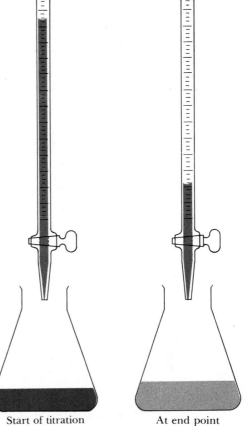

Figure 4–1. An acid–base titration. The solution in the flask contains an unknown number of equivalents of base (or acid). The burette is calibrated to show volume to the nearest 0.001 cm³. It is filled with a solution of strong acid (or base) whose concentration is known. Small increments of acid are added until, at the end point, one drop or less changes the indicator color permanently. (An indication that the end point is being approached is the transient appearance of the color that the indicator assumes in acid media.) At the end point, the total amount of acid is recorded from the burette readings. The number of equivalents of added acid equals the initial number of equivalents of base.

Start of titration   At end point

For a titration of acid and base, the number of equivalents of each is the same, and

$$N_A V_A = N_B V_B = w_A/W_A = w_B/W_B \qquad (4\text{–}9)$$

in which w is the weight in grams, N the normality, V the volume, W the equivalent weight, and the subscripts A and B represent acid and base, respectively. From a titration such as the one in Figure 4–1, we can determine the following:

1) The relative normalities of a solution of an acid and a solution of base.
2) The amount of an acid (or base) in a solution by titration with a standard solution of base (or acid).
3) The equivalent weight of an unknown acid (or base) by titration of a weighed sample with a standard solution of base (or acid).

Example. Twenty-five (25.00) milliliters (ml) of phosphoric acid (H_3PO_4) are neutralized by 30.25 ml of sodium hydroxide (NaOH). What is the ratio of the normalities of the two solutions? What is the ratio of their molarities?

Solution. When working problems dealing with reactions, you should always first write the equation for the reaction:

$$3NaOH + H_3PO_4 \rightarrow Na_3PO_4 + 3H_2O$$

Equivalents of base = equivalents of acid

$$\frac{30.25}{1000} \text{liter} \times N_B \frac{\text{equiv}}{\text{liter}} = \frac{25.00}{1000} \text{liter} \times N_A \frac{\text{equiv}}{\text{liter}}$$

$$\frac{N_A}{N_B} = \frac{30.25}{25.00} = 1.210$$

Notice that for these two reagents, the normality of A is three times the molarity, and that of B is equal to its molarity. Therefore, the molarity ratio is

$$\frac{\text{Molarity of } A}{\text{Molarity of } B} = \frac{N_A/3}{N_B} = \frac{1.210}{3} = 0.403$$

Example. In a titration, 25.00 ml of a saturated solution of $Ca(OH)_2$ required 10.81 ml of 0.100-normal HCl for neutralization. What is the normality of $Ca(OH)_2$? Translate the normality into solubility of $Ca(OH)_2$ in terms of molarity and in terms of grams per liter.

Solution

$$Ca(OH)_2 + 2HCl \rightarrow CaCl_2 + 2H_2O$$

$$0.01081 \text{ liter HCl} \times 0.100 \text{ equiv liter}^{-1} = 0.001081 \text{ equiv HCl}$$
$$= 0.001081 \text{ equiv } Ca(OH)_2$$

$$\text{Concentration of } Ca(OH)_2 = \frac{0.001081 \text{ equiv}}{0.0250 \text{ liter}}$$
$$= 0.0432 \text{ normal}$$

There are 2 gram-equivalents per mole of $Ca(OH)_2$.

$$\text{Concentration of saturated } Ca(OH)_2 = 0.0216 \text{ molar}$$

$$0.0216 \text{ mole liter}^{-1} \times 74 \text{ g mole}^{-1} = 1.60 \text{ g liter}^{-1}$$

Example. An organic chemist has synthesized a new compound that is an acid. He dissolves 0.500 g in a convenient volume of water and finds that it requires 15.73 ml of 0.437-normal NaOH for neutralization. What is the gram equivalent weight of the new compound as an acid?

Solution

$$0.01573 \text{ liter} \times 0.437 \text{ equiv liter}^{-1} = 0.00687 \text{ equiv NaOH}$$
$$0.500 \text{ g acid} = 0.00687 \text{ equiv}$$
$$72.8 \text{ g acid} = 1 \text{ equiv}$$

Thus, the equivalent weight of the acid is 72.8 g.

4–5   QUANTITIES OF ENERGY IN CHEMICAL CHANGES

When hydrogen peroxide in solution reacts to form oxygen gas and liquid water, heat is liberated. The amount of heat will vary somewhat with the temperature at which the reaction occurs, but at 298°K (the commonly accepted standard "room temperature" for tabulating heat data), each mole of H_2O_2 that reacts produces 22.64 kcal of heat. [A kilocalorie (kcal) is 1000 calories (cal); 1 calorie of heat is that amount required to raise the temperature of 1 g of water from 14.5°C to 15.5°C.] The heat involved in a chemical reaction carried out at constant pressure is referred to as the change in the system's enthalpy, ΔH. [As we shall see in Chapter 15, the change in the system's energy, ΔE, is the heat of a reaction at constant volume, as in a bomb calorimeter (Figure 4–2). The difference between these terms is small but significant.]

If heat is released, the enthalpy of the system falls, and ΔH is negative. Such a reaction is called *exothermic*. In an *endothermic* reaction, heat is absorbed and the enthalpy of the reacting system rises. For the hydrogen peroxide system, we can write the heat relationship as

$$H_2O_2(aq) \rightarrow H_2O(l) + \tfrac{1}{2}O_2(g) \qquad \Delta H_{298} = -22.64 \text{ kcal} \qquad (4\text{--}10)$$

The heat of reaction as quoted is for one *stoichiometric unit* of the reaction as written. One stoichiometric unit of Reaction 4–10 would occur when 1 mole of H_2O_2 has decomposed to 1 mole of water and $\tfrac{1}{2}$ mole of oxygen.

The heat of a reaction depends on how the reaction is written. If all the coefficients in Equation 4–10 are doubled to represent the decomposition of 2 moles of H_2O_2 to 2 moles of H_2O and one of O_2, then the heat of reaction also must be doubled:

$$2H_2O_2(aq) \rightarrow 2H_2O(l) + O_2(g) \qquad \Delta H_{298} = -45.28 \text{ kcal} \qquad (4\text{--}11)$$

The symbol (*aq*) for "aqueous" indicates what is called an "ideally dilute solution"—a solution so dilute that encounters of solute molecules with each other can be neglected. Each solute molecule is then effectively alone in a sea of solvent.

The heat of a reaction also depends on the state of the reactants and products. If hydrogen peroxide were decomposed to oxygen gas and water vapor instead of liquid, part of the 22.64 kcal would be diverted to evaporating

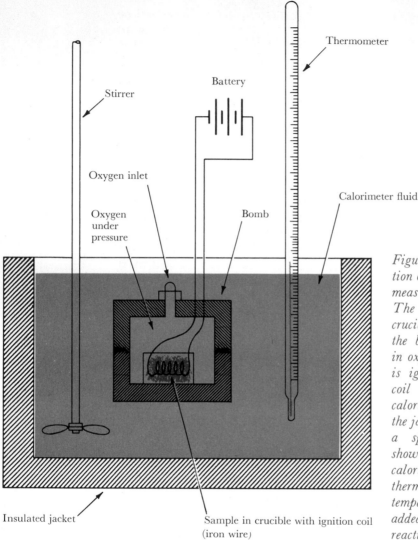

Figure 4–2. A schematic representation of a bomb calorimeter used for the measurement of heats of combustion. The weighed sample is placed in a crucible (which in turn is placed in the bomb) and is burned completely in oxygen under pressure. The sample is ignited by an iron wire ignition coil that glows when heated. The calorimeter is insulated by means of the jacket, which can be equipped with a special thermal regulator (not shown). The temperature of the calorimeter fluid is measured with the thermometer. From the change in temperature and the amount of heat added through the wire, the heat of reaction can be calculated.

water, and less heat would be evolved:

$$H_2O_2(aq) \rightarrow H_2O(g) + \tfrac{1}{2}O_2(g) \qquad \Delta H_{298} = -12.12 \text{ kcal} \qquad (4\text{–}12)$$

Furthermore, if pure liquid H_2O_2 were the reactant instead of a dilute aqueous solution, slightly *more* heat would be liberated:

$$H_2O_2(l) \rightarrow H_2O(l) + \tfrac{1}{2}O_2(g) \qquad \Delta H_{298} = -23.48 \text{ kcal} \qquad (4\text{–}13)$$

More heat is liberated because H_2O_2 produces 0.84 kcal of heat per mole when it dissolves in water, and the solution therefore has less enthalpy to emit during the reaction.

There is a hidden assumption in the foregoing: that heats of reaction or enthalpy changes are additive. We derived Equation 4–12 by adding to Equation 4–10 another equation representing the vaporization of water; the heat terms associated with the two reactions also were added:

$$H_2O_2(aq) \rightarrow H_2O(l) + \tfrac{1}{2}O_2(g) \qquad \Delta H_{298} = -22.64 \text{ kcal} \qquad (4\text{–}10)$$

$$\underline{H_2O(l) \quad \rightarrow H_2O(g) \qquad\qquad\qquad \Delta H_{298} = +10.52 \text{ kcal}}$$

$$H_2O_2(aq) \rightarrow H_2O(g) + \tfrac{1}{2}O_2(g) \qquad \Delta H_{298} = -12.12 \text{ kcal} \qquad (4\text{–}12)$$

The statement that this addition is a proper procedure is one form of the first law of thermodynamics, to which we shall return in Chapter 15. A collection of chemicals in a given state has a certain energy and a certain enthalpy, neither of which depend in any way on how the chemicals were brought to that state. Hence, the difference between enthalpies of reactants and products of a reaction, which is the heat of the reaction, depends only on the starting and ending states and not on the exact path by which the reaction was carried out. This application of the first law of thermodynamics to chemical reactions sometimes is called *Hess' law of heat summation:* If a chemical reaction can be expressed as the algebraic sum of two other reactions, the enthalpy change of the given reaction will be the algebraic sum of the enthalpy changes of the other two.

This fact means a tremendous saving in experimentation in thermochemistry. The enthalpy of every chemical reaction need not be measured and its results tabulated. If we know the heat of vaporization of liquid water, and also the heat of reaction of hydrogen peroxide to form liquid water, then we do not need to make any further measurements to find out how much heat is liberated in the reaction to produce water vapor in the decomposition of H_2O_2. We simply can calculate it. If a reaction is inconvenient to accomplish, there may be a series of easier ones whose sum is the reaction in question. After these individual experiments are done, the enthalpy changes can be added in the same way as the chemical equations to find the difficult-to-measure heat.

For example, suppose that someone proposes to you a scheme for making diamonds (abbreviated "dia") by oxidizing methane (CH_4):

$$CH_4(g) + O_2(g) \rightarrow C(dia) + 2H_2O(l) \qquad\qquad (4\text{–}14)$$

and asks you whether the reaction is likely to produce heat that must be allowed for in the design of the reaction vessel. The reaction of Equation 4–14 never has been carried out and probably never will be, yet you can give your misguided friend his answer from a knowledge of the heats of combustion of methane and of diamond. The *heat of combustion* of a substance containing C, N, O, and H is the heat (per mole of substance) of the reaction with oxygen to produce nitrogen, carbon dioxide, and water. These were among the first reactions that chemists measured (Figure 4–2), and extensive tables are available in books such as the *CRC Handbook of Chemistry and Physics* and *Lange's Handbook of Chemistry*. The two combustion reactions and their heats are

$$CH_4(g) + 2O_2(g) \rightarrow CO_2(g) + 2H_2O(l) \qquad \Delta H_{298} = -212.8 \text{ kcal}$$
$$C(\text{dia}) + O_2(g) \rightarrow CO_2(g) \qquad\qquad \Delta H_{298} = -94.5 \text{ kcal}$$

The desired diamond-synthesizing equation is produced by adding the first reaction to the reverse of the second (for which the heat has the reverse sign). The enthalpy change that would be measured *if* the diamond synthesis could be induced is

$$CH_4(g) + 2O_2(g) \rightarrow CO_2(g) + 2H_2O(l) \qquad \Delta H_{298} = -212.8 \text{ kcal}$$
$$\underline{\qquad\qquad CO_2(g) \rightarrow O_2(g) + C(\text{dia}) \qquad \Delta H_{298} = +94.5 \text{ kcal}}$$
$$CH_4(g) + O_2(g) \rightarrow C(\text{dia}) + 2H_2O(l) \qquad \Delta H_{298} = -118.3 \text{ kcal}$$

Therefore, because of the first law of thermodynamics (or Hess' law), some reactions do not have to be tabulated; we require only the ones from which all others can be obtained by the proper combination. The ones that have been agreed upon by scientists and engineers are the heats of formation of compounds from elements in standard states. The heat of formation reactions for the compounds in the diamond-synthesis equation are

$$C(\text{gr}) + 2H_2(g) \rightarrow CH_4(g) \qquad \Delta H^0_{298} = -17.89 \text{ kcal} \qquad (4\text{-}15)$$
$$C(\text{gr}) \rightarrow C(\text{dia}) \qquad\qquad \Delta H^0_{298} = +0.45 \text{ kcal} \qquad (4\text{-}16)$$
$$H_2(g) + \tfrac{1}{2}O_2(g) \rightarrow H_2O(l) \qquad \Delta H^0_{298} = -68.32 \text{ kcal} \qquad (4\text{-}17)$$

The symbol "gr" on carbon represents graphite. The superscript zero on the enthalpy denotes that all reactants and products are in standard states. The *standard state* for a solid or liquid is the pure substance at 1 atm external pressure and at 25°C, or 298°K. The standard state for a gaseous element at 25°C is the gas at 1 atm partial pressure. The standard heat of formation of an element in its standard state is zero, by definition. But the standard state of solid carbon was chosen to be graphite, C(gr), and not diamond, and the conversion of graphite to diamond requires 0.45 kcal of heat per mole. (In an emergency, diamonds are a slightly better source of heat than graphite.)

A table of standard heats of formation is given in Appendix 2. We can derive Equation 4–14 from the three heat of formation equations by a process that can be represented symbolically by

$$(4\text{-}14) = (4\text{-}16) + 2(4\text{-}17) - (4\text{-}15)$$

In exactly the same way we find the heat of the reaction from the heats of formation:

$$\Delta H^0_{298} = (+0.45) + 2(-68.32) - (-17.89) = -118.30 \text{ kcal}$$

As you can see, the consequences of not keeping strict track of signs can be disastrous. (Unfortunately, tables of heats of formation are given at 25°C, whereas heats of combustion tables commonly are at 20°C. If you were to repeat these calculations at 20°C, your final value would differ by 2 kcal.)

The foregoing method of writing heat of formation equations in full, and canceling equivalent terms on either side to obtain the desired equation (thereby treating heats of formation exactly like the equations are treated) is a foolproof method. A convenient but sometimes risky shortcut is to think of a heat of formation of a compound as if it were, in some sense, the enthalpy of the compound. Then the heat of a reaction is the sum of the products' heats of formation, *minus* the sum of the reactants' heats of formation. Of course, each heat of formation must be multiplied by the coefficient of the compound as written in the equation.

Example. What is the standard enthalpy change of the reaction by which ferric oxide is reduced in a blast furnace?

Solution

$$Fe_2O_3(s) + 3C(gr) \rightarrow 2Fe(s) + 3CO(g)$$
$$\Delta H^0_{298} = -196.5 \quad 0.0 \quad 2(0.0) \quad 3(-26.4) \text{ kcal}$$

For the reaction as written,

$$\Delta H^0_{298} = 2(0.0) + 3(-26.4) - (-196.5) - (0.0) = +117.3 \text{ kcal}$$

which is consistent with the fact that much heat must be supplied to reduce the ore to iron. Note, however, that this is the net heat which would be absorbed if the reaction were run at 298°K (25°C); it may or may not approximate the situation at 1800°K in a blast furnace. Yet this calculated figure also is the heat absorbed if ferric oxide and carbon are heated from 298°K to 1800°K, allowed to react, and the products cooled again to room temperature. The enthalpy change in the reaction depends only on the initial and final states of the chemical system, and not on whether the temperature was raised to blast-furnace levels during the reaction; all that is important is that the temperature is lowered at the end. Whether a mixture of H_2 and O_2 explodes violently and the resultant water is cooled back to 298°K, or whether a mixture of hydrogen and oxygen reacts slowly with a platinum black catalyst at 298°K, the net heat evolved will be the same. So, in referring to heats of reaction, when we state that the values are correct for the reaction conducted "at 1 atm pressure and 298°K," we require only that the reactants begin at these conditions, and that the products end there. This is why tables of heats of formation under standard conditions such as the ones in Appendix 2 are of practical use in laboratory chemistry.

Example. What is the standard heat of combustion of liquid benzene?

Solution

$$2C_6H_6(l) + 15O_2(g) \rightarrow 12CO_2(g) + 6H_2O(l)$$
$$\Delta H^0_{298} = 2(+11.72) \quad 0.0 \quad 12(-94.05) \quad 6(-68.32)$$

And for the reaction as written (the combustion of 2 moles of benzene),

$$\Delta H^0_{298} = 12(-94.05) + 6(-68.32) - 2(+11.72)$$
$$= -1561.96 \text{ kcal}$$

Thus, the standard heat of combustion of benzene at 25°C is -780.98 kcal mole^{-1}. (Check the heat of combustion tables in a handbook to see if the heat of combustion at 20° is greater or less, and by how much. Can you think of any reasons for the trend you see?)

Before you go on.   You may feel that you need more drill to help you to understand the material in this section. You will find further discussion of the methods for calculating the amount of energy evolved or absorbed in chemical reactions and problems illustrating these methods in Section 4–5 of Lassila's book.

4–6 SUMMARY

This entire chapter has been concerned with measurements and numbers. In a sense, the task of a scientist is not to create new theories—there are always plenty of those—but to destroy inadequate ones. The tools of this demolition process are careful measurements. Above the rubble will be the ideas that best agree with reality.

We have seen the importance of the mole concept in placing chemistry on a quantitative basis. Also, we have examined the information in chemical equations about moles, and masses and volumes of reactants and products. You have seen how to balance chemical equations: simple equations using conservation of mass, and acid–base equations using proton balancing. Energy and enthalpy, too, are conserved in a chemical reaction. That is, the energy liberated as heat is always the same in a reaction between a given initial and final state, regardless of how the reaction is performed. Aristotle would have been pleased to see how faithfully present-day chemists adhere to his doctrine of "ex nihilo nihil fit"—nothing can come from nothing. In a properly balanced chemical equation that can represent the behavior of real substances, neither mass, nor kinds of atoms, nor protons, nor electrons, nor energy can be created or destroyed.

A large part of science is a search for conservation principles: for quantities such as mass, charge, and energy that do not change during a process. Out of delicate experiments on the conservation of symmetry of subnuclear particles, for example, there may come answers to broadly relevant questions such as: Are antiworlds possible? Did our material universe have a beginning? Does time, itself, have a preferred direction? To those who protest that scientists have not really explained things but have only succeeded in keeping track of them, we reply: As products and parts of the universe that we study, what more can we be expected to do?

SUGGESTED READING

S. W. Benson, *Chemical Calculations*, Wiley, New York, 1971, 3rd ed.

W. F. Kieffer, *The Mole Concept in Chemistry*, Reinhold, New York, 1962.

L. K. Nash, *Stoichiometry*, Addison-Wesley, Reading, Mass., 1966.

M. J. Sienko, *Chemistry Problems*, W. A. Benjamin, Menlo Park, Calif., 1972, 2nd ed. Chapters 2 through 5 are especially relevant here.

C. A. VanderWerf, *Acids, Bases, and the Chemistry of the Covalent Bond*, Reinhold, New York, 1961.

QUESTIONS

1 How many gram-atoms of gold are there in a kilogram-atom of gold? How many atoms of gold are there in a kilo-gram-atom of gold?

2 What is the difference between a molecular formula and an empirical formula of a compound?

3 What is the molecular formula for water? The empirical formula? What is the molecular formula for hydrogen peroxide? The empirical formula?

4 What is a stoichiometric unit of a reaction? In the reaction $H_2 + Cl_2 \rightarrow 2HCl$, how many moles of chlorine form one stoichiometric unit? How many grams? In the reaction $\frac{1}{2}H_2 + \frac{1}{2}Cl_2 \rightarrow HCl$, how many moles of chlorine form one stoichiometric unit? How many grams?

5 How does Brønsted's definition of acids and bases differ from that of Arrhenius? What is a conjugate acid and a conjugate base?

6 What is the principle of balancing acid–base neutralization reactions?

7 Under what conditions is a 1-*molal* solution also 1 *molar*, or effectively so?

8 Why are molality and molarity not identical in solvents with a density of 1000 g liter^{-1}?

9 What is the difference between molarity and normality? In an acid–base titration, what is the normality of a 0.10-molar solution of carbonic acid?

10 In an acid–base titration, what is the equivalent weight of calcium hydroxide?

11 Does the enthalpy of a reacting system rise, or fall, in an exothermic reaction?

12 Why does the heat of a reaction depend on the state of the reactants and products?

13 What are heats of reaction, heats of combustion, and heats of formation?

14 Why is it possible to find the heat of a reaction from the heats of formation of all of the reactants and products?

15 In the symbol ΔH_{298}^0, what do the superscript zero and the subscript 298 signify?

PROBLEMS

1 The following is a balanced chemical equation:

$$Zn + H_2SO_4 \rightarrow H_2 + ZnSO_4$$

State the qualitative meaning of the equation. What is the meaning in terms of the relative weights of compounds involved? What do we mean by saying that the equation is balanced?

2 Balance the following equations:

a) $Fe_2O_3 + Al \rightarrow Fe + Al_2O_3$

b) $Na_2SO_3 + HCl \rightarrow$
$$NaCl + SO_2 + H_2O$$

c) $Mg_3N_2 + H_2O \rightarrow$
$$Mg(OH)_2 + NH_3$$

d) $Pb + PbO_2 + H_2SO_4 \rightarrow$
$$PbSO_4 + H_2O$$

State what each equation means in terms of the appearance and disappearance of moles of reactants and products.

3 Write a balanced equation for each of the following: (a) the reaction of sodium with water to produce hydrogen and sodium hydroxide; (b) the reaction of calcium hydroxide and carbon dioxide to form calcium carbonate and water; (c) the reaction of carbon monoxide with hydrogen to form methane and water; (d) the reaction of aluminum nitrate with ammonium hydroxide to produce aluminum hydroxide and ammonium nitrate.

4 Write balanced equations for each of the following: (a) the reaction of aluminum with hydrogen chloride to produce aluminum chloride and hydrogen; (b) the reaction of ammonia with oxygen to produce nitric oxide (NO) and water; (c) the reaction of zinc with phosphorus to produce zinc phosphide; (d) the reaction of nitric acid with zinc hydroxide to produce zinc nitrate and water.

5 Carbon burns in air to produce carbon dioxide. What weight of carbon dioxide results from burning 100 g of carbon? What is the volume of the carbon dioxide at STP?

6 Laughing gas, N_2O, causes hysteria and unconsciousness when inhaled. It is made from ammonium nitrate by the reaction

$$NH_4NO_3 \rightarrow H_2O + N_2O$$

Balance the equation, and calculate the volume of N_2O produced at 25°C and 1 atm from 7.50 g of ammonium nitrate.

7 Potassium dichromate, $K_2Cr_2O_7$, reacts with oxalic acid, $H_2C_2O_4$, and sulfuric acid, H_2SO_4, according to the following equation:

$$3H_2C_2O_4 + K_2Cr_2O_7 + 5H_2SO_4 \rightarrow$$
$$2KHSO_4 + Cr_2(SO_4)_3$$
$$+ 6CO_2 + 7H_2O$$

(a) Is this equation balanced? If not, balance it. (b) If 450 ml of 0.2-molar potassium dichromate solution react with excess oxalic and sulfuric acids, how many moles of CO_2 will form? How many grams? How many liters at STP?

8 Sodium sulfide reacts with sulfuric acid to produce sodium sulfate and hydrogen sulfide. Assume that excess sulfuric acid is allowed to react with 10.0 g of sodium sulfide and calculate:

a) the moles of sodium sulfide used.

b) the moles of hydrogen sulfide liberated.

c) the grams of hydrogen sulfide liberated.

d) the volume of hydrogen sulfide liberated at STP.

9 Calcium phosphide, Ca_3P_2, reacts with water to produce phosphine gas (PH_3) and calcium hydroxide. Write a balanced equation for this process. How many liters of PH_3 gas will 1.75 g of calcium phosphide produce, measured at 300°K and 2.00 atm?

10 When vanadium oxide, VO, reacts with iron oxide, Fe_2O_3, the products are V_2O_5 and FeO. If no other reactants or products are involved, write a balanced equation for the reaction. If 6.50 g of vanadium oxide react with an excess of iron oxide, how many grams of V_2O_5 are produced? How many grams of V_2O_5 can be formed from 2.00 g of VO and 5.75 g of Fe_2O_3?

11 When 2.81 g of an unknown metal are treated with dilute sulfuric acid, 560 ml of H_2 gas evolve when measured at 1 atm and 273°K. (a) How many moles of hydrogen gas are evolved? (b) What is the combining weight of the metal? (c) The atomic weight of the metal is about 100. What is the correct atomic weight? What is the chemical formula of the resulting sulfate? (Use the symbol M for the metal.) (d) What is the metal?

12 Ammonia, NH_3, dissolves in water and forms NH_4^+ and OH^- ions. In the system of ammonia and water, which of the two compounds is the acid, and which is the base?

13 Write a balanced equation for the reaction of magnesium hydroxide with H_3PO_4 that produces $Mg_3(PO_4)_2$ and water.

14 Potassium perchlorate, $KClO_4$, has a solubility of about 7.5 g liter^{-1} of solution in water at 0°C. What is the molarity of a saturated solution at 0°C?

15 An amount of 50.0 ml of ether, C_2H_5—O—C_2H_5, which has a density of 0.714 g ml^{-1}, is dissolved in enough ethyl alcohol to make 100 ml of solution. Calculate the molarity of ether in the solution.

16 A solution of sulfuric acid was prepared from 95.94 g of H_2O and 10.66 g of H_2SO_4. The volume of the solution that resulted was 100.00 ml. Calculate the molality, molarity, normality, and density of the solution.

17 How many grams of solute are required to prepare the indicated amounts of the following solutions: (a) 2 liters of 2.5-molar sulfuric acid; (b) 0.5 liter of 1.0-normal H_3PO_4. Assume that the dissociation produces HPO_4^{2-} ions, and that the solution is to be used in acid–base neutralization reactions; (c) 1.0 liter of sodium hydroxide solution that will titrate ml-for-ml with 0.5-normal HCl.

18 How many milliliters of water must be *added* to 200 ml of 5.00-molar HNO_3 to make a solution that is 2 molar in HNO_3?

19 What volume of a solution of 1.53-molar sulfuric acid must be diluted with water to obtain 25 ml of 0.0500-molar sulfuric acid?

20 Calculate the volume of: (a) 2.10-molar KOH needed to make 500 ml of 0.0100-molar KOH; (b) 18-molar sulfuric acid needed to make 2.0 liters of 0.100 molar H_2SO_4; (c) the solution made by diluting 2.0 ml of 6-normal HNO_3 to give 0.01-normal HNO_3.

21 When 25.0 ml of 0.400-molar H_2SO_4 and 50.0 ml of 0.850-molar H_2SO_4 are mixed, what is the molarity of sulfuric acid in the final solution?

22 Calculate the molarity and normality of a solution that contains 0.0156 g of $Ba(OH)_2$ in 245 ml of solution.

23 What volume of $0.200M$ H_2SO_4 will neutralize 20.0 ml of $0.120N$ NaOH?

24 Equal volumes of $0.050M$ $Ba(OH)_2$ and $0.040M$ HCl are allowed to react. Calculate the molarity of each of the ions present after the reaction.

25 The density of 65% nitric acid is 1.40 g ml^{-1}. How many milliliters of nitric acid would be required to prepare 500 ml of a $0.50N$ solution?

26 How much water must be added to 100.0 ml of concentrated hydrochloric acid to make a $0.1000N$ solution? Concentrated hydrochloric acid contains 37.00% by weight of HCl and has a density of 1.190 g ml^{-1}

27 Calculate the weight in grams of $Mg(OH)_2$ required to react with 20.0 ml of 0.103-normal H_3PO_4 and convert it completely to PO_4^{3-}.

28 Sodium carbonate, Na_2CO_3, dissolves in water and forms carbonate ions, CO_3^{2-}, each of which can accept two protons to form carbonic acid, H_2CO_3. What is the number of gram-equivalents of base per mole of sodium carbonate? What is the normality of a solution made by adding 1.35 g of sodium carbonate to enough water to make 50.0 ml of solution?

29 What volume of 0.2-normal HCl is needed to titrate 20 ml of 0.35-normal NaOH?

30 If 0.350 g of calcium hydroxide is dissolved in water, how many milliliters of 0.100-molar HCl are required to titrate the solution to neutrality?

31 A volume of 35.8 ml of 0.1-molar sodium hydroxide is needed to titrate 20.0 ml of sulfuric acid solution. What are the normality and molarity of the sulfuric acid?

32 Find the number of grams of lysergic acid, $C_{15}H_{15}N_2COOH$, in a solution that requires 8.6 ml of $0.10M$ NaOH solution to titrate to a phenol-red end point. (Only one H per lysergic acid molecule reacts with NaOH.)

33 A strong base will react with an ammonium salt to liberate ammonia. In addition to providing a qualitative test for NH_4^+, this reaction may be used to determine the molarity of an NH_4Cl solution. To 20.0 ml of an NH_4Cl solution, 50.0 ml of $0.500M$ NaOH were added. The ammonia was expelled by heating the solution. The remaining solution was titrated to a methyl-orange end point with 15.0 ml of $0.500M$ HCl. Write balanced equations for the reactions that occurred, and calculate the molarity of the NH_4Cl solution. What weight of NH_3 was liberated?

34 A student was given an unknown acid, which was either acetic (CH_3-COOH), pyruvic ($CH_3COCOOH$), or propionic (CH_3CH_2COOH). A solution of the unknown acid was prepared by dissolving 0.100 g of the acid in 50.0 ml of water. The solution was titrated to a phenolphthalein end point with 11.3 ml

of $0.100M$ NaOH. Identify the unknown acid. (Only one H per acid molecule reacts with NaOH.)

35 The molecular weight of potassium acid phthalate (HOOC—C_6H_4—COOK, abbreviated KHP) is 204.2. If 1.673 g of KHP are dissolved in 80 ml of water, and this solution requires 34.50 ml of a NaOH solution to give phenolphthalein indicator a slight pink color, what is the molarity of the NaOH?

36 Formic acid, HCOOH, can be produced by distilling ants. What weight of formic acid dissolved in 4.32 ml of aqueous solution will require 3.72 ml of 0.0173-normal NaOH for complete neutralization? Assume that only one of the hydrogen atoms in formic acid dissociates.

37 Oxalic acid, HOOC—COOH, is a moderately poisonous constituent of rhubarb leaves. Calculate the volume of 0.114-normal NaOH required to react completely with 0.273 g of the acid if the gram equivalent weight of oxalic acid is 45.0 g. If this equivalent weight is correct, what does this indicate about the number of hydrogen atoms that dissociate per molecule of acid?

38 When dissolved in water a 0.375-g sample of a weak acid requires 28.8 ml of 0.1250-molar sodium hydroxide to cause phenolphthalein to turn a pale pink. What is the equivalent weight of the acid?

39 A chemist dissolves 0.300 g of an unknown acid in a convenient volume of water. He finds that 14.60 ml of 0.426-normal NaOH are required to neutralize the acid. What is the equivalent weight of the acid?

40 A 0.162-g sample of an unknown acid requires 12.7 ml of 0.0943-molar solution of NaOH for neutralization. What is the equivalent weight of the acid? The acid is either CH_3—C_6H_4—COOH or CH_3—CH_2—C_6H_4—COOH. Which possibility is more likely?

41 When titrating 0.15-molar HCl with a magnesium hydroxide solution of unknown concentration, 35.0 ml of the acid are required to neutralize 25.0 ml of the base. Calculate the molarity of the base.

42 In a laboratory exercise, the ammonia in 0.250 g of a compound Cu(NH$_3$)$_x$SO$_4$ is neutralized by 25.37 ml of 0.201-normal HCl when titrated to a methyl-orange end point. (a) What is the percentage of NH$_3$ in the compound by weight? (b) What is the value of the subscript, x, in the formula?

43 Calculate the enthalpy change or the heat of the following reactions at 298°C by using data in Appendix 2:

a) $2HI(g) \rightarrow H_2(g) + I_2(s)$
b) $HI(g) \rightarrow \frac{1}{2}H_2(g) + \frac{1}{2}I_2(s)$
c) $2HI(g) \rightarrow H_2(g) + I_2(g)$
d) $H_2(g) + I_2(s) \rightarrow 2HI(g)$
e) $2HI(g) \rightarrow H_2(g) + 2I(g)$
f) $2HI(g) \rightarrow 2H(g) + 2I(g)$
g) $3HI(g) \rightarrow H_2(g) + I_2(s) + HI(g)$

What is the heat of sublimation of I_2?

44 Nitrogen trichloride, NCl$_3$, is an unstable yellow oil that explodes at 95°C with the release of N$_2$, Cl$_2$, and 55 kcal mole^{-1} of heat. Write a balanced equation for the reaction, including ΔH. Calculate the amount of heat released by the decomposition of 10.0 g of NCl$_3$.

45 Sulfur has two crystalline forms, rhombic and monoclinic. With the data in Appendix 2, calculate the heat of conversion from rhombic to monoclinic sulfur. What is the heat of formation of SO$_2(g)$ from O$_2(g)$ and monoclinic S?

46 Is the conversion of rutile, TiO$_2$, to Ti$_3$O$_5$ in the open atmosphere exothermic or endothermic? What is the heat of the reaction? Write a balanced equation for it.

47 Calculate the heat of the following reaction at 298°K; ΔH_f^0 for PbO$_2(s)$ is -66.12 kcal.

$$4Al(s) + 3PbO_2(s) \rightarrow$$
$$3Pb(s) + 2Al_2O_3(s)$$

What does the algebraic sign in your answer indicate?

48 What is the heat of the reaction that produces sodium carbonate from Na$_2$O and carbon dioxide gas at 298°K?

49 Given the following reactions:

$$2P(s) + 3Cl_2(g) \rightarrow 2PCl_3(l)$$
$$\Delta H^0 = -151.8 \text{ kcal}$$

$$PCl_3(l) + Cl_2(g) \rightarrow PCl_5(s)$$
$$\Delta H^0 = -32.81 \text{ kcal}$$

calculate the heat of formation of solid PCl$_5$ at 25°C. What is the heat of vaporization of PCl$_3(l)$ at this temperature?

50 The heat of combustion of CH$_3$OH(l) to carbon dioxide gas and liquid water at 298°K is -170.9 kcal mole^{-1}, whereas that of formic acid, HCOOH(l), is -62.8 kcal mole^{-1}. Calculate the heat of the reaction

$$CH_3OH(l) + O_2(g) \rightarrow$$
$$HCOOH(l) + H_2O(l)$$

51 A combustion bomb containing 5.40 g of aluminum metal and 15.97 g of Fe$_2$O$_3$ is placed in an ice calorimeter that initially contained 8.000 kg of ice and 8.000 kg of liquid water. The reaction

$$2Al(s) + Fe_2O_3(s) \rightarrow$$
$$Al_2O_3(s) + 2Fe(s)$$

is begun by remote control. It is observed that the calorimeter contains 7.746 kg of ice and 8.254 kg of water. The heat of fusion of ice is 80 cal g^{-1}. What is the enthalpy change for the reaction?

52 Nitrogen dioxide, NO$_2$, is one of the atmospheric pollutants produced by automobiles. It is formed when the high temperature in the internal combustion engine causes atmospheric N$_2$ and O$_2$ to combine to produce NO, which then reacts with more O_2 to give NO_2. This pollutant ultimately is converted to HNO_3. A proposed equation for this conversion is

Calculate the change in enthalpy for the conversion reaction. Necessary data can be obtained from Appendix 2.

$$O_3(g) + 2NO_2(g) + H_2O(g) \rightarrow$$
$$2HNO_3(aq) + O_2(g)$$

Challenge. There are several problems in *Relevant Problems for Chemical Principles*, by Butler and Grosser, that apply to topics in this chapter. You may want to see whether you can determine the correct formula for a recently prepared xenon-fluorine compound (Problem 4–5); try problems that may be considered by a county medical examiner (Problems 4–13 and 4–14) and a water pollution control engineer (Problem 4–15); work problems that apply to industrial design, biomedical analysis, and city planning (Problems 4–16 to 4–19); apply thermochemical principles to nutrition, rocket propulsion, and architectural design (Problems 4–22, 4–24, and 4–25).

When you can measure what you are speaking about, and express it in numbers, you know something about it; but when you cannot measure it, when you cannot express it in numbers your knowledge is of a meagre and unsatisfactory kind; it may be the beginning of knowledge, but you have scarcely, in your thoughts, advanced to the stage of science.
William Thomson, Lord Kelvin (1891)

5 WILL IT REACT? AN INTRODUCTION TO CHEMICAL EQUILIBRIUM

The main question asked in the preceding chapter was "If a given set of substances will react to give a desired product, how much of each substance would be needed?" Our basic physical assumptions were that a chemist cannot arbitrarily create or destroy matter, and that atoms going into a reaction must come out again as products. This is chemical bookkeeping in the tradition of Lavoisier (Chapter 1).

In this chapter we will ask a second question: "Will a reaction occur, eventually?" That is, is there a tendency or a drive for a given reaction to take place, and if we wait long enough will we find that reactants have been converted spontaneously into products? This question leads to the ideas of *spontaneity* and of *chemical equilibrium*. The third question, "Will a reaction occur in a reasonably short time?", involves chemical kinetics, which will be discussed in Chapter 18. For the moment, we will be satisfied if we can predict which way a chemical reaction will go by itself, ignoring the time factor.

5–1 SPONTANEOUS REACTIONS

A chemical reaction that will occur on its own, given enough time, is said to be spontaneous. In the open air, and under the conditions inside an automobile engine, the combustion of gasoline is spontaneous:

$$C_7H_{16} + 11O_2 \rightarrow 7CO_2 + 8H_2O$$

(The reaction is exothermic, or heat-emitting. The enthalpy change, which was defined in Chapter 4, is large and negative: $\Delta H_{298} = -1150$ kcal mole^{-1}. The heat emitted causes the product gases to expand, and it is the pressure from these expanding gases that drives the car.) In contrast, the reverse reaction under the same conditions is not spontaneous:

$$7CO_2 + 8H_2O \not\rightarrow C_7H_{16} + 11O_2$$

No one seriously proposes that gasoline can be obtained spontaneously from a mixture of water vapor and carbon dioxide.

Explosions are examples of rapid, spontaneous reactions, but a reaction need not be as rapid as an explosion to be spontaneous. It is important to understand clearly the difference between the two ideas. If you mix oxygen and hydrogen gases at room temperature, they will remain together without appreciable reaction for years. Yet the reaction to produce water is genuinely spontaneous:

$$2H_2 + O_2 \rightarrow 2H_2O$$

We know that this is the case because we can trigger the reaction with a match, or with a catalyst of finely divided platinum metal.

The preceding sentence is the key to why a chemist is interested in whether a reaction is spontaneous, that is, whether it has a natural tendency to occur. If a desirable chemical reaction is spontaneous but slow, it may be possible to find a means to speed up the process. Raising the temperature often will do the trick, or a catalyst may work. We will discuss the function of a catalyst in detail in Chapter 18. But in brief, we can say now that a catalyst is a substance that helps a naturally spontaneous reaction to go faster by providing an easier pathway for it. Gasoline will burn in air at a high enough temperature, thus the role of a spark plug in an automobile engine is to provide this initial temperature. The heat produced by the reaction maintains the high temperature needed to keep it going thereafter. Gasoline also will combine with oxygen at room temperature if the proper catalyst is used, because the reaction is naturally spontaneous. But no catalyst ever will make carbon dioxide and oxygen gases recombine to produce gasoline at room temperature and moderate pressures, and it would be a foolish chemist who spent time trying to find such a catalyst. Therefore, the idea of spontaneous and nonspontaneous reactions helps a chemist to see the limits on what is *possible*. If a reaction is possible but not currently realizable, it may be worthwhile to look for ways to carry it out. If the process is inherently impossible, then it is time to study something else.

5–2 EQUILIBRIUM AND THE EQUILIBRIUM CONSTANT

The speed with which a reaction takes place ordinarily depends on the concentrations of the reacting substances. This is common sense, since most reactions take place by collisions of molecules, and the more molecules there are per unit of volume, the more often collisions will occur.

The industrial fixation of atmospheric nitrogen is of great importance in the manufacture of agricultural fertilizers (and explosives). One of the steps in nitrogen fixation, in the presence of a catalyst, is

$$N_2 + O_2 \rightarrow 2NO \tag{5-1}$$

If this reaction took place by simple collision of one molecule of N_2 and one molecule of O_2, then we would expect the rate of collision (and hence the rate of reaction) to be proportional to both the concentration of N_2 and that of O_2:

Rate of NO production $\propto [N_2][O_2]$

or

$$R_1 = k_1[N_2][O_2] \tag{5-2}$$

in which k_1 is the forward-reaction rate constant and the bracketed terms $[N_2]$ and $[O_2]$ represent concentrations in moles per liter. This rate constant, which we will discuss in more detail in Chapter 18, usually varies with temperature. Most reactions go faster at higher temperatures, so k_1 is larger at higher temperatures. But k_1 does *not* depend on the concentrations of nitrogen and oxygen gases present. All of the concentration dependence of the overall forward reaction rate, R_1, is contained in the terms $[N_2]$ and $[O_2]$. If this reaction began rapidly in a sealed tank with high starting concentrations of both gases, then as more N_2 and O_2 were consumed, the forward reaction would become progressively slower. The rate of reaction would decrease because the frequency of collision of molecules would diminish as the number of N_2 and O_2 molecules left in the tank decreased.

The reverse reaction also can occur. If this reaction occurred by the collision of two molecules of NO to make a molecule each of the starting gases,

$$2NO \rightarrow N_2 + O_2 \tag{5-3}$$

then the rate of reaction again would be proportional to the concentration of the colliding molecules. Since these molecules are of the same compound, NO, the rate would be proportional to the *square* of the NO concentration:

Rate of NO removal $\propto [NO][NO]$

or

$$R_2 = k_2[NO]^2 \tag{5-4}$$

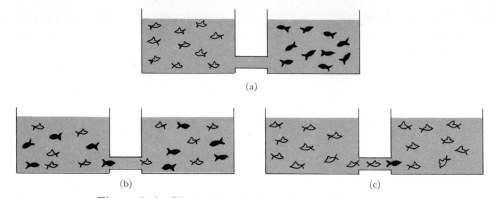

Figure 5–1. Illustration of dynamic aqualibrium: two fish tanks connected by a channel. (a) Start of experiment, with ten goldfish in the left tank and ten guppies in the right. (b) Equilibrium state, with five of each kind of fish in each tank. (c) If we were to observe one single fish (here a guppy among goldfish), we would find that it spends half its time in each tank. The equilibrium of the tank in (b) is a dynamic, averaged state and not a static condition. The fish do not stop swimming when they have become evenly mixed.

If little NO is present when the experiment begins, this reaction will occur at a negligible rate. But as more NO accumulates by the forward reaction, the faster it will be broken down by the reverse reaction. Thus as the forward rate, R_1, decreases, the reverse rate, R_2, increases. Eventually the point will be reached at which the forward and reverse reactions exactly balance:

$$R_1 = R_2$$
$$[N_2][O_2]k_1 = k_2[NO]^2 \qquad (5\text{–}5)$$

This is the condition of *equilibrium*. Had you been monitoring the concentrations of the three gases, you would have found that the composition of the reacting mixture had reached an equilibrium state and thereafter ceased to change with time. This would not mean that the individual reactions had stopped, only that they were proceeding at equal rates; that is, they had arrived at, and thereafter maintained, a condition of balance or equilibrium.

The condition of equilibrium can be illustrated by imagining two large fish tanks, connected by a channel (Figure 5–1). One tank initially contains ten goldfish, and the other contains ten guppies. If you watch the fish swimming aimlessly long enough, you eventually will find that approximately five of each type of fish are present in each tank. Each fish has the same chance of blundering through the channel into the other tank. But as long as there are more goldfish in the left tank [Figure 5–1(a)], there is a greater probability that a goldfish will swim from left to right than the reverse. Similarly, as long as the number of guppies in the right tank exceeds that in the left, there will be a net flow of guppies to the left, even though there is nothing in the left tank to make the guppies prefer it. Thus the rate of flow of guppies is proportional to the concentration of guppies present. A similar statement can be made for the goldfish.

At equilibrium [Figure 5–1(b)], on an average there will be five guppies and five goldfish in each tank. But they will not always be the same five of each fish. If one guppy wanders from the left tank into the right, then it or a different guppy may wander back a little later. Thus at equilibrium we find that the fish have not stopped swimming, only that over a period of time the total number of guppies and goldfish in each tank remains constant. If we were to fill each tank with nine goldfish and throw in one guppy, we would see that, in its aimless swimming, it would spend half its time in one tank and half in another [Figure 5–1(c)].

In the NO reaction we considered, there will be a constant concentration of NO molecules at equilibrium, but they will not always be the same NO molecules. Individual NO molecules will react to re-form N_2 and O_2, and other reactant molecules will make more NO. As with the goldfish, only on a head-count or concentration basis have changes ceased at equilibrium.

The equilibrium condition for the NO reaction can be rewritten in a more useful form:

$$\frac{[NO]^2}{[N_2][O_2]} = \frac{k_1}{k_2} = K_{eq} \tag{5–6}$$

in which the ratio of forward and reverse rate constants can be expressed as a simple constant, the *equilibrium constant*, K_{eq}. This equilibrium constant will vary as the temperature varies, but is independent of the concentrations of the reactants and products. It tells us the ratio of products to reactants at equilibrium, and is an extremely useful quantity for determining whether a desired reaction will take place spontaneously.

We derived the equilibrium-constant expression for the NO reaction by assuming that we knew the molecular mechanisms of the forward and reverse steps. If the NO reaction proceeded by simple collision of two molecules, our derivation would be perfectly correct. In fact, the actual mechanism of this reaction is more complicated. But it is important, and fortunate for chemists, that we do not have to know the reaction mechanism to write the proper equilibrium-constant expression. (We shall prove this in Chapter 15.) The NO forward reaction actually takes place by a series of complicated chain steps. The reverse reaction takes place by a complementary set of reactions, so that these complications cancel one another in the final ratio of concentrations that gives us the equilibrium constant. Therefore, to write the equilibrium-constant expression, we need know only the overall reaction, and not the details of its mechanism.

If a general chemical reaction is written as

$$aA + bB \rightleftharpoons cC + dD \qquad \text{(\textit{Note:} The two arrows indicate a} \tag{5–7}$$
state of equilibrium.)

in which A, B, C, and D are chemical compounds and a, b, c, and d are the numbers of the moles of the compounds in the reaction, then the equilibrium constant for that reaction is given by

$$K_{eq} = \frac{[C]^c [D]^d}{[A]^a [B]^b} \tag{5-8}$$

This is called the *law of mass action*.

We derived the equilibrium-constant expression for the NO reaction by assuming that we knew the molecular mechanisms for the forward and reverse reactions. However, as far as equilibrium constants are concerned, the reaction can be treated as *if* it were a simple one-step process, with reaction rates depending on overall amounts of substances present in a given volume. As we shall see, if we know the equilibrium constant for a given reaction, we then can predict whether the reaction will be spontaneous in the forward or the reverse direction, and what the concentrations of reactants and products will be when equilibrium is reached. If we mix several compounds and observe no immediate reaction, an equilibrium-constant calculation can tell us whether it will be useful or futile to search for a catalyst to make the reaction go.

Example 1.   In the "contact process" for making sulfuric acid, H_2SO_4, the following reaction occurs:

$$2SO_2 + O_2 \rightleftarrows 2SO_3$$

What is the equilibrium-constant expression?

Solution

$$K_{eq} = \frac{[SO_3]^2}{[SO_2]^2 [O_2]}$$

Example 2.   The sulfuric acid reaction also could have been written as

$$SO_2 + \tfrac{1}{2}O_2 \rightleftarrows SO_3$$

What is the equilibrium-constant expression for this version of the reaction?

Solution

$$K'_{eq} = \frac{[SO_3]}{[SO_2][O_2]^{1/2}}$$

Notice that this equilibrium constant is simply the square root of the preceding one because each concentration term on the right side is the square root of the corresponding term in the first example. It is important to write the equilibrium-constant expression properly for the chemical reaction *as given*, and when using an equilibrium constant from some reference, to be very sure exactly what reaction this constant applies to.

Example 3. The Deacon process for producing chlorine gas involves the reaction

$$\tfrac{1}{2}O_2 + 2HCl \rightleftharpoons H_2O + Cl_2$$

What is the expression for the equilibrium constant?

Solution

$$K_{eq} = \frac{[H_2O][Cl_2]}{[O_2]^{1/2}[HCl]^2}$$

Again, note that there is nothing wrong with fractional exponents for the concentration terms. However, it is usual to square K_{eq} or to raise it to whatever power is necessary to make all the exponents integers.

Example 4. If the Deacon process reaction were written

$$O_2 + 4HCl \rightleftharpoons 2H_2O + 2Cl_2$$

what relationship would this new equilibrium constant have to the preceding one?

Solution. The new equilibrium constant would be the square of the first one. Verify this by writing both equilibrium expressions and comparing them.

Example 5. The dissociation of sulfur trioxide can be written

$$SO_3 \rightleftharpoons SO_2 + \tfrac{1}{2}O_2$$

How does the equilibrium-constant expression for this reaction compare with that for the reverse reaction discussed previously?

Solution. The new equilibrium constant is the inverse or reciprocal of the previous one. Verify this by writing both expressions and comparing them.

5–3 USING EQUILIBRIUM CONSTANTS

The most convincing way to demonstrate the validity of a theory is to test it with real data. One reaction that has been studied extensively is that between hydrogen and iodine:

$$H_2(g) + I_2(g) \rightleftharpoons 2HI(g) \tag{5–9}$$

If we mix hydrogen and iodine in a sealed flask and observe the reaction, we can detect the production of hydrogen iodide by the gradual fading of the purple iodine vapor. This reaction was studied first by Max Bodenstein in

Table 5–1.   Experimental Measurements of Equilibrium Concentrations[a]

Experiment				Calculations from experiment		
$[H_2]$ (a)	$[I_2]$ (b)	$[HI]$ (c)	$\dfrac{c}{a\ b}$	$\dfrac{c^2}{a\ b} = K_{eq}$	Deviation from average K_{eq}
18.14	0.41	19.38	2.60	50.50	−0.03
10.96	1.89	32.61	1.57	51.34	+0.71
4.57	8.69	46.28	1.16	53.93	+3.40
2.23	23.95	51.30	0.96	49.27	−1.26
0.86	67.90	53.40	0.91	48.83	−1.70
0.65	87.29	52.92	0.93	49.35	−1.18
				6)303.22	6)8.28
				50.53	1.38

Average $K_{eq} = 50.53$ $\dfrac{1.38}{50.53} \times 100 = 2.7\%$ mean deviation

[a] For the reaction $H_2(g) + I_2(g) \rightleftarrows 2HI(g)$, at 448°C in a sulfur vapor constant-temperature bath. Concentrations are in mole per liter $\times 10^{+3}$ (e.g., the first hydrogen concentration is 18.14×10^{-3} mole liter^{-1}).

1893. Table 5–1 contains the data from Bodenstein's experiments. The experimental data are in the first three columns. In the fourth column, we have calculated the simple ratio of product and reactant concentrations, $[HI]/[H_2][I_2]$, to see if it is constant. It clearly is not, for as the hydrogen concentration is decreased and the iodine concentration is increased, this ratio varies from 2.60 to less than 1. The law of mass action dictates that the equilibrium-constant expression should contain the *square* of the HI concentration, since the reaction involves two moles of HI for every mole of H_2 and I_2. The fifth column shows that the ratio $[HI]^2/[H_2][I_2]$ is constant within a mean deviation of approximately 3%. Therefore, this ratio is the proper equilibrium expression, and the average value of K_{eq} for these six runs is 50.53.

The equilibrium constant can be used to determine whether a reaction under specified conditions will go spontaneously in the forward or the reverse direction. The ratio of product-to-reactant concentrations, analogous to the equilibrium constant but not necessarily for equilibrium conditions, is called the *reaction quotient*, Q:

$$Q = \frac{[HI]^2}{[H_2][I_2]} \quad \text{(not necessarily at equilibrium)} \tag{5–10}$$

If there are too many reactant molecules present for equilibrium to exist, then the concentration terms in the denominator will make the reaction quotient smaller than K_{eq}. The reaction will go forward spontaneously to make more product. However, if an experiment is set up so that the reaction quotient is greater than K_{eq}, then too many product molecules are present for equilibrium and the reverse reaction will proceed spontaneously. Therefore, a comparison of the actual concentration ratio or reaction quotient with the equilibrium constant allows us to predict which direction a reaction will go spontaneously under a given set of circumstances.

Example 6. If 1.0×10^{-2} mole each of hydrogen and iodine gases are placed in a 1-liter flask at 448°C with 2.0×10^{-3} mole of HI, will more HI be produced?

Solution. The reaction quotient under these conditions is

$$Q = \frac{(2.0 \times 10^{-3})^2}{(1.0 \times 10^{-2})^2} = 0.040$$

This is smaller than the equilibrium value of 50.53 in Table 5–1, which tells us that excess reactants are present. Hence, collisions between reactant molecules will occur more often than collisions between product molecules, and equilibrium will not be reached until more HI has been formed.

Example 7. If only 1.0×10^{-3} mole each of H_2 and I_2 had been used, together with 2.0×10^{-3} mole of HI, would more HI be produced spontaneously?

Solution. You can verify that the reaction quotient is $Q = 4.0$. Because this is less than K_{eq}, the forward reaction is still spontaneous.

Example 8. If the conditions of Example 7 are changed so that the HI concentration is increased to 2.0×10^{-2} mole liter^{-1}, what happens to the reaction?

Solution. The reaction quotient now is $Q = 400$. This is greater than K_{eq}. There now are too many product molecules and too few reactant molecules for equilibrium to exist. Thus the reverse reaction occurs more rapidly than the forward reaction. Equilibrium is reached only by converting some of the HI to H_2 and I_2, so the reverse reaction is spontaneous.

Example 9. If a 1-liter flask contains 1.0×10^{-3} mole each of H_2 and I_2 at 448°C, what amount of HI is present if the gas mixture is at equilibrium?

Solution. The solution now is

$$\frac{[HI]^2}{(1.0 \times 10^{-3})^2} = K_{eq} = 50.53$$

$$[HI]^2 = 50.53 \times 1.0 \times 10^{-6}$$

$$[HI] = 7.1 \times 10^{-3} \text{ mole liter}^{-1}$$

You can verify that in Example 7 the HI concentration was less than this equilibrium value, and in Example 8 it was more.

Example 10.   One tenth mole of hydrogen iodide is placed in a 5-liter flask at 448°C. When the contents have come to equilibrium, how much hydrogen and iodine will be in the flask?

Solution.   From the stoichiometry of the equation, the concentrations of H_2 and I_2 must be the same. For every mole of H_2 and I_2 formed, two moles of HI decompose. Let y equal the number of moles of H_2 or I_2 *per liter* present at equilibrium. The initial concentration of HI before any dissociation has occurred is

$$[HI]_0 = \frac{0.10 \text{ mole}}{5 \text{ liters}} = 0.020 \text{ mole liter}^{-1}$$

Begin by writing a balanced equation for the reaction, then make a table of concentrations at the start and at equilibrium:

$$H_2 + I_2 \rightleftarrows 2HI$$

Start (moles liter^{-1}): 0	0	0.020
Equilibrium: y	y	$0.020 - 2y$

The HI concentration of 0.020 mole liter^{-1} has been decreased by $2y$ for every y molecules of H_2 and I_2 that are formed. The equilibrium-constant expression is

$$50.53 = \frac{(0.020 - 2y)^2}{y^2}$$

We immediately see that we can take a shortcut by taking the square root of both sides:

$$7.11 = \frac{0.020 - 2y}{y}$$

$$9.11y = 0.020$$

$$y = 0.0022 \text{ mole liter}^{-1}$$

For five liters, $5 \times 0.0022 = 0.011$ mole of H_2 and of I_2 will be present at equilibrium. Only $(0.020 - 0.0044) \times 5 = 0.078$ mole of HI will be left in the 5-liter tank, and the fraction of HI dissociated at equilibrium is

$$\frac{2y}{[HI]_0} = \frac{0.0044}{0.020} = 0.22, \text{ or } 22\% \text{ dissociation}$$

Shortcuts such as taking the square root in the preceding example are not always possible, yet part of the skill of solving equilibrium problems lies in recognizing shortcuts when they occur and then using them. The key to this is often a good physical intuition about what quantities are large and small relative to one another, and this intuition comes from thoughtful practice and understanding of the chemistry involved. You should remember that these are chemical problems, not mathematical ones.

In many cases the quadratic equation is useful.

Example 11. If 0.00500 mole of hydrogen gas and 0.0100 mole of iodine gas are placed in a 5-liter tank at 448°C, how much HI will be present at equilibrium?

Solution. The initial concentrations of H_2 and I_2 are

$$[H_2]_0 = \frac{0.00500 \text{ mole}}{5 \text{ liters}} = 0.00100 \text{ mole liter}^{-1}$$

$$[I_2]_0 = \frac{0.0100 \text{ mole}}{5 \text{ liters}} = 0.00200 \text{ mole liter}^{-1}$$

This time, let the unknown variable y be the moles per liter of H_2 or I_2 that have reacted at equilibrium:

	H_2	$+$	I_2	$\rightleftarrows 2HI$
Start (moles liter^{-1}):	0.00100		0.00200	0.0
Equilibrium:	$0.00100 - y$		$0.00200 - y$	$2y$

The equilibrium expression is

$$50.53 = \frac{(2y)^2}{(0.00100 - y)(0.00200 - y)}$$

The square-root shortcut now is impossible because the starting concentrations of H_2 and I_2 are unequal. Instead we must reduce the equation to a quadratic expression:

$$46.53y^2 - 0.1516y + 1.011 \times 10^{-4} = 0$$

A general quadratic equation of the form $ay^2 + by + c = 0$ can be solved by the quadratic formula,

$$y = \frac{-b \pm \sqrt{b^2 - 4ac}}{2a}$$

Thus for this problem

$$y = \frac{0.1516 \pm \sqrt{0.02298 - 0.01881}}{93.06}$$

$$y = 2.32 \times 10^{-3} \text{ and } 0.935 \times 10^{-3} \text{ mole liter}^{-1}$$

The first solution is physically impossible since it shows more H_2 reacting than was originally present. The second solution is the correct answer:

$y = 0.935 \times 10^{-3}$ mole liter^{-1}. Therefore, the equilibrium concentrations are

$$[H_2] = 0.00100 - 0.000935 = 0.065 \times 10^{-3} \text{ mole liter}^{-1}$$
$$[I_2] = 0.00200 - 0.000935 = 1.065 \times 10^{-3} \text{ mole liter}^{-1}$$
$$[HI] = 2(0.935 \times 10^{-3}) \quad\quad = 1.87 \;\times 10^{-3} \text{ mole liter}^{-1}$$

5–4   FACTORS AFFECTING EQUILIBRIUM: LE CHATELIER'S PRINCIPLE

Equilibrium represents a balance between two opposing reactions. How sensitive is this balance to changes in the conditions of a reaction? What can be done to change the equilibrium state? These are very practical questions if, for example, one is trying to increase the yield of a useful product in a reaction.

Under specified conditions, the equilibrium-constant expression tells us the ratio of product to reactants when the forward and backward reactions are in balance. This equilibrium constant is not affected by changes in concentration of reactants or products. However, if products can be withdrawn continuously, then the reacting system can be kept constantly off-balance and short of equilibrium. More reactants will be used and a continuous stream of new products will be formed. This method is useful when one product of the reaction can escape as a gas, be condensed or frozen out of a gas phase as a liquid or solid, washed out of the gas mixture by a spray of a liquid in which it is especially soluble, or precipitated from gas or solution.

For example, when solid lime and coke are heated in an electric furnace to make calcium carbide,

$$CaO(s) + 3C(s) \rightleftharpoons CaC_2(s) + CO(g)\uparrow \tag{5–11}$$

the reaction, which at 2000–3000°C has an equilibrium constant of close to 1.00, is tipped toward calcium carbide formation by the continuous removal of carbon monoxide gas. In the industrial manufacture of titanium dioxide for pigments, $TiCl_4$ and O_2 react as gases:

$$TiCl_4(g) + O_2(g) \rightleftharpoons TiO_2(s)\downarrow + 2Cl_2(g) \tag{5–12}$$

The product separates from the reacting gases as a fine powder of solid TiO_2, and the reaction thus is kept moving in the forward direction. When ethyl acetate or other esters used as solvents and flavorings are synthesized from carboxylic acids and alcohols,

$$\underset{\text{acetic acid}}{CH_3COOH} + \underset{\text{ethanol}}{HOCH_2CH_3} \rightleftharpoons \underset{\text{ethyl acetate}}{CH_3COOCH_2CH_3} + H_2O \tag{5–13}$$

the reaction is kept constantly off-balance by removing the water as fast as it is formed. This can be done with a drying agent such as Drierite ($CaSO_4$), by running the reaction in benzene and boiling off a constant-boiling benzene–water mixture, or by running the reaction in a solvent in which the water is completely immiscible and separates as droplets in a second phase. Finally, since ammonia is far more soluble in water than either hydrogen or nitrogen is, the yield of ammonia in the reaction

$$N_2(g) + 3H_2(g) \rightleftarrows 2NH_3(g) \tag{5–14}$$

can be raised to well over 90% by washing the ammonia out of the equilibrium mixture of gases with a stream of water and recycling the nitrogen and hydrogen.

Temperature

All of the preceding techniques are methods of upsetting an equilibrium (in our examples, in favor of desired products) without altering the equilibrium constant. A chemist often can enhance yields of desired products by increasing the equilibrium constant so that the ratio of products to reactants at equilibrium is larger. The equilibrium constant usually is temperature dependent. In general, both forward and reverse reactions are speeded up by increasing the temperature, because the molecules move faster and collide more often. If the increase in the rate of the forward reaction is greater than that of the reverse, then K_{eq} increases with temperature and more products are formed at equilibrium. If the reverse reaction is favored, then K_{eq} decreases. Thus K_{eq} for the hydrogen–iodine reaction at 448°C is 50.53, but at 425°C it is 54.4, and at 357°C, it increases to 66.9. Production of HI is favored to some extent by an increase in temperature, but its dissociation to hydrogen and iodine is favored much more.

The hydrogen iodide-producing reaction is exothermic or heat-emitting:

$$H_2(g) + I_2(g) \rightleftarrows 2HI(g) \tag{5–15}$$
$$\Delta H_{298} = -2.5 \text{ kcal per 2 moles of HI}$$

(If you check this figure against the tables in Appendix 2, remember that this reaction involves gaseous iodine and not solid.) If the external temperature of this reaction is lowered, the equilibrium is shifted in favor of the heat-emitting or forward reaction; if the temperature is raised, the reverse reaction producing H_2 and I_2 is favored. The equilibrium shifts so as to counteract to some extent the effect of adding heat externally (raising the temperature) or removing it (lowering the temperature).

The temperature dependence of the equilibrium point is one example of a more general principle. If an external stress is applied to a system at chemical equilibrium, then the equilibrium point will change in such a way as to counteract the effects of that stress. This is *Le Chatelier's principle*. If the forward half of an equilibrium reaction is exothermic, then K_{eq} will decrease as the temperature increases; if it is endothermic, K_{eq} will increase. Only for a heat-absorbing reaction can the equilibrium yield of products be improved by increasing the temperature. A good way to remember this is to write the reaction explicitly with a heat term:

$$H_2(g) + I_2(g) \rightleftharpoons 2HI(g) + \text{heat (given off)} \qquad (5\text{–}16)$$

Then adding heat, just like adding HI, shifts the reaction to the left.

Pressure

Le Chatelier's principle is true for other kinds of stress, such as pressure changes. The equilibrium constant, K_{eq}, is not altered by a pressure change at constant temperature. However, the relative amounts of reactants and products will change in a way that can be predicted from Le Chatelier's principle.

The hydrogen–iodine reaction involves an equal number (two) of moles of reactants and product. Therefore, if we double the pressure at constant temperature, the volume of the mixture of gases will be halved. All concentrations in moles per liter will be doubled, but their *ratio* will be the same. In Example 11, doubling the concentrations of the reactants and product does not change the equilibrium constant:

$$
\begin{aligned}
K_{eq} &= \frac{(1.87 \times 10^{-3} \text{ mole liter}^{-1})^2}{(0.065 \times 10^{-3} \text{ mole liter}^{-1})(1.065 \times 10^{-3} \text{ mole liter}^{-1})} \\
&= \frac{(3.74 \times 10^{-3})^2}{(0.13 \times 10^{-3})(2.13 \times 10^{-3})} = 50.51
\end{aligned}
\qquad (5\text{–}17)
$$

Thus the hydrogen–iodine equilibrium is not sensitive to pressure changes. Notice that in this case K_{eq} does not have units, since the concentration units in the numerator and denominator cancel.

In contrast, the dissociation of ammonia is affected by changes in pressure because the number of moles (two) of reactant does not equal the total number of moles (four) of products:

$$2NH_3(g) \rightleftharpoons N_2(g) + 3H_2(g) \qquad (5\text{–}18)$$

The equilibrium constant for this reaction at 25°C is

$$K_{eq} = \frac{[N_2][H_2]^3}{[NH_3]^2} = 2.5 \times 10^{-9} \text{ mole}^2 \text{ liter}^{-2} \qquad (5\text{–}19)$$

(Note that K_{eq} is no longer a unitless number, since the number of concentration terms in numerator and denominator is not the same. In this reaction, K_{eq} has the units of the square of the concentration.) One set of equilibrium conditions is

$$N_2 = 3.28 \times 10^{-3} \text{ mole liter}^{-1}$$
$$H_2 = 2.05 \times 10^{-3} \text{ mole liter}^{-1}$$
$$NH_3 = 0.106 \text{ mole liter}^{-1}$$

(Can you verify that these concentrations satisfy the equilibrium condition?)

If we now double the pressure at constant temperature, thereby halving the volume and doubling each concentration:

$$N_2 = 6.56 \times 10^{-3} \text{ mole liter}^{-1}$$
$$H_2 = 4.10 \times 10^{-3} \text{ mole liter}^{-1}$$
$$NH_3 = 0.212 \text{ mole liter}^{-1}$$

the ratio of products to reactants, the reaction quotient, is no longer equal to K_{eq}:

$$Q = \frac{(6.56 \times 10^{-3})(4.10 \times 10^{-3})^3}{(0.212)^2} = 1.0 \times 10^{-8} \text{ mole}^2 \text{ liter}^{-2} \quad (5\text{--}20)$$

Since Q is greater than K_{eq}, too many product molecules are present for equilibrium. The reverse reaction will run spontaneously, thereby forming more NH_3 and decreasing the amounts of H_2 and N_2. Consequently, part of the increased pressure is offset when the reaction shifts in the direction that lowers the total number of moles of gas present. In general, a reaction that reduces the number of moles of gas will be favored by an increase in pressure, and one that produces more gas will be disfavored.

Example 12. Had the hydrogen–iodine reaction been run at a lower temperature at which the iodine was a solid, would an increase in pressure shift the equilibrium reaction toward more HI or less? What would be the effect of pressure on K_{eq}?

Solution. Since the reaction of two moles of gaseous HI now yields one mole of gaseous H_2 and another of solid I_2, the stress of increased pressure is relieved by dissociating HI to H_2 and I_2. However, K_{eq} will be unchanged by the pressure increase.

Catalysis

What effect does a catalyst have on a reaction *at equilibrium?* The answer is: none. A catalyst cannot change the value of K_{eq}, but it can increase the speed with which equilibrium is reached. This is the main function of a catalyst. It cannot take the reaction anywhere else except to the same equilibrium state that would be reached eventually without the catalyst.

Catalysts are useful, nevertheless. Many desirable reactions, although spontaneous, occur under ordinary conditions at extremely slow rates. The main smog-producing reaction in automobile engines involving oxides of nitrogen is

$$N_2 + O_2 \rightleftarrows 2NO \quad (5\text{--}21)$$

(Once NO is present it reacts readily with more oxygen to make brown NO_2.) At the high temperature of an automobile engine, K_{eq} for this reaction is large enough that appreciable amounts of NO are formed. However, at

25°C, $K_{eq} = 10^{-30}$. (From only the previous two bits of information and Le Chatelier's principle, predict whether the reaction as written is endothermic or exothermic. Check your answer using data from Appendix 2.) The amount of NO present in the atmosphere at equilibrium at 25°C should be negligible. NO should decompose spontaneously to N_2 and O_2 as the exhaust gases cool. But any Southern Californian can verify that this is not true. NO and NO_2 are indeed present, because the gases of the atmosphere are not at equilibrium.

The rate of decomposition of NO is extremely slow, although spontaneous. One approach to the smog problem has been to search for a catalyst for the reaction

$$2NO \rightleftarrows N_2 + O_2 \tag{5-22}$$

which could be housed in an exhaust system and could break down NO in the exhaust gases as they cool. Finding a catalyst is possible; a practical problem arises from the gradual poisoning of the catalyst by additives in the gasoline, such as lead compounds. This is one reason for the current interest in low-lead and lead-free fuels.

A proof of the assertion that a catalyst cannot change the equilibrium constant is illustrated in Figure 5–2. If a catalyst *could* shift the equilibrium point of a reacting gas mixture and produce a volume change, then this expansion and contraction could be harnessed by some such mechanical means and made to do work. We would have a true perpetual-motion machine that would deliver power without an energy source. From common sense and experience we know this to be impossible. This "common sense" is stated scientifically as the first law of thermodynamics, which will be discussed in Chapter 15. A mathematician would call this a *proof by contradiction*: If we assume that a catalyst can alter K_{eq}, then we must assume the existence of a perpetual-motion machine. However, a perpetual-motion machine cannot exist; therefore our initial assumption was wrong, and we must conclude that a catalyst cannot alter K_{eq}.

In summary, K_{eq} is a function of temperature, but is not a function of reactant or product concentrations, total pressure, or the presence or absence of catalysts. The relative amounts of substances at equilibrium can be changed by applying an external stress to the equilibrium mixture of reactants and products, and the change is one that will relieve this stress. This last statement, Le Chatelier's principle, enables us to predict what will happen to a reaction when external factors are changed, without having to make exact calculations.

Challenge. If you want to apply Le Chatelier's principle to the technique of pouring champagne and manufacturing explosives, work Problems 5–10 and 5–13 in Butler and Grosser.

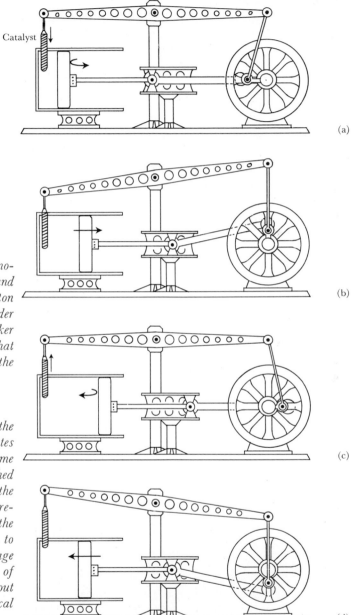

Figure 5–2. *The ammonia perpetual-motion engine. A mixture of* NH_3, H_2, *and* N_2 *is contained in a chamber by the piston at the left, and the hatched cylinder suspended from the left end of the rocker arm contains a mythical catalyst that would shift the equilibrium point of the reaction*

$$2NH_3(g) \rightleftarrows N_2(g) + 3H_2(g)$$

to the right. In Steps (a) and (b), as the catalyst is introduced, ammonia dissociates to nitrogen and hydrogen. The total volume of gas increases and the piston is pushed to the right. In Steps (c) and (d), as the catalyst is withdrawn, N_2 and H_2 re-associate to form ammonia; hence the volume shrinks and the piston is driven to the left. This self-contained, two-stage process provides an unlimited supply of power at the flywheel on the right, without an external input of energy. For practical difficulties, see the text.

5-5 EQUILIBRIA IN AQUEOUS SOLUTION: ACIDS AND BASES

When a solid dissolves in a liquid, it does so because the forces between solute (the dissolved matter) and solvent (the dissolving medium) molecules are stronger than those between the undissolved solute molecules or ions. In some solutions, such as organic-dye molecules dissolved in ethyl ether (C_2H_5—O—C_2H_5), the solute molecules are dispersed through the solvent intact. Water molecules are polar, and water is an especially good solvent for molecules that also are polar (see Figure 5–3). Thus water is a better solvent for polar methanol molecules (CH_3OH) than for nonpolar methane molecules (CH_4), and even better for ionic compounds. The remainder of this chapter will be devoted to aqueous solutions of ionized substances, and ionic equilibria.

Ionization of Water and the pH Scale

Water itself ionizes to a small extent:

$$H_2O(l) \rightleftharpoons H^+(aq) + OH^-(aq) \tag{5-23}$$

Each ion is surrounded with polar water molecules [like Na^+ and OH^- are in Figure 5–3(d)], with the oxygen atoms of water molecules closest to hydrogen ions, and the hydrogen atoms of other water molecules surrounding each hydroxide ion. Ions that interact electrostatically in this way with surrounding water molecules are said to be *hydrated*. The hydrated state of the proton, H^+, sometimes is represented as H_3O^+, meaning $H^+ \cdot H_2O$. But this is an unnecessary and even misleading notation. A more accurate representation of a hydrated proton would be $H_9O_4{}^+$, or $H^+ \cdot (H_2O)_4$, to represent the cluster:

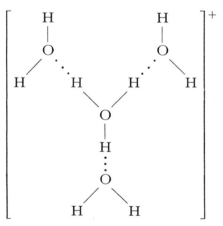

A convenient shorthand notation is $H^+(aq)$, in which (aq) signifies a hydrated proton in a large amount of water as solvent. We will assume that H^+ and OH^-, like every other ion, are hydrated in aqueous solution, and will represent them simply as H^+ and OH^-.

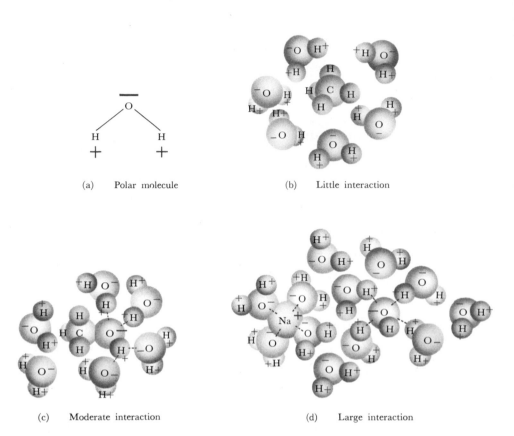

(a)   Polar molecule

(b)   Little interaction

(c)   Moderate interaction

(d)   Large interaction

Figure 5–3. (*a*) *Water is a polar molecule with excess electrons and a small negative charge on the oxygen atom; and an electron deficiency and a small positive charge on each hydrogen atom.* (*b*) *The methane molecule,* CH_4*, is nonpolar; its electrons are distributed evenly over the molecule. It has no local regions of positive and negative charge to attract water molecules, so water is a poor solvent for methane.* (*c*) *Methanol,* CH_3OH*, is polar, although less so than water. It has an excess of electrons and a small negative charge on the oxygen atom, and a small positive charge on the attached hydrogen atom. Methanol interacts well with water molecules by electrostatic forces, making it water soluble.* (*d*) *Sodium hydroxide,* $NaOH$*, dissociates into positive and negative ions. These ions interact strongly with the polar water molecules, so* $NaOH$ *is extremely water soluble. Each* Na^+ *and* OH^- *ion has a cluster of water molecules surrounding it, with their negative charges closest to the sodium ions and their positive charges closest to the hydroxide ions. The ions are said to be hydrated.*

Table 5–2. Measured Temperature Dependence of the Ion-Product Constant for Water, K_w

$$K_w = [H^+][OH^-]$$

T, °C:	0	25	40	60
K_w:	0.115×10^{-14}	1.008×10^{-14}	2.95×10^{-14}	9.5×10^{-14}

The equilibrium-constant expression for the dissociation of water is

$$K_{eq} = \frac{[H^+][OH^-]}{[H_2O]} \tag{5-24}$$

In reasonably dilute solutions, the amount of water that is used up or formed during a chemical reaction is small in comparison with the total amount of water present. The concentration of water is unaffected significantly by production or use of H_2O in chemical reactions, and is virtually the same as the concentration of water in its pure state:

$$[H_2O] = \frac{1000 \text{ g liter}^{-1}}{18.0 \text{ g mole}^{-1}} = 55.5 \text{ moles liter}^{-1} \tag{5-25}$$

It is more convenient to eliminate this constant from the denominator of the equilibrium-constant expression by bringing it to the left side of the equation and incorporating it into the constant:

$$K_w = 55.5 K_{eq} = [H^+][OH^-] \tag{5-26}$$

This new equilibrium constant, K_w, is called the *ion-product constant* for water. Like most equilibrium constants K_w varies with temperature. Some experimental values of the ion-product constant are given in Table 5–2.

Exercise. From the data in Table 5–2 and Le Chatelier's principle, predict whether the dissociation of water liberates or absorbs heat.

(*Answer:* Since a higher temperature favors dissociation, dissociation is an endothermic or heat-absorbing process. From Appendix 2, $\Delta H_{(diss\,of\,H_2O)} = +13.36$ kcal mole^{-1}. This is the energy required to break one O—H bond, thereby leaving both electrons with the oxygen atom.)

It is customary to take $K_w = 10^{-14}$ as being accurate enough for room-temperature equilibrium calculations. This means that in pure water, where the concentrations of hydrogen and hydroxide ions are equal,

$$[H^+] = [OH^-] = 10^{-7} \text{ mole liter}^{-1} \tag{5-27}$$

Since large powers of ten are clumsy to deal with, a logarithmic notation has been devised, called the *pH scale*. (The symbol pH stands for "negative power of hydrogen ion concentration.") The pH is the negative logarithm of $[H^+]$:

$$pH = -\log_{10} [H^+] \tag{5–28}$$

If the hydrogen ion concentration is 10^{-7} mole liter^{-1}, then

$$pH = -\log_{10} (10^{-7}) = +7$$

By an analogous definition,

$$pOH = -\log_{10} [OH^-] \tag{5–29}$$

and the pOH of pure water is also $+7$. The equilibrium constant K_w also can be expressed in logarithmic terms:

$$pK_w = -\log_{10} K_w = +14 \tag{5–30}$$

Finally, the equilibrium expression for dissociation of water,

$$[H^+][OH^-] = K_w = 10^{-14} \tag{5–31}$$

can be written

$$pH + pOH = 14 \tag{5–32}$$

In an acid solution, $[H^+]$ is greater than 10^{-7}, thus the pH is less than 7. The ion-product equilibrium still holds, and $[OH^-]$ can be found from the expression

$$[OH^-] = \frac{K_w}{[H^+]} = \frac{10^{-14}}{[H^+]} \tag{5–33}$$

or

$$pOH = pK_w - pH = 14 - pH \tag{5–34}$$

The approximate pH values of several common solutions are given in Table 5–3.

Example 13. From Table 5–3, what is the hydrogen ion concentration of orange juice? What is the hydroxide ion concentration?

Solution. Since the pH is 2.8, the hydrogen ion concentration is

$$[H^+] = 10^{-2.8} = 10^{+0.2} \times 10^{-3} = 1.6 \times 10^{-3}$$
$$= 0.0016 \text{ mole liter}^{-1}$$

(Logarithms and antilogarithms can be read from a slide rule. The logarithm of 1.6 is 0.2, and the antilogarithm of 0.2 is 1.6. A useful conversion to remember as insurance that you are using the log scale properly on a slide rule is that the logarithm of 2 is 0.30; that is, $2 = 10^{0.30}$ or $\log_{10} 2 = 0.30$.)

Table 5–3.   Acidity (expressed as pH) of Some Common Solutions

Substance	pH
Commercial concentrated HCl (37% by weight)	~ −1.1
1-molar HCl solution	0.0
Gastric juice	1.4
Lemon juice	2.1
Orange juice	2.8
Wine	3.5
Tomato juice	4.1
Black coffee	5.0
Urine	6.0
Rainwater	6.5
Milk	6.9
Pure water at 24°C	7.0
Blood	7.4
Baking soda solution	8.5
Borax solution	9.2
Limewater	10.5
Household ammonia	11.9
1-molar NaOH solution	14.0
Saturated NaOH solution	~15.0

The hydroxide ion concentration can be obtained by either of two equivalent methods:

$$[OH^-] = \frac{10^{-14}}{1.6 \times 10^{-3}} = 6.3 \times 10^{-12} \text{ mole liter}^{-1}$$

or

$$pOH = 14 - pH = 11.2$$
$$[OH^-] = 10^{-11.2} = 10^{+0.8} \times 10^{-12} = 6.3 \times 10^{-12} \text{ mole liter}^{-1}$$

Example 14.   What is the ratio of hydrogen ions to hydroxide ions in pure water? In orange juice?

Solution.   In pure water the ratio is $10^{-7}/10^{-7}$ or 1/1. In orange juice, from Table 5–3, the ratio is $1.6 \times 10^{-3}/6.3 \times 10^{-12}$ or 250,000,000/1. To maintain equilibrium, the added H^+ ions from the juice have pushed the water dissociation reaction in the direction of undissociated H_2O, thereby removing OH^- ions from the solution. Orange juice is not a particularly strong acid, and the enormous fluctuation of ionic ratios even in this example illustrates the usefulness of power-of-ten and logarithmic (pH, pOH, pK) notation.

Strong and Weak Acids

Acids are substances that increase the hydrogen ion concentration in a solution, and bases are substances that decrease $[H^+]$. In aqueous solution, acids are classified as either strong or weak. Strong acids are completely dissociated or ionized, and they include hydrogen acids such as hydrochloric acid (HCl) and hydroiodic acid (HI), and oxyacids such as nitric acid (HNO_3), sulfuric acid (H_2SO_4), and perchloric acid ($HClO_4$). Each of these acids loses one proton in solution, and the acid-dissociation constant, K_a, is so large ($> 10^3$) that too little undissociated acid remains to be measured. (HSO_4^- loses a second proton and is a weak acid.)

Weak acids have measurable ionization constants in aqueous solution, because they do not dissociate completely. Examples (at 25°C) are

Sulfuric: $HSO_4^- \rightleftharpoons H^+ + SO_4^{2-}$ $K_a = \dfrac{[H^+][SO_4^{2-}]}{[HSO_4^-]}$
(2nd ionization)

$$= 1.2 \times 10^{-2} \qquad (5\text{--}35)$$

Hydrofluoric: $HF \rightleftharpoons H^+ + F^-$ $K_a = \dfrac{[H^+][F^-]}{[HF]}$

$$= 3.5 \times 10^{-4} \qquad (5\text{--}36)$$

Acetic: $CH_3COOH \rightleftharpoons CH_3COO^- + H^+$ $K_a = \dfrac{[H^+][CH_3COO^-]}{[CH_3COOH]}$

$$= 1.76 \times 10^{-5} \qquad (5\text{--}37)$$

Hydrocyanic: $HCN \rightleftharpoons H^+ + CN^-$ $K_a = \dfrac{[H^+][CN^-]}{[HCN]}$

$$= 4.9 \times 10^{-10} \qquad (5\text{--}38)$$

The distinction between strong and weak acids is somewhat artificial. The ionization of HCl is not simply a dissociation, rather it is the result of successful competition of H_2O molecules with Cl^- ions for the proton, H^+:

$$HCl + xH_2O \rightleftharpoons H^+ \cdot (H_2O)_x + Cl^- \qquad (5\text{--}39)$$

In the Brønsted–Lowry theory of acids and bases, any *proton donor* is an acid, and any *proton acceptor* is a base. Therefore, HCl is an acid, and Cl^- is its *conjugate base*. Since HCl loses a proton readily it is a strong acid, and because Cl^- has so little affinity for the proton it is a weak base. In contrast, HCN is a very weak acid, because it loses its proton only to a small extent. Its conjugate base, CN^-, is a strong base by virtue of its high affinity for a proton.

Water is a somewhat stronger base than Cl^-, and when it is present in excess, as in an aqueous solution of HCl, it takes virtually all of the protons from HCl, thereby leaving it completely ionized. CN^- is a much stronger base than H_2O, thus only a small fraction of the protons from HCN become bound to the water molecules. In other words, HCN is only slightly ionized

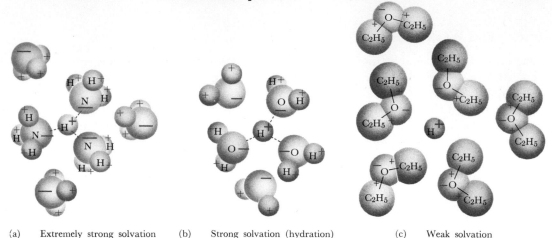

(a)   Extremely strong solvation   (b)   Strong solvation (hydration)   (c)   Weak solvation

Figure 5–4. Comparison of relative strengths of solvation of a hydrogen ion in (a) liquid ammonia, (b) water, and (c) ethyl ether. The binding between proton and solvent ammonia molecules is extremely strong, and liquid ammonia will take protons from and make strong acids of substances that in aqueous solution are only weak acids. In contrast, ethyl ether is such an ineffectual proton-solvating molecule that many substances that are strong acids in water can retain their proton and be only partially dissociated weak acids in ethyl ether.

in aqueous solution, as its K_a of 4.9×10^{-10} indicates. [We shall be able to explain the relative attractions of Cl^-, H_2O, and CN^- for protons after we know more about the arrangement of electrons around ions (Chapter 10). For the moment, we shall accept these data as experimentally true and use them.]

Because water is present in great excess, any acid whose conjugate base is weaker than H_2O (i.e., has a lesser affinity for protons than has H_2O) will be ionized essentially completely in aqueous solution. We cannot distinguish between the behavior of HCl and $HClO_4$ (perchloric acid) in water solution. Both are completely dissociated, and thus are strong acids. However, for a solvent with a lesser attraction for protons than water, we find differences between HCl and $HClO_4$ (Figure 5–4). With diethyl ether as a solvent, perchloric acid is still a strong acid, but HCl is only partially ionized and hence is a weak acid. Diethyl ether does not solvate a proton as strongly as water does. ("Solvation" is a generalization of the concept of hydration, which applies to solvents other than water.) The equilibrium point in the reaction

$$HCl + xC_2H_5OC_2H_5 \rightleftarrows H^+ \cdot (C_2H_5OC_2H_5)_x + Cl^- \qquad (5-40)$$

lies far to the left, thus HCl is only partially dissociated in ether. Only in an extremely strong acid, such as perchloric acid, does the anion have so little attraction for the proton that it will release it to methanol as an acceptor

Table 5–4.   Dissociation Constants of Some Acids^a at 25°C

Acid	HA	A$^-$	K_a	pK_a
Perchloric	$HClO_4$	ClO_4^-	$\sim 10^{+8}$	~ -8
Permanganic	$HMnO_4$	MnO_4^-	$\sim 10^{+8}$	~ -8
Chloric	$HClO_3$	ClO_3^-	$\sim 10^{+3}$	~ -3
Nitric	HNO_3	NO_3^-		
Hydrobromic	HBr	Br^-		
Hydrochloric	HCl	Cl^-		
Sulfuric (1)^b	H_2SO_4	HSO_4^-		
Hydrated proton or protonated solvent	$H^+(aq)$	H_2O(solvent)	1.00	0.00
Trichloroacetic	CCl_3COOH	CCl_3COO^-	2×10^{-1}	0.70
Oxalic (1)	$HOOC—COOH$	$HOOC—COO^-$	5.9×10^{-2}	1.23
Dichloroacetic	$CHCl_2COOH$	$CHCl_2COO^-$	3.32×10^{-2}	1.48
Sulfurous (1)	H_2SO_3	HSO_3^-	1.54×10^{-2}	1.81
Sulfuric (2)	HSO_4^-	SO_4^{2-}	1.20×10^{-2}	1.92
Phosphoric (1)	H_3PO_4	$H_2PO_4^-$	7.52×10^{-3}	2.12
Bromoacetic	$CH_2BrCOOH$	CH_2BrCOO^-	2.05×10^{-3}	2.69
Malonic (1)	$HOOC—CH_2—COOH$	$HOOC—CH_2—COO^-$	1.49×10^{-3}	2.83
Chloroacetic	$CH_2ClCOOH$	CH_2ClCOO^-	1.40×10^{-3}	2.85
Nitrous	HNO_2	NO_2^-	4.6×10^{-4}	3.34
Hydrofluoric	HF	F^-	3.53×10^{-4}	3.45
Formic	$HCOOH$	$HCOO^-$	1.77×10^{-4}	3.75
Benzoic	C_6H_5COOH	$C_6H_5COO^-$	6.46×10^{-5}	4.19
Oxalic (2)	$HOOC—COO^-$	$^-OOC—COO^-$	6.4×10^{-5}	4.19
Acetic	CH_3COOH	CH_3COO^-	1.76×10^{-5}	4.75
Propionic	CH_3CH_2COOH	$CH_3CH_2COO^-$	1.34×10^{-5}	4.87
Malonic (2)	$HOOC—CH_2—COO^-$	$^-OCC—CH_2—COO^-$	2.03×10^{-6}	5.69
Carbonic (1)	$CO_2 + H_2O$	HCO_3^-	4.3×10^{-7}	6.37
Sulfurous (2)	HSO_3^-	SO_3^{2-}	1.02×10^{-7}	6.91
Hydrogen sulfide (1)	H_2S	HS^-	9.1×10^{-8}	7.04
Phosphoric (2)	$H_2PO_4^-$	HPO_4^{2-}	6.23×10^{-8}	7.21
Ammonium ion	NH_4^+	NH_3	5.6×10^{-10}	9.25
Hydrocyanic	HCN	CN^-	4.93×10^{-10}	9.31
Silver ion	$Ag^+ + H_2O$	$AgOH$	9.1×10^{-11}	10.04
Carbonic (2)	HCO_3^-	CO_3^{2-}	5.61×10^{-11}	10.25
Hydrogen peroxide	H_2O_2	HO_2^-	2.4×10^{-12}	11.62
Hydrogen sulfide (2)	HS^-	S^{2-}	1.1×10^{-12}	11.96
Phosphoric (3)	HPO_4^{2-}	PO_4^{3-}	2.2×10^{-13}	12.67
Water^c	H_2O	OH^-	1.8×10^{-16}	15.76

^{a} HA is the acid form, with acid strength decreasing down the table. A$^-$ is the conjugate base, with base strength increasing down the table. The equilibrium is HA $\rightleftarrows$ H$^+$(aq) + A$^-$(aq), and the equilibrium-constant expression is

$$K_a = \frac{[H^+][A^-]}{[HA]} \qquad pK_a = -\log_{10} K_a$$

^{b} (1) is a first dissociation or proton-transfer reaction; (2) is a second dissociation; (3) is a third dissociation.

^{c} Note that this K_a value for water explicitly uses $[H_2O] = 55.5$ moles liter^{-1} in the denominator, for the sake of consistency with the other entries in the table, and that $55.5 \times 1.8 \times 10^{-16} = 1.0 \times 10^{-14} = K_w$.

solvent. Thus, by using solvents other than water, we can see differences in acidity (or proton affinity) that are masked in aqueous solution. This masking of relative acid strengths by basic solvents such as water is known as the "leveling effect."

The dissociation constants for several acids in aqueous solution are listed in Table 5–4, with estimates of the K_a for strong acids that are "leveled" by the solvent. The dissociation of protonated solvent H_2O, into hydrated protons and H_2O, represents merely a shuffling of protons from one set of water molecules to another, and must have a K_{eq} of 1.00. In liquid ammonia as a solvent, all acids whose conjugate bases are weaker than NH_3 would be leveled by the solvent and would be totally ionized strong acids. Thus hydrofluoric acid and acetic acid are strong acids in liquid ammonia.

The leveling effect of solvent, and the origin of strong and weak acids are summarized in Figure 5–4. The distinction between strong and weak acids depends upon the solvent as much as it does upon the inherent properties of the acids themselves. Nevertheless, in aqueous solution the distinction is real. As long as the discussion is confined to aqueous solutions (as ours will be from now on), we shall find it useful to think about and to treat the two classes of acids separately.

Strong and Weak Bases

In ordinary terminology a base is a substance that decreases the hydrogen ion concentration of a solution. Sodium hydroxide, potassium hydroxide, and similar compounds, are bases because they dissolve and dissociate completely in aqueous solution to yield hydroxide ions:

$$NaOH \rightleftharpoons Na^+ + OH^-$$
$$KOH \rightleftharpoons K^+ + OH^- \tag{5–41}$$

These excess hydroxide ions then perturb the water dissociation equilibrium, and combine with some of the protons normally found in pure water:

$$H^+ + OH^- \rightleftharpoons H_2O \qquad [H^+] = \frac{K_w}{[OH^-]} < 10^{-7} \tag{5–42}$$

In the more generalized Brønsted–Lowry definition of a base, the hydroxide ion itself is the Brønsted–Lowry base, because it is the substance that combines with the proton. The Na^+ and K^+ ions merely provide the positive ions that are necessary for overall electrical neutrality for the chemical compound.

The commonly encountered hydroxides of alkali metals dissolve and dissociate completely to produce the same Brønsted–Lowry base, OH^-. These hydroxides all are *strong bases*, analogous to strong acids such as HCl and HNO_3. Other substances such as ammonia and many organic nitrogen compounds also can combine with protons in solution and act as Brønsted–Lowry bases. These compounds are *weaker* bases than the hydroxide ion, because they have a smaller attraction for protons. For example, when ammonia competes with OH^- for protons in an aqueous solution, it is only partially successful. It can combine with only a portion of the H^+ ions, thus

$$NH_3 + H^+ \rightleftarrows NH_4^+ \tag{5–43}$$

will have a measurable equilibrium constant.

There is no logical reason why this reaction cannot be described by an acid-dissociation constant, as in Table 5–4. The ammonium ion, NH_4^+, is the Brønsted–Lowry *conjugate acid* of the base NH_3. There is no reason why, in an acid-base pair, it is the acid that must be neutral and the base charged, as in HCl/Cl^- and HCN/CN^-. The NH_4^+ ion is just as respectable an acid as HCl or HCN, and although weaker than HCl, it actually is stronger than HCN. Thus, we can describe the ammonia reaction as an acid dissociation:

$$NH_4^+ \rightleftarrows NH_3 + H^+ \qquad K_a = 5.6 \times 10^{-10} \tag{5–44}$$
$$\text{(from Table 5–4)}$$

or if we want to focus on the basic behavior of NH_3:

$$NH_3 + H^+ \rightleftarrows NH_4^+ \qquad K_{eq} = \frac{1}{K_a} = 1.79 \times 10^{+9} \tag{5–45}$$

In spite of this, chemical language has become trapped by the older acid-base terminology introduced by Arrhenius, and you should be aware of this. Arrhenius thought of a base as a substance that releases OH^- ions into aqueous solution. For alkali metal hydroxides the process was straightforward:

$$NaOH \rightleftarrows Na^+ + OH^- \tag{5–46}$$

But what about NH_3? Where do the hydroxide ions come from? Arrhenius assumed that when ammonia dissolved in water the reaction was

$$NH_3 + H_2O \rightleftarrows NH_4OH \rightleftarrows NH_4^+ + OH^- \tag{5–47}$$

This brought NH_3 into line by postulating an intermediate—ammonium hydroxide—that dissociated like any other hydroxide. Sodium hydroxide is a strong base that dissociates completely; ammonium hydroxide would be a weak base that dissociates only partially. Arrhenius defined a base dissociation constant, K_b, as

$$BOH \rightleftarrows B^+ + OH^- \qquad K_b = \frac{[B^+][OH^-]}{[BOH]} \tag{5–48}$$

For ammonia, K_a and K_b would be related by

$$K_b = \frac{[NH_4^+][OH^-]}{[NH_3]} = \frac{[NH_4^+][OH^-][H^+]}{[NH_3][H^+]} = \frac{K_w}{K_a} \tag{5–49}$$

$$K_b = \frac{10^{-14}}{5.6 \times 10^{-10}} = 1.79 \times 10^{-5} \tag{5–50}$$

Unfortunately for Arrhenius' theory, there is no evidence that ammonium

Table 5–5.   Dissociation Constants of Some Weak Bases^a at 25°C

Base	B	BH$^+$	K_b	pK_b
Aniline	(phenyl)—NH$_2$	(phenyl)—NH$_3{}^+$	4.3×10^{-10}	9.37
Pyridine	(pyridine ring) N	(pyridinium ring) N$^+$—H	1.8×10^{-9}	8.75
Imidazol	(imidazole ring)	(imidazolium ring)	9.1×10^{-8}	7.05
Hydrazine	N$_2$H$_4$	N$_2$H$_5{}^+$	9.8×10^{-7}	6.01
Ammonia	NH$_3$	NH$_4{}^+$	1.79×10^{-5}	4.75
Trimethylamine	(CH$_3$)$_3$N	(CH$_3$)$_3$NH$^+$	6.4×10^{-5}	4.19
Methylamine	CH$_3$—NH$_2$	CH$_3$—NH$_3{}^+$	3.7×10^{-4}	3.34
Dimethylamine	(CH$_3$)$_2$NH	(CH$_3$)$_2$NH$_2{}^+$	5.4×10^{-4}	3.27

^{a} If B represents the base, the equilibrium equation is B + H$_2$O $\rightleftarrows$ BH$^+$ + OH$^-$, in which BH$^+$ is the conjugate acid. Base strengths increase down the table, and conjugate acid strengths decrease. The equilibrium-constant expression is

$$K_b = \frac{[BH^+][OH^-]}{[B]} \qquad pK_b = -\log_{10} K_b$$

hydroxide, NH$_4$OH, really exists as a stoichiometric compound. It is more accurate to say that the polar ammonia molecule is hydrated like any other polar molecule: NH$_3 \cdot$(H$_2$O)$_x$ or NH$_3$(aq). Ammonia, NH$_3$, combines directly with a proton and with water molecules:

$$
\begin{aligned}
&\text{NH}_3 + \text{H}^+ + x\text{H}_2\text{O} \rightleftarrows \text{NH}_4{}^+(aq) && \text{(in acid solutions)} \\
&\text{NH}_3 + x\text{H}_2\text{O} \qquad\quad \rightleftarrows \text{NH}_4{}^+(aq) + \text{OH}^- && \text{(in basic solutions)}
\end{aligned}
\qquad (5\text{--}51)
$$

Nevertheless, Arrhenius' notation is too deeply embedded in the fabric of chemistry to dislodge, and we often will use K_b for weak bases rather than K_a for their conjugate acids. In general, the completely dissociated strong bases that we shall encounter will be hydroxide compounds, and the weak bases will be ammonia and organic nitrogen compounds such as those listed in Table 5–5. K_b always can be found from K_a and K_w and the expression

$$K_a \times K_b = K_w \qquad (5\text{--}52)$$

5–6 SOLUTIONS OF STRONG ACIDS AND BASES: NEUTRALIZATION AND TITRATION

When an amount of strong acid is added to water, the effect is that of adding the same amount of hydrogen ions, since the acid is totally dissociated.

Example 15. What is the hydrogen ion concentration of a 0.01-molar nitric acid solution? What is the pH?

Solution

$$[H^+] = 0.01 \text{ mole liter}^{-1}$$
$$pH = -\log_{10}(10^{-2}) = 2.0$$

The solution is quite acidic.

Example 16. What are the hydrogen ion concentration and pH of a 0.005-molar sodium hydroxide solution?

Solution. The hydroxide ion contribution from completely dissociated NaOH is

$$[OH^-] = 0.005 \text{ mole liter}^{-1}$$

This large amount of hydroxide ion will repress the normal dissociation of water and enhance the reaction to the left:

$$H_2O \rightleftharpoons H^+ + OH^-$$

The hydrogen ion concentration is found from the water equilibirum expression:

$$[H^+] = \frac{K_w}{[OH^-]} = \frac{10^{-14}}{0.005} = 2 \times 10^{-12} \text{ mole liter}^{-1}$$

$$pH = -\log_{10}(2) - \log_{10}(10^{-12}) = -0.30 + 12.0 = 11.7$$

The solution is strongly basic.

Example 17. What will be the pH if we mix equal volumes of the solutions of the previous two examples?

Solution. If equal volumes are mixed, then the concentration of each solute will be halved, since the final volume is twice the volume of each starting solution. The final solution would be 0.0050 molar in nitric acid and 0.0025 molar in sodium hydroxide. But acid and base will react and neutralize one another until one or the other is used up:

$$H^+ + NO_3^- + Na^+ + OH^- \rightleftharpoons H_2O + NO_3^- + Na^+$$

or simply

$$H^+ + OH^- \rightleftharpoons H_2O$$

since sodium and nitrate ions take no part in the neutralization reaction. In this case, sodium hydroxide is in shorter supply. When all of the base has been neutralized, we still have

$$0.0050 - 0.0025 \quad = 0.0025 \text{ mole liter}^{-1} \text{ excess nitric acid}$$
$$[H^+] \quad = 0.0025 \quad = 2.5 \times 10^{-3} \text{ mole liter}^{-1}$$
$$pH \quad = -\log_{10}(2.5) + 3.0 = 2.6$$

Example 18. How many milliliters of 0.10-molar HCl must we add to 200 ml of 0.005-molar KOH to bring the pH down to 10?

Solution. Without HCl, the pH of the potassium hydroxide solution would be 11.7, as in Example 16. Let y equal the number of milliliters of HCl solution needed to yield a pH of 10. Since 0.005 mole liter^{-1} is the same as 0.005 millimoles ml^{-1}, the total number of millimoles of KOH is

$$n_{KOH} = 0.005 \text{ millimole ml}^{-1} \times 200 \text{ ml} = 1.0 \text{ millimole}$$

The total number of millimoles of HCl that must be added is

$$n_{HCl} = 0.10 \text{ millimole ml}^{-1} \times y \text{ ml} = 0.10y \text{ millimole}$$

Since the final solution is basic, $n_{KOH} > n_{HCl}$. The net amount of hydroxide ion left over after partial neutralization by HCl is

$$n_{base} = n_{KOH} - n_{HCl} = 1.0 - 0.10y$$

The final volume is

$$V = 200 + y \text{ ml}$$

and therefore the final hydroxide ion concentration is

$$[OH^-] = \frac{n_{base}}{V} = \frac{1.0 - 0.10y}{200 + y}$$

A pH of 10 means a pOH of 4 and $[OH^-] = 10^{-4}$ mole liter^{-1}, thus

$$\frac{1.0 - 0.10y}{200 + y} = 10^{-4}$$

and

$$y = 9.8 \text{ ml of 0.10-molar HCl to be added.}$$

Titration and Titration Curves

If we add equal numbers of equivalents of a strong acid and a strong base, they will neutralize one another completely, and the pH will be 7.0. This leads to a way of measuring the amount of an acid present in an unknown solution. Simply add a measured amount of base solution of known concentration to the point of neutralization, or end point, and calculate how many equivalents of acid had been present from the number of equivalents of base used. This process is called *titration*, and is a standard analytical technique. We have discussed it already in Chapter 4.

Example 19. One hundred fifty milliliters of HCl solution of unknown concentration are titrated with 0.10-molar NaOH. Eighty milliliters of base solution are required to neutralize the acid. How many moles of HCl were present originally, and what was the acid-solution concentration?

Solution. The number of millimoles of base used is

$$n_{NaOH} = 0.10 \text{ mmole ml}^{-1} \times 80 \text{ ml} = 8.0 \text{ mmoles}$$

This must be the same as the number of millimoles of acid originally present, if neutralization was complete.

Thus the original concentration of HCl was

$$[HCl]_0 = \frac{8.0 \text{ mmoles}}{150 \text{ ml}} = 0.053 \text{ mmole ml}^{-1} \text{ or mole liter}^{-1}$$

A common way of determining the end point of titration is with an acid–base indicator. Indicators are weak organic acids or bases, which have different colors in their ionized and neutral states (or in two ionized states). If their color change occurs in the neighborhood of pH 7, and if we add a few drops of indicator solution to the solution being titrated, we see this color change at the end point of the titration. We will discuss some common indicators in the section on weak acids. The matching of indicator color-change point and the end point of a titration does not have to be very exact, because the pH swings drastically through several units as neutralization becomes complete. This can make life easy for the analytical chemist, and it is worth looking more closely at the behavior of pH during titration. To illustrate what we have just said, let us calculate the titration curve for a typical strong acid and strong base.

Example 20. Fifty milliliters of 0.10-molar nitric acid are titrated with 0.10-molar KOH, in an experimental arrangement such as that shown in Figure 4–1. Calculate the pH of the solution as a function of the volume of KOH solution added (v, in ml).

Solution. It is easiest to treat this calculation in three parts: before neutralization, at neutralization (end point), and after neutralization. Before the end point, calculate how much base has been added, assume that all of this base was used to neutralize some of the acid, and calculate how much acid would remain unneutralized, as a function of the volume of base solution added.

Original: $n_{HNO_3} = 50 \text{ ml} \times 0.10 \text{ mmole ml}^{-1} = 5.0 \text{ mmoles}$

Added: $n_{KOH} = v \text{ ml} \times 0.10 \text{ mmole ml}^{-1}$

Net acid: $n_{acid} = 5.0 - 0.10v \text{ mmole}$

Total volume: $V = 50 + v \text{ ml}$

$$[H^+]_{net} = \frac{5.0 - 0.10v}{50 + v} = \frac{50 - v}{50 + v}(0.10) \text{ mmole ml}^{-1}$$

The calculation of $[H^+]$ for various values of v is shown in Table 5–6(a), and these calculations are plotted with open circles at the left of Figure 5–5. At the end point, the amounts of acid and base are equal and the pH is 7.0.

Table 5–6.   Titration of 50 ml of 0.10M Nitric Acid by 0.10M Potassium Hydroxide ($v = ml$ of base solution added)

a) Before equivalence point:

v (ml)	$\dfrac{50 - v}{50 + v}$	[H$^+$]	pH
0	1.00	0.100	1.00
10	$\frac{40}{60}$	0.067	1.18
20	$\frac{30}{70}$	0.043	1.37
30	$\frac{20}{80}$	0.025	1.60
40	$\frac{10}{90}$	0.011	1.95
45	$\frac{5}{95}$	0.0053	2.28
48	$\frac{2}{98}$	0.0020	2.69
49	$\frac{1}{99}$	0.0010	3.00
49.9	$\frac{0.1}{99.9}$	0.0001	4.00
49.99	$\frac{0.01}{99.99}$	0.00001	5.00

b) After equivalence point:

v (ml)	$\dfrac{v - 50}{v + 50}$	[OH$^-$]	pOH	pH
50.01	$\frac{0.01}{100.01}$	0.00001	5.00	9.00
50.1	$\frac{0.1}{100.1}$	0.0001	4.00	10.00
51	$\frac{1}{100}$	0.0010	3.00	11.00
52	$\frac{2}{102}$	0.0020	2.71	11.29
55	$\frac{5}{105}$	0.0048	2.32	11.68
60	$\frac{10}{110}$	0.0091	2.04	11.96
70	$\frac{20}{120}$	0.0167	1.78	12.22
80	$\frac{30}{130}$	0.023	1.64	12.36
90	$\frac{40}{140}$	0.029	1.54	12.46
100	$\frac{50}{150}$	0.033	1.48	12.52

After the end point, we only need to calculate how much base was added in excess of that required to neutralize the acid, and use this to find [OH$^-$], pOH, and pH:

Original: $n_{HNO_3} = 5.0$ mmoles (as before)

Added: $n_{KOH} = v$ ml $\times$ 0.10 mmole ml^{-1}

Net base: $n_{base} = 0.10v - 5.0$ mmoles

Final volume: $V = 50 + v$ ml

Hydroxide ion concentration:

$$[OH^-] = \frac{0.10v - 5.0}{50 + v} = \frac{v - 50}{v + 50} (0.10) \text{ mmole ml}^{-1}$$

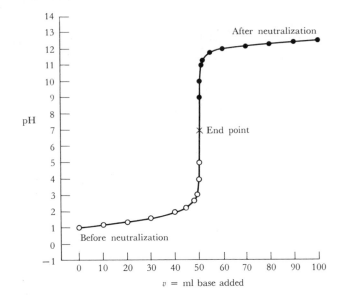

Figure 5–5. Titration curve for typical strong acid and base. Fifty milliliters of 0.10M HNO₃ *are titrated with increasing amounts of* 0.10M KOH. *Data are given in Table 5–6. Notice how rapidly the* pH *changes in the region of the end point, or of exact neutralization of acid by base. Any acid–base indicator that changes color in the region of* pH 4 *to* 10 *could be used to detect the end point in this titration.*

This calculation for several values of v and the corresponding pH values are listed in Table 5–6(b) and are plotted with solid circles on the right of Figure 5–5. It now is obvious why the choice of an indicator is not too critical in such a titration. Any indicator that changes color in the pH range between 3 and 11 will do.

Titrating a weak acid with a strong base, or a weak base with a strong acid, is more complicated because the weak component is only partially dissociated. Dissociation equilibria of the type discussed in the next section must be used. We will not be concerned in this chapter with such titrations, but they are treated in Appendix 3.

5–7 WEAK ACIDS AND BASES

As we discussed previously, weak acids and bases are only partially dissociated in water. Therefore, the contribution of acetic acid to hydrogen ion concentration is less than the total concentration of added acid.

Example 21. What is the pH of a solution of 0.0100-molar acetic acid? Compare this with the pH of the same concentration of nitric acid (Example 15).

Solution. It is common to represent the acetate ion, CH_3COO^-, by Ac^-, and to write HAc for acetic acid instead of CH_3COOH. (OAc^- and HOAc also are used, to indicate that this is an oxyacid with the dissociating proton attached to an oxygen atom.) The dissociation of HAc is incomplete:

$$HAc \rightleftharpoons H^+ + Ac^-$$

and the equilibrium expression describing dissociation is

$$K_a = \frac{[H^+][Ac^-]}{[HAc]} = 1.76 \times 10^{-5} \quad \text{(from Table 5–4)}$$

We know the initial overall concentration of acetic acid:

$$c_0 = 0.0100 \text{ mole liter}^{-1}$$

and we know that at equilibrium some of this acetic acid remains undissociated and some of it has ionized to acetate ions, Ac^-:

$$c_0 = [HAc] + [Ac^-]$$

This is called a *mass-balance* equation, because it states that *total* acetate is neither created nor destroyed during dissociation. We also know that the concentrations of hydrogen ions and acetate ions are equal, since dissociation of HAc is the only source of H^+. (It is legitimate to neglect H^+ from the dissociation of water, since acetic acid represses water dissociation even below its normal small extent.) Thus

$$[H^+] = [Ac^-]$$

This is known as a *charge-balance* equation, because it states that the total positive charge in the solution must equal the total negative charge. We now can use these data about conservation of acetate and neutrality of the solution to simplify the equilibrium-constant expression. Let the hydrogen ion concentration that we are seeking be $[H^+] = y$, and eliminate $[Ac^-]$ at once using the charge-balance equation:

$$K_a = \frac{y^2}{[HAc]} \qquad \text{(equilibrium equation)}$$

$$c_0 = [HAc] + y \qquad \text{(mass-balance equation)}$$

The second equation tells us that the concentration of undissociated HAc equals the original overall concentration, c_0, minus the amount that has dissociated, y:

$$[HAc] = c_0 - y$$

The equilibrium expression then is

$$K_a = \frac{y^2}{c_0 - y}$$

Substituting the value of K_a from Table 5–4, we get

$$1.76 \times 10^{-5} = \frac{y^2}{0.0100 - y}$$

or

$$y^2 + 1.76 \times 10^{-5}y - 1.76 \times 10^{-7} = 0$$

This is a quadratic equation, which can be solved with the quadratic formula:

$$y = \frac{-1.76 \times 10^{-5} \pm \sqrt{3.10 \times 10^{-10} + 7.04 \times 10^{-7}}}{2}$$

or

$$y = \frac{-1.76 \times 10^{-5} \pm 8.39 \times 10^{-4}}{2}$$

Only the positive answer is reasonable, because one cannot have a negative concentration. Thus the answer is

$$y = 4.11 \times 10^{-4} \text{ mole liter}^{-1}$$

Under certain physical conditions you can take a shortcut to avoid the quadratic formula. In this example, since you know that the acid is only slightly dissociated, you can try neglecting y in the denominator of the equilibrium expression for K_a, thereby assuming that it is small in comparison with 0.0100 mole liter^{-1}, and that the concentration of undissociated acetic acid is virtually the same as the total acetic acid present. This assumption gives

$$1.76 \times 10^{-5} = \frac{y^2}{0.0100}$$

and an approximate answer of

$$y = 4.2 \times 10^{-4} = 0.00042 \text{ mole liter}^{-1}$$

This is close to the correct answer of 0.000411 mole liter^{-1}. You can make a quick improvement by using this approximate value in the undissociated acetate concentration in the denominator:

$$1.76 \times 10^{-5} = \frac{y^2}{0.0100 - 0.00042}$$

$$y = 4.11 \times 10^{-4} \text{ mole liter}^{-1}$$

If your physical intuition for how much dissociation the acid undergoes is good enough, you often can solve an equilibrium problem by an approximate solution and a quick correction in less time than it takes to solve the quadratic formula.

As our results show, acetic acid indeed is only slightly dissociated at 0.0100-molar concentration. Of the initial 0.0100 mole per liter, 0.000411 mole has dissociated, and 0.0096 mole remains as dissolved but undissociated HAc molecules. The percent dissociation is

$$\frac{4.11 \times 10^{-4} \text{ mole}}{0.0100 \text{ mole}} \times 100 = 4.11\%$$

The pH of this solution is 3.39.

What happens if we dilute the acetic acid solution? Does a greater or lesser percent of the acetic acid then dissociate? Does the pH increase or decrease?

Example 22. What are the pH and percent dissociation in a solution of 0.00100-molar acetic acid?

Solution. The equilibrium expression is as before:

$$K_a = \frac{y^2}{c_0 - y}$$

$$1.76 \times 10^{-5} = \frac{y^2}{0.00100 - y}$$

Neglecting y in comparison with c_0, the approximate solution is

$$1.76 \times 10^{-5} = \frac{y^2}{0.00100}$$

$$y = 1.33 \times 10^{-4} \text{ mole liter}^{-1}$$

and the solution obtained by using this value to correct the undissociated HAc concentration is

$$y = 1.24 \times 10^{-4} \text{ mole liter}^{-1}$$

Now the pH is 3.91 instead of 3.39, and the percent dissociation is

$$\frac{1.24 \times 10^{-4} \text{ mole}}{0.00100 \text{ mole}} \times 100 = 12.4\%$$

Although the actual hydrogen ion concentration is lower (witness the larger pH), a greater fraction of the HAc present is dissociated into ions. This is Le Chatelier's principle again. If a solution containing HAc, H^+, and Ac^-

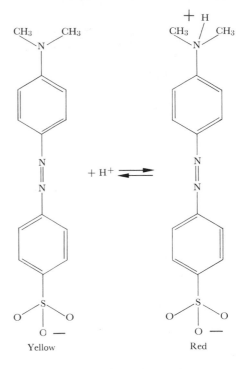

Figure 5–6. The basic form (left) and acid form (right) of the indicator methyl orange. The different colors of the two structures, yellow and red, give methyl orange its usefulness in displaying the pH of a solution into which it has been introduced. The complex structure can be symbolized by an ion, In⁻, which can combine with a proton as shown at the bottom of the figure.

is diluted, thereby lowering the overall concentration of all ions and molecules, the equilibrium will attempt to reestablish itself as reactions change in the direction that will increase the overall concentration of solute particles of one kind or another. Compare this behavior with the effect of increasing the pressure on the ammonia gas equilibrium in Section 5–4.

Indicators

An indicator is a weak acid (or weak base) with sharply different colors in its dissociated and undissociated state. Methyl orange (Figure 5–6) is a complex organic compound that is red in its neutral form and yellow when ionized. It can be represented as the weak acid HIn:

$$\underset{\text{red}}{\text{HIn}} \rightleftarrows \text{H}^+ + \underset{\text{yellow}}{\text{In}^-} \tag{5–53}$$

The intensity of color from indicators such as methyl orange is so great that the colors can be seen easily even when the amount added to a solution is too small to have an appreciable influence on the pH of the solution. Nevertheless, the ratio of dissociated to undissociated indicator will depend on the hydrogen ion concentration

$$K_a = \frac{[\text{H}^+][\text{In}^-]}{[\text{HIn}]} \tag{5–54}$$

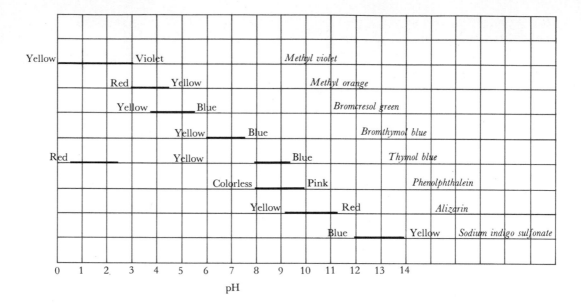

Figure 5–7. Some common acid–base indicators, with the pH *ranges in which their color changes occur. The choice of an indicator for an acid–base titration depends on the expected* pH *at the end point of the titration and the width of swing of* pH *values as the end point is passed.*

and

$$\frac{[\text{In}^-]}{[\text{HIn}]} = \frac{K_a}{[\text{H}^+]} \tag{5-55}$$

$$\log_{10}\left(\frac{[\text{In}^-]}{[\text{HIn}]}\right) = \text{pH} - pK_a \tag{5-56}$$

For methyl orange, $K_a = 1.6 \times 10^{-4}$ and $pK_a = 3.8$. The neutral (red) and dissociated (yellow) forms of the indicator are present at equal concentrations when the pH = 3.8. The eye is sensitive to color changes over a range of concentration ratios of approximately 100, or over two pH units. Below pH 2.8, a solution containing methyl orange is red, and above approximately 4.8 it is clearly yellow. As you can see from Figure 5–5, an indicator change over two pH units is quite satisfactory for strong acid–base titrations.

Methyl orange could be used for the titration in Figure 5–5 even though its pK_a is far from the titration end point of 7.0 only because the change in pH at the end point is so large. For weak acid titrations, this would not be true, and it would be better to pick an indicator with a pK_a closer to the expected end point. Other indicators are shown in Figure 5–7, along with the pH range in which their color changes occur. Phenolphthalein is a particularly convenient and common indicator that changes from colorless to pink in the range of pH 8 to 10.

Before you go on. If you still are having difficulty with acid–base calculations, work through Section 5–2 of *Programed Reviews of Chemical Principles.*

5–8   WEAK ACIDS AND THEIR SALTS

What will happen to a weak acid such as acetic acid if we add some sodium acetate, the salt of a strong base (NaOH) and acetic acid? The salt will dissolve and dissociate completely to sodium and acetate ions. From Le Chatelier's principle, we would expect these added acetate ions to force the weak acetic acid equilibrium system in the direction of less dissociation. This is exactly what happens. The acid-equilibrium expression is the same:

$$K_a = \frac{[H^+][Ac^-]}{[HAc]} \tag{5-57}$$

However, two sources of acetate ions now exist: NaAc and HAc. The acetate ion supplied by sodium acetate is measured by c_s, the total molarity of the salt, since dissociation is complete. Acetate concentration from acetic acid is measured by the hydrogen ion concentration, since every dissociation of HAc to produce Ac^- also produces a proton. Therefore, the total acetate ion concentration is

$$[Ac^-]_{total} = [Ac^-]_{NaAc} + [Ac^-]_{HAc} = c_s + [H^+] \tag{5-58}$$

(Again, we have neglected any protons from the dissociation of water.) The concentration of un-ionized acetic acid is the overall acid concentration, c_a, less the acetate from dissociation:

$$[HAc] = c_a - [Ac^-]_{HAc} = c_a - [H^+] \tag{5-59}$$

If we represent the hydrogen ion concentration by y, we have

$$K_a = \frac{y(c_s + y)}{(c_a - y)} \tag{5-60}$$

When the added salt concentration, c_s, is zero, this is the simple weak acid-dissociation equilibrium expression that we have seen previously.

Example 23. What are the pH and percent dissociation of a solution of 0.010-molar acetic acid in the presence of no NaAc, 0.0050-molar NaAc, 0.010- and 0.020-molar NaAc?

Solution. From Le Chatelier's principle, we would expect that as more NaAc is added, the dissociation of HAc is repressed. The pH should increase and the percent dissociation should decrease. The problem without NaAc was solved in Section 5–7, yielding pH 3.39 and 4.11% dissociation. For $c_s = 0.0050$ mole liter^{-1}:

$$1.76 \times 10^{-5} = \frac{y(0.0050 + y)}{0.010 - y} \quad \text{(from Equation 5–60)}$$

Table 5-7.   Effect of Adding Sodium Acetate to 0.10M Acetic Acid Solution

c_s = concentration of sodium acetate in moles liter^{-1}						
c_s:	0.0	0.001	0.002	0.005	0.010	0.020
pH:	3.4	3.8	4.1	4.5	4.8	5.1
Percent dissociation of acetic acid:	4.1	1.5	0.84	0.35	0.18	0.09

As a first approximation we can assume that y will be smaller than 0.0050 or 0.010, and neglect it when added to or subtracted from these quantities:

$$y_1 = 1.76 \times 10^{-5} \times \frac{0.010}{0.0050} = 3.52 \times 10^{-5} = 0.000035 \text{ mole liter}^{-1}$$

As a second approximation, we can use this trial value of y to "correct" 0.0050 to 0.005035, and 0.010 to 0.009965, and solve the equation again:

$$y_2 = 1.76 \times 10^{-5} \times \frac{0.009965}{0.005035} = 3.48 \times 10^{-5} \text{ mole liter}^{-1}$$

A third approximation is unnecessary, and the answer should be rounded to 3.5×10^{-5} mole liter^{-1}:

$$pH = 5 - \log_{10} 3.5 = 5 - 0.54 = 4.46$$

$$\text{Percent dissociation} = \frac{3.5 \times 10^{-5}}{0.010} \times 100 = 0.35\%$$

For c_s = 0.010 mole liter^{-1},

$$y = [H^+] = 1.76 \times 10^{-5} \text{ mole liter}^{-1}$$

$$pH = 4.75$$

$$\text{Percent dissociation} = 0.18\%$$

Notice that the acetic acid now dissociates so little that the first approximation is adequate.

Results for these and a few other sodium acetate concentrations are listed in Table 5-7 and are plotted in Figure 5-8. The first salt added has a large effect on the degree of dissociation and pH; later additions of salt cause less change. When acid and salt are present in equal concentrations, the pH is equal to the pK_a of the acid, as shown in Example 24.

Buffers

If the concentrations of a solution of a weak acid and a salt of the acid anion are reasonably high, then the solution is resistant to changes in hydrogen ion concentration.

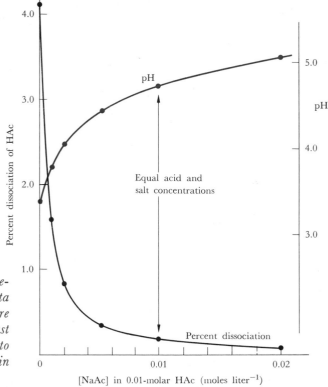

Figure 5–8. The effect of added sodium acetate on the dissociation of acetic acid. Data plotted here are listed in Table 5–7, and were calculated as explained in the text. The first salt added represses acetic acid dissociation to a great extent and causes a rapid increase in pH. Later additions are not as effective.

Example 24. A solution is 0.050-molar in HAc and 0.050-molar in NaAc. Calculate the change in pH when 0.0010-molar hydrochloric acid (HCl) is added. Compare this to the pH of a 0.0010-molar HCl solution without HAc and NaAc.

Solution. Before adding HCl the acetic acid equilibrium is

$$K_a = \frac{[H^+][Ac^-]}{[HAc]} = \frac{y(0.050)}{(0.050)}$$

Thus

$$y = K_a = 1.76 \times 10^{-5} \text{ mole liter}^{-1}$$
$$pH = pK_a = 4.75$$

(Again, we were justified in ignoring y in the $[Ac^-]$ and $[HAc]$ terms because the value is small compared to 0.050.)

The added protons from HCl combine with acetate ions to form more acetic acid:

$$Ac^- + H^+ \text{ (from HCl)} \rightarrow HAc$$

Thus to a good approximation, all of the added protons are used up, and the new acetic acid and acetate concentrations are

$$[HAc] = 0.050 + [H^+]_{HCl} = 0.051 \text{ mole liter}^{-1}$$
$$[Ac^-] = 0.050 - [H^+]_{HCl} = 0.049 \text{ mole liter}^{-1}$$
$$K_a = \frac{y(0.049)}{(0.051)}$$
$$y = 1.76 \times 10^{-5} \times \frac{0.051}{0.049} = 1.83 \times 10^{-5} \text{ mole liter}^{-1}$$
$$pH = 5 - 0.26 = 4.74$$

The pH changes from 4.75 to 4.74, a difference of only 0.01 unit. In the absence of HAc and NaAc, the same concentration of HCl would produce a pH of 3.0.

This resistance to pH change is called *buffering action*, and the solution of HAc and NaAc is an acetate buffer. Buffers are used widely for pH control in laboratory chemistry, in the chemical industry, and in living organisms. A carbonate buffer system in your bloodstream, involving the reaction

$$H^+ + HCO_3^- \rightleftarrows H_2CO_3 \rightleftarrows CO_2 + H_2O \tag{5-61}$$

maintains the blood pH around 7.4. When a biochemist studies enzyme activity *in vitro*, he must use a buffer system to maintain a constant pH during the experiments, otherwise his results may have little meaning. One of the sillier disputes in commercial advertising is that between two pharmaceutical companies as to whether buffers added to aspirin to combat an acid reaction in the stomach are a benefit or an adulterant.

In general, if the concentration of acid added to a buffer solution is x moles liter^{-1}, the equilibrium equation becomes

$$K_a = \frac{[H^+][A^-]}{[HA]} = \frac{[H^+](c_s - x)}{c_a + x} \tag{5-62}$$

in which c_s and c_a are the salt and buffering acid concentrations, respectively. After addition of the foreign acid the hydrogen ion concentration is

$$[H^+] = K_a \frac{(c_a + x)}{(c_s - x)} \tag{5-63}$$

and the pH is

$$pH = pK_a + \log_{10} \frac{(c_s - x)}{(c_a + x)} \tag{5-64}$$

If base is added, hydrogen ions are removed, and the same expressions can be used with a negative value of x.

Exercise. A formic acid buffer is prepared with 0.010 mole liter^{-1} each of formic acid (HCOOH) and sodium formate (HCOONa). What is the pH of the solution? What is the pH if 0.0020 mole liter^{-1} of sodium hydroxide is added? What would be the pH of the sodium hydroxide solution without buffer? What would the pH have been after adding sodium hydroxide if the buffer concentrations had been 0.10 mole liter^{-1} instead of 0.010?

(*Answer*:

Buffer:	pH = 3.75
After adding NaOH:	pH = 3.92
Without buffer:	pH = 11.30
With stronger buffer:	pH = 3.77)

In the preceding exercise, you can see the dramatic effect of the formate buffer in keeping the solution acidic in spite of the added base, and the importance of reasonably high buffer concentrations if the buffering capacity of the solution is not to be exceeded.

Challenge. The carbonate system is one of the most important buffer systems in humans. To learn how this buffer works, try Problems 5–38 through 5–41 in *Relevant Problems for Chemical Principles.*

5–9 SALTS OF WEAK ACIDS AND STRONG BASES: HYDROLYSIS

A sodium chloride solution is neutral, with a pH of 7.0. This is reasonable, because sodium hydroxide is a strong base and hydrochloric acid is a strong acid, and if equal amounts of each were added neutralization would be complete. In contrast, sodium acetate is the salt of a strong base and a weak acid. We intuitively would expect a sodium acetate solution to be somewhat basic, and it is. Some of the acetate ions from the salt combine with water to form undissociated acetic acid and hydroxide ions:

$$Ac^- + H_2O \rightleftarrows HAc + OH^- \tag{5–65}$$

This sometimes is called a *hydrolysis reaction*, the implication being that H_2O breaks up crystals of sodium acetate (hydrolysis means "breaking up with water"). It does, when the salt crystal dissolves in water, but this is not the point. In solution the acetate ion acts as a base. It is equally as good a Brønsted base as ammonia:

$$NH_3 + H_2O \rightleftarrows NH_4^+ + OH^-$$

We should not let the different charges on acetate ion (-1) and ammonia (0) obscure the similarity of their acid–base behavior.

The equilibrium constant for acetate hydrolysis is

$$K_b = \frac{[HAc][OH^-]}{[Ac^-]} \tag{5–66}$$

where, as usual, the virtually unchanging water concentration is incorporated into the equilibrium constant. This constant sometimes is written K_h for "hydrolysis constant," but the added nomenclature is unnecessary. It is a simple base-equilibrium constant of the kind we have seen before, except that acetate ion is the base.

As always, K_b is related to the acid-dissociation constant of acetic acid, K_a, by

$$K_b = \frac{[HAc][OH^-]}{[Ac^-]} = \frac{[HAc][OH^-][H^+]}{[Ac^-][H^+]} = \frac{K_w}{K_a} \tag{5-67}$$

Thus

$$K_b = \frac{10^{-14}}{1.76 \times 10^{-5}} = 5.68 \times 10^{-10} \tag{5-68}$$

(Recall the ammonia–water equilibrium expressions at the end of Section 5–5.) This value is all we need to calculate the pH of a sodium acetate solution.

Example 25.   What is the pH of a solution of 0.010-molar NaAc?

Solution.   The equilibrium expression is

$$5.68 \times 10^{-10} = \frac{[HAc][OH^-]}{[Ac^-]}$$

Let the hydroxide ion concentration be z. Since every reaction of an acetate ion with water produces one hydroxide ion and one undissociated HAc molecule, the concentrations of these latter two species both must be z. The remaining acetate ions are those originally present from NaAc minus those that have combined with water:

$$[Ac^-] = 0.010 - z$$

and we arrive at the familiar expression

$$K_b = \frac{z^2}{0.010 - z} = 5.68 \times 10^{-10}$$

This is even easier to solve than the weak-acid problems. Since the equilibrium constant is so small, z will be correspondingly small and can be neglected in the denominator in comparison to 0.010. The result is

$$z^2 = 0.010 \times 5.68 \times 10^{-10} = 5.68 \times 10^{-12}$$

$$z = 2.38 \times 10^{-6} \text{ mole liter}^{-1} = [OH^-]$$

$$[H^+] = \frac{K_w}{[OH^-]} = \frac{10^{-14}}{2.38 \times 10^{-6}} = 4.20 \times 10^{-9}$$

$$pH = 9 - 0.62 = 8.38$$

5–10 EQUILIBRIA WITH SLIGHTLY SOLUBLE SALTS

When most solid salts dissolve in water, they dissociate completely into hydrated positive and negative ions. The solubility of a salt in water represents a balance between the attraction of the ions in the crystal lattice and the attraction between these ions and the polar water molecules. This balance may be a delicate one, easily changed in going from one compound to an apparently similar one, or from one temperature to another. It is not possible to give hard-and-fast rules as to whether a compound is soluble, or even to account for all observed behavior.

One important factor certainly is the electrostatic attraction between ions. Crystals with small ions that can be packed closely together generally are harder to pull apart than crystals with large ions. Therefore, for a given cation, fluorides (F^-) and hydroxides (OH^-) are less soluble than nitrates (NO_3^-) and perchlorates (ClO_4^-). Chlorides are intermediate in size, and their behavior is difficult to predict from general principles.

The charge on the ions also is important. More highly charged ions such as phosphates (PO_4^{3-}) and carbonates (CO_3^{2-}) interact strongly with cations and are less soluble than are the singly charged nitrates and perchlorates.

The terms "soluble" and "insoluble" are relative, and the degree of solubility can be related to an equilibrium constant. For a "slightly soluble" salt such as silver chloride, an equilibrium exists between the dissociated ions and the solid compound:

$$AgCl(s) \rightleftarrows Ag^+(aq) + Cl^-(aq) \qquad (5\text{–}69)$$

[The (aq) is a reminder that each ion is hydrated by water molecules. This will be understood from now on and the (aq) symbol will be dropped.] The equilibrium expression for this reaction is

$$K_{eq} = \frac{[Ag^+][Cl^-]}{[AgCl]_{solid}} \qquad (5\text{–}70)$$

As long as solid AgCl remains, its effect on the equilibrium does not change. As with the H_2O concentration in the water dissociation equilibrium, the concentration of the solid salt can be incorporated into the equilibrium constant:

$$K_{sp} = K_{eq}[AgCl]_{solid} = [Ag^+][Cl^-] \qquad (5\text{–}71)$$

This new equilibrium constant, K_{sp}, is called the *solubility-product constant*. For substances in which the ions are not in a 1:1 ratio, the form of the solubility-product expression is analogous to our previous equilibrium expressions:

$$PbCl_2 \rightleftarrows Pb^{2+} + 2Cl^- \qquad K_{sp} = [Pb^{2+}][Cl^-]^2$$
$$Al(OH)_3 \rightleftarrows Al^{3+} + 3OH^- \qquad K_{sp} = [Al^{3+}][OH^-]^3$$
$$Ag_2CrO_4 \rightleftarrows 2Ag^+ + CrO_4^{2-} \qquad K_{sp} = [Ag^+]^2[CrO_4^{2-}]$$
$$Ba_3(PO_4)_2 \rightleftarrows 3Ba^{2+} + 2PO_4^{3-} \qquad K_{sp} = [Ba^{2+}]^3[PO_4^{3-}]^2$$

Solubility equilibria are useful in predicting whether a precipitate will form under specified conditions, and in choosing conditions under which two chemical substances in solution can be separated by selective precipitation.

The solubility-product constant of a slightly soluble compound can be calculated from its solubility in moles liter^{-1}.

Example 26.   The solubility of AgCl in water is 0.000013 mole liter^{-1} at 25°C. What is its solubility-product constant, K_{sp}?

Solution.   The equilibrium expression is

$$AgCl \rightleftharpoons Ag^+ + Cl^-$$

The concentrations of Ag^+ and Cl^- are equal because for each mole of solid AgCl that dissolves, one mole each of Ag^+ and Cl^- ions would be produced. Hence the concentration of each ion is equal to the overall solubility of the solid, s, in moles per liter:

$$[Ag^+] = [Cl^-] = s = 1.3 \times 10^{-5} \text{ mole liter}^{-1}$$
$$K_{sp} = [Ag^+][Cl^-] = s^2 = 1.7 \times 10^{-10}$$

Example 27.   At a certain temperature the solubility of $Fe(OH)_2$ in water is 7.7×10^{-6} mole liter^{-1}. Calculate its K_{sp} at that temperature.

Solution.   The equilibrium equation is

$$Fe(OH)_2 \rightleftharpoons Fe^{2+} + 2OH^-$$

and the solubility-product expression is

$$K_{sp} = [Fe^{2+}][OH^-]^2$$

Since one mole of dissolved $Fe(OH)_2$ produces one mole of Fe^{2+} and *two* moles of OH^-,

$$[Fe^{2+}] = s = 7.7 \times 10^{-6} \text{ mole liter}^{-1}$$
$$[OH^-] = 2s = 1.54 \times 10^{-5} \text{ mole liter}^{-1}$$
$$K_{sp} = 7.7 \times 10^{-6} \times (1.54 \times 10^{-5})^2 = 1.8 \times 10^{-15}$$

The solubility-product constants of several substances are listed in Table 5–8, grouped according to the approximate decreasing solubility of anions, and for a given anion, according to decreasing solubility product. Once the solubility-product constant is known, it can be used to calculate the solubility of a compound at a specified temperature.

Example 28.   What is the solubility of $PbSO_4$ in water at 25°C?

Solution.   The dissociation reaction is

$$PbSO_4 \rightleftharpoons Pb^{2+} + SO_4{}^{2-}$$

Table 5–8.   Solubility-Product Constants, K_{sp}, at 25°C

Fluorides		Chromates (continued)		Hydroxides (continued)	
BaF_2	2.4×10^{-5}	$BaCrO_4$	8.5×10^{-11}	$Ni(OH)_2$	1.6×10^{-16}
MgF_2	8×10^{-8}	Ag_2CrO_4	1.9×10^{-12}	$Zn(OH)_2$	4.5×10^{-17}
PbF_2	4×10^{-8}	$PbCrO_4$	2×10^{-16}	$Cu(OH)_2$	1.6×10^{-19}
SrF_2	7.9×10^{-10}			$Hg(OH)_2$	3×10^{-26}
CaF_2	3.9×10^{-11}	Carbonates		$Sn(OH)_2$	3×10^{-27}
		$NiCO_3$	1.4×10^{-7}	$Cr(OH)_3$	6.7×10^{-31}
Chlorides		$CaCO_3$	4.7×10^{-9}	$Al(OH)_3$	5×10^{-33}
$PbCl_2$	1.6×10^{-5}	$BaCO_3$	1.6×10^{-9}	$Fe(OH)_3$	6×10^{-38}
$AgCl$	1.7×10^{-10}	$SrCO_3$	7×10^{-10}	$Co(OH)_3$	2.5×10^{-43}
$Hg_2Cl_2{}^a$	1.1×10^{-18}	$CuCO_3$	2.5×10^{-10}		
		$ZnCO_3$	2×10^{-10}	Sulfides	
Bromides		$MnCO_3$	8.8×10^{-11}	MnS	7×10^{-16}
$PbBr_2$	4.6×10^{-6}	$FeCO_3$	2.1×10^{-11}	FeS	4×10^{-19}
$AgBr$	5.0×10^{-13}	Ag_2CO_3	8.2×10^{-12}	NiS	3×10^{-21}
$Hg_2Br_2{}^a$	1.3×10^{-22}	$CdCO_3$	5.2×10^{-12}	CoS	5×10^{-22}
		$PbCO_3$	1.5×10^{-15}	ZnS	2.5×10^{-22}
Iodides		$MgCO_3$	1×10^{-15}	SnS	1×10^{-26}
PbI_2	8.3×10^{-9}	$Hg_2CO_3{}^a$	9.0×10^{-15}	CdS	1.0×10^{-28}
AgI	8.5×10^{-17}			PbS	7×10^{-29}
$Hg_2I_2{}^a$	4.5×10^{-29}			CuS	8×10^{-37}
		Hydroxides		Ag_2S	5.5×10^{-51}
Sulfates		$Ba(OH)_2$	5.0×10^{-3}	HgS	1.6×10^{-54}
$CaSO_4$	2.4×10^{-5}	$Sr(OH)_2$	3.2×10^{-4}	Bi_2S_3	1.6×10^{-72}
Ag_2SO_4	1.2×10^{-5}	$Ca(OH)_2$	1.3×10^{-6}		
$SrSO_4$	7.6×10^{-7}	$AgOH$	2.0×10^{-8}		
$PbSO_4$	1.3×10^{-8}	$Mg(OH)_2$	8.9×10^{-12}	Phosphates	
$BaSO_4$	1.5×10^{-9}	$Mn(OH)_2$	2×10^{-13}	Ag_3PO_4	1.8×10^{-18}
		$Cd(OH)_2$	2.0×10^{-14}	$Sr_3(PO_4)_2$	1×10^{-31}
Chromates		$Pb(OH)_2$	4.2×10^{-15}	$Ca_3(PO_4)_2$	1.3×10^{-32}
$SrCrO_4$	3.6×10^{-5}	$Fe(OH)_2$	1.8×10^{-15}	$Ba_3(PO_4)_2$	6×10^{-39}
$Hg_2CrO_4{}^a$	2×10^{-9}	$Co(OH)_2$	2.5×10^{-16}	$Pb_3(PO_4)_2$	1×10^{-54}

^{a} As $Hg_2{}^{2+}$ ion. $K_{sp} = [Hg_2{}^{2+}][X^-]^2$

Let the unknown solubility be s moles liter^{-1}. Then since each mole of dissolved $PbSO_4$ produces one mole of each ion,

$$[Pb^{2+}] = [SO_4{}^{2-}] = s$$

The solubility-product equation is

$$K_{sp} = [Pb^{2+}][SO_4{}^{2-}] = s^2 = 1.3 \times 10^{-8} \quad \text{(from Table 5–8)}$$
$$s = 1.14 \times 10^{-4} \text{ mole liter}^{-1}$$

Example 29.   In Table 5–8 we see that $CdCO_3$ and Ag_2CO_3 have approximately the same solubility-product constants. Compare their molar solubilities in water (at 25°C).

Solution. For cadmium carbonate:

$$K_{sp} = [Cd^{2+}][CO_3^{2-}] = s^2 = 5.2 \times 10^{-12}$$
$$s = 2.3 \times 10^{-6} \text{ mole liter}^{-1}$$

For Ag_2CO_3 the expression is slightly different. If the solubility again is s moles liter^{-1}, since each mole of salt produces two moles of Ag^+ ions,

$$[Ag^+] = 2s$$
$$[CO_3^{2-}] = s$$
$$K_{sp} = [Ag^+]^2[CO_3^{2-}] = (2s)^2 \times s = 4s^3 = 8.2 \times 10^{-12}$$
$$s = 1.3 \times 10^{-4} \text{ mole liter}^{-1}$$

Although cadmium carbonate and silver carbonate have nearly the same solubility-product constants, their solubilities in moles per liter differ by a factor of 100 because the form of the solubility-product expression is different. The solubility of Ag_2CO_3 is sensitive to the square of the metal-ion concentration, because two silver ions per carbonate ion are necessary to build the solid crystal.

Common-Ion Effect

In the preceding example, the solubility of silver carbonate in pure water was calculated to be 1.3×10^{-4} mole liter^{-1}. Will silver carbonate be more soluble or less soluble in silver nitrate solution? Le Chatelier's principle leads us to predict that a new, outside source of silver ions would shift the silver carbonate equilibrium reaction in the direction of less dissociation:

$$Ag_2CO_3 \rightleftharpoons 2Ag^+ + CO_3^{2-} \tag{5-72}$$

or that silver carbonate would be less soluble in a silver nitrate solution than in pure water. This decrease in the solubility of one salt in a solution of another salt that has a common cation or anion is called the *common-ion effect*.

Example 30. What is the solubility at 25°C of CaF_2 (a) in pure water, (b) in 0.10-molar $CaCl_2$, and (c) in 0.10-molar NaF?

Solution. a) If the solubility in pure water is s, then

$$[Ca^{2+}] = s$$
$$[F^-] = 2s$$
$$K_{sp} = s \times 4s^2 = 4s^3 = 3.9 \times 10^{-11}$$
$$s = 2.1 \times 10^{-4} \text{ mole liter}^{-1}$$

b) In 0.10-molar $CaCl_2$, the calcium ion concentration is the sum of the concentration of calcium ions from calcium chloride and from calcium fluoride, whose solubility we are seeking:

$$[Ca^{2+}] = 0.10 + s$$
$$[F^-] = 2s$$
$$K_{sp} = (0.10 + s)(2s)^2 = 3.9 \times 10^{-11}$$

This is a cubic equation, but a moment's thought about the chemistry involved will eliminate the need to solve it as such. With such a small solubility-product constant, you can predict that the solubility of calcium fluoride will be very small in comparison with 0.10 mole liter^{-1}. (You already should realize from Part (a) and Le Chatelier's principle that in this problem, s will be less than 2.1×10^{-4} mole liter^{-1}.) If our prediction is valid, we can simplify the solubility-product equation and calculate the approximate solubility:

$$0.10 \times (2s)^2 = 3.9 \times 10^{-11}$$

$$s^2 = \frac{3.9 \times 10^{-11}}{4 \times 0.10} = 9.75 \times 10^{-11}$$

$$s = 0.99 \times 10^{-5} = 9.9 \times 10^{-6} \text{ mole liter}^{-1}$$

Therefore the approximation is justified. Only five percent as much CaF_2 will dissolve in 0.10-molar $CaCl_2$ as in pure water:

$$\frac{9.9 \times 10^{-6}}{2.1 \times 10^{-4}} \times 100 = 5\%$$

c) In 0.10-molar NaF,

$$[Ca^{2+}] = s$$

and

$$[F^-] = 0.10 + 2s$$

since fluoride ions come from NaF as well as from CaF_2. The solubility-product equation is

$$K_{sp} = s(2s + 0.10)^2 = 3.9 \times 10^{-11}$$

Again, thinking about the chemical meaning will avoid the necessity of solving a cubic equation. The $2s$ term will be very small compared to 0.10 mole liter^{-1}, therefore,

$$s(0.10)^2 = 3.9 \times 10^{-11}$$

$$s = 3.9 \times 10^{-9} \text{ mole liter}^{-1}$$

This approximation is even more valid than the previous one, since from the calculation

$$\frac{3.9 \times 10^{-9}}{2.1 \times 10^{-4}} \times 100 = 0.002\%$$

only 0.002% as much CaF_2 will dissolve in 0.10-molar NaF as in pure water. Fluoride is more effective than calcium as a common ion because it has a second-power effect on the solubility equilibrium.

The common-ion method of controlling solubility often is used with solutions of sulfide ion, S^{2-}, because many metals form insoluble sulfides, and the sulfide ion concentration can be controlled by adjusting the pH.

Example 31. What is the maximum possible concentration of Ni^{2+} ion in water at 25°C that is saturated with H_2S and maintained at pH 3.0 with HCl?

Solution. From the solubility-product equilibrium equation we predict that too much nickel ion will cause the precipitation of NiS:

$$K_{sp} = [Ni^{2+}][S^{2-}] = 3 \times 10^{-21}$$

The only new twist to this problem is finding the sulfide ion concentration from the H_2S equilibrium. Hydrogen sulfide dissociates in two steps, each with an equilibrium constant[1]:

$$H_2S \rightleftharpoons H^+ + HS^- \qquad K_{a_1} = 9.1 \times 10^{-8}$$
$$\underline{HS^- \rightleftharpoons H^+ + S^{2-} \qquad K_{a_2} = 1.1 \times 10^{-12}}$$
$$H_2S \rightleftharpoons 2H^+ + S^{2-} \qquad K_{a_{12}} = K_{a_1} \times K_{a_2}$$

Because the overall dissociation is the sum of two dissociation steps, the overall equilibrium constant, $K_{a_{12}}$, is the product of K_{a_1} and K_{a_2}:

$$K_{a_{12}} = \frac{[H^+][HS^-]}{[H_2S]} \times \frac{[H^+][S^{2-}]}{[HS^-]} = \frac{[H^+]^2[S^{2-}]}{[H_2S]}$$

$$K_{a_{12}} = 9.1 \times 10^{-8} \times 1.1 \times 10^{-12} = 1.0 \times 10^{-19}$$

Saturated H_2S is approximately 0.10 molar at 25°C, and the very small value of $K_{a_{12}}$ means that dissociation of H_2S is very slight. Hence we can write

$$[H_2S] = 0.10 \text{ mole liter}^{-1} \qquad \text{and} \qquad [H^+]^2[S^{2-}] = 1.0 \times 10^{-20}$$

in a saturated H_2S solution. This "ion product" for saturated H_2S is a useful relationship to remember.

In this problem, the pH has been adjusted to 3.0 with hydrochloric acid so

$$[H^+] = 10^{-3} \text{ mole liter}^{-1}$$

Therefore, the sulfide ion concentration can be calculated from

$$[S^{2-}] = K_{a_{12}} \times \frac{[H_2S]}{[H^+]^2} = 1.0 \times 10^{-19} \times \frac{0.10}{(10^{-3})^2}$$

which gives

$$[S^{2-}] = 1.0 \times 10^{-14} \text{ mole liter}^{-1}$$

Since NiS will precipitate if the solubility product is exceeded, the highest

---

[1] Polyprotic acids, or acids that yield more than one proton upon dissociation, are discussed in more detail in Appendix 3.

value that the nickel ion concentration can have is

$$[Ni^{2+}] = \frac{K_{sp}}{[S^{2-}]} = \frac{3 \times 10^{-21}}{1 \times 10^{-14}} = 3 \times 10^{-7} \text{ mole liter}^{-1}$$

Separation of Compounds by Precipitation

Solubility-product constants can be used to devise methods for separating ions in solution by selective precipitation. The entire, traditional qualitative-analysis scheme is based on the use of these equilibrium constants to determine the correct precipitating ions and the correct strategy.

Example 32. A solution is 0.010-molar in $BaCl_2$ and 0.020-molar in $SrCl_2$. Can either Ba^{2+} or Sr^{2+} be precipitated selectively with concentrated sodium sulfate solution? Which ion will precipitate first? When the second ion just begins to precipitate, what is the residual concentration of the first ion, and what fraction of the original amount of the first ion is left in solution? (For simplicity, assume that the Na_2SO_4 solution is so concentrated that the volume change in the Ba–Sr solution can be neglected.)

Solution. The upper limit on barium sulfate solubility is given by

$$K_{sp} = [Ba^{2+}][SO_4^{2-}] = 1.5 \times 10^{-9}$$

With 0.010 mole liter^{-1} of Ba^{2+}, precipitation of barium sulfate will not occur until the sulfate ion concentration increases to

$$[SO_4^{2-}] = \frac{1.5 \times 10^{-9}}{0.010} = 1.5 \times 10^{-7} \text{ mole liter}^{-1}$$

Strontium sulfate will precipitate when the sulfate concentration is

$$[SO_4^{2-}] = \frac{K_{sp(SrSO_4)}}{[Sr^{2+}]} = \frac{7.6 \times 10^{-7}}{0.020} = 3.8 \times 10^{-5} \text{ mole liter}^{-1}$$

Therefore, barium will precipitate first. When the sulfate concentration has risen to 3.8×10^{-5} mole liter^{-1} and strontium sulfate just begins to precipitate, the residual barium concentration left in solution will be

$$[Ba^{2+}] = \frac{1.5 \times 10^{-9}}{3.8 \times 10^{-5}} = 3.9 \times 10^{-5} \text{ mole liter}^{-1}$$

This quantity is

$$\frac{3.9 \times 10^{-5}}{0.010} \times 100 = 0.4\%$$

or four tenths percent of the original Ba^{2+} present. Thus 99.6% of the barium has been precipitated before any strontium begins to precipitate.

Before you go on. For more practice doing solubility-product calculations, work the problems in Section 5–3 of Lassila.

5–11 SUMMARY

In this chapter we have encountered two of the most fundamental ideas of chemistry, namely, spontaneity and chemical equilibrium. They are fundamental because they tell us when a reaction has an inherent tendency to occur (which is not to say that it will occur rapidly without help). If the forward and backward reactions of a chemical process are occurring at the same rate, this condition of balance is *equilibrium*. A reaction that is not at equilibrium but is moving in the direction of equilibrium is *spontaneous*. A catalyst accelerates the movement of a spontaneous reaction toward equilibrium, but does not affect the final equilibrium state.

The higher the concentrations of reacting substances, the greater will be their tendency to react to form products. Conversely, as the concentration of products increases, the more the reverse reaction is favored over the forward reaction. At equilibrium the ratio of products to reactants has a characteristic value for the reaction, known as the *equilibrium constant*, K_{eq}. For the general reaction

$$aA + bB \rightleftarrows cC + dD$$

the form of the equilibrium-constant expression is

$$K_{eq} = \frac{[C]^c[D]^d}{[A]^a[B]^b}$$

in which each concentration term is raised to a power corresponding to its stoichiometry in the overall reaction. Although the simple mass action derivation of K_{eq} is defective, the results are correct. The value of K_{eq} can be calculated from the overall stoichiometric equation of the reaction, independently of the reaction mechanism. Correspondingly, the equilibrium constant can tell us nothing about the actual mechanism by which the reaction occurs.

Le Chatelier's principle is an important summary of how equilibria behave. It says that if a stress is applied to a chemical system at equilibrium, the system will shift in such a way as to reduce that stress. If the stress is a change in concentrations of reactants or products, in pressure, or in volume, the equilibrium amounts of the chemical participants may change but their concentration ratio—the equilibrium constant, K_{eq}—will *not*. However, if the temperature is changed, then K_{eq} usually will have a new value. In agreement with Le Chatelier's principle, the equilibrium constant for a reaction that absorbs heat will be greater at higher temperatures.

Dissociation of acids, bases, and salts in solution represents a competition between the mutual attraction of ions of the substance and the attraction of water (or other solvent) molecules for these ions. With acids and bases, the important factor is the relative attraction of neutral molecules or ions for the proton, H^+. According to the Brønsted–Lowry theory, any substance that gives up a proton is an *acid*, and any substance that can combine with a proton and remove it from solution is a *base*. As an acid loses its proton, it becomes the *conjugate base*. A strong acid such as HCl has a weak conjugate base, Cl^-, and a weak acid, such as HAc or NH_4^+, has a strong conjugate base, Ac^- or NH_3. Any acid whose conjugate base is sufficiently weaker than H_2O (less affinity for H^+) will be dissociated completely in aqueous solution, and hence is classed as a *strong acid*. Acids that dissociate only partially in aqueous solution are *weak acids*.

The dissociation of weak acids and bases, and of water itself, can be described by the equilibrium constants, K_a, K_b, and K_w. For the same reasons of numerical convenience that lead us to describe the hydrogen ion concentration logarithmically as $pH = -\log_{10} [H^+]$, these equilibrium constants can be written as pK_a, pK_b, and pK_w. Solving acid–base equilibrium problems requires that one keep track of equilibria that may exist between chemical species, that one allow for the impossibility of destroying or creating matter during the reactions, and that overall charge neutrality be maintained. When first set up, an equilibrium problem may appear more involved than it really is. Chemical common sense about when one can make approximations usually reduces the complexity of equilibrium computations to slide-rule dimensions.

If a dissociation constant is smaller than 10^{-4} or 10^{-5}, one usually can neglect the amount of a substance that dissociates in comparison with the amount remaining, and approximate the concentration of undissociated substance at equilibrium by the initial overall concentration. Even when a check of the answer shows that this value is incorrect, it usually is the best starting guess for a second cycle of calculations. It also helps to realize how inexact most equilibrium constants are. Refining an answer to less than 5% accuracy is a waste of time.

Most of the general comments just made about solving acid–base equilibrium problems are applicable to solubility equilibria. Solubility-product calculations are most useful as guides for indicating whether precipitation will occur under certain conditions, what the upper limits on concentration of an ion in solution may be, and whether two ions can be separated in solution by selective precipitation.

William Thomson made a very perceptive observation about numbers and science, which was quoted at the beginning of this chapter. The two areas where chemists first began to apply numbers to their experiments, after Lavoisier's early comments about the importance of mass, were in heats of reaction—thermochemistry—and equilibrium experiments. In Chapter 15 we shall see that these are two sides of the same phenomenon. What we can learn from measuring thermodynamic properties will permit us to calculate equilibrium constants directly. In this chapter we have either accepted them as known or considered them to be the results of equilibrium experiments. We also have avoided any questions about *how fast* reactions take place. Reaction rates will be the subject of Chapter 18. For the moment, it is a large step forward to be able to predict ahead of time whether a reaction, if we wait patiently, eventually will take place.

SUGGESTED READING

A. J. Bard, *Chemical Equilibrium*, Harper and Row, New York, 1966.

J. N. Butler, *Ionic Equilibrium*, Addison-Wesley, Reading, Mass., 1964.

J. N. Butler, *Solubility and pH Calculations*, Addison-Wesley, Reading, Mass., 1964.

A. F. Clifford, *Inorganic Chemistry of Qualitative Analysis*, Prentice-Hall, Englewood Cliffs, N.J., 1961.

E. S. Gould, *Inorganic Reactions and Structure*, Holt, Rinehart and Winston, New York, 1962, 2nd ed.

K. B. Harvey and G. B. Porter, *Introduction to Physical Inorganic Chemistry*, Addison-Wesley, Reading, Mass., 1963.

E. J. King, *Acid–Base Equilibria*, Macmillan, New York, 1965.

E. J. King, *Qualitative Analysis and Electrolyte Solutions*, Harcourt, Brace & World, New York, 1959.

T. Moeller and R. O'Conner, *Ions in Aqueous Systems*, McGraw-Hill, New York, 1972.

L. Pauling, *The Nature of the Chemical Bond*, Cornell University Press, Ithaca, N.Y., 1960, 3rd ed.

M. J. Sienko, *Chemistry Problems*, W. A. Benjamin, Menlo Park, Calif., 1972, 2nd ed.

M. J. Sienko and R. A. Plane, *Physical Inorganic Chemistry*, W. A. Benjamin, Menlo Park, Calif., 1963.

QUESTIONS

1 What is a spontaneous reaction? Must a spontaneous reaction be rapid? Illustrate with an example other than those given in this chapter.

2 How does a catalyst affect a spontaneous reaction? What does it do to the point of equilibrium in a reaction?

3 What is meant by the rate constant for a chemical reaction? What is an equilibrium constant? How does the equilibrium constant depend on the concentrations of reactants and products? How does it depend on the relative numbers of molecules that are involved in a chemical reaction?

4 How is the reaction quotient related to the equilibrium constant? Are they ever equal? If the reaction quotient under a given set of conditions is greater than the equilibrium constant, what does this tell you about the spontaneity of the reaction as written under these conditions?

5 What is Le Chatelier's principle? What would it indicate about the effect on an equilibrium constant of raising the temperature at which the reactions occurred? How could it predict the effect on equilibrium concentrations of a change in overall pressure?

6 How does a catalyst affect the conditions of equilibrium? Justify your answer in terms of the nonexistence of perpetual-motion machines. What makes a catalyst useful?

7 Why is water a better solvent for methanol than for methane? Why is table salt more soluble in water than in ethyl ether?

8 What is the ion-product constant for the dissociation of water? Since it is an equilibrium constant, why does the concentration of water not appear in the ion-product expression?

9 What is the pH scale, and why is it useful? Does a strong acid have a high or a low pH? Does a high pH indicate a high or a low hydroxide ion concentration? If the pH is 3, what are the hydrogen ion and hydroxide ion concentrations? When the pH changes by two units, by what factor does the hydrogen ion concentration change?

10 What is an acid, and what is a base? What is the distinction between strong and weak acids?

11 In the Brønsted–Lowry theory, what are conjugate acids and bases? Give two examples of conjugate acid-base pairs, in one of which the acid is charged and the base is neutral, and in the other of which the acid is neutral and the base is charged.

12 How does the nature of the solvent affect whether an acid is classified as strong or weak? What is the "leveling effect"?

13 How are K_a and K_b defined? Why is it always true that their product is equal to the ion product for water, K_w? Illustrate with an example other than the one used in this chapter.

14 What is meant by an "equivalent" of an acid or a base in titration? (See Chapter 4 if you have forgotten.) How many equivalent weights of acid are there per mole of hydrochloric acid? Phosphoric acid? Sulfuric acid? How many moles of sodium hydroxide would be required to neutralize one mole of sulfuric acid?

15 Why is the choice of an indicator relatively uncritical in the titration of a strong acid by a strong base?

16 Is the pH of a solution of a weak base greater or smaller than the pH of a solution of the same concentration of a strong base, assuming the same number of equivalents of base per mole? Why?

17 How does the weak acid or weak base character of an indicator make it useful in determining when a given pH has been reached during a titration?

18 What do "charge balance" and "mass balance" equations represent, and how are they useful in dealing with the equilibrium expression for a weak acid?

19 Will a solution of ammonium chloride be acidic, neutral, or basic? As the concentration of the solution is increased, what will happen to the pH?

20 What will be the effect on the degree of dissociation of an aqueous ammonia solution if we add some ammonium chloride? What do you think would happen to the degree of dissociation if instead we added a substance that formed a complex ion with ammonia molecules such as $Cu(NH_3)_4^{2+}$?

21 How does a buffer counteract attempts to change the pH of a solution? What are the two components of a typical buffer solution?

22 How is the hydrolysis constant related to acid and base equilibrium constants?

23 What is the relationship between solubility and solubility-product constants for $CaCO_3$? For CaF_2? For $Ca_3(PO_4)_2$?

24 What is the common-ion effect, and how does it influence solubility equilibria?

25 How can a knowledge of solubility-product constants be used to make analytical separations of ions in solution?

26 How can pH be used to control the concentration of sulfide ion, S^{2-}, in solution? As the pH is increased, does the sulfide ion concentration increase or decrease? Give a physical explanation for your answer.

PROBLEMS

1 The equilibrium constant for the reaction

$$A_2(g) + B_2(g) \overset{k_1}{\underset{k_2}{\rightleftarrows}} 2AB(g)$$

is 2.5×10^{-6}, at a certain temperature. The rate constant (k_2) for the reverse reaction is 151 atm^{-1} sec^{-1}. Calculate the rate constant for the forward reaction.

2 The equilibrium constant for the reaction

$$N_2(g) + O_2(g) \rightleftarrows 2NO(g)$$

at 2130°C is 2.5×10^{-3}. Calculate the equilibrium constant for the reaction

$$NO(g) \rightleftarrows \tfrac{1}{2}O_2(g) + \tfrac{1}{2}N_2(g)$$

3 The reaction of Problem 2 proceeds with the absorption of heat. For the following conditions, determine whether a net reaction will occur, and if so, in which direction:

a) A 1-liter box contains 0.02 mole of NO, 0.01 mole of O_2, and 0.02 mole of N_2 at 2130°C.

b) A 20-liter box contains 1×10^{-2} mole of N_2, 1×10^{-3} mole of O_2, and 2×10^{-2} mole of NO at 2130°C.

c) A 1-liter box contains 1.00 mole of N_2, 16 moles of O_2, and 0.2 mole of NO at 2500°C.

4 Into a 1-liter tank at 448°C are placed 0.10 mole of HI, 1.5 moles of I_2, and 1.0 mole of H_2. Calculate the equilibrium concentrations of HI, H_2, and I_2.

5 At 25°C and 20 atm, the reaction

$$N_2(g) + 3H_2(g) \rightleftarrows 2NH_3(g)$$

has a ΔH of -22.1 kcal. If the temperature is raised to 300°C while the pressure is held at 20 atm, will more or less ammonia be present at equilibrium? If the pressure is increased to 30 atm while the temperature remains at 25°C, will more or less ammonia be present compared with the initial conditions? If half the ammonia is removed and the system allowed to come to equilibrium again, will the amount of nitrogen gas present increase or decrease? What will be the effect on the original equilibrium mixture if a catalyst for ammonia synthesis is added?

6 What is the pH of a 0.01-molar NaOH solution?

7 What is the pH of a 10^{-10}-molar HCl solution?

8 If a 0.10-molar acetic acid solution is 1.3% ionized, what is the pH of the solution? What is K_a for acetic acid? Compare your value with that in Table 5–4.

9 If a 0.10-molar HF solution is 5.75% ionized, what is the pH of the solution? What is K_a for HF? Compare your value with that in Table 5–4.

10 From the data in Table 5–4, calculate the dissociation constant for ammonium hydroxide. Is undissociated NH_4OH really present in the solution? If not, what is the reaction for the production of ammonium ion and OH^-? What is the pH of a 0.0100-molar solution of ammonia?

11 A detergent box must bear a warning label if its contents will form a solution that has a pH greater than 11 because strong base degrades protein structure. Should a box bear such a label if the H^+ concentration of a solution of its contents is found to be 2.5×10^{-12} mole liter^{-1}?

12 The ionization constant for arsenous acid ($HAsO_2$) is 6.0×10^{-10}. What is the pH of a 0.10-molar solution of arsenous acid? What is the pH of a 0.10-molar solution of $NaAsO_2$?

13 A solution of ammonia has a hydrogen ion concentration of 8.0×10^{-9} mole liter^{-1}. What is the pOH of this solution?

14 What is the CN^- ion concentration and the pOH in a 1.00-molar aqueous solution of HCN?

15 Pyridine is an organic base that reacts with water as follows:

$$C_5H_5N + H_2O \rightleftarrows C_5H_5NH^+ + OH^-$$

The base-dissociation constant for this reaction, K_b, is 1.58×10^{-8}. What is the concentration of $C_5H_5NH^+$ ion in a solution that was initially 0.10 molar in pyridine? What is the pH of the solution?

16 What is the equilibrium concentration of NO_2^- ion in a 0.25-molar aqueous solution of nitrous acid? What is the pH? What is the percent ionization of HNO_2?

17 Hydrazine is a weak base that dissociates in water according to the equation

$$N_2H_4 + H_2O \rightleftarrows N_2H_5^+ + OH^-$$

The equilibrium constant for this dissociation at 25°C is 2.0×10^{-6}. Write the equilibrium-constant expression for this reaction. If the initial hydrazine concentration is 0.010 molar, what is the concentration of hydrazinium ion, $N_2H_5^+$? What is the pH?

18 What is the pH of a 0.18-molar solution of ammonium chloride?

19 What is the pH of a 0.025-molar solution of sodium acetate?

20 The hypobromous ion, OBr^-, is the conjugate base of the weak hypobromous acid, HOBr. When 0.100-molar sodium hypobromite is dissolved in water, the pH of the solution is 10.85. Write the equation for the hydrolysis of OBr^- and the equilibrium-constant expression for the reaction. Calculate the value of the hydrolysis constant and of the acid-dissociation constant for HOBr.

21 The phenolate ion, $C_6H_5O^-$, is the anion of the weak acid phenol, C_6H_5OH. The anion undergoes hydrolysis according to the equation

$$C_6H_5O^- + H_2O \rightleftarrows C_6H_5OH + OH^-$$

A 0.0100-molar solution of sodium phenolate has a pH of 11.0. Write the expression for the hydrolysis constant. Calculate the numerical values of the hydrolysis constant and the acid-dissociation constant for phenol.

22 The pH of a 0.100-molar sodium nitrite solution is 8.15. Calculate the hydrolysis constant, K_b, for NO_2^-. Calculate the dissociation constant for nitrous acid.

23 What is the pH of a 1.0-molar solution of sodium cyanide?

24 A buffer solution is made with 0.30-molar sodium cyanide and 0.30-molar HCN. What is the pH of the buffer solution?

25 What is the pH of a buffer prepared to be 0.20 molar in NH_3 and 0.40 molar in NH_4Cl?

26 A buffer solution is made from equal volumes of 0.10-molar acetic acid and 0.10-molar sodium acetate. What is the pH of the buffer?

27 What is the pH of a solution made from equal volumes of 0.20-molar propionic acid and 0.20-molar sodium propionate?

28 A solution is 0.10 molar in formic acid and 0.010 molar in sodium formate. What is the pH of the solution?

29 If 0.010 mole of HCl gas is dissolved in 1 liter of pure water, what is the final pH? If the same amount of HCl is dissolved instead in 1 liter of the buffer solution of Problem 27, what is the final pH?

30 If 20 ml of a solution of 0.6-molar ammonia are mixed with 10 ml of a 1.8-molar ammonium chloride solution, what is the final pH? If 1 ml of a 1.0-molar HCl solution is added, what will the pH become? If the buffer solution had been prepared from 0.06-molar ammonia and 0.18-molar ammonium chloride, would the same HCl solution change the pH more or less than in the first situation? Why?

31 A student titrated a spoonful of an unknown monoprotic acid with NaOH solution of unknown concentration. After the addition of 5.00 ml of base the pH of the solution was found to be 6.00. The end point was reached when 7.00 ml of additional base were added. Calculate the dissociation constant of the acid.

32 Novocain (Nvc) is a weak organic base that reacts with water as follows:

$$Nvc + H_2O \rightleftarrows NvcH^+ + OH^-$$

The base-equilibrium constant for this reaction is $K_b = 9.0 \times 10^{-6}$. Suppose that a 0.010-molar solution of Novocain is titrated with nitric acid. (a) What is the pH of the Novocain solution at the beginning of titration, before any acid has been added? (b) At the end point of the titration, the solution behaves just like a solution of 0.010-molar $NvcH^+$-NO_3^-. What is the pH of this solution? (c) The indicator bromcresol green has a pK_a of 5.0. Is this indicator suitable for the titration?

33 If 0.10-molar pyridine solution $(K_b = 1.58 \times 10^{-8})$ is titrated with HCl, what is the pH of the solution when the ratio of equivalents of H^+ added to initial equivalents of pyridine is 0.50? What is the pH when this ratio is 1.00? From the information in Figure 5–7, which indicator would be most suitable for this titration: methyl violet, methyl orange, bromthymol blue, or alizarin?

34 Determine equilibrium constants for the following reactions:

a) $NO_2^- + HF \rightleftarrows HNO_2 + F^-$

b) $CH_3COOH + F^- \rightleftarrows$
$$HF + CH_3COO^-$$

c) $CH_3COOH + SO_3^{2-} \rightleftarrows$
$$HSO_3^- + CH_3COO^-$$

d) $NH_3 + HSO_3^- \rightleftarrows SO_3^{2-} + NH_4^+$

Arrange the Brønsted acids in the preceding equations in order of their increasing acid strength.

35 The solubility of silver phosphate, Ag_3PO_4, in water is 0.0065 g liter^{-1} at 20°C. What is the solubility product (K_{sp}) for this salt? What is the solubility of silver phosphate, in moles liter^{-1}, in a solution that contains a total of 0.10 mole liter^{-1} of Ag^+?

36 If a solution containing 0.16 mole liter^{-1} of Pb^{2+} is made 0.10 molar in chloride ion, 99.0% of the Pb^{2+} is removed as $PbCl_2$. What is K_{sp} for $PbCl_2$?

37 With data in Table 5–8, calculate the solubility in moles liter^{-1} of MgF$_2$ in pure water. What is the solubility in 0.050-molar NaF?

38 What is the solubility of CoS in pure water, in moles liter^{-1}? What is the solubility of CoS in 0.10-molar sodium sulfide solution?

39 What is the silver ion concentration in a solution of silver chromate in pure water? In 0.10-molar chromate solution?

40 Calculate the calcium ion concentration in a saturated solution of calcium fluoride.

41 A solution is made 0.10 molar in Mg^{2+}, 0.10 molar in NH$_3$, and 1.0 molar in NH$_4$Cl. Will Mg(OH)$_2$ precipitate?

42 How many grams of ammonium chloride must be added to 100 ml of 0.050-molar ammonium hydroxide to prevent the precipitation of ferrous hydroxide, Fe(OH)$_2$, when the NH$_4$Cl–NH$_4$OH mixture is added to 100 ml of 0.020-molar FeCl$_2$? Assume that the addition of solid NH$_4$Cl produces no volume change.

43 The K_{sp} of calcium phosphate, Ca$_3$(PO$_4$)$_2$, is 1.3×10^{-32}, and the third ionization constant of phosphoric acid is 2.2×10^{-13}. Suppose that 0.31 g of calcium phosphate are added to 100 ml of water, and the pH of the solution is adjusted until all of the calcium phosphate dissolves. What is this pH? (Assume that HPO$_4{}^{2-}$ is the only other species formed in the solution and that CaHPO$_4$ is soluble.)

44 In the precipitation of metal sulfides, selective precipitation can be achieved by adjusting the hydrogen ion concentration. At what pH does ZnS begin to precipitate from a 0.077-molar solution of H$_2$S containing 0.08-molar Zn^{2+}? (Necessary data are in Tables 5–4 and 5–8.)

45 What is the solubility of AgOH in a buffer at pH 13?

46 In a water solution saturated with H$_2$S, $[H^+]^2[S^{2-}] = 1.3 \times 10^{-21}$. Calculate the solubility of FeS at pH 9 and at pH 2. Can you see how this behavior might be useful in analytical separations?

47 Calculate the solubility of Mg(OH)$_2$ in aqueous solution at pH 2 and pH 12. How is this behavior useful in chemical separations?

48 Frequently, communities partially soften their water by adding slaked lime, Ca(OH)$_2$, to the water supply. The slaked lime reacts with HCO$_3{}^-$,

$$Ca(OH)_2(s) + 2HCO_3{}^-(aq) \rightarrow$$
$$CaCO_3(s) + CO_3{}^{2-}(aq) + 2H_2O(l)$$

to produce a mole of CO$_3{}^{2-}$, which further reacts with Ca^{2+} originally in the water to precipitate CaCO$_3$. Thus Ca^{2+} ion is added to remove Ca^{2+} ion. A malfunctioning water-softening plant recently delivered saturated Ca(OH)$_2$ solution to the home owners' taps in Charleston, Illinois. Calculate the pH of a saturated solution of Ca(OH)$_2$ at 0°C. Is it unsafe for human consumption? (See Problem 11.) The solubility of Ca(OH)$_2$ can be found in a handbook.

49 Three suggestions are made for ways to remove silver ions from solution. (a) Make the solution 0.010 molar in NaI. (b) Buffer the solution at pH 13. (c) Make the solution 0.0010 molar in Na$_2$S. What will be the equilibrium silver ion concentration in each case? Which course of action is most effective in removing Ag$^+$ ions?

It often matters vastly with what others,
In what arrangements the primordial germs
Are bound together, and what motions, too,
They exchange among themselves,
   for these same atoms
Do put together sky, and sea, and lands,
Rivers and suns, grains, trees and
   breathing things.
But yet they are commixed in different ways
With different things, with motions each its own.
Lucretius (55 B.C.)

6 CLASSIFICATION OF THE ELEMENTS AND PERIODIC PROPERTIES

From the properties of solutions of electrolytes, from the chemical changes produced by electric current, and from the direct study of gaseous ions we have seen that atoms are electrical in nature. Moreover, an atom appears to be made of a positive core that contains most of the mass of the atom, and one or more negatively charged and relatively loosely held electrons. These electrons, each weighing only 1/1836 as much as a hydrogen atom, can be removed to form positive ions. Other electrons can be added to neutral atoms to form negative ions. This ability to add or lose electrons varies from one element to another, and does so in a systematic way that correlates with the chemical behavior of the element. The gain and loss of electrons is fundametal to chemical reactivity.

In this chapter we shall examine correlations of atomic, electrical, and chemical properties, and how they lead directly to a fundamental classification scheme for matter known as the periodic table. To

Rutherford the physicist, the periodic table would be the ultimate stamp album. It would only confirm his impression of chemistry, if this were the final chapter. But we organize the elements of the universe into the periodic table so chemistry can *begin*, not end. Once the classification scheme is set up, it must be explained in terms of electrons and other subatomic particles from which atoms are constructed. This explanation is the task of Part 3. But before we begin to theorize about the world, let us see what it is really like.

6–1   EARLY CLASSIFICATION SCHEMES

Very early in the development of chemistry, chemists recognized that certain elements have similar properties. The earliest classification scheme of the elements consisted of only two divisions, metals and nonmetals. Metallic elements have a certain lustrous appearance, they are malleable and ductile, they conduct heat and electricity, and they form compounds with oxygen that are basic. Nonmetallic elements have no one characteristic appearance, they generally do not conduct heat and electricity, and they form acidic oxides.

Döbereiner's Triads

In 1829, the German chemist Johann Döbereiner observed several groups of three elements (*triads*) with similar chemical properties. In every case the atomic weight of one element in the triad was nearly the average of the other two. An example of one of Döbereiner's triads is chlorine, bromine, and iodine. Each element in this triad forms colored vapors containing diatomic molecules. Each element combines with metals and exhibits a combining weight equal to its atomic weight. Also, each forms ions with oxygen that have a single negative charge as, for example, ClO^-, ClO_3^-, BrO_3^-, and IO_3^-. The atomic weight of bromine (80) is approximately the average of those of chlorine (35.5) and iodine (127). In Table 6–1 the similar properties of elements in this and other triads are given.

In addition to recognizing the triads given in Table 6–1, Döbereiner observed a peculiar triad of the metals iron, cobalt, and nickel, all of which have similar properties and almost the same atomic weights. The metals are used in structural materials (steel) and may be ferromagnetic like iron; in their $+2$ and $+3$ states they form complex ions that are colored.

This discovery of families of elements (the number 3 per family proved to be insignificant) provided an incentive to those who were attempting to find a rational means of classifying the elements.

Newlands' Law of Octaves

During 1850–1865 many new elements were discovered. Furthermore, in this period chemists made considerable progress in the determination of

Table 6–1. Description of Döbereiner's Triads

Triad elements and atomic weights	Elementary form	Principal compounds	Special properties
(I) Cl, Br, I; 35.5, 80, 127	Colored diatomic molecules: Cl_2 (yellow), Br_2 (brown), I_2 (violet)	Form simple salts containing (-1) ions: Cl^-, Br^-, I^-. Form oxyions of (-1) charge containing one to four oxygen atoms: ClO_4^-, ClO_3^-, BrO_3^-, IO_3^-, ClO^-, IO_4^-. Hydrogen compounds are molecular: HCl, HBr, HI.	Free elements react vigorously with electron donors to form negative ions Cl^-, Br^-, I^-: $$2Na + Cl_2 \rightarrow 2Na^+ + 2Cl^-$$ $$I_2 + S^{2-} \rightarrow 2I^- + S$$ Salts (like NaCl) are very soluble in water. Halide salts of Li, Na, and K give neutral solutions. Hydrogen compounds are strong acids and ionize completely in water: $$HBr + H_2O \rightarrow H_3O^+ + Br^-$$
(II) S, Se, Te; 32, 79, 127.6	Colored crystalline nonmetals (Te somewhat metallic): S_8 (yellow), Se_8 (red)	Form simple salts with (-2) ions: S^{2-}, Se^{2-}, Te^{2-}, and very smelly compounds with hydrogen: H_2S, H_2Se, H_2Te. Form oxyions of (-2) charge with up to four oxygen atoms: SO_3^{2-},[a] SO_4^{2-}, SeO_4^{2-}. Form dioxides and trioxides: SO_2, SO_3,[a] SeO_2, TeO_2, TeO_3.	Salts, except with triads (III) and (IV), are slightly soluble in water: CuS, ZnS, HgS. Soluble salts (Na_2S) give basic solutions: $$S^{2-} + H_2O \rightarrow HS^- + OH^-$$ Hydrogen compounds are weak acids.
(III) Ca, Sr, Ba; 40, 88, 137	Reactive metals	Form salts containing $(+2)$ ions: Ca^{2+}, Sr^{2+}, Ba^{2+} in $BaSO_4$, $CaCO_3$, $SrCl_2$, etc.	Salts give bright colors in flame: Ca (orange), Sr (red), Ba (green). Sulfates and carbonates are insoluble. Metals replace hydrogen slowly from water.
(IV) Li, Na, K; 7, 23, 39	Very reactive metals	Form salts containing $(+1)$ ions: Li^+, Na^+, K^+ in Li_2CO_3, NaCl, K_3PO_4, etc.	Almost all salts are soluble; metals and salts give brightly colored flames: Li (red), Na (yellow), K (purple). Metals react violently with water to produce hydrogen and soluble ionic hydroxides: $$2Na + 2H_2O \rightarrow H_2 + 2Na^+ + 2OH^-$$

[a] Note the importance of charge: SO_3^2 is very different from SO_3 (no charge).

atomic weights. Thus, more accurate atomic weights were made available for old elements, and reasonably accurate values were presented for new elements. In 1865, the English chemist John Newlands (1839–1898) explored the problem of the periodic recurrence of similar behavior of elements. He arranged the lightest of the known elements in order of increasing atomic weight as follows:

H	Li	Be	B	C	N	O
F	Na	Mg	Al	Si	P	S
Cl	K	Ca	Cr	Ti	Mn	Fe

Newlands noticed that the eighth element (fluorine) resembled the first (hydrogen), the ninth resembled the second, and so forth. His observation that every eighth element had similar properties led him to compare his chemical octaves with musical octaves, and he himself called it his *law of octaves*. Periodicity by octaves in chemistry suggested to him a fundamental chemical harmony like the one in music. The comparison, although appealing, is invalid. Had Newlands known of the noble gases, his periodicity of properties would have been by nines rather than by eights. He never would have used his musical analogy, and might have been spared some of the ridicule and indifference that he suffered. (See Section 6–7 for more on Newlands.)

Newlands' effort was admittedly a step in the right direction. However, three serious criticisms can be directed at his classification scheme:

1) There were no places in his table for new elements, which were being discovered rapidly. Moreover, in the later parts of the table, there were several places where two elements were forced into the same position in the table (Section 6–7).

2) There was no scholarly evaluation of the work on atomic weights and no selection of probable best values.

3) Certain elements did not seem to belong where they were placed in the scheme. For example, chromium (Cr) is not sufficiently similar to aluminum (Al), nor manganese metal (Mn) to phosphorus (P), a nonmetal. Iron metal (Fe) and sulfur (S), a nonmetal, do not resemble each other either.

6–2 THE BASIS FOR PERIODIC CLASSIFICATION

The development of the periodic table as we now know it is credited mainly to the Russian chemist Dmitri Mendeleev (1834–1907), although the German chemist Lothar Meyer worked out essentially the same system independently and almost simultaneously. So far as we know, neither man was aware of Newlands' work. Mendeleev's periodic table, presented in 1869, followed Newlands' plan of arranging the elements in order of increasing atomic weights (Figure 6–1), but with the following substantial improvements:

1) Long periods were instituted for the elements now known as transition metals. These long periods are shown folded in half in his original table, with each full period taking two lines. This innovation removed the necessity of placing metals such as vanadium (V), chromium, and manganese under nonmetals such as phosphorus, sulfur, and chlorine.

2) If the properties of an element suggested that it did not fit in the arrangement according to atomic weight, a space was left for a new element. For example, no element existed that would fit in the space below silicon (Si). Thus, a space was left for a new element, which was named *ekasilicon*.

Row	Group I R₂O	Group II RO	Group III R₂O₃	Group IV RH₄ RO₂	Group V RH₃ R₂O₅	Group VI RH₂ RO₃	Group VII RH R₂O₇	Group VIII RO₄
1	H = 1							
2	Li = 7	Be = 9.4	B = 11	C = 12	N = 14	O = 16	F = 19	
3	Na = 23	Mg = 24	Al = 27.3	Si = 28	P = 31	S = 32	Cl = 35.5	
4	K = 39	Ca = 40	— = 44	Ti = 48	V = 51	Cr = 52	Mn = 55	Fe = 56, Co = 59, Ni = 59, Cu = 63
5	(Cu = 63)	Zn = 65	— = 68	— = 72	As = 75	Se = 78	Br = 80	
6	Rb = 85	Sr = 87	?Yt = 88	Zr = 90	Nb = 94	Mo = 96	— = 100	Ru = 104, Rh = 104, Pd = 106, Ag = 108
7	(Ag = 108)	Cd = 112	In = 113	Sn = 118	Sb = 122	Te = 125	I = 127	
8	Cs = 133	Ba = 137	?Di = 138	?Ce = 140				
9								
10			?Er = 178	?La = 180	Ta = 182	W = 184		Os = 195, Ir = 197, Pt = 198, Au = 199
11	(Au = 199)	Hg = 200	Tl = 204	Pb = 207	Bi = 208			
12				Th = 231		U = 240		

Figure 6–1. The periodic table of Mendeleev as it appeared when published in English, in 1871. The elements appear in order of increasing atomic weight. Note the empty space under Si for an unknown (at that time) element of atomic weight 72, and the incorrect atomic weights (for example, In). "R" in the column headings is the general symbol for an element in the table.

3) A scholarly evaluation of atomic weight data was made. For example, as a result of this work the combining capacity of chromium in its highest oxide was changed from five to the correct value of six. The combining weight of Cr was known to be 8.66 g. Hence, instead of 43.3 (5 × 8.66), the revised atomic weight of Cr became 52.0 (6 × 8.66).

Indium (In), with a combining weight of 38.5, had been assigned a combining capacity of two and therefore an atomic weight of 77, and had been placed between arsenic (As) and selenium (Se). Since their properties were consistent with placement below adjacent phosphorus and sulfur, arsenic and selenium had to be adjacent in Mendeleev's scheme. A reevaluation showed that indium has an atomic weight of 114.8 and a combining capacity of three, which is consistent with its position below aluminum and gallium (Ga) in the present table.

The atomic weight of platinum (Pt) was thought to be greater than gold (Au). Mendeleev thought otherwise, because of the chemistry of the two metals and the places that they should occupy in his table. New

Table 6–2. Mendeleev's Predictions for the Element Ekasilicon (Germanium)

Properties	Silicon and its compounds	Mendeleev's predictions for ekasilicon	Winkler's reports for germanium	Tin and its compounds
Atomic weight	28	72	72.6	118
Appearance	Gray, diamondlike	Gray metal	Gray metal	White metal or gray nonmetal
Melting point, °C	1410	High	958	232
Density, g cm^{-3}	2.32	5.5	5.36	7.28 or 5.75
Action of acid and alkali	Acid resistant; slow attack by alkali	Acid and alkali resistant	Not attacked by HCl or lye (NaOH); attacked by HNO$_3$	Slow attack by conc. HCl; attacked by HNO$_3$; inert to lye (NaOH)
Oxide formula and density, g cm^{-3}	SiO$_2$, 2.65	EsO$_2$, 4.7	GeO$_2$, 4.70	SnO$_2$, 7.0
Sulfide formula and properties	SiS$_2$, decomposes in water	EsS$_2$, insoluble in water, soluble in ammonium sulfide	GeS$_2$, insoluble in water, soluble in ammonium sulfide	SnS$_2$, insoluble in water, soluble in ammonium sulfide solution
Chloride formula	SiCl$_4$	EsCl$_4$	GeCl$_4$	SnCl$_4$
Boiling point of chloride, °C	57.6	100	83	114
Density of chloride, g cm^{-3}	1.50	1.9	1.88	2.23
Preparation of element	Reduction of K$_2$SiF$_6$ with sodium	Reduction of EsO$_2$ or K$_2$EsF$_6$ with sodium	Reduction of K$_2$GeF$_6$ with sodium	Reduction of SnO$_2$ with carbon

determinations inspired by Mendeleev showed 198 for platinum and 199 for gold, thereby placing platinum ahead of gold and under palladium (Pd), which of all the other elements most resembles platinum.

4) On the basis of the known periodic behavior summarized in the table, predictions of the properties of the undiscovered elements were made. These predictions later proved to be amazingly accurate. A good example is the comparison of predicted properties of the element "ekasilicon" and the properties (as reported by Winkler) of the element called germanium (Ge), which now occupies the ekasilicon space. This comparison is given in Table 6–2.

From the table it is evident how Mendeleev was able to predict accurately the physical and chemical properties of the missing element. Its position in the periodic table was below silicon and above tin. The physical properties of germanium are just about the average between those observed for silicon and tin. Predicting the chemical properties required information from the known relative properties of phosphorus, arsenic, and antimony (Sb) in the column to the right in the periodic chart.

Correlations such as this guided the search for new elements and compounds and stimulated investigation when known data did not conform with other correlations. One consequence of this research was that we gained improved values for atomic weights and densities.

The Periodic Law

Mendeleev summarized his discoveries by stating the *periodic law: The properties of chemical elements are not arbitrary, but vary with the atomic weight in a systematic way.*

After most of the elements had been discovered and their atomic weights carefully determined, several discrepancies still persisted. For example, the order of increasing atomic weight within Mendeleev's Group VIII (Figure 6–1) was found to be Fe, Ni, Co, Cu in the fourth period, Ru, Rh, Pd, Ag in the fifth, and Os, Ir, Pt, Au in the sixth. Yet Ni resembles Pd and Pt more than Co does. Again, Te has a higher atomic weight than I, but I clearly belongs with Br and Cl, and Te resembles Se and S in chemical properties. When the noble gases were discovered, it was revealed that Ar had a higher atomic weight than K, whereas all the other noble gases had lower atomic weights than the adjacent alkali metals. In these three instances, increasing atomic weight clearly is *not* acceptable as a means of placing elements in the periodic table. Therefore, the elements were assigned *atomic numbers* from 1 to 92 (now 105). (The atomic numbers of the elements *approximately* increase with their atomic weights.) When the elements are arranged according to increasing atomic number, chemically similar elements lie in vertical columns (families) of the periodic table.

In 1912, Henry G. J. Moseley (1888–1915) observed that the frequencies of x rays emitted from elements could be correlated better with atomic numbers than with atomic weights. The relationship between an element's atomic number and the frequency (or energy) of x rays emitted from the element is a consequence of atomic structure. As we shall see in Chapter 8, the electrons around an atom are arranged in *energy levels*. When an element is bombarded by a powerful beam of electrons, electrons from the innermost levels or shells (closest to the nucleus) can be ejected from the atoms. When outer electrons drop into these shells to fill the vacancies, energy is emitted as x radiation. The x-ray spectrum of an element (the collection of frequencies of x rays emitted) contains information about the electronic energy levels of the atom. The important point for the present purposes is that the energy of a level varies correspondingly with the charge on the nucleus of the atom. The greater the nuclear charge, the more tightly the innermost electrons are bound. More energy is required to knock off one of these electrons; consequently, there is more energy emitted when an electron falls back into a vacancy in the shells. Moseley discovered that the frequency of x rays emitted, v, varies with atomic number, Z, according to

$$v = c(Z - b)^2$$

in which c and b are characteristic of a given x-ray line and are the same for all elements.

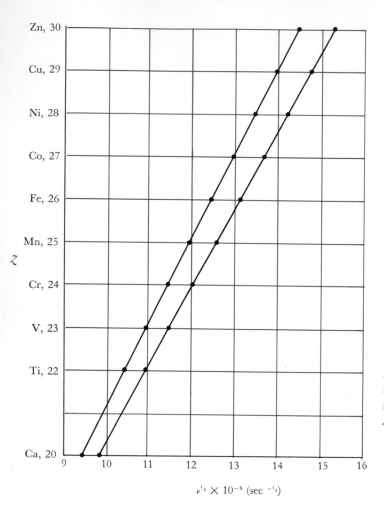

z

Ca, 20

9   10   11   12   13   14   15   16

$\nu^{1_2} \times 10^{-8}$ (sec $^{-1_2}$)

Figure 6–2. Moseley's plot of the square root of x-ray frequency against atomic numbers for the elements calcium through zinc. The two lines come from two different, identifiable frequencies in each atom's spectrum.

In April 1914, Moseley published the results of his work on 39 elements from $_{13}$Al to $_{79}$Au. (The atomic number is indicated by a subscript before the symbol of an element.) A portion of his data is plotted in Figure 6–2. Moseley wrote the following:

The spectra of the elements are arranged on horizontal lines spaced at equal distances. The order chosen for the elements is the order of the atomic weights, except in the cases of Ar, Co, and Te, where this clashes with the order of the chemical properties. Vacant lines have been left for an element between Mo and Ru, an element between Nd and Sm, and an element between W and Os, none of which are yet known. . . . This is equivalent to assigning to successive elements a series of successive characteristic integers. . . . Now if either the elements were not characterized by these integers, or any mistake had been made in the order chosen or in the number of places left for unknown elements, these regularities (the straight lines) would at once disappear. We can therefore conclude from the evidence of the x-ray spectra alone, without using any theory of atomic structure, that these integers are really characteristic of the elements. . . . Now Rutherford has proved that the most important constituent of an atom is its central positively charged nucleus, and van den Broek has put forward the view that the charge carried by this nucleus is in all cases an integral multiple of the charge on the hydrogen nucleus. There is every reason to suppose that the integer which controls the x-ray spectrum is the same as the number of electrical units in the nucleus, and these experiments therefore give the strongest possible support to the hypothesis of van den Broek.[1]

The three undiscovered elements mentioned by Moseley later were found to be elements 43 (technetium, Tc), 61 (promethium, Pm), and 75 (rhenium, Re). A confusing "double element" was cleared up in 1923, when D. Coster and G. Hevesy showed that one of the unoccupied horizontal lines on Moseley's chart belonged to the new element hafnium (Hf, 72). Moseley's work was perhaps the most fundamental single step in the development of the periodic table. It proved that atomic number (or the charge on the nucleus), and not atomic weight, was the essential property in explaining chemical behavior.

6–3   THE   MODERN   PERIODIC   TABLE

The easiest way to understand the periodic table is to build it. Although this may seem to be a difficult task, surprisingly little knowledge of chemistry is required to understand that the form on the inside front cover of this book is inevitable. If we arrange elements by atomic number, as Moseley did, then certain chemical properties repeat at definite intervals (Figure 6–3, top). The chemically inert gases (at least thought to be inert until 1962, when chemists produced xenon tetrafluoride), He, Ne, Ar, Kr, Xe, and Rn, have atomic numbers 2, 10, 18, 36, 54, and 86, or numerical intervals of 2, 8, 8, 18, 18, and 32. Each of these gases precedes an extremely reactive, soft metal that tends to form a +1 ion: the *alkali metals* Li, Na, K, Rb, Cs, and Fr. And each gas is preceded by a reactive element that can gain an electron to form a −1 ion: hydrogen and the *halogens* F, Cl, Br, I, and At. These key elements are shown in color in the row at the top of Figure 6–3.

These chemical similarities are best represented by folding the list of 105 elements into seven rows or *periods* (illustrated at the bottom of Figure 6–3). However, the first period has only two elements, the next two have eight, the next two, 18, and the sixth and probably the seventh periods have 32. How can we align eight entries over 18, and 18 over 32?

The *alkaline earth metals*, Be, Mg, Ca, Sr, and Ba, are so similar in chemical properties that we need little imagination to place them as shown.

---

[1] By the time these lines were published, Moseley was in the British army, and less than a year later he was dead, at the age of 27, on a hillside on Gallipoli.

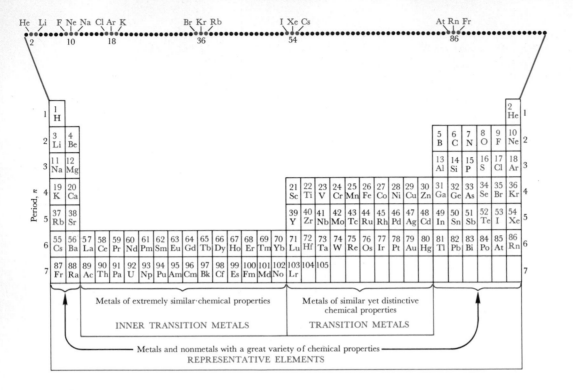

Figure 6–3. When elements are listed in order of increasing atomic number, as in the strip at the top, the recurrence of similar chemical properties suggests the folding into the "superlong" form of the periodic table shown below the strip. Elements can be classified into three categories, based on the extent to which chemical and physical properties change from one position in the table to the next.

Nonmetals are at the right end of each period, and O, S, Se, and Te constitute a series of elements with a combining capacity of two and an increase in metallic behavior from O to Te: O is a nonmetal; Te exists in the unspecific intermediate zone known as the *semimetals* or *metalloids*. Elements N, P, As, Sb, and Bi compose a group whose characteristics are the gain of three electrons in certain compounds and a gradation from the nonmetallic N and P, to the semimetallic As, to the metallic Sb and Bi. The elements C, Si, Ge, Sn, and Pb all have a combining capacity of four. For these latter elements, the border zone between metals and nonmetals is located at an earlier period; C is a nonmetal, Si and Ge are semimetals, and Sn and Pb are metals. Finally, the series B, Al, Ga, In, and Tl form +3 ions. B is semimetallic and the others are metallic. Al and Ga have more similar properties than do Al and Sc. To bring Al above Ga, it is necessary to shift the eight-element periods to the extreme right above the 18-element period below.

The "superfluous" elements in Periods 4 and 5 ($_{21}$Sc to $_{30}$Zn, and $_{39}$Y to $_{48}$Cd) constitute a series of metals, all of which exhibit a great variety of ionic states. The $+2$ and $+3$ states appear to be the most common. Their properties do not change from one element to another nearly as much as the properties in the series B, C, N, O, and F change; we call these "super-fluous" elements *transition metals*. (We defer the question of *what* is in transition to Chapter 9.) When we look for chemical parallels between the fifth and sixth periods, we find that $_{40}$Zr and $_{72}$Hf are virtually identical in behavior. Again, our preferred arrangement is to place the elements in Period 5 beyond $_{38}$Sr as far as possible to the right atop Periods 6 and 7. The extra elements in Period 6, $_{57}$La to $_{70}$Yb, are practically identical in chemical behavior. These elements are called the *rare earths*, or *lanthanides*. Their partners in the seventh period ($_{89}$Ac to $_{102}$No) are known as the *actinides*. Because the lanthanides are so similar in chemical properties, they are found together in nature and are extremely difficult to separate.

In summary, the elements can be classified into three groups (Figure 6–3): the *representative elements*, with diverse properties, the *transition metals*, more similar but yet clearly distinguishable, and the *inner transition metals* (lanthanides and actinides), with closely similar properties. The representative elements are called representative because they show a broader range of properties than are found in the other elements, and because they are the elements with which we are most familiar.

(The radioactivity and nuclear instability of the actinides, especially uranium, have given them an historical significance that their chemical properties perhaps would not have justified. An old-time chemist is a man who still thinks of uranium chiefly as an obscure heavy element used in yellow pottery glazes and stained glass. It is ironic that a nuclear war would be fought with the raw material of stained-glass windows.)

There is a more compact form of the periodic table that indicates more clearly the relative variability of properties of neighboring elements (Figure 6–4). Trends in chemical behavior often are easier to understand if only the representative elements are examined, with the transition metals set to one side as a special case and the inner transition metals virtually ignored. In this table, the vertical columns are called *groups*, and those of the representative elements are numbered IA through VIIA and 0. The groups of the transition elements are numbered in a way to remind you that they should be inserted in the representative element table. The numbering includes Groups IIIB to VIIB, then three columns all labeled collectively Group VIIIB, then Groups IB and IIB. Group IIIB follows Group IIA in the representative elements, and Group IIB precedes Group IIIA. This kind of numbering is clearer in the standard, "long form" of the periodic table on the inside front cover. We can see that the standard form is a compromise between the compactness of Figure 6–4 and the completeness of Figure 6–3. The lanthanides and actinides have been of so little relative importance that they have not been given group numbers.

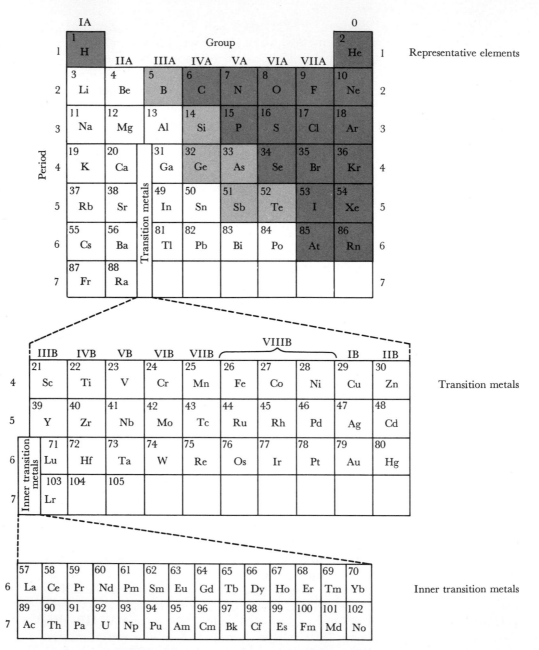

Figure 6–4. This compact, folded form of the periodic table emphasizes the natural division of elements into three categories: the extremely variable representative elements, the more similar transition metals, and the quite similar inner transition metals. Nonmetals are in color, and semimetals in a lighter color. The standard long version of the periodic table on the inside front cover is a compromise between this table and the one in Figure 6–3.

6-4   TRENDS IN PHYSICAL PROPERTIES

A regularity in certain physical properties of the elements is reflected in a corresponding regularity in chemical behavior. The most important such properties are the ease of losing or adding electrons and the size of atoms and ions.

First Ionization Energies

Since chemical reactivity is related to the gain, loss, or sharing of electrons, one of the most interesting properties of an element is the ease with which a gain or loss can occur. In Figure 6–5 we plot the *first ionization energy*, or the energy required to remove one electron from a neutral, gaseous atom: $M(g) \rightarrow M^+(g) + e^-$.

Figure 6–5. First ionization energy, or the energy required to remove one electron from a neutral, gaseous ion. This is a three-dimensional graph, in which the horizontal base a–b–c is the standard long form of the periodic table, and the vertical distance above this base represents the ionization energy of each element. An energy scale is given at the front right. Periods 2, 4, and 6 are in black, and 3 and 5 are in color. The three lines between each pair of period lines are only to help give shape to the surface of the graph, and have no other significance. The gray stripe across the graph from B to Sb has its lower boundary at 180 kcal mole^{-1}, and 225 kcal mole^{-1} is its upper boundary. This is roughly the region of transition between metallic and nonmetallic behavior.

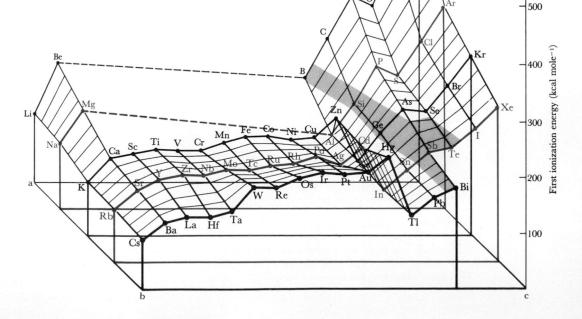

It is relatively easy to remove an electron from an alkali metal such as potassium, but it becomes increasingly difficult for elements across a period through the transition metals. Beyond the transition metals, at Ga, In, and Tl, the removal of an electron becomes suddenly easier. Then the difficulty increases again and reaches a peak with the halogens and noble gases. Successive elements in the periodic table have more and more electrons. The atomic number, as Moseley suggested, gives the charge on the nucleus in a neutral atom. Each of the seven periods approximately represents the filling of a *shell* of electrons. The innermost shell contains two electrons, and the successive ones reach a point of stability at intervals of 8, 8, 18, 18, and 32 more electrons. The noble gases in Group 0 possess such stable electron shells. Elements in the same group (vertical column) have the same number of electrons outside of the last completed shell.[2]

These outer electrons, referred to as *valence electrons*, are primarily responsible for chemical behavior. In the representative elements at least, the number of outer valence electrons is given by the group number; it ranges from one in the alkali metals to a stable, unreactive eight in the noble gases. It is easier to remove the single valence electron from Cs than from Li because in Cs the electron is in a shell farther from the influence of the positive nucleus. This trend in ionization energy is generally valid within a group. In contrast, it becomes generally more difficult to remove one electron as we move to higher atomic numbers across a period. This is because the nuclear charge is steadily increasing. The other valence electrons in the outermost shell contribute relatively little toward shielding a valence electron from the nucleus, so the attraction of the nucleus for a valence electron increases within the period. Explanations for the troughs and ridges of Figure 6–5 will have to wait for a more exact picture of electronic structure in Chapter 9. But we can see already that, among the representative elements, the third valence electron added (Group IIIA: B and Al) is less tightly bound than the second (Group IIA: Be and Mg), and that the sixth electron (O, S, Se, Te) is less tightly bound than the fifth (N, P, As, Sb).

Second and Higher Ionization Energies: The Formation of Ions

Figure 6–5 only shows the energy required to remove the *first* electron from a neutral atom. If there are several valence electrons, they all can be removed (at least in metals) without using very high energies. However, to remove electrons from a stable inner shell requires more energy than is normally available in chemical reactions (Figure 6–6). Therefore, the maximum

---

[2] This statement is an oversimplification that will be corrected in Chapter 9. The transition metals represent an interruption while a previously unused part of the immediately prior shell is being filled. The similarity of the transition metals is a reflection of the fact that a *buried* shell and not the outermost shell is changing.

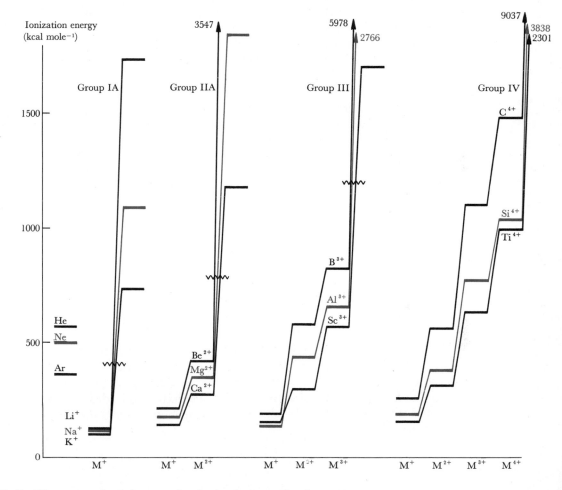

Figure 6–6. First, second, and successive ionization energies for representative elements in Groups I–IV. For the alkali metals in Group I, only about 100 kcal mole^{-1} are needed to remove the valence electron. But to dip into the completed shell beneath and strip off one of its electrons requires an additional 750 to 1750 kcal. Therefore, alkali metals do not form +2 ions. Similarly, removing only two electrons in Group II, or three in Group III, is much easier than removing one more and disrupting the inert shell below. As the group number increases, it is progressively more difficult to take all the outer valence electrons, and there is a decrease in the tendency to form cations having a charge equal to the group number.

positive charge that a representative element can have is equal to its group number.

But if it takes more than 100 kcal mole^{-1} of energy to pull electrons from an alkali metal, why should the atom ionize at all? The answer is that the energy required to remove an electron from an isolated gaseous atom is not the whole story. The alkali metals are solids, not gases. Moreover, when an ion goes into solution, it attracts water molecules to it. In H_2O molecules, the oxygen atoms are slightly negative, and the hydrogen atoms have compensating positive charges. Hence, the water molecules orient themselves around a cation, and the resulting *hydrated* ion [symbolized "(aq)"] is more stable than an isolated cation and water molecules by themselves. This saving in energy is the *energy of hydration* of the ion.

The overall energy involved in producing a hydrated Ca^{2+} ion from calcium metal is shown diagrammatically in Figure 6–7. This energy is the net result of four steps:

		Energy required
1. Sublimation:	$Ca(s) \rightarrow Ca(g)$	$+38$ kcal mole^{-1}
2. Ionization: 1st I.E.	$Ca(g) \rightarrow Ca^+(g) + e^-$	$+141$ kcal mole^{-1}
2nd I.E.	$Ca^+(g) \rightarrow Ca^{2+}(g) + e^-$	$+274$ kcal mole^{-1}
3. Hydration of ion:	$Ca^{2+}(g) \rightarrow Ca^{2+}(aq)$	-358 kcal mole^{-1}
4. Reduction of H^+:	$2e^- + 2H^+(aq) \rightarrow H_2(g)$	-230 kcal mole^{-1}
Overall reaction:	$Ca(s) + 2H^+(aq) \rightarrow Ca^{2+}(aq) + H_2(g)$	-135 kcal mole^{-1}

These energies are all *free energies*, ΔG, rather than enthalpies, ΔH. The difference between these two energies is explained in Chapter 15. But the difference here is small and need not concern us now.

This reaction converting metallic calcium to hydrated Ca^{2+} ion liberates 135 kcal of free energy per mole. The *oxidation energy* for the overall reaction then is

$$135 \text{ kcal mole}^{-1} \times \frac{1 \text{ electron volt ion}^{-1}}{23.06 \text{ kcal mole}^{-1}} = 5.85 \text{ eV ion}^{-1}$$

or as it is more commonly tabulated, 2.92 eV equivalent^{-1}. The *oxidation potential* is 2.92 V.

Step 4 is necessary because we cannot oxidize one substance in solution without simultaneously reducing something else. Steps 1 to 3 alone would leave free electrons in the solution, which is an impossible situation in water. The standard reaction to which other oxidation and reduction reactions are compared in tabulations of oxidation and reduction potentials is that of Step 4. The number 2.87 from Table 6–3 is an accurate value obtained from the potentials of electrolytic cells. Our calculated value here is inaccurate because it was derived as a small difference between large numbers. Oxidation potentials ordinarily are measured directly from cell voltages, as we shall see in Chapter 17.

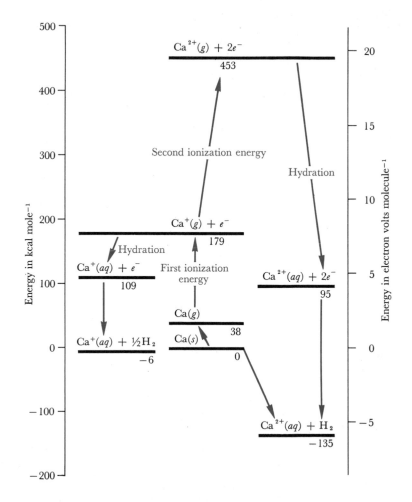

Figure 6–7. The oxidation energy of a reaction in water is the energy needed to change a solid metal into solution as the hydrated ion, with the electrons used to reduce H^+ ions. We show oxidation energies here in kilocalories per mole, but it is more common to tabulate oxidation potentials in volts per equivalent. [If the reaction involves n electrons, then 1 electron volt (eV) per equivalent equals 23.06n kcal per mole.] The overall reaction is the sum of the energies of sublimation, ionization, hydration of the gaseous cation, and use of the electrons to reduce H^+. Note that $Ca^+(aq)$ is slightly more stable than $Ca(s)$, but $Ca^{2+}(aq)$ is so much more stable that any monovalent calcium present would change spontaneously into equal amounts of divalent calcium and calcium metal. Hence, monovalent calcium does not exist in solution.

Table 6–3.   Oxidation Potentials[a] for the Production of Hydrated
Cations from the Metal

Group IA		Group IIA	
Li	3.02	Be	1.70
Na	2.71	Mg	2.34
K	2.92	Ca	2.87
Rb	2.99	Sr	2.89
Cs	3.02	Ba	2.90

[a] Measured in volts.

A high, positive oxidation potential means that there is a strong tendency for the oxidation reaction to occur, and that the products are thermodynamically quite stable relative to the reactants. The oxidation potentials in Table 6–3 indicate a steady increase in stability of the hydrated ion with increasing atomic number in Group IIA. This situation is a result of the lower atomic ionization energies of the heavier elements because of the greater distances between valence electrons and nuclei. In Group IA, however, the lithium ion is unexpectedly stable. The reason for this is its small size. Water molecules can come closer to the small charged ion and can form a more stable hydrated complex. This situation more than compensates for the slightly higher first ionization energy of Li (Figure 6–6). The hydrogen ion, H^+, is even smaller and more firmly hydrated. The number of hydrating water molecules is uncertain, so it is common practice to represent the hydrated proton by the symbol H_3O^+ and call it the *hydronium ion*.

Challenge. To apply the principle of first ionization energy to pressure measurement in a high vacuum system, and to see if you understand the quantitative aspects of second ionization energies, try Problems 6–9 and 6–13 in *Relevant Problems for Chemical Principles* by Butler and Grosser.

Electron Affinity

If energy is required to remove an electron from a neutral atom, it should not be too surprising to discover that energy can be given off when an additional electron is bound to a neutral atom. The reason lies in the imperfect shielding of the nucleus by the outer valence electrons. If the extra electron is added to the outer valence shell, the other electrons in this shell do not counteract the charge on the nucleus as well as those electrons in lower shells do. The extra electron has a slight attraction to the nucleus. The *electron affinity* is the energy released when the following reaction occurs:

$$M(g) + e^- \rightarrow M^-(g)$$

It is more difficult to measure electron affinities than ionization energies,

Figure 6–8. Electron affinity is the energy released when an electron binds itself to a gaseous neutral atom. The base of this graph is the representative elements portion of the periodic table, and electron affinity is plotted vertically. Halogens have unusually high electron affinities because the addition of one electron to the uncharged atom produces the eight-electron stable configuration of the noble gases. Note that the electron affinity of F *is less than that of* Cl, *although the perspective of the drawing masks the fact. This anomaly has not yet been explained.*

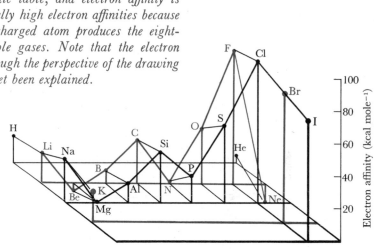

but the measured values for some of the representative elements are presented in Figure 6–8.

Halogens have very strong electron affinities. Although it requires 400 kcal mole^{-1} to remove one electron from F, 80 kcal mole^{-1} are released when an extra electron is added to form F$^-$. The electron affinities of Cl, Br, and I are comparable. There is a strong tendency to add the eighth valence electron per atom and thus form a stable closed shell like the neighboring noble gas. O and S have only half the electron affinity of their neighboring halogens, and elements to the left of these have even less. It is easier to add an electron at the right end of a period for the same reason that it is more difficult to remove one: Electrons are attracted more strongly by the greater charge on the nucleus. The variation in electron affinity across a period is consistent with that of ionization energy. Eight electrons in a shell form the most stable arrangement, but five or two electrons also are relatively stable. There is a stronger tendency for C or Si to add the fifth electron than for N or P to add a sixth. Similarly, Be and Mg are particularly stable; it requires 4 kcal mole^{-1} of energy to force another electron onto Be.

The Sizes of Atoms and Ions

There are three main measurements of sizes of atoms: van der Waals (or nonbonded) radii, covalent radii, and ionic radii. We determine van der Waals radii by seeing how close nonbonded atoms can come to one another in the solid state. This radius is the effective packing size of an atom. We determine covalent radii by examining bond lengths between atoms that

Table 6–4.   Van der Waals, Ionic, and Covalent Radii for Some Representative Elements[a]

	H			He	Li	Be
van der Waals	1.2 Å			0.93	—	—
Ionic	1.54			(0.93)[b]	0.60	0.31
Covalent	0.37			—	1.35	0.90
	N	O	F	Ne	Na	Mg
van der Waals	1.5	1.4	1.25	1.12	—	—
Ionic	1.71	1.40	1.36	(1.12)	0.95	0.65
Covalent	0.70	0.66	0.64	—	1.54	1.30
	P	S	Cl	Ar	K	Ca
van der Waals	1.9	1.85	1.80	1.54	—	—
Ionic	2.12	1.84	1.81	(1.54)	1.33	0.99
Covalent	1.10	1.04	0.99	—	1.96	1.74
	As	Se	Br	Kr	Rb	Sr
van der Waals	2.0	2.00	1.95	1.69	—	—
Ionic	2.22	1.98	1.95	(1.69)	1.48	1.13
Covalent	1.21	1.17	1.14	—	2.11	1.92
	Sb	Te	I	Xe	Cs	Ba
van der Waals	2.2	2.20	2.15	1.90	—	—
Ionic	2.45	2.21	2.16	(1.90)	1.69	1.35
Covalent	1.41	1.37	1.33	—	2.25	1.98
Charge on the ion for ionic radii (H is −1)	−3	−2	−1	0	+1	+2

[a] From L. Pauling, *The Nature of the Chemical Bond,* Cornell University Press, Ithaca, N.Y., 1960, 3rd ed.

[b] Values in parentheses are repetitions of the van der Waals radii.

share a single electron pair (Chapter 10). These are smaller than van der Waals radii, because the formation of a bond permits closer approach of the atoms. Ionic radii are a measure of the separations between ions of opposite charge in crystals.

These three types of radii are given for some representative elements in Table 6–4 and are illustrated in Figure 6–9. Van der Waals radii are measured between nonbonded, neutral atoms and do not apply to metals. Instead, we find their covalent radii from bond lengths in gaseous dimers (two-atom molecules). These radii are virtually the same as the metallic radii found for ions of identical charge in metals. Neither ionic nor covalent radii have meaning for atoms that do not form compounds, but the van der

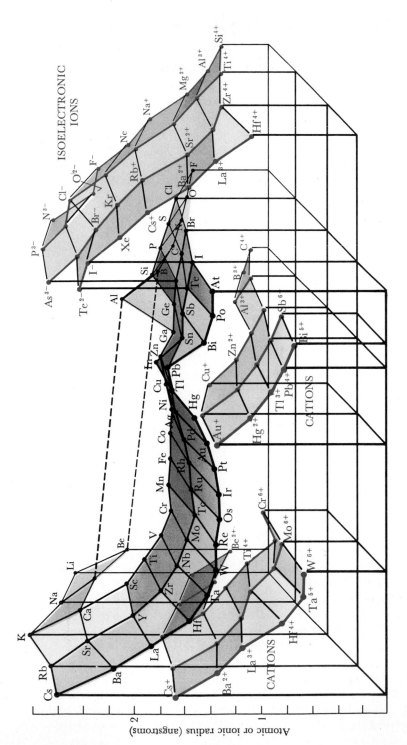

Figure 6–9. Single-bond covalent radii (black) and ionic radii (color) for the elements. The base of the graph is the long form of the periodic table; however, the table has been extended to the right to include the first four isoelectronic ions of each succeeding period. Each ion has gained or lost enough electrons to produce the stable eight-electron shell of the neighboring noble gas. Note the smaller size of the cations and the larger size of the anions. Note also that the isoelectronic ions form a smooth progression, in which radii decrease as nuclear charges increase.

Waals packing radii for the noble gases can be thought of as ionic radii of "ions" with zero charge.

The ions N^{3-}, O^{2-}, F^-, Ne^0, Na^+, and Mg^{2+} are *isoelectronic;* that is, they all have the same number of electrons and differ only in the charge on their nuclei. The greater the nuclear charge, the greater is the attraction for electrons. Hence, these ions show a steady decrease in size from N^{3-} to Mg^{2+} (Figure 6–9). The radii of neutral atoms show a less marked decrease to the right across any one period. Although successive atoms have more electrons, they also have a greater nuclear charge. This charge pulls all of the electronic shells more strongly and causes a small decrease in size. The transition elements exhibit slight irregularities in this overall trend; although the nuclear charge is continuing its steady increase, the new electrons are being added to an inner shell rather than to the outer valence shell.

These three factors—ionization energy, electron affinity, and size—are the most important in predicting chemical behavior. Several features of these quantities have been noted without explanation, and in fact cannot be explained without a more precise description of the arrangement of electrons around an atom. Meanwhile, the last aspect of periodicity that we will examine at this level is periodicity in chemical behavior.

6–5   TRENDS IN CHEMICAL PROPERTIES

In this section we will learn how the periodic table enables us to predict chemical properties, and will discuss trends in those properties that occur horizontally across a period, vertically within a group, and in some cases, diagonally from one corner to the other.

Types of Bonds

Bond formation involves the gain, loss, or sharing of electrons. One extreme type of bonding exists between an element with a low ionization energy and one with a high electron affinity, such as Na and Cl. When atoms of these elements react there is a complete transfer of an electron from Na to Cl and the formation of a nondirectional electrostatic, or ionic, bond between Na^+ and Cl^-. The other extreme is when two atoms have the same attraction for electrons. An example of this occurs in the central groups of the representative elements where the energy of either removing or adding many electrons is too high. Carbon does not satisfy its shortage of four valence electrons by accepting four from a donor, rather, it shares its four with other atoms in the form of four electron pairs. These electron pairs are called *covalent* bonds. Each shared electron pair is called a single bond; two pairs of electrons shared between the same two atoms are a double bond. Covalent bonds, unlike ionic bonds, are highly directional. One of the tests of any theory of covalent bonds will be its success in explaining why covalent bonds are only in certain relative directions around an atom. The shapes of molecules are dictated by the ways in which their component atoms form covalent bonds. (In Chapter 10 we shall consider bonding in more detail; at the moment we wish only to define briefly these two types of bonds.)

The physical states of the pure elements reflect these differences in bonding (Figure 6–10). For example, Group 0 elements form bonds only with the two small atoms F and O, which have a strong attraction for electrons. Group 0 elements exist as monatomic gases at STP. The nonmetals to the upper right of the table—the halogens, O, and N—share their electrons in pairs and form diatomic molecules. The lighter of these nonmetals (N_2, O_2, F_2, and Cl_2) are gases at STP. The heavier Br_2 is a liquid. The heaviest, I_2 and At_2, are low-melting solids. The slightly less nonmetallic elements, C, Si, P, As, S, Se, and Te, do not share all of their valence electrons with a single, similar atom. Instead, S, Se, and Te form rings of eight molecules that pack into a solid; Se and Te also have alternative solid structures in which endless parallel chains are packed. The elements P and As each form molecules having four atoms at the corners of a tetrahedron, with three bonds per atom. But at the lower end of the group the more metallic Bi and Sb make layers or sheets with three bonds per atom within the sheet. The more stable form of As has this sheet structure as well.

Group IVA elements form four bonds per atom. They do this by crystallizing in an endless array, in which each atom is bonded to four others at the corners of a tetrahedron. We obtain this *diamond structure* from ZnS (see Figure 3–18) by placing carbon atoms at both the Zn and the S positions. Carbon exists in a second variety, graphite, in which each atom is bonded to three others in a *hexagonal* pattern in a plane. The fourth bond is spread out to give all C—C bonds in a sheet a partial double bond character. The sheets are held together only by weak packing forces. This is the reason why graphite feels slippery to the touch and is a good lubricant.

This partial double bond character of graphite is reminiscent of the ability of the gaseous nonmetals such as N_2 to form multiple bonds between like atoms. The atoms in N_2 are held together by a triple bond. However, this structure is denied to the semimetals Si and Ge, both of which use only the diamond structure. Sn exists in two different crystal forms, or *allotropes*: gray tin, which has a diamond structure and nonmetallic properties, and white tin, which has the structure and properties of a metal. The heavier elements in Group IIIA and the elements in Group IIB form distorted metal structures. All of the elements to the left of Group IIB in the periodic table have metal structures of one type or another: ccp, hcp, bcc, or some variation of these.

In summary, starting from F in the periodic table, we see that metallic character increases in a direction that is diagonally downwards and to the left. The most nonmetallic elements have directional, multiple bonds and form small molecules. As elements approach metallic characteristics, their ability to form multiple bonds disappears. The result is networks of directional, single bonds in a rigid, nonionic solid. As the directionality of the

He, Ne, Ar, Kr, Xe

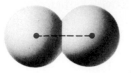

H_2, N_2, O_2, F_2, Cl_2, Br_2, I_2

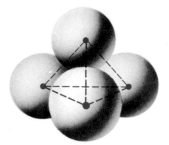

P_4, As_4

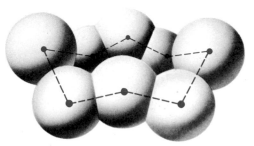

S_8, Se_8, Te_8

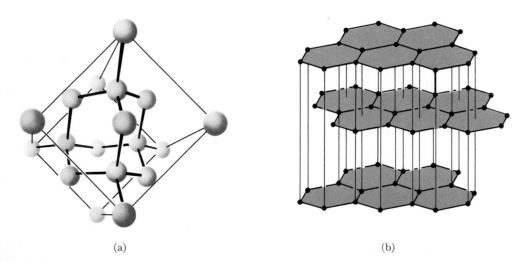

(a)   (b)

Figure 6–10. Bonding and molecular structure in nonmetallic elements. Noble gases exist as single, gaseous atoms. Halogens exist as diatomic molecules with single bonds, and O_2 and N_2 are small diatomic molecules with double and triple bonds, respectively. Sulfur and the larger elements in Group VIA use two bonds per atom to build rings or chains. Phosphorus and arsenic use three bonds, like nitrogen, but in this case they bond to three different neighbors in a tetrahedral configuration. Carbon and the other elements in Group IVA use four bonds with four different neighbors in a three-dimensional tetrahedral framework. Only carbon, within this group, also can form a layer structure in which a given atom shares four bonds with three neighbors within a layer.

Figure 6–11. Heats of fusion, in kilocalories per mole, for the solid elements. The base of the graph is the long form of the periodic table. Note the low heats of fusion of gaseous nonmetals at the right, and the extremely high heats of fusion of nonmetals such as C and Si, which form bonded covalent networks in a crystal. No contrast could be more striking than the one between crystalline diamond and atmospheric nitrogen and oxygen gases, elements that are adjacent in the periodic table. Note also the relatively high heats of fusion of the transition metals, which have free electrons for binding in the metal.

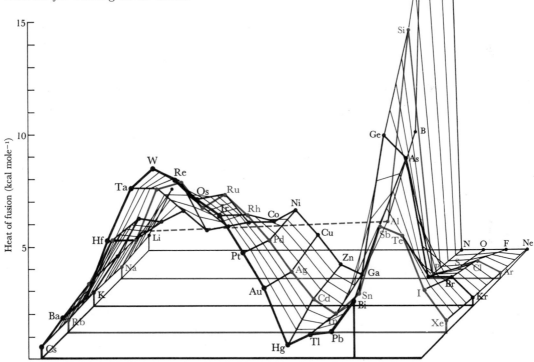

bonds diminishes and metallic behavior increases, the dominant factor in solid structure becomes the way in which the atoms are packed.

These differences in bonding are reflected in the melting points, boiling points, and heats of fusion (the heat required to melt the solid, in kcal $mole^{-1}$) of the elements. All three properties show the same general behavior. As an example, the heats of fusion of solids are shown in Figure 6–11. There is a sharp and striking contrast between the small amounts of heat needed to melt crystals of the dimeric nonmetals (N_2, O_2, Cl_2) and the large energies required to break lattices in which the covalent bonds form an endless framework (C, Si, Ge) or internally bonded sheets (C, As, Sb). Elements P, S, and Se have abnormally low heats of fusion since the P_4 tetrahedra or S_8

rings persist into the liquid phase. The disintegration of such solids is more like that of crystals of diatomic molecules than of an extended framework. As we shall see in Chapter 9, the strength of bonding in the transition metals depends on the number of free, unpaired electrons. The number increases to a maximum in the middle of the transition series and then diminishes again. The heats of fusion and melting points of the transition elements both reflect this factor.

Combining Ratios with Hydrogen

The number of hydrogen atoms that combine with one atom of a given representative element in the first three periods of the table varies (as shown in Figure 6–12) from one to four and back to one again across each period. This number is equal to either the number of valence electrons in the outermost shell or the number required to complete a stable octet, whichever is smaller. This fact alone offers a clue to the way in which H is bonded in each hydrogen compound.

Compounds of metals with hydrogen—called *hydrides*—are mostly ionic. In alkali hydrides such as KH or NaH, there is a transfer of approximately half a negative charge to each hydrogen atom. In alkaline earth hydrides (MgH_2 or CaH_2), there is about a quarter charge transferred. Alkali hydrides have the NaCl crystal structure, but BeH_2, MgH_2, and AlH_3 manifest a new phenomenon, "divalent" hydrogen. In this arrangement each H atom in the crystal is equidistant between two metal atoms and appears to form a hydrogen bridge between them. Whenever H has a net negative charge, this extra charge apparently can be used to make a second bond to another atom,

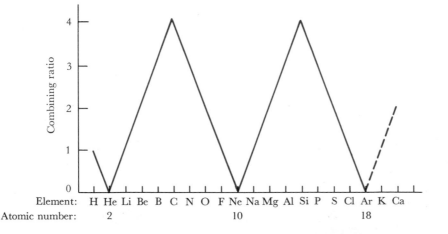

Figure 6–12. Periodicity of combining ratios of the lightest elements in compounds with hydrogen. The number of H atoms combined with one atom of these elements usually is the group number or eight minus the group number, whichever is smaller.

provided that there is enough potential bonding capacity in the other atoms. The negatively charged H is present in NaH, but not the capacity for multiple bonds. However, Be and Mg, and Al, satisfy both demands, and bridge structures are formed. The boron–hydrogen compound B_2H_6 (Figure 6–13) is an example of hydrogen bridging *within* a molecule, and the other known boron hydrides (such as B_4H_{10}, B_5H_9, B_5H_{11}, B_6H_{10}, and $B_{10}H_{14}$) all make extensive use of such hydrogen bridges (Section 12–1).

Compounds of hydrogen with elements in the right half of the periods are small molecular compounds in which a hydrogen atom *shares* its lone electron in a covalent bond. The number of hydrogen atoms in a molecule is dictated by the number of covalent bonds that the other atom can form. Molecules of such compounds are held together in crystals only by weak forces between molecules; thus the melting and boiling points are very low (Figure 6–13).

Figure 6–13. Hydrogen compounds of the elements in the first three rows of the periodic table. Combining ratios of the elements (with hydrogen) increase to four (in CH_4 and SiH_4) and then decrease. The hydrides of Li, Be, Na, Mg, and Al are solids at room temperature (25°C) and have infinitely extended network structures. The simple molecules LiH, BeH_2, NaH, and MgH_2 exist only at low pressures and high temperatures in the gas phase. AlH_3 is not an isolated molecule; it can exist only in the polymeric form $(AlH_3)_x$. The remaining hydrogen compounds are liquids or gases at room temperature. These compounds consist of discrete molecules having the composition and structure depicted schematically in the figure. The structure of the interesting B_2H_6 molecule is discussed in detail in Chapter 12.

Hydrogen $Z = 1$ H/H = 1/1 mp = −259°C bp = −252°C						
(LiH)$_x$ Lithium $Z = 3$ Li/H = 1/1 mp = 680°C bp = (decom.)	(BeH$_2$)$_x$ Beryllium $Z = 4$ Be/H = 1/2 mp = (decom.)	Boron $Z = 5$ B/H = 1/3 mp = −165°C bp = −92°C	Carbon $Z = 6$ C/H = 1/4 mp = −182°C bp = −161°C	Nitrogen $Z = 7$ N/H = 1/3 mp = −78°C bp = −33°C	Oxygen $Z = 8$ O/H = 1/2 mp = 0°C bp = 100°C	Fluorine $Z = 9$ F/H = 1/1 mp = −83°C bp = 20°C
(NaH)$_x$ Sodium $Z = 11$ Na/H = 1/1 mp = 700–800°C (decom.)	(MgH$_2$)$_x$ Magnesium $Z = 12$ Mg/H = 1/2 mp = (decom.)	(AlH$_3$)$_x$ Aluminum $Z = 13$ Al/H = 1/3 mp = (decom.)	Silicon $Z = 14$ Si/H = 1/4 mp = −185°C bp = −111°C	Phosphorus $Z = 15$ P/H = 1/3 mp = −134°C bp = −88°C	Sulfur $Z = 16$ S/H = 1/2 mp = −86°C bp = −60°C	Chlorine $Z = 17$ Cl/H = 1/1 mp = −114°C bp = −85°C

Ionic hydrides react with water to produce basic solutions:

$$Na^+H^- + H_2O \rightarrow H_2 + OH^- + Na^+$$

Conversely, the halogen compounds at the other end of the period are acidic:

$$HCl + H_2O \rightarrow H_3O^+ + Cl^-$$

Combinations with Oxygen: Binary Oxides

The representative elements form oxides with the formulas expected from the elements' positions in the periodic table; in the third period these oxides are Na_2O, MgO, Al_2O_3, SiO_2, P_2O_5, SO_3, and Cl_2O_7. Oxides at the lower left of the table are strong bases. They have a large negative charge on the O atom, and are ionic. The melting points of these ionic oxides are typically around 2000°C, and many decompose before melting. They react with water to make basic solutions:

$$Na_2O + H_2O \rightarrow 2Na^+ + 2OH^-$$

At the other extreme, oxides of elements at the upper right of the table are strong acids:

$$Cl_2O_7 + 3H_2O \rightarrow 2H_3O^+ + 2ClO_4^- \qquad \text{(perchloric acid)}$$
$$SO_3 + 3H_2O \rightarrow 2H_3O^+ + SO_4^{2-} \qquad \text{(sulfuric acid)}$$
$$P_2O_5 + 9H_2O \rightarrow 6H_3O^+ + 2PO_4^{3-} \qquad \text{(phosphoric acid)}$$

Cl_2O_7 is explosively unstable, and SO_3 and P_2O_5 react vigorously with water to produce acid solutions. The acids have been represented here as completely ionized or dissociated, but this can be as misleading as writing them in their undissociated forms: $HClO_4$, H_2SO_4, and H_3PO_4. As we saw in Chapter 5, H_2SO_4 and H_3PO_4 are partially dissociated in water.

Between acidic and basic oxides lies a diagonal band of oxides that are *amphoteric*: BeO, Al_2O_3, and Ga_2O_3; GeO_2 through PbO_2; and Sb_2O_5 and Bi_2O_5. These amphoteric oxides show both acidic and basic behavior. They are virtually insoluble in water but can be dissolved by either acids or bases:

$$BeO + 2H_3O^+ \rightarrow Be^{2+} + 3H_2O$$
$$BeO + 2OH^- + H_2O \rightarrow Be(OH)_4^{2-}$$

The notation in the first equation above is conventional but inconsistent. The hydration of the proton is denoted by the symbol H_3O^+. However, the Be^{2+} cation also is very strongly hydrated, especially so because of its small size. It should be written as $Be(H_2O)_n^{2+}$, or at least $Be(aq)^{2+}$. But so long as the hydration of cations is understood, it need not be spelled out every time.

The amphoteric and basic oxides are solids with high melting points. For instance, Al_2O_3 is the abrasive known as corundum, or emery; SiO_2 is quartz. Only the oxides of C, N, S, and the halogens are normally liquids or gases. The contrast between C and N in diamond and nitrogen gas is analogous to the contrast between C and Si in carbon dioxide and quartz. The difference between C and Si arises because C can make double bonds to O and therefore form a molecular compound of limited size. However, Si must make single bonds with four different O atoms; hence, it must assume a three-dimensional network structure in which tetrahedrally arranged Si atoms are connected by bridging oxygen atoms.

In all of the oxides discussed so far, the chemical formula can be predicted from the group number. But there are other oxides whose formulas cannot be predicted from the group numbers. For example, C can form CO as well as CO_2. N_2O_5 is not the only nitrogen oxide: NO_2, N_2O_3, and NO are others. Sulfur can form SO_2, S_2O_3, S_2O, and S_2O_7, as well as SO_3. But in these compounds the element does not make full use of its potential combining capacity. Thus, the general trends in properties are best illustrated by the oxides that we have been examining.

Before you go on. The topics discussed in this chapter—ionization energy, electron affinity, bonding, ionic charge, chemical properties of the elements, and the relationship of these to the periodic table—all are important to an understanding of chemistry. A summary of these relationships for the representative elements that should help you to remember them is in Review 6 of Lassila *et al.*, *Programed Reviews of Chemical Principles*.

6–6   SUMMARY

Physical and chemical properties of elements are periodic functions, not of atomic weight, but of *atomic number*. Moseley suggested, and it later was verified, that the atomic number is the total positive charge on the nucleus. It is also of necessity equal to the total number of electrons around the nucleus in a neutral atom.

Particularly stable, inert elements occur at intervals of 2, 8, 8, 18, 18, and 32 in atomic number. These intervals, and only the most basic knowledge of similarities among elements, lead to a *periodic table*, in which similar elements are in vertical columns or *groups*, and in which chemical properties change in an orderly manner along horizontal rows or *periods*. The full, extended periodic table can be folded into a compact form that illustrates the division of the elements into three categories: the diversified *representative elements*, the more similar *transition metals*, and the virtually identical *inner transition metals*.

Chemical bonding is a function of the gain, loss, or sharing of the outer electrons, usually referred to as the *valence electrons*. The particularly stable noble gases all have eight such outer electrons, and among the representative elements the number of valence electrons is given by the *group number*. A strong tendency to form a cation by losing electrons is indicated by a low *ionization energy;* a strong tendency to gain electrons and to become an anion is identifiable by a high *electron affinity*. The values of these two quantities also suggest that, for reasons we cannot yet explain, *two* or *five* valence electrons provide a somewhat more stable electronic arrangement, although nothing like the inertness associated with eight electrons in the valence shell. Two extreme forms of bonding have been introduced: *ionic*, in which electrons have been transferred from one atom to another, and *covalent*, in which electrons are shared between atoms. Ionic bonds are electrostatic and are not oriented in a specific direction; covalent bonds are highly directional.

The combining capacity of an element is related to the number of outer valence electrons that it possesses. The correlation between combining capacity and group number in the periodic table has been illustrated by the hydrides and oxides of the representative elements.

Elements at the lower left of the periodic table are metals, with low ionization energies and low electron affinities. They form solids made of closely packed spherical cations, which are held together by a sea of mobile valence electrons. Their hydrides and oxides are ionic, and aqueous solutions of these compounds are basic.

Elements at the upper right corner of the table are nonmetals, with high ionization energies and high electron affinities. Bonds in the pure element of these nonmetals are covalent, and the full combining capacity of the element is used in multiple bonds between pairs of atoms to form small molecules. These pure elements are therefore gases at STP. Their hydrogen compounds and oxides are likewise small molecules with covalent bonds, are gases or liquids, and are acidic.

Between these two extremes at upper right and lower left there is a gradation of properties. As the elements pass from nonmetals through semimetals to metals, they first lose their ability to form multiple bonds between like atoms, and then lose their directional covalent bonding completely. Correspondingly, the hydrogen compounds go from acidic, to neutral or inert, to basic (although there are many complications to this overall trend), and the oxides proceed in a more regular manner from acidic, to amphoteric, to basic.

With this outline of the general behavior of the elements in mind, it is now time to look more closely at the process of chemical bonding.

6–7 POSTSCRIPT TO THE CLASSIFICATION OF THE ELEMENTS

The story of John A. R. Newlands is a melancholy illustration of the fact that, in science, a good idea alone is not enough. The idea must be substantiated with enough evidence to gain acceptance. It is also an example of the dangers of poor nomenclature.

Newlands was the son of a Scottish minister and a graduate of Glasgow University. From his mother, who was of Italian descent, he inherited a love of music and the fervor to join Garibaldi in the struggle for Italian independence in 1860. On his return to England he completed his chemical studies and established himself as a private analytical chemist for industry.

His reputation in chemistry was based on his expertise in sugar chemistry, but his lifelong hobby was chemical periodicity.

The high point of his work on periodicity was to occur on March 1, 1866, when he presented his "law of octaves" before The Chemical Society in London. He expected acclaim, but received only indifference and heavy-handed humor. The paper on which his talk was based was rejected by the *Journal of the Chemical Society.* The account of the meeting was reported in *Chemical News* [**13**, 113 (1866)] as follows:

Mr. John A. R. Newlands read a paper entitled "The Law of Octaves and the Causes of Numerical Relations among the Atomic Weights." The author claims the discovery of a law according to which the elements analogous in their properties exhibit peculiar relationships, similar to those subsisting in music between a note and its octave. Starting from the atomic weights on Cannizzaro's system, the author arranges the known elements in order of succession, beginning with the lowest atomic weight (hydrogen) and ending with thorium (=231.5); placing, however, nickel and cobalt, platinum and iridium, cerium and lanthanum, etc., in positions of absolute equality or in the same line. The fifty-six elements so arranged are said to form the compass of eight octaves, and the author finds that chlorine, bromine, iodine, and fluorine are thus brought into the same line, or occupy corresponding places in his scale. Nitrogen and phosphorus, oxygen and sulfur, etc., are also considered as forming true octaves. The author's supposition will be exemplified in Table II, shown to the meeting, and here subjoined:

Table II — Elements Arranged in Octaves

H	1	F	8	Cl	15	Co & Ni	22	Br		29	Pd	36	I		43	Pt & Ir	50
Li	2	Na	9	K	16	Cu	23	Rb		30	Ag	37	Cs		44	Os	51
G	3	Mg	10	Ca	17	Zn	24	Sr		31	Cd	38	Ba & V		45	Hg	52
B	4	Al	11	Cr	18	Y	25	Ce & La		32	U	39	Ta		46	Tl	53
C	5	Si	12	Ti	19	In	26	Zr		33	Sn	40	W		47	Pb	54
N	6	P	13	Mn	20	As	27	Di & Mo		34	Sb	41	Nb		48	Bi	55
O	7	S	14	Fe	21	Se	28	Ro & Ru		35	Te	42	Au		49	Th	56

Dr. Gladstone made objection on the score of its having been assumed that no elements remain to be discovered. The last few years had brought forth thallium, indium, cesium, and rubidium, and now the finding of one more would throw out the whole system. The speaker believed there was as close an analogy subsisting between the metals named in the last vertical column as in any of the elements standing on the same horizontal line.

Professor G. F. Foster humorously inquired of Mr. Newlands whether he had ever examined the elements according to the order of their initial letters? For he believed that any arrangement would present occasional coincidences, but he condemned one which placed so far apart manganese and chromium, or iron from nickel and cobalt.

Mr. Newlands said that he had tried several other schemes before arriving at that now proposed. One founded upon the specific gravity of the elements had altogether failed, and no relation could be worked out of the atomic weights under any other system than that of Cannizzaro.

And so dies a good story. The questioner did not ask about "chords and arpeggios," as is sometimes said, but only about alphabetical order. The disbelief was apparent, however, and the unfortunate musical analogy made Newlands' ideas look even more like numerology instead of science. The lack of space for new elements and the crowding of two elements into one space were serious flaws. Perhaps the main feature that made Mendeleev's scheme superior was the introduction of the long periods after the first two eight-element ones. Mendeleev buttressed his table with a host of chemical evidence, including his famous predictions for new elements and their chemistry. He clearly deserves his reputation as the creator of the periodic table.

Yet we should not forget Newlands, struggling to have his contribution recognized. He published note after note in *Chemical News*, first elaborating on his table, and then welcoming Mendeleev's table in 1869 as the vindication of his own. Seven years after the *Journal of the Chemical Society* rejected his 1866 paper, he was given a reason, of sorts, by the Society president, Dr. Odling. The paper had not been published, he said, because they ". . . had made it a rule not to publish papers of a purely theoretical nature, since it was likely to lead to correspondence of a controversial character."

Newlands collected all of his papers and published them as a book in 1884, and documented his claims to priority in the pages of *Chemical News* and in an account to the German Chemical Society. Perhaps in an outburst of conscience, the Royal Society of Great Britain awarded him the Davy medal in 1887, five years after it had presented the same award to Mendeleev.

SUGGESTED READING

J. L. Hall and D. A. Keyworth, *Brief Chemistry of the Elements*, W. A. Benjamin, Menlo Park, Calif., 1971.

J. W. Mellor, *Comprehensive Treatise on Inorganic and Theoretical Chemistry*, Macmillan, New York, 1922. See especially Chapter VI.

R. L. Rich, *Periodic Correlations*, W. A. Benjamin, Menlo Park, Calif., 1965.

R. T. Sanderson, *Chemical Periodicity*, Reinhold, New York, 1960.

M. E. Weeks and H. M. Leicester, *Discovery of the Elements*, Chemical Education Publishing Co., Easton, Pa., 1968, 7th ed.

QUESTIONS

1 In what ways was Mendeleev's classification of the elements superior to Newlands'?

2 Why did Mendeleev's period classification lead to a reexamination of combining capacities?

3 How did Mendeleev predict the properties of "ekasilicon"?

4 What is incorrect about Mendeleev's periodic law?

5 How did Moseley deduce the existence of undiscovered elements?

6 What are the identifying characteristics of the following groups of elements: halogens, alkali metals, noble gases, alkaline earths?

7 What is a group in the periodic table? What is a period? How many elements are in each of the first six periods?

8 What is the difference between the elements in the three categories of representative elements, transition metals, and inner transition metals?

9 What does the letter "A" or "B" after the group number signify?

10 How does metallic character vary within Groups IIIA, IVA, or VA? How does metallic character vary across a period?

11 How does the first ionization energy correlate with metallic properties?

12 If energy is required to ionize an atom, why do ions form at all?

13 Why is the fourth ionization energy of aluminum so much higher than the first three?

14 Why does the first ionization energy of an element decrease within a group from the first to the seventh period?

15 What is the difference between ionization energy and electron affinity?

16 Why should the electron affinity decrease in the series Cl, Br, and I?

17 What evidence is there that two or five electrons of a possible eight-electron outer valence shell form a particularly stable configuration?

18 How is an oxidation potential related to an ionization energy? What other energy information is needed to calculate oxidation potentials from ionization energies?

19 What are isoelectronic ions? Why should the series Te^{2-}, I^-, Xe, Cs^+, Ba^{2+}, La^{3+}, and Hf^{4+} show a steady decrease in size?

20 What is an allotrope? Can you think of examples other than white and gray tin?

21 How are the main features on the right half of Figure 6–11, Heats of Fusion, accounted for by the bonding in the nonmetallic elements? Why the tremendous difference between C and N?

22 How does the combining capacity of the representative elements vary with group number in the hydrogen compounds and the oxides?

23 What is the difference in bonding in the hydrogen compounds NaH, MgH_2, and NH_3?

24 Why is there such a difference between the melting points or boiling points of CO_2 and SiO_2? Is there any similarity between this phenomenon and the difference in melting points of carbon and nitrogen?

25 How do the chemical properties of the oxides change from left to right across a period of the table?

26 Which elements are out of their proper sequence in Newlands' table in Section 6–7? Why do you think that they are misplaced as they are? (Glucinium, G, was an early name for beryllium, Be.)

PROBLEMS

1 Efforts presently are being made to synthesize or discover new elements of very high atomic number [G. T. Seaborg, "From Mendeleev to Mendelevium and Beyond," *Chemistry* **43**, 6 (1970)]. Which existing element would be most like Element 111? Like 112? Like 118? How would the ionization energy of Element 118 compare with the other elements of its group? Predict the empirical formulas of chlorides of Elements 111, 112, and 118.

2 Account for the two straight lines plotted in Figure 6–2 rather than one. Could three lines have been plotted?

3 How many valence electrons are present in (a) N; (b) Al; (c) Cl; (d) Rb?

4 Moseley demonstrated that the energy levels of an atom are dependent upon the charge of the nucleus by showing that plots of atomic number against the square roots of the frequencies of the emitted x rays give straight lines. For an isoelectronic series, a plot of atomic number against the square roots of the ionization energies also gives a straight line. Predict the ionization energy of N^{4+} from the following data:

Species	Ionization energy (kcal mole^{-1})
Li	124
Be$^+$	420
B^{2+}	874
C^{3+}	1487

5 Consider the following series of bromides: $MgBr_2$, $AlBr_3$, $SiBr_4$, PBr_5. Does the ionic character increase or decrease from first to last?

6 In the elements Si, Ge, Sn, and Pb, do the nonmetallic properties increase or decrease in the series?

7 Which has the largest first ionization energy: Li, Na, K, or Rb? Why?

8 Which of the following statements correctly describes trends in the first ionization energies of atoms: (a) Ionization energies decrease regularly from left to right across a row of the periodic table. (b) Ionization energies increase regularly from left to right across a row of the table. (c) Ionization energies decrease from left to right across a row, but there are irregularities for atoms with three or six valence electrons. (d) Ionization energies increase regularly from left to right across a row, but there are irregularities for atoms with three or six valence electrons. (e) Ioniza-

tion energies increase downward in a column of the table.

9 Which of the ions, Be^{2+}, Mg^{2+}, Ca^{2+}, or Sr^{2+}, has the largest ionic radius?

10 Which of the following statements correctly describes the observed trends in atomic radii: (a) Atomic radii decrease with increasing atomic number from left to right across a row of the periodic table, but increase with increasing atomic number (Z) down a column. (b) Radii increase with increasing Z across a row, but do not change in a column. (c) Radii decrease with increasing Z across a row, and decrease with increasing Z downward in a column. (d) Radii increase with increasing Z across a row, but do not change in a column. (e) Radii increase with increasing Z across a row and also increase with increasing Z in a column.

11 Write a balanced equation for the reaction of hydrogen iodide with water.

12 Write a balanced equation for the reaction of calcium hydride, CaH_2, with water.

13 Imagine that you were taking chemistry prior to the discovery of strontium ($Z = 38$). Considering its position in the periodic table, predict the following properties of Sr: (a) chemical formula of its most common oxide; (b) chemical formula of its most common chloride; (c) chemical formula of its most common hydride; (d) solubility of its hydride in water, and the acidity or basicity of the resulting solution; (e) principal ion formed in aqueous solution.

14 What would you predict as the formulas of the hydrogen compounds of the following elements: Ca, Te, Ge, S, W? Which compounds will be ionic? In which will the hydrogens behave as cations? In which will they be anions? Which aqueous solution will be most basic?

When our views are sufficiently extended as to enable us to reason with precision concerning the proportions of elemental atoms, we shall find the arithmetical relation will not be sufficient to explain their mutual action and we shall be obliged to acquire a geometrical conception of their relative arrangement in all three dimensions. . . . When the number of particles (combined with one particle) exists in the proportion of 4/1, stable equilibrium may take place if the four particles are situated at the angles of the four equilateral triangles composing a regular tetrahedron. . . . It is perhaps too much to hope that the geometrical arrangement of primary particles will ever be perfectly known.

W. H. Wollaston (1808)

7 OXIDATION, COORDINATION, AND COVALENCE

The systematic, periodic variation in properties from one element to another is reflected in the ways in which elements form bonds. We have seen how the combining ratios of elements with hydrogen and oxygen vary systematically with group number in the periodic table. One of the first concepts invented to keep track of how elements form compounds was that of combining weights and combining capacities. So far we have used combining capacities (or combining numbers) as a description of observations, with no theoretical justification. In this chapter we shall look more closely at the different kinds of combining capacities of atoms with other atoms and molecules. We shall explore in more depth how chemical bonding can be related to the periodicity of the elements. We shall look at more complex substances such as $Co(NH_3)_6Cl_3$, in which three different kinds of chemical combination appear to be present. And finally we shall learn of the ingenuity of the late nineteenth- and early twentieth-century chemists, who deduced rather precise structural information about molecules and ions, primarily from a simple knowledge of atomic ratios.

7–1   TYPES OF CHEMICAL BONDING

The most fundamental observation about bonding is that bonding occurs as interactions between relatively small numbers of discrete atoms; hence, we can define large units of matter that represent the atomic process with a constant multiplication or "scale-up" factor. We use this idea in our concept of the mole. The idea of units of reaction is implicit in the explanation of Dalton's law of multiple proportions. Let us imagine ourselves in the position of a late nineteenth-century chemist, well-grounded in Dalton's atomic theory, with a good set of atomic weights provided by Cannizzaro, with a convenient, if not well-understood, periodic table (thanks to Mendeleev), and with a large body of experimental observations about chemical combination. How would we explain chemical bonding? The facts available to us could be summarized as follows:

1) Two kinds of elements are known, metals and nonmetals. (The noble gases were recognized just before 1900.)

2) Of the physical forces of attraction, only the one between opposite electric charges (positive and negative) appears to be of a magnitude and character suitable for holding atoms together. (Gravity is too weak; nuclear forces were unknown; and magnetic poles cannot be separated into isolated particles as electrostatic charges can.)

3) Metals combine chemically only with nonmetals. Both are in a very real sense opposites in terms of properties and behavior. (Alloys between metals were recognized as solutions of variable composition.)

4) Nonmetals combine not only with metals, but with themselves (H_2, O_2, S_8) and with each other (CH_4, CO_2, SO_3, ICl, HF).

5) The combining ratios among atoms in molecules are ratios of small whole numbers (4/1 in CH_4, 7/2 in Cl_2O_7).

6) Some combinations of atoms pass unchanged through many chemical reactions as if they were atomic units in their own right [SO_4^{2-}, OH^-, NH_4^+, $Pt(CN)_6^{2-}$].

7) Some metals can combine in two different ways with neutral molecules and with nonmetallic elements in a single, complex compound [$Co(NH_3)_6Cl_3$].

8) Some compounds have identical formulas, yet different chemical properties (ethyl alcohol, C_2H_5OH, and dimethyl ether, CH_3—O—CH_3; the complex molecules [$Co(NH_3)_5SO_4$]$^+$ Br^- and [$Co(NH_3)_5Br$]$^{2+}$ SO_4^{2-}).

Oxidation Numbers

As aids in writing chemical formulas, chemists sought to assign combining numbers to elements. The *combining number*, or combining capacity, was defined as the ratio of the atomic weight of the element to the combining weight in the compound in question. The combining number of a metal in its compound with oxygen was named the *oxidation number*. This oxidation number is equal to the positive charge on the metal ion when the oxide is brought into solution; oxidation numbers for metals were given a positive sign. Oxidation numbers were extended to the nonmetals by assigning a negative charge to the combining capacity of a nonmetal. A principle was recognized in writing formulas: The sum of the oxidation numbers of all of the atoms in a properly written formula for an uncharged compound must be zero, and the sum of the oxidation numbers of all of the atoms in an ion must equal the charge on the ion.

The oxidation number was given more than a purely bookkeeping significance. This simple theory of bonding proposed that, when zinc and oxygen react to form ZnO, the zinc atom loses two electrons and is oxidized to Zn^{2+}, while the oxygen atom gains two electrons and is reduced to O^{2-}. The relative number of atoms in the compound then is dictated by the necessity of balancing positive and negative charges. Since antimony has an oxidation state of $+3$ in many compounds, and oxygen has a customary oxidation state of -2, it could be predicted (correctly) that antimony would have an oxide with the formula Sb_2O_3. The six positive charges on the two metal atoms counterbalance the six negative charges on the three oxygen atoms.

Chemists observed that, among the representative elements, the metals often showed positive oxidation numbers equal to their group numbers in the periodic table, and that the nonmetals often showed negative oxidation numbers equal to eight *minus* their group number. Thus, sodium can have an oxidation number of $+1$, calcium $+2$, the halogens -1, oxygen and sulfur -2, and nitrogen and phosphorus -3. However, many elements required several oxidation numbers, depending on the compound in which they were found. For instance, sulfur has an oxidation number of $+4$ in SO_2 and $+6$ in SO_3. In $HClO_3$, if H is $+1$ and O is -2, the oxidation number of chlorine must be $+5$ for the compound's net oxidation number to be zero. Does this imply that an atom of chlorine actually loses five of its electrons to an oxygen atom?

The answer to the last question is *No*. Chemists realized that oxidation numbers cannot be taken as indicators of actual electron transfers. However, the numbers are still useful as a means of keeping track of chemical reactions and balancing chemical equations. The rules for assigning oxidation numbers and using them in balancing equations are given in Section 7–2. Yet something better is needed to explain what really happens in chemical bonding.

Covalence

The ionic attraction theory failed completely to explain how nonmetals could combine with one another. Another quite different bonding theory was proposed for nonmetals, first on a purely empirical basis and later with a certain amount of theoretical justification. Each nonmetal atom was considered to have a certain number of potential bonds, of which it could use some or all in forming a molecule. A popular but unrealistic picture of this bonding was that an atom has a certain number of "hooks," which could form bonds by attaching to the hooks of another atom. The combining capacity of an element was represented symbolically by the number of hooks on its atoms. Hydrogen was considered to have only one hook, and oxygen two. Two hydrogen atoms could join hooks and form one *covalent* bond between them in a diatomic molecule:

$$H \underline{\quad \mathbf{\mathfrak{X}} \quad} H$$

Or, the two hydrogen atoms each could combine with one of the hooks on an oxygen atom to form water, H_2O:

$$H \underline{\quad \mathbf{\mathfrak{X}} \quad} O \underline{\quad \mathbf{\mathfrak{X}} \quad} H$$

Of course, no one regarded the hooks seriously as real physical entities; they were convenient bookkeeping devices to represent bonding behavior that lacked a good explanation. They were simply a graphic representation of combining capacity.

The old name for the oxidation–reduction concept of chemical bonding was *electrovalence;* the bonding we have just been discussing was called *covalence* since it involved cooperation between similar elements rather than electrostatic interaction. Covalent bonding accounted rather well for many observed chemical compounds. The covalence, or combining capacity, of a nonmetal was often equal to eight minus the group number. For example, the halogens had a covalence of one, as in HCl and HF; elements in Group VIA showed a covalence of two (H_2O, H_2S, OCl_2); those in Group VA had a covalence of three (NH_3, PCl_3); and carbon, with its immense variety of organic compounds, exhibited a covalence of four (CH_4, CH_3—CH_3, CH_3—OH). If we represent a covalent bond by a straight line rather than a pair of hooks, we can write these chemical formulas as

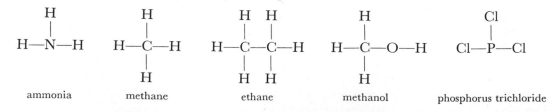

ammonia   methane   ethane   methanol   phosphorus trichloride

But some compounds could not be represented so simply, and chemists had to presume that atoms need not use all of their hooks or covalence. Sulfur shows its maximum covalence of six in SO_3, but only four bonds are used in SO_2 (we assume in both compounds that each oxygen atom uses its full covalence of two to form two bonds to the sulfur):

$$
\begin{matrix}
 & O & \\
 & \parallel & \\
O\!=\!\!S\!=\!\!O & & O\!=\!\!S\!=\!\!O
\end{matrix}
$$

In sulfuric acid, sulfur uses all six bonds, but in sulfurous acid it uses only four:

$$
\begin{matrix}
 & O & & & & O & \\
 & \parallel & & & & \parallel & \\
H\!-\!O\!-\!\!S\!-\!O\!-\!H & & & & H\!-\!O\!-\!\!S\!-\!O\!-\!H \\
 & \parallel & & & & & \\
 & O & & & &
\end{matrix}
$$

There appeared to be some validity to the idea of variable covalence. The compounds in which sulfur had a covalence of six all seemed closely related (H_2SO_4, SO_3), whereas those with covalences of four and two formed different self-contained sets (H_2SO_3 and SO_2; H_2S and CS_2). The most common covalence of an element was $8 - gn$, where gn is the group number of a representative element; however, other particularly frequent covalences were gn and $gn - 2$. There was obviously some sort of relationship between covalence among nonmetals and electrovalence between nonmetals and metals. The mathematics of oxidation numbers was useful for both. However, the nature of this relationship was unclear.

Coordination and Coordination Number

Compounds such as $Co(NH_3)_6Cl_3$ clearly involve more than one type of bonding. In solution, this compound produces Cl^- ions that can be precipitated by silver salts, and it has a limiting van't Hoff i factor of four, which suggests a dissociation into three Cl^- ions and one complex cation: $Co(NH_3)_6^{3+}$. This complex chemical group combines with other anions, such as SO_4^{2-}, in the correct proportions as if it were a cation of charge $+3$: $[Co(NH_3)_6]_2 [SO_4]_3$. The bonding between the cation $Co(NH_3)_6^{3+}$ and anions is purely electrovalent or, in modern terminology, *ionic*. But what about the bonding within the cation?

Each cobalt ion, Co^{3+}, has six neutral molecules of ammonia, NH_3, coordinated to it. Other neutral molecules or ions also can coordinate to cobalt; the preferred number appears to be six. The total charge on the complex ion is the charge on the central metal ion plus the sum of the charges on the coordinating groups. If one or more of the neutral NH_3 molecules are replaced by Cl^-, F^-, or NO_2^-, we obtain ions such as $Co(NH_3)_5Cl^{2+}$, $Co(NH_3)_5NO_2^{2+}$, $Co(NH_3)_2(NO_2)_4^-$, and CoF_6^{3-}.

The number of neutral molecules or groups associated closely with a metal ion in this way is known as its *coordination number*. The platinum(IV) ion

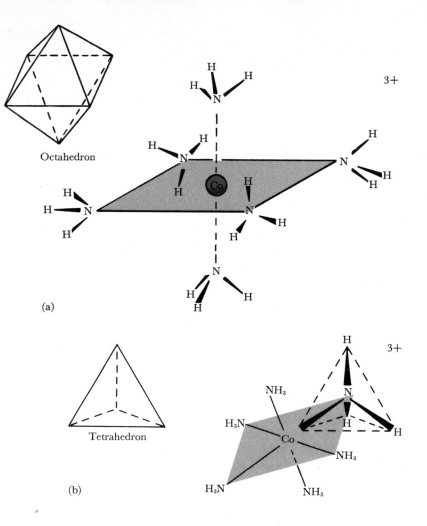

Octahedron

(a)

Tetrahedron

(b)

3+

3+

Figure 7–1. The structure of the complex ion Co(NH₃)₆³⁺. *(a) The six* NH₃ *molecules are around the* Co *at the six vertices of an octahedron. (b) Each nitrogen atom is tetrahedrally coordinated, with* Co *and the three* H *atoms at the four vertices of a tetrahedron. The geometry of such coordination in many cases was worked out solely by an ingenious analysis of data on combining ratios, before methods such as x-ray diffraction were available.*

has a coordination number of six, like Co^{3+}, and can form complex ions such as

$$Pt(NH_3)_6^{4+} \qquad Pt(NH_3)_5Cl^{3+} \qquad Pt(NH_3)_3Cl_3^{+} \qquad PtCl_6^{2-}$$

In contrast, the platinum(II) ion, Pt^{2+}, has a coordination number of only

Figure 7–2. Three-dimensional structures of some coordination compounds built with Benjamin/Maruzen models. It is often easier to understand such structures if you build them first, even if from toothpicks and clay. (a) Tetrahedral coordination: ammonium ion. (b) Square planar coordination: tetrachloroplatinate(II) ion. (c) and (d) Octahedral coordination: hexachloroplatinate(IV) and hexaamminecobalt(III) ions.

four and forms complex groups such as

$$Pt(NH_3)_4{}^{2+} \qquad Pt(NH_3)_2Cl_2 \qquad Pt(NH_3)Cl_3{}^{-} \qquad PtCl_4{}^{2-}$$
(neutral molecule)

This formation of complex ions by the coordination of neutral molecules or ions is most characteristic of the transition metals. The most common coordination numbers are six and four, although others exist. Coordination chemists suggested quite early in the development of the theory that, in six-fold coordination, the six groups lie at equal distances on either side of the central metal ion, along mutually perpendicular x, y, and z axes. This arrangement places the coordinating groups at the vertices of an octahedron and is called *octahedral coordination*. The geometrical relationship between coordinating groups and metal ions is identical to the one between metal atoms and an atom in an octahedral hole in an interstitial alloy (Figure 3–11). The arrangement of NH_3 molecules around the Co^{3+} ion in $Co(NH_3)_6^{3+}$ is shown in Figures 7–1(a) and 7–2(d).

The term "coordination number" has been extended to signify the number of nearest neighbors to any atom in a structure. For example, each nitrogen atom in the cobalt complex is bound to three hydrogen atoms and to the cobalt atom. The nitrogen atom has a coordination number of four; and since we know from x-ray crystal-structure analysis that the hydrogen atoms and the cobalt atom lie at the corners of a tetrahedron about the nitrogen atom, we call this *tetrahedral coordination* [Figure 7–1(b)].

In summary, there are three types of bonding in the compound $Co(NH_3)_6Cl_3$. Hydrogen atoms are bound to nitrogen atoms by covalent bonds, NH_3 groups are coordinated to the cobalt atom, and the entire complex cation is associated with Cl^- ions by ionic forces. Within each bond type, the combining units interact in the ratios of small whole numbers:

$$N/H = 1/3 \text{ in } NH_3$$
$$Co^{3+}/NH_3 = 1/6 \text{ in } Co(NH_3)_6^{3+}$$
$$Co(NH_3)_6^{3+}/Cl^- = 1/3 \text{ in } Co(NH_3)_6Cl_3$$

Let us now look at each of these bonding ideas in more detail.

7–2 OXIDATION NUMBERS

An element in any compound can be assigned an oxidation number by the following simple rules:

1) The oxidation number for any free element is zero; thus, H_2, O_2, Fe, Cl_2, and Na have zero oxidation numbers.

2) The oxidation number for any simple one-atom ion is equal to its charge; thus, Na^+ has an oxidation number of $+1$, Ca^{2+} of $+2$, and Cl^- of -1.

3) The oxidation number of hydrogen in any *nonionic* compound is $+1$. This is true for the great majority of hydrogen compounds such as H_2O, NH_3, HCl, and CH_4. For the ionic metal hydrides such as NaH, the oxidation number of hydrogen is -1.

4) The oxidation number of oxygen is -2 in all compounds in which it does not form an O—O covalent bond. Thus, its oxidation number is -2 in H_2O, H_2SO_4, NO, CO_2, and CH_3OH, but in hydrogen peroxide, H_2O_2, it is -1. Another exception to the rule that oxygen has an oxidation number of -2 is OF_2, in which O is $+2$ and F is -1.

5) In combinations of nonmetals *not* involving hydrogen or oxygen, the nonmetal that is either above or to the right of the other in the periodic

table is considered negative. Its oxidation number is given the same value as the charge on its most commonly encountered negative ion. This rule can be paraphrased: The least metallic element is assigned the negative oxidation number. For instance, in CCl_4, the oxidation number for chlorine is -1 and carbon is $+4$; whereas in CH_4, hydrogen is $+1$ and carbon is -4. In SF_6, fluorine is -1 and sulfur is $+6$; but in CS_2, S is -2 and C is $+4$. The last molecule is a borderline case. However, the assignment is made because S is farther to the right of C than C is above S. In completely molecular compounds such as N_4S_4, for which this rule is indefinite, the concept of empirical oxidation number loses usefulness.

6) The algebraic sum of the oxidation numbers of all atoms in the formula for a neutral compound must be zero. Hence, in NH_4Cl, the total oxidation number for the four hydrogen atoms is $4(+1) = +4$, and the number for Cl is -1, so the number for N must be -3 in order that the sum $+4 - 1 - 3$ equal zero.

7) The algebraic sum of oxidation numbers of all atoms in an ion must equal the charge on the ion. Thus, in NH_4^+, the oxidation number of N again must be -3, so $-3 + 4 = +1$. In SO_4^{2-}, since the four oxygen atoms have a total oxidation number of -8, the number for sulfur must be $+6$ in order that $+6 - 8 = -2$.

8) In chemical reactions, the *total oxidation number is conserved*. It is this last rule that makes oxidation numbers useful in modern chemistry. If the oxidation number of an element increases during a chemical reaction, the element is *oxidized;* if the number decreases, the element is *reduced*. This last principle can be paraphrased: In a balanced chemical reaction, *oxidations and reductions must exactly compensate one another*.

Calculating Oxidation Numbers

From the preceding rules, we can calculate the oxidation number of the elements in most compounds. Certain oxidation numbers are characteristic of a given element, and these can be related to the position of the element in the periodic table. Figure 7–3 shows the variation of oxidation numbers with atomic number. The *maximum* oxidation number generally increases across a period from $+1$ to $+7$.

Representative metals.   Metals in Groups I–III in the periodic table form ions with positive charges numerically equal to the numbers of their respective groups; that is, their oxidation numbers are the same as their group numbers.

Nonmetals.   Nonmetals often assume either of two characteristic oxidation numbers. Their minimum oxidation number is usually $-(8 - gn)$, where gn is the number of the group in the periodic table; thus, each atom can

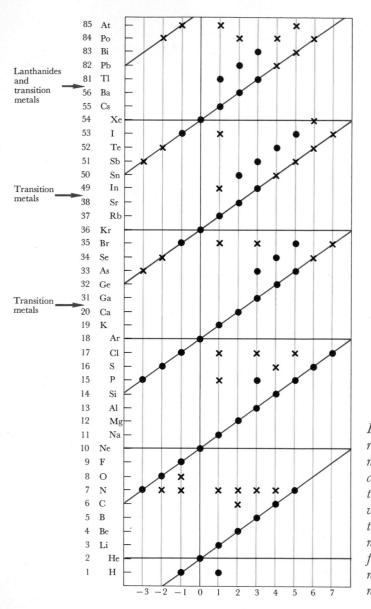

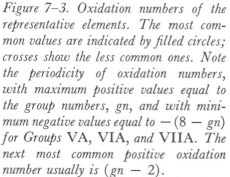

Figure 7–3. Oxidation numbers of the representative elements. The most common values are indicated by filled circles; crosses show the less common ones. Note the periodicity of oxidation numbers, with maximum positive values equal to the group numbers, gn, and with minimum negative values equal to $-(8 - gn)$ *for Groups VA, VIA, and VIIA. The next most common positive oxidation number usually is* $(gn - 2)$.

combine with $8 - gn$ hydrogen atoms. For example, sulfur atoms in Group VI each combine with two hydrogen atoms since sulfur has an oxidation number of -2. Their maximum oxidation number is commonly $+gn$, especially in oxygen compounds. Examples are SO_3 and H_2SO_4, in which the oxidation number of sulfur is $+6$. Most nonmetals also exhibit intermediate oxidation numbers (see Table 7–1).

Table 7–1. Oxidation Numbers of Nonmetals

Element	Oxidation number	Representative compounds
F	−1	Fluorides: HF, Na^+F^-
O	−2	H_2O, OH^-, O^{2-}, SO_2
	−1	Peroxides: H_2O_2, O_2^{2-}
N	−3	NH_3, NH_4^+, N^{3-}
	+5	HNO_3, NO_3^-, N_2O_5
	All intermediate values	N_2H_4, NH_2OH, N_2O, NO, NO_2^-, NO_2
C	+4	CO_2, CCl_4, CF_4
	−4	CH_4
	Complicated by chain formation	C_2H_6, C_4H_{10}, C_2H_6O
Cl	−1	HCl, Cl^-
	+7	$HClO_4$, ClO_4^-
	Intermediate values	ClO^-, ClO_2^-, ClO_2, ClO_3^-
S	−2	H_2S, S^{2-}
	+4	H_2SO_3, SO_2, HSO_3^-, SO_3^{2-}
	+6	H_2SO_4, SO_3, SO_4^{2-}, SF_6
	Intermediate values	$S_2O_3^{2-}$, $S_2O_4^{2-}$, $S_5O_6^{2-}$
P	−3	PH_3, PH_4^+, P^{3-}
	+5	H_3PO_4, P_4O_{10}, PO_4^{3-}, PCl_5
	Intermediate values	H_3PO_3, H_3PO_2
Si	+4	SiO_2, SiO_4^{4-}
	−4 unstable	
Br	−1	HBr, Br^-
	+5	$HBrO_3$, BrO_3^-, BrF_5
	Intermediate values	BrF, BrF_3
I	−1	HI, I^-
	+5, +3, +1	IO_3^-, ICl_4^-, ICl
	+7	HIO_4, H_5IO_6, IF_7
Se, Te	−2	H_2Se, H_2Te
	+4	SeO_2, TeO_2
	+6	H_2SeO_4, $Te(OH)_6$
As, Sb	−3	AsH_3, SbH_3
	+3	$AsCl_3$, $SbCl_3$
	+5	AsO_4^{3-}, $Sb(OH)_6^-$

Transition metals. Oxidation numbers follow trends among transition metals, as illustrated in Table 7–2. Early members of the series of transition metals exhibit maximum oxidation numbers of increasing magnitude, up to +7 for manganese in MnO_4^-, which correspond to the group numbers. Thereafter, the maximum oxidation number usually falls again by *one* number for each step to the right across the second half of the transition metals. [Can you

Table 7–2.   Oxidation Numbers of First-Row Transition Metals[a]

Oxidation number

	IIIB	IVB	VB	VIB	VIIB	VIII			IB	IIB
7					MnO_4^-					
6				CrO_4^{2-}	MnO_4^{2-}	FeO_4^{2-}				
5			VO_4^{3-}	$CrOCl_5^{2-}$	MnO_4^{3-}	*				
4		TiO_2	VO^{2+}	*	MnO_2	*	CoO_2	NiO_2		
3	$\underline{Sc^{3+}}$	Ti^{3+}	V^{3+}	Cr^{3+}	Mn^{3+}	Fe^{3+}	Co^{3+}	Ni_2O_3	Cu^{3+}	
2	*	TiO	V^{2+}	Cr^{2+}	$\underline{Mn^{2+}}$	$\underline{Fe^{2+}}$	$\underline{Co^{2+}}$	$\underline{Ni^{2+}}$	$\underline{Cu^{2+}}$	$\underline{Zn^{2+}}$
1		*	*	*	$Mn(CN)_6^{5-}$	*	*	$Ni_2(CN)_6^{4-}$	$\underline{Cu^+}$	
0		*	*	$Cr(CO)_6$	$Mn_2(CO)_{10}$	$Fe(CO)_5$	$Co_2(CO)_8$	$Ni(CO)_4$		

[a]Underlined species are those most commonly encountered under ordinary conditions in solids and in aqueous solutions. The asterisk indicates that oxidation numbers have been observed only in rare complex ions or unstable compounds.

anticipate later theories and imagine any connection between this behavior of the maximum oxidation number and the trend in melting points (Figure 9–5) and heats of fusion (Figure 6–11)]?

Inner transition metals.   Lanthanide and actinide elements (Figure 6–3) compose another type of transition series, in which adjacent elements have very similar properties. The oxidation number +3 is common to all lanthanides and actinides in their compounds. Other oxidation numbers are possible and in some cases preferred (e.g., Eu^{2+}, Ce^{4+}, and U^{6+}).

Most compounds containing transition and inner transition metals are colored (Figure 7–4). This phenomenon and the similarity of horizontally adjacent elements in these transition series can be explained in terms of modern theories of atomic and molecular structure.

Challenge.   For application of oxidation numbers to "hypo" (sodium thiosulfate) and the uranium ore pitchblende, try Problems 7–7 and 7–8 in *Relevant Problems for Chemical Principles* by Butler and Grosser.

7–3 OXIDATION–REDUCTION REACTIONS

We often refer to an element with a particular oxidation number as being in the *oxidation state* of that number; thus, in H_2O, H is in the +1 oxidation state and O is in the −2 oxidation state. Reactions in which the oxidation states of component atoms change are called *oxidation–reduction (redox) reactions.* If an atom's oxidation number increases, the atom is *oxidized;* if its oxidation number decreases, it is *reduced.*

We would like to be able to deduce, from various types of information, the detailed *mechanism* or steps involved during a chemical change. Proton transfer (Section 4–3) is an example of such a mechanism for acid–base reactions. Chemists have deduced two general mechanisms by which oxida-

Sc_2O_3 white	TiO_2 white	V_2O_5 orange	CrO_3 red	Mn_2O_7 green	Fe_2O_3 red-brown	CoO green-brown	NiO green-black	Cu_2O red	ZnO white	Ga_2O_3 white	GeO_2 white	As_2O_5 white	SeO_3 white
Y_2O_3 white	ZrO_2 white	Nb_2O_5 white	MoO_3 white	Tc_2O_7 yellow	RuO_4 yellow	RhO_2 brown	PdO green-blue	Ag_2O black	CdO brown	In_2O_3 yellow	SnO_2 white	Sb_2O_5 yellow	TeO_3 white
La_2O_3 white	HfO_2 white	Ta_2O_5 white	WO_3 yellow	Re_2O_7 yellow	OsO_4 yellow	IrO_2 black-blue	PtO violet-black	Au_2O dark	HgO red-yellow	TlO_3 brown	PbO_2 brown	Bi_2O_5 red-brown	

CeO_2 white	PrO_2 brown-black	Nd_5O_3 blue	Pm	Sm_2O_3 yellow	Eu_2O_3 pale red	Gd_2O_3 white	Tb_2O_3 white	Dy_2O_3 white	Ho_2O_3 tan	Er_2O_3 red	Tm_2O_3 green-white	Yb_2O_3 white	Lu_2O_3 white
ThO_2 white	Pa_2O_5 white	UO_3 orange	NpO_2^+ green	PuO_2^+ red-violet	AmO_2^+ green	Cm	Bk	Cf	Es	Fm	Md	No	Lr

Figure 7–4. Oxides of many transition and inner transition metals are colored.

tion–reduction can occur: electron transfer and atom transfer. Many complicated reactions probably include steps involving both mechanisms. (We must emphasize that it is impossible to deduce from the balanced equation alone what the mechanism of a chemical reaction is. The equation describes the stoichiometry of a reaction; kinetic experiments are required to supply data that will help us to know precisely how substances react with each other. This is the subject of Chapter 18.)

Electron Transfer

We already have encountered examples of electron transfer in the reactions occurring at electrodes during electrolysis (Section 3–2). At the cathode, reduction occurs; at the anode, oxidation occurs. It seems reasonable, then, that electrons might be exchanged directly among molecules. Electron-deficient molecules can accept electrons from *reducing agents* (electron donors), and electron-rich molecules can donate electrons to *oxidizing agents* (electron acceptors). (Therefore, the terminology is reciprocal.) Some common oxidizing and reducing agents are listed in Table 7–3. Acceptance of electrons by a substance must cause a decrease in oxidation number, whereas loss of electrons must involve an increase. It follows that any electron-transfer reaction is an oxidation–reduction in which the reducing agent gives electrons to the oxidizing agent. Thus, in rusting, iron reduces oxygen, and oxygen oxidizes (the origin of the term) iron:

$$4Fe + 3O_2 \rightarrow 4Fe^{3+} + 6O^{2-} \qquad \text{(in the form of } 2Fe_2O_3)$$

$$\text{12}e^-$$

By donating electrons, copper reduces silver ions, and the silver ions oxidize the metallic copper in the reaction

$$Cu + 2Ag^+ \rightarrow Cu^{2+} + 2Ag$$

$$2e^-$$

If an element can have several oxidation states, the intermediate oxidation states can be either oxidizing or reducing agents. Mn^{3+} can act as an oxidizing agent and be reduced to Mn^{2+}, or as a reducing agent and be oxidized to Mn^{4+}. In fact, Mn^{3+} in solution is unstable and spontaneously *disproportionates* with self oxidation–reduction to give the $+2$ and $+4$ oxidation states:

$$2Mn^{3+} + 2H_2O \rightarrow Mn^{2+} + MnO_2 \downarrow + 4H^+$$

Furthermore, oxidizing and reducing agents occur in pairs such that when each oxidizing agent reacts it *becomes* a potential reducing agent, and vice versa. This process is similar to the Brønsted acid–base theory (Chapter 4), in which every acid, by giving up a proton, becomes a base, and every base, by accepting a proton, becomes an acid.

Since electron loss and gain correspond directly to oxidation number gain and loss, it is convenient to think of all redox reactions as if they were electron-transfer reactions. However, after careful investigation, chemists have determined a second mechanism for altering oxidation numbers.

Atom Transfer

In the reaction

$$ClO_3^- + 3SO_3^{2-} \rightarrow Cl^- + 3SO_4^{2-}$$

sulfur has an oxidation number of $+4$ in SO_3^{2-} and $+6$ in SO_4^{2-}. (If you do not see why, look again at Section 7–2.) Chlorine has an oxidation number of $+5$ in ClO_3^-, and -1 in Cl^-. So three units of S are required for each unit of Cl. By using $KClO_3$ enriched with the isotope oxygen-18, we find that even in water solution the ^{18}O is transferred directly from ClO_3^- to

Table 7–3.   Common Oxidizing and Reducing Agents

Oxidizing agents

1. Free (elemental) nonmetals become negative ions:

fluorine	$F_2 + 2e^- \rightarrow 2F^-$
oxygen	$O_2 + 4e^- \rightarrow 2O^{2-}$
chlorine	$Cl_2 + 2e^- \rightarrow 2Cl^-$
bromine	$Br_2 + 2e^- \rightarrow 2Br^-$
iodine	$I_2 + 2e^- \rightarrow 2I^-$
sulfur	$S + 2e^- \rightarrow S^{2-}$

2. Positive (usually metal) ions become neutral:

$$Ag^+ + e^- \rightarrow Ag$$
$$2H^+ + 2e^- \rightarrow H_2$$

3. Higher oxidation states become lower:

$$8H^+ + MnO_4^- + 5e^- \rightarrow Mn^{2+} + 4H_2O$$
$$Cu^{2+} + e^- \rightarrow Cu^+ \quad \text{(often written as } Cu^{2+}|Cu^+)$$
$$Fe^{3+} + e^- \rightarrow Fe^{2+} \quad \text{(or } Fe^{3+}|Fe^{2+})$$
$$Cr_2O_7^{2-}|Cr^{3+}$$
$$ClO_3^-|Cl^-$$
$$NO_3^-|(NO_2, NO, N_2O, NH_4^+, \text{etc.})$$
$$Ce^{4+}|Ce^{3+}$$

Reducing agents

1. Metals yield ions plus electrons:

$$Zn \rightarrow Zn^{2+} + 2e^-$$
$$Na \rightarrow Na^+ + e^-$$

All metals yielding their common ions may be included here.

2. Nonmetals combine with other nonmetals, such as O and F, which they take from compounds with metals:

$$C + [O^{2-}] \rightarrow CO + 2e^-$$

Here $[O^{2-}]$ represents oxygen in a -2 oxidation state in combination with a metal such as Fe in the following total equation:

$$3C + Fe_2O_3 \rightarrow 3CO + 2Fe$$

3. Lower oxidation states become higher:

$Fe^{2+} \rightarrow Fe^{3+} + e^-$	(or $Fe^{2+}|Fe^{3+}$)
$SO_3^{2-} + H_2O \rightarrow SO_4^{2-} + 2H^+ + 2e^-$	(or $SO_3^{2-}|SO_4^{2-}$)
$NO + 2H_2O \rightarrow NO_3^- + 4H^+ + 3e^-$	(or $NO|NO_3^-$)

$SO_3{}^{2-}$, since most of the ^{18}O is present in the product, $SO_4{}^{2-}$:

$$Cl^{18}O_3{}^- + 3SO_3{}^{2-} \rightarrow Cl^- + 3[SO_3\,{}^{18}O]^{2-}$$

The transfer of oxygen alters the oxidation number of Cl and S, but not the charge on either ionic species. Even though atoms and not electrons are transferred, the oxidation numbers of chlorine and sulfur change because the number of oxygen atoms in the ions alters without a change in ionic charge. But since an overall equation does not specify the exact mechanism, all oxidation–reduction equations still can be balanced as if they were electron-transfer reactions.

7–4   BALANCING OXIDATION–REDUCTION EQUATIONS

Let us look again at the reaction involving $K_2Cr_2O_7$ and HI in Chapter 4 (Equation 4–4). If we assume that the reactants and products are known,[1] then the problem is how to find the mole ratios and to balance the following equation properly:

$$K_2Cr_2O_7 + HI + HClO_4 \rightarrow KClO_4 + Cr(ClO_4)_3 + I_2 + H_2O$$

$$(7\text{–}1)$$

Two methods have been developed for balancing redox equations systematically. With the oxidation-number method we use the fact that the amount of oxidation must equal the amount of reduction in the total chemical reaction. With the ion–electron method we consider a redox reaction to be the formal sum of two half-reactions, one that donates electrons and the other that accepts them.

Oxidation-Number Method

1) Identify the elements that change oxidation number during the reaction. It is helpful to write the oxidation numbers of these elements above their symbol on both sides of the reaction. In Equation 7–1, chromium (Cr) goes from $+6$ in $K_2Cr_2O_7$ to $+3$ in $Cr^{3+}(ClO_4{}^-)_3$. Imagine that each Cr atom accepts three electrons to change its oxidation state from $+6$ to $+3$. Iodine goes from -1 in HI to 0 in I_2 and loses one electron per atom in the process.

2) Now choose enough of the reductant and oxidant so the electrons lost by one are used completely by the other. There must be three times as many I atoms involved as Cr, and since $K_2Cr_2O_7$ has two Cr atoms, the

---

[1] At this point you should not feel that you should be able to predict products of reactions. As you gain experience, especially in the laboratory, you will be able to make more and more predictions.

reaction requires six HI molecules:

$$\overset{+6}{K_2Cr_2O_7} + \overset{-1}{6HI} + HClO_4 \rightarrow$$
$$\overset{+3}{KClO_4} + 2\overset{+3}{Cr}(ClO_4)_3 + \overset{0}{3I_2} + H_2O$$

3) Balance the other metals that do not change oxidation number (K^+ in this case):

$$\overset{+6}{K_2Cr_2O_7} + \overset{-1}{6HI} + HClO_4 \rightarrow$$
$$\overset{+3}{2KClO_4} + 2\overset{+3}{Cr}(ClO_4)_3 + \overset{0}{3I_2} + H_2O$$

4) Balance the anions that do not change (ClO_4^- in this case):

$$K_2Cr_2O_7 + 6HI + 8HClO_4 \rightarrow$$
$$2KClO_4 + 2Cr(ClO_4)_3 + 3I_2 + H_2O$$

5) Balance the hydrogens, and make sure that oxygen is also balanced:

$$K_2Cr_2O_7 + 6HI + 8HClO_4 \rightarrow$$
$$2KClO_4 + 2Cr(ClO_4)_3 + 3I_2 + 7H_2O$$

The balancing process is thus completed. The sequence of balancing steps can be summarized as oxidation numbers–cations–anions–hydrogens–oxygens. In what follows we shall try the same equation by another method.

Challenge. Air pollution monitoring and hair bleaching involve redox processes. Try Problems 7–11 and 7–13 in Butler and Grosser, which also uses the oxidation-number method.

Ion–Electron (Half-Reaction) Method

It is often useful to pretend that oxidation and reduction are occurring separately, and then to combine enough of each half-reaction to cancel all the free electrons. Chemical reactions occurring at electrodes in batteries or electrolysis cells (Chapter 3) are examples of half-reactions that actually occur. For example,

$$Cu^{2+} + 2e^- \rightarrow Cu \qquad \text{(cathode)}$$
$$2H_2O \rightarrow O_2 + 4H^+ + 4e^- \qquad \text{(anode)}$$

Redox reactions that occur in solution can be considered as the sum of two such half-reactions that proceed without the addition of an external driving force (the battery). In all electron-transfer reactions *the number of electrons donated by the reducing agent must equal the number of electrons accepted by the oxidizing agent.*

The $K_2Cr_2O_7$ reaction can be balanced by half-reactions as follows:

1) First, simplify the reaction by eliminating all "spectator" ions, such as K^+ or ClO_4^-, which do not really participate in the reaction:

$$Cr_2O_7{}^{2-} + I^- + H^+ \rightarrow Cr^{3+} + I_2 + H_2O$$

2) Now construct two balanced half-reactions, one involving Cr and one involving I.

   a) The unbalanced reactions are

$$Cr_2O_7{}^{2-} \rightarrow 2Cr^{3+}$$
$$2I^- \rightarrow I_2$$

   b) Balance the atoms in each half-reaction by adding H^+ and H_2O if the reactions occur in an acid medium, or H_2O and OH^- if in a basic one:

$$Cr_2O_7{}^{2-} + 14H^+ \rightarrow 2Cr^{3+} + 7H_2O \qquad \text{(Cr, O, and H atoms balanced)}$$

$$2I^- \rightarrow I_2 \qquad \text{(no } H^+ \text{ or } OH^- \text{ on either side; no atoms needed)}$$

   c) Balance the charge by adding electrons:

$$6e^- + Cr_2O_7{}^{2-} + 14H^+ \rightarrow 2Cr^{3+} + 7H_2O$$
$$2I^- \rightarrow I_2 + 2e^-$$

If the half-reaction is balanced properly, the number of electrons will indicate exactly the change in oxidation number. The two Cr require six electrons, and the two I^- produce two electrons.

3) Multiply the half-reactions by coefficients that will make the number of electrons transferred in each half-reaction the same:

$$Cr_2O_7{}^{2-} + 14H^+ + 6e^- \rightarrow 2Cr^{3+} + 7H_2O$$
$$6I^- \rightarrow 3I_2 + 6e^-$$

4) Add the two half-reactions and cancel species that appear on both sides of the overall reaction:

$$Cr_2O_7{}^{2-} + 14H^+ + 6I^- \rightarrow 2Cr^{3+} + 3I_2 + 7H_2O$$

As a precaution, make sure that the *number of atoms* on both sides is the same, that the *charges* balance, and that there are *no net electrons left*.

5) Complete the equation by restoring the "uninvolved" species and by grouping ions to form known species:

$$Cr_2O_7{}^{2-} + 8H^+ + 6HI \rightarrow 2Cr^{3+} + 3I_2 + 7H_2O$$
$$K_2Cr_2O_7 + 8HClO_4 + 6HI \rightarrow 2Cr(ClO_4)_3 + 2KClO_4 + 3I_2$$
$$+ 7H_2O$$

This process can be summarized as: half-reactions–whole reaction–uninvolved ions.

After you have had some practice, the oxidation-number method is faster, but the ion–electron method is safer and more foolproof.

As a second example, balance the equation representing the reaction between potassium permanganate ($KMnO_4$) and ammonia (NH_3) that produces potassium nitrate (KNO_3), manganese dioxide (MnO_2), potassium hydroxide (KOH), and water.

1) *Oxidation-number method.* The unbalanced reaction is

$$KMnO_4 + NH_3 \rightarrow KNO_3 + MnO_2 + KOH + H_2O$$

In this reaction manganese and nitrogen change oxidation number:

$$Mn^{7+} \rightarrow Mn^{4+} \qquad \text{(change of } -3\text{)}$$
$$N^{3-} \rightarrow N^{5+} \qquad \text{(change of } +8\text{)}$$

To conserve overall oxidation numbers, we need eight manganese atoms for three nitrogen atoms:

$$\overset{+7}{8KMnO_4} + \overset{-3}{3NH_3} \rightarrow \overset{+5}{3KNO_3} + \overset{+4}{8MnO_2} + KOH + H_2O$$

Potassium (K^+) is the cation that does not change oxidation number; it now must be balanced:

$$\overset{+7}{8KMnO_4} + \overset{-3}{3NH_3} \rightarrow \overset{+5}{3KNO_3} + \overset{+4}{8MnO_2} + 5KOH + H_2O$$

The hydrogen atoms must be balanced:

$$8KMnO_4 + 3NH_3 \rightarrow 3KNO_3 + 8MnO_2 + 5KOH + 2H_2O$$

The oxygen atoms must balance; there are 32 on each side, and the process is complete.

2) *Half-reaction method.* Begin by simplifying the reaction. The K^+ ion does not change, so it is omitted:

$$MnO_4^- + NH_3 \rightarrow NO_3^- + MnO_2 + OH^- + H_2O$$

In this reaction, MnO_4^- is reduced and NH_3 is oxidized:

$$MnO_4^- \rightarrow MnO_2$$
$$NH_3 \rightarrow NO_3^-$$

Since OH^- is involved in the reaction, H_2O and OH^- are used to balance the atoms in each half-reaction:

$$MnO_4^- + 2H_2O \rightarrow MnO_2 + 4OH^-$$
$$NH_3 + 9OH^- \rightarrow NO_3^- + 6H_2O$$

Electrons are added to balance the charge for each half-reaction:

$$3e^- + MnO_4^- + 2H_2O \rightarrow MnO_2 + 4OH^-$$

$$NH_3 + 9OH^- \rightarrow NO_3^- + 6H_2O + 8e^-$$

The half-reactions are multiplied by eight and three, respectively, and then added:

$$\cancel{24}e^- + 8MnO_4^- + \cancel{16H_2O} \overset{5}{\rightarrow} 8MnO_2 + \cancel{32}OH^-$$

$$3NH_3 + \overset{2}{\cancel{27OH^-}} \rightarrow 3NO_3^- + \cancel{18}H_2O + \cancel{24}e^-$$

$$8MnO_4^- + 3NH_3 \rightarrow 8MnO_2 + 3NO_3^- + 5OH^- + 2H_2O$$

Before you go on. The ability to balance chemical equations is obviously an essential part of your study of chemistry because an initial understanding of a reaction and its stoichiometry is dependent upon a balanced equation. You will find a step-by-step development of the ion–electron method of balancing redox equations in Review 7 of Lassila, *Programed Reviews of Chemical Principles.* You may want to practice balancing such equations.

7–5   REDOX TITRATIONS

The *gram equivalent weight* of an acid or base in a neutralization reaction is the quantity of acid or base that will release or take up one mole of protons. In a similar way, the gram equivalent weight of an oxidizing or reducing agent in a redox reaction is defined as the amount of compound that will produce one mole of oxidation number change. In the reaction

$$Na \rightarrow Na^+ + e^-$$

the sodium undergoes a change in oxidation number of one unit, so the equivalent weight of sodium *in this reaction* is equal to its atomic weight. In the oxidation of a divalent metal

$$Mg \rightarrow Mg^{2+} + 2e^-$$

each atom of magnesium changes by two oxidation number units, and each mole of magnesium metal furnishes two gram-equivalents of reducing ability. Therefore, the equivalent weight of Mg in this reaction is half its atomic weight.

   The equivalent weight of HCl in an acid–base neutralization reaction is equal to its molecular weight. The equivalent weight of HCl in a redox reaction depends on the change in oxidation number of chlorine during the reaction. If a chloride ion is oxidized to Cl_2,

$$\overset{-1}{Cl^-} \rightarrow \overset{0}{\tfrac{1}{2}Cl_2} + e^-$$

then there is one redox equivalent per Cl, and the equivalent weight and molecular weight of HCl are identical. But if the reaction is

$$\overset{-1}{Cl^-} + 3H_2O \rightarrow \overset{+5}{ClO_3^-} + 6H^+ + 6e^-$$

then each HCl furnishes six equivalents of reducing power, and the equivalent weight is one sixth the molecular weight.

In titrations using solutions of oxidants or reductants as reagents, it is convenient to use equivalents; for when all of an oxidizing agent in a sample has reacted with a solution's reducing agent in the burette, the number of equivalents of oxidant and reductant is the same. As with neutralization reactions, the *normality* of a solution is the *number of gram equivalent weights per liter of solution.*

Example. An amount of 50 ml of solution containing 1.00 g of $KMnO_4$ is to be used in titrating a reducing agent. During the reaction, MnO_4^- will be reduced to Mn^{2+}. What is the molarity of the solution? What is the normality?

Solution. The molarity is

$$1.00 \text{ g } \cancel{KMnO_4} \times \frac{1 \text{ mole}}{158 \text{ g } \cancel{KMnO_4}} \times \frac{1}{0.050 \text{ liter}}$$

$$= 0.127 \text{ mole liter}^{-1}$$

$$= 0.127 \text{ molar}$$

The reduction of MnO_4^- to Mn^{2+} is

$$\overset{+7}{MnO_4^-} + 8H^+ + 5e^- \rightarrow \overset{+2}{Mn^{2+}} + 4H_2O$$

Since Mn changes by five oxidation units in this reaction, the equivalent weight of $KMnO_4$ is one fifth the molecular weight, and the normality is five times the molarity:

$$0.127 \text{ mole liter}^{-1} \times 5.00 \text{ equiv mole}^{-1}$$

$$= 0.634 \text{ equiv liter}^{-1}$$

$$= 0.634 \text{ normal}$$

Example. A 31.25-ml solution of 0.100-molar $Na_2C_2O_4$ (sodium oxalate) in acid is titrated with 17.38 ml of $KMnO_4$ solution of unknown strength. What is the normality of the $Na_2C_2O_4$ and of the $KMnO_4$, and the molarity of the $KMnO_4$?

Solution. The reaction is

$$\overset{+7}{2MnO_4^-} + \overset{+3}{5C_2O_4^{2-}} + 16H^+ \rightarrow \overset{+2}{2Mn^{2+}} + \overset{+4}{10CO_2} + 8H_2O$$

Manganese goes from $+7$ to $+2$, so each MnO_4^- provides 5 equivalents of oxidizing power. Carbon goes from $+3$ to $+4$, so each $C_2O_4^{2-}$, with *two* carbon atoms, provides 2 equivalents of reducing power. Another way of understanding this is to write the two half-reactions:

$$MnO_4^- + 8H^+ + 5e^- \rightarrow Mn^{2+} + 4H_2O$$
$$C_2O_4^{2-} \rightarrow 2CO_2 + 2e^-$$

For this reaction the 0.100-molar $Na_2C_2O_4$ solution is 0.200 normal. The number of milliequivalents of oxidant and reductant are equal at neutralization. [One milliequivalent (meq) is 10^{-3} equivalent.] So

$$meq\ Na_2C_2O_4 = meq\ KMnO_4$$
$$31.25\ ml \times \frac{0.200\ meq}{1\ ml} = 17.38\ ml \times \frac{X\ meq}{1\ ml}$$

(Note that $1\ meq\ ml^{-1} = 1$ equivalent $liter^{-1}$.)

$$\text{Normality of } KMnO_4 = X = 0.360\ normal$$
$$\text{Molarity of } KMnO_4 = \frac{0.360\ \cancel{equiv}\ liter^{-1}}{5\ \cancel{equiv}\ mole^{-1}}$$
$$= 0.072\ mole\ liter^{-1}$$
$$= 0.072\ molar$$

The importance of writing the equations or half-reactions when dealing with equivalents is illustrated by the fact that MnO_4^- can be reduced in various circumstances in the following ways:

1) $MnO_4^- + e^- \qquad\qquad \rightarrow MnO_4^{2-}$
2) $MnO_4^- + 2H_2O + 3e^- \rightarrow MnO_2 + 4OH^-$
3) $MnO_4^- + 8H^+ \quad + 4e^- \rightarrow Mn^{3+} + 4H_2O$
4) $MnO_4^- + 8H^+ \quad + 5e^- \rightarrow Mn^{2+} + 4H_2O$

The number of equivalent weights per mole of $KMnO_4$ in these examples is 1, 3, 4, and 5. The last reaction is the most frequently encountered, but the others also occur. The normality of any $KMnO_4$ solution thus will depend on how we use it.

Challenge. Calcium is essential for the growth of bones and teeth; it is the fifth most abundant element in man. To see if you can determine the calcium content of a blood sample, try Problem 7–22 in the Butler and Grosser book.

7–6 COVALENCE

Oxidation numbers are helpful for predicting the proportions in which elements will react to form compounds, but are of little value for explaining why they do so or how the atoms in a molecule will be arranged in space.

Table 7–4. Common Covalences

Element	Covalence	Examples
H	1	H_2, H_2O, HCl
F	1	F_2, HF, ClF
O	2	H_2O, Cl_2O, CO_2
N	3	NH_3, NF_3, N_2H_4
C	4	CH_4, CO_2, CCl_4
B	3	BCl_3, BF_3
Cl	1	Cl_2, CCl_4, Cl_2O
S	2, 6	H_2S, SF_6, SO_3
P	3, 5	PH_3, P_4O_{10}, PCl_5, PCl_3
Br	1	Br_2, HBr, PBr_3
I	1	I_2, HI, ICl

The simple covalency ideas introduced in Section 7–1 are a beginning for these explanations.

Covalent bonds are formed primarily between nonmetals. Each element has a certain combining capacity or covalency, and bonds can be written between atoms by matching their covalences. Common covalences for nonmetals are given in Table 7–4, along with illustrative compounds. There can be two or even three covalent bonds between the same two atoms: These are called *double* and *triple* bonds. Some examples of covalent bond diagrams of simple molecules are

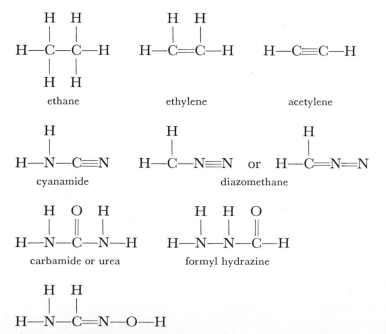

ethane   ethylene   acetylene

cyanamide   diazomethane

carbamide or urea   formyl hydrazine

formamidoxime

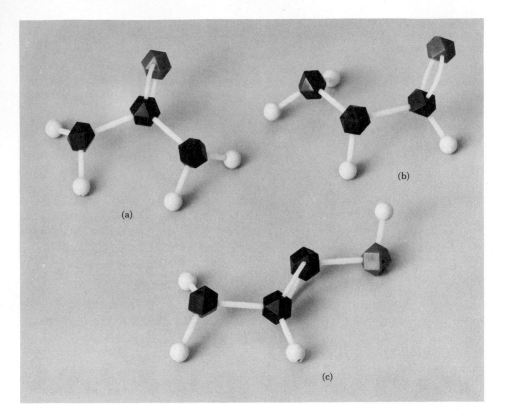

Figure 7–5. Three isomers of CH_4N_2O: (*a*) *Carbamide or urea:*
$H_2N—CO—NH_2$. (*b*) *Formyl hydrazine:* $H_2N—NH—CO—H$.
(*c*) *Formamidoxime:* $H_2N—CH=N—OH$. (*The urea molecule
should be planar, with H atoms in the same plane as N, C, and O.*)
*Only one of these three isomers can have stereoisomers as well. Can you
find it?*

Note that cyanamide and diazomethane are *isomers:* They both have the
molecular formula CH_2N_2, but have dissimilar molecular structure and properties. The last three compounds also are isomers (Figure 7–5). All three
are solids at room temperature, yet their melting points vary: 133°C for
urea, 54°C for formyl hydrazine, and 105°C for formamidoxime. The chemistry of the amides, hydrazines, and oximes is quite different.

Covalency alone cannot explain the shapes of molecules. Are the four
C—H bonds in methane, CH_4, directed to the corners of a square, all in the
same plane with the carbon, or are they directed to the vertices of a tetrahedron? Or is a less regular arrangement possible?

The Dutch chemist Jacobus van't Hoff (1852–1911) proposed, in 1874,
that all four hydrogen atoms in methane were structurally equivalent and
were at the four corners of a tetrahedron with the carbon atom in the center.

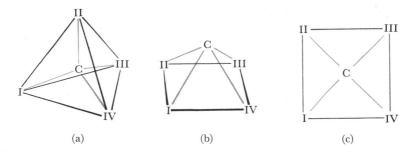

Figure 7–6. The three possible geometrical arrangements for methane,
CH_4: (a) *Tetrahedral,* (b) *square pyramidal, and* (c) *square planar.*
I–IV *represent hydrogen atoms.*

His reasoning exemplifies the deductive analysis by which much of the stereochemistry of molecules was inferred long *before* the development of x-ray crystallography.

When methane reacts with bromine at high temperatures, bromomethane is produced:

$$CH_4 \quad + \quad Br_2 \quad \rightarrow \quad CH_3Br \quad + \quad HBr$$

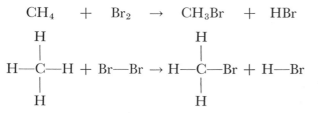

Only *one* form of bromomethane is ever produced. Therefore, unless there is some reason why only one of the four hydrogen atoms in methane can react, all four must be in equivalent and symmetrical positions. What sort of symmetrical arrangements of four H's around one C could be imagined? (Note the hidden assumption in this reasoning that the geometry around the carbon atom does not change during the bromination reaction. This is true for carbon in this reaction; in reactions of coordination complexes such an assumption is not always valid.)

Figure 7–6 illustrates the three possible structures for CH_4 in which all four H's are equivalent. The choice among them was made possible by the knowledge that there are no isomers of CH_2Br_2, dibromomethane. Only one arrangement of two Br's around the four H positions is possible. As you can see from Figure 7–6, if methane had either structures (b) or (c), pyramidal or planar, there would be two *stereoisomers:* one with bromine atoms at adjacent positions (I and II; II and III, etc.) and one with bromine atoms across the diagonal of the square (I and III; II and IV). (Stereoisomers are isomers that have the same bonds between the same pairs of atoms, and differ only in the spatial arrangement of bonds around the atoms.) Hence, (a) is the only

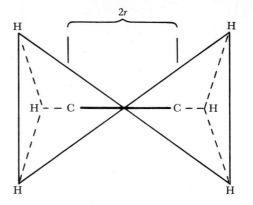

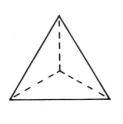

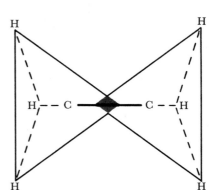

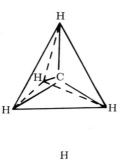

(a)

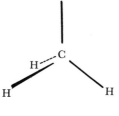

(b)

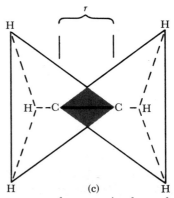

(c)

Figure 7–7. *Geometric structures of some simple carbon compounds.* (a) *A tetrahedron.* (b) *Structure of methane* (CH_4). (c) *Structure of ethane* (C_2H_6). *In ethane, one "corner" of the tetrahedral structure of one carbon atom is shared with the corner of the tetrahedron of the other carbon atom. It is not possible to predict from this tetrahedral model the distance between carbon atoms, because it might be anything from r to 2r as shown.* (d) *The structure of ethylene* (C_2H_4). *We imagine that this*

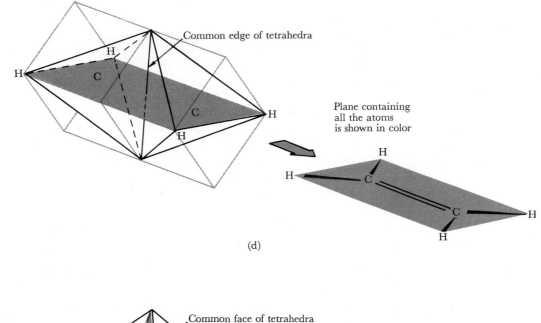

Common edge of tetrahedra

Plane containing
all the atoms
is shown in color

(d)

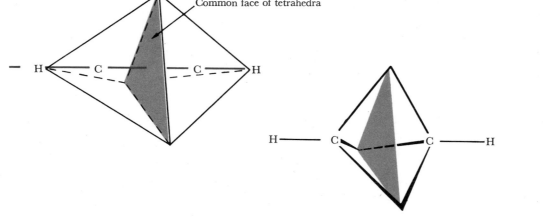

Common face of tetrahedra

(e)

structure has two carbon tetrahedra joined by sharing an edge. By inscribing the two tetrahedra sharing an edge inside two cubes sharing a face, we see that the two carbon atoms and four hydrogen atoms must lie in a common plane. The model is consistent with all physical and chemical evidence that the ethylene molecule is, in fact, planar. (e) Structure of acetylene (C_2H_2). The two carbon atoms share a tetrahedral face. The distance from one carbon atom to another is r, and the carbon atoms are triply bonded to each other. A line through the two free apexes of the tetrahedra also passes through the two carbon atoms. Thus, C_2H_2 must be linear. These arguments, although simple and incomplete, were the starting point for all structural organic chemistry.

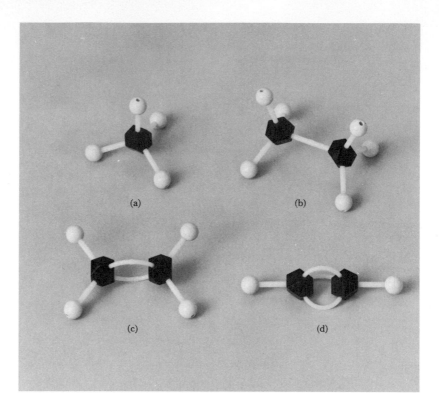

Figure 7–8. Benjamin/Maruzen models of molecules illustrating single and multiple bonds. (a) Methane, CH₄. (b) Ethane, C₂H₆. (c) Ethylene, C₂H₄, with a double bond. (d) Acetylene, C₂H₂, with a triple bond.

possible arrangement. So two simple observations on the lack of isomers of CH_3Br and CH_2Br_2 are sufficient to prove that all hydrogen atoms in CH_4 are equivalent and that they are arranged tetrahedrally around the carbon atom.

The proposal of the tetrahedral geometry around a carbon atom was the beginning of structural chemistry. Chemists used similar symmetry arguments to prove that ethylene, C_2H_4, is planar:

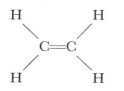

and that acetylene, C_2H_2, is linear:

$$H—C≡C—H$$

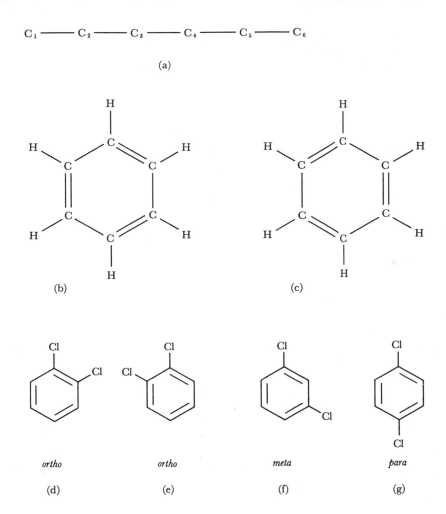

$$C_1 \text{——} C_2 \text{——} C_3 \text{——} C_4 \text{——} C_5 \text{——} C_6$$

(a)

(b)   (c)

ortho   *ortho*   *meta*   *para*

(d)   (e)   (f)   (g)

Figure 7–9. Possible benzene structures. (a) Linear; (b), (c) Kekulé's hexagonal ring; (d), (e), (f), and (g) possible isomers of dichlorobenzene. The prefix "ortho" indicates adjacent substitution, and "para" means substitution across the ring as shown. (The etymology is completely nonsensical. "Ortho" really means "straight," and "para" means "alongside of." But then, neither did the Greeks know chemistry.)

Double and triple bonds could be understood as a sharing of edges or faces of tetrahedra (Figures 7–7 and 7–8). We now could construct models of all of the carbon compounds shown so far in this section; and from the structures of these compounds, we could propose mechanisms for their reactions.

Shortly before van't Hoff postulated the tetrahedral carbon atom, August Kekulé conceived of a structure for the puzzling carbon compound benzene,

C_6H_6. Six C—H single bonds were known to be present, and all six hydrogen atoms could be replaced by Cl. Dichlorobenzene, $C_6H_4Cl_2$, existed in *only three isomers: ortho*-dichlorobenzene (o-$C_6H_4Cl_2$), melting at $-17.2°C$; *meta*-dichlorobenzene (m-$C_6H_4Cl_2$), melting at $-24.8°C$; and *para*-dichlorobenzene (p-$C_6H_4Cl_2$), melting at $53.1°C$. *If* benzene were linear, as in Figure 7–9(a), then there could be nine isomers in all. The pairs of hydrogen atoms replaced by chlorine would be given by (1, 2), (1, 3), (1, 4), (1, 5), (1, 6), (2, 3), (2, 4), (2, 5), and (3, 4). [All other combinations are equivalent to these, with a reversal of the chain: (5, 6) is the same as (1, 2) and (3, 5) is the same as (2, 4).] Kekulé's structure for benzene was a ring of six carbon atoms, with single and double bonds alternating between carbon atoms, and with one C—H single bond from each carbon [Figure 7–9(b, c)].[2] The two Cl atoms could be substituted on the six-membered ring on adjacent carbon atoms (*ortho*-dichlorobenzene), separated by one carbon atom (*meta*-dichlorobenzene), or across the ring from one another (*para*-dichlorobenzene). There was still one flaw. By Kekulé's structure, there should be two variants of the ortho form, with the chlorine atoms added at the ends of a single bond [Figure 7–9(e)] or a double bond (d). No such variants were found. The original explanation was that the two Kekulé structures, (b) and (c), "resonated" back and forth, averaging out their differences. This is quite poor and misleading terminology, although the word "resonance" is too firmly established in scientific literature to abandon. What we now realize is happening is that *neither* Kekulé structure is fully correct. It is impossible to draw any simple structure with single and double bonds that expresses accurately the bonding in benzene. Each of the six C—C bonds is more than single and less than double. X-ray diffraction has shown that all six bonds are the same length and are intermediate between the single C—C bond length in ethane and the double C=C bond length in ethylene. This partial double bond character gives such flat ring compounds their distinctive chemical properties. The formation of such structures will have to be explained by modern bonding theory in Part 3.

7–7   COORDINATION NUMBER

Chemists originally applied the term "coordination number" only to the central atom in a complex ion to indicate how many direct links it has to other atoms or molecules. In our earlier example, $Co(NH_3)_6{}^{3+}$, cobalt has a coordination number of six. This same coordination number of cobalt is maintained in the compounds $Co(NH_3)_6Cl_3$, $Co(NH_3)_5Cl_3$, and $Co(NH_3)_4Cl_3$; for the octahedrally coordinated cation in each of these cases is $Co(NH_3)_6{}^{3+}$, $Co(NH_3)_5Cl^{2+}$, and $Co(NH_3)_4Cl_2{}^+$, with Cl^- substituting for NH_3.

---

[2] Kekulé's own explanation of how he finally stumbled upon the hexagonal structure for benzene involves a dazed sleep and a dream of six snakes that coiled themselves into a hexagon. Chemical euphoria did not begin with LSD.

Table 7–5.   Variation of Coordination Number with Periods in the Periodic Table

Ion or molecule	Row in periodic table	Coordination number
Complex ions in which the element is coordinated to oxygen:		
BO_3^{3-}, CO_3^{2-}, NO_3^-	2	3
SiO_4^{4-}, PO_4^{3-}, SO_4^{2-}, ClO_4^-	3	4
AsO_4^{3-}, SeO_4^{2-}	4	4
$Sn(OH)_6^{2-}$, $Sb(OH)_6^-$, $Te(OH)_6$, IO_6^{5-}	5	6
Molecules and complex ions in which fluorine or hydrogen is attached to the central atom:		
BF_4^-, CF_4, CH_4, NH_4^+	2	4
AlF_6^{3-}, SiF_6^{2-}, PF_6^-, SF_6	3	6
SnF_6^{2-}, SbF_6^-, TeF_6, IF_7	5	6, 7

The terminology has been broadened with time; we now speak of the Na^+ ion as having a coordination number of six in solid NaCl because it has six nearest neighbor Cl^- ions. The Cs^+ ion in CsCl is given a coordination number of eight because of its eight neighboring Cl^- ions at the corners of the cube (Figure 3–17). Finally, the metal atoms in a cubic close-packed structure are described as being in twelvefold coordination because each atom has 12 similar atoms packed around it (see Figure 3–10).

In complex anions composed of oxygen and a nonmetal (SO_4^{2-} or NO_3^-), the coordination number for the nonmetal is the number of oxygen atoms to which the nonmetal is bonded. The coordination number of chlorine varies from one to four as the number of bonded oxygen atoms increases from one to four in ClO^-, ClO_2^-, ClO_3^-, and ClO_4^-. The maximum coordination number that an element commonly shows increases with the size of the central atom within a group (Table 7–5). Thus, later rows in the periodic table contain elements that tend to have larger coordination numbers. The reasons for this situation are the same as the ones for the radius ratio arguments of Section 3–7. Tin has more coordinating groups than carbon because there is more room for the groups to approach the larger tin ion without touching one another.

Nearly all transition metals form ions of charge +2 or +3, and nearly all of these ions form complex ions with a coordination number of six. Exceptions to this rule are the ions nickel(II), palladium(II), platinum(II), and gold(III). These ions generally are found in square planar structures with a coordination number of four. Examples of coordination complexes are in Table 7–6.

Table 7–6.   Common Coordination Numbers

Element	Coordination number	Examples
Fe	6	$Fe(CN)_6^{4-}$
Fe	6	$Fe(CN)_6^{3-}$
Co	6	$Co(NH_3)_6^{3+}$
Co	4, 6	$CoCl_4^{2-}$, $Co(H_2O)_6^{2+}$
Ni	4, 6	$Ni(CN)_4^{2-}$, $Ni(NH_3)_6^{2+}$
Cu	4, 6	$CuCl_4^{2-}$, $Cu(H_2O)_6^{2+}$
Zn	4	$Zn(CN)_4^{2-}$
Pt	4	$PtCl_4^{2-}$
Pt	6	$PtCl_6^{2-}$
B	3, 4	BO_3^{3-}, BF_4^-
C	3, 4	CO_3^{2-}, CH_4, CF_4
N	3, 4	NO_3^-, NH_4^+
Si	4, 6	SiO_4^{2-}, SiF_6^{2-}
S	4, 6	SO_4^{2-}, SF_6
Cl	1, 2, 3, 4	ClO^-, ClO_2^-, ClO_3^-, ClO_4^-
As	3, 4	AsO_3^{3-}, AsO_4^{3-}
Sb	6	$Sb(OH)_6^-$, $SbCl_6^-$
I	3, 4, 6	IO_3^-, IO_4^-, IO_6^{5-}

7–8   SUMMARY

We have discussed two important numbers in this chapter: oxidation number and coordination number. The first is a formal representation of the charge that is transferred (or supposedly transferred) during a chemical reaction. It is extremely useful for keeping track of quantities of matter in chemical reactions, even though it is not an accurate picture of what is happening on the atomic level. The oxidation number is an aid in keeping track of electrons during chemical reactions. Balancing a redox equation is equivalent to requiring that electrons be neither created nor destroyed. Therefore, electrons join numbers of atoms and charges on ions as quantities that must balance in a chemical equation.

The coordination number indicates how many other atoms are surrounding a given atom. This number is of most obvious importance in the chemistry of inorganic complex ions, but the coordination chemistry of iron, copper, and a few other transition metals is vital to biochemistry. For example, the familiar protein hemoglobin is an elaborate package for carrying iron in octahedral coordination.

The transition metals are the most common elements that form complex ions which most frequently have octahedral, tetrahedral, or square planar geometry. Both neutral molecules such as H_2O or NH_3 and ions such as Cl^-, SCN^-, or NO_2^- can coordinate to metal ions. The charge on the complex ion is the sum of the charge on the metal ion and the charge on the groups coordinated to it.

Just after World War I, two different kinds of bonds between atoms were distinguished: ionic bonds between metals and nonmetals (or between complex ions) and covalent bonds between nonmetals. Walter Kossel in Germany was the great proponent of ionic bonds, and much of what was known about covalence was the contribution of G. N. Lewis at the University of California. But in 1926, Erwin Schrödinger, at the University of Zurich, proposed his theory of *quantum mechanics* in a form that was to be extremely valuable to chemists. In the next few chapters we shall see how the bonding ideas of this chapter are the extreme forms of a more general and inclusive theory of chemical bonding. All of the ideas of Part 2 are *necessary*, but by themselves they are not *sufficient*. We are now ready to improve on this simple picture.

7–9 POSTSCRIPT TO OXIDATION, COORDINATION, AND COVALENCE

It is difficult now to realize the state of utter confusion in chemistry before the advent of quantum mechanics. The facts were known—even the facts about chemical bonding. Yet chemists were at a loss to explain what they saw. The following passage is quoted from J. W. Mellor's *Comprehensive Treatise on Inorganic and Theoretical Chemistry*, published in 1922 (Volume I, p. 225), only to suggest to you how confused and uncertain matters were. As you read it, remember that the beginnings of quantum theory were a generation earlier than the book, and that only four years after its publication Schrödinger presented his *wave mechanics* in the form that would prove so useful to chemistry.

Attempts to Explain Valency

The composition of all chemical compounds, says H. von Euler (1903) can be regarded as a function of a valency force—*Valenzkraft*—which is probably of an electric nature, and dependent on the temperature, pressure, and the nature of the solvent. Numerous attempts have been made to invent some peculiarities in the structure of the atoms which will explain that strange power manifesting itself as valency. Even Lucretius attributed the differences in the behavior of his atoms to differences in their shape, size, and mode of motion. The subject has rather lent itself to hypotheses established by the absence of a knowledge of contradictory facts. A brief *résumé* of the more striking forms of these hypotheses may act as a danger beacon.

I. Differences in the valency of different elements have been explained by supposing that *an atom of an n-valent element is compounded of n units, each of which is capable of attracting one other unit.* A constant quantity of one element, said E. Erlenmeyer (1862), never binds itself to more or to less than a constant quantity of another element—this he called *the law of constant affinivalencies.* W. Odling (1855) called these attracting units *subatoms;* G. Ensrud (1907), *Kernen* or *nuclei;* L. Knorr (1894), *Valenzkorper* or *valency bodies;* E. Erlenmeyer (1867), *affinivalencies;* A. W. von Hofmann (1865), *minimum atom-binding quantities* of an element; and J. Wislicenus (1888), *primitive atoms*, which are located in certain parts of the atom and from which they exert their influence. W. Lossen, in an important paper *Ueber die Vertheilung der Atome in der Molekul* (1880), pointed out that this hypothesis cannot be sound, for if a constant mass of, say, carbon binds itself to a constant mass of oxygen in the molecule of carbon dioxide, CO_2, the same mass of carbon is bound to half the same constant mass of oxygen in carbon monoxide, CO. Hence the assumed constant mass must be variable. G. Ensrud (1907) supposed an atom to be compounded of an enveloping shell of a substance of small density with a nucleus of great density and eccentric shape. The envelopes of different atoms repel one another, the nuclei attract one another in the direction along which valency acts. An atom of an *n*-valent element has *n* nuclei. . . . Some of these hypotheses appear to have arisen by confusing the fractional parts of an atom with the fractional parts of its weight, and assuming that the former are equal to the latter. There is nothing to show that if the atom were divided up into a number of attracting portions, each would be the same fractional part of the weight of the atom. The modern electron hypothesis of valency is one form of this hypothesis.

II. Other hypotheses assume that *valency is an attracting force localized at certain parts of the atom*. The atoms are supposed to be joined together at these attracting points; in other words, some parts of the atom are less active than others. E. Erlenmeyer (1867) and A. Michaelis (1872) suggested that the attractive forces are not exerted uniformly in all directions as is the case with gravitation, but are specially strong in certain definite directions so that a straight line joining two atoms directly bound together expresses the direction of the mutually exerted force. . . . A. C. Brown (1861) assumed that each atom possesses two kinds of attractive forces—positive and negative—and the point toward which these forces act was called a *pole* or active point. He made no assumption as to the nature of the attractive or repellent forces. . . . In order to support the assumption that valency is due to centers of attraction localized on the atom, subsidiary, hypotheses have to be invented. For instance, it has been assumed (i) that the atoms are bound to one another through the attraction of electric or magnetic charges localized on the atoms; and also (ii) that the intensity of the attractive force is modified by the shape of the atom.

(i) *Electric charges localized on the atom.* —The idea that the reacting units are polarized, and carry definite electric charges, each charge representing one valency, naturally grew from Davy's and Berzelius' electrochemical hypothesis, and Faraday's work. There are many modified forms of the hypothesis. For example, V. Meyer and E. Riecke (1888) assumed that the carbon atom is surrounded by an aethereal envelope which, in the case of isolated atoms, has a spherical shape like that supposed to be possessed by the atoms themselves. The atom in the core carries the specific affinities; the aethereal envelope is the seat of the valencies. Each valency is determined by the presence of two opposite electrical poles—called double or *di-poles*—situated at the ends of a straight line which is small in comparison with the diameter of the aethereal shell. The four valencies of carbon are represented by four such di-poles each of which is able to move freely within the aethereal shell, and to turn freely about its middle point. The carbon atom attaches other atoms to its surface by the attractions of the di-poles. . . .

(ii) *The shape of the atom.* —J. H. van't Hoff, in his *Ansichten uber die Organische Chemie* (Braunschweig, 1881), showed that the attractive forces emanating from an atom will be uniform in all directions if the atom is spherical, but if the shape be not spherical the intensity of the force, at short distances, will be more concentrated in certain spots than in others. Thus, if the atom were shaped like a regular tetrahedron, it would behave as if it were quadrivalent, for the centers of the four bounding faces would represent maximal attractions. Given the number of maximal points on the atom, it would be possible to deduce the valency, and conversely . . . J. Wislicenus (1888) has expressed a similar idea; he said:

> "It is not impossible that the carbon atom more or less resembles—perhaps very closely—the form of a regular tetrahedron; and further, that the causes of those attractions which are exhibited by the so-called units of affinity or bonds are concentrated at the apices of this tetrahedral structure, so that where there is least matter there is most force. These attractions are possibly analogous to the electrical state of a metal tetrahedron charged with electricity."

III. Another set of hypotheses has assumed that *valency is due to the need for harmonizing the motions of the combining atoms so as to form complexes whose parts move in stable equilibrium.* . . . According to L. Meyer (1884), the atoms in a molecule are not in a state of rest, but they move rotationally about a centre of equilibrium; the orbits of similar atoms have the same paths, but the orbits of different atoms are greater, the greater the valency of the atom. E. Molinari (1893) suggested a modification of this hypothesis in a paper entitled *Motochemistry* (*moto*, motion). The valency of an atom in a molecule is determined by the nature or energy of its oscillatory motion; and he claims that the constitution of compounds is dependent upon the intramolecular movements rather than on the relative positions of the atoms in space. F. A. Kekulé (1872) considered that valency is determined by the relative number of impacts which an atom receives from other atoms in unit time; each of the univalent atoms in a diatomic molecule impinges once, while the bivalent atoms impinge twice in unit time. It is not very clear how this explains valency. . . . F. M. Flavitzky (1896), following N. N. Beketoff (1880), supposed that the atoms move in curves which lie in planes parallel to one another; the atoms of different elements move in planes which are inclined at definite angles to one another; the motion of the atoms of one element can be completely counteracted by the motions of the atoms of another element only when the two planes of motion are parallel; otherwise, according to the size of the angle between the planes of motion, an atom of one element may require two, three, or more atoms of another element to balance it; and only those components come into action which are parallel to the plane of motion of another atom. Accordingly, F. M. Flavitzky refers the valency of an element to the difference in the angles between the planes of the orbits of the different rotating atoms. . . .

The sole impression that you should retain from the foregoing is of a state of almost riotous confusion. Most chemistry texts select from this profusion only the ideas that were proved to be correct by later developments. We remember Kekulé's benzene ring but conveniently forget his bouncing-ball theory of covalent bonding. We remember van't Hoff's tetrahedral carbon bonding but forget that it was associated with an atom that was a real, solid tetrahedron. We remember the electrical suggestion for bonding but forget the pivoting dipoles in the "aether" around the nucleus. All of this editorial selection by hindsight creates the erroneous impression that chemistry progresses in a straight line.

Buried in the confusion of Mellor's summary are suggestions of what was to come: the comment about the "modern electron hypothesis of valency" as a subheading of Hypothesis I. G. N. Lewis had proposed a relationship between the eight electrons in a stable, noble-gas valence shell and the four vertices of the tetrahedron formed by the bonds of carbon. Each covalent bond, Lewis suggested, is constructed by the sharing of an electron from each of the bound atoms to create an electron pair. This idea of the *electron-pair bond* developed, with the aid of quantum mechanics, into the bonding theory that we believe best explains the facts.

As you struggle through the intricacies of Part 3, come back and reread this postscript from time to time to remind yourself of how simple quantum mechanics is.

SUGGESTED READING

F. Basolo and R. C. Johnson, *Coordination Chemistry*, W. A. Benjamin, Menlo Park, Calif., 1964. Too advanced overall, but good in its earlier sections.

W. Herz, *The Shape of Carbon Compounds*, W. A. Benjamin, Menlo Park, Calif., 1963. Parts of Chapters 1–4 are useful now; others will be later.

L. Holliday, "Early Views on Forces Between Atoms," *Scientific American*, May, 1970.

J. W. Mellor, *A Comprehensive Treatise on Inorganic and Theoretical Chemistry*, Macmillan, New York, 1922. A good summary of prequantum attempts to explain bonding, in Chapter 5.

R. T. Sanderson, *Chemical Periodicity*, Reinhold, New York, 1960. Chapters 4, "Principles of Coordination Chemistry," and 17, "Survey of Coordination Chemistry by Periodic Groups," are particularly relevant.

QUESTIONS

1 What is the oxidation number of Co in $Co(NH_3)_6Cl_3$? In K_3CoF_6? In K_2CoI_4? In $Co(NH_3)_6Cl_2$? What is the coordination number of Co in each compound?

2 How are the suffixes "ic" and "ous" associated with oxidation state? Match the following formulas and names:

Sulfuric and sulfurous acids:
$$H_2SO_3 \text{ and } H_2SO_4$$

Nitric and nitrous acids:
$$HNO_2 \text{ and } HNO_3$$

3 Oxidation numbers are a convenient bookkeeping device for keeping track of electrons; they are useful even though in the actual reaction electrons are not totally removed from one atom and entirely given to another. The principle of conservation behind balancing redox equations is this: In a chemical reaction, electrons are neither created nor destroyed. How does this principle lead inevitably to Rule 8 (Section 7-2): In chemical reactions, the total oxidation number is conserved?

4 How does Rule 5, dealing with the relative oxidation numbers of nonmetals, follow from what you know about ionization energies and electron affinities in Chapter 6?

5 How are the most common oxidation states of the representative elements related to their group number?

6 What pattern of maximum oxidation number can be seen across a period in the transition metals?

7 Is a substance that gives up electrons in a reaction an oxidizing agent or a reducing agent? Is it oxidized or reduced? Does its oxidation number increase or decrease in the process?

8 What is the equivalent weight of sulfuric acid, H_2SO_4, in each of the following processes:

a) An acid–base titration.

b) A redox reaction in which the sulfate ion goes to sulfite.

c) A redox reaction in which the sulfate ion goes to sulfide.

9 What is the normality of a 0.1-molar solution of sulfuric acid in each of the three processes of Question 8?

PROBLEMS

1 In the compound C_3H_8 carbon exhibits a covalence of four and hydrogen a covalence of one. Write a chemical formula for the compound that is consistent with these covalences. Use a straight line to represent a covalent bond. Do the same exercise with C_4H_{10}.

10 What is the difference between covalency and oxidation number? In water, the oxidation numbers of the elements are numerically equal to their covalency, one for hydrogen and two for oxygen. This is also true for compounds such as methane, sulfuric acid, and ethyl alcohol. Yet for chlorine gas, the oxidation number is zero but the covalency is one, and for oxygen gas, the oxidation number is zero but the covalency is two. Why the difference?

11 What are isomers? What are stereoisomers? Are the isomers carbamide, formyl hydrazine, and formamidoxime also stereoisomers?

12 How many different isomers can there be of chloroform, $CHCl_3$?

13 How many different isomers can there be of chloroethane, CH_3CH_2Cl? Of dichloroethane, $C_2H_4Cl_2$?

14 What would be the answers to Question 13 if ethane had a flat structure in one plane such as

Remember in this question that turning the molecule around or tumbling it in space does not produce separate isomers.

15 If $[Co(NH_3)_5SO_4]Br$ and $[Co(NH_3)_5Br]SO_4$ are isomers, each with an octahedrally coordinated cobalt, sketch the coordination about the cobalt atom in each case. What is the charge on the cation in each compound?

2 The nitrogen atom in NH_4^+ is bound to four hydrogen atoms and the ion is said to have tetrahedral geometry. The cobalt atom in $Co(NH_3)_6^{3+}$ is bound to six ammonia molecules and the ion has octahedral geometry. Would it be more correct to describe the geometry of the

$Co(NH_3)_6{}^{3+}$ ion as hexahedral? Explain.

3 Suppose that you have just prepared a compound in the laboratory and want to determine whether it is ionic or covalent. Give two methods by which you could experimentally decide this question.

4 What is the coordination number of the central atom in each of the following ions: $SiF_6{}^{2-}$, $BO_3{}^{3-}$, $NH_4{}^+$, $Ni(CN)_4{}^{2-}$, $[Co(NH_3)_5Cl]^+$? What is the oxidation number of each central atom?

5 Xenon forms several nonionic compounds with F and O. Give the coordination number and the oxidation number of the central Xe atom in XeO_4, XeF_2, XeO_3, XeF_4, and XeF_6.

6 Xenon also forms several ionic compounds, including $CsXeF_7$ and $CsXeF_8$. What are the ions in each compound? What are the ionic charge, coordination number, and oxidation number of Xe in each ion containing Xe?

7 In $[Cd(NH_3)_3Cl]^+Cl^-$, what are the coordination number and ionic charge of cadmium? Using only the cadmium ion, ammonia molecules, and chloride ions, write the formula of a compound that will (a) produce three ions in aqueous solution and (b) produce no ions in aqueous solution.

8 What is the coordination number of the central atom in each of the following ions or molecules: $Co(CN)_5{}^{3-}$, $PtCl_6{}^{2-}$, $CO_3{}^{2-}$, SF_4, $[Mn(H_2O)_3Br_3]^{2+}$?

9 What is the oxidation number of nitrogen in each of the following ions or molecules: NH_3, N_2H_4, NO, NO_2, $NO_2{}^-$, $NO_3{}^-$?

10 What are the coordination number and oxidation number of platinum in the complex ion $PtCl_4{}^{2-}$?

11 Assign oxidation numbers to the atoms in the following ions and molecules: (a) gold, Au; (b) iodine, I_2;

(c) barium chloride, $BaCl_2$; (d) ethane, C_2H_6; (e) stannous oxide, SnO; (f) stannic oxide, SnO_2; (g) nitrous oxide, N_2O; (h) phosphorus pentoxide P_2O_5; (i) calcium hydride, CaH_2; (j) magnesium hydroxide, $Mg(OH)_2$; (k) sulfurous acid, H_2SO_3; (l) telluric acid, H_6TeO_6; (m) hypochlorous acid, $HClO$; (n) perchloric acid, $HClO_4$; (o) dichromate, $Cr_2O_7{}^{2-}$; (p) cyanide, CN^-.

12 What is the oxidation number of the underlined element in each ion or molecule: $\underline{V}O_2{}^+$, $\underline{P}_2O_7{}^{4-}$, $\underline{P}H_3$, $K\underline{N}O_2$, $H_2\underline{O}_2$, $Li\underline{H}$, $Mg_3\underline{N}_2$, $\underline{N}F_3$, $\underline{I}Cl_5$, $Ag(\underline{N}H_3)_2{}^+$?

13 When the equation

$$MnO_2 + I^- + H^+ \rightarrow Mn^{2+} + I_2 + H_2O$$

is balanced, what is the net charge on each side of the equation? What element is oxidized? What element is reduced?

14 For each of the following reactions list (1) the substance reduced, (2) the substance oxidized, (3) the reducing agent, and (4) the oxidizing agent if the reaction is an oxidation–reduction type.

a) $6H^+ + 2MnO_4{}^- + 5SO_3{}^{2-} \rightarrow 5SO_4{}^{2-} + 2Mn^{2+} + 3H_2O$

b) $Cl^- + Ag^+ \rightarrow AgCl$

c) $3Cl_2 + 6OH^- \rightarrow ClO_3{}^- + 5Cl^- + 3H_2O$

d) $MgSO_3 \rightarrow MgO + SO_2$

15 Phosphine, PH_3, is a colorless, highly toxic gas that smells like rotten fish and is produced in small amounts when animal and vegetable matter decay in moist situations such as damp graveyards. Traces of P_2H_4 are produced simultaneously and cause the PH_3 to ignite in air to give pale, flickering lights commonly called "corpse candles" or "will-o'-the-wisps." In the laboratory, the gas can be prepared by adding water to calcium phosphide. Write a balanced equation for the reaction. Assign oxidation numbers to each of the elements present.

16 Balance the reaction

$$MnO_4^- + H^+ + H_2S \rightarrow$$
$$Mn^{2+} + H_2O + S$$

What is the oxidation number of Mn in MnO_4^-? What element is oxidized? What element is the oxidizing agent? Is this latter element oxidized or reduced?

17 Balance the following equations by the oxidation-number method:

a) $H_2S + Cr_2O_7^{2-} \rightarrow S + Cr^{3+}$
(acidic solution)

b) $NH_3 + O_2 \rightarrow NO + H_2O$

c) $NO_2^- + MnO_4^- \rightarrow$
$$NO_3^- + MnO_2$$
(basic solution)

d) $NH_3 + OCl^- \rightarrow N_2H_4 + Cl^-$
(basic solution)

e) $H_2 + OF_2 \rightarrow H_2O + HF$

f) $MnO_2 + Al \rightarrow Al_2O_3 + Mn$

18 Balance the following equations by the half-reaction method:

a) $MnO_2 + KOH + O_2 \rightarrow$
$$K_2MnO_4 + H_2O$$

b) $CuCl_4^{2-} + Cu \rightarrow CuCl_2^-$
(acidic solution)

c) $NO_3^- + Zn \rightarrow NH_4^+ + Zn^{2+}$
(acidic solution)

d) $ClO_2 \rightarrow ClO_2^- + ClO_3^-$
(basic solution)

e) $Fe^{2+} + Cr_2O_7^{2-} \rightarrow Fe^{3+} + Cr^{3+}$
(acidic solution)

f) $Cu + NO_3^- \rightarrow Cu^{2+} + NO$
(acidic solution)

19 Balance the following equations by any method you choose:

a) $H_3PO_4 + CO_3^{2-} \rightarrow$
$$PO_4^{3-} + CO_2 + H_2O$$
(neutral solution)

b) $MnO_2 + SO_3^{2-} \rightarrow$
$$Mn(OH)_2 + SO_4^{2-}$$
(basic solution)

c) $H^+ + Cr_2O_7^{2-} + H_2SO_3 \rightarrow$
$$Cr^{3+} + HSO_4^-$$
(acidic solution)

d) $MnO_4^- + V^{2+} \rightarrow VO_2^+ + Mn^{2+}$
(acidic solution)

e) $FeSO_4 + NaClO_2 \rightarrow$
$$Fe_2(SO_4)_3 + NaCl$$
(in sulfuric acid solution)

f) $KNO_2 + KMnO_4 \rightarrow$
$$KNO_3 + MnO_2$$
(in KOH solution)

g) $MnO_4^- + OH^- + I^- \rightarrow$
$$MnO_4^{2-} + IO_3^- + H_2O$$

h) $KMnO_4 + NH_3 \rightarrow$
$$KNO_3 + MnO_2 + KOH + H_2O$$

20 Twelve grams of $KMnO_4$ were dissolved in sufficient water to make a liter of solution. Calculate the molarity of the solution. The solution was divided into parts and used in four different reactions. Reaction 1 was carried out in basic solution, thereby producing MnO_2 as a product. Reaction 2 was carried out in very strong base to give MnO_4^{2-}. Reaction 3 gave Mn^{3+} in acid solution, and Reaction 4 yielded Mn^{2+}. Calculate the equivalent weight of $KMnO_4$ and the normality of the potassium permanganate solution for each reaction.

21 A solution of potassium dichromate, $K_2Cr_2O_7$, is made by adding enough water to 3.52 g of the salt to fill a 100-ml volumetric flask. (a) What is the molarity of the solution? (b) The solution will be used in a titration in which the dichromate ion is reduced to Cr^{3+}. What is the normality of the solution?

22 Find the equivalent weight of an oxidizing agent that oxidizes Fe^{2+} to Fe^{3+} if 0.664 g of the compound requires 23.5 ml of 0.540-molar Fe^{2+} solution.

23 (a) When H_3PO_4 reacts with NaOH to produce NaH_2PO_4, how many gram equivalent weights of H_3PO_4 are there per mole? (b) When H_3PO_4 is reduced to H_3PO_2, how many gram equivalent weights of H_3PO_4 are there per mole?

24 Exactly 6.40 g of gaseous SO_2 are absorbed in 95 ml of water. When absorption is complete, water is added so the final volume of solution is 100 ml. Calculate for each of the following reactions the volume of solution *of this composition* that contains one gram equivalent weight of H_2SO_3: (a) neutralization to SO_3^{2-} with sodium hydroxide, (b) oxidation to SO_4^{2-}, (c) reduction to S^{2-}, (d) reduction to elemental sulfur.

25 An acidic solution is 0.10 molar in TiO^{2+}. What is its normality when reacted with: (a) dilute $NaOH$ to produce TiO_2, (b) dilute $FeSO_4$ to produce Ti^{3+}, (c) concentrated H_2SO_4 to produce Ti^{4+}?

26 A slightly acidic solution is 0.01 molar in Cl_2. What is its normality when reacted with: (a) dilute $FeSO_4$ to produce Fe^{3+} and (b) dilute H_2O_2 to produce $HClO$?

27 If chlorine gas is bubbled through a basic solution of potassium iodide, the following reaction occurs:

$$KOH + Cl_2 + KI \rightarrow$$
$$KCl + KIO_3 + H_2O$$

Balance the equation. If chlorine gas is bubbled through 25 ml of a 0.10-normal solution of KI in aqueous KOH at STP, what volume of Cl_2 gas will be required to react completely with the KI? (KI normality is based on its reducing action.)

28 Given the equation

$$H_2SO_4 + HI \rightarrow H_2S + I_2 + H_2O$$

calculate the number of moles of sulfuric acid consumed by reaction with 25.00 ml of 0.100-normal HI. (Check first to see if the reaction is balanced.)

29 When H_2S reacts with $KMnO_4$ in acidic solution, the following reaction can be written:

$$MnO_4^- + H_2S + H^+ \rightarrow$$
$$Mn^{2+} + S + H_2O$$

Balance the equation. If a 0.05-normal solution of H_2S is used to titrate 50 ml of permanganate solution, 70 ml are required to reach the end point. What is the normality of the original permanganate solution? What is the molarity?

30 A 25.00-ml sample of an unknown copper solution is treated with excess potassium iodide in acidic solution, and the liberated iodine is titrated with 0.0250-molar sodium thiosulfate. If 12.50 ml of the thiosulfate solution are required to reach the end point, what is the molarity of the unknown copper solution? The unbalanced reactions are

$$Cu^{2+} + I^- \rightarrow CuI + I_3^-$$
$$I_3^- + S_2O_3^{2-} \rightarrow I^- + S_4O_6^{2-}$$

31 In oxidizing H_2SO_3 to SO_4^{2-}, IO_3^- is reduced to I_2 in acidic solution. Write a balanced chemical equation for this redox reaction. What volume of 0.25-normal KIO_3 is required for the complete oxidation to sulfate of 125 ml of 0.10-normal H_2SO_3?

32 In acidic aqueous solution, permanganate ion will oxidize oxalic acid $(COOH)_2$ to carbon dioxide while being reduced to manganous ion, Mn^{2+}. (a) What are the oxidation numbers of C and Mn in the reactants and products? (None of the substances is a peroxide.) (b) Write a balanced equation for the reaction. (c) What are the molecular weights and equivalent weights of the reactants $KMnO_4$ and $(COOH)_2$? (d) How many moles of permanganate can be reduced by 0.01 mole of oxalic acid? (e) How many equivalents of oxalic acid can be oxidized by 0.04 equivalent of potassium permanganate? (f) What is the normality of a solution that contains 15.8 g of potassium permanganate per liter? (g) What is the molarity of a solution that contains 4.5 g of oxalic acid per liter? (h) What weight of oxalic acid can be oxidized to carbon dioxide by 150.0 ml of a solution containing 6.25 g of potassium permanganate per liter?

PART THREE

THE QUANTUM REVOLUTION

In 1922, the British chemist J. W. Mellor introduced his *Comprehensive Treatise*, from which we already have quoted, with the gloomy words below. He had been commenting on P. J. Macquer's *Dictionnaire de chymie* (Paris, 1766), in which the chemical compounds were arranged alphabetically:

> We now flatter ourselves that the periodic law has given inorganic chemistry a scheme of classification which enables the facts to be arranged and grouped in a scientific manner. The appearance of order imparted by that guide is superficial and illusory. Allowing for certain lacunae in the knowledge of the scarcer elements prior to the appearance of that law, the arrangements employed by the earlier chemists were just as satisfactory, and in some cases, indeed, more satisfactory than those based on the periodic law.

Such pessimism originated in the inability of chemists to explain why the periodic table should be as it is. Why do the numbers' of elements in the periods vary from 2, to 8, to 18, to 32? How could properties of elements be predicted from their positions in the table? Some progress had been made, but chemists such as Mellor still considered the periodic table only a scheme for classification. The momentum derived from the concept of atoms was slackening, and new impetus was needed.

The seeds of the next advance in chemistry took root at the turn of the century, when physicists began to be concerned about their inability to explain how light or other forms of energy interact with matter. The specific heat of solids, the spectrum of radiation from a glowing hot object, the emission of electrons by metals when irradiated by light, and the absorption of light by atoms were "wrong" according to classical physics. Physicists were compelled by experiments more than by inclination to conclude that energy behaved in many circumstances as if it came in packets with a minimum size, or in *quanta.* Quantum mechanics was developed to account for this behavior. As is often true, time was required before the implications of this new theory were appreciated in other fields. The period of the "quantization" of chemistry was in the 1920's and 1930's. Even now we are far from knowing all there is to know from the new chemistry. Many of the great chemists of this period, R. S. Mulliken, Linus Pauling, and C. A. Coulson (to name only a few and to risk offending the partisans of the others), are still alive and working. Although it is always chancy to write memoirs in the heat of battle, it is already clear that the change in chemistry in this century is as great as the one that took place in Dalton's time, and that we may accurately call the events of the past generation the "quantum revolution."

In the next seven chapters we shall see how this quantum revolution arose, how it was applied to chemistry, and how we can derive from it an understanding of chemical behavior. In these chapters we shall cover some of the same ground that we covered in Chapters 3 through 7. In our initial treatment, we were more concerned with observing how elements and compounds behaved. Now we shall try to explain this behavior by using the ideas of structure and bonding from the quantum theory of matter. We should not think that this new procedure gives the statements about chemical behavior any more validity than they previously had. When we compare data and theory, we are verifying the theory and not the data. But the theory makes the data easier to understand and easier to remember.

*The continuity of all dynamical effects was formerly taken for granted as the basis of all physical theories and, in close correspondence with Aristotle, was condensed in the well-known dogma—*Natura non facit saltus—*nature makes no leaps. However, present-day investigation has made a considerable breach even in this venerable stronghold of physical science. This time it is the principle of thermodynamics with which that theorem has been brought into collision by new facts, and unless all signs are misleading, the days of its validity are numbered. Nature does indeed seem to make jumps—and very extraordinary ones.*
Max Planck (1914)

8 QUANTUM THEORY AND ATOMIC STRUCTURE

Physics seemed to be settling down quite satisfactorily in the late nineteenth century. A clerk in the U.S. Patent Office wrote a now-famous letter of resignation in which he expressed a desire to leave a dying agency, an agency that would have less and less to do in the future since most inventions already had been made. The famous physicist A. A. Michelson suggested at the dedication of a physics laboratory in Chicago, in 1894, that the more important physical laws all had been discovered, and "Our future discoveries must be looked for in the sixth decimal place." Thermodynamics, statistical mechanics, and electromagnetic theory had been brilliantly successful in explaining the behavior of matter. Atoms themselves had been found to be electrical, and undoubtedly would follow Maxwell's electromagnetic laws.

Then came x rays and radioactivity. In 1895, Wilhelm Röntgen (1845–1923) evacuated a Crookes tube (Figure 3–6) so the cathode rays struck the anode without being blocked by gas molecules. He discovered that a new and penetrating form of radiation was emitted by the anode. This radiation, which he called *x rays*, traveled with ease through paper, wood, and flesh but was absorbed by heavier substances such as bone and metal. Röntgen demonstrated that x rays were not deflected by electric or magnetic fields and therefore were not beams of charged particles. Other scientists suggested that the rays might be electromagnetic radiation like light, but of a shorter wavelength. Max von Laue proved this hypothesis 18 years later when he diffracted x rays with crystals (Section 3–7).

In 1896, Henri Becquerel (1852–1928) observed that uranium salts emitted radiation that penetrated the black paper coverings of photographic plates and exposed the photographic emulsion. He named this behavior *radioactivity*. In the next few years, Pierre and Marie Curie isolated two entirely new, and radioactive, elements from uranium ore and named them *polonium* and *radium*. Radioactivity, even more than x rays, was a shock to physicists of the time. They gradually realized that radiation occurred during the breakdown of atoms, and that atoms were not indestructible but could decompose and decay into other kinds of atoms. The old certainties, and the hopes for impending certainties, began to fall away.

The radiation most commonly observed was of three kinds, designated alpha (α), beta (β), and gamma (γ). γ Radiation proved to be electromagnetic radiation of even higher frequency than x rays. β Rays, like cathode rays, were beams of electrons. Electric and magnetic deflection experiments showed that α radiation has a mass of 4 amu and a charge of $+2$; α particles were simply nuclei of helium, $^4_2$He.

The next certainty to slip away was the quite satisfying model of the atom that had been proposed by J. J. Thomson.

8–1 RUTHERFORD AND THE NUCLEAR ATOM

Thomson had proposed a model of the atom in which all of the mass and positive charge were distributed uniformly throughout the atom, with electrons embedded in the atom like raisins in a pudding. Mutual repulsion of electrons separated them uniformly. The resulting close association of positive and negative charges was reasonable. Ionization could be explained as a stripping away of some of the electrons from the pudding, thereby leaving a massive, solid atom with a positive charge.

In 1910, Ernest Rutherford (1871–1937) disproved the Thomson model, more or less by accident, while measuring the scattering of a beam of α particles by extremely thin metal foils. (His experimental arrangement is shown in Figure 8–1.) He expected to find a relatively small deflection of particles, as would occur if the positive charge and mass of the atoms were distributed throughout a large volume in a uniform way [Figure 8–2(a)]. What he

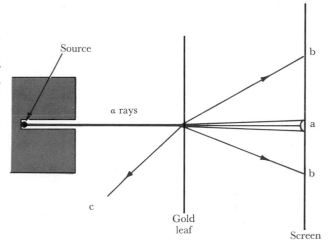

Figure 8–1. The experimental arrangement for Rutherford's measurement of the scattering of α particles by very thin metal foils. The source of the α particles was radioactive polonium, encased in a lead block that protected the surroundings from radiation and confined the α particles to a beam. The gold foil used was about 6×10^{-5} cm thick. Most of the α particles passed through the gold leaf with little or no deflection, a. A few were deflected at wide angles, b, and occasionally a particle rebounded from the foil, c, and was detected by a screen or counter placed on the same side of the foil as the source.

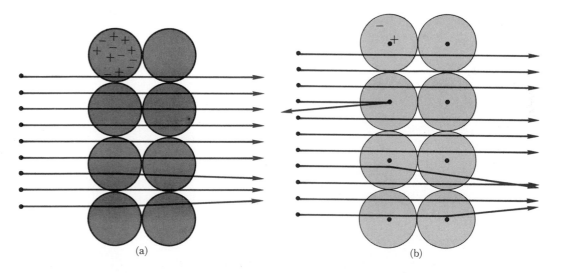

(a)   (b)

Figure 8–2. The expected outcome of the Rutherford scattering experiment, if one assumes (a) the Thomson model of the atom, and (b) the model deduced by Rutherford. In the Thomson model, mass is spread throughout the atom, and the negative electrons are embedded in the positive mass in a uniform manner. There would be little deflection of the beam of positively charged α particles. In the Rutherford model, all the positive charge and virtually all the mass is concentrated in a very small nucleus. Most α particles pass through undeflected. But close approach to a nucleus will produce a strong swerve in the path of the particle, and head-on collision with the nucleus will lead to the rebound of the α particle in the direction from which it came.

observed was quite different, and wholly unexpected. In his own words:

> In the early days I had observed the scattering of α particles, and Dr. Geiger in my laboratory had examined it in detail. He found in thin pieces of heavy metal that the scattering was usually small, of the order of one degree. One day Geiger came to me and said, "Don't you think that young Marsden, whom I am training in radioactive methods, ought to begin a small research?" Now I had thought that too, so I said, "Why not let him see if any α particles can be scattered through a large angle?" I may tell you in confidence that I did not believe they would be, since we knew that the α particle was a very fast massive particle, with a great deal of energy, and you could show that if the scattering was due to the accumulated effect of a number of small scatterings, the chance of an α particle's being scattered backwards was very small. Then I remember two or three days later Geiger coming to me in great excitement and saying, "We have been able to get some of the α particles coming backwards." . . . . It was quite the most incredible event that has ever happened to me in my life. It was almost as incredible as if you fired a 15-inch shell at a piece of tissue paper and it came back and hit you.

Rutherford, Geiger, and Marsden calculated that this observed backscattering was precisely what would be expected if virtually all of the mass and positive charge of the atom were concentrated in a dense nucleus at the center of the atom [Figure 8–2(b)]. They also calculated the charge on the gold nucleus as 100 $\pm$20 (actually 79), and the radius of the nucleus as something less than 10^{-12} cm (actually nearer to 10^{-13} cm).

The picture of the atom that emerged from these scattering experiments was of an extremely dense, positively charged nucleus surrounded by negative charges—electrons. These electrons inhabited a region with a radius 100,000 times that of the nucleus. The majority of the α particles passing through the metal foil were not deflected because they never encountered the nucleus. However, particles passing close to such a great concentration of charge would be deflected; and those few particles that happened to collide with the small target would be bounced back in the direction from which they had come.

The validity of Rutherford's model has been borne out by later investigations. An atom's nucleus is comprised of protons and neutrons (Figure 8–3). Just enough electrons are around this nucleus to balance the nuclear charge. But this model of an atom is unexplainable by classical physics. What keeps the positive and negative charges apart? If the electrons were stationary, electrostatic attraction would pull them toward the nucleus to form a miniature version of Thomson's atom. Conversely, if the electrons were moving in orbits around the nucleus, things would be no better. An electron moving in a circle around a positive nucleus is an oscillating dipole when the atom is viewed in the plane of the orbit; the negative charge appears to oscillate up and down relative to the positive charge. By all the laws of classical electromagnetic theory, such an oscillator should broadcast energy as electromagnetic waves. But if this happened, the atom would lose energy, and the electron would spiral into the nucleus. By the laws of classical physics, the Rutherford model of the atom could not be valid. Where was the flaw?

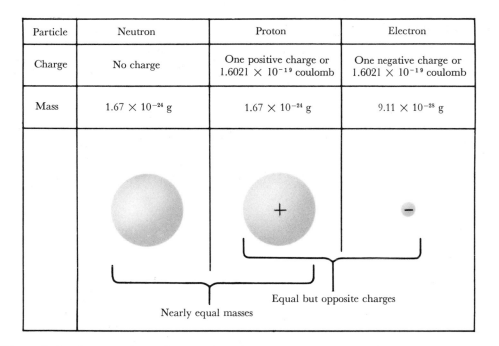

Particle	Neutron	Proton	Electron
Charge	No charge	One positive charge or 1.6021×10^{-19} coulomb	One negative charge or 1.6021×10^{-19} coulomb
Mass	1.67×10^{-24} g	1.67×10^{-24} g	9.11×10^{-28} g

Equal but opposite charges

Nearly equal masses

Figure 8–3. A comparison of properties of a neutron, proton, and an electron. The mass of a proton is 1836 times as great as that of an electron. However, the electrostatic force of attraction between two particles is independent of their masses and dependent only on their charges and how far they are apart.

8–2   THE QUANTIZATION OF ENERGY

Other flaws were appearing in physics at this time, which were equally as disturbing as Rutherford's impossibly stable atoms. By the turn of the century it was realized that radio waves, infrared, visible light, and ultraviolet radiation (and x rays and γ rays a few years later) were *electromagnetic waves* with different wavelengths. These waves all travel at the same speed, c, which is 2.9979×10^{10} cm sec^{-1} or 186,000 miles sec^{-1}. (This speed seems almost instantaneous until you recall that the slowness of light is responsible for the 1.3-sec delay each way in radio messages between the earth and the moon.) Waves such as these are described by their wavelength (λ), amplitude, and frequency (ν), which is the number of cycles of a moving wave that passes a given point per unit of time (Figure 8–4). The speed of the wave, c, which is constant for all these kinds of electromagnetic radiation, is the product of the frequency (the number of cycles per second) and the length of each cycle (the wavelength):

$$c = \nu\lambda \tag{8–1}$$

The reciprocal of the wavelength is called the wave number, $\bar{\nu}$:

$$\bar{\nu} = 1/\lambda$$

Its units are commonly waves per centimeter, or cm^{-1}.

The electromagnetic spectrum as we know it is shown in Figure 8–5(a). The scale is logarithmic rather than linear in wavelength; that is, it is in

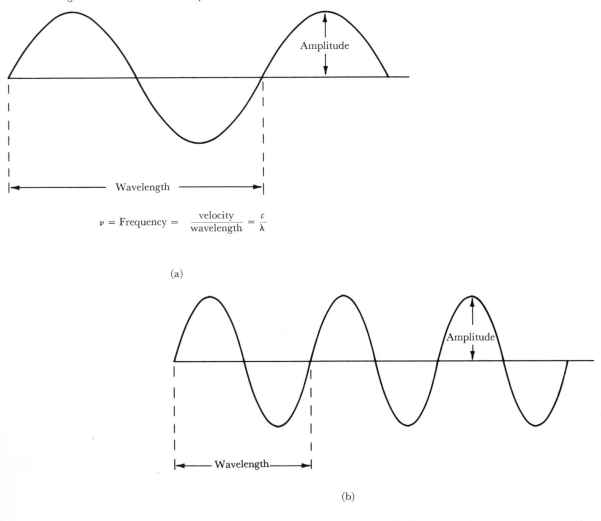

Light wave travels at velocity $c = 2.9979 \times 10^{10}$ cm sec^{-1}

Amplitude

Wavelength

$$\nu = \text{Frequency} = \frac{\text{velocity}}{\text{wavelength}} = \frac{c}{\lambda}$$

(a)

Amplitude

Wavelength

(b)

Figure 8–4. An electromagnetic wave such as x rays, light, microwaves, or radio waves. (a) The profile of a traveling wave at an instant of time that shows amplitude, wavelength (λ), velocity (c), and frequency (ν). The wave number, $\bar{\nu}$, measured in waves per centimeter, or cm^{-1}, is the reciprocal of the wavelength. (b) A wave of shorter wavelength and hence higher frequency since the product, $\lambda\nu$, is constant.

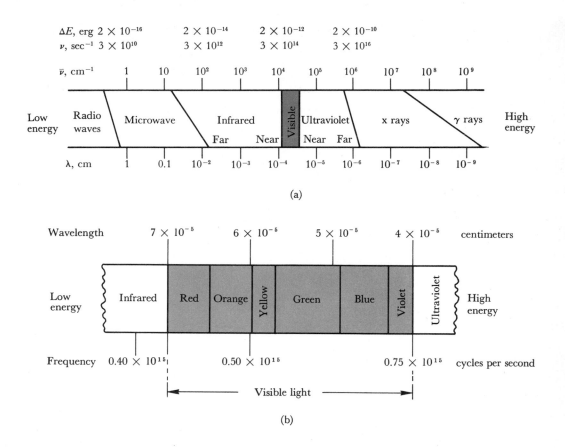

Figure 8–5. *The spectrum of electromagnetic radiation. The visible region is only a small part of the entire spectrum. (a) Overall spectrum. (b) Visible region.*

increasing powers of 10. On this logarithmic scale, the portion of the electromagnetic radiation that our eyes can see is only a small sector halfway between radio waves and gamma rays. The visible part of the spectrum is shown in Figure 8–5(b).

Challenge. You now should be able to solve problems involving wave motion and frequency in applications such as earth–Mars conversation, grand opera, and fireworks. Try Problems 8–4 to 8–6 in *Relevant Problems for Chemical Principles* by Butler and Grosser.

The Ultraviolet Catastrophe

Classical physics gave serious trouble even when used to try to explain why a red-hot iron bar is red. Solids emit radiation when heated. The ideal radiation from a perfect absorber and emitter of radiation is called *blackbody*

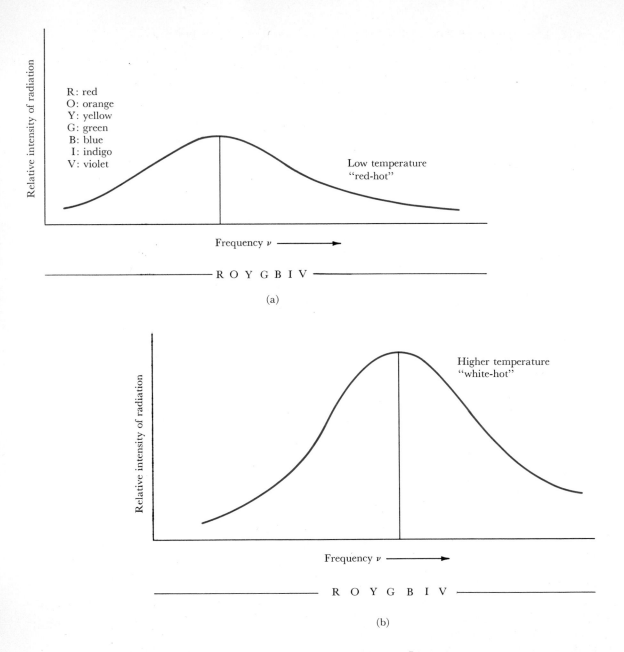

R: red
O: orange
Y: yellow
G: green
B: blue
I: indigo
V: violet

Low temperature
"red-hot"

Frequency ν ⟶

—— R O Y G B I V ——

(a)

Higher temperature
"white-hot"

Frequency ν ⟶

—— R O Y G B I V ——

(b)

Figure 8–6. The radiation from a hot object is referred to as blackbody radiation. (a) For a moderately hot object, most of the radiation is in the red region of the spectrum, and the object is said to be "red hot." (b) As the temperature is increased, the object glows orange, then yellow as the maximum of the radiation curve moves to higher frequencies, and finally white when the radiation occurs at all visible wavelengths in significant amounts. At even higher temperatures, there is less radiation in the red region, and the glow assumes a blueish tinge.

radiation. The *spectrum*, or plot of relative intensity against frequency, of radiation from a red-hot solid is shown in Figure 8–6(a). Since most of the radiation is in the red and infrared frequency regions, we see the color of the object as red. As the temperature is raised, the peak of the spectrum moves to higher frequencies, and we see the hot object as orange, then yellow, and finally white when enough energy is radiated through the entire visible spectrum.

The difficulty in this observation is that classical physics predicts that the distribution curve should keep rising to the right rather than falling after a maximum. Thus there should be much more blue and ultraviolet radiation emitted than actually is observed, and all heated objects should appear blue to our eyes. This complete contradiction of theory by facts was called the "ultraviolet catastrophe" by physicists of the time.

In 1900, Max Planck provided an answer for this paradox by discarding a hallowed tenet of science: Variables in nature change in a continuous way, or nature does not make jumps. According to classical theory, light of a certain frequency is emitted because charged objects in the solid vibrate with that frequency. The intensity curve of the spectrum is calculable if the relative number of tiny oscillators—atoms or groups of atoms—that oscillate with a given frequency is known. It is assumed that all frequencies are possible; the energy associated with a frequency depends only on how many oscillators are vibrating with that frequency. There should be no lack of high-frequency oscillators in the blue and ultraviolet regions.

Planck made the revolutionary suggestion that the energy of electromagnetic radiation comes in packages, or *quanta*. The energy of one package of radiation is proportional to the frequency of the radiation:

$$E = h\nu \tag{8–2}$$

The proportionality constant, h, is known as Planck's constant and has the value 6.6262×10^{-27} erg sec. By Planck's theory, a group of atoms cannot be emitting a *small* amount of energy at a *high* frequency; high frequencies can be emitted only by oscillators with a *large* amount of energy, as given by $E = h\nu$. The probability of finding oscillators with high frequencies is therefore slight because the probability of finding groups of atoms with such unusually large vibrational energies is low. Instead of rising in an ultraviolet catastrophe, the spectral curve falls at high frequencies, as in Figure 8–6.

Was Planck's theory correct, or was it only an *ad hoc* explanation to account for one isolated phenomenon? Science is plagued with theories that explain the phenomenon for which they were invented, and thereafter never explain another phenomenon correctly. Was the idea that electromagnetic energy comes in bundles of fixed energy that is proportional to frequency only another one-shot explanation?

The Photoelectric Effect

Albert Einstein (1879–1955) provided another example of the quantization of energy, in 1905, when he successfully explained the photoelectric effect.

Light striking a metal surface can cause electrons to be given off; this phenomenon is called the *photoelectric effect*. (Photocells in automatic doors use the photoelectric effect to generate the electrons that operate the door-opening circuits.) It was observed that, for a given metal, there is a minimum frequency of light below which no electrons are emitted, no matter how intense the beam of light. To classical physicists it seemed nonsensical that for some metals the most intense beam of red light could not drive off electrons that could be ejected by a faint beam of blue light.

Einstein showed that Planck's hypothesis explained these kinds of phenomena beautifully. The energy of the quanta of light striking the metal, he said, is greater for blue light than for red. As an analogy, imagine that the low-frequency red light is a beam of ping pong balls and that the high-frequency blue light is a beam of steel balls with the same velocity. Each impact of a quantum of energy of red light is too small to dislodge an electron; and in our analogy, a steady stream of ping pong balls cannot do what one rapidly moving steel ball can. These quanta of light were named *photons*. Because of the successful explanation of both the blackbody and photoelectric effects, physicists began recognizing that light behaves like particles as well as like waves.

Challenge. You may want to apply these ideas to burglar alarms and TV cameras; try Problems 8–10 and 8–11 in the Butler and Grosser book.

The Spectrum of the Hydrogen Atom

The most striking example of the quantization of light, to a chemist, appears in the search for an explanation of atomic spectra. Isaac Newton (1642–1727) was one of the first men to demonstrate with a prism that white light is a spectrum of many colors, from red at one end to violet at the other. We know now that the electromagnetic spectrum continues on both sides of the small region to which our eyes are sensitive; it includes the infrared at low frequencies and the ultraviolet at high ones.

All atoms and molecules absorb light of certain characteristic frequencies. The pattern of absorption frequencies is called an *absorption spectrum* and is an identifying property of any particular atom or molecule. The absorption spectrum of hydrogen atoms is shown in Figure 8–7. The lowest-energy absorption corresponds to the line at 82,259 cm^{-1}. Notice that the absorption lines are crowded closer together as the limit of 109,678 cm^{-1} is approached. Above this limit absorption is continuous.

If atoms and molecules are heated to high temperatures, light of certain frequencies is emitted. For example, hydrogen atoms emit red light when heated. An atom that possesses excess energy (e.g., an atom that has been heated) emits light in a pattern known as its *emission spectrum*. A portion of the emission spectrum of atomic hydrogen is shown in Figure 8–8. Note that the lines occur at the same wave numbers in the two types of spectra.

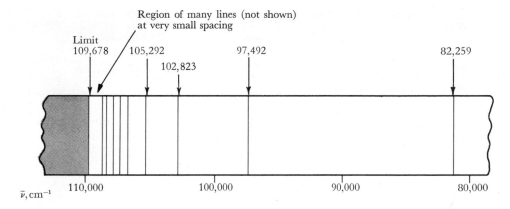

Figure 8-7. The electromagnetic absorption spectrum of hydrogen atoms in the ultraviolet region. The scale is in wave numbers (cm^{-1}). The lines in this spectrum represent ultraviolet radiation that is absorbed by hydrogen atoms as a mixture of all wavelengths is passed through a gas sample.

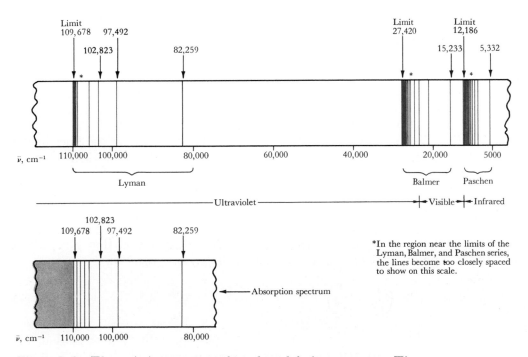

*In the region near the limits of the Lyman, Balmer, and Paschen series, the lines become too closely spaced to show on this scale.

Figure 8-8. The emission spectrum from heated hydrogen atoms. The emission lines occur in series, named for their discoverers: Lyman, Balmer, Paschen; the Brackett and Pfund series are farther to the right in the infrared region. The lines become more closely spaced to the left in each series, until they finally merge at the series limit.

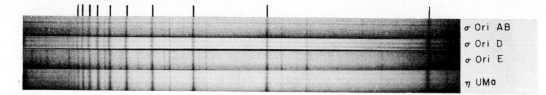

Figure 8–9. Balmer hydrogen-atom spectra from several stars. The three σ Ori spectra are from a group of stars, σ Orionis, located just below the belt in the constellation Orion. η UMa is η Ursa Majoris, the end of the handle in the Big Dipper. Note the universality of the atomic hydrogen spectrum. Hydrogen is hydrogen, no matter where you find it. Balmer lines are marked above the spectra. The other lines are primarily from helium. (Courtesy John Oke, California Institute of Technology.)

If we look more closely at the emission spectrum in Figure 8–8, we see that there are three distinct groups of lines. These three groups or series are named after the scientists who discovered them. The series that starts at 82,259 cm^{-1} and continues to 109,678 cm^{-1} is called the *Lyman series* and is in the ultraviolet portion of the spectrum. The series that starts at 15,233 cm^{-1} and continues to 27,420 cm^{-1} is called the *Balmer series* and covers a large portion of the visible and a small part of the ultraviolet spectrum. The lines between 5,332 cm^{-1} and 12,186 cm^{-1} are called the *Paschen series* and fall in the near-infrared region. The Balmer spectra of hydrogen from several stars are shown in Figure 8–9.

J. J. Balmer proved, in 1885, that the wave numbers of the lines in the Balmer spectrum of the hydrogen atom are given by the empirical relationship

$$\bar{\nu} = R_{\mathrm{H}} \times \left(\frac{1}{4} - \frac{1}{n^2} \right) \qquad n = 3, 4, 5, \ldots \tag{8-3}$$

Later, Johannes Rydberg formulated a general expression that gives all of the line positions. This expression, called the *Rydberg equation*, is

$$\bar{\nu} = R_{\mathrm{H}} \times \left(\frac{1}{n_1^2} - \frac{1}{n_2^2} \right) \tag{8-4}$$

In the Rydberg equation n_1 and n_2 are integers, with n_2 greater than n_1; R_{H} is called the *Rydberg constant* and is known accurately from experiment to be 109,677.581 cm^{-1}.

Example.   Calculate $\bar{\nu}$ for the lines with $n_1 = 1$ and $n_2 = 2$, 3, and 4.

Solution

$n_1 = 1$, $n_2 = 2$ line:

$$\bar{\nu} = 109{,}678 \left(\frac{1}{1^2} - \frac{1}{2^2}\right) = 109{,}678 \left(1 - \frac{1}{4}\right) = 82{,}259 \text{ cm}^{-1}$$

$n_1 = 1$, $n_2 = 3$ line:

$$\bar{\nu} = 109{,}678 \left(\frac{1}{1^2} - \frac{1}{3^2}\right) = 109{,}678 \left(1 - \frac{1}{9}\right) = 97{,}492 \text{ cm}^{-1}$$

$n_1 = 1$, $n_2 = 4$ line:

$$\bar{\nu} = 109{,}678 \left(\frac{1}{1^2} - \frac{1}{4^2}\right) = 109{,}678 \left(1 - \frac{1}{16}\right) = 102{,}823 \text{ cm}^{-1}$$

We see that the preceding wave numbers correspond to the first three lines in the Lyman series. Thus we expect that the Lyman series corresponds to lines calculated with $n_1 = 1$ and $n_2 = 2, 3, 4, 5, \ldots$. Let us check this by calculating the wave number for the line with $n_1 = 1$ and $n_2 = \infty$.

$n_1 = 1$, $n_2 = \infty$ line:
$$\bar{\nu} = 109{,}678(1 - 0) = 109{,}678 \text{ cm}^{-1}$$

The wave number $109{,}678 \text{ cm}^{-1}$ corresponds to the highest emission line in the Lyman series.

The wave number for $n_1 = 2$ and $n_2 = 3$ is

$$\bar{\nu} = 109{,}678 \left(\frac{1}{4} - \frac{1}{9}\right) = 15{,}233 \text{ cm}^{-1}$$

This corresponds to the first line in the Balmer series. Thus, the Balmer series corresponds to the $n_1 = 2$, $n_2 = 3, 4, 5, 6 \ldots$ lines. You probably would expect the lines in the Paschen series to correspond to $n_1 = 3$, $n_2 = 4, 5, 6, 7 \ldots$; and they do. Now you should wonder where the lines are with $n_1 = 4$, $n_2 = 5, 6, 7, 8, \ldots$, and $n_1 = 5$, $n_2 = 6, 7, 8, 9 \ldots$. They are exactly where the Rydberg equation predicts they should be. The $n = 4$ series was discovered by Brackett and the $n = 5$ series was discovered by Pfund. The series with $n = 6$ and higher are located at very low frequencies and are not given special names.

The Rydberg formula, Equation 8–4, is a summary of observed facts about hydrogen atomic spectra. It states that the wave number of a spectral line is the difference between two numbers, each inversely proportional to the square of an integer. If we draw a set of horizontal lines at a distance R_H/n^2 down from a baseline, with $n = 1, 2, 3, 4, \ldots$, then each spectral line in any of the hydrogen series is observed to correspond to the distance between two such horizontal lines in the diagram (Figure 8–10). The Lyman series occurs between line $n = 1$ and those above it; the Balmer series occurs

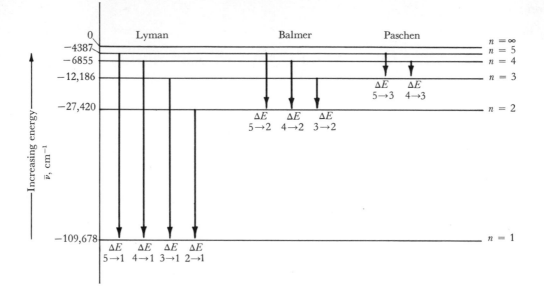

Figure 8–10. An energy-level diagram that accounts for the observed hydrogen spectrum. This diagram can be considered as a graphic representation of the Rydberg equation: $\bar{\nu} = R_{\mathrm{H}}[(1/n_1^2) - (1/n_2^2)]$. However, Bohr attributed more meaning to it. He proposed that these levels represent the only possible energy states of a hydrogen atom: $E \propto 1/n^2$. He also postulated that an atom generates a spectral line when it goes from one energy state to another of lower energy, and that the wave number of the emitted line is determined by the change in energy: $\Delta E = hc\bar{\nu}$. Only lines with $n = 1, 2, 3, 4,$ and 5 and the limit at $n = \infty$ are shown.

between line $n = 2$ and those above it; the Paschen series occurs between line $n = 3$ and those above it; and the higher series are based on lines $n = 4$, 5, and so on. Is the agreement between this simple diagram and the observed wave numbers of spectral lines only a coincidence? Does the idea of a wave number of an emitted line being the difference between two "wave-number levels" have any physical significance, or is this just a convenient graphical representation of the Rydberg equation?

8–3   BOHR'S THEORY OF THE HYDROGEN ATOM

In 1913, Niels Bohr (1885–1962) proposed a theory of the hydrogen atom that, in one blow, did away with the problem of Rutherford's unstable atom and gave a perfect explanation of the spectra we have just discussed.

There are two ways of proposing a new theory in science, and Bohr's work illustrates the less obvious one. One way is to amass such an amount of data that the new theory becomes obvious and self-evident to any observer.

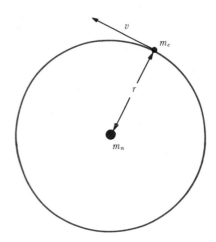

Figure 8–11. Bohr's picture of the hydrogen atom. A single electron of mass m_e moves in a circular orbit with velocity v at a distance r from a nucleus of mass m_n. To explain the spectrum of Figure 8–8, or the Rydberg equation diagram of Figure 8–10, Bohr had to postulate that the angular momentum of the electron, $m_e v r$, was restricted to integral multiples of the quantity $h/2\pi$. The integers are the numbers n of Figure 8–10.

The theory then is almost a summary of the behavior of the data. This is essentially the way Dalton reasoned from combining weights to atoms. The other way is to make a bold new assertion that initially does not seem to follow from the data, and then to demonstrate that the consequences of this assertion, when worked out, explain many observations. With this method, a theorist says, "You may not see why, yet, but please suspend judgment on my hypothesis until I show you what I can do with it." Bohr's theory is more of this type.

Bohr answered the question of why the electron does not spiral into the nucleus by simply postulating that *it does not*. In effect, he said to classical physicists: "You have been misled by your physics to expect that the electron would radiate energy and spiral into the nucleus. Let us assume that it does not, and see if we can account for more observations than by assuming that it does." The observations that he explained so well are the wavelengths of lines in the atomic spectrum of hydrogen.

Bohr's model of the hydrogen atom is illustrated in Figure 8–11: an electron of mass m_e moving in a circular orbit at a distance r from a nucleus. If the electron has a velocity of v, it will have an *angular momentum of $m_e v r$.* (You can appreciate what angular momentum means if you think of an ice skater spinning on one blade like a top. The skater begins spinning with his arms extended. As he brings his arms to his sides, he spins faster and faster. This is because, in the absence of any external forces, angular momentum is conserved. As the mass of the skater's arms comes closer to the axis of rotation, or as r decreases, the velocity of his arms must increase in order that the product mvr remain constant.) Bohr postulated as the first basic assumption of his theory that, in a hydrogen atom, there only could be orbits for which *the angular momentum is an integral multiple of Planck's constant divided by 2π:*

$$m_e v r = n\left(\frac{h}{2\pi}\right)$$

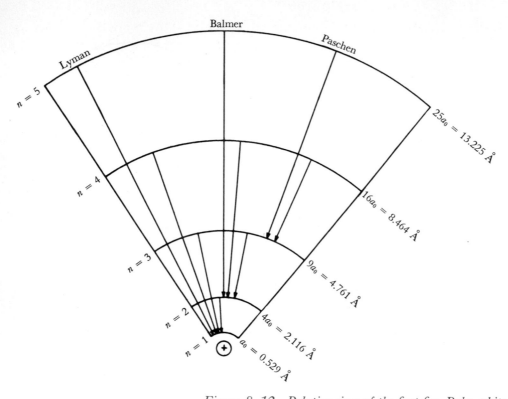

Figure 8–12. Relative sizes of the first five Bohr orbits for atomic hydrogen. The transitions from one orbit to a lower one are indicated as in Figure 8–10. Each arc of a circle represents part of a circular orbit for an electron around the positive nucleus at the bottom of the diagram. The radius of the nth orbit is calculated from $r = n^2 a_0$, in which a_0 is the first Bohr radius; $a_0 = (h^2/4\pi^2 m_e e^2) = 0.529$ Å.

There is no obvious justification for such an assumption; it will be accepted only if it leads to the successful explanation of other phenomena. Bohr then showed that, with no more new assumptions, and with the laws of classical mechanics and electrostatics, his principle leads to the restriction of the energy of an electron in a hydrogen atom to the values

$$E = -\frac{k}{n^2} \qquad n = 1, 2, 3, 4, \ldots \tag{8–5}$$

The integer n is the same integer as in the angular momentum assumption, $m_e v r = n(h/2\pi)$; k is a constant that depends only on Planck's constant, h, the mass of an electron, m_e, and the charge on an electron, e:

$$k = \frac{2\pi^2 m_e e^4}{h^2} = 13.595 \text{ electron volts (eV) atom}^{-1}$$
$$= 313.5 \text{ kcal mole}^{-1}$$

The radius of the electron's orbit also is determined by the integer n:

$$r = n^2 a_0 \tag{8-6}$$

The constant, a_0, is called the *first Bohr radius* and is given in Bohr's theory by

$$a_0 = \frac{h^2}{4\pi^2 m_e e^2} = 0.529 \text{ Å}$$

The first Bohr radius often is used as a measure of length called the *atomic unit*, a.u.

The energy that an electron in a hydrogen atom can have is *quantized*, or limited to certain values, by Equation 8–5. The integer, n, that determines these energy values is called the *quantum number*. An electron that is removed completely (dissociated) from an atom is described as exciting the electron to the quantum state $n = \infty$. From Equation 8–5, we see that as n approaches ∞, E approaches zero. Thus, the energy of a completely dissociated electron has been chosen as the zero energy level. Because energy is required to remove an electron from an atom, an electron that is bound to an atom must have less energy than this, and hence a negative energy. The relative sizes of the first five hydrogen-atom orbits are compared in Figure 8–12.

Exercise. For a hydrogen atom, what is the energy, relative to the dissociated atom, of the *ground state*, for which $n = 1$? How far is the electron from the nucleus in this state? What are the energy and radius of orbit of an electron in the *first excited state*, for which $n = 2$?

(*Answer:*

$$E_1 = -\frac{k}{1^2} = -313.5 \text{ kcal mole}^{-1}$$

$$E_2 = -\frac{k}{2^2} = -78.4 \text{ kcal mole}^{-1}$$

$$r_1 = 1^2 \times 0.529 \text{ Å} = 0.529 \text{ Å}$$

$$r_2 = 2^2 \times 0.529 \text{ Å} = 2.116 \text{ Å})$$

Example. Using the Bohr theory, calculate the ionization energy of the hydrogen atom.

Solution. The ionization energy is that energy required to remove the electron, or to go from quantum state $n = 1$ to $n = \infty$. This energy is

$$\Delta E = E_\infty - E_1 = 0.00 - (-313.5 \text{ kcal mole}^{-1})$$
$$= +313.5 \text{ kcal mole}^{-1}$$

Exercise. Diagram the energies available to the hydrogen atom as a series of horizontal lines. Plot the energies in units of k for simplicity. Include at least the first eight quantum levels and the ionization limit. Compare your result with Figures 8–10 and 8–13.

In the second part of his theory, Bohr postulated that absorption and emission of energy occur when an electron moves from one quantum state to another. The energy emitted when an electron drops from state n_2 to a lower quantum state n_1 is the difference between energies of the two states:

$$\Delta E = E_1 - E_2 = -k \left(\frac{1}{n_1^2} - \frac{1}{n_2^2} \right) \tag{8-7}$$

The light emitted is assumed to be quantized in exactly the way predicted from the blackbody or photoelectric experiments:

$$|\Delta E| = h\nu = hc\bar{\nu} \tag{8-8}$$

If we divide Equation 8–7 by hc to convert from energy to wave number units, we obtain the Rydberg equation,

$$\bar{\nu} = \frac{k}{hc} \left(\frac{1}{n_1^2} - \frac{1}{n_2^2} \right) \tag{8-9}$$

With the Bohr theory, we now can calculate the Rydberg constant from first principles:

$$R_H = \frac{k}{hc} = \frac{2\pi^2 m_e e^4}{h^3 c} = 109{,}737.3 \text{ cm}^{-1}$$

Recall that the experimental value of R_H is $109{,}677.581 \text{ cm}^{-1}$.

The graphic representation of the Rydberg equation, Figure 8–10, now is seen to be an energy-level diagram of the possible quantum states of the hydrogen atom. We can see why light is absorbed or emitted only at specific wave numbers. The absorption of light, or the heating of a gas, provides the energy for an electron to move to a higher orbit. Then the *excited* hydrogen atom can emit energy in the form of light quanta when the electron falls back to a lower-energy orbit. From this emission come the different series of spectral lines:

1) The Lyman series of lines arises from transitions from the $n = 2, 3, 4, \ldots$ levels to the ground state $(n = 1)$.

2) The Balmer series arises from transitions from the $n = 3, 4, 5, \ldots$ levels to the $n = 2$ level.

3) The Paschen series arises from transitions from the $n = 4, 5, 6, \ldots$ levels to the $n = 3$ level.

An excited hydrogen atom in quantum state $n = 8$ may drop directly to the ground state and emit a photon in the Lyman series; or it may drop first to $n = 3$, emit a photon in the Paschen series, and then drop to $n = 1$ and emit a photon in the Lyman series. The frequency of each photon depends on the energy difference between levels:

$$\Delta E = E_a - E_b = h\nu$$

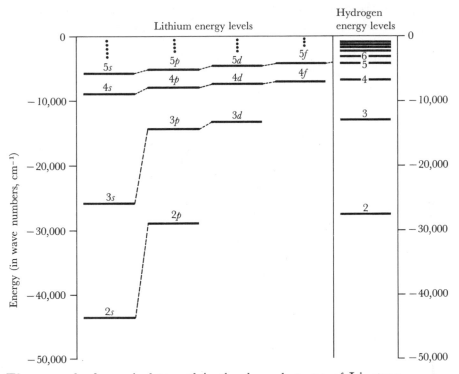

Figure 8–13. The energy levels required to explain the observed spectra of Li *atoms compared with the hydrogen levels at the right. Levels* n = 1 *are off scale at the bottom. For quantum number n, there are n levels, traditionally identified by the letters s, p, d, f, g, ... The levels farthest to the right for each quantum number (2p, 3d, 4f, . . .) approach the corresponding hydrogen levels, whereas all other levels of the same quantum number are more stable. Sommerfeld accounted for this stability by postulating elliptical orbits, in which the s orbits were most elliptical and the 2p, 3d, 4f, . . . orbits were nearly circular (Figure 8–14).*

By cascading down the energy levels, the electron in one excited hydrogen atom can successively emit photons in several series. Therefore, all series are present in the emission spectra from hot hydrogen. However, when measuring the absorption spectrum of hydrogen gas at lower temperatures we find virtually all of the hydrogen atoms in the ground state. Therefore, almost all of the absorption will involve transitions from $n = 1$ to higher states, and only the Lyman series will be observed.

The Need for a Better Theory

The Bohr theory of the hydrogen atom suffered from the fatal weakness that it explained nothing *except* the hydrogen atom and any other combination of

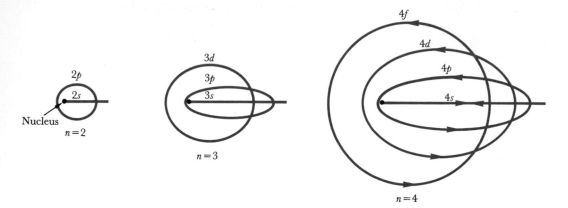

Figure 8–14. The Sommerfeld orbits for hydrogen. For a point nucleus, all orbits of the same principal quantum number, n, will have the same energy. For a nucleus surrounded by a shielding cloud of electrons, the more elliptical orbits that penetrate this cloud will experience (for part of their path) a stronger attraction from the nucleus and will be more stable. Hence, the 4p level, for example, is lower in energy than the 4f (Figure 8–13).

a nucleus and one electron. For example, it could account for the spectra of He^+ and Li^{2+}, but it did not provide a general explanation for atomic spectra. Even the alkali metals, which have a single valence electron outside a closed shell of inner electrons, produce spectra that are at variance with the Bohr theory. The lines observed in the spectrum of Li could be accounted for only by assuming that each of the Bohr levels beyond the first was really a collection of levels of different energies as in Figure 8–13: two levels for $n = 2$, three levels for $n = 3$, four for $n = 4$, and so on. The levels for a specific n were given letter symbols based on the appearance of the spectra involving these levels: s for "sharp," p for "principal," d for "diffuse," and f for "fundamental."

Arnold Sommerfeld (1868–1951) proposed an ingenious way of saving the Bohr theory. He suggested that orbits might be elliptical as well as circular. Further, he explained the differences in stability of levels with the same principal quantum number, n, in terms of the ability of the highly elliptical orbits to bring the electron closer to the nucleus (Figure 8–14). For a point nucleus of charge $+1$ in hydrogen, the energies of all levels with the same n were identical. But for a nucleus of $+3$ screened by an inner shell of two electrons in Li, an electron in an outer circular orbit would experience a net attraction of $+1$, whereas one in a highly elliptical orbit would penetrate the screening shell and feel a charge approaching $+3$ for part of its traverse. Thus, the highly elliptical orbits would have the additional stability illustrated in Figure 8–13. The s orbits, being the most elliptical of all in Sommerfeld's model, would be much more stable than the others in the set of common n.

The Sommerfeld scheme led no further than the alkali metals: Li, Na, K, Rb, Cs. Again an impasse was reached, and an entirely fresh approach was needed.

8–4   PARTICLES OF LIGHT AND WAVES OF MATTER

At the beginning of this century, scientists generally believed that all physical phenomena could be divided into two distinct and exclusive classes. The first class included all phenomena that could be described by laws of classical, or Newtonian, mechanics of motion of discrete particles. The second class included all phenomena showing the continuous properties of waves.

One outstanding property of matter, apparent since the time of Dalton, is that it is built of discrete particles. Most material substances appear to be continuous: water, mercury, salt crystals, gases. But if our eyes could see the nuclei and electrons that constitute atoms, and the fundamental particles that make up nuclei, we would discover quickly that every material substance in the universe is composed of a certain number of these basic units and therefore is quantized. *Objects appear continuous only because of the minuteness of the individual units.*

In contrast, light was considered to be a collection of waves traveling through space at a constant speed; any combination of energies and frequencies was possible. However, Planck, Einstein, and Bohr showed that, when observed under the right conditions, light also behaves as though it occurs in particles, or quanta.

In 1924, the French physicist Louis de Broglie (1892–   ) advanced the complementary hypothesis that all matter possesses wave properties. De Broglie pondered the Bohr atom, and asked himself where, in nature, does quantization of energy occur most naturally. An obvious answer is in the vibration of a string with fixed ends. A violin string can vibrate with only a selected set of frequencies: a fundamental tone with the entire string vibrating as a unit, and overtones of shorter wavelengths. A wavelength in which the vibration fails to come to a *node* (a place of zero amplitude) at both ends of the string would be an impossible mode of vibration (Figure 8–15). The vibration of a string with fixed ends is quantized by the *boundary conditions* that the ends cannot move.

Can the idea of standing waves be carried over to the theory of the Bohr atom? Standing waves in a circular orbit can exist only if the circumference of the orbit is an integral number of wavelengths [Figure 8–15(c), (d)]. If it is not, waves from successive turns around the orbit will be out of phase and will cancel. The value of the wave amplitude at 10° around the orbit from a chosen point will not be the same as at 370° or 730°, yet all of these represent the same point in the orbit. Such ill-behaved waves are not *single-*

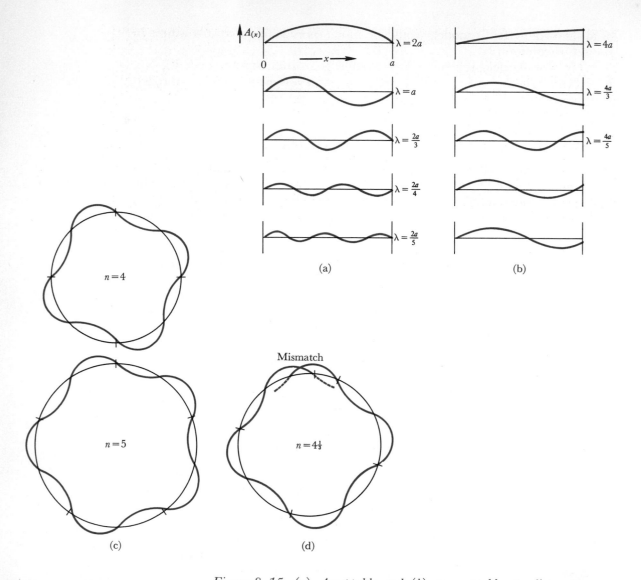

Figure 8–15. (a) *Acceptable and* (b) *unacceptable standing waves or modes of vibration in a violin string, as determined by the boundary conditions that the ends of the string be motionless. The boundary conditions limit the possible wavelengths of vibration to* $\lambda = 2a/n$, *in which a is the length of the string and n is an integer* $n = 1, 2, 3, \ldots$. (c) *Acceptable and* (d) *unacceptable electron waves in a Bohr orbit. The boundary conditions for a standing wave in the circular orbit are that the circumference be an integral number of wavelengths:* $2\pi r = n\lambda$. *This requirement and Bohr's angular momentum postulate,* $mvr = n(h/2\pi)$, *lead directly to de Broglie's relationship between the mass of a particle, its velocity, and its wavelength:* $\lambda = h/mv$.

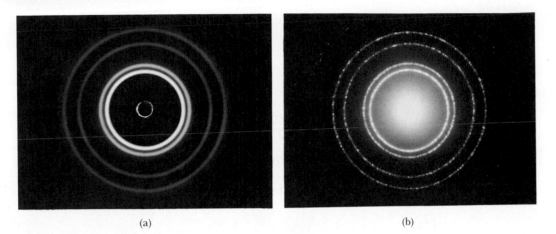

(a)   (b)

Figure 8–16. Diffraction of waves by aluminum foil. (a) X rays of wavelength 0.71 Å. (b) Electrons of energy 600 eV, or wavelength 0.50 Å. The similarity of these two patterns is strong evidence for the wave properties of particles. (Courtesy Film Studio, Education Development Center.)

valued at any point on the orbit: Single-valuedness is a boundary condition on acceptable waves.

For single-valued standing waves around the orbit, the circumference is an integer, n, times the wavelength:

$$2\pi r = n\lambda$$

But from Bohr's original assumption about angular momentum,

$$2\pi r = n\left(\frac{h}{mv}\right)$$

Therefore, the idea of standing waves leads to the following relationship between the mass of the electron, m, its velocity, v, and its wavelength, λ:

$$\lambda = \frac{h}{mv} \tag{8–10}$$

De Broglie proposed this relationship as a general one. With every particle, he said, there is associated a wave. The wavelength depends on the mass of the particle and how fast it is moving. If this is so, the same sort of diffraction from crystals that von Laue observed with x rays should be produced with electrons.

In 1927, C. Davisson and L. H. Germer demonstrated that metal foils diffract a beam of electrons in exactly the same way as they diffract an x-ray beam, and that the wavelength of a beam of electrons is given correctly by de Broglie's relationship (Figure 8–16). Electron diffraction now is a standard technique for determining molecular structure.

Example. A typical electron diffraction experiment is conducted with electrons accelerated through a potential drop of 40,000 volts, or with 40,000 eV of energy. What is the wavelength of the electrons?

Solution. First convert the energy, E, from electron volts to ergs:

$$E = 40,000 \text{ eV} \times \frac{1.6022 \times 10^{-12} \text{ erg}}{1 \text{ eV}} = 6.408 \times 10^{-8} \text{ erg}$$

(This and several other useful conversion factors, plus a table of the values of frequently used physical constants, are in Appendix 1.) Since the energy is $E = \frac{1}{2}mv^2$, the velocity of the electrons is

$$v = \left(\frac{2E}{m}\right)^{1/2} = \left(\frac{2 \times 6.408 \times 10^{-8} \text{ g cm}^2 \text{ sec}^{-2}}{9.109 \times 10^{-28} \text{ g}}\right)^{1/2}$$

$$v = (1.410 \times 10^{20} \text{ cm}^2 \text{ sec}^{-2})^{1/2} = 1.186 \times 10^{10} \text{ cm sec}^{-1}$$

(In the expression $E = \frac{1}{2}mv^2$, if the mass is in grams and the velocity is in cm sec^{-1}, then the energy is in ergs: 1 erg equals 1 g cm^2 sec^{-2} of energy. We used this conversion of units in the preceding step. The mass of the electron, $m = 9.109 \times 10^{-28}$ g, is found in Appendix 1.) The momentum of the electron, mv, is

$$mv = 9.109 \times 10^{-28} \text{ g} \times 1.186 \times 10^{10} \text{ cm sec}^{-1}$$

$$mv = 10.81 \times 10^{-18} \text{ g cm sec}^{-1}$$

Finally, the wavelength of the electron is obtained from the de Broglie relationship:

$$\lambda = \frac{h}{mv} = \frac{6.6262 \times 10^{-27} \text{ erg sec}}{10.81 \times 10^{-18} \text{ g cm sec}^{-1}}$$

$$\lambda = 0.0613 \times 10^{-8} \frac{\text{g cm}^2 \text{ sec}^{-2} \text{ sec}}{\text{g cm sec}^{-1}} = 0.0613 \times 10^{-8} \text{ cm}$$

$$\lambda = 0.0613 \text{ Å}$$

So 40-kilovolt (kV) electrons produce the diffraction effects expected from waves with a wavelength of six hundredths of an angstrom.

Such calculations are all very well, but the question remains: Are electrons waves or are they particles? Are light rays waves or particles? Scientists worried about these questions for years, until they gradually realized that they were arguing about language and not about science. Most things in our everyday experience behave either as what we would call "waves" or as what we would call "particles," and we have created idealized categories and used the words "wave" and "particle" to identify them. The behavior of matter as small as electrons cannot be described accurately by these large-scale categories. Electrons, protons, neutrons, and photons are not waves, and they are not particles. They sometimes act as if they were what we commonly call waves, and in other circumstances, as if they were what we call particles. But to demand, "Is an electron a wave or a particle?" is as pointless as asking a Swiss friend, "Are you a Republican or a Democrat?" The categories do not apply.

This wave–particle duality is present in all objects; it is only because of the scale of certain objects that one behavior predominates and the other is neglected. For example, a thrown baseball has wave properties, but with a wavelength so short that we cannot detect it.

Exercise. A 200-g baseball is thrown with a speed of 3.0×10^3 cm sec^{-1}. Calculate its de Broglie wavelength.

(*Answer:* $\lambda = 1.10 \times 10^{-32}$ cm $= 1.10 \times 10^{-24}$ Å.)

Example. How fast (or rather, how slowly) must a 200-g baseball travel to have the same de Broglie wavelength as a 40-kV electron?

Solution. The wavelength of a 40-kV electron is 0.0613 Å.

$$v = \frac{h}{\lambda m} = \frac{6.6262 \times 10^{-27} \text{ g cm}^2 \text{ sec}^{-2} \text{ sec}}{0.0613 \times 10^{-8} \text{ cm} \times 200 \text{ g}}$$

$$v = 0.540 \times 10^{-19} \text{ cm sec}^{-1} = 1.70 \times 10^{-4} \text{ Å year}^{-1}$$

Such a baseball would take over ten thousand years to travel the length of a carbon–carbon bond, 1.54 Å. This sort of motion is completely outside our experience with baseballs; thus we never regard baseballs as having wave properties.

8-5 THE UNCERTAINTY PRINCIPLE

One of the most important consequences of the dual nature of matter is the uncertainty principle, proposed in 1927 by Werner Heisenberg (1901–). This principle states that you cannot know simultaneously both the position and the momentum of any particle with absolute accuracy. The product of the uncertainty in position, Δx, and momentum, $\Delta(mv)$, will be equal to or greater than Planck's constant divided by 4π:

$$[\Delta x][\Delta(mv_x)] \geq \frac{h}{4\pi} \tag{8-11}$$

We can understand this principle by considering how we determine the position of a particle. If the particle is large, we can touch it without disturbing it seriously. If the particle is small, a more delicate means of locating it is to shine a beam of light on it and observe the scattered rays. Yet light acts as if it were made of particles—photons—with energy proportional to frequency: $E = h\nu$. When we illuminate the object, we are pouring energy onto it. If the object is large, it will become warmer; if the object is small

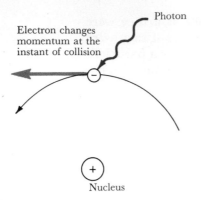

Electron changes
momentum at the
instant of collision

Photon

Nucleus

Figure 8–17. The position of an electron e^- at an instant of time should be determinable by a "super microscope" using light of small wavelength, λ (x rays or γ rays). However, photons of light of small λ have great energy and therefore very large momentum. A collision of one of these photons with an electron instantly changes the electronic momentum. Thus, as the position is better resolved, the momentum becomes more and more uncertain.

enough, it will be pushed away and its momentum will become uncertain. The least interference that we can cause is to bounce a single photon off the object and watch where the photon goes. But we now are caught in a dilemma. The detail in an image of an object depends on the fineness of the wavelength of the light used to observe the object. To make the photon of sufficiently low energy that the momentum of an atom is unaltered, we must choose such a long wavelength that the position of the atom is unclear. Conversely, if we try to locate the atom accurately by using a short-wavelength photon, the energy of the photon sends the atom ricocheting away with an uncertain momentum (Figure 8–17). We can design an experiment to obtain an accurate value of either an atom's momentum or its position, but the product of the errors in these quantities is limited by Equation 8–11.

Example. Suppose that we want to locate an electron whose velocity is 10^8 cm sec^{-1} by using a beam of green light whose frequency is $\nu = 0.60 \times 10^{15}$ sec^{-1}. How does the energy of one photon of such light compare with the energy of the electron to be located?

Solution. The energy of the electron is

$$E = \tfrac{1}{2}mv^2 = \frac{9.109 \times 10^{-28}\,\text{g} \times 10^{16}\,\text{cm}^2\,\text{sec}^{-2}}{2}$$

$$E = 4.55 \times 10^{-12}\,\text{erg}$$

But the energy of the photon is almost as large:

$$E_p = h\nu = 6.6262 \times 10^{-27}\,\text{erg sec} \times 0.60 \times 10^{15}\,\text{sec}^{-1}$$

$$E_p = 3.97 \times 10^{-12}\,\text{erg}$$

Finding the position and momentum of such an electron with green light is as questionable a procedure as finding the position and momentum of one billiard ball by striking it with another. In either case, you detect the particle at the price of disturbing its momentum. As a final difficulty, green light is a hopelessly coarse yardstick for finding objects of atomic dimensions. An atom is about an angstrom in radius, whereas the wavelength of green light is around 5000 Å. Shorter wavelengths make the energy quandary worse.

We do not see the uncertainty limitations in large objects because of the sizes of the masses and velocities involved. Compare the following two problems.

Example. An electron is moving with a velocity of 10^8 cm sec^{-1}. Assume that we can measure its position to 0.01 Å, or 1% of a typical atomic radius. Compare the uncertainty in its momentum, p, with the momentum of the electron itself.

Solution. The uncertainty in position is $\Delta x \simeq 0.01$ Å $= 0.01 \times 10^{-8}$ cm. The momentum of the electron is approximately

$$p = mv \simeq 10^{-27} \text{ g} \times 10^8 \text{ cm sec}^{-1} = 10^{-19} \text{ g cm sec}^{-1}$$

By the Heisenberg uncertainty principle, the uncertainty in the knowledge of the momentum is

$$\Delta p = \frac{h/4\pi}{\Delta x} \simeq \frac{0.5 \times 10^{-27} \text{ g cm}^2 \text{ sec}^{-1}}{0.01 \times 10^{-8} \text{ cm}}$$

$$\simeq 0.5 \times 10^{-17} \text{ g cm sec}^{-1}$$

The *uncertainty* in the momentum of the electron is 50 times as great as the momentum itself!

Exercise. A baseball of mass 200 g is moving with a velocity of 3.0×10^3 cm sec^{-1}. If we can locate the baseball with an error equal in magnitude to the wavelength of light used (e.g., 5000 Å), how will the uncertainty in momentum compare with the total momentum of the baseball?

(*Answer:* $p = 6 \times 10^5$ g cm sec^{-1}, and $\Delta p = 1 \times 10^{-23}$ g cm sec^{-1}. The intrinsic uncertainty in the momentum is only one part in 10^{28}, far below any possibility of detection in an experiment.)

8–6 WAVE EQUATIONS

The two usual approaches to the Schrödinger wave equation in introductory courses are either to pick its solutions out of thin air with little explanation of where they came from, or else to work through the difficult mathematics of solving the differential equations in detail.[1] We shall try to take a middle

---

[1] Equations of motion are always *differential equations* because they relate the change in one quantity to the change in another, such as the change in position with change in time.

road. The mathematics of the Schrödinger equation is beyond us, but the mode of attack, or the strategy of finding its solution, is not. If you can see how physicists go about solving the Schrödinger equation, even though you cannot solve it yourself, then the appearance of quantization and quantum numbers may be a little less mysterious. This section is an attempt to explain the method of solving a differential equation of motion of the type that we encounter in quantum mechanics. We shall explain the strategy with the simpler analogy of the equation of a vibrating string.

The de Broglie wave relationship and the Heisenberg uncertainty principle should prepare you for the two main features of quantum mechanics that contrast it with classical mechanics:

1) Information about a particle is obtained by solving an equation for a wave.

2) The information obtained about the particle is not its position; rather, it is the *probability* of finding the particle in a given region of space.

We are not able to say whether an electron is in a certain place around an atom, but we are able to measure the probability that it is there rather than somewhere else.

Wave equations are familiar in mechanics. For instance, the problem of the vibration of a violin string is solved in three steps:

1) Set up the equation of motion of a vibrating string. This equation will involve the displacement or amplitude of vibration, $A_{(x)}$, as a function of position along the string, x.

2) Solve the differential equation to obtain a general expression for amplitude. For a vibrating string with fixed ends, this general expression is a sine wave. As yet, there are no restrictions on wavelength or frequency of vibration.

3) Eliminate all solutions except those that leave the ends of the string stationary. This restriction on acceptable solutions of the wave equation is a boundary condition. Figure 8–15(a) shows solutions that fit this boundary condition of fixed ends of the string; Figure 8–15(b) shows solutions that fail. The only acceptable vibrations are those with $\lambda = 2a/n$, or $\bar{\nu} = n/2a$, in which $n = 1, 2, 3, 4, \ldots$. *The boundary conditions and not the wave equation are responsible for the quantization of the wavelengths of string vibration.*

Exactly the same procedure is followed in quantum mechanics:

1) Set up a general wave equation for a particle. Such an equation was proposed, in 1926, by Erwin Schrödinger (1887–1961). The Schrödinger equation is written in terms of the function $\psi_{(x,y,z)}$, which is analogous to the amplitude, $A_{(x)}$, in our violin-string analogy. *The square of this amplitude, $|\psi|^2$, is the relative probability density of the particle at position (x, y, z).*

That is, if a small element of volume, dv, is located at (x, y, z), the probability of finding an electron within that element of volume is $|\psi|^2 \, dv$.

2) Solve the Schrödinger equation to obtain the most general expression for $\psi_{(x,y,z)}$.

3) Apply the boundary conditions of the particular physical situation. If the particle is an electron in an atom, the boundary conditions are that $|\psi|^2$ must be *continuous*, *single-valued*, and *finite* everywhere. All these conditions are only common sense. First, probability functions do not fluctuate radically from one place to another; the probability of finding an electron a few thousandths of an angstrom from a given position will not be radically different from the probability at the original position. Second, the probability of finding an electron in a given place cannot have two different values simultaneously. Third, since the probability of finding an electron somewhere must be 100% or 1.000, if the electron really exists, the probability at any one point cannot be infinite.

We now shall compare the wave equation for a vibrating string and the Schrödinger wave equation for a particle. In this text you will not be expected to do anything with either equation, but you should note the similarities between them.

Vibrating string. The amplitude of vibration at a distance x along the string is $A_{(x)}$. The differential equation of motion is

$$\frac{d^2 A_{(x)}}{dx^2} + 4\pi^2 \bar{\nu}^2 A_{(x)} = 0 \tag{8–12}$$

The general solution to this equation is a sine function

$$A_{(x)} = A_{\max} \sin (2\pi\bar{\nu}x + \alpha)$$

and the only acceptable solutions [Figure 8–15(a)] are those for which $\bar{\nu} = n/2a$, where $n = 1, 2, 3, 4, \ldots$, and for which the phase shift, α, is zero:

$$A_{(x)} = A_{\max} \sin n \left(\frac{\pi}{a}\right) x$$

Schrödinger equation. The square of the amplitude $|\psi_{(x,y,z)}|^2$ is the probability density of the particle at (x, y, z). The differential equation is

$$\frac{\partial^2 \psi}{\partial x^2} + \frac{\partial^2 \psi}{\partial y^2} + \frac{\partial^2 \psi}{\partial z^2} + \frac{8\pi^2 m}{h} (E - V_{(x,y,z)})\psi_{(x,y,z)} = 0 \tag{8–13}$$

V is the potential energy function at (x, y, z), and m is the mass of the electron.

Although solving Equation 8–13 is not a simple process, it is purely a mathematical operation; there is nothing in the least mysterious about it. The energy, E, is the variable that is restricted or quantized by the boundary conditions on $|\psi|^2$. Our next task is to determine what the possible energy states are.

8–7   THE HYDROGEN ATOM

The sine function that is the solution of the equation for the vibrating string is characterized by one integral quantum number: $n = 1, 2, 3, 4, \ldots$. The first few acceptable sine functions are

$$
\left.
\begin{aligned}
A_{1(x)} &= A_0 \sin \left(\frac{\pi}{a}\right) x \\[2mm]
A_{2(x)} &= A_0 \sin 2 \left(\frac{\pi}{a}\right) x \\[2mm]
A_{3(x)} &= A_0 \sin 3 \left(\frac{\pi}{a}\right) x \\[2mm]
A_{4(x)} &= A_0 \sin 4 \left(\frac{\pi}{a}\right) x
\end{aligned}
\right\}
\quad A_{n(x)} = A_0 \sin n \left(\frac{\pi}{a}\right) x
\tag{8–14}
$$

. . . etc.

These are the first four curves in Figure 8–15(a).

An atom is three-dimensional, whereas the string has only length. The solutions of the Schrödinger equation for the hydrogen atom are characterized by three integer quantum numbers: n, l, and m. These arise when solving the equation for the wave function, ψ, which is analogous to the function, $A_{n(x)}$, in the vibrating string analogy. In solving the Schrödinger equation, we divide it into three parts. The solution of the *radial* part describes how the wave function, ψ, varies with distance from the center of the atom. If we borrow the customary coordinate system of the earth, an *azimuthal* part produces a function that reveals how ψ varies with north or south latitude, or distance up or down from the equator of the atom. Finally, an *angular* part is a third function that suggests how the wave function varies with east–west longitude around the atom. The total wave function, ψ, is the product of these three functions.

In the process of separating these parts, a constant, n, appears in the radial expression, another constant, l, occurs in the radial and azimuthal expressions, and m appears in the azimuthal and angular expressions. The boundary conditions that give physically sensible solutions to these three equations are that each function (radial, azimuthal, and angular) be continuous, single-valued, and finite at all points. These conditions will not be met unless n, l, and m all are integers, and furthermore that l is zero or a positive integer less than n, and that m has a value from $-l$ to $+l$. From a one-dimensional problem (the vibrating string) we obtained one quantum number. Now with a three-dimensional problem we obtain three quantum numbers.

The *principal quantum number*, n, can be any positive integer: $n = 1, 2, 3, 4, 5, \ldots$. The *azimuthal quantum number*, l, can have any integral value from 0 to $n - 1$. The *magnetic quantum number*, m, can have any integral value from

Table 8-1.   Quantum States for the Hydrogen Atom through n = 4

n	l	m	s	Common name	Number of states
1	0	0	$\pm\frac{1}{2}$	$1s$	2
2	0	0	$\pm\frac{1}{2}$	$2s$	2
2	1	-1	$\pm\frac{1}{2}$		
		0	$\pm\frac{1}{2}$	$2p$	6
		$+1$	$\pm\frac{1}{2}$		
3	0	0	$\pm\frac{1}{2}$	$3s$	2
3	1	-1	$\pm\frac{1}{2}$		
		0	$\pm\frac{1}{2}$	$3p$	6
		$+1$	$\pm\frac{1}{2}$		
3	2	-2	$\pm\frac{1}{2}$		
		-1	$\pm\frac{1}{2}$		
		0	$\pm\frac{1}{2}$	$3d$	10
		$+1$	$\pm\frac{1}{2}$		
		$+2$	$\pm\frac{1}{2}$		
4	0	0	$\pm\frac{1}{2}$	$4s$	2
4	1	-1	$\pm\frac{1}{2}$		
		0	$\pm\frac{1}{2}$	$4p$	6
		$+1$	$\pm\frac{1}{2}$		
4	2	-2	$\pm\frac{1}{2}$		
		-1	$\pm\frac{1}{2}$		
		0	$\pm\frac{1}{2}$	$4d$	10
		$+1$	$\pm\frac{1}{2}$		
		$+2$	$\pm\frac{1}{2}$		
4	3	-3	$\pm\frac{1}{2}$		
		-2	$\pm\frac{1}{2}$		
		-1	$\pm\frac{1}{2}$		
		0	$\pm\frac{1}{2}$	$4f$	14
		$+1$	$\pm\frac{1}{2}$		
		$+2$	$\pm\frac{1}{2}$		
		$+3$	$\pm\frac{1}{2}$		
5	0	...	etc.		

$-l$ to $+l$. The different quantum states that the electron can have are listed in Table 8–1. For an electron around the nucleus in a hydrogen atom, the energy depends only on n. Moreover, the energy expression is exactly the same as in the Bohr theory:

$$E_n = -\frac{k}{n^2} \qquad k = \frac{2\pi^2 m_e e^4}{h^2} \qquad (8\text{–}5)$$

Quantum states, with $l = 0, 1, 2, 3, 4, 5, \ldots$ are called the $s, p, d, f, g, h, \ldots$

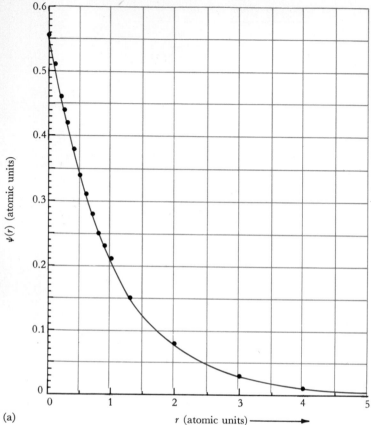

(a)

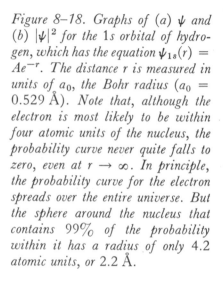

Figure 8–18. Graphs of (a) ψ and (b) $|\psi|^2$ for the 1s orbital of hydrogen, which has the equation $\psi_{1s}(r) = Ae^{-r}$. The distance r is measured in units of a_0, the Bohr radius ($a_0 = 0.529$ Å). Note that, although the electron is most likely to be within four atomic units of the nucleus, the probability curve never quite falls to zero, even at $r \to \infty$. In principle, the probability curve for the electron spreads over the entire universe. But the sphere around the nucleus that contains 99% of the probability within it has a radius of only 4.2 atomic units, or 2.2 Å.

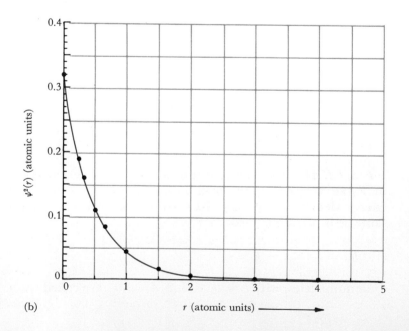

(b)

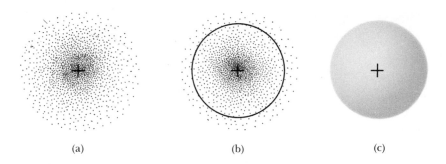

(a)   (b)   (c)

*Figure 8–19. Three ways of representing the spherical electron prob-
ability-density function of the 1s orbital of hydrogen. (a) $|\psi|^2$ represented
by the density of stippling; (b) a black circle representing a cross section
through the spherical shell that encloses 90% of the probability (radius
2.7 atomic units or 1.4 Å); (c) the 90% probability shell portrayed as a
surface.*

states in an extension of the old spectroscopic notation (Figure 8–13). All
of the l states for the same n have the *same* energy in the hydrogen atom; the
energy-level diagram is as in Figure 8–10.

Each of the quantum states differentiated by n, l, and m in Table 8–1
corresponds to a different probability distribution function for the electron
in space. The simplest such probability functions, for s states ($l = 0$), are
spherically symmetrical. The probability of finding the electron is the same
in all directions, but varies with distance from the nucleus. The dependence
of ψ and of the probability density $|\psi|^2$ on the distance of the electron from
the nucleus in the 1s state is plotted in Figure 8–18. You can see the spherical
symmetry of this state more clearly in Figure 8–19. The quantity $|\psi|^2 \, dv$ can
be thought of either as the probability of finding an electron in the volume
element dv in one atom, or as the *average* electron density within the corresponding volume element in a great many different hydrogen atoms. The
electron is no longer an orbit in the Bohr–Sommerfeld sense; rather, it is in
an electron probability cloud, which commonly is called an *orbital*.

The 2s orbital is also spherically symmetrical, but its radial distribution
function has a node, that is, zero probability, at $r = 2$ atomic units (one
atomic unit is $a_0 = 0.529$ Å). The probability density has a crest at four
atomic units, which is the radius of the Bohr orbit for $n = 2$. There is a high
probability of finding an electron in the 2s orbital closer to or farther from
the nucleus than $r = 2$, but there is no probability of ever finding it in the
spherical shell at a distance $r = 2$ from the nucleus (Figure 8–20). The 3s
orbital has two such spherical nodes, and the 4s has three. However, these
details are not as important in explaining bonding as are the general observations that s orbitals are spherically symmetrical and that they increase in
size as n increases.

There are three 2p orbitals: $2p_x$, $2p_y$, and $2p_z$. Each orbital is cylindrically
symmetrical with respect to rotation around one of the three principal axes

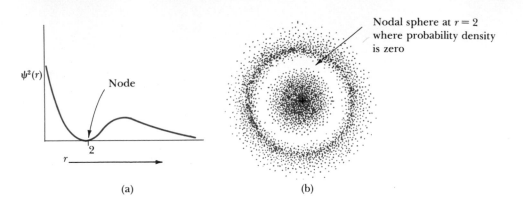

Figure 8–20. The 2s hydrogen orbital. (a) The graph of $|\psi|^2$ against r. (b) A cross section through the probability function plotted in three dimensions. Probability density is represented by stippling.

X, Y, or Z, as identified by the subscript. Each $2p$ orbital has two lobes of high electron density separated by a nodal plane of zero density (Figures 8–21 and 8–22). The sign of the wave function, ψ, is positive in one lobe and negative in the other. The $3p$, $4p$, and higher p orbitals have one, two, or more additional nodal shells around the nucleus (Figure 8–23); again, these details are of secondary importance. The significant facts are that the three p orbitals are mutually perpendicular, strongly directional, and of increasing size as n increases.

The five d orbitals first appear for $n = 3$. For $n = 3$, l can be zero, one, or two, thus s, p, and d orbitals are possible. The $3d$ orbitals are shown in Figure 8–24. Three of them, d_{xy}, d_{yz}, and d_{xz}, are identical in shape but different in orientation. Each has four lobes of electron density bisecting the angles between principal axes. The remaining two are somewhat unusual: The $d_{x^2-y^2}$ orbital has lobes of density along the X and Y axes, and the d_{z^2} orbital has lobes along the Z axis, with a small doughnut or ring in the XY plane. However, there is nothing sacrosanct about the Z axis. The proper combination of wave functions of these five d orbitals will give us another set of five d orbitals in which the d_{z^2}-like orbital points along the X axis, or the Y axis. We could even combine the wave functions to produce a set of orbitals, all of which were alike but differently oriented. However, the set that we have described, d_{xy}, d_{yz}, d_{xz}, $d_{x^2-y^2}$, and d_{z^2}, is convenient and is used conventionally in chemistry. The sign of the wave function, ψ, changes from lobe to lobe, as indicated in Figure 8–24.

The azimuthal quantum number l is related to the shape of the orbital, and is referred to as the *orbital-shape quantum number:* s orbitals with $l = 0$ are spherically symmetrical, p orbitals with $l = 1$ have plus and minus extensions along one axis, and d orbitals with $l = 2$ have extensions along two mutually perpendicular directions (Figure 8–25). The third quantum num-

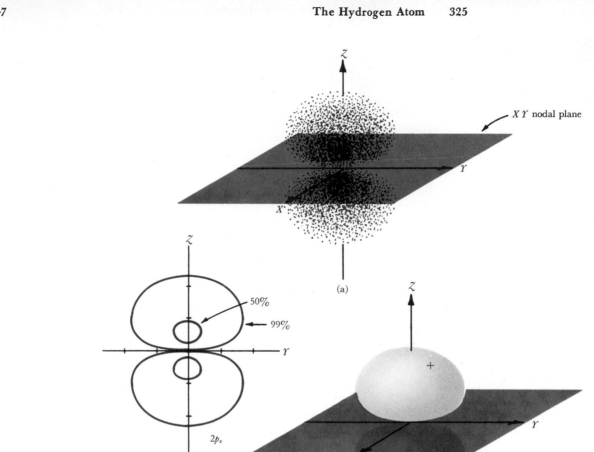

Figure 8–21. Three ways of representing the $2p_z$ atomic orbital of hydrogen. (a) $|\psi|^2$ represented by stippling. (b) Contour diagram of the $2p_z$ orbital. The contours represent lines of constant $|\psi|^2$ in the YZ plane and have been chosen so that, in three dimensions, they enclose 50% or 99% of the total probability density. The $2p_z$ orbital is symmetrical around the Z axis. (c) The 99% probability shell portrayed as a surface. The plus and minus signs on the two lobes represent the relative signs of ψ and should not be confused with electric charge. Note that there is no probability of finding the electron on the XY plane. Such a surface, which need not be planar, is called a nodal surface.

ber, m, describes the orientation of the orbital in space. It sometimes is called the magnetic quantum number because the usual way of distinguishing between orbitals with different spatial orientations is to place the atoms in a magnetic field and to note the differences in energy produced in the orbitals. We will use the more descriptive term, *orbital-orientation quantum number*.

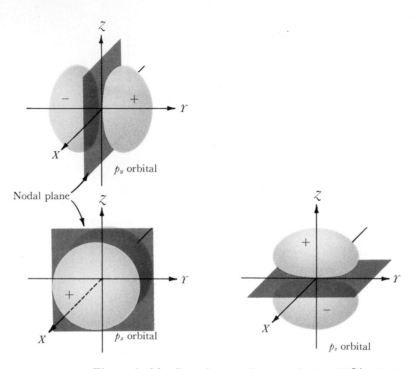

Figure 8–22. Boundary surfaces enclosing 99% of the probability for $2p_y$, $2p_x$, and $2p_z$ orbitals of hydrogen. Note the nodal plane of zero probability density in each orbital.

There is a fourth quantum number that has not been mentioned. Atomic spectra, and more direct experiments as well, indicate that an electron behaves as if it were spinning around an axis and has a small magnetic moment. Each electron has a choice of two spin states with spin quantum numbers $s = +\frac{1}{2}$ or $-\frac{1}{2}$. A complete description of the state of an electron in a hydrogen atom requires the specification of all four quantum numbers: n, l, m, and s.

Figure 8–23. Contour diagrams in the XZ plane for hydrogen wave ▶
functions that show the 50% and 99% contours. X and Z axes are marked in intervals of five atomic units. All orbitals shown except the $3d_{xz}$ have rotational symmetry around the Z axis. The $3p_z$ orbital differs from the $2p_z$ in having another nodal surface as a spherical shell around the nucleus at a distance of approximately six atomic units. But so far as chemical bonding is concerned, these inner details do not matter; the important difference between a 2p and a 3p orbital is size.

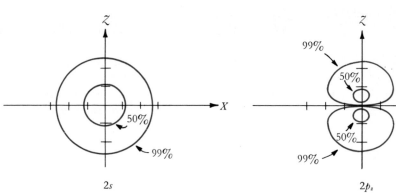

2s

2p_z

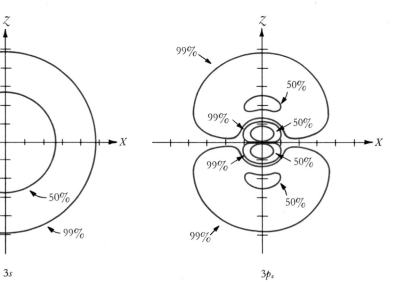

3s

3p_z

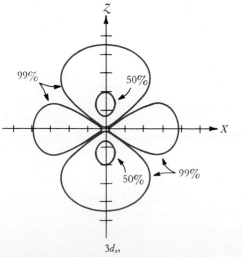

3d_{z^2}

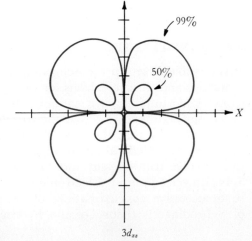

3d_{xz}

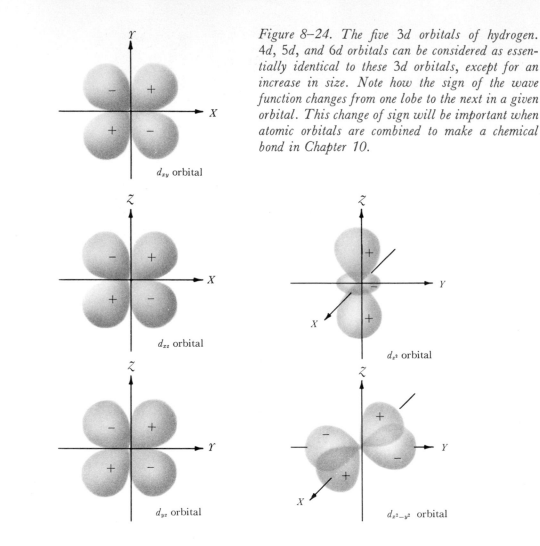

Figure 8–24. The five 3d orbitals of hydrogen. 4d, 5d, and 6d orbitals can be considered as essentially identical to these 3d orbitals, except for an increase in size. Note how the sign of the wave function changes from one lobe to the next in a given orbital. This change of sign will be important when atomic orbitals are combined to make a chemical bond in Chapter 10.

8–8   MANY-ELECTRON ATOMS

It is possible to set up the Schrödinger wave equation for lithium, which has a nucleus and three electrons, or uranium, which has a nucleus and 92 electrons. Unfortunately, we cannot solve the differential equations. There is little comfort in knowing that the structure of the uranium atom is calculable *in principle*, and that the fault lies with mathematics and not with physics. Physicists and physical chemists have developed many approximate methods that involve trial guesses and successive approximations to solutions of the Schrödinger equation. Electronic computers have been of immense value in such successive approximations or iterative calculations. But the advantage of Schrödinger's theory of the hydrogen atom is that it gives us a clear

Figure 8–25. Summary of the most important aspects of the hydrogen orbitals. (a) The principal quantum number, n, indicates the approximate relative size of the orbital. (b) The orbital-shape quantum number, l, indicates the shape or the degree of asymmetry of the orbital.

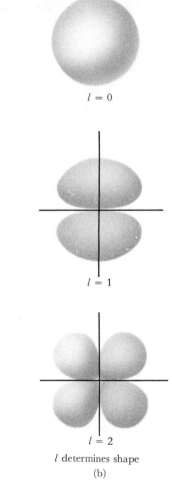

$l = 0$

$n = 1$

$n = 2$

$l = 1$

$n = 3$

$l = 2$

n determines effective volume

(a)

l determines shape

(b)

qualitative picture of the electronic structure of many-electron atoms without such additional calculations. Bohr's theory was too simple and could not do this, even with Sommerfeld's help.

The extension of the hydrogen-atom picture to many-electron atoms is one of the most important steps in understanding chemistry, and we shall reserve it for the next chapter. We shall begin by assuming that electronic orbitals for other atoms are *similar* to the orbitals for hydrogen and that they can be described by the same four quantum numbers and have analogous probability distributions. If the energy levels deviate from the ones for hydrogen (which they do), then we shall have to provide a persuasive argument, in terms of the hydrogenlike orbitals, for these changes. The next chapter will be devoted to explaining the structures of many-electron atoms in the language of the hydrogen-atom model.

Before you go on.   You will be able to appreciate better the next few chapters if you understand thoroughly how quantum numbers, energy levels, and orbitals are used to describe the structures of atoms. Review 8 of Lassila, *Programed Reviews of Chemical Principles*, examines these topics, and you may wish to check your proficiency with them before you begin new chapters.

8–9   SUMMARY

The quantum revolution is now history. Having been born into the post-revolutionary era, the new generation of chemists readily accepts the innovations that quantum mechanics brought to chemistry and finds it difficult to see anything startling in the ideas that all matter is both wavelike and particle-like, that light comes in bundles, that there are limits to how much we can know about a particle, and that energy is quantized. These ideas were certainly shocking when they were proposed. Yet each new step was forced upon us, not by mathematicians enamored of a new differential equation, but by puzzled experimentalists who could not reconcile what they saw with what they thought they knew. Planck abandoned the idea that *nature makes no jumps* with reluctance. To Rutherford, his α-particle scattering results were ". . . quite the most incredible event that has ever happened to me in my life." The entire edifice of quantum mechanics seemed alien, confusing, and improbable to many good scientists. It is accepted now for the best of all reasons: It explains more observations in physics and chemistry than any other theory yet conceived.

Bohr proposed that electrons move in orbits, and de Broglie suggested that stable orbits were those that could accommodate standing waves. We know now that electrons do have wave properties, but we think in terms of electron probability distributions, or orbitals, instead of fixed orbits. In the hydrogen atom, several physically sensible boundary conditions on acceptable probability distributions lead to the quantization of the energy of the atom, and to the selection of certain ones of the many possible wave functions. The acceptable quantum states are described in terms of four quantum numbers: n, l, m, and s. The energy depends only on n, which also is a measure of the size of the orbital. Quantum number l is related to the shape of the orbital, and m to its orientation in space. The last number, s, describes the two possible spin orientations of an electron in an orbital.

Although we cannot solve the complete Schrödinger equation for more complex atoms, we can use the hydrogen-atom results as a point of departure in explaining many of the physical and chemical properties of the heavier elements. This is the subject of the next chapter.

SUGGESTED READING

A. W. Adamson, "Domain Representations of Orbitals," *J. Chem. Educ.* **42,** 141 (1965).

R. S. Berry, Advisory Council on College Chemistry Resource Paper on "Atomic Orbitals," *J. Chem. Educ.* **43,** 283 (1966).

J. B. Birks, Ed., *Rutherford at Manchester*, W. A. Benjamin, Menlo Park, Calif., 1963. A good account of what it was like to be a scientist in Britain in the early part of the century. Written by and about one of the sharpest minds, finest men, and best writers in physics.

I. Cohen and T. Bustard, "Atomic Orbitals: Limitations and Variations," *J. Chem. Educ.* **43,** 187 (1966).

U. Fano and L. Fano, *Physics of Atoms and Molecules*, University of Chicago Press, Chicago, 1972.

R. P. Feynman, R. B. Leighton, and M. Sands, *The Feynman Lectures on Physics: Quantum Mechanics*, Addison-Wesley, Reading, Mass., 1965. Exhilarating book. Ostensibly written for a beginning physics course, but excellent reading even when you cannot follow it all.

G. Gamow, *Mr. Tomkins in Wonderland*, Cambridge University Press, New York, 1939. What would our world be like if the speed of light were ten miles per hour? If Planck's constant were 27 orders of magnitude larger? A collection of stories answering such questions, with each story based on a changed value for one important physical constant in quantum mechanics. The quantized world as seen through the eyes of a middle-class bank clerk with a taste for free public lectures. Highly recommended.

G. Gamow, *Thirty Years that Shook Physics: The Story of Quantum Theory*, Doubleday Anchor, New York, 1966.

W. Heisenberg, *The Physical Principles of Quantum Theory*, Dover, New York, 1930. An early statement by one of the pioneers in quantum mechanics.

C. N. Hinshelwood, *The Structure of Physical Chemistry*, Oxford University Press, New York, 1951.

R. M. Hochstrasser, *The Behavior of Electrons in Atoms*, W. A. Benjamin, Menlo Park, Calif., 1964. A supplement written for general chemistry students. Goes into more details than this chapter but at a similar level.

R. C. Johnson and R. R. Rettew, "Shapes of Atoms," *J. Chem. Educ.* **42,** 145 (1965).

E. A. Ogryzlo and G. B. Porter, "Contour Surfaces for Atomic and Molecular Orbitals," *J. Chem. Educ.* **40,** 256 (1963).

B. Perlmutter-Hayman, "The Graphical Representation of Hydrogen-Like Wave Functions," *J. Chem. Educ.* **46,** 428 (1969).

R. E. Powell, "The Five Equivalent *d* Orbitals," *J. Chem. Educ.* **45,** 1 (1968).

H. H. Sisler, *Electronic Structure, Properties and the Periodic Law*, Reinhold, New York, 1963.

G. Thomson, *The Atom*, Oxford University Press, New York, 1962.

QUESTIONS

1 What are α, β, and γ rays? Which of them are composed of particles? Which are waves? Why is this an unfair question?

2 How are Thomson's and Rutherford's models of the atom different, and how does the scattering of α particles differentiate between them?

3 What is intolerable about Rutherford's atom, according to classical physics?

4 Which of the following is proportional to energy in electromagnetic radiation: speed, wave number, or wavelength?

5 If wavelengths are normally the measured quantities in spectroscopy, why are wave numbers preferable to frequencies when a quantity proportional to energy is desired?

6 What is the ultraviolet catastrophe, and how did Planck resolve it?

7 How did Planck's central assumption in explaining the ultraviolet catastrophe also account for the photoelectric effect?

8 What was the empirical formula for obtaining wave numbers of spectral lines in atomic hydrogen that was obtained by Balmer and others? How was it explained by Bohr? Can you think of any possible explanations for such a formula, other than ones of the type that Bohr proposed?

9 What was Bohr's basic assumption for obtaining the spectral formula from his model? What justification was there for that assumption? (Do not confuse Bohr's contribution with de Broglie's.)

10 What is *quantization* of energy? How does it arise in

a) the Bohr atom?
b) de Broglie's interpretation of the Bohr atom?
c) a vibrating violin string?
d) the Schrödinger equation for the hydrogen atom?
e) an organ pipe with closed ends?
f) an organ pipe with the top end open?

11 Can you see any logical connection between the quantization of energy in the hydrogen atom and Caruso shattering a wineglass with a high note?

12 How can the same atom of hydrogen, in quick succession, emit a photon in the Pfund, Brackett, Paschen, Balmer, and Lyman series? Can it emit them in the reverse order? Why, or why not?

13 How does the Bohr theory fail for the lithium atom? How did Sommerfeld try to overcome the difficulty? Where did his theory fail?

14 Why are energies of electrons in atoms always negative?

15 Why is wave behavior not seen in automatic rifle fire, although we can see it in a beam of neutrons?

16 What is the uncertainty principle? Why can we neglect it in everyday life?

17 What is a boundary condition in the solution of a wave equation? What is its physical meaning? What boundary conditions are imposed on the solution of the vibration of a violin string? What are the boundary conditions on the solution of the Schrödinger equation for an electron in a hydrogen atom?

18 What are the three quantum numbers that are found in the solution of the Schrödinger equation for the hydrogen atom? What values can each quantum number have, and what do they signify?

19 What is an atomic orbital? How does it differ from an orbit?

20 What is the difference between a probability density and a probability? Why is it incorrect to talk about a probability of an electron being at a particular *point* in space?

21 How many d levels are there in a quantum level? How do the shapes of the d orbitals, in their common chemical representation, differ from those of the p orbitals? How can the d orbitals be given different energies?

22 What is the spin quantum number? What numerical values can it have?

23 Consider two hydrogen atoms. The electron in the *first* hydrogen atom is in the $n = 1$ Bohr orbit. The electron in the *second* hydrogen atom is in the $n = 4$ Bohr orbit. (a) Which atom has the ground-state electronic configuration? (b) In which atom is the electron moving faster? (c) Which orbit has the larger radius? (d) Which atom has the lower potential energy? (e) Which atom has the higher ionization energy?

24 In each of the following statements, choose one of the four possibilities (1), (2), (3), and (4), that most accurately completes the statement. Only one answer should be given for each statement. Read the statements carefully.

a) Rutherford, Geiger, and Marsden carried out experiments in which a beam of helium nuclei (α particles) was directed at a thin piece of gold foil. They found that the gold foil (1) severely deflected most of the particles of the beam directed at it. (2) deflected very few of the particles of the beam, and deflected these only very slightly. (3) deflected most of the particles of the beam, but deflected these only very slightly. (4) deflected very few of the particles of the beam, but deflected these severely.

b) From the results in (a), Rutherford concluded that (1) electrons are massive particles. (2) the positively charged parts of atoms are extremely small and extremely heavy particles. (3) the positively charged parts of atoms are moving about with a velocity approaching that of light. (4) the diameter of an electron is approximately equal to that of the nucleus.

c) Max Planck was led to his formulation of the quantum theory in attempting to provide a theoretical explanation of the fact that (1) electrons are emitted from a metal when light of sufficiently short wavelength falls upon it. (2) the thermal (or "blackbody") radiation from a hot object contains a relatively large amount of ultraviolet light, contrary to the prediction of classical mechanics. (3) the thermal radiation from a hot object contains a relatively small amount of ultraviolet light, contrary to the prediction of classical mechanics. (4) the thermal radiation from a hot object can occur at all frequencies, contrary to the prediction of classical mechanics.

d) Which one of the following statements concerning the photoelectric effect is *not* true? (1) No electrons are emitted from the surface of a metal until the frequency of the light directed on the surface exceeds a certain "threshold" value. (2) Above the threshold frequency, the greater the intensity of the light, the greater will be the velocity of the emitted electrons. (3) Above the threshold frequency, the smaller the wavelength of the light, the greater will be the velocity of the emitted electrons. (4) Above the threshold frequency, the greater the intensity of the light, the greater will be the number of electrons emitted per second.

e) Which one of the following statements concerning the Bohr theory of the hydrogen atom is *not* true? (1) The theory successfully explained the observed emission and absorption spectra of the hydrogen atom. (2) The theory requires that the greater the energy of the electron in the hydrogen atom, the greater its velocity. (3) The theory requires that the energy of the electron in the hydrogen atom have only certain discrete values. (4) The theory requires that the distance of the electron from the nucleus in the hydrogen atom can have only certain discrete values.

PROBLEMS

1 Spectroscopists often express the energy of light in units of reciprocal centimeters (cm^{-1}). Calculate the frequency of a photon of green light that has an energy of 20,000 cm^{-1}.

2 Calculate the wavelength of a photon that has a frequency of 1.2×10^{15} sec^{-1}. What is the energy of the photon in ergs $photon^{-1}$? What is the energy in kcal $mole^{-1}$ of photons? What do we usually call such radiation?

3 X rays typically have wavelengths of 1 Å to 10 Å. Calculate the energy in ergs of photons with a 2-Å wavelength. Calculate the energy in kcal $mole^{-1}$ of such 2-Å photons, and compare this with the bond energy of a carbon–carbon single bond, 83 kcal $mole^{-1}$. Would you expect x rays to be able to produce chemical reactions?

4 Calculate the energy of photons, in ergs $photon^{-1}$ and in kcal $mole^{-1}$, for 1000-kilocycle broadcast-band radio waves. (One kilocycle is a frequency of 10^3 sec^{-1}.) What is the wavelength of such photons? How does the energy compare with that for a carbon–carbon single bond? Would you expect radio waves to be able to produce chemical reactions?

5 The first ionization energy (see Chapter 6) of Cs is 89.8 kcal $mole^{-1}$. Calculate the first ionization energy (in kcal and eV) for one atom of Cs. Calculate the wavelength of light that would be just sufficient to ionize a Cs atom.

6 In 1914, Moseley discovered that $\nu = c(Z - b)^2$ (see Chapter 6), in which ν is the frequency of x rays emitted when an element is bombarded by a beam of electrons. Use the Bohr theory to explain the dependence of Z on the square root of ν in this expression.

7 What is the wavelength of photons that have an energy of 83 kcal $mole^{-1}$?

What do we call such radiation? [See Figure 8–5(a).]

8 Einstein interpreted the photoelectric effect in terms of which of the following ideas: (a) the particle nature of light, (b) the wave nature of light, (c) the wave nature of matter, (d) the uncertainty principle?

9 When a photon strikes a metal surface, a certain minimum energy is required to eject an electron from the metal. This minimum or threshold energy is known as the work function of the metal. Any energy in the original photon above this minimum is translated into kinetic energy for the ejected electron. The threshold wavelength for photoelectric emission from Li, above which no electrons are emitted, is 5200 Å. Calculate the velocity of electrons emitted as the result of absorption of light at 3600 Å.

10 Which of the following best describes the emission spectrum of atomic hydrogen: (a) a continuous emission of light at all frequencies, (b) discrete series of lines, each line within a series being equidistant from the next, (c) discrete lines occurring in pairs, each pair equidistant from the next pair, (d) only two lines observed over the entire spectrum, (e) discrete series of lines whose spacing within the series decreases at higher wave numbers?

11 The Lyman series of lines results from transitions from more energetic orbits to which lower-energy Bohr orbit? A spectral line is found at 103,000 cm^{-1}. What is the quantum number of the initial orbit of the electron undergoing this transition?

12 Which of the following describes the process responsible for the emission spectrum of hydrogen, according to the Bohr theory? (a) Electrons are excited into higher-energy orbits and emit light when they fall back into lower-energy orbits. (b) The hydrogen atom emits light when electrons are excited into higher-energy orbits. (c) The hydrogen atom absorbs light when electrons are excited into higher-energy orbits. (d) Electrons are deexcited into lower-energy orbits and emit light when they return to higher-energy orbits.

13 Calculate the wave number of the photons emitted when a hydrogen atom decays from a state with $n = 3$ to one with $n = 2$. What is the name of the series containing such emission?

14 In the Bohr theory of the atom, which of the following concepts was introduced arbitrarily by Bohr? (a) The electron is attracted to the nucleus by coulombic forces. (b) The electron moves in circular orbits about the nucleus. (c) The angular momentum of the electron is restricted to discrete values. (d) The kinetic energy of the electron is given by $\frac{1}{2}m_e v^2$. (e) The mass of the electron is restricted to discrete or quantized values.

15 Which of the following was not explained by the simple Bohr theory: (a) the ionization energy of hydrogen, (b) the details of atomic spectra of atoms with many electrons, (c) the locations of the lines in the hydrogen spectrum, (d) the spectra of hydrogenlike atoms such as He^+ and Li^{2+}, (e) the energy levels of the hydrogen atom?

16 If the energy associated with the first Bohr orbit is -13.60 eV atom^{-1}, what is the energy associated with the fourth Bohr orbit?

17 If the second Bohr orbit has a radius of 2.12 Å, what is the radius of the fourth Bohr orbit?

18 Which of the following experiments most directly supports de Broglie's hypothesis of the wave nature of matter: (a) x-ray diffraction, (b) photoelectric effect, (c) α-particle scattering by a metal foil, (d) the blackbody effect, (e) electron diffraction?

19 Which of the following aspects of the Bohr theory is not allowed by Heisenberg's uncertainty principle: (a) discrete atomic energy levels, (b) simple circular orbits, (c) quantum numbers, (d) electron orbitals, (e) electron waves? Why does the aspect that you chose clash with the uncertainty principle?

20 An electron in a hydrogen atom has a principal quantum number of 4. List the values of the second quantum number, l, that the electron can have.

21 With which of the following is the quantum number, m, associated: (a) the spatial orientation of the orbital, (b) the shape of the orbital, (c) the energy of the orbital in the absence of a magnetic field, (d) the effective volume of the orbital?

22 If an electron has an orbital-shape quantum number of $l = 3$, what values of m can it have? What do we call such an $l = 3$ electron?

23 An electron is in a $4f$ orbital. What possible values for the quantum numbers, n, l, m, and s, can it have?

24 Calculate the de Broglie wavelength of an average helium atom at 27°C (see Chapter 2). Assume that the position of an average He atom can be measured to 0.10 Å. Compare the uncertainty in the momentum of the He atom to its actual momentum.

25 The probability of finding a p-orbital electron at the nucleus of an atom is zero. A contradiction arises when the two lobes of a p orbital are described as touching each other. What is the contradiction? [*J. Chem. Educ.* **38**, 20 (1961).]

*The electron has conquered physics, and many
worship the new idol rather blindly.*
H. Poincaré (1907)

9 ELECTRONIC STRUCTURE AND CHEMICAL PROPERTIES

We now know the wave functions and energy levels for a hydrogen atom. We cannot solve the Schrödinger equation exactly for atoms with more than one electron, but with one new principle we can construct a picture of many-electron atoms that will explain the periodic table and account for the chemical behavior of the elements. This new principle is the Pauli exclusion principle. The application of this principle to many-electron atoms is the *aufbau*, or buildup, process. Using this process we shall determine the electronic structures of all the elements.

These electronic structures will lead directly to the periodic table in the form in which we have seen it in Figures 6–3 and 6–4. They will explain the stability of eight-electron shells in noble gases and the oxidation numbers of the representative elements. Finally, electronic structure will elucidate many of the chemical and physical properties of the class of elements that has been explained poorly so far, the transition metals.

9-1   BUILDUP OF MANY-ELECTRON ATOMS

Although we cannot solve the Schrödinger equation exactly for many-electron atoms, we can show that no radical new features are expected as the atomic number increases. There are the same quantum states, the same four quantum numbers (n, l, m, and s), and virtually the same electronic probability functions or electron-density clouds. The energies of the quantum levels are not identical for all elements, but vary in a regular fashion from one element to the next.

In studying the electronic structure of a many-electron atom, we first shall assume a nucleus and the required number of electrons. We shall assume that the possible electronic orbitals are hydrogenlike, if not identical to the hydrogen orbitals. Then we shall build the atom by adding electrons one at a time, by placing each new electron in the lowest-energy orbital available. In this way we shall build a model of an atom in its *ground state*, or the state of lowest electronic energy. Wolfgang Pauli (1900–1958) first suggested this treatment of many-electron atoms, and called it the *aufbau*, or buildup, process.

The *aufbau* process involves three principles:

1) No two electrons in the same atom can be in the same quantum state. This principle is known as the *Pauli exclusion principle*. It means that no two electrons can have the same n, l, m, and s values. Therefore, one atomic orbital, described by n, l, and m, can hold a maximum of two electrons: one of spin $+\frac{1}{2}$ and one of spin $-\frac{1}{2}$. We often shall represent an atomic orbital by a circle

and indicate electrons by arrows

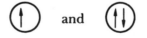

When two electrons occupy one orbital with spins $+\frac{1}{2}$ and $-\frac{1}{2}$, we say that their spins are *paired*.

2) Orbitals are filled with electrons in order of increasing energies. The s orbital can hold a maximum of two electrons. The three p orbitals can hold six electrons, the five d orbitals can hold 10 electrons, and the seven f orbitals can hold 14. We must decide on the order of increasing energies of the levels before the buildup process can begin. For atoms with more than one electron, in the absence of an external electric or magnetic field, energy depends on n and l (the size and shape quantum numbers) and not on m, the orbital-orientation quantum number.

3) When electrons are added to orbitals of the same energy (such as the five $3d$ orbitals), they will fill each of the available orbitals with one

electron before pairing electrons in any one orbital. This is *Hund's rule*, which states that in orbitals of identical energy, electron spins remain unpaired if possible. This behavior is understandable in terms of electron–electron repulsion. Two electrons, one in a p_x orbital and one in a p_y orbital, remain farther apart than two electrons paired in the same p_x orbital (Figure 8–22). A consequence of Hund's rule is that a half-filled set of orbitals, each containing a single electron, is a particularly stable arrangement. The sixth electron in a set of five d orbitals is forced to pair with another electron in a previously occupied orbital. The mutual repulsion of negatively charged electrons means that less energy is required to remove this sixth electron than to remove one of the five in a set of five half-filled d orbitals. Similarly, the fourth electron in a set of three p orbitals is held less tightly than the third.

Relative Energies of Atomic Orbitals

The $3s$, $3p$, and $3d$ orbitals in the hydrogen atom have the same energy but differ in the closeness of approach of the electron to the nucleus. An electron in a $3d$ orbital has a low probability of being close to the nucleus (Figure 9–1). In contrast, a $3p$ electron has a high probability of being found far from the nucleus, a node of no probability near $r = 6$ atomic units, and a low probability near $r = 3$. An electron in a $3s$ orbital has two inner nodes and two inner probability peaks, one at only 1 Bohr radius (atomic unit) from the nucleus.

The energy of an electron in an orbital depends on the attraction exerted on it by the positively charged nucleus. Electrons with low principal quantum numbers will lie close to the nucleus and will screen some of this electrostatic attraction from electrons with higher principal quantum numbers. In the Li^+ ion, the *effective nuclear charge* beyond 1 or 2 atomic units (a.u.) from the nucleus is not the true nuclear charge of $+3$, but a *net* charge of $+1$ produced by the nucleus plus the two $1s$ electrons. Similarly, the lone $n = 3$ electron in sodium experiences a net nuclear charge of approximately $+1$ rather than the full nuclear charge of $+11$. Outer (valence) electrons with the same principal quantum number are approximately the same distance from the nucleus and have little screening effect upon one another. Thus the net nuclear charge felt by the valence electrons of elements in the third period of the table is approximately the atomic number less 10, because the full nuclear charge is counteracted by the ten electrons in the filled $n = 1$ and $n = 2$ orbitals.

If the net charge from the nucleus and the filled inner orbitals were concentrated at a point at the nucleus, then the energies of $3s$, $3p$, and $3d$ orbitals would be the same. But the screening electrons extend over an appreciable volume of space. The net attraction that an electron with a principal quantum number of 3 experiences depends on how close it comes to the nucleus, and whether it penetrates the lower screening electron clouds. As

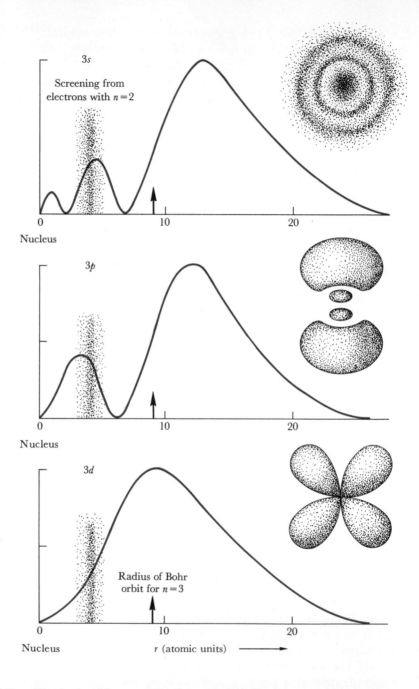

in Sommerfeld's elliptical-orbit model, the *s* orbital comes closer to the nucleus and is somewhat more stable than the *p*, and the *p* is more stable than the *d*. This is the reason for the variation of the *l* energy levels in the Li energy-level diagram in Figure 8–13.

◄ *Figure 9–1. Radial distribution functions for electrons in the 3s, 3p, and 3d atomic orbitals of hydrogen. These curves are obtained by spinning the orbital in all directions around the nucleus to smear out all details that depend on direction away from the nucleus, and then by measuring the smeared electron probability as a function of distance from the nucleus. The 3s orbital, which is already spherically symmetrical without the smearing operation, has a most probable radius at 13 atomic units and two minor peaks close to the nucleus. The 3p orbital has a maximum density near r = 12 atomic units, one spherical node at r = 6 atomic units and a density peak close to the nucleus. The 3d orbital has only one density peak, which occurs very close to the Bohr orbit radius of 9 atomic units. The shapes of the three orbitals before the spherical smearing process are to the right of each curve. An electron in the hydrogen atom with n = 2 will be in the neighborhood of r = 4 atomic units. The scale of distances changes in many-electron atoms, but relative distances in different orbitals in the atom are the same as in* H. *An electron in a 3s orbital is more stable than one in a 3p or 3d orbital because it has a greater probability of being inside the orbital of n = 2 electrons, in which it experiences a greater attraction from the nucleus. The 3p orbital is similarly more stable than the 3d.*

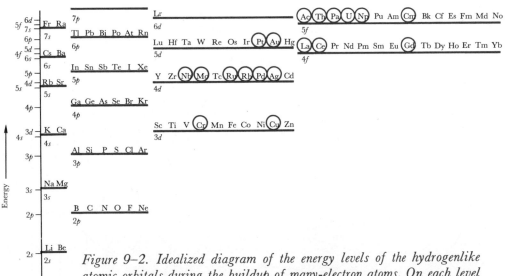

Figure 9–2. Idealized diagram of the energy levels of the hydrogenlike atomic orbitals during the buildup of many-electron atoms. On each level are written the symbols of those elements that are completed with the addition of electrons on that level. Note the nearly equal energies of 4s and 3d levels, of 5s and 4d, of 6s, 4f, and 5d, and finally of 7s, 5f, and 6d. The near equivalence of energies is reflected in some irregularity in the order of filling levels in the transition metals and inner transition metals. Elements with such irregularities are circled. For example, the first electron after the 6s and 7s orbitals fill, in La *and* Ac, *goes into a d orbital rather than an f. See Figure 9–4 for details.*

For a given value of the principal quantum number, n, the order of increasing energy is s, p, d, f, g, ... It is less easy to decide whether and when the high l-value orbitals of one n overtake the low-l orbitals of the next: for example, whether a $4f$ orbital has a higher energy than a $5s$, or a $3d$ a higher energy than a $4s$. The question originally was settled empirically by choosing the order of overlap that accounted for the observed structure of the periodic table. The energies since have been calculated theoretically, and (fortunately for quantum mechanics) they agree with the observed order of levels. The sequence of energy levels is shown in Figure 9–2.

The Buildup Process

We now can build up all of the atoms in the periodic table, one after the other, in order of increasing atomic number. This is accomplished by

Figure 9–3. The map of first ionization energies that appeared earlier as Figure 6–5. The first ionization energy is the energy required to pull the outermost electron from the neutral, gaseous atom. The fifth electron in boron must be placed in the higher-energy 2p level because the 2s orbital is full. This 2p electron is bound less tightly, so the first ionization energy of B is less than that of Be. Similarly, the eighth electron in O must pair with one of the three electrons in the three half-filled 2p orbitals of N. Because of electron–electron repulsion, this fourth p electron is held less tightly; thus the first ionization energy of O is less than that of N.

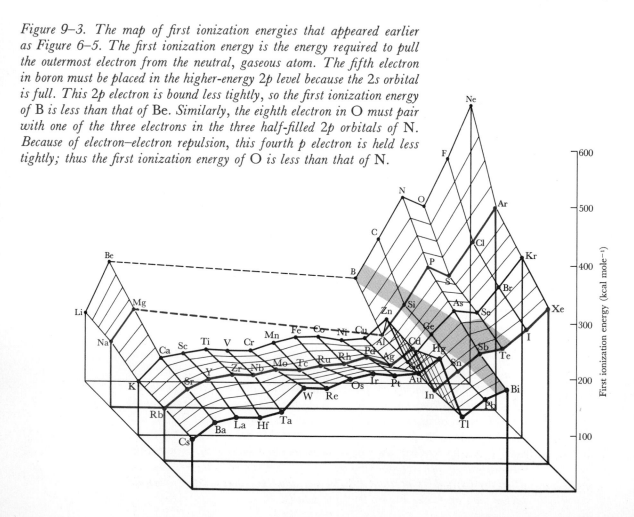

adding electrons to the hydrogenlike orbitals in order of increasing energy, and by increasing the nuclear charge each time by one. We can monitor this process by watching the experimental first ionization energies to see if they are consistent with the filling of orbitals (Figure 9–3). The periodic table in Figure 9–4 is helpful in following this building process.

A hydrogen atom has only one electron, which in the ground state obviously must go in the 1s orbital. The electronic configuration is represented as

H: $1s^1$

In He, the second electron also can be in the 1s orbital if its spin is paired with the first electron. In spite of electron–electron repulsion, this electron is more stable in the 1s orbital than in the 2s orbital. Helium is represented as

He: $1s^2$

Two electrons fill the 1s orbital; the third electron in Li must, by the Pauli exclusion principle, occupy the next lowest-energy orbital, namely, the 2s:

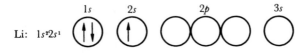

Li: $1s^2 2s^1$

The fourth electron in Be fills the 2s orbital, and the fifth electron in B must occupy one of the higher-energy 2p orbitals:

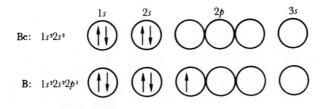

Be: $1s^2 2s^2$

B: $1s^2 2s^2 2p^1$

The first ionization energy of B is less than that of Be because its outermost electron is in a less stable orbital (higher energy). In C, two of the three 2p orbitals contain an electron. As Hund's rule predicts, in N the three p elec-

s block

n	s¹	s²
1	1 H	2 He
2	3 Li	4 Be
3	11 Na	12 Mg
4	19 K	20 Ca
5	37 Rb	38 Sr
6	55 Cs	56 Ba
7	87 Fr	88 Ra

Inner transition metals (n−2) f

	f^1	f^2	f^3	f^4	f^5	f^6	f^7	f^8	f^9	f^{10}	f^{11}	f^{12}	f^{13}	f^{14}
6	57 La (d^1)	58 Ce (f^1d^1)	59 Pr	60 Nd	61 Pm	62 Sm	63 Eu	64 Gd (f^7d^1)	65 Tb	66 Dy	67 Ho	68 Er	69 Tm	70 Yb
7	89 Ac (d^1)	90 Th (d^2)	91 Pa (f^2d^1)	92 U (f^3d^1)	93 Np (f^4d^1)	94 Pu	95 Am	96 Cm (f^7d^1)	97 Bk (f^9d^1)	98 Cf	99 Es	100 Fm	101 Md	102 No

Transition metals (n−1) d

	d^1	d^2	d^3	d^4	d^5	d^6	d^7	d^8	d^9	d^{10}
4	21 Sc	22 Ti	23 V	24 Cr (d^5s^1)	25 Mn	26 Fe	27 Co	28 Ni	29 Cu ($d^{10}s^1$)	30 Zn
5	39 Y	40 Zr	41 Nb (d^4s^1)	42 Mo (d^5s^1)	43 Tc	44 Ru (d^7s^1)	45 Rh (d^8s^1)	46 Pd (d^{10})	47 Ag ($d^{10}s^1$)	48 Cd
6	71 Lu	72 Hf	73 Ta	74 W	75 Re	76 Os	77 Ir	78 Pt (d^9s^1)	79 Au ($d^{10}s^1$)	80 Hg
7	103 Lr	104	105							

np

	p^1	p^2	p^3	p^4	p^5	p^6
2	5 B	6 C	7 N	8 O	9 F	10 Ne
3	13 Al	14 Si	15 P	16 S	17 Cl	18 Ar
4	31 Ga	32 Ge	33 As	34 Se	35 Br	36 Kr
5	49 In	50 Sn	51 Sb	52 Te	53 I	54 Xe
6	81 Tl	82 Pb	83 Bi	84 Po	85 At	86 Rn
7						

Figure 9–4. The "superlong" form of the periodic table, with an indication at the head of each column of the last electron to be added in the Pauli buildup process. Those elements whose electronic structure in the ground state differs from this simple buildup model are shown in color. In Gd, Cm, Cr, Mo, Cu, Ag, and Au, this difference arises from the extra stability of half-filled (f^7, d^5) or completely filled (d^{10}) shells. The other deviations arise from the extremely small energy differences between d and s, or d and f levels. These deviations are less important to us now than the overall patterns of buildup and the way in which they account for the structure of the periodic table. In this table, He is placed over Be in Group IIA since the second electron is added to complete the s orbital in each of these elements. In the usual periodic table (inside front cover), He is placed over Ne, Ar, and the other noble gases to indicate that the entire valence shell is filled in these elements.

trons are found in all three $2p$ orbitals, instead of two being paired in one:

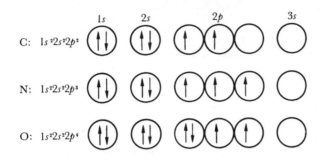

The fourth $2p$ electron in an oxygen atom is held less tightly than the first three because of the electron–electron repulsion with the other electron in one of the $2p$ orbitals. The first ionization energy of O is accordingly low.

The general trend across this period is for each new electron to be held *more* tightly because of the increased charge on the nucleus. Because the other $2s$ and $2p$ electrons are approximately the same distance from the nucleus, they do not shield the new electron from the steadily increasing charge. This increased charge overcomes the electron repulsion as the fifth $2p$ electron is added in F. Therefore, the fifth electron is held very tightly in F, and the first ionization energy increases again. The most stable configuration results when the sixth $2p$ electron is added to complete the $n = 2$ shell with the noble gas Ne:

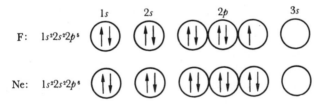

The complete $n = 1$ shell of two electrons often is given the symbol K, and the complete $n = 2$ shell of eight electrons is given the symbol L. Briefer representation of the Ne atom then is

Ne: KL

The buildup of the next period of the periodic table proceeds in exactly the same way. Each new electron is bound more firmly because of the increasing nuclear charge, except for the fluctuations at Al and S produced by the filling of $3s$ in Mg and the half-filling of $3p$ in P:

Na: $KL\ 3s^1$ KL

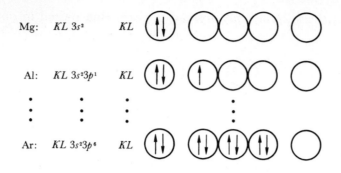

The outermost electron for each element in this period is bound *less* firmly than the outermost electron in the corresponding element of the previous period because the $n = 3$ electrons are farther from the nucleus. Therefore, the first ionization energies for the $n = 3$ elements are smaller than for the corresponding $n = 2$ elements. With the completion of the 3s and 3p orbitals, we again have reached a particularly stable electronic configuration with the noble gas Ar.

Something unusual happens in the fourth period. The 4s orbital penetrates closer to the nucleus than does the 3d orbital, and at this point in the buildup process the 4s has slightly lower energy than the 3d. Hence, the one and two electrons that are added to form K and Ca go into the 4s orbital before the 3d orbital is filled in the elements Sc through Zn. If we assume a constant inner electronic configuration of $KL\ 3s^2 3p^6$, the valence electronic configurations for the 4s and 3d elements are

K:	$3d^0 4s^1$	Mn:	$3d^5 4s^2$
Ca:	$3d^0 4s^2$	Fe:	$3d^6 4s^2$
Sc:	$3d^1 4s^2$	Co:	$3d^7 4s^2$
Ti:	$3d^2 4s^2$	Ni:	$3d^8 4s^2$
V:	$3d^3 4s^2$	Cu:	$3d^{10} 4s^1$
Cr:	$3d^5 4s^1$	Zn:	$3d^{10} 4s^2$

There are two anomalies in this order of filling. The half-filled (d^5) and filled (d^{10}) levels are particularly stable, therefore the Cr and Cu atoms have only one 4s electron each.

Although the 4s orbital penetrates closer to the nucleus than the 3d and therefore has a lower energy, the *majority* of the probability density of the 4s orbital is farther from the nucleus than in the 3d. An electron in a 4s orbital is simultaneously farther from the nucleus, on the average, than a 3d electron and is more stable because of the small but not negligible probability that it will be very close to the nucleus. In chemical bonding, the energies of electrons in such closely spaced levels in atoms are not as significant as distances of the electrons from the nucleus. Therefore, the 4s electrons have more of an effect on chemical properties than the relatively buried 3d electrons. With the exception of Cr and Cu, all of the elements from Ca through Zn have the same outer electronic structure: two 4s electrons. The chemical properties of this series of elements will vary less rapidly than in a series in which s or p electrons are being added. This is the reason for the relatively unchanging properties of the *transition metals*.

After the $3d$ orbitals are filled, the $4p$ orbitals fill, in a straightforward manner, to form the representative elements from Ga ($3d^{10}4s^24p^1$) to the noble gas Kr ($3d^{10}4s^24p^6$). The first ionization energy, which had risen with increasing nuclear charge in the transition metals, plummets at Ga when the next electron is placed in the less stable $4p$ orbital.

The fifth period repeats the same pattern: first the filling of the $5s$ orbitals, then an interruption while the buried $4d$ orbitals are filled in another series of transition metals, and finally the filling of the $5p$ orbitals, ending with the noble gas Xe ($4d^{10}5s^25p^6$). The common feature of all noble gases is the outermost electronic arrangement s^2p^6. This is the origin of the stable eight-electron shells that we mentioned in Chapter 6. The late filling of the d orbitals (and f orbitals) produces the observed lengths of the periods of the periodic table: first 2, then 8, then only 8 instead of 18 for $n = 3$, then only 18 instead of 32 for $n = 4$.

Exercise. If the order of energy levels were in strict numerical order ($1s$, $2s$, $2p$, $3s$, $3p$, $3d$, $4s$, $4p$, $4d$, $4f$, $5s$, $5p$, etc.) and if the stable elements that we call noble gases occurred when the last electron of a given n value was added, what would be the atomic numbers of the noble gases? Compare them with the true atomic numbers.

According to the energy diagram in Figure 9–2, the $6s$ orbital is more stable than the $5d$, which is not surprising since we have seen the same behavior in the two previous periods. However, the $4f$ orbitals also are generally more stable than the $5d$, although the difference is small and there are exceptions. The *idealized* filling pattern is for the $6s$ orbital to fill in Cs and Ba, followed by the deeply buried $4f$ orbitals in the 14 inner transition elements La through Yb. There are minor deviations from this pattern, as shown in Figure 9–4. The most important of these deviations is that the first electron after Ba goes into the $5d$ orbital in La and not into the $4f$. Lanthanum is more properly a transition metal than an inner transition metal. It is more relevant to understand the idealized filling pattern, however, than to worry about the individual exceptions to it.

The chemical properties of the inner transition metals from Ce to Lu vary even less than the properties of the transition metals, because successive electrons are in the deeply buried $4f$ orbitals. After the $4f$ orbitals are filled, the balance of the third transition-metal series, Hf to Hg, occurs with the filling of the $5d$ orbitals. The representative elements Tl through Rn are formed as the $6p$ orbitals fill.

The seventh and last period begins in the same way. First the $7s$ orbital fills, then the inner transition metals from Ac to No (with the irregularities shown in Figure 9–4) and finally the beginning of a fourth transition-metal series with Lr. There are more deviations from this simple *f*-first, *d*-next filling pattern in the actinides than in the lanthanides (Figure 9–4), and consequently the first few actinide elements show a greater diversity of chemical properties than do the lanthanides.

In summary, the idealized sequence of filling of orbitals across a period is as follows:

1) For period *n*, the *ns* orbital is filled first with two electrons. These elements are the alkali metals and the alkaline earths, and are classed with the *representative elements*.

2) The very deeply buried $(n - 2)f$ orbitals are filled next. They exist only for $(n - 2)$ greater than three, or for Periods 6 and 7. These elements, which have virtually identical outer electronic structure and therefore which have virtually identical chemical properties, are the *inner transition metals*.

3) The less deeply buried $(n - 1)d$ orbitals then are filled if they exist. They exist only for $(n - 1)$ greater than two, or for Period 4 and greater. These elements are similar to one another, but not as similar as the inner transition metals. They are called the *transition metals*.

4) Finally, the three *np* orbitals are filled to form the remaining *representative elements* and to conclude in each period with the outermost s^2p^6 configuration of the noble gases.

Challenge.   To see if you can predict the ground-state electronic configurations of a semiconductor, a conductor, a nuclear reactor fuel, and a metal cation in a biologically important molecule, try Problems 9–6 to 9–10 in *Relevant Problems for Chemical Principles* by Butler and Grosser.

9–2 ATOMIC PROPERTIES

We now can explain many of the facts that we observed in Chapter 6. The structure of the periodic table, with its groups and periods, can be seen to be a consequence of the order of energy levels (Figure 9–2). Elements in the same group have similar chemical properties because they have the same outer electronic structure in the *s* and *p* orbitals. The outer valence electrons that are so important in chemistry are these *s* and *p* electrons. The closed, inert shell of the noble gases is the completely filled s^2p^6 configuration. We can understand the mechanism of formation of the transition metals and the inner transition metals in terms of the filling of inner *d* and *f* orbitals. We can see the reasons for the features of the plot of first ionization energies, for general trends across a period or down a group, and for local fluctuations within a period.

We also can explain the details of the plot of electron affinities (Figure 6–8). The halogens have the highest electron affinity within a period because the net nuclear charge, after the effect of screening electrons in lower quantum levels has been accounted for, is greater for a halogen than for any other element in the period. The noble gases have low electron affinities because the new electron must be added to the next higher principal quantum level in each atom. Not only would the added electron be farther from the nucleus than the other electrons, it also would receive the full screening effect from all the others. The net nuclear charge felt by an added electron in S is $+6$; in Cl it is $+7$; in Ar it is 0.

Lithium and sodium have moderate electron affinities; beryllium and magnesium have smaller electron affinities because their s orbitals are full, and the added electron must go into the higher-energy p orbitals. Nitrogen and phosphorus have low electron affinities because an added electron must pair with an electron in one of the half-filled p orbitals.

The atomic and ionic radii shown in Figure 6–9 also make sense now. The various series of isoelectronic ions (As^{3-}, Se^{2-}, Br^-, Kr^0, Rb^+, Sr^{2+}, Y^{3+}, Zr^{4+}; or Au^+, Hg^{2+}, Tl^{3+}, Pb^{4+}, Bi^{5+}; and others) have the same electronic configuration but a steadily increasing positive charge on the nucleus. The ions show a steady decrease in size because of the increase in electrostatic attraction. Ionic radii within a vertical group (F^-, Cl^-, Br^-, I^-) show a steady increase with atomic number because the principal quantum number is greater and the outer electrons are farther from the nucleus. Nonionic bonding radii (metallic and covalent) show a similar but less pronounced decrease from left to right across a horizontal period as the increased nuclear charge attracts the electrons more strongly. The minimum in the transition-metal, nonionic radii arises because these values were obtained from interatomic distances in the metal. The tightness of metal bonding in these elements is a function of the number of unpaired d electrons. This number is a maximum at d^5: Mn, Tc, and Re. Hence, these metals have the strongest metallic bonds and closest approach of atoms.

The bonding effect of the d electrons is also evident in the high melting and boiling points in the middle of the transition metal series (Figure 9–5). The stronger the metallic bonds, the more thermal energy is needed to break some of them to form a liquid, and to break all of them to form a gas. Those transition metals with no unpaired d electrons (Zn, Cd, and Hg with $d^{10}s^2$) have relatively low melting and boiling points. The high boiling points of the elements of Group IVA (Si, Ge, and Sn) are caused by the presence, even in the liquid, of covalent bonds that must be broken during the vaporization process. Carbon does not melt at all. When enough thermal energy is supplied to break the many tetrahedral covalent bonds of diamond, the carbon atoms separate as a gas.

The irregularities in filling d orbitals in the transition metals, shown in color in Figure 9–4, are caused by the extra stability of a completely filled or half-filled d orbital. These irregularities have interesting consequences in the electrical conductivities of metals. A solid conducts electricity if the number of easily available electronic states is greater than the number of electrons. An electron in such a solid can be excited to a slightly higher state in which it has the energy to move through the metal. If all of the ground states are full and there is an energy gap before the next higher state, then the substance

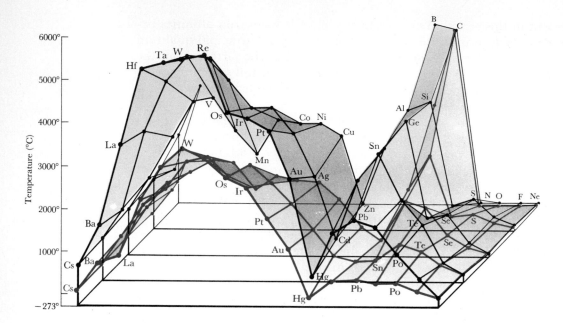

Figure 9–5. Boiling points (black dots) and melting points (colored dots) for the elements. Melting and boiling points of the transition metals are highest when there are a maximum of five unpaired d electrons (d^5s^1 and d^5s^2) and a maximum amount of metal bonding in the solid or liquid. The final transition metals with no unpaired d electrons Zn, Cd, Hg (with $d^{10}s^2$) are easily melted and vaporized. The high melting and boiling points of the nonmetals such as C, Si, *and* Ge *arise from the rigid framework of covalent bonds in the solid, some of which persist in the liquid state.*

is an insulator. A metal with filled outer *s* orbitals probably will be a poorer conductor than one with the configuration s^1, although conductivity is sufficiently complicated so this rule is not inviolable. The best conductors of all the elements are the *coinage metals:* Cu, Ag, and Au (Figure 9–6). All of these elements have the atypical electronic arrangement $d^{10}s^1$ instead of d^9s^2, thus there are twice as many *s* orbitals available as there are *s* electrons to fill them. Mercury, cadmium, and zinc, with the $d^{10}s^2$ configuration, are much poorer conductors. Similarly, s^1 above a half-filled *d* shell also produces high conductivity: W (d^5s^1 in the solid), Mo, and Cr are much better conductors than the neighboring Re, Tc, and Mn. [Note in Figure 9–6 the sharp distinction between metals and nonmetals (e.g., between Al and Si).]

Electronegativity

When atoms are joined, two extreme bond types are observed: ionic and covalent. If an element such as potassium, with low ionization energy and low electron affinity, combines with an element such as chlorine, with high

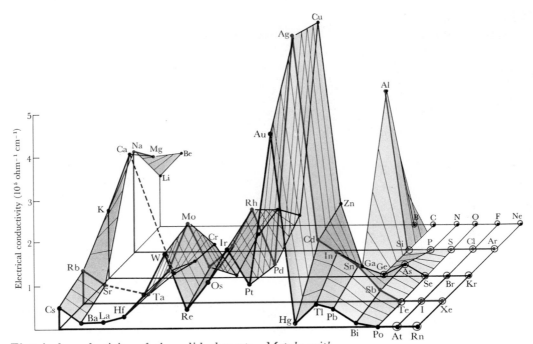

Figure 9–6. Electrical conductivity of the solid elements. Metals with one outer s electron above a half-filled or filled d shell are especially good conductors of electricity. Those with a filled s orbital conduct less well. Thus the coinage metals, Cu, Ag, *and* Au, *which have the outer electronic structure* $d^{10}s^1$, *are better conductors than* Zn, Cd, *and* Hg, *which have the* $d^{10}s^2$ *structure.* Cr, Mo, *and* W, *which have the* d^5s^1 *structure in the solid, are better conductors than* Mn, Tc, *and* Re, *which have the* d^5s^2 *structure.*

ionization energy and high electron affinity, an ionic compound will be formed in which there is a transfer of an electron from K to Cl. In contrast, elements with similar ionization energies and electron affinities form covalent bonds. Linus Pauling (1900–) defined a quantity called the *electronegativity*, X, in 1932, and R. S. Mulliken (1896–) showed two years later that it could be related to the average of the electron affinity and ionization energy. The electronegativity is thus a measure of the tendency of an element to hold electrons when it forms a compound. The difficulty with Mulliken's calculations was that reliable electron affinities were known for relatively few elements.

Pauling obtained electronegativity values by comparing the bond energy of a bond between unlike atoms with the *average* energies of the bonds between each of the two elements; for example, by comparing the energy of the HF bond with the energies of the bonds in H_2 and F_2. If HF formed a covalent bond like H_2 and F_2, then we would expect the bond energy in HF to be close to the average of the bond energies in H_2 and F_2. However, in molecules such as HF, the bonds are stronger than such averages. We can see

Figure 9–7. Electronegativities of the elements, on the Pauling scale. Electronegativity is a measure of the relative tendency of an atom to bind electrons in a chemical compound. Compounds of substances with large differences in electronegativity, such as sodium and chlorine, will be ionic. Compounds of substances of roughly equal electronegativities, such as carbon and hydrogen, will be covalent. Most chemical bonds must be considered to have partly covalent character and partly ionic; the relative contributions of covalent and ionic bonding depend on the electronegativity differences of the atoms.

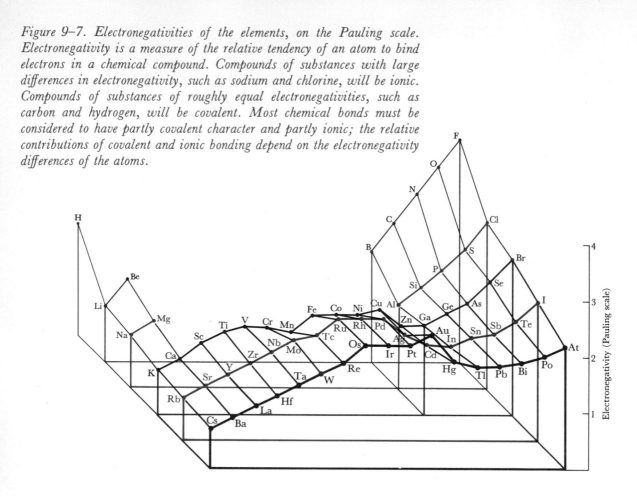

this phenomenon in the heats of formation of the hydrogen halides from their gaseous dimer elements. When HI is made from H_2 and I_2, H—I bonds are formed, and H—H and I—I bonds are broken. Therefore, the heat of formation of HI is a measure of the extra stability of the bonds between unlike atoms:

$$\tfrac{1}{2}H_2(g) + \tfrac{1}{2}I_2(g) \rightarrow HI(g) \qquad \Delta H^0_{298} = -1.2 \text{ kcal mole}^{-1}$$
$$\tfrac{1}{2}H_2(g) + \tfrac{1}{2}Br_2(g) \rightarrow HBr(g) \qquad \Delta H^0_{298} = -12.3 \text{ kcal mole}^{-1}$$
$$\tfrac{1}{2}H_2(g) + \tfrac{1}{2}Cl_2(g) \rightarrow HCl(g) \qquad \Delta H^0_{298} = -22.1 \text{ kcal mole}^{-1}$$
$$\tfrac{1}{2}H_2(g) + \tfrac{1}{2}F_2(g) \rightarrow HF(g) \qquad \Delta H^0_{298} = -64.2 \text{ kcal mole}^{-1}$$

Hydrogen and iodine have similar electronegativities and form a bond with much covalent character; hydrogen and fluorine differ in electronegativity and form a more ionic molecule. This heat of formation from diatomic gases, ΔH^0_{298}, is used by Pauling to calculate electronegativities.

If the bond energy of a compound AB, or the heat of formation of the bond from the gaseous atoms A and B, is D_{AB}, and the bond energy for the

Table 9–1. *First Ionization Energies, Electron Affinities, and Electronegativities of the Elements*

	1 H	2 He																
IE^a	313	567																
EA^b	+16.1	−14.4																
X^c	2.20	−																

	3 Li	4 Be										5 B	6 C	7 N	8 O	9 F	10 Ne
IE	124	215										191	260	336	314	402	497
EA	+18.4	−4.4										+7.6	+28.6	+1.1	+33.9	+83.5	−13.1
X	0.98	1.57										2.04	2.55	3.04	3.44	3.98	—

	11 Na	12 Mg										13 Al	14 Si	15 P	16 S	17 Cl	18 Ar
IE	119	176										138	188	254	239	300	363
EA	+28.0	0										+12.0	+33.0	+18.0	+47.7	+85.1	—
X	0.93	1.31										1.61	1.90	2.19	2.58	3.16	—

	19 K	20 Ca	21 Sc	22 Ti	23 V	24 Cr	25 Mn	26 Fe	27 Co	28 Ni	29 Cu	30 Zn	31 Ga	32 Ge	33 As	34 Se	35 Br	36 Kr
IE	100	141	151	158	156	156	171	182	181	176	178	216	138	187	231	225	273	323
EA	+19.0																+79.6	
X	0.82	1.00	1.36	1.54	1.63	1.66	1.55	1.8	1.88	1.91	1.90	1.65	1.81	2.01	2.18	2.55	2.96	—

	37 Rb	38 Sr	39 Y	40 Zr	41 Nb	42 Mo	43 Tc	44 Ru	45 Rh	46 Pd	47 Ag	48 Cd	49 In	50 Sn	51 Sb	52 Te	53 I	54 Xe
IE	96	131	152	160	156	166	167	173	178	192	175	207	133	169	199	208	241	280
EA																	+72.7	
X	0.82	0.95	1.22	1.33	1.6	2.16	1.9	2.28	2.2	2.20	1.93	1.69	1.78	1.96	2.05	2.1	2.66	—

	55 Cs	56 Ba	57 La	72 Hf	73 Ta	74 W	75 Re	76 Os	77 Ir	78 Pt	79 Au	80 Hg	81 Tl	82 Pb	83 Bi	84 Po	85 At	86 Rn
IE	90	120	129	127	138	184	182	201	212	207	213	241	141	171	185	—	—	248
EA																		
X	0.79	0.89	1.10	1.3	1.5	2.36	1.9	2.2	2.2	2.28	2.54	2.00	2.04	2.33	2.02	2.0	2.2	—

^{a} IE is the first ionization energy in kcal mole^{-1} of atoms.
^{b} EA is electron affinity in kcal mole^{-1} of atoms.
^{c} X is Pauling electronegativity as recalculated by A. L. Allred, *J. Inorg. Nucl. Chem.* **17**, 215 (1961).

two covalently bonded elements is D_{AA} and D_{BB}, then the added stability produced by the ionic bond is

$$\Delta_{AB} = D_{AB} - \frac{D_{AA} + D_{BB}}{2}$$

This Δ_{AB} is equal to $-\Delta H^0_{298}$ for reactions of the type shown previously. Pauling defined the difference in electronegativities of atoms A and B as

$$X_A - X_B = 0.208\sqrt{\Delta_{AB}}$$

with X for hydrogen set at 2.1. Thus, by Pauling's method electronegativities can be calculated for many more elements. These values appear in Figure 9–7 and in Table 9–1.

Electronegativity is a measure of the attraction that an atom has for electrons in a bond it has formed with another atom. Consequently, from the data in Table 9–1, you can predict that CsF will be ionic and CH_4 will be covalent. Bonds between atoms have varying degrees of covalent and ionic character, depending on the difference in electron-binding ability (electro-negativity) of the atoms involved.

Oxidation and Reduction Potentials

The oxidation potential of a reaction is a measure of the tendency for a reaction of the type

(reduced substance) $\rightarrow$ (oxidized substance) + (electrons)

to take place, in comparison with the reaction

$$\tfrac{1}{2}H_2 \rightarrow H^+ + e^-$$

which is *assigned* an oxidation potential of zero. If the oxidation potential of a reaction is positive, the reaction has a stronger tendency to occur than does the oxidation of H_2. This is true for sodium metal; its standard oxidation potential at 25°C, $\mathscr{E}^0$, is

$$Na \rightarrow Na^+ + e^- \qquad \mathscr{E}^0 = +2.71 \text{ V}$$

The drive toward the oxidized state of Na in water is so strong that water itself is decomposed, and hydrogen ions are reduced to H_2 gas.

If the oxidation potential for a reaction is negative, the favored drive is toward the reduced rather than the oxidized state:

$$Ag \rightarrow Ag^+ + e^- \qquad \mathscr{E}^0 = -0.80 \text{ V}$$

Thus the reverse of this reaction will occur. We will return to a systematic study of oxidation potentials in Chapter 17, where we will see how they are measured in electrolytic cells. At the moment, we want to use them only as measures of the relative tendency of elements to exist in different oxidation states in solution.

The qualification, "in solution," in the preceding sentence is an important one. The first ionization energy of sodium measures the tendency of a *gaseous atom* of Na to lose an electron and to form a *gaseous ion*. In contrast, the oxidation potential measures the tendency of *solid* Na to lose an electron and to form a *hydrated* sodium ion in aqueous solution. This is a much more useful quantity in most chemical applications. In Section 6–4 we noted how the oxidation energy and the oxidation potential of Ca can be obtained from the heat of sublimation of solid Ca, the first and second ionization energies of calcium, and the heat of hydration of the calcium ion. Sometimes the result of oxidation of a metal in solution is not a hydrated cation but an oxide complex:

$$Mn + 4H_2O \rightarrow MnO_4^- + 8H^+ + 7e^- \qquad \mathscr{E}^0 = -0.771 \text{ V}$$

The oxidation potential of Mn to MnO_4^- is much more meaningful in solution chemistry than is the energy required to strip seven electrons from a Mn atom in the gas phase.

Although it is more common to work with oxidation potentials in the United States, most of the rest of the world uses reduction potentials. (The United States is switching gradually to reduction potentials also.) If the oxidation potential for the oxidation of a sodium atom is +2.71 V, the reduction potential for the reduction of a sodium ion to the neutral atom is −2.71 V:

$$Na^+ + e^- \rightarrow Na \qquad \mathscr{E}^0 = -2.71 \text{ V}$$

Reduction potentials are oxidation potentials with the signs reversed, and you must be aware of which one you are dealing with in a tabulation of electrode potentials. Since we are comparing the relative ease of oxidation of metals, we will use oxidation potentials in this chapter, but in Chapter 17 we will adopt the international convention and use reduction potentials. You should be familiar with both.

9-3   CHEMICAL PROPERTIES: THE s-ORBITAL METALS

The next four sections contain descriptive chemistry of the Group IA and IIA metals, the transition metals, and, briefly, the representative elements of Groups IIIA and beyond. We shall return to the chemistry of the non-metals several times after the chapters on chemical bonding, so they are not covered in depth in this chapter. Knowing the electronic structures of atoms, we can interpret the chemical properties of the metals in a reasonable way. You should not attempt to memorize all of the facts in these four sections. Instead, you should try to pick out of the descriptive material those properties that show regular trends across the table, and that can be explained by electronic structure. Not every chemical property becomes absolutely clear once we know the electronic structure of the element; we are not as far from Mellor's harsh judgment in the introduction to Part 3 as we would like to be. But much of what we observe *does* make sense now, and it is this sense that we shall look for in the mass of chemical data.

Group IA. Alkali Metals: Li, Na, K, Rb, and Cs

All of these metals have an s^1 outer electronic configuration. The electron is lost easily; thus these elements have low ionization energies and low electro-negatives. Ionization energy and electronegativity decrease from Li to Cs as the distance of the outer shells from the nucleus increases.

These metals are the most reactive known. They never occur naturally in the metallic state but always in combination with oxygen, chlorine, or other elements, and always in the +1 oxidation state. All of their compounds are ionic, even the hydrides. Virtually any substance capable of being reduced will be, in the presence of an alkali metal. The oxidation potentials of the alkali metals, from Li to Cs, are

$$Li(s) \rightarrow Li(aq)^+ + e^- \qquad \mathscr{E}^0 = +3.05 \text{ V}$$
$$Na(s) \rightarrow Na(aq)^+ + e^- \qquad \mathscr{E}^0 = +2.71 \text{ V}$$
$$K(s) \rightarrow K(aq)^+ + e^- \qquad \mathscr{E}^0 = +2.92 \text{ V}$$
$$Rb(s) \rightarrow Rb(aq)^+ + e^- \qquad \mathscr{E}^0 = +2.93 \text{ V}$$
$$Cs(s) \rightarrow Cs(aq)^+ + e^- \qquad \mathscr{E}^0 = +2.92 \text{ V}$$

Each of these metals has a strong tendency to lose electrons and to become oxidized in solution. In contrast, it is difficult to reduce their ions; potassium ions have a reduction potential of -2.92 V. Lithium loses electrons in solution more readily than Cs, in spite of the higher ionization energy of Li, because the small size of a Li^+ ion permits water molecules to approach the center of the ion more closely, which makes the hydrated ion quite stable.

Water attacks all of the alkali metals, and the reaction with all of these metals is violent and exothermic. A typical reaction is

$$Na + H_2O \rightarrow Na^+ + OH^- + \tfrac{1}{2}H_2 \qquad \Delta H^0_{298} = -40 \text{ kcal}$$

The hydrogen gas evolved is ignited by the heat of the reaction and burns spontaneously in air. Sodium metal carelessly thrown into a sink is one of the hazards of a beginning chemistry laboratory. The alkali metals ordinarily are stored in kerosene or some other unreactive hydrocarbon.

Because the alkali metals are the strongest reducing agents known, the free metals cannot be prepared conveniently by reduction of their compounds with another substance:

$$Li^+ + (\text{reducing substance}) \rightarrow Li + (\text{oxidized substance})$$

Instead, the metals usually are prepared by electrolysis of their molten compounds.

The easily lost valence electron is responsible for the metallic properties of the alkali metals. With only one mobile electron per atom, metallic bonds are weak. The bonds become weaker with increasing atomic number as the valence electrons become more distant from the nucleus. The metals have low melting and boiling points, and are soft, malleable (can be hammered into thin sheets), and ductile (can be drawn into wires). Lithium can be cut with a knife with difficulty, but Cs is as soft as cheese. Cesium has the lowest ionization energy of any element, and its valence electron can be ejected most easily by light in a photoelectric cell. The photoelectric effect is used in photocells and in television cameras such as the iconoscope, in which the optical image falling on cesium-coated cathodes is converted to electrical impulses.

Almost all of the compounds of the alkali metals are soluble in water. The alkali metal ions in solution are colorless. Color is produced when an electron in an atom is excited from one energy level to another, and when the difference in energy of these levels is in the visible portion of the spectrum. The alkali metal ions have no free electrons to be excited by energies in the visible region. The oxides of the alkali metals all are basic, and all react with water to form basic hydroxides that are soluble and completely dissociated.

Alkali metals have the interesting property of being soluble in liquid ammonia and forming intensely blue solutions that leave behind the original metal when the ammonia is evaporated. The atoms dissociate into positive ions and electrons, and the electrons associate with the NH_3 solvent molecules. Such electrons are known as *solvated electrons*. The intense color has been shown to arise from the solvated electrons and not from the metal ions; the same color can be produced by introducing electrons into the ammonia from a platinum electrode.

Group IIA. Alkaline Earth Metals: Be, Mg, Ca, Sr, and Ba

The chemistry of the alkaline earth metals is the chemistry of atoms with two easily lost electrons. All are typical metals and strong reducing agents, although not quite as strong as the alkali metals. The nuclear charge has increased by one from the alkali metals in a given period, but the screening of the nucleus by electrons in inner orbitals is similar for both groups, so the net nuclear charge is greater. Thus, the alkaline earth atoms are smaller and have higher first ionization energies than those of alkali metals in corresponding periods. Their oxidation potentials in aqueous solution are

$$Be(s) \rightarrow Be(aq)^{2+} + 2e^- \qquad \mathscr{E}^0 = +1.85 \text{ V}$$
$$Mg(s) \rightarrow Mg(aq)^{2+} + 2e^- \qquad \mathscr{E}^0 = +2.37 \text{ V}$$
$$Ca(s) \rightarrow Ca(aq)^{2+} + 2e^- \qquad \mathscr{E}^0 = +2.76 \text{ V}$$
$$Sr(s) \rightarrow Sr(aq)^{2+} + 2e^- \qquad \mathscr{E}^0 = +2.89 \text{ V}$$
$$Ba(s) \rightarrow Ba(aq)^{2+} + 2e^- \qquad \mathscr{E}^0 = +2.90 \text{ V}$$

They are more electronegative than the alkali metals, but all of their compounds, with the exception of some Be compounds, are ionic. Beryllium is the first example of the general observation that, within a group, elements with lower principal quantum number will be less metallic because their outer electrons are closer to the nucleus and are held more tightly. This behavior is reflected in the greater electronegativities of the smaller atoms within a group (Figure 9–7). Beryllium has a lower oxidation potential, or lower tendency to lose an electron in solution, for the same reason that it has a higher first ionization energy than other elements in the group (Figure 9–3). It is true that Be, like Li, has a high hydration energy because of its small size. This leads us to expect a strong tendency to oxidize in aqueous solution and a large positive oxidation potential. However, Be has an extraordinarily high ionization energy and energy of vaporization (Table 9–2), and these two effects combine to dominate in the oxidation of Be to $Be(aq)^{2+}$ and the oxidation potential is somewhat lower than might be expected.

The second ionization energies of these metals usually are double their first ionization energies; thus, we might expect $+1$ ions to form and the $+1$ oxidation state to exist in solution. But this is not the case. The hydration of

Table 9–2. Properties of the Alkaline Earth Metals

Element	Be	Mg	Ca	Sr	Ba
Electronegativity	1.6	1.3	1.0	0.95	0.89
Metallic radius (Å)	0.89	1.36	1.74	1.91	1.98
Melting point(°C)	1278	651	842	769	725
Boiling point (°C)	2970	1107	1487	1384	1140
Heat of fusion (kcal mole^{-1})	2.8	2.2	2.2	2.2	1.8
Heat of vaporization (kcal mole^{-1})	70.4	30.8	35.8	33.2	36.1
MCl_2 (melting point, °C)	405	708	772	873	963
MCl_2 (boiling point, °C)	520	1412	1600	1250	1560
Equivalent conductivity of MCl_2 (ohm^{-1} mole^{-1})	0.086	29.0	52.0	—	—

the doubly charged cation gives it enough extra stability to overcome the energy required to remove the second electron (Section 6–4). Any solution of Ca^+ ions would disproportionate spontaneously to Ca metal and Ca^{2+} ions (Figure 6–7):

$$2Ca(aq)^+ \rightarrow Ca(s) + Ca(aq)^{2+}$$

The solution chemistry of the alkaline earth metals is exclusively that of the +2 oxidation state.

The free metals do not occur in nature because they are too reactive. Beryllium and magnesium are found in complex silicate minerals such as beryl ($Be_3Al_2Si_6O_{18}$) and asbestos ($CaMg_3Si_4O_{12}$) (Chapter 14). Emerald is impure beryl, colored with a trace of Cr. Magnesium, calcium, strontium, and barium occur as the relatively insoluble carbonates, sulfates, and phosphates. Calcium and magnesium are much more common than the other elements in the group. Calcium carbonate, $CaCO_3$, is found as chalk, limestone, and marble, usually from deposits of shells and skeletons of marine organisms. Like the alkali metals, the pure alkaline earth metals commonly are prepared from molten compounds by electrolysis because of the difficulty of finding anything with a higher oxidation potential with which to reduce them chemically.

The pure metals have higher melting and boiling points than do alkali metals because they have two electrons per atom for forming metallic bonds. For the same reason, they are harder, although they still can be cut with a sharp steel knife. Beryllium and magnesium are the only elements that commonly are used as structural metals; they are used pure or in alloys for aircraft and spacecraft, in which weight is an important factor.

Alkaline earth compounds generally are less soluble in water than the compounds of the alkali metals. The hydrides of Ca, Sr, and Ba (CaH_2, SrH_2, and BaH_2) all are ionic and are white powders that release H_2 gas upon reaction with water:

$$CaH_2 + 2H_2O \rightarrow Ca^{2+} + 2OH^- + 2H_2$$

The oxides, all with the expected 1/1 atomic ratio (BeO, CaO, etc.), are hard, relatively insoluble in water, and basic with the exception of BeO. In water, the basic oxides form hydroxides, which also are only slightly soluble:

$$CaO(s) + H_2O \rightarrow Ca(aq)^{2+} + 2OH^- \rightarrow Ca(OH)_2(s)$$

$Ba(OH)_2$ is strongly basic; $Mg(OH)_2$ is weakly basic.

Beryllium is definitely the odd-man-out in Group IIA. Its oxide is _amphoteric_, showing both acidic and basic properties. It is virtually insoluble in water, but in strong acid it acts as if it were basic:

$$BeO + 2H^+ + 3H_2O \rightarrow Be(H_2O)_4{}^{2+}$$

and in strong base it acts as if it were acidic:

$$BeO + 2OH^- + H_2O \rightarrow Be(OH)_4{}^{2-}$$

In both cases, the cation is so small that only a coordination number of four is possible. The coordinating groups are arranged tetrahedrally around Be. The amphoteric behavior of BeO arises because Be is so small and electronegative. Be^{2+} attracts electrons from neighboring water molecules and makes it easier for them to lose a proton to the surroundings:

$$Be^{2+} + 4H_2O \rightarrow Be(OH)_4{}^{2-} + 4H^+$$

Beryllium shows many other signs of nonmetallic behavior in addition to the amphoterism of its oxide. Its melting and boiling points and heats of vaporization all are unusually high, which is a reversal of the trends within the rest of Group IIA (Table 9–2). All of these facts suggest that the covalent bonds in Be, like those in diamond, persist in the liquid. Solid $BeCl_2$ is composed of covalently bonded chains that are held together only by weak intermolecular forces. $BeCl_2$ has the low melting point expected of a molecular, covalent compound instead of an ionic solid such as $CaCl_2$. Finally, liquid $BeCl_2$ does not conduct electricity, which indicates the absence of ions.

9–4   THE FILLING OF THE _d_ ORBITALS: TRANSITION METALS

The transition metals are hard metals with high melting points, boiling points, and heats of fusion. All of these properties depend on the number of unpaired _d_ electrons, which begins at one in Sc–Y–Lu, increases to a maximum of five at Mn–Tc–Re, and decreases to zero at Zn–Cd–Hg (Figure 9–4). These properties are illustrated in Figures 6–11 and 9–5. The atoms tend to become smaller with increasing atomic number across a period because of the increased nuclear charge. Atoms in the second transition-metal series, Y to Cd, are larger than those in the first, Sc to Zn. But atoms in the third transition-metal series, Lu to Hg, are not as much larger than the atoms in the second series as would be expected. The reason is that the first _inner_ transition-metal series, the lanthanides, is interposed after La. There is a steady decrease in size from La to Lu because of increasing nuclear charge, which produces the *lanthanide contraction*. Therefore, hafnium is not as large as it would have been had it followed directly after La. The nuclear charge in Zr is 18 greater than in Ti, but that in Hf is 32 greater than in Zr. The result is that the second- and third-series transition metals have not only the same outer electronic configurations in corresponding groups, but almost the same size as well. Thus, the second and third series are more similar in properties than either is to the first. Titanium resembles Zr and Hf less than Zr and Hf resemble one another. Vanadium is distinct from Nb and Ta, but the very names of Ta and Nb reflect the difficulty in separating them. Tantalum and niobium were discovered in 1801 and 1802, but for nearly half a century many chemists thought that they were the same element. Because of the difficulty in isolating it, Ta was named after Tantalus, the Greek mythological figure who was doomed to an eternity of frustrating labor. Niobium, in turn, was named for Niobe, the daughter of Tantalus. The smaller increase in radius that accompanies the third transition series is obscured by the perspective in Figure 6–9, but it is there if you look closely.

The Structure of Transition-Metal Ions

In the K^+ and Ca^{2+} ions, the 4s orbital is slightly more stable than the 3d, and added electrons go into the 4s orbital. In contrast, at Sc^{3+}, the 3d orbital energy dips below the 4s, and it remains there for all higher atomic numbers. The lone electron in Sc^{2+} is in a 3d orbital and not the 4s. This behavior is typical of all transition metals. The crossover of s- and d-orbital energies occurs at the beginning of a transition-metal series. Although the s orbitals fill first in Groups IA and IIA, it is the d orbitals that are occupied in transition-metal ions. The outer electronic configuration of Ti^{2+} is $3d^2$, and not $4s^2$.

The lowest oxidation state in all the 3d transition metals, with the exception of Cu and a few rare compounds of other metals, is $+2$, with both s electrons lost. Other higher oxidation states occur with the loss of more electrons from the d orbitals, up to a maximum equal to the number of *unpaired* electrons in the d orbitals. This is why the maximum oxidation number increases from $+3$ in Sc to $+7$ in Mn (five d plus two s), and thereafter falls by one per group to $+2$ in Zn (loss only of the two s electrons). The most common oxidation states are $+2$ and $+3$. In the first half of the series, the maximum oxidation state for each element—Sc(III), Ti(IV), V(V), Cr(VI), and Mn(VII)—is also common (Table 7–2). (Oxidation states of metals often are shown by Roman numerals in parentheses.)

These generalizations are true for the first transition series. There are some higher oxidation states observed in the second and third series, as in RuO_4 and OsO_4. It is more important that you know the behavior of the first transition series than that you remember the exceptions in the heavier metals.

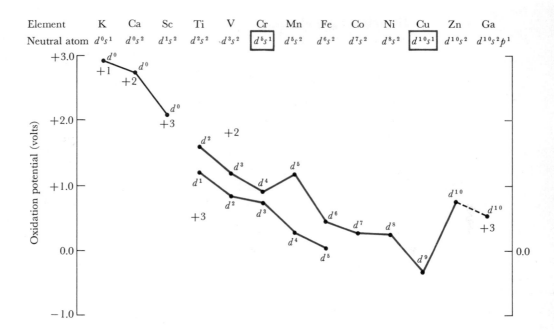

Figure 9–8. Oxidation potentials for the fourth-period metals, including the first transition-metal series. Potentials are for the production of simple cations in solution from the solid metals. K, Ca, and Sc are given for the production of +1, +2, and +3 ions with the Ar noble-gas electronic structure. Transition-metal potentials are shown for +2 and +3 ions. By each point is given the outer electronic configuration of the ion. For an explanation of the irregularities in the +2 ion curve, see text.

Oxidation Potentials

The oxidation potentials for the production of ions from neutral atoms are plotted in Figure 9–8 for the first transition series. The driving force for the production of ions decreases as the atomic number increases because the electrons are held more tightly. The +2-ion curve represents the removal of the two *s* electrons and the retention of the original *d* configuration. It is particularly difficult to remove two electrons in Cu because only one is in the *s* orbital and the second would have to come from the filled d^{10} orbitals. In contrast, it is quite easy to remove the two *s* electrons in Zn and leave the stable, filled d^{10} orbitals untouched. This same effect is observed to a lesser extent in Cr and Mn, in which a half-filled, d^5 configuration is present instead of a filled d^{10}. The *d* electrons always must be disturbed when a +3 ion is produced, consequently we do not have the local fluctuations observed in the +2 curve. The elements preceding the transition series become cations by losing all of their electrons outside the inner noble-gas shells; the representative elements following a transition series achieve their maximum oxidation number by using all of their electrons outside the filled d^{10} shells.

Chemical Properties of Individual Groups: Sc and Ti Groups

The Sc–Y–Lu triad has the outer electronic configuration $d^1 s^2$, and shows only +3 oxidation states. The properties of these elements are similar to those of Al in Group IIIA. All react with water as does Al. But Sc_2O_3 is a basic oxide rather than amphoteric as Al_2O_3 because Sc^{3+} is larger than Al^{3+}. The difference in behavior resembles that between CaO and BeO.

In the Ti–Zr–Hf triad, which has the $d^2 s^2$ electronic configuration, Ti and Zr show +2, +3, and +4 oxidation states, whereas Hf has only +4. This is an example of a general trend in the transition metals: The lower oxidation numbers are less important for the second and third transition series because the electrons are farther from the nucleus. If they lose some electrons, they are likely to lose all of them. The lower oxidation states of Ti are ionic, and the +4 state is more covalent and nonmetallic. Titanium(II) oxide, TiO, is basic, ionic, and has the NaCl crystal structure. In contrast, the dioxide, TiO_2, is a white, insoluble pigment that has both basic and acidic properties. The chlorides are a particularly good illustration of this progression of properties. The dichloride, $TiCl_2$, is a strong reducing agent and oxidizes spontaneously in air. It is an ionic solid that decomposes at 475°C in a vacuum. Since it reduces water to H_2, there is no aqueous chemistry of Ti^{2+}. Titanium trichloride, $TiCl_3$, is another strong reducing agent and is an ionic solid that decomposes at 440°C. In contrast, the tetrachloride, $TiCl_4$, is a stable liquid that freezes at $-25°C$ and boils at $+136°C$. It boils before $TiCl_3$ melts (decomposes) and is a molecular compound with covalent bonds.

Vanadium Group and the Colors of Ions

The chemistry of the V–Nb–Ta elements is similar to that of the previous triad. V and Ta have the $d^3 s^2$ electronic configuration and Nb has the $d^4 s^1$ configuration. Vanadium has oxidation states of +2, +3, +4, and +5, whereas only the +5 state is important in Nb and Ta (although some +3 and +4 compounds are known). Like Ti, Zr, and Hf, these metals react easily with N, C, and O at high temperatures, and it is difficult to prepare them by the high-temperature reduction processes used with Fe and other metals. At low temperatures an oxide coating protects them; consequently, the metals are more inert than their oxidation potentials would suggest. At the top of the group, V_2O_5 is amphoteric like TiO_2. It dissolves in both acids and bases to form complex and poorly characterized polymers. The +4 oxidation state of V is also on the borderline between ionic and covalent character; VCl_4 is a molecular liquid with a boiling point of 154°C. In contrast, the V(III) compounds all are ionic.

The vanadium ions are good examples of the colors that are typical of transition-metal compounds. Vanadium(V) as VO_4^{3-} is colorless. The vanadyl ion, VO^{2+}, is deep blue, the V^{3+} ion is green, and the V^{2+} ion is violet. Only these colors are seen from the entire visible spectrum because the three solutions absorb orange light (ca. 6100 Å), red light (ca. 6800 Å), and yellow light (ca. 5600 Å), respectively. The colors we see are the complementary colors to those absorbed (Table 11–3). Most electronic energy levels are so far apart that the radiation absorbed in an electronic excitation is in the ultraviolet. But when ions or molecules coordinate with transition metals to form complexes, the different *d* orbitals are given slightly different energies. (These complexes will be examined in more detail in Chapter 11.) The splitting of *d*-orbital energies is so small that the frequencies of radiation required for transitions are in the visible region; hence, colors are produced. The smallest energy absorption, in the red region, would leave unabsorbed the wavelengths that produce a complementary blue-green color in the solution or compound. Larger and larger energy absorptions produce blue-green, blue, violet, purple, red-orange, and finally yellow; these are the complementary colors to red, orange, yellow, green, blue, and violet. Therefore, color is an approximate guide to the electronic energy differences in metal complexes, as we shall see in Chapter 11.

Chromium Group and the Chromate Ion

The elements Cr, Mo, and W have high melting and boiling points (Figure 9–5) and are hard metals. They are relatively inert to corrosion because films of oxides formed on the surface adhere and protect the metal beneath. The thin layer of Cr_2O_3 on chromium metal makes chrome plating an efficient protection for the more easily attacked metals such as iron. Along with V, these three metals are used mainly as alloying agents in steels. Vanadium gives steels ductility, tensile strength, and shock resistance. Chromium makes stainless steels corrosion resistant, Mo acts as a toughening agent, and W is used in steel cutting tools that remain hard even at red heat.

Chromium(III) is the most prevalent oxidation state of chromium. Chromium(II) is a good reducing agent, and Cr(VI) is a good oxidizing agent. As we would expect, the acidity of the oxides varies with the oxidation state: CrO_3 is acidic, Cr_2O_3 is amphoteric, and CrO and $Cr(OH)_2$ are basic. A common anion is the yellow chromate ion, CrO_4^{2-}, which dimerizes in acid to form the orange dichromate ion:

$$2CrO_4^{2-} + 2H^+ \rightarrow Cr_2O_7^{2-} + H_2O$$

The dichromate ion is a powerful oxidizing agent, and the reaction by which dichromate ion is reduced to Cr^{3+} has a large positive reduction potential:

$$Cr_2O_7^{2-} + 14H^+ + 6e^- \rightarrow 2Cr^{3+} + 7H_2O \qquad \mathscr{E}^0 = +1.33 \text{ V}$$

We could express the same chemical fact by saying that the reaction by which Cr^{3+} is oxidized to dichromate ion has a low negative oxidation potential, and that the reaction tends to go in the other direction:

$$2Cr^{3+} + 7H_2O \rightarrow Cr_2O_7^{2-} + 14H^+ + 6e^- \qquad \mathscr{E}^0 = -1.33 \text{ V}$$

Manganese Group and the Permanganate Ion

Of the Mn–Tc–Re triad, only Mn has any real importance. Rhenium was discovered in 1925, and Tc was the first element produced artificially. Technetium was discovered by Perrier and Segré, in 1937, in a sample of Mo that had been irradiated by deuterons ($^2_1$H particles) in the Berkeley cyclotron by Ernest Lawrence (for whom, incidentally, element 103 was named). The new element was named technetium from the Greek *technetos*, artificial.

The chief use of Mn metal is to make hard and tough manganese steels. Oxidation states of $+2$ to $+7$ are known; the two extremes are the most important. Unlike Ti^{2+}, V^{2+}, and Cr^{2+}, Mn^{2+} shows little tendency to go to higher oxidation states. It is strongly resistant to oxidation and is not a good reducing agent. Manganese(II) in water forms the pink $Mn(H_2O)_6^{2+}$ octahedral complex, and the $MnSO_4$ and $MnCl_2$ salts are also pink. The oxidation states Mn(III) through Mn(VI) are rare, except for the chief natural ore, MnO_2. Mn(VI) does exist as the manganate ion, MnO_4^{2-}. The Mn(VII) state is chiefly important for the deep purple *permanganate* ion, MnO_4^-. It is one of the most powerful common oxidizing agents, with a reduction potential of $+1.49$ V:

$$MnO_4^- + 8H^+ + 5e^- \rightarrow Mn^{2+} + 4H_2O \qquad \mathscr{e}^0 = +1.49 \text{ V}$$

Note again that a compound with a high positive reduction potential or a low negative oxidation potential will be a good oxidizing agent because the compound itself will have a strong tendency to go to the reduced form.

Solutions of the permanganate ion are used as disinfectants. (One of our black comedies on the World War II theme has this bitter remark about the medical services available to enlisted men: "The first time you come in, they give you two aspirins; the second time they paint your gums purple. If you show up again, they arrest you for impersonating an officer." The purple, of course, is $KMnO_4$.) Manganese is an excellent example of the dependence of chemical properties on oxidation state. Manganese(II) exists in solution as a cation and has a basic oxide, MnO, and hydroxide, $Mn(OH)_2$. At the other extreme, the $+6$ and $+7$ states exist as anions: MnO_4^{2-} and MnO_4^-, corresponding to the acidic oxides MnO_3 and Mn_2O_7.

The Iron Triad and the Platinum Metals

In Group VIII, the horizontal similarity between Fe, Co, and Ni is greater than between these and the corresponding elements in the second and third transition series. These nine elements usually are separated into the iron triad, Fe–Co–Ni, and the light and heavy platinum triads, Ru–Rh–Pd and Os–Ir–Pt. Iron, cobalt, and nickel have the electronic configurations d^6s^2, d^7s^2, and d^8s^2, respectively. They all are ferromagnetic (Section 10–3), and all show chiefly the $+2$ and $+3$ oxidation states. (The $+3$ state is very rare for Ni.) Iron is one of the most important structural metals. Many of the other transition metals are important chiefly as alloying agents with iron. Iron is found in three main oxide ores: FeO, Fe_2O_3, and the magnetic mixed oxide magnetite, Fe_3O_4 or $FeO \cdot Fe_2O_3$. Iron is produced by high-temperature reduction with the CO from coke in a blast furnace. The result is cast iron with 3–4% carbon. The open hearth and Bessemer processes are means of burning out most of this carbon with streams of oxygen to obtain steels with 0.1–1.5% carbon.

Iron is no more intrinsically reactive than the other transition metals that we have been discussing. Unfortunately, however, the iron oxides do not have crystal-lattice dimensions comparable to those of metallic iron and therefore do not adhere to the surface. Rust (iron oxide) flakes off as it is formed and exposes fresh metal to attack (see Section 17–7). Chrome steel or stainless steel is more corrosion resistant, but the customary protection is an added surface layer such as chromium, tin, nickel, or paint. Ferrous or iron(II) compounds usually are green, and the hydrated ferric ion, $Fe(H_2O)_6^{3+}$, is a pale violet. Both +2 and +3 states form octahedral complexes with cyanide: $Fe(CN)_6^{4-}$ and $Fe(CN)_6^{3-}$. The traditional names for these anions are "ferrocyanide" and "ferricyanide" ions. In the modern, systematic nomenclature, they are "hexacyanoferrate(II)" and "hexacyanoferrate(III)." The nomenclature of complex ions is given in Chapter 11.

Cobalt in solution exists mainly as the +2 cation, since Co^{3+} is an excellent oxidizing agent and has a strong tendency to be reduced to Co^{2+}:

$$Co^{3+} + e^- \rightarrow Co^{2+} \qquad \mathscr{E}^0 = +1.84 \text{ V}$$

But in many octahedral complexes of Co(III) the ligands (the ions or molecules attached to Co) stabilize it against reduction. Nickel forms octahedral and square planar complexes in the Ni(II) state. Most of its octahedral salts, as well as the hydrated cation, are green. The square planar complexes usually are red or yellow.

The light and heavy platinum triads can be passed over quickly. The metals are relatively rare, and much work remains to be done on their reactions. All are relatively unreactive metals, and are found naturally as the pure metals. The oxidation states +2, +3, and +4 are most important, and the metals form octahedral or square planar complex ions in solution. Complex ions of Pt(IV) and Ir(III) are octahedral. The square planar $PtCl_4^{2-}$, the tetrachloroplatinate(II) ion, shows a strong tendency to bind to sulfur in proteins and has been useful in preparing heavy-atom derivatives of proteins for x-ray crystallographic analysis.

The Coinage Metals

Copper, silver, and gold have the slightly irregular outer electronic configuration $d^{10}s^1$. They have lower melting and boiling points than the preceding transition metals and are moderately soft. These properties are part of a downward trend that began with Group VIB (Cr–Mo–W), which arises from the decreasing number of unpaired *d* electrons. They are excellent conductors of electricity and heat since their electronic arrangement makes the s electrons extremely mobile. They are malleable, ductile, inert, and can be found naturally in the metallic state. Although rare enough to be prized, they are much less scarce than the platinum metals. This relative abundance plus their occurrence as uncombined metals meant that they were the first metals to be collected and worked by man. The first metal to be reduced from its ore probably was copper. Metallurgy began when it was discovered that an alloy of copper with tin (a naturally occurring impurity) produced the much harder bronze. Copper artifacts have been unearthed in some of the earliest farming communities in the Middle East, dating from 7000–6000 B.C. Bronze was known in the Sumerian cities of Ur and Eridu from 3500 B.C., during the era that also included the invention of writing.

These three metals have been the source of more strife and trouble than any other elements. Until a century ago, they were used mainly for their symbolic and decorative qualities. More recently, the physical properties of Ag and Au—electrical and thermal conductivity and corrosion resistance—have become so valuable that the metals can be spared no longer for their traditional coinage roles. Gold now is used for plating external surfaces of delicate components in satellites and space probes.

Copper, silver, and gold have little resemblance to the alkali metals, with which they are associated in the short form of the periodic table derived from Mendeleev's table (Figure 6–1). Copper shows mainly the $+2$ oxidation state in solution, and $+1$ to a lesser extent. The reverse is true for Ag: The $+1$ state is common, and the $+2$ and $+3$ can be obtained only under extreme oxidizing conditions. Gold occurs in the $+3$ state and less frequently in the $+1$. The metals have low negative oxidation potentials, thereby indicating their inertness and reluctance to oxidize:

$$\begin{aligned}
\text{Cu} &\rightarrow \text{Cu}^{2+} + 2e^- & \mathscr{E}^0 &= -0.34 \text{ V} \\
\text{Ag} &\rightarrow \text{Ag}^+ + e^- & \mathscr{E}^0 &= -0.80 \text{ V} \\
\text{Au} &\rightarrow \text{Au}^{3+} + 3e^- & \mathscr{E}^0 &= -1.42 \text{ V}
\end{aligned}$$

Copper(I) is unstable in solution and disproportionates spontaneously to Cu and Cu^{2+}. However, it can be stabilized by complexes such as $CuCl_2^-$. Copper(I) exists as the solid and extremely insoluble Cu_2O and Cu_2S, which are the principal ores of copper. The chemistry of Cu(II) is similar to that of other transition metals in the $+2$ oxidation state. The hydrated Cu(II) ion has a characteristic blue color, and tetraamminecopper(II), $Cu(NH_3)_4^{2+}$, is an intense blue. The complex is square planar. Silver(I) forms complexes such as $AgCl_2^-$, $Ag(NH_3)_2^+$, and $Ag(S_2O_3)_2^{3-}$, and Au(III) forms the very stable $AuCl_4^-$ complex.

The Chemistry of Photography

All of the silver halides except AgF are sensitive to light and are the basis of the photographic process. In making photographic film, fine crystals of AgBr

are spread in gelatin on a film backing. Light from the camera image interacts with the crystalline AgBr in a poorly understood process that appears
to involve defects in the crystal structure, and makes the grains, or crystals,
more sensitive to reduction. The sensitized AgBr is reduced in the developer
by a mild organic reducing agent such as hydroquinone:

$$AgBr + e^- \text{ (reducing agent)} \rightarrow Ag + Br^-$$

Then the unsensitized AgBr grains are dissolved and washed away in a
sodium thiosulfate solution, one of the few solutions in which silver halides
are soluble:

$$AgBr + 2S_2O_3{}^{2-} \rightarrow Ag(S_2O_3)_2{}^{3-} + Br^-$$

The old name for sodium thiosulfate was sodium hyposulfite; hence, the
synonym "hypo" for fixer.

The Low-Melting Transition Metals

The most distinguishing characteristic of Zn, Cd, and Hg is their weak
coherence as metals. They have low melting points, low boiling points, and
are soft. Mercury is the only liquid metal at room temperature. Zinc and Cd
resemble the alkaline earth metals in their chemical behavior. Mercury is
more inert and resembles Cu, Ag, and Au. All three elements have a $+2$
oxidation state. Mercury also has a $+1$ state in compounds such as Hg_2Cl_2.
But mercury(I) always appears as the dimeric ion, $Hg_2{}^{2+}$, and x-ray and
magnetic measurements show that the two Hg atoms are held together by a
covalent bond. Therefore, mercury has a $+1$ oxidation number in Hg_2Cl_2
only in the same formal sense that oxygen is -1 in hydrogen peroxide,
H—O—O—H. The metal is still divalent. The chemistry of the $+2$ state in
this group is as expected. The oxides ZnO, CdO, and HgO are only slightly
soluble in water but quite soluble in strong acids, as basic oxides should be.
Yet ZnO is also soluble in strong bases; hence it is amphoteric. Again, this
behavior arises because of the small size of the Zn^{2+} cation and the ease
with which it can pull electrons from water molecules and cause them to
release protons.

With these elements, the *d* orbitals are filled and the transition series
are closed. The next electrons to be added must go into the higher-energy *p*
orbitals, and the more rapidly varying representative elements are begun.

Trends in the Transition Metals

What systematic trends in behavior of the transition metals can we see from
the foregoing material? Some of the trends in properties are as follows.

1) The transition metals can lose, at most, the two *s* electrons and all
   unpaired *d* electrons in the outer shell. Therefore, the maximum oxidation number is three for Sc, and increases by one per group to a maxi-

mum of seven for Mn. Thereafter it falls by one per vertical column, through Fe, Co, Ni, and Cu, to two for Zn. The only exception to this rule is the absence of oxidation number five for Co. We shall explain this Co anomaly in Chapter 11, when we discuss transition-metal complexes.

2) Oxidation states lower than this maximum are found, with $+2$ and $+3$ being especially common.

3) The first transition-metal series, Sc to Zn, shows the full range of oxidation states. The second, and especially the third series, Lu to Hg, show only the higher oxidation states.

4) For a given element with a range of oxidation numbers, the behavior of the lowest oxidation state will be the most metallic, and that of the highest oxidation state, the least metallic. For example, compounds of V(III) are ionic, whereas V(V) has many compounds with covalent bonds. Among oxides of Cr, CrO is basic, Cr_2O_3 is amphoteric, and CrO_3 is acidic. Titanium dichloride, $TiCl_2$, and $TiCl_3$ are ionic solids, whereas $TiCl_4$ is a molecular liquid.

5) In higher oxidation states, the isolated cation is not stable, even when coordinated to water molecules. Such high oxidation states can be stabilized by coordination to oxide ions. Thus, Sc^{3+} exists as a hydrated ion, $Sc(H_2O)_6{}^{3+}$; Ti(IV) requires the stabilizing influence of coordinating groups such as hydroxide in $Ti(OH)_2(H_2O)_4{}^{2+}$; and V(V), Cr(VI), and Mn(VII) are coordinated to oxide ions in $VO_2{}^+$, $CrO_4{}^{2-}$, and $MnO_4{}^-$. Oxidation states that are not stable in solution sometimes can be stabilized by the formation of complexes such as $CuCl_2{}^-$.

6) Oxidation potentials for the oxidation of transition metals to the $+2$ or $+3$ cation in aqueous solution generally decrease as the atomic number increases, thereby reflecting the greater difficulty in removing electrons from the atoms. Ionization energies increase in a corresponding manner.

7) The physical properties of the transition metals (melting point, boiling point, heats of fusion and vaporization, and hardness) all reflect the number of unpaired d electrons in the atoms. All of these properties increase to a maximum in the Mn group and then decrease with increasing atomic number.

Before you go on. You may want more practice in using oxidation numbers and nomenclature of the more important transition metals. If you do, try Review 9 of *Programed Reviews of Chemical Principles* by Lassila *et al.*

9–5 THE FILLING OF f ORBITALS: LANTHANIDES AND ACTINIDES

After the atomic number reaches 57, the energy of the $4f$ orbitals is sufficiently low to allow those orbitals to be utilized. Consequently, after barium in Period 6 the seven $4f$ orbitals can fill successively with electrons, thereby producing the 14 lanthanide metals. Likewise in Period 7 after $Z = 89$, when the $5f$ and the $6d$ orbitals have virtually the same energy, there are 14 actinide metals, which are formed when the seven $5f$ orbitals are filled successively with electrons. The electronic configurations of these inner transition elements are shown in Figure 9–4. As with d orbitals in the transition metals, there are irregularities in the f-orbital utilization, more so for the actinides than for the lanthanides. But again, it is sufficient now to know the trend and leave the irregularities until later. (Incidentally, it is because the first element of each series—La and Ac—have d^1 configurations instead of f^1 that these series sometimes are shown to begin with Ce and Th, as in the periodic table on the inside front cover.)

All the lanthanides and actinides are typical metals with high luster and conductivity. These metals are chemically reactive, with oxidation potentials in the range 2–3 V. Because of their high oxidation potentials (and also low first ionization energies) the metals tarnish readily in air and react vigorously with water to displace hydrogen.

The most important characteristic of the lanthanides is their close similarity to each other. This similarity is due primarily to the successive electrons going into the low-lying f orbitals, which results in small changes in atomic and ionic radii ($\sim$0.01 Å) from one element to the next. The predominant oxidation state for the lanthanides, and to a lesser extent for the actinides, is $+3$; nearly all compounds of these elements are ionic salts with discrete M^{3+} ions. Because of their great similarity, lanthanide compounds are found together in nature and are difficult to separate from each other.

The actinides are distinguished from the lanthanides in that they all are radioactive. (Promethium, $Z = 61$, is the only radioactive lanthanide.) The transuranium actinides ($Z = 93$ to $Z = 103$) are man-made elements.

9–6   THE *p*-ORBITAL OR REPRESENTATIVE ELEMENTS

After the interruptions for the transition and inner transition metals, the filling of the outermost p orbitals begins again (as with B and Al in Periods 2 and 3) and goes to completion. The alkali metals and alkaline earth metals are remarkable for their smooth gradation of properties within a group. The transition metals also vary gradually from one element to the next. But beginning with Group IIIA, we shall see sharp variations within a group, although these variations are systematic and fall into a pattern that can be followed across the remainder of the periodic table. The sharp variations occur at the transition between metallic and nonmetallic properties. Some of the trends are shown in Tables 9–3 and 9–4.

Table 9–3 shows the melting points of the halides of Li, Be, B, and C. The Li halides all are ionic. Their melting points decrease with the heavier anions because these large anions cannot approach the Li^+ ion as closely and are not bound as strongly by electrostatic forces. The carbon tetrahalides

Table 9–3.   Melting Points of Halides of Second-Period Elements (°C)

X =	F	Cl	Br	I	
LiX	842	614	547	450	} Ionic
BeX$_2$	800^a	405	490	510	
BX$_3$	−127	−107	−46	50	} Covalent
CX$_4$	−184	− 23	90	decomposes	

^{a} Extended chains held together by ionic forces.

exhibit the opposite behavior. They all are covalent molecular compounds, and the heavier molecules have a higher melting point because more thermal energy is required to move them. By this criterion, the boron trihalides are also covalent. Beryllium constitutes the borderline case mentioned previously. With elements of similar electronegativity, it forms covalent compounds. With the strongly electronegative F, Be forms long chains similar to those in BeCl$_2$. But because F is so electronegative, Be acquires a positive charge and F a negative charge. The covalently bonded chains are held together by ionic forces.

The border between metallic and nonmetallic behavior also is shown in Table 9–4. Note the diagonal character of the border area: The amphoteric properties of Be in Group IIA appear one period lower in the neighboring group, in Al instead of B. Boron is a nonmetal and makes covalent bonds in compounds in which it has a +3 oxidation state. Aluminum also can make covalent bonds but is definitely metallic. The acidic oxide B$_2$O$_3$ forms boric acid in water:

$$\tfrac{1}{2}B_2O_3 + \tfrac{5}{2}H_2O \rightarrow B(OH)_3 + H_2O \rightarrow B(OH)_4^- + H^+$$

Aluminum oxide, Al$_2$O$_3$, is amphoteric, and the oxides of Ga, In, and Tl are basic. Except boron, the elements in Group IIIA are metals. Gallium has only the +3 oxidation state and a chemistry quite similar to Al; In exhibits both +3 and +1; Tl shows both states, but +1 is more common.

Group resemblances are even less apparent in Group IVA. Carbon is a nonmetal that almost always makes four covalent bonds with other elements. It can polymerize with itself in a chain to form what are classified as organic compounds, and can construct multiple covalent bonds with the same atom. Silicon is a nonmetal with several metallic properties, including a silvery sheen. It can form a limited number of hydrides, called *silanes*, analogous to the hydrocarbons, with the general formula Si$_x$H$_{2x+2}$. But the chains break beyond $x = 6$, and even these low-molecular-weight silanes are explosively reactive with halogens and oxygen. Silicon also can form another

Table 9–4. *Properties of Elements on the Metal–Nonmetal Border of the Periodic Table*

BeCl₂ Covalent molecular chains held together in solid and liquid by weak intermolecular forces. Narrow liquid range. mp 400°C, bp 520°C. Forms a dimer, Be₂Cl₄.	**BCl₃** Molecular gas: mp −107°C, bp 13°C. Hydrolyzes completely in solution.	**CCl₄** Molecular gas above 76.8°C. Inert in water.
BeF₂ Covalent molecular chains held together in solid by ionic forces. mp 800°C. BeF₄²⁻ complex in solution.	**BF₃** Molecular gas: mp −127°C. BF₄⁻ complex in solution.	**CF₄** Molecular gas: mp −184°C. Inert in water.
Be(OH)₂ Amphoteric hydroxide.	**B(OH)₃** Boric acid. Forms polymeric ions.	**CO₂** Gaseous, acidic oxide.
MgCl₂ Ionic solid: mp 708°C, bp 1412°C.	**AlCl₃** Covalent network solid: Sublimes 178°C. Forms dimer, Al₂Cl₆.	**SiCl₄** Volatile molecular liquid: mp −70°C, bp 57.6°C. Hydrolyzes completely. in water.
MgF₂ Ionic solid: mp 1266°C, bp 2239°C.	**AlF₃** Ionic solid: mp 1040°C. AlF₆³⁻ complex in solution.	**SiF₄** Molecular gas: mp −90°C. SiF₆²⁻ complex in solution.
Mg(OH)₂ Basic hydroxide.	**Al(OH)₃** Amphoteric hydroxide.	**SiO₂** Solid acidic oxide. Forms polymeric anions.

class of polymers, the *silicones*, in which the Si atoms are bridged with oxygen atoms:

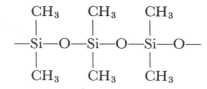

These silicones are inert, water repellent, electrically insulating, and stable to heat. Silicon, in spite of the science fiction writers, is not a suitable alternative to carbon for life forms, at least under terrestrial conditions.

Germanium is a semimetal, and tin and lead both are metals. Carbon and silicon show the $+4$ oxidation state almost exclusively. (By picking the right compound, we can find almost any oxidation number we please for carbon in compounds with chains of carbon–carbon bonds. Thus, following the rules of Chapter 7, carbon has a formal oxidation number of 4 in CH_4, 3 in C_2H_6, and $2\frac{2}{3}$ in C_3H_8. But such formal oxidation numbers in chain compounds have little meaning. In compounds without carbon–carbon bonds, carbon almost always has an oxidation number of 4.) Germanium and tin have both $+4$ and $+2$ states, and the chemistry of lead is almost wholly that of the $+2$ state.

The same behavior occurs in Group VA, but the break between metals and nonmetals is lower in the group. Nitrogen and phosphorus are nonmetals whose covalent chemistry and oxidation states are governed by the presence of five valence electrons: s^2p^3. Nitrogen and phosphorus most commonly have oxidation states -3, $+3$, or $+5$. Arsenic and antimony are semimetals with amphoteric oxides, and only Bi is metallic. For As and Sb, the $+3$ state is the most important. For Bi it is the only state, except under extraordinary conditions. Bismuth cannot lose all five valence electrons; the energy required is too high. However, it does lose the three $6p$ electrons to produce Bi^{3+}.

The trend in Group VIA is similar to that in the N group. Oxygen and sulfur both are nonmetals. Oxygen is strongly electronegative and has only the -2 oxidation state, except in OF_2 and the peroxides. Sulfur has the -2 state and several positive states as well, especially $+4$ and $+6$. Selenium and tellurium both are semimetals, but with a chemistry that resembles sulfur. Polonium, a rare, radioactive element, has the electrical conductivity of a metal.

With Group VIIA, all metallic properties have been lost; thus the halogens are nonmetals. They lack only one electron of possessing a noble-gas electronic arrangement and are reduced easily to anions with the s^2p^6 electronic configuration. Their *reduction potentials* are

$$F_2 + 2e^- \rightarrow 2F^- \qquad \mathscr{E}^0 = +2.87 \text{ V}$$
$$Cl_2 + 2e^- \rightarrow 2Cl^- \qquad \mathscr{E}^0 = +1.36 \text{ V}$$
$$Br_2 + 2e^- \rightarrow 2Br^- \qquad \mathscr{E}^0 = +1.06 \text{ V}$$
$$I_2 + 2e^- \rightarrow 2I^- \qquad \mathscr{E}^0 = +0.54 \text{ V}$$

For Cl, Br, and I, all odd oxidation numbers between -1 and $+7$ are known. But F, like O, is too electronegative to exhibit the positive states and occurs only in the -1 state.

9–7 SUMMARY

With hydrogenlike atomic wave functions and their energies, and with the Pauli buildup principle, we now can write the ground-state electronic structure for every element. The stable shell of eight electrons that was proposed on chemical grounds is the set of eight electrons in the outermost s, p_x, p_y, and p_z orbitals. The detailed structure of the periodic table is a result of the order of energies of the levels, and the delayed filling of d and of f orbitals.

The outermost s and p electrons are responsible for most chemical properties. They are what traditionally are meant by the valence electrons. The d and f orbitals are buried more deeply, and differences in occupancy of these levels in the transition metals and inner transition metals have less effect on differences in chemical behavior. Elements that have one or two electrons per atom beyond a closed noble-gas shell lose them easily to form metallic cations. Elements at the right of the table that need only one or two electrons per atom to complete a $s^2 p^6$ shell gain them easily to form anions. Within a group, electrons are lost most easily from the elements of larger atomic radii, and gained most easily by the smaller elements. All of these trends are summed up in the electronegativity scale. Substances that differ greatly in electronegativity will transfer electrons and form ionic compounds; elements of approximately equal electronegativity will combine by sharing electrons in covalent bonds.

The border between metals and nonmetals sweeps diagonally across the table in an ill-defined band that goes approximately from Be and B to Po and At. The borderline occurs near an electronegativity of 2.0 on a scale in which the most electronegative, F, is 3.98 and the least electronegative, Cs, is 0.79. Yet the high electronegativities of the platinum metals are a reminder that electronegativity is not the only factor in determining non-metallic behavior.

The transition metals undergo a smooth variation in melting point, boiling point, heats of fusion and vaporization, hardness, and common oxidation number, all of which can be related to the number of unpaired d electrons. Because of the lanthanide contraction, elements in the second and third transition series are more similar in size and in chemical properties than are those in the first and second periods.

We have spent less time on the nonmetals in this survey because they will be covered more thoroughly later. Ionic bonds are easy to understand, but covalent bonds are less self-evident. The next step in understanding chemistry is to find out what quantum mechanics can tell us about the process of covalent bonding.

SUGGESTED READING

M. J. Bigelow, *The Representative Elements*, Bogden and Quigley, New York, 1970.

J. L. Dye, "The Solvated Electron," *Scientific American*, February, 1967.

J. L. Hall and D. A. Keyworth, *Brief Chemistry of the Elements*, W. A. Benjamin, Menlo Park, Calif., 1971.

K. J. Laidler and M. H. Ford-Smith, *The Chemical Elements*, Bogden and Quigley, New York, 1970.

E. M. Larsen, *Transitional Elements*, W. A. Benjamin, Menlo Park, Calif., 1965. Written as a supplement to a general chemistry course. More extensive than this chapter, but at a comparable level.

G. Oster, "The Chemical Effects of Light," *Scientific American*, September, 1968.

R. L. Rich, *Periodic Correlations*, W. A. Benjamin, Menlo Park, Calif., 1965. The first line of defense if you are unclear on the relationship between the periodic table and chemical properties after reading this chapter.

R. T. Sanderson, *Chemical Periodicity*, Reinhold, New York, 1960. A quite readable text, slightly more advanced than Rich, with a vast amount of comparative data.

H. H. Sisler, *Electronic Structure, Properties and the Periodic Law*, Reinhold, New York, 1963. Another systematic treatment of periodicity similar to Rich and Sanderson.

V. F. Weisskopf, "How Light Interacts with Matter," *Scientific American*, September, 1968.

QUESTIONS

1 What is the Pauli exclusion principle, and how does it permit us to construct models for the electronic configurations of atoms with more electrons than hydrogen has?

2 What would the periodic table look like if the Pauli exclusion principle did not hold?

3 What is Hund's rule, and what role does it play in the buildup of electronic configurations of atoms? What is the physical justification for Hund's rule?

4 Why do the $4s$, $4p$, $4d$, and $4f$ orbitals have the same energy in the hydrogen atom, but different energies in a many-electron atom?

5 What is "screening" by electrons in an atom?

6 Why is the first ionization energy of sulfur less than that of phosphorus?

7 Why is the first ionization energy of thallium less than that of mercury?

8 What evidence do we have that the relative order of energy levels in Figure 9–2 is correct?

9 Why are the chemical properties of the transition metals less varied than those of the representative elements?

10 What are the "inner transition metal" series, and how are they explained in terms of electronic configurations?

11 Why does the electron affinity, which had been rising steadily from N to O to F in Figure 6–8, drop so abruptly at Ne?

12 What are isoelectronic ions? Why do the radii of the isoelectronic ions As^{3-},

Se^{2-}, Br^-, Kr^0, Rb^+, Sr^{2+}, and Y^{3+} decrease as they do in Figure 6–9?

13 Why do the radii of the neutral atoms decrease from K to Ca to Sc to Ti as they do in Figure 6–9?

14 Why do Ta, W, and Re have such high melting and boiling points?

15 Why do C, Si, and Ge have higher melting and boiling points than their neighbors N, P, and As?

16 How is electronegativity related to first ionization energy and electron affinity? How is it related to the bond energies of molecules?

17 In what part of the periodic table are elements with the highest electronegativity found? What three elements have highest electronegativities? Which element has the lowest?

18 Would you expect rubidium chloride to be ionic or covalent? Why? What about an osmium or iridium chloride? On what would you base your predictions?

19 The imaginary element turbidium (Tu) has the following oxidation potential data:

$$Tu \rightarrow Tu^{3+} + 3e^- \qquad \mathscr{E}^0 = -3.00 \text{ V}$$

Is turbidium a good oxidizing agent or reducing agent, or neither? Is the Tu^{3+} ion a good oxidizing agent or reducing agent, or neither?

20 What is the difference between oxidation potentials and reduction potentials?

21 Why are Li and Na not found free in nature, whereas Ag and Au are?

22 From which metal can electrons be ejected with a longer wavelength, Na or Cs? Why?

23 Which element is more metallic, Be or Ba? What evidence permits you to say this? How can you explain this in terms of electronic configuration?

24 Why is calcium harder than potassium?

25 Why is BeO amphoteric? What does this term mean in relation to chemical behavior?

26 What is the lanthanide contraction? Why does it occur? What detectable chemical effects does it produce?

27 If the first electrons after those in the Kr noble-gas shells go into the $5s$ orbital for Rb and Sr, why is the outer electronic configuration of Zr^{2+} $4d^2$, rather than $5s^2$ as in Sr?

28 How does the maximum oxidation number change with atomic number in the first transition-metal series, Sc to Zn?

29 In a transition metal with several oxidation states, which state is usually most metal-like in its compounds? Can you give an example?

30 As the principal quantum number of the element increases within a transition-metal group (vertical column), do the higher or lower oxidation states become more important? Can you give a reason for this behavior?

31 How is the color of a chemical related to electronic transitions between energy levels?

32 Why are transition-metal compounds more often colored than those of the representative elements?

33 How do the oxides of Cr, Ni, Cu, Al, and many other metals differ from iron oxide in a way that has great economic importance? Try to imagine a world in which iron oxide behaved the same way. What important chemical industry would be badly hurt if this were so?

34 Why do Zn, Cd, and Hg have such low melting points compared with Cr, Mo, and W?

35 Why do amphoteric oxides appear for those elements which lie on a diagonal across the periodic table (Be, Al, Ge,

Sb) rather than in one group separating the metals from the nonmetals?

PROBLEMS

1 Which of the following configurations of electrons represent ground states, which represent excited states, and which are impossible? Why are these latter ones unacceptable? What neutral atom can have each permissible configuration?

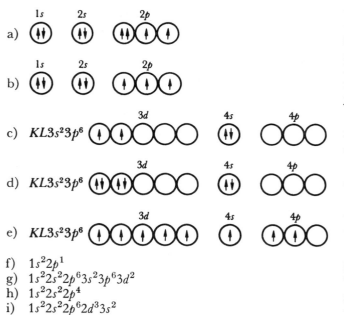

f) $1s^2 2p^1$
g) $1s^2 2s^2 2p^6 3s^2 3p^6 3d^2$
h) $1s^2 2s^2 2p^4$
i) $1s^2 2s^2 2p^6 2d^3 3s^2$

2 What are the ground-state electronic configurations of the following atoms or ions [use the s-p-d-f notation of Problem 1(f–i)]: (a) As, (b) Co^{2+}, (c) Cu, (d) S^{2-}, (e) Kr, (f) C, (g) W, (h) H^+, (i) H^-, (j) Cl^-.

3 Write the ground-state electronic configurations of the two atoms $^{18}_{8}O$ and $^{16}_{8}O$.

4 Ten atoms and their electronic configurations are given below. For each, decide whether a neutral atom, a positive ion, or a negative ion is represented.

36 How does metallic character change within one group of the table?

In addition, specify whether the electronic state represented is a ground state, an excited state, or impossible.

a) $_3$Li: $1s^2 2p^1$
b) $_1$H: $1s^2$
c) $_{16}$S: $1s^2 2s^2 2p^6 3s^2 3p^4$
d) $_6$C: $1s^2 2s^2 2p^1 2d^1$
e) $_{10}$Ne: $1s^2 2s^1 2p^7$
f) $_7$N: $1s^2 2s^1 2p^3$
g) $_9$F: $1s^2 2s^2 2p^5 3s^1$
h) $_2$He: $1p^1$
i) $_{21}$Sc: $1s^2 2s^2 2p^6 3s^2 3p^6 3d^1 4s^2$
j) $_8$O: $1s^2 2s^2 2p^3$

5 Write ground-state electronic configurations for Li, Lu, La, and Lr and also for Li^+, Lu^{3+}, La^{3+}, and Lr^{3+}.

6 Compute the electronegativity of a chlorine atom by the Pauling method.

7 Would you predict an actinide contraction analogous to that for the lanthanides? Explain.

8 Consider an excited hydrogen atom in which the electron is in the 3s orbital. The energy required per mole to remove such an electron is 34.8 kcal. However, to remove the 3s electron of Na requires 119 kcal mole^{-1}. Why this difference?

9 Arrange the following metals in a series of increasing reducing ability: Ca, Na, Ba, K, Ag.

10 Explain the fact that the second ionization energy of Mg is larger than the first, but not as large as the second ionization energy of Na. How does this explain the observed chemical behavior of Mg and Na?

11 In Mulliken's first simple definition, the electronegativity of an element was proportional to the sum of its first ionization energy and its electron affinity. This relationship is not strictly true for the numerical values given in Table 9–1 since the ionization energies, electron affinities, and electronegativities in this table have been calculated by different people using different methods. Nevertheless, the proportionality is approximately valid. From the data in Table 9–1, plot a graph of the sum of ionization energy and electron affinity, against electronegativity on the Pauling scale, for the elements in the second and third periods of the table. (a) Draw the best straight line that you can through these data points and the origin. (b) Use this plot to estimate the electronegativity of Ne. If Ne—F bonds existed, would you expect them to be ionic or covalent? (c) Work backward from your plot to calculate the electron affinities of the fifth-period elements Rb through In. Plot these values as a function of atomic number. In terms of electronic configurations of the atoms, explain the general trend of electron affinities across the transition metals in this period, and the striking behavior at Ag–Cd–In.

12 Why should the melting points of the alkaline earth metal chlorides have the smooth trend that you find in Table 9–2?

13 Which atom has the lowest first ionization energy: Li, F, Cs, or Xe? Why?

14 Which atom has the greatest electron affinity: Cl, I, O, or Na? Why?

15 Which hydrogen compound is most ionic: Li—H, Cs—H, F—H, or I—H? In which compound will H have the largest positive charge? The largest negative charge?

16 Which element will show the largest oxidation number in its chloride: Bi, Mg, P, or Si?

17 If you suspected that compounds of element 120 might occur naturally in small quantities, which of the following minerals would be the most sensible starting point in looking for the new element: KCl, $BaSO_4$, Al_2O_3, UO_3, or Gd_2O_3?

18 Suppose that you have found element 120, and that it is stable. What will be the most probable formula of its most common hydride? Its most common oxide? What will be its most common oxidation state? Estimate its electronegativity. What might be the melting point of its chloride? Will the solid chloride conduct electricity? Will the liquid chloride conduct electricity? Why, or why not?

19 A contestant in the American Chemical Society Laboratory Technician Beauty Contest was described in the following terms: "Her hair was like spun $[Ar]3d^64s^2$, her skin was as soft as $[Xe]4f^{14}5d^46s^2$, her perfume was the hydride of $[Ne]3s^23p^4$, and her personality was completely $[He]2s^22p^1$." Describe the lady in more conventional terms. What convention about notation of inner filled orbitals do you have to recognize in order to detect the lady's qualities?

A theory is not an ultimate goal; its object is physical rather than metaphysical. From the point of view of the physicist, a theory is a matter of policy rather than a creed.

J. J. Thomson

10 COVALENT BONDING

Ionic bonding is easy to understand but often is inadequate as an explanation of chemical behavior. The nonmetals were slighted in the preceding chapter because much of their chemistry is incomprehensible without the theory of covalent bonding. In this chapter we look at a simple method of representing covalent bonding, Lewis structures. Then we go more deeply into the actual formation of covalent bonds by using molecular orbital theory. We see how to write Lewis electron-dot structures for familiar compounds, and we interpret them in terms of electron-pair sharing and the completion of noble gas valence shells. We explain oxidation number in terms of electron-pair sharing, and look at oxidation states as environments of different electronegativities. We also note how the electron-dot picture begins to fail even with so simple a molecule as SO_3.

Turning to a better theory, we see how molecular orbitals can be constructed for diatomic molecules with the same type of atoms and with different atoms. These models are tested against measured bond energies, bond lengths, and paramagnetism. Molecular orbital theory can account for the shapes of molecules such as NH_3 and H_2O and for the difference in structure of two supposedly similar molecules such as H_2O and H_2S. Hybridization is introduced to explain the tetrahedral structure of carbon compounds. And finally, we examine the idea of bonds between pairs of atoms, and see when such an idea can be used and when (as with benzene, C_6H_6) it cannot.

10–1 LEWIS STRUCTURES

Electron-dot formulas for chemical compounds were developed by G. N. Lewis, in 1916, as an attempt to comprehend covalent bonding. Our understanding of bonding rests on firmer ground now, but the dot formulas are still a convenient notation. Each valence electron (i.e., an electron in the outermost s and p orbitals) is represented by a dot placed beside the chemical symbol: H·, He:. In modern terminology, each of the four compass points of the symbol represents one of the s, p_x, p_y, and p_z atomic orbitals. For example, atoms of the second-period elements are written as

$$\text{Li·} \qquad \text{Be} \qquad \text{·B} \qquad \text{·C} \qquad \text{·N·} \qquad \text{:O·} \qquad \text{:F:} \qquad \text{:Ne:}$$

The loss and gain of electrons in the formation of ions can be illustrated by the formation of sodium chloride from sodium and chlorine atoms:

$$\text{Na·} + \text{:Cl:} \rightarrow \text{Na}^+\text{:Cl:}^-$$

Each ion now has the outer electronic configuration of a noble gas: The sodium ion has the configuration of Ne, and the chloride ion, of Ar. This transfer of an electron occurs because Cl is more electronegative than Na (3.16 for Cl versus 0.93 for Na). What happens in HI, in which the electronegativities are nearly equal (2.20 and 2.66, respectively)?

According to Lewis' theory of covalence, each atom completes a noble-gas configuration, not by transferring, but by *sharing* an electron:

$$\text{H·} + \text{·I:} \rightarrow \text{H:I:}$$

The H atom now has two electrons in its outer valence orbital, as in He, and I has eight electrons, as in Xe. Lewis set forth the principle: *Atoms form bonds by losing, gaining, or sharing enough electrons to achieve the outer electronic configurations of noble gases.* The type of bond, ionic or covalent, depends on whether electrons are transferred or shared. The combining capacities of atoms are a consequence of the proportions in which they must associate to achieve noble-gas configurations. Lewis' theory explains bond type and the pattern of connections of atoms within a molecule. However, it is not able to explain the geometry of molecules.

The Lewis theory made bonding between like atoms, as it occurs in H_2, F_2, or N_2, understandable for the first time. Two hydrogen atoms share their electrons to provide each atom with the He closed-shell structure:

$$H \cdot + H \cdot \rightarrow H : H \qquad \text{or} \qquad H—H$$

A straight-line bond symbol often is used, as here, in the special sense of a symbol for a Lewis electron pair. Two fluorine atoms share one pair of electrons; thus, each F atom has the Ne structure:

$$:\!\ddot{F}\cdot \; + \; :\!\ddot{F}\cdot \; \rightarrow \; :\!\ddot{F}\!:\!\ddot{F}\!: \qquad \text{or} \qquad :\!\ddot{F}\!—\!\ddot{F}\!:$$

The unshared pairs of electrons on F are called *lone pairs;* we now would interpret them as spin-paired electrons in atomic orbitals that are not involved in bonding. The *bond energy*, or the energy required to break the diatomic molecule into two infinitely separated atoms, is 103 kcal mole^{-1} for H_2 and only 33 kcal mole^{-1} for F_2. Part of this relative instability of the F_2 molecule may arise from electrostatic repulsion between lone pairs of electrons on the two F atoms.

Multiple Bonds

If we try to construct O_2 in a manner similar to F_2, we end with unpaired electrons and only seven electrons in the neighborhood of each O atom:

$$:\!\ddot{O}\cdot \; + \; :\!\ddot{O}\cdot \; \rightarrow \; :\!\ddot{O}\!:\!\ddot{O}\!: \qquad \text{or} \qquad :\!\ddot{O}\!—\!\ddot{O}\!:$$

This defect can be eliminated by assuming that the oxygen atoms share *two* pairs of electrons (without regard to the geometry of the process):

$$:\!\ddot{O}\!:\!:\!\ddot{O}\!: \qquad \text{or} \qquad :\!\ddot{O}\!=\!\ddot{O}\!:$$

Thus, there is a double bond between two oxygen atoms. A triple bond must be assumed in N_2 to give each nitrogen atom a noble-gas configuration:

$$:\!N\!:\!:\!:\!N\!: \qquad \text{or} \qquad :\!N\!\equiv\!N\!:$$

This concept of multiple bonds is not all imagination; bond energies and bond lengths both support the idea of a single bond in F_2, a double bond in O_2, and a triple bond in N_2:

Molecule:	N_2	O_2	F_2
Bond energy (kcal mole^{-1}):	225	118	33
Bond length (Å):	1.10	1.21	1.42

A molecule such as N_2, with a triple bond between two atoms, is said to have a *bond order* of three. (The bond order is the number of electron-pair bonds.) The oxygen molecule has a bond order of two, and in the F_2 molecule, of one. The higher the bond order, the more tightly the atoms will be held, the greater will be the bond energy, and the shorter will be the bond.

The three covalent hydrogen compounds water, ammonia, and methane have the Lewis structures

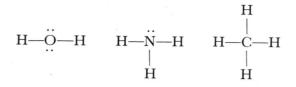

The Lewis structures show the connections between atoms within the molecule correctly, which the simple molecular formulas do not do. The Lewis structures also indicate the presence of lone pairs of electrons, which often are important in the chemical behavior of the compounds.

The simple hydrocarbons ethylene and acetylene illustrate multiple bonding; ethane is shown for comparison:

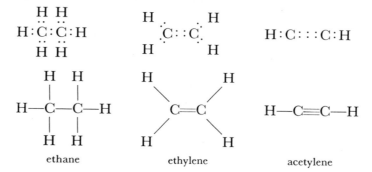

ethane   ethylene   acetylene

The lone electron-pair notation is useful when atoms other than C and H are involved:

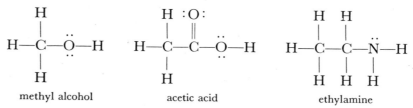

methyl alcohol   acetic acid   ethylamine

The ammonium ion, NH_4^+, has the Lewis structure

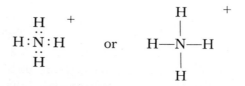

Note that the ammonium ion is *isoelectronic* with methane. That is, it has the same number of electrons as does CH_4. If we could reach into the nucleus of nitrogen in an ammonium ion and remove one proton, we would have a molecule of methane (although with ^{13}C rather than the common isotope

^{12}C). Yet the chemical properties of ammonium ion and methane are very different, precisely because of the additional positive charge in the nitrogen nucleus in NH_4^+.

Challenge. To see if you can draw Lewis structures for "wood alcohol," an antiseptic, a bleach, a dry-cleaning solvent, a rocket fuel, and a source of silicone rubber, try Problems 10–2 to 10–5 in *Relevant Problems for Chemical Principles* by Butler and Grosser.

Formal Charges

Carbon dioxide is easy to represent by Lewis structures, but carbon monoxide raises a problem. Each O needs two electrons to achieve the stable eight-electron (*octet*) structure; thus it should share two electron pairs with C. Yet the carbon atom needs four electrons and should share four pairs. The only satisfactory Lewis structure for CO is obtained by letting three pairs be shared, and by distributing the other four valence electrons in such a way as to complete an eight-electron shell around each atom.

Carbon dioxide: $:\overset{..}{O}::C::\overset{..}{O}:$ or $:\overset{..}{O}\!=\!C\!=\!\overset{..}{O}:$

Carbon monoxide: $:C:::O:$ or $:\overset{\ominus}{C}\!\equiv\!\overset{\oplus}{O}:$

Carbon monoxide is isoelectronic with N_2, so we might expect the triple bond in CO to be a completely satisfactory representation. (We can imagine some hypothetical "Schrödinger's Demon" making a molecule of carbon monoxide from a molecule of N_2 by removing a proton from one nitrogen nucleus and adding it to the other.) Nevertheless, there is a difficulty in such a structure for CO. If we assume that each shared electron pair is shared equally between atoms, then carbon has three of the six electrons from the triple bond, plus the two in the lone pair. It has five electrons but a nuclear charge that will counterbalance only four of them. Similarly, oxygen has five valence electrons but a nuclear charge designed for six. Therefore, carbon has a *formal charge* of -1, and oxygen, of $+1$. This statement does not mean that these charges are fully present, but only that the demands of bonding lead to a nonuniform distribution of charge.

When calculating the formal charge on the atoms in a molecule, we assign to each atom *one* electron for each covalent electron-pair bond that it makes, plus all of its lone-pair electrons. The formal charge on the atom is then the charge that it would have if it were an isolated ion with the same number of valence electrons:

$$\text{Formal charge} = Z - (N_{\text{bonds}} + N_{\text{nonbonding}})$$

Here Z is the atomic number, N_{bonds} is the number of covalent electron-pair bonds that the atom makes with other atoms, and $N_{\text{nonbonding}}$ is the total number of electrons possessed by the atom that are *not* involved in covalent bonds. You should verify for yourself that, in every uncharged molecule discussed so far in this section except CO, the formal charge on each atom is zero.

Electron Donor and Acceptor Compounds

The colorless liquid BF_3 is represented by the Lewis structure

It is unusual because it does not have four electron pairs around the B atom. It reacts with ammonia to form the *addition compound* BF_3NH_3 as follows:

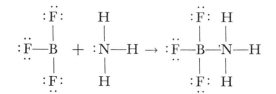

In this compound, nitrogen, with a lone pair, donates both the electrons of the covalent bond. Such a donor–acceptor bond sometimes is called a co-ordinate covalent bond; however, the distinction is pointless, because once the bond is formed it is like any other covalent bond.

BF_3NH_3 is isoelectronic with CF_3CH_3, and differs from it only by the charges on the nuclei of the central atoms. The charge is zero for each atom in the carbon compound. But if you work out the formal assignment of charge for the boron compound, you can see that B has a formal charge of -1, and N, of $+1$. Since formal charges arise from the way in which electrons are distributed in a molecule or ion, the total formal charge of all the atoms must be equal to the total charge on the ion, or zero for a neutral molecule.

Lewis Acids and Bases

A compound such as BF_3, which can accept an electron pair, is called a *Lewis acid*, and an electron-pair donor is a *Lewis base*. This terminology, like that of Brønsted (see Chapter 5), is an extension of the simple Arrhenius acid–base theory. By the Arrhenius theory, an acid is a substance that produces hydrogen ions or protons in aqueous solution, and a base is a substance that produces hydroxide ions. Bronsted's terminology is more general: An acid is any substance that can donate protons, and a base is any substance that can accept protons. And now, in Lewis' theory, an acid is a substance that can accept electrons in a reaction, and a base is a substance that can donate them. To illustrate the difference in the three definitions, consider the neutralization of HCl and NaOH,

$$HCl + NaOH \rightarrow H_2O + NaCl$$

In terms of the species present in aqueous solution, the reaction should be written as

$$H_3O^+ + Cl^- + Na^+ + OH^- \rightarrow Na^+ + Cl^- + 2H_2O$$

To Arrhenius, HCl is the acid, and NaOH is the base. To Brønsted, H_3O^+ is the acid, and the hydroxide ion is the base since it is the species that combines with the proton. To Lewis, the proton is the acid, because it will combine with the lone pair on the hydroxide ion; the hydroxide ion is the electron-pair donor and hence the base:

$$H^+ + :\overset{..}{\underset{..}{O}}\!-\!H^- \rightarrow H\!-\!\overset{..}{\underset{..}{O}}\!-\!H$$

The Brønsted and Lewis theories both are applicable to nonaqueous solutions, whereas Arrhenius' is not. Both Brønsted's and Lewis' theories will be useful later. These more general definitions of acids and bases are helpful because they include compounds that do not contain hydrogen, and which we might not recognize as having properties of acids with Arrhenius' theory. For example, BF_3, because it is an electron acceptor, often will catalyze organic reactions that are catalyzed by protons.

The Meaning of Oxidation Numbers

Chlorine is found in a series of oxyanions, ClO^-, ClO_2^-, ClO_3^-, and ClO_4^-, that illustrate its entire range of positive oxidation states. The chloride ion has the Ar noble-gas structure with four pairs of valence electrons. The four oxyanions can be thought of as the products of the reaction of this Cl^- ion as a Lewis base, in which one to four oxygen atoms act as electron-pair acceptors and Lewis acids. Following are the four reactions and the oxidation number of Cl in each oxyanion:

$$:\overset{..}{\underset{..}{Cl}}:^- \quad + \overset{..}{O}: \rightarrow :\overset{..}{\underset{..}{Cl}}:\overset{..}{\underset{..}{O}}:^- \qquad \text{Oxidation number} +1$$

$$:\overset{..}{\underset{..}{Cl}}:\overset{..}{\underset{..}{O}}:^- \quad + \overset{..}{O}: \rightarrow :\overset{..}{O}:\overset{..}{\underset{..}{Cl}}:\overset{..}{\underset{..}{O}}:^- \qquad \text{Oxidation number} +3$$

$$:\overset{..}{O}:\overset{..}{\underset{..}{Cl}}:\overset{..}{\underset{..}{O}}:^- + \overset{..}{O}: \rightarrow :\overset{..}{O}:\overset{\overset{\textstyle :\overset{..}{O}:}{..}}{\underset{..}{Cl}}:\overset{..}{\underset{..}{O}}:^- \qquad \text{Oxidation number} +5$$

$$:\overset{\overset{\textstyle :\overset{..}{O}:}{..}}{\underset{..}{O}}:\overset{..}{\underset{..}{Cl}}:\overset{..}{O}: \quad + \overset{..}{O}: \rightarrow :\overset{\overset{\textstyle :\overset{..}{O}:}{..}}{\underset{\underset{\textstyle :\overset{..}{O}:}{..}}{O}}:\overset{..}{\underset{..}{Cl}}:\overset{..}{O}:^- \qquad \text{Oxidation number} +7$$

There are no chlorine oxyanions with more than four O's because there are no more valence electron pairs on Cl.

The formal charge on each atom is determined by assigning one electron in a bond to each participating atom. In contrast, the oxidation number is found by assigning *both* electrons in a bond to the more electronegative of the two atoms. (This is the meaning of Rule 5 in Section 7–2.) Thus the oxidation number is the charge that the atom would have if it were an isolated ion with the assigned number of electrons:

$$\text{Oxidation number} = Z - (N_{\text{assigned}} + N_{\text{nonbonding}})$$

Here Z is the atomic number, N_{assigned} is the total number of electrons in bonds between the given atom and atoms that are *less* electronegative than it is, and $N_{\text{nonbonding}}$ is the total number of electrons possessed by the atom that are *not* involved in covalent bonds. In calculating oxidation numbers, we always pretend that both electrons in a bond belong to the more electronegative of two bonded atoms. Fluorine, the most electronegative element, always has an oxidation number of -1. Oxygen always has an oxidation number of -2, except in peroxides and compounds of fluorine. In the Cl oxyanions, since Cl has an electronegativity of 3.16, and O, of 3.44, both electrons in a bond are assigned to oxygen. In Cl^-, chlorine has a net charge of -1 and an oxidation number of -1. In ClO^-, chlorine has six assigned electrons and an oxidation number of $+1$, since the bonding lone pair has been assigned to oxygen because of oxygen's greater electronegativity. In ClO_4^-, all four electron pairs have been "abducted" by the more electronegative oxygen atoms, and Cl has an oxidation number of $+7$, *as if* it really had lost all seven of its valence electrons. Oxidation numbers are couched in the language of electron loss and gain. What they really measure, however, is the extent of combination of an atom with other atoms more electronegative than itself.

Resonance Structures

For the examples given so far, we have been able to write single Lewis structures that are compatible with experimental data for the compounds and that are otherwise reasonable. In this section we will discuss a few examples of familiar compounds for which single Lewis structures are not completely satisfying.

In writing the "best" Lewis structure for a substance, we would like to select the one that puts the minimum formal charge on the atoms involved. But because of other considerations this is not always the most reasonable way to proceed. Consider the following exercise.

Exercise.   Calculate the formal charge on the oxygen and chlorine atoms in the chlorine oxyanions, as shown in the preceding section.

(*Answer:* Each O has a formal charge of -1, and that on Cl varies from 0 in ClO^- to $+3$ in ClO_4^-.)

The formal charge on Cl in ClO_4^- is somewhat disturbing. It is possible to write Lewis structures with no formal charge on Cl, but only at the price of abandoning the equivalence of the Cl—O bonds and of placing more than eight electrons around Cl (or of making more than four covalent bonds).

Now let us consider even more troublesome examples.

Example. Write Lewis structures for SO_2 and SO_3. How can you accommodate three oxygen atoms around one S in SO_3? Can you write a Lewis structure for SO_3 in which all three S—O are equivalent? What is the formal charge on S then?

Solution. An SO_2 structure with equivalent bonds is

$$:\ddot{O}\!=\!\ddot{S}\!=\!\ddot{O}:$$

This SO_2 structure is satisfactory in that both S—O bonds are equivalent and there is no formal charge on the sulfur atom; however, it places ten valence electrons around the S atom. But as we already have seen in Chapter 7, in octahedral complexes as many as twelve valence electrons can be placed around a central atom if it is large enough. The Lewis octet of electrons is really an oversimplification, and is most applicable to small atoms in the first two periods of the periodic table. If we insist on no more than eight electrons around each atom, we cannot draw a Lewis structure with equivalent bonds. However, we can draw *two* Lewis structures for SO_2 in which the S—O bonds are not equivalent:

$$:\ddot{O}\!-\!\ddot{S}\!=\!\ddot{O}: \qquad :\ddot{O}\!=\!\ddot{S}\!-\!\ddot{O}:$$

The SO_3 molecule involves more difficulties. There are two Lewis structures in which the three S—O bonds are equivalent in each:

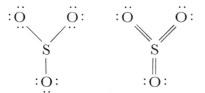

The first structure places a formal charge of +3 on S and has only three electron pairs around the central atom. The second structure avoids formal charges, but at the expense of surrounding the S atom with six electron pairs. Three arrangements can be drawn with an octet of electrons around S, but which give it a +2 formal charge:

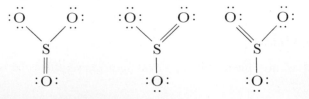

Three additional structures place a $+1$ formal charge on S:

All of these last six structures are inadequate because they suggest that the S—O bonds are of different lengths. It is impossible to write a structure that simultaneously avoids formal charges, places a Lewis octet of electrons around the S, and makes all S—O bonds equivalent.

X-ray and spectroscopic data reveal that all three bonds in SO_3 are identical, and that their length is shorter than that expected for a single bond but longer than for a double bond. Therefore, we cannot describe accurately the actual molecule with any one Lewis structure, and our failure illustrates the inadequacy of such a simple bond model. We can compromise and say that the structure of the SO_3 molecule has something of the character of the two equivalent-bond extremes, with a little of the other six structures as well. These eight Lewis-structure alternatives are called *resonance structures*, which is an unfortunate choice of terminology because it creates the erroneous impression that the molecule flips back and forth between the Lewis structures. We shall find the idea of resonance structures useful again in Section 10–10 when we examine the benzene molecule.

Exercise. Write Lewis structures for H_2S, $SO_3{}^{2-}$, and $SO_4{}^{2-}$ by analogy with the chlorine oxyanions. What are the oxidation number and the formal charge on the S atom in each species?

Before you go on. Lewis structures are simple yet useful for depicting the bonding in both ionic and covalent compounds. If you need additional practice in drawing and using these diagrams, see Section 10–1 of *Programed Reviews of Chemical Principles* by Lassila *et al.*

10–2 ACIDITY OF OXYACIDS

The simple Lewis model of bonding, with all its defects, does give us a physical understanding of the relative acidity of compounds that contain a central atom bonded to oxygen atoms alone, or to the oxygen atoms in hydroxide ion and water. Oxides of nonmetals dissolve in water to form acids. An example is

$$SO_3 + H_2O \rightarrow H_2SO_4$$

(However, sulfuric acid is not prepared this way commercially.) In such oxyacids the protons are bound to oxygen:

$$
\begin{array}{c}
\ddot{\text{O}}: \\
\parallel \\
:\!\ddot{\text{O}}\!=\!\text{S}\!-\!\ddot{\text{O}}\!-\!\text{H} \\
| \\
:\!\ddot{\text{O}}: \\
| \\
\text{H}
\end{array}
$$

Consider a series of compounds containing hydroxide groups bound to positive ions ranging from Na^+ to Cl^{7+}: $NaOH$, $Mg(OH)_2$, $Al(OH)_3$, $Si(OH)_4$, H_3PO_4, $O_2S(OH)_2$ or H_2SO_4, and (O_3ClOH) or $HClO_4$. Now consider the bonds in the structure

$$M\overset{1}{-}O\overset{2}{-}H$$

in which M is the central atom. If the bond breaks at position 1 to form M^+ and OH^-, the compound is basic. If it breaks at position 2 to form $M—O^-$ and H^+, the compound is acidic. For the preceding series of compounds, Na^+OH^- is an ionic compound that is very soluble in water; thus, OH^- and Na^+ are separated easily. The smaller, more highly charged Mg^{2+} ion binds more tightly to OH^-, thereby making magnesium hydroxide, $Mg(OH)_2$, less soluble than $NaOH$ and a weaker base. Aluminum hydroxide is virtually insoluble in water, but loses OH^- to strong acids and H^+ to strong bases:

$$Al(OH)_3 + 3H^+ \rightarrow Al^{3+} + 3H_2O$$
$$Al(OH)_3 + OH^- \rightarrow AlO(OH)_2^- + H_2O$$

A compound such as aluminum hydroxide, which is able to react either as a base with an acid, or as an acid with a base, is called *amphoteric*. Hydrogen species such as HCO_3^- that can either lose a proton (giving CO_3^{2-}) or gain a proton (forming H_2CO_3) are called *amphiprotic*.

The exact formula of the ion represented by $AlO(OH)_2^-$ is uncertain. Recent work has shown that the hydrated aluminum ion is $Al(H_2O)_6^{3+}$. Removal of three protons from this would give neutral, insoluble $Al(OH)_3(H_2O)_3$, which could lose another proton and form $Al(OH)_4(H_2O)_2^-$. This formula probably most nearly represents the actual situation.

Silicic acid, $Si(OH)_4$, easily gives up water molecules to form SiO_2. It is a weak acid and reacts with $NaOH$. However, it does not react with HCl; thus, the compound is not amphoteric:

$$Si(OH)_4 + 2NaOH \rightarrow Na_2SiO_3 + 3H_2O$$
$$Si(OH)_4 + HCl \qquad \rightarrow \text{no reaction}$$

The compounds H_3PO_4, H_2SO_4, and $HClO_4$ become progressively more acidic; $HClO_4$ is the strongest oxyacid known. It appears that the O—H

Table 10–1.   Effect of Formal Charge on Acid Strength

Substance	Formal charge on O	pK
H_3O^+	+1	−1.75
H_2O	0	+15.7
OH^-	−1	25

Substance	Formal charge on N	pK
NH_4^+	+1	9.25
NH_3	0	35

Substance	Formal charge on S	pK
H_2S	0	7.04
HS^-	−1	11.96

bonds in this series become easier to break as the oxidation state of the central atom becomes more positive. However, in HNO_3, in which N is in the +5 oxidation state, the acid is much stronger than $Te(OH)_6$, in which Te is in the +6 oxidation state. A better correlation can be made between acidity of oxyacids and the formal charge on the central atom (Section 10–1). The formal charge on the central atom in several oxyacids is:

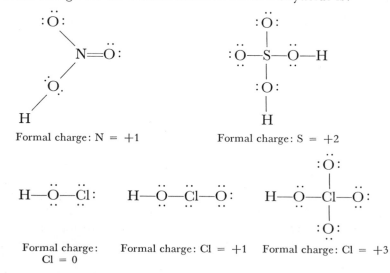

Formal charge: N = +1   Formal charge: S = +2

Formal charge: Formal charge: Cl = +1   Formal charge: Cl = +3
    Cl = 0

In general, acidity increases with increasing formal charge on the central atom, provided that all structures are written to be consistent with the octet rule. The higher the formal positive charge on a central atom, the more it will attract electrons from attached oxygen atoms. This results in a weakening of the O—H bond, thereby allowing easier removal of the H^+ and therefore an increase in acid strength. The effect of formal charge on acidity for similar

Table 10–2.   pK Values for Inorganic Oxyacids^a

$X(OH)_m$ (very weak)		$XO(OH)_m$ (weak)		$XO_2(OH)_m$ (strong)		$XO_3(OH)_m$ (very strong)	
$Cl(OH)$	7.5	$ClO(OH)$	2.0	$ClO_2(OH)$	(−3)	ClO_3OH	(−8)
$Br(OH)$	8.7	$NO(OH)$	3.4	$NO_2(OH)$	−1.4	$MnO_3(OH)$	(−8)
$I(OH)$	10.6	$IO(OH)$	1.6	$IO_2(OH)$	0.8		
$B(OH)_3$	9.2	$SO(OH)_2$	1.8	$SO_2(OH)_2$	(−3)		
$Sb(OH)_3$	11.0	$SeO(OH)_2$	2.5	$SeO_2(OH)_2$	(−3)		
$Si(OH)_4$	9.7	$TeO(OH)_2$	2.5				
$Ge(OH)_4$	8.6	$CO(OH)_2$	6.4				
$Te(OH)_6$	7.7	$PO(OH)_3$	2.1				
$As(OH)_3$	9.2	$AsO(OH)_3$	2.3				
		$HPO(OH)_2$	1.8				
		$H_2PO(OH)$	2.0				

^{a}Values given within parentheses are estimated values.

species is shown by the data in Table 10–1; a small increase in negative charge on an atom to which a proton is attached enormously decreases the acidity of the species.

There is another, more simple correlation between the acidity of oxyacids and their structures. Acidity increases as the number of oxygen atoms without hydrogen atoms attached increases. Table 10–2 shows such a correlation in which the pK's of oxyacids are clearly functions of n in the formula $XO_n(OH)_m$ and are much less influenced by the value of m.

The correlations of acidity with structure are approximate but have led to interesting discoveries. For instance, the pK of H_3PO_3 is 1.8 for the first hydrogen ionization. A chemist would instinctively write the structure as

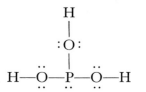

for this acid, which would suggest a pK of about 7 to 9 (see the "very weak" group in Table 10–2). However, the pK_1 of 1.8 suggests that phosphorus has one oxygen bound to it, which is not bound to a hydrogen atom. Structural studies show that only two protons can be ionized and that the third proton is attached to phosphorus. Thus, the structure of H_3PO_3 should be written as

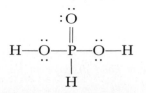

This structure is clearly consistent with the correlation in Table 10–2.

10–3   MOLECULAR ORBITALS

In Chapter 8, we used atomic orbitals to explain the properties of atoms. From the Schrödinger equation we obtained a set of wave functions, $\psi_{(x,y,z)}$, of such a nature that $|\psi|^2_{(x,y,z)}$ at any point is the electron probability density at that point. If the electron is in the quantum state described by n, l, and m, the probability of finding the electron within a small element of volume dv at (x, y, z) is

$$|\psi_{n,l,m,(x,y,z)}|^2 \, dv$$

It was helpful in building a picture of many-electron atoms to imagine the probability functions or orbitals as if they had a shadowy existence of their own, and then to fill these orbitals by adding electrons like peas dropped into cups. In the same way, for molecules we first shall find a set of *molecular orbitals* for a given arrangement of atoms, and then fill them with the available electrons, no more than two to an orbital as before. But before the procedure becomes so formal, let us look at what happens when two hydrogen atoms come together to make a molecule.

Bonding in the H_2 Molecule

If two hydrogen atoms are far apart, they have no effect on one another. As they are brought closer together, they begin to exert an effect. The two nuclei, having the same positive charge, repel one another, and the two electron clouds also repel one another. However, most important of all is the attraction between the nucleus of one atom and the electron cloud of the other atom. As the atoms approach, the electron clouds are pulled toward the region between the nuclei [Figure 10–1(d)]. The combination of two nuclei and two electrons is more stable (has lower energy) than two isolated nuclei, each with its electron. The closer the nuclei come together, the more electron density is attracted between them, the lower the energy falls, and the more stable the assembly (which now can be called a molecule) becomes. However, there is a limit to this process. If the nuclei come too close together, the repulsion between them begins to dominate. Beyond this limit, the nuclear repulsion is greater than the nucleus–electron cloud attraction.

Figure 10–1. Bonding in the H_2 molecule. (a) Probability density in the 1s atomic ▶ *orbital of hydrogen. (b) The spherical surface that encloses 99% of the probability density. (c) Two hydrogen atoms sufficiently far apart will exert no effect on one another. (d) As the atoms are brought together, each electron cloud begins to respond to the attraction of the nucleus of the other atom. The electron clouds become distorted, and electron density increases in the region between the nuclei. (e) At closer proximity, repulsion between the nuclei becomes significant. The equilibrium bond distance in the H_2 molecule is the point of balance between this attraction and repulsion.*

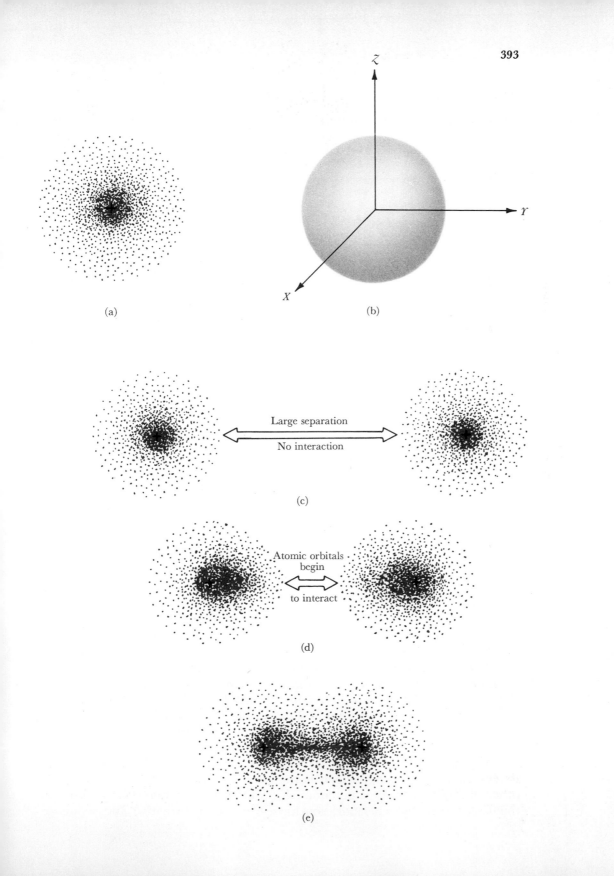

(a)

(b)

Large separation

No interaction

(c)

Atomic orbitals
begin

to interact

(d)

(e)

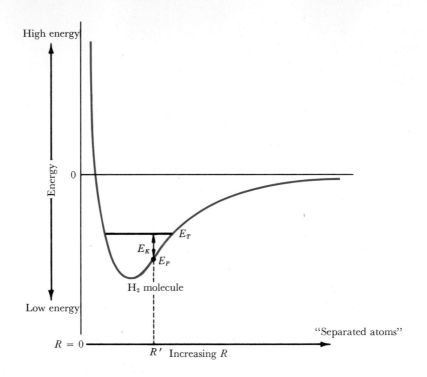

Figure 10–2. Potential energy curve for the H_2 molecule. As the distance between nuclei decreases, the potential energy decreases because of electron cloud–nucleus attraction, and then increases because of nucleus–nucleus repulsion. The horizontal line marked E_T is the total energy of a vibrating molecule. At the extremes of vibration, where the E_T line touches the potential energy curve, kinetic energy is zero and all energy is the potential energy of the extended or compressed bond. In the center of the vibration, kinetic energy of motion, E_K, is at a maximum; potential energy, E_P, is at a minimum. At any other point in the vibration, the sum of kinetic and potential energies is constant: $E_T = E_K + E_P$. The "equilibrium bond length" is the bond length at the minimum E_P.

   There is an intermediate *equilibrium* distance at which both forces are balanced. Pull the atoms apart, and attractive forces pull them back again. Push them together, and repulsive forces push back. The two atoms act very much as if they were tied together by a spring. This condition of balance, or equilibrium distance, is what we normally mean when we speak of the *bond length* [Figure 10–1(e)].

   The energy of a vibrating H_2 molecule is shown in Figure 10–2. Part of its energy is kinetic, E_K, which is the energy of motion of the atoms. The other part of its energy is potential, E_P, which is the energy that motionless molecules with a given separation have because of attractive and repulsive

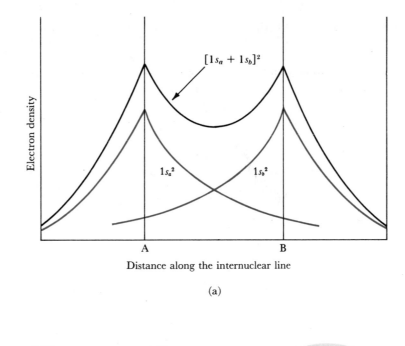

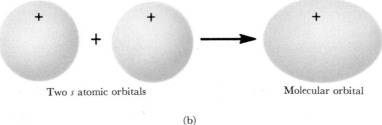

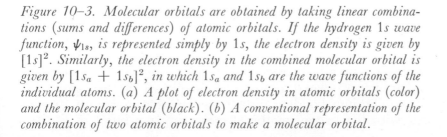

(b)

Figure 10–3. Molecular orbitals are obtained by taking linear combinations (sums and differences) of atomic orbitals. If the hydrogen $1s$ wave function, ψ_{1s}, is represented simply by $1s$, the electron density is given by $[1s]^2$. Similarly, the electron density in the combined molecular orbital is given by $[1s_a + 1s_b]^2$, in which $1s_a$ and $1s_b$ are the wave functions of the individual atoms. (a) A plot of electron density in atomic orbitals (color) and the molecular orbital (black). (b) A conventional representation of the combination of two atomic orbitals to make a molecular orbital.

forces. At the extreme limits of stretch or compression, the atoms are motionless at the instant of turnaround, but the forces on them are greatest. In the middle of one vibration, the potential energy is least, but the atoms are moving most rapidly and the kinetic energy is greatest. So long as the molecule is not disturbed, the total energy, E_T, is constant. The point representing the molecule in the energy diagram of Figure 10–2 moves back and forth from one end of the horizontal E_T line to the other, but at all points $E_T = E_K + E_P$.

The attraction that makes the molecule stable is the attraction of the nuclei for the electron density concentrated between them. We can think of this concentration as an *overlap* of the 1s atomic orbitals. If, for convenience, we represent the atomic wave function ψ_{1s} simply by the symbol 1s, then electron density in the atom is represented by $[1s]^2$. [Square brackets are used here to avoid confusion with the notation for electronic configuration in atoms: $(1s)^2(2s)^1$, etc.] We can construct a molecular orbital by adding the two atomic wave functions from atoms *a* and *b* to produce the molecular wave function $1s_a + 1s_b$. The electron probability density in such a molecular state is given by the square of the molecular wave function: $[1s_a + 1s_b]^2$. As you can see in Figure 10–3(a), such a combination of atomic orbitals does produce the pileup of electron density that we have been using to explain bonding. In the hydrogen molecule, this molecular orbital is filled with the two electrons having opposite spins (paired), and a single covalent bond is formed. This type of molecular orbital is a *bonding orbital*.

There is more than one way of combining two atomic wave functions, $1s_a$ and $1s_b$. What if they were subtracted instead of added? Expressed differently, what if the atomic wave functions were combined with opposite sign, or were out of phase? The results are compared in Figure 10–4(a) and (b). The first drawing shows the addition of atomic wave functions to make the molecular orbital $[1s_a + 1s_b]^2$. The second shows the subtraction of one from the other to make the molecular orbital $[1s_a - 1s_b]^2$. The wave function, $1s_a - 1s_b$, changes sign halfway from one nucleus to the other, so its square falls to zero at the *nodal plane*. If electrons are in this molecular orbital in the molecule, there is *no* probability of finding them on a plane halfway between the nuclei. In fact, most of the electron density is concentrated outside the two nuclei. Rather than being pulled together, the nuclei are pulled apart. This type of molecular orbital is an *antibonding* orbital.

The potential energies of the bonding and antibonding orbitals are shown in Figure 10–5(a). The closer the nuclei come in the antibonding state, the more they are held back by the drag of their electron clouds and the higher is the energy of the molecule. At every point, the energy of the molecule is greater than that of two isolated atoms. The energies of the two molecular orbitals at the equilibrium bonding distance are plotted in Figure 10–5(b) and compared with the energy of the electrons in 1s orbitals of isolated atoms.

In summary, the two atomic 1s orbitals can be combined in two different ways to produce two molecular orbitals, one bonding and one antibonding. The bonding orbital concentrates electron density between the nuclei; the antibonding orbital concentrates it outside the two nuclei and has no density at all on a plane halfway between them. Both these molecular orbitals are symmetrical with respect to rotation around the line joining the nuclei; that is, when the orbital is spun around this line, neither the appearance of the electron density cloud nor the sign of the wave function composing it is

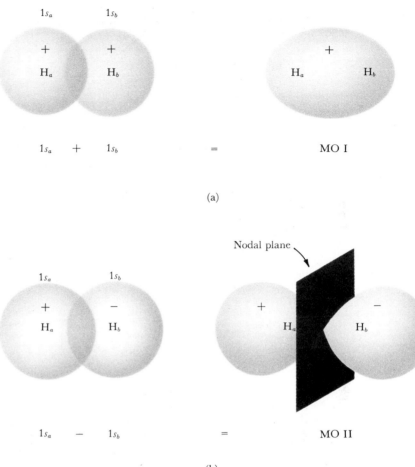

$1s_a$ $1s_b$

$1s_a$ + $1s_b$ =   MO I

(a)

$1s_a$ $1s_b$

Nodal plane

$1s_a$ − $1s_b$ =   MO II

(b)

*Figure 10–4. Two atomic 1s orbitals give rise to two molecular orbitals.
(a) If the two atomic wave functions are added, or combined with the same
sign, the resulting molecular orbital has high electron density between the
nuclei. Electrons in such an orbital hold the molecule together, and it is
called a* bonding *orbital. (b) If the two atomic functions are subtracted,
or combined with opposite signs, the electron density in the molecular
orbital is concentrated away from the internuclear region. There is zero
probability of finding the electron in a nodal plane halfway between the
nuclei. Electrons in such a molecular orbital pull the molecule apart, so
it is called an* antibonding *orbital.*

altered. Orbitals with such symmetry are called sigma, σ, orbitals. The bonding orbital is given the superscript, b, and the antibonding orbital, the superscript, *. (The types of molecular orbitals are described by the symbols σ, π,
δ, . . . , by analogy with s, p, d, . . . for atomic orbitals.)

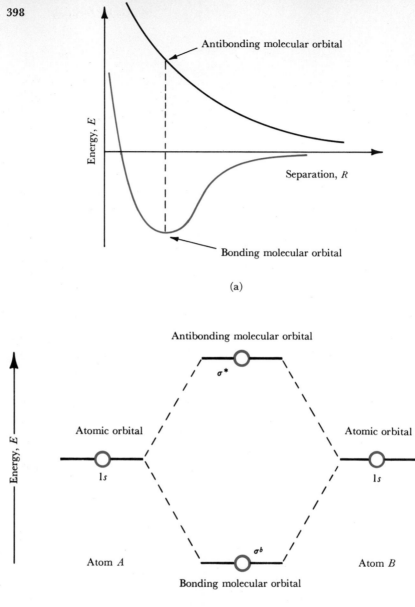

Figure 10–5. (a) The energy of a molecule with electrons in the bonding orbital falls to a minimum at the observed interatomic distance. The energy of a molecule with electrons in the antibonding orbital is always greater than the energy of completely separated atoms; it increases steadily as the atoms are brought closer together. (b) The two lowest molecular orbitals for the hydrogen molecule, and the atomic 1s orbitals from which they came. The symbol σ (sigma) indicates that the orbital is symmetrical around the line between the nuclei, and that the orbital could be spun around that line as an axis without changing the orbital's appearance. The superscript b indicates bonding character, and * indicates antibonding character.

The Pauli Buildup Process in Molecules

Now we can use an *aufbau* process to explain the occurrence or nonoccurrence of the molecules H_2^+, H_2, He_2^+, and He_2. The hydrogen molecule-ion, H_2^+, has two nuclei but only one electron. By Pauli's reasoning, this electron will be in the lowest-energy molecular orbital, which Figure 10–5(b) indicates is the bonding σ^b orbital. The H_2^+ molecule-ion should be weakly stable.

The hydrogen molecule, H_2, has two nuclei and two electrons. Both electrons can be accommodated in the σ^b orbital if their spins are paired, and a covalent electron-pair bond is created. The bond energy (the energy needed to pull the atoms apart) should be substantially larger than that of the hydrogen molecule-ion.

The helium molecule-ion He_2^+, has two helium nuclei and three electrons. Although the energies of the helium orbitals, atomic and molecular, are different from those of hydrogen because of the different nuclear charge, the relative arrangement of atomic and molecular energy levels is similar. We can use Figure 10–5(b) for He as well as for H if we make the proper adjustments to the energy scale on the left.

The first two electrons in He_2^+ pair their spins and fill the σ^b bonding orbital. But what happens to the third electron? By the Pauli exclusion principle, it cannot occupy the σ^b state, so it must go into the next lowest energy level, which is the antibonding σ^* orbital. This third electron is pushed away from the region between the nuclei by the presence of the first two and is forced into the region outside the two nuclei. This electron is a disruptive influence; it pulls the nuclei apart. The molecule would be more stable if the third electron were not there. The electron effectively counteracts the contribution of one of the bonding electrons, thereby leaving a *net* bonding action of one electron, or half a covalent bond. The bond energy of He_2^+ should be less than that of H_2.

In He_2, the fourth electron also must go into the antibonding orbital. Now there are two bonding electrons and two antibonding electrons. The molecule is no more stable than the isolated atoms and falls apart. We should not expect to find a He_2 molecule.

Table 10–3.   Comparison of Predicted and Observed Bonding in Simple Diatomic Molecules

	Molecular orbital theory predictions		Experimental observations		
Molecule	Bonding electrons	Antibonding electrons	Net bonding electrons	Bond length (Å)	Bond energy (kcal mole⁻¹)
H_2^+	1	0	1	1.06	61
H_2	2	0	2	0.74	103
He_2^+	2	1	1	1.08	60
He_2	2	2	0	none	none

Enough of theory for a moment. What actually happens? Table 10–3 lists the observed bond energies and bond lengths for H_2^+, H_2, and He_2^+; as predicted, He_2 does not exist. Moreover, the measured bond energies are consistent with the number of net bonding electrons given by molecular orbital theory. Bond lengths, too, are consistent. The more bonding electrons, the tighter the interaction and the shorter the bond length. Thus far, molecular orbital theory explains the data well. How can we extend this process to more complicated molecules?

The process that we shall use to explain first the diatomic molecules of heavier atoms and then more complicated molecules can be summarized as follows:

1) Combine atomic orbitals in a suitable way to obtain a set of molecular orbitals. The total number of molecular orbitals obtained always will be equal to the number of atomic orbitals that we began with.

2) Decide the order of energies of these molecular orbitals.

3) Feed all of the electrons in the molecule into these molecular orbitals. Start from the lowest and work up; place no more than two electrons in any one orbital.

4) Examine the filled bonding and antibonding orbitals to determine the net number of bonding electrons. (Some antibonding orbitals will have lower energy than other bonding orbitals and will be filled before these bonding orbitals are. The criterion for a bonding orbital is not that it have a low energy, but that it have a *minimum* in energy, as in Figure 10–5(a), at some interatomic distance.) Two net bonding electrons correspond to what we have called a single bond in the Lewis model.

10–4 DIATOMIC MOLECULES WITH ONE TYPE OF ATOM

The O_2, N_2, and Cl_2 molecules, with only one type of atom, are called *homonuclear* molecules. In contrast, HCl, CO, and HI are *heteronuclear*. We want to extend the simple molecular orbital treatment of H_2 and He_2 to homonuclear diatomic molecules of elements in the second period of the periodic table. Some of these molecules, such as N_2, O_2, and F_2, are stable at STP. Others, such as C_2 and Li_2, are found only at high temperatures. Some do not exist at all. What are the predictions of molecular orbital theory?

The first step in the treatment is to construct molecular orbitals. The atomic orbitals available are the $2s$ and three $2p$ orbitals from each of the two atoms. Their energies are diagramed in Figure 10–6, and the molecular orbitals that result from their combination are shown in Figure 10–7.

The two $2s$ atomic orbitals can be combined into a bonding σ_s^b and an antibonding σ_s^* orbital in the same manner as for the $1s$. If the line joining the nuclei is the Z axis, then there are two kinds of $2p$ orbitals: the $2p_z$ orbital, which is parallel to the internuclear axis, and the $2p_x$ and $2p_y$ orbitals, which are perpendicular to it. The two $2p_z$ orbitals from the two atoms can

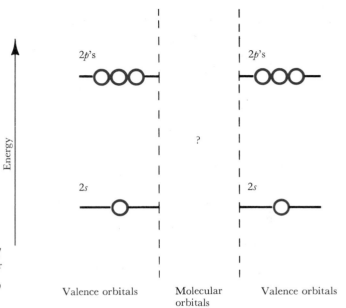

Figure 10–6. The 2s and 2p atomic orbital energies of elements in the second period of the periodic table, before combination into molecular orbitals.

be combined with the same signs in the internuclear region, thereby producing a concentration of electron density between the nuclei. They also can be combined with opposing signs, thus producing a deficiency of electron density between the nuclei. These molecular orbitals are labeled σ_z^b and σ_z^*, and are at the left and right center in Figure 10–7. They are still σ orbitals because they are rotationally symmetrical around the Z axis.

The $2p_x$ orbitals on the two atoms can be combined as sums (lower left, Figure 10–7) or differences (lower right). The first combination, $[2p_x + 2p_x]^2$, produces a molecular orbital that looks like an exaggerated version of the original $2p$ orbitals. Maximum electron density occurs in two watermelon-shaped lobes, above and below a nodal plane that was also the nodal plane of the original atomic orbitals. The wave function itself has opposite signs in the two lobes. The other combination, $[2p_x - 2p_x]^2$, leads to a molecular orbital with a second nodal plane, and four lobes whose electron densities lie mostly outside the internuclear region (lower right, Figure 10–7). The two-lobed orbital is bonding; the four-lobed orbital is antibonding. They are called π orbitals and have a different kind of symmetry around the Z axis. If either orbital is rotated 180° around the Z axis, the electron-density cloud has the same appearance, but the signs of the wave function in the different lobes all are reversed. The two orbitals are labeled π_x^b and π_x^*. A corresponding pair, π_y^b and π_y^*, results from the two $2p_y$ atomic orbitals.

From eight atomic orbitals we have obtained eight molecular orbitals, four of them bonding (σ_s^b, σ_z^b, π_x^b, π_y^b) and the other four antibonding (σ_s^*, σ_z^*, π_x^*, π_y^*). What is the order of energy of these orbitals?

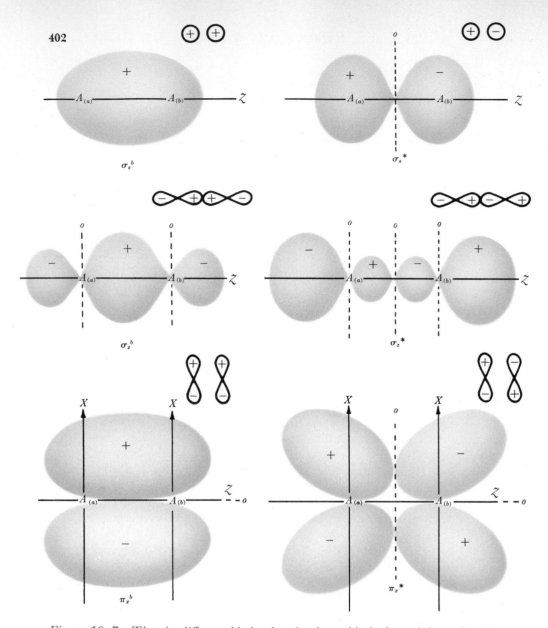

Figure 10–7. The six different kinds of molecular orbitals formed from the s, p_x, p_y, and p_z orbitals of two similar atoms in a diatomic molecule. The line drawn through the two nuclei is chosen as the Z axis. The symbol π indicates that, if the molecular orbital is rotated 180° around the axis, the electron distribution is unchanged. The only effect is to reverse the signs of the parts of the wave function. Plus and minus signs represent only the signs on the wave function, and not electric charge. The atomic orbitals from which these are obtained are shown, with their appropriate signs, at the upper right of each molecular orbital. The atomic orbitals used are s (top row), p_z (middle row), and p_x (bottom row), which is equivalent to p_y. Bonding orbitals are in the left column; antibonding orbitals are in the right one. Dashed lines denoted by o are nodal planes of zero electron density.

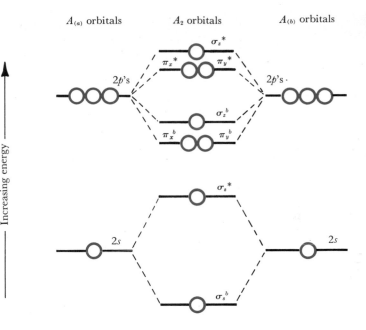

Figure 10–8. Energy levels for the molecular orbitals shown in Figure 10–7. Among the orbitals that come from either s or p atomic orbitals, bonding molecular orbitals are more stable than antibonding orbitals. The π_x^b and π_y^b orbitals are more stable than the σ_z^b because they permit the electrons to remain farther away from the filled σ_s^b orbital.

The orbitals derived from the s atomic orbitals will have lower energy than those from the p orbitals. Moreover, of two orbitals derived from the same atomic orbitals, the bonding orbital will lie lower than the antibonding orbital. Therefore, the first two most stable levels are the σ_s^b and σ_s^*. The most stable of the bonding orbitals obtained from $2p$ orbitals are π_x^b and π_y^b rather than σ_z^b, which is contrary to earlier ideas and to the diagrams in many older texts. This order of levels has been found from recent, careful, spectroscopic and magnetic studies of B_2 and N_2^+. It is reasonable, because electrons added to π_x^b are farther removed in space from those in filled σ_s^b and σ_s^* than they would be if they were in the σ_z^b orbital. (There is a crossover of energy levels at O, so in O_2 and F_2 the σ_z^b is more stable than π_x^b and π_y^b. This really does not matter in our discussion since all three levels are filled in O_2 and F_2 anyway.) The π_x^b and π_y^b orbitals have the same energy, and are said to be *degenerate* energy levels. Above these orbitals lies the σ_z^b, then the two antibonding π_x^* and π_y^*, and last of all the antibonding σ_z^*. The complete energy diagram of the molecular orbitals from $2s$ and $2p$ atomic orbitals appears in Figure 10–8.

Paramagnetism and Unpaired Electrons

Substances whose molecules and ions have electrons with unpaired spins tend to be drawn into magnetic fields. The magnetic field aligns the spins and magnetizes the substance. If the substance retains its aligned spins and its magnetic properties after the magnetic field is removed, it is called *ferromagnetic*. Iron is the most familiar ferromagnetic substance, but Co and Ni

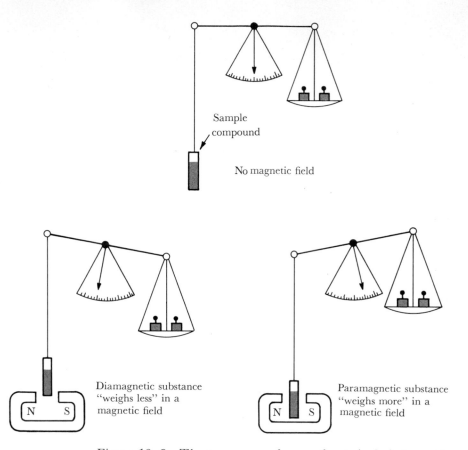

Sample compound

No magnetic field

Diamagnetic substance "weighs less" in a magnetic field

N   S

Paramagnetic substance "weighs more" in a magnetic field

N   S

Figure 10–9. The presence or absence of unpaired electron spins can be determined by a magnetic or Gouy balance. A diamagnetic substance, with no unpaired electrons, is slightly repelled by a magnetic field (bottom left). A paramagnetic substance, with unpaired electron spins, is attracted into the magnetic field (bottom right).

also are ferromagnetic. The powerful *Alnico* magnets are made of an alloy of aluminum, nickel, and cobalt with iron. Many more substances, called *paramagnetic* substances, lose their magnetism when removed from the magnetic field. These materials also have unpaired electrons, and the strength of the attraction by a magnetic field can be used to determine how many such unpaired electrons there are per mole of substance. If a molecule has no unpaired electrons, it is *diamagnetic* and is slightly repelled by a magnetic field because of the small opposing magnetic moments induced in it by the field. The number of unpaired electrons in a molecule of a substance can be determined with a magnetic balance as shown schematically in Figure 10–9.

The three types of experimental data that we shall use to test the predictions of molecular orbital theory are bond energy, bond length, and the number of unpaired electrons.

Buildup of Diatomic Molecules

We now are ready to feed electrons into molecular orbitals, two electrons to an orbital, and to build the diatomic molecules from Li_2 through Ne_2. There always will be four electrons from the lower $n = 1$ atomic orbitals. In the diatomic molecule, two of these inner-shell electrons will be in the $\sigma_{1s}{}^b$ bonding molecular orbital and two in the $\sigma_{1s}{}^*$ antibonding orbital. However, it makes no difference to the net bonding whether we think of them as in $1s$ atomic orbitals or the molecular orbitals obtained from $1s$ orbitals. The bonding properties of the molecule arise only from the outer shell of $n = 2$ electrons, and we need to consider only the molecular orbitals derived from $2s$ and $2p$ atomic orbitals.

Lithium. The Li atom has one valence electron, so the Li_2 molecule has two potential bonding electrons. These are paired in the lowest available molecular orbital, $\sigma_s{}^b$. Therefore, the Li_2 molecule contains a single covalent bond. The bond length is longer than in H_2, 2.67 Å compared to 0.74 Å, because the larger $n = 2$ orbitals are involved, rather than the $n = 1$. For the same reason the bond is weaker: 26.3 kcal mole^{-1} rather than 103 kcal mole^{-1} as in H_2. The nuclei are farther apart, the electron cloud is spread over a greater volume, and the overall attractive forces are weaker.

Beryllium. There are four valence electrons available in the Be_2 molecule. Two are paired in the bonding $\sigma_s{}^b$ molecular orbital, and two are paired in the antibonding $\sigma_s{}^*$. This configuration gives no net bonds, which is consistent with the absence of Be_2 from the family of stable second-row diatomic molecules.

Boron. The two additional valence electrons of B_2 go into the next lowest unfilled molecular orbitals, $\pi_x{}^b$ and $\pi_y{}^b$. By Hund's rule, electron–electron repulsion ensures that one electron occupies each orbital rather than having them both spin-paired in one. Whether the electrons are paired or not, the effect of two bonding electrons is a single covalent bond. The electronic configuration for B_2 is

$$KK(\sigma_s{}^b)^2(\sigma_s{}^*)^2(\pi_x{}^b)^1(\pi_y{}^b)^1$$

The symbol KK represents the four electrons in the inner $n = 1$ shells that have no effect on bonding. The experimental bond length in B_2, 1.59 Å, is less than that in Li_2, 2.67 Å. The bond energy is greater; it is 65 kcal mole^{-1} rather than 26 kcal mole^{-1}. Both effects arise from the greater positive charge on the B nucleus and the tightness with which the electrons are held. Perhaps the most satisfying test of the molecular orbital theory is the finding of two unpaired electrons in B_2 from magnetic measurements. This is a direct confirmation of the order of $\sigma_z{}^b$ and $\pi_x{}^b$ orbital energies in Figure 10–8; if the order were reversed, both electrons would be paired in $\sigma_z{}^b$ and the molecule would have no unpaired spins. (As a matter of historical fact, the unpaired electrons in B_2 were not predicted in advance. The existence of the

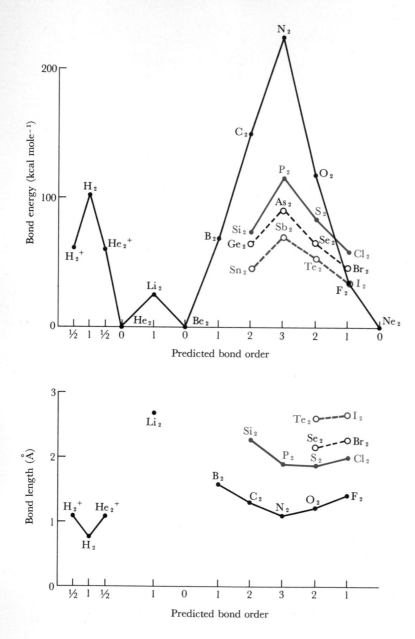

Figure 10–10. Plot of bond energies and bond lengths against predicted bond order for homonuclear diatomic molecules. Bond energies increase with increasing bond order, and bond lengths decrease.

unpaired electrons compelled scientists to revise their original order of orbital energies to that of Figure 10–8.)

Carbon. The two additional electrons in carbon, C_2, complete the π_x^b and π_y^b molecular orbitals. There are four net bonding electrons, and hence two covalent bonds in Lewis' terminology. There should be no magnetic moment or unpaired spins in the ground electronic state. True to predictions, the bond energy of C_2 is twice that of B_2 (144 kcal mole^{-1} and 65 kcal mole^{-1}), and the bond length is less (1.24 Å and 1.59 Å). Moreover, C_2 is not paramagnetic.

Nitrogen. With nitrogen, all of the bonding orbitals in Figure 10–8 are filled. The N_2 molecule has the electronic configuration

$$KK(\sigma_s{}^b)^2(\sigma_s{}^*)^2(\pi_x{}^b)^2(\pi_y{}^b)^2(\sigma_z{}^b)^2 \qquad \text{or} \qquad KK(\sigma_s{}^b)^2(\sigma_s{}^*)^2(\pi_{x,y}^b)^4(\sigma_z{}^b)^2$$

There are six net bonding electrons, so the N_2 molecule contains a triple bond. Since there are no unpaired electrons, no paramagnetism is expected.

Nitrogen has the greatest bond energy and the shortest bond length of any element in the second period, 225 kcal mole^{-1} and 1.10 Å. The increase in bond energy with theoretical bond order (single, double, or triple bonds) in Figure 10–10 is remarkably constant. As predicted, N_2 has no magnetic moment.

We now can interpret the Lewis structure of N_2,

$$\overset{\cdot\cdot}{N}\!\equiv\!\overset{\cdot\cdot}{N}$$

The three bonds involve the $\pi_x{}^b$, $\pi_y{}^b$, and $\sigma_z{}^b$ orbitals. The two lone pairs correspond, at least formally, to the self-canceling pair of orbitals: $(\sigma_s{}^b)^2(\sigma_s{}^*)^2$.

We did not try to write a Lewis structure for C_2. Nothing in Lewis' covalence theory suggested that it should exist. Now, by analogy with N_2, we would write the C_2 molecule as

$$\overset{\cdot\cdot}{C}\!=\!\overset{\cdot\cdot}{C}$$

in which the bonds correspond to the filled $\pi_x{}^b$ and $\pi_y{}^b$ orbitals, and the lone pairs are as in N_2. But C_2, like BF_3, is electron-deficient; there are only six valence electrons around each carbon atom. We might expect that C_2 would accept electron pairs from a donor in the way that BF_3 accepts them from NH_3 to make the addition compound BF_3NH_3. But C_2 also has a lone pair and can play the role of donor as well. The C_2 molecule is found only at high temperatures. At lower temperatures, each atom in C_2 accepts electrons from one new C and donates electrons to another. The result is a network in which each C is covalently linked to at least three others (graphite) or, alternatively, to four (diamond).

Oxygen. In oxygen, the next two electrons must go into the two antibonding orbitals $\pi_x{}^*$ and $\pi_y{}^*$, one in each by Hund's rule. Of the 12 valence electrons in O_2, a total of 8 are in bonding orbitals and 4 are in antibonding orbitals. There are 4 *net* bonding electrons, thus the molecule has a double bond. The two additional electrons, which go into antibonding orbitals, cancel the effects of two of the electrons in the orbitals that gave N_2 a triple bond. Both bond length and bond energy agree well with theory (Figure 10–10). The electronic configuration for O_2 is

$$KK\,(\sigma_s{}^b)^2(\sigma_s{}^*)^2(\sigma_z{}^b)^2(\pi_{x,y}^b)^4(\pi_{x,y}^*)^2$$

Notice that the relative order of energy levels $\sigma_z{}^b$ and $\pi_{x,y}^b$ has changed, as was mentioned previously.

Molecular orbital theory explains why O_2 is paramagnetic, indicating two unpaired electrons, whereas the Lewis theory fails. The Lewis structure for O_2 has no unpaired electrons:

$$:\overset{..}{O}=\overset{..}{O}:$$

The only possible Lewis structures with a double bond and two unpaired electrons violate the symmetry of the molecule by making the oxygen atoms different, and make it appear that the unpaired electrons are both associated with a particular atom:

$$\cdot\overset{..}{\underset{\cdot}{O}}=\overset{..}{O}: \qquad :\overset{..}{O}=\overset{..}{\underset{\cdot}{O}}\cdot$$

You can partially redeem the Lewis structures by saying that these two structures are the two *resonance structures* for O_2, and that the true structure is unrepresentable but has the character of both resonance structures in equal amounts. But this treatment hardly seems worth the effort. It is easier to abandon Lewis structures and to think in molecular orbital terms.

Fluorine. In F_2, all of the molecular orbitals in Figure 10–8 are occupied except the highest one. The molecule has one net covalent bond from its two net bonding electrons, and the electronic structure is

$$KK(\sigma_s{}^b)^2(\sigma_s{}^*)^2(\sigma_z{}^b)^2(\pi_{x,y}^b)^4(\pi_{x,y}^*)^4$$

Bond energy and bond length are as expected for a single bond, and the F_2 molecule has no magnetic moment.

Neon. The Ne_2 molecule would have all of the molecular orbitals in the center of Figure 10–8 filled, and an equal number of bonding and antibonding electrons. There would be no net bonding electrons and no reason for the atoms to remain together. As predicted, there is no Ne_2 molecule.

Later periods in the table. Experimental data on several diatomic molecules and molecule-ions are given in Table 10–4. Some of these data have been plotted in Figure 10–10 as well. The trends for the nonmetals are regular and understandable in terms of larger orbitals (in which $n = 3$, 4, and 5) and weaker forces holding the electrons. The unexpected weakness of the F_2 bond is odd. The lone electron pairs in F_2 are considerably closer together than they are in the larger halogens, and we think that such close lone-pair repulsion may be at least part of the reason for the weak F_2 bond.

Example. Write the molecular orbital electronic structure of the molecule-ion $O_2{}^-$. What is the bond order, and how many unpaired electrons are there?

Table 10-4. Bond Properties of Some Homonuclear Diatomic Molecules and Ions[a]

Molecule	Bond length (Å)	Bond dissociation energy (kcal mole⁻¹)
Ag_2	—	38.7 ± 2.2
As_2	2.288	91.3
Au_2	2.472	53.9 ± 2.2
B_2	1.589	65.5 ± 5
Bi_2	—	46.6 ± 1.5
Br_2	2.2809	45.440 ± 0.003
C_2	1.2425	144
Cl_2	1.988	57.18 ± 0.006
Cl_2^+	1.8917	99.2
Cs_2	—	10.4
Cu_2	2.2195	47.3 ± 2.2
F_2	1.417	33.2 ± 1.6
Ge_2	—	65
H_2	0.74116	103.24
H_2^+	1.06	61.06
He_2^+	1.080	77.0
I_2	2.6666	35.55
K_2	3.923	11.8
Li_2	2.672	26.3
N_2	1.0976	225.07
N_2^+	1.116	201.28
Na_2	3.078	17.3
O_2	1.20741	117.96
O_2^+	1.1227	—
O_2^-	1.26	93.9
O_2^{2-}	1.49	—
P_2	1.8937	114
Pb_2	—	23
Rb_2	—	11.3
S_2	1.889	100.69
Sb_2	2.21	71.3
Se_2	2.1663	77.6
Si_2	2.246	75
Sn_2	—	46
Te_2	2.5574	62.3

[a] The values of bond dissociation energy (*DE*) given generally refer to the ΔE^0 for the process $A_2(g) \rightarrow A(g) + A(g)$.

Solution. The ion has six valence electrons from each oxygen atom plus one extra for the -1 charge, or 13 valence electrons. Filling orbitals in Figure 10–8 from the bottom, we find an electronic structure of

$$KK(\sigma_s^{\,b})^2(\sigma_s^{\,*})^2(\sigma_z^{\,b})^2(\pi_{x,y}^b)^4(\pi_{x,y}^*)^3$$

There are three net bonding electrons; hence, a bond order of $1\frac{1}{2}$. The molecule has one unpaired electron.

Exercise. With what neutral molecule is $O_2{}^{2-}$ isoelectronic? Explain the observed bond lengths of $O_2{}^{+}$, O_2, $O_2{}^{-}$, and $O_2{}^{2-}$ in Table 10–2.

(*Answer:* F_2; bond orders $2\frac{1}{2}$, 2, $1\frac{1}{2}$, and 1.)

Exercise. Plot the bond energies and bond lengths against period numbers for the diatomic alkali metal molecules, and account for trends you see.

Before you go on. You can study the molecular orbital theory for diatomic homonuclear molecules in greater detail and more slowly by reading Section 10–2 of Lassila, *Programed Reviews of Chemical Principles.*

10–5 DIATOMIC MOLECULES WITH DIFFERENT ATOMS

With the methods used for homonuclear diatomic molecules in mind, let us examine the molecular orbital treatment of molecules with two different atoms.

Hydrogen Fluoride and Potassium Chloride

When we carry out the mathematical operations that are behind the expression, "combining two atomic orbitals to produce an antibonding and a bonding molecular orbital," we find that the two atomic orbitals should be reasonably close in energy. In the H_2 molecule, each of the two molecular orbitals has a 50% contribution from each of the two hydrogen $1s$ atomic orbitals. At the other extreme, if in a molecule of the type AB we combined an orbital from A having extremely high energy with one from B of quite low energy, we would find at the end of the mathematical analysis that the antibonding molecular orbital was almost pure A, and the bonding orbital was almost pure B. Then a pair of electrons in this "bonding" orbital would not be a true covalent bonding orbital at all. It would be a lone electron pair in a B orbital. The interaction of these two atomic orbitals would be negligible. We shall see, for the HF molecule, what this means in terms of partial ionic character in a bond.

   In HF, the energies of the hydrogen $1s$ and fluorine $1s$ atomic orbitals are so different that there is effectively no interaction between them. The fluorine $2s$ orbital has too low an energy as well. Only the $2p$ orbitals are close enough in energy to the hydrogen $1s$ that there is appreciable combination into molecular orbitals. Moreover, the $2p_x$ and $2p_y$ orbitals have the wrong symmetry for combining with hydrogen $1s$, as Figure 10–11 shows. The total overlap of either p with the $1s$ is zero if proper account is taken of the signs of the wave functions. The molecular orbitals in HF are obtained by a combination of hydrogen $1s$ and fluorine $2p_z$ atomic orbitals. This combination

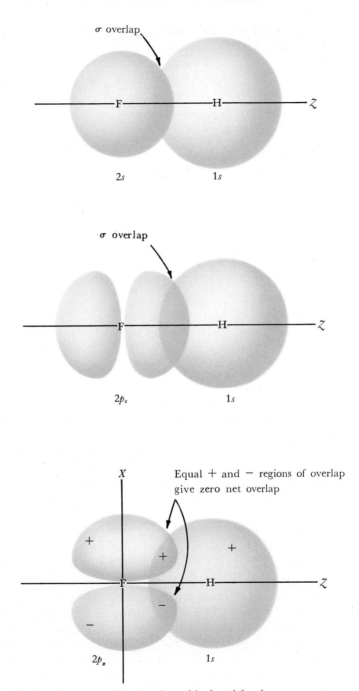

Figure 10–11. Overlap of the hydrogen 1s atomic orbital with the valence orbitals of fluorine. The net overlap of a $2p_x$ or $2p_y$ orbital of fluorine with the hydrogen 1s orbital is zero, and these two p orbitals cannot be used in forming molecular orbitals.

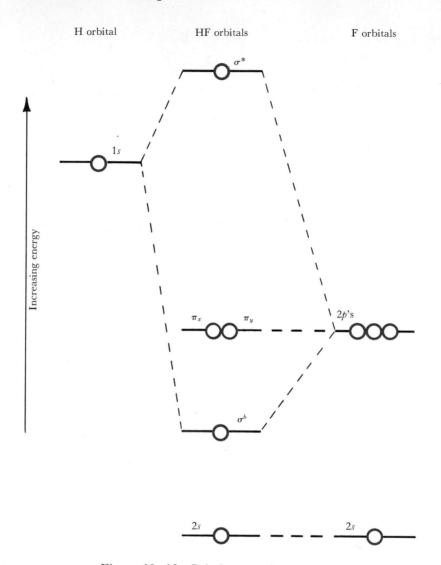

Figure 10–12. Relative energies of atomic and molecular orbitals in HF. The energy of an electron in the hydrogen atom $1s$ orbital is -313 kcal mole^{-1} (the first ionization energy of H is $+313$ kcal mole^{-1}), and the energy in the $2p$ orbitals in F is -402 kcal mole^{-1} (first ionization energy of $+402$ kcal mole^{-1}).

produces two orbitals with σ symmetry, one bonding (σ^b) and one anti-bonding (σ^*).

The energy levels for HF appear in Figure 10–12. The π_x and π_y orbitals are essentially lone-pair orbitals on fluorine and might just as well be designated $2p_x$ and $2p_y$. A third lone pair occupies the fluorine $2s$ orbital. There are eight valence electrons in HF; seven from F and one from H. These electrons fill all of the HF orbitals except the highest antibonding σ^*. In the HF molecule this assignment produces one covalent bond and three lone pairs on F. The Lewis structure, H—$\overset{\cdot\cdot}{\underset{\cdot\cdot}{F}}:$, is accurate.

The energies of the $1s$ atomic orbital at the left of Figure 10–12 and the $2p$ orbitals at the right are obtainable from the first ionization energies of H and F. If 313 kcal mole^{-1} are required to remove the electron from H, the energy of the electron before removal is -313 kcal mole^{-1}. Similarly, the first ionization energy of F is 402 kcal mole^{-1}, so the energy of the $2p$ levels is -402 kcal. The two atomic levels differ by 89 kcal. The σ^b molecular orbital is closer in energy to the fluorine $2p$ than to the hydrogen $1s$. This means that there will be more of a fluorine $2p$ character to the σ^b orbital. The covalent bond is not perfectly symmetrical; there is a small inequality in charge distribution and a partial ionic character to the bond. Electrons in the σ^b orbital will have a greater probability of being near the F atom. A small charge displacement is represented by a lower case delta, δ. We can show the partial ionic character of the HF molecule by $H^{\delta+} F^{\delta-}$.

Imagine what would happen to the levels if the hydrogen $1s$ atomic orbital energy were to fall slowly. The energy separations between the σ^b molecular orbital and the two atomic orbitals from which it came would equalize; σ^b would assume an equal contribution from each. The charge displacement would diminish, and the bond would approach the perfectly symmetrical covalent bond of F_2 or H_2. This is more nearly the situation for HCl, in which the first ionization energies of H and Cl are close: 313 kcal mole^{-1} and 300 kcal mole^{-1}. In HCl, HBr, and HI the bonds are much more covalent and the charge separation in the molecule much less than in HF.

In the HCl example, the numbers just given make it appear that the electrons would be more attracted to H than to Cl since the first ionization energy of H (313 kcal mole^{-1}) is larger than that of Cl (300 kcal mole^{-1}). But ionization energies are only part of the story; relative electron affinities also must be considered. The electron affinity of Cl (85 kcal mole^{-1}) is so much larger than that of H (16 kcal mole^{-1}) that the prediction based on ionization energies alone is reversed. The combination of ionization energy and electron affinity, or the *electronegativity* of each atom, is the true deciding factor in determining charge distribution in the bond.

Now imagine instead that the H orbital at the left of Figure 10–12 is raised from its present position of -313 kcal toward an eventual limit of zero energy. As this happens, the σ^b molecular orbital becomes even more like the original $2p_z$. The limit of this trend is for the hydrogen $1s$ orbital to go to zero energy (which means complete dissociation of the electron) and for the σ^b containing the two bonding electrons to become the $2p_z$ of F (which means the formation of a F$^-$ anion). This behavior is approached in KF. Here the first ionization energy of K is only 100 kcal mole^{-1}, and the energy of the potassium $3s$ level is -100 kcal.

Table 10–5.   Percent Ionic Character of Bonds, from Dipole Moments

Molecule	Bond length, r (Å)	Calculated dipole moment, D_c	Observed dipole moment, D_o	Percent ionic character, $D_o/D_c \times 100$
H_2	0.74	3.5	0.00	0
F_2	1.42	6.8	0.00	0
HI	1.60	7.65	0.38	5
BrCl	2.14	10.4	0.57	5
ICl	2.32	11.1	0.65	6
FCl	1.63	7.9	0.88	11
HBr	1.41	6.75	0.79	12
FBr	1.76	8.45	1.29	15
HCl	1.27	6.1	1.07	17
HF	0.92	4.4	1.82	41
KI	3.05	14.6	9.24	63
LiH	1.60	7.65	5.88	77
KF	2.17	10.3	8.60	83

Dipole Moments

A heteronuclear diatomic molecule such as HF possesses an *electric dipole moment* caused by the separation of positive and negative charges. If a positive and a negative charge of magnitude q are separated by a distance r, the dipole moment, D, is

$$D = qr$$

The measured experimental dipole moment of HF is 1.82 debye units [1 debye (D) unit is 10^{-18} esu cm. Since the charge on an electron in the electrostatic system of units is 4.8×10^{-10} esu, two unit charges of opposite sign separated by 1 Å will have a dipole moment of $4.8 \times 10^{-10} \times 10^{-8}$ esu cm $= 4.8 \times 10^{-18}$ esu cm $= 4.8$ debyes]. If H really had a full $+1$ charge and F had a full -1 charge, and if these charges were separated by the true bond length of 0.92 Å, the dipole moment of HF would be calculated as 4.4 debyes from the formula above. The separated partial charges on the HF molecule are given by the ratio of true to calculated dipole moments: $1.82/4.4 = 0.41$. We say that the HF bond has 41% ionic character.

The percent ionic characters of several other diatomic molecules are listed in Table 10–5. The HCl bond has only 17% ionic character, and the KF bond is 83% ionic by the dipole moment criterion.

This treatment of HF indicates that no bond is purely ionic or purely covalent. These are not two different mechanisms of bonding but only two extreme examples of a continuous range of polarity. What matters in molecular orbital theory is the degree of match or mismatch of energy levels from the two atoms. This match or mismatch is related to the electronegativities of the atoms.

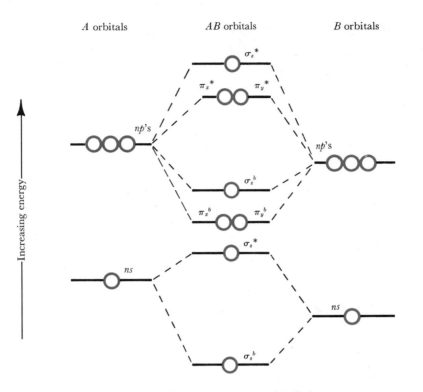

Figure 10–13. Energy levels for a general AB molecule, in which B is more electronegative than A. Compare with the homonuclear AA molecular levels in Figure 10–8. As atom B becomes more electronegative, its atomic energy levels decrease in energy and the bonding molecular orbitals assume more B-atom character.

A General *AB* Type Diatomic Molecule

The treatment of heteronuclear diatomic molecules of the general *AB* type is similar to that of homonuclear molecules. The energy-level diagram is similar, except that the atomic levels of the more electronegative atom are lower than those of the more electropositive atom (Figure 10–13). Therefore, bonding orbitals have more of the character of the electronegative atom, and antibonding orbitals, of the electropositive atom. The molecular orbitals are skewed toward one or the other atom, as shown in Figure 10–14.

Filling of orbitals with electrons occurs exactly as before. The BN molecule is isoelectronic with C_2, except that the $\pi_{x,y}^b$ and σ_z^b levels are so close together that the energy required to promote one electron to the σ_z^b orbital can be provided by the energy gained in unpairing two electrons. The electronic configuration of BN is

$$KK(\sigma_s^b)^2(\sigma_s^*)^2(\pi_{x,y}^b)^3(\sigma_z^b)^1$$

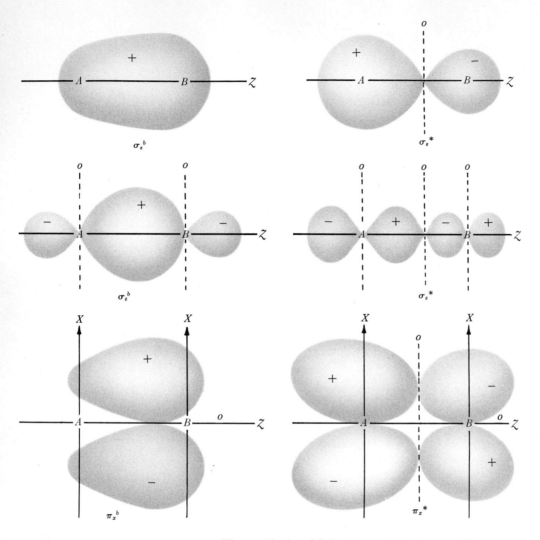

Figure 10–14. Molecular orbitals in an AB molecule, in which B is more electronegative than A. Compare with Figure 10–7. Note the increased electron probability near the more electronegative atom in bonding orbitals, and the opposite trend in antibonding orbitals.

The BN bond energy of 92 kcal mole^{-1} is suspiciously low in comparison with 144 kcal mole^{-1} for C_2. Further experimental work is necessary to verify the BN bond energy.

The species BO, CN, and CO$^+$ have nine valence electrons. From molecular orbital theory we can predict a bond order of $2\frac{1}{2}$ for them. The ions and molecules NO$^+$, CO, and CN$^-$ have 10 valence electrons and are isoelectronic with N_2. NO has 11 electrons and is one of the few common

Table 10–6. Bond Properties of Some Heteronuclear Diatomic Molecules and Ions

Molecule	Bond length (Å)	Bond dissociation energy (kcal mole⁻¹)
AsN	1.620	115
AsO	1.623	113
BF	1.262	131
BH	1.2325	70
BN	1.281	92
BO	1.2043	191.2 ± 2.3
BaO	1.940	130.4 ± 6
BeF	1.3614	135.9 ± 2.3
BeH	1.297	53
BeO	1.3308	106.1 ± 2.3
BrCl	2.138	52.1
BrF	1.7555	55
CF	1.2718	106
CH	1.1202	80
CN	1.1719	188
CN⁺	1.1727	—
CN⁻	1.14	—
CO	1.1283	255.8
CO⁺	1.1152	192.4
CP	1.5583	122.1 ± 5
CS	1.5349	173.6 ± 3.5
CSe	1.66	138 ± 5
CaO	1.822	91.32 ± 1.4
ClF	1.6281	60.3
CsBr	3.072	91.5
CsCl	2.9062	101.7
CsF	2.345	122
CsH	2.494	42
CsI	3.315	75.4
GeO	1.650	157
HBr	1.4145	86.5
HBr⁺	1.459	—
HCl	1.2744	102.2
HCl⁺	1.3153	108.3
HF	0.91680	135.1
HI	1.6090	70.5
HS	1.3503	81.4
IBr	2.485	41.90
ICl	2.32070	49.63
IF	1.908	45.7
KBr	2.8207	91.4
KCl	2.6666	100.8
KF	2.1715	118.9

Table 10–6 continued

Molecule	Bond length (Å)	Bond dissociation energy (kcal mole^{-1})
KH	2.244	43
KI	3.0478	77.2
LiBr	2.1704	101
LiCl	2.018	113.25
LiF	1.5639	135.8
LiH	1.5953	56
LiI	2.3919	81
MgO	1.749	81
NH	1.045	85
NH$^+$	1.081	—
NO	1.1508	162
NO$^+$	1.0619	—
NP	1.4910	—
NS	1.495	115
NS$^+$	1.25	—
NaBr	2.502	88
NaCl	2.3606	98.5
NaF	1.9260	113.9
NaH	1.8873	47
NaI	2.7115	69
NaK	—	14.3
NaRb	—	13.8
OH	0.9706	101.5
OH$^+$	1.0289	101.0
PH	1.4328	—
PN	1.4869	174.6
PO	1.473	124
RbBr	2.9448	90.9
RbCl	2.7868	102.8
RbF	2.2704	119.5
RbH	2.367	39
RbI	3.1769	77.7
SO	1.4810	123.66
SbO	1.848	74
SiF	1.6008	129.5
SiH	1.5201	74
SiN	1.575	104
SiO	1.5097	182.8
SiS	1.929	148
SnH	1.785	74
SnO	1.838	126.5
SnS	2.209	110.3
SrO	1.9199	99.2

gases with an odd number of electrons. The electronic configuration of NO is

$$KK(\sigma_s{}^b)^2(\sigma_s{}^*)^2(\pi_{x,y}^b)^4(\sigma_z{}^b)^2(\pi_{x,y}^*)^1$$

It has a bond order of $2\frac{1}{2}$, and both its bond energy and its bond length are intermediate between those of N_2 and O_2. Data on other AB diatomic molecules are in Table 10–6.

Exercise. What is the electronic configuration of CF? What is its bond order? Does it have unpaired electrons? (The answers to this exercise are in the preceding paragraph.)

Exercise. Add the data on second-period AB molecules in Table 10–6 to a plot such as Figure 10–10. Do they correlate well with the data on homonuclear molecules? Can you account for any differences?

Challenge. Try Problem 10–20 in the Butler and Grosser book to see whether you can use bond order to predict the most abundant species in the earth's ionosphere.

10–6 MOLECULES WITH MORE THAN TWO ATOMS: LOCALIZED BONDS

For a polyatomic molecule such as water, H_2O, the most thorough procedure would be to combine the $2s$, $2p_x$, $2p_y$, and $2p_z$ atomic orbitals of oxygen and the two $1s$ hydrogen orbitals into six molecular orbitals, each extending over the entire molecule. Then the four molecular orbitals of lowest energy would be filled by the eight valence electrons, and the three atoms would be bonded.

It is usually easier to consider the atoms two at a time and to think of *localized* bonds rather than filled bonding orbitals covering the entire molecule. In this picture, the oxygen atom and one hydrogen atom are held together by one localized bond, and the oxygen atom and the other hydrogen atom are held together by another. Two lone pairs of electrons on the oxygen atom are not involved in bonding, and the Lewis structure is

$$H\!-\!\overset{\displaystyle ..}{\underset{\displaystyle ..}{O}}\!-\!H$$

What leads us to think that we have any right to simplify the bonding picture in this way? Why can we speak of bonds between *pairs* of atoms in polyatomic molecules? There are many kinds of evidence that make this localized-bond picture reasonable; one of the simplest is outlined in Table 10–7. Carbon and hydrogen occur in a series of compounds with the general formula C_nH_{2n+2}. The simple picture in which each carbon atom makes four bonds, either to another carbon atom or to hydrogen atoms, results in the pattern of bonds shown in the second column of the table. In this localized-bond model, each compound has one more C—C bond and two more C—H

Table 10–7.   Standard Heats of Formation and Bond Energies of Some Carbon–Hydrogen Compounds

Compound (all as gases)	Bond sketch	ΔH^0_{298} (kcal mole^{-1})	Change in ΔH^0_{298} (kcal mole^{-1})
CH_4	H | H—C—H | H	−17.89	
			−2.35
C_2H_6	H H | | H—C—C—H | | H H	−20.24	
			−4.58
C_3H_8	H H H | | | H—C—C—C—H | | | H H H	−24.82	
			−4.99
C_4H_{10}	H H H H | | | | H—C—C—C—C—H | | | | H H H H	−29.81	
			−5.19
C_5H_{12}	H H H H H | | | | | H—C—C—C—C—C—H | | | | | H H H H H	−35.00	
			−4.96
C_6H_{14}	—C—C—C—C—C—C—	−39.96	
			−4.93
C_7H_{16}	—C—C—C—C—C—C—C—	−44.89	
			−4.93
C_8H_{18}	—C—C—C—C—C—C—C—C—	−49.82	

bonds than its predecessor in the table. The standard heats of formation of the gaseous compounds from gaseous hydrogen and solid carbon are given in the third column of the table. The increments in these heats are in the last column. Each compound (after C_2H_6) is approximately 5 kcal mole^{-1} more stable than its predecessor. The increase in the number of bonds is paralleled exactly by an increase in the magnitude of the heat of formation.

The simplest explanation is that each bond in the molecule, as represented by the bond sketch, has a bond energy that is virtually independent of how many other atoms are in the molecule.

We had been using this idea consistently when we spoke of bonds and of bond energies, but now we see some evidence that our preconceived idea was justified. Perhaps it is more thorough to work with full molecular orbitals or delocalized molecular orbitals, but it is far easier and generally almost as valid to employ localized molecular orbitals instead. From here on we always shall use localized orbitals unless they are clearly unacceptable. Benzene, which is discussed in Section 10–10, is an example of a compound for which localized molecular orbitals cannot be utilized, and we must rely on at least partially delocalized orbitals instead.

Before you go on. You can find emphasis on orbital diagrams for predicting the shapes of polyatomic molecules in Section 10–3 of Lassila, *Programed Reviews of Chemical Principles.*

10–7 HYBRID ORBITALS

The most outstanding feature of bonding in many carbon compounds is that each carbon atom makes four bonds to neighboring atoms. How can this be, if atomic carbon has the electronic configuration

$$(1s)^2(2s)^2(2p_x)^1(2p_y)^1$$

and only *two* unpaired electrons? One answer might be that the two electrons in the 2s orbital become unpaired and one of them is promoted to the third 2p orbital:

$$(1s)^2(2s)^1(2p_x)^1(2p_y)^1(2p_z)^1$$

This promotion would require 96 kcal mole^{-1} of energy, but the promotion energy would be compensated by the added stability of the four covalent bonds in which these four unpaired electrons could participate.

What would be the geometry of methane, CH_4, if each hydrogen atom made a bond with the *s* or one of the three *p* orbitals of a carbon atom? The three *p* orbitals could make three bonds at right angles, but where could the bond to the *s* orbital go? Moreover, we know that all four C—H bonds in methane are equivalent and that there are no isomers of CH_3Cl.

Linus Pauling solved the problem in 1931, when he showed that the appropriate linear combination of these four atomic orbitals produced another set of atomic orbitals that were all *equivalent* and oriented to the corners of a tetrahedron. The four combinations are

$$t_1 = \tfrac{1}{2}(s + p_x + p_y + p_z)$$
$$t_2 = \tfrac{1}{2}(s - p_x - p_y + p_z)$$
$$t_3 = \tfrac{1}{2}(s + p_x - p_y - p_z)$$
$$t_4 = \tfrac{1}{2}(s - p_x + p_y - p_z)$$

(*text continues on page 424*)

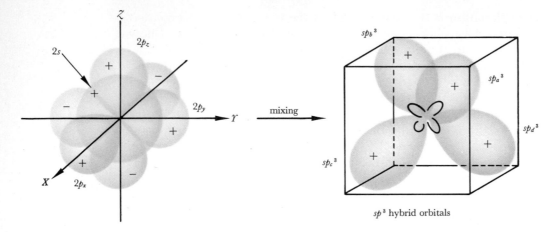

Figure 10–15. *The formation of four sp³ hybrid atomic orbitals from the s, p_x, p_y, and p_z orbitals.*

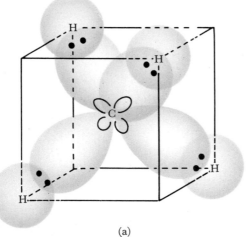

(a)

Figure 10–16. *The CH₄ molecule is held together by electrons in bonding molecular orbitals that are combinations of 1s hydrogen orbitals and the four tetrahedral sp³ orbitals of carbon shown in (a). Each of the four bonding molecular orbitals is occupied by a pair of electrons. (a) The atomic orbitals. (b) The geometry of the tetrahedral bonds.*

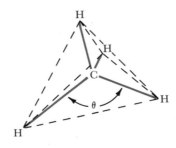
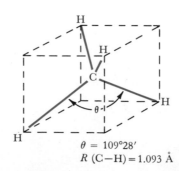

$\theta = 109°28'$
$R (C-H) = 1.093$ Å

(b)

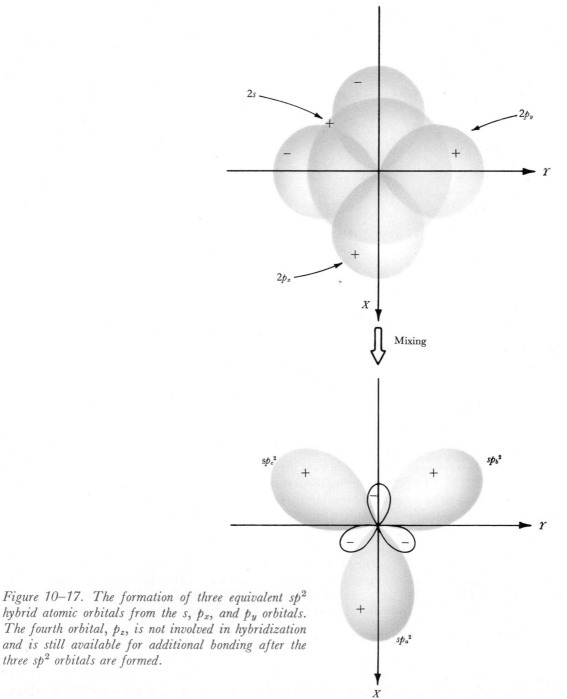

Figure 10–17. The formation of three equivalent sp^2 hybrid atomic orbitals from the s, p_x, and p_y orbitals. The fourth orbital, p_z, is not involved in hybridization and is still available for additional bonding after the three sp^2 orbitals are formed.

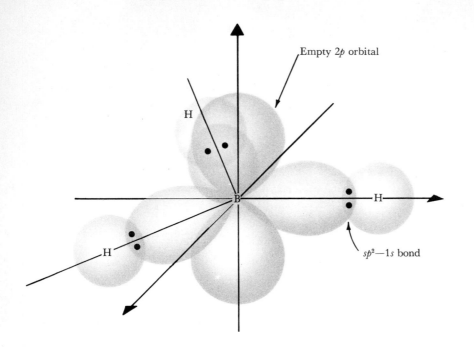

Figure 10–18. Bonding in BH$_3$. *This molecule has been observed in a mass spectrometer as a decomposition product of* B$_2$H$_6$. *The molecular orbitals are obtained from combinations of the hydrogen 1s orbitals and the three sp^2 hybrid boron orbitals shown here. The remaining unhybridized p orbital is empty.*

These tetrahedral orbitals are called *sp^3 hybrid atomic orbitals* and are designated by the subscripts *a* through *d*, respectively, in Figure 10–15. Each of these hybrid atomic orbitals can be combined with a 1s hydrogen orbital to produce two localized molecular orbitals, one bonding and one antibonding. If each of these four bonding molecular orbitals is filled with an electron pair, the methane molecule, CH$_4$, results (Figure 10–16).

Two of the three *p* orbitals also can be combined with the *s* orbital to produce three equivalent *sp^2* hybrid orbitals directed at 120° angles in a plane (Figure 10–17). The unused *p* orbital is unaffected, and its nodal plane is the plane containing the three equivalent *sp^2* orbitals. Such orbitals would be used in the BH$_3$ molecule (Figure 10–18), which is not stable under normal conditions, and the BF$_3$ molecule, which is stable. The unhybridized *p* orbital is vacant.

These orbitals look like possible candidates for explaining the ammonia molecule, NH$_3$, but they are not. The ammonia molecule is not a flat triangle of H atoms around a central N atom; instead, it is a pyramid with N at the apex. We can think of the lone pair of electrons in the unhybridized *p* orbital pushing the three N—H bonds beneath the plane in Figure 10–18 by electrostatic repulsion. However, if this occurs, another set of hybrid orbitals is a better first approximation. We will come back to NH$_3$ in the next section.

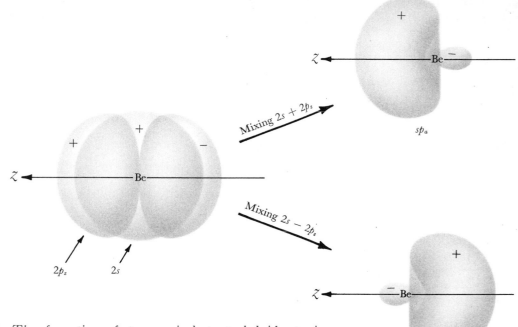

Figure 10–19. The formation of two equivalent sp hybrid atomic orbitals from the s and p_z orbitals of Be. The p_x and p_y orbitals, not shown here, are unaffected. These sp orbitals have their true shape, whereas the sp^2 and sp^3 hybrid orbitals in the previous four figures have been slimmed down for the sake of clarity.

Finally, one p orbital can hybridize with an s orbital on the same atom to produce two equivalent sp hybrid orbitals directed 180° apart (Figure 10–19). These are the hybrid orbitals used for the structure of BeH_2 (Figure 10–20).

In all of these hybridizations, the final hybrid set is of higher energy and is less stable than the initial unhybridized state. An amount of 96 kcal mole^{-1} of energy is required to go from the carbon $2s^2 2p^2$ state to the unpaired $2s^1 2p^3$ state. Another 75 kcal mole^{-1} would be required to produce atoms with each of the four electrons in one of the hybrid sp^3 tetrahedral orbitals. This energy is regained in the formation of bonds. The hybrid $(sp^3)^4$ state of the carbon atom never really occurs; rather, it is a hypothetical intermediate state in the process of achieving the stable state of a tetrahedral molecule such as CH_4.

10–8   ELECTRON REPULSION AND MOLECULAR STRUCTURE

When attempting to determine molecular structures, it is helpful to remember that electrons repel one another. We can rationalize the tetrahedral structure

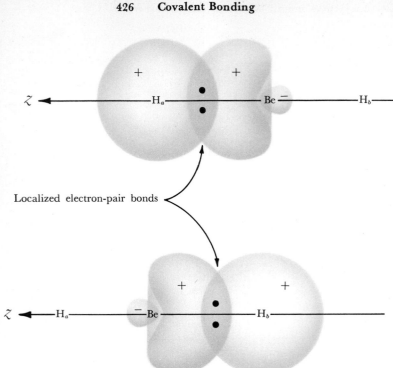

Localized electron-pair bonds

Figure 10–20. Bonding in BeH_2 using the two sp hybrid orbitals from Be. Each Be sp orbital forms a localized bonding molecular orbital with a hydrogen 1s orbital.

Figure 10–21.   The tetrahedral placement of four electron pairs minimizes their electrostatic repulsion.

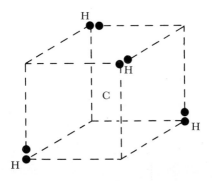

of methane in terms of interelectronic repulsions of the four bonding electron pairs. If four equal charges are confined to the surface of a sphere, they will adopt a tetrahedral arrangement, because this is the arrangement with the lowest energy. The most stable arrangement for the four strongly repelling C—H electron pairs, too, is the tetrahedral array shown in Figure 10–21.

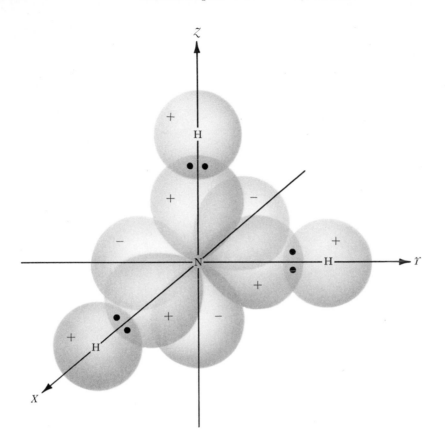

Figure 10–22. A model of the NH_3 *molecule can be constructed by using the three* $2p$ *nitrogen orbitals for bonding, and the nitrogen* $2s$ *(not shown) for the lone pair of electrons.*

Ammonia

As an example of interelectronic repulsion, let us return to the ammonia molecule. The Lewis structure is

with three single bonds and one lone pair of electrons. One simple bond model assumes that each of the three nitrogen $2p$ orbitals combines with a hydrogen $1s$ orbital, while the lone pair goes into the nitrogen $2s$ orbital. If this were so, the H—N—H bond angles would be 90° (Figure 10–22). As we discussed in the preceding section, a second simple model makes three N—H

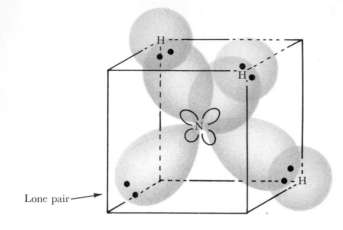

Lone pair——▶

Figure 10–23. An alternative model for NH_3 *uses four* sp^3 *hybrid orbitals for the three bonds and the lone electron pair.*

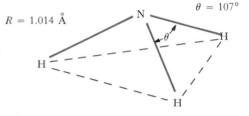

Figure 10–24. The NH_3 *molecule actually is a trigonal pyramid with N at the apex.*

bonds with the three sp^2 hybrid orbitals and places the lone pair in the unused nitrogen $2p$ orbital. In this model, the H—N—H bond angle is 120°.

A third model for ammonia assumes that the dominant factor in determining shape is the mutual repulsion of electron pairs, whether in a bond or a lone pair. In this model, the four electron pairs extend to the corners of a tetrahedron, as in Figure 10–23. Three electron pairs bond to H atoms and the fourth is a lone pair. In this electron repulsion model, the H—N—H angle is the tetrahedral angle of 109°28′. It is natural to think of these electron pairs as using the four equivalent tetrahedral sp^3 orbitals, thereby allowing minimum electron repulsion.

From all three bonding models we can predict that NH_3 will have threefold symmetry, in which all N—H bonds are the same but differ in the details of bond angles. The real ammonia molecule is a trigonal pyramid with a H—N—H bond angle of 107° (Figure 10–24). The electron-pair repulsion model, although not perfect, is a more accurate description of NH_3 than the other two.

This third model for ammonia is an example of the use of the *valence-shell electron-pair repulsion* (VSEPR) theory to predict the geometry of bonding in a molecule. This is a simple but surprisingly accurate theory. It states that covalent bonds can be considered as localized electron pairs, and that the bonds orient themselves around the central atom to minimize the electrostatic repulsion between electron pairs. Lone pairs repel in the way that bond-

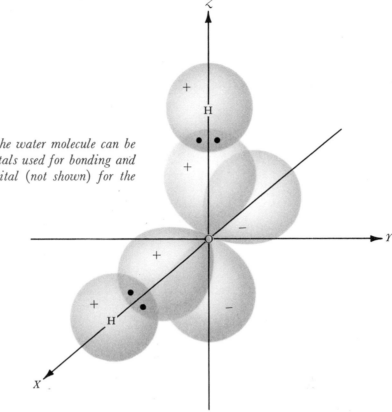

Figure 10–25. A simple model of the water molecule can be built with two of the oxygen 2p orbitals used for bonding and the other 2p orbital and the 2s orbital (not shown) for the two lone electron pairs.

ing pairs do. And when the distinction must be made, lone pairs repel one another and bonding pairs slightly more strongly than bonding pairs repel each other. For four bonds or lone pairs around a central atom, as in ammonia, it is customary to regard the electron pairs as using the tetrahedral hybrid orbitals, although the simple VSEPR theory suggests nothing about what orbitals the repelling electron pairs occupy.

Water

Water is another example of the power of the electron-pair repulsion model. The Lewis structure of water is

$$H—\overset{..}{\underset{..}{O}}—H$$

with two bonds and two lone pairs. We can imagine that the bonding in water involves two of the three p orbitals of oxygen, with the two lone pairs in the third p orbital and in the s orbital. In this model, the H—O—H bond angle is 90° (Figure 10–25).

The electron-pair repulsion model again predicts that the two bonds and two lone pairs will be directed to the corners of a tetrahedron (Figure 10–26).

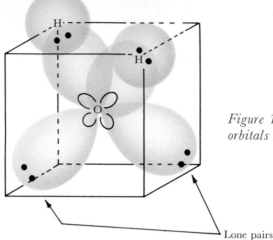

Figure 10–26. The tetrahedral sp³ hybrid orbitals also can be used as a model for H₂O.

Lone pairs

In this model, the H—O—H bond angle is 109°28′. The observed bond angle in water is 105°; thus, again the electron repulsion model is quite close to the truth.

   As the size of the central atom increases, the electrons in valence orbitals are farther from each other, and interelectronic repulsions are less important in determining molecular shapes. For example, sulfur is larger than oxygen, and it is known from atomic spectra that interelectronic repulsions in the S valence orbitals are substantially smaller than in O. This probably is the reason why the H—S—H bond angle in H_2S is 92°, which is much closer to the 90° value that would be expected if bonding occurred with the use of p orbitals (Figure 10–25).

10–9   SINGLE AND MULTIPLE BONDS IN CARBON COMPOUNDS

The tetrahedral sp^3 hybrid atomic orbitals explain the bonding in methane quite well. They also explain the structures of ethane, C_2H_6, and many other organic compounds in which carbon atoms are linked in chains by single bonds. In ethane, each of the carbon atoms has three hydrogen atoms linked to it by covalent bonds that use three of the four sp^3 hybrid orbitals. The fourth sp^3 orbital on each carbon atom links the carbon atoms together in a covalent bond. In forming the bond, the two sp^3 atomic orbitals combine to produce a stable bonding molecular orbital and an unstable antibonding orbital. The bonding orbital, which is symmetrical around the C—C axis and is thus a σ^b orbital, is filled by two spin-paired electrons, and the bond is complete. The disposition of sp^3 orbitals is shown in Figure 10–27(a); the measured bond lengths and bond angles are in Figure 10–27(b). Propane (CH₃—CH₂—CH₃), butane (CH₃—CH₂—CH₂—CH₃), and all of the large array of straight-chain and branched-chain hydrocarbons, including the

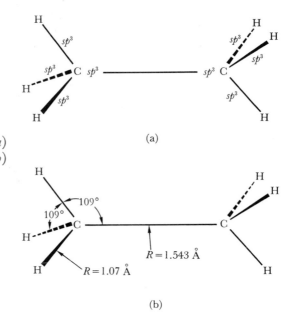

Figure 10–27. Bonding in ethane, C_2H_6. (a) The atomic orbitals contributed by carbon. (b) Observed structural parameters in ethane.

components of kerosene, gasoline, and paraffin wax, can be constructed from tetrahedrally hybridized orbitals of carbon atoms that combine with one another and with hydrogen atomic orbitals.

The Lewis structure for ethylene, C_2H_4, assumes a double bond between carbon atoms:

```
H  H
|  |
C==C
|  |
H  H
```

In the best model of this molecule, each carbon atom uses trigonal, sp^2 hybridization. Two of the three equivalent hybrid orbitals on each carbon atom are used to bond to two hydrogen atoms; the third orbital from each carbon atom participates in a single C—C bond with σ symmetry [Figure 10–28(a)]. The second bond of the double bond arises from a combination of the two $2p$ atomic orbitals that were not involved in hybridization. A π molecular orbital is produced exactly like the π_x^b and π_y^b orbitals in homonuclear diatomic molecules [Figure 10–28(c)]. There are 12 valence electrons in ethylene: four each from the two carbon atoms, and one each from the four hydrogen atoms. Eight of these electrons are used in the four electron-pair bonds to H, and two more in the σ C—C bond. The last two valence electrons occupy the π C—C bond and complete the expected double bond.

In this model of ethylene, all six atoms must lie in a plane, because if one $-CH_2$ group were to be twisted relative to the other around the C—C

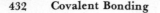

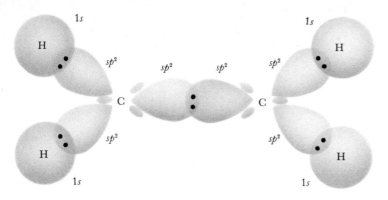

5 σ-bonding pairs = 10 electrons

(a)

Figure 10–28. Bonding in ethylene, C_2H_4. (a) The σ-bonding structure. (b) The Lewis structure. (c) The π-orbital contribution to the double bond.

(b)

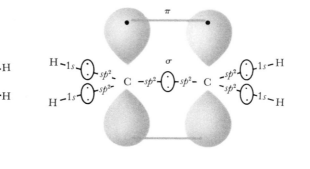

(c)

line, the overlap between $2p$ orbitals in the π^b molecular orbital would be weakened, and the bond would be reduced to something approaching a single σ^b bond. The bond energy of the single C—C bond in ethane is 83 kcal mole^{-1}, and that of the double bond in ethylene is 125 kcal mole^{-1}. The energy required to twist the ethylene molecule by 90° should be the difference between these two numbers, or 42 kcal mole^{-1}. This is a formidable amount of energy, so the ethylene molecule should be planar.

X-ray analysis shows that ethylene *is* planar, and that its bond angles in the plane agree closely with the 120° predicted from sp^2 hybridization: 117° for each H—C—H angle and 121°31′ for each H—C—C angle. The molecular structure of C_2H_4 is in close agreement with the molecular orbital picture, and we have a good example of the structure of a double bond.

There is another way in which the double bond in ethylene could be explained by molecular orbitals: tetrahedral sp^3 hybridization. In this model, two of the four sp^3 orbitals on one carbon atom overlap with two similar orbitals on the other carbon atom. The two carbon tetrahedra share an edge, as was described in Figure 7–7(d). However, the total overlap of atomic

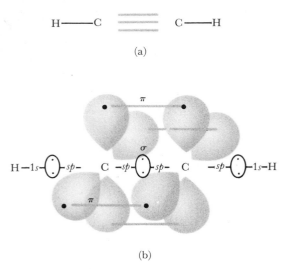

Figure 10–29. Bonding in acetylene, C_2H_2. (a) The Lewis structure. (b) The σ bonding from sp carbon orbitals and the two π bonds from carbon p orbitals.

orbitals is less in this model than in the sp^2 hybridization model, which means that the bond is not as strong. In addition, the tetrahedral model with two bent bonds predicts that the H—C—H angle is close to the tetrahedral value of 109° rather than the sp^2 value of 120°. The observed value of 117° is a strong argument for the model of the double bond in Figure 10–28 rather than the bent-bond sp^3 hybrid model.

In C_2H_2, there is only one hydrogen atom attached to each carbon atom. We can construct a localized bonding model for C_2H_2 as follows. The s and one p orbital on each carbon atom combine to produce two sp hybrid orbitals that are oriented 180° to each other. One sp orbital is used for the σ bond to a hydrogen atom and the other for the σ C—C single bond. Each C has two unused $2p$ orbitals. These combine to form two π^b molecular orbitals. This model is shown in Figure 10–29.

In summary, according to molecular orbital theory, the double bond in C_2H_4 consists of one σ bond and one π bond. The triple bond in C_2H_2 is made up of one σ bond and two π bonds. The relationship of bond order, bond distance, and bond energy is clearly illustrated by experimental data for these three compounds. As the C—C bond order increases, the bond length decreases, and the energy required to break the bond increases, as shown in Table 10–8.

Table 10–8.   Effect of Bond Order on Bond Length and Bond Energy

Molecule	C—C bond order	C—C bond length (Å)	$H_nC—CH_n$ energy (kcal mole^{-1})
C_2H_6	1	1.54	83
C_2H_4	2	1.35	125
C_2H_2	3	1.21	230

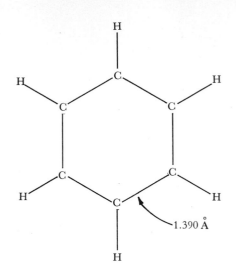

Figure 10–30. *The skeleton of the benzene molecule. The C—C bond length has been determined by x-ray crystal structure analysis.*

~1.390 Å

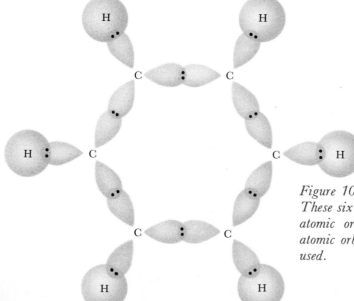

Figure 10–31. *The pattern of σ bonding in benzene. These six C—C bonds and six C—H bonds use 24 atomic orbitals and 24 electrons. Six 2p carbon atomic orbitals and six valence electrons remain unused.*

10–10   BENZENE AND DELOCALIZED ORBITALS

With benzene, the localized molecular orbital treatment fails. Just as we could not produce a satisfactory Lewis structure for O_2, so we cannot produce a satisfactory localized molecular orbital structure for C_6H_6.

Benzene has the planar hexagonal skeleton shown in Figure 10–30. Each carbon atom in the hexagon is attached to one hydrogen atom and two

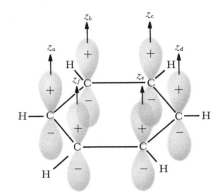

Figure 10–32. The unused 2p atomic orbitals in benzene. These orbitals combine to make molecular orbitals that can accommodate the six valence electrons unused in the σ bonding.

other carbon atoms with bond angles of 120°. The bond angles alone suggest sp^2 hybridization. If we use sp^2 hybridization for carbon, we form the σ bonding network shown in Figure 10–31. Each carbon atom is joined to one H atom and two other C atoms by single covalent bonds.

This cannot be the whole story, because the observed C—C bond length in benzene, 1.390 Å, is too short for a single bond (1.54 Å). Each carbon atom would have an unused $2p$ orbital perpendicular to the plane of the hexagonal ring (Figure 10–32). There are 30 valence electrons in benzene: four each from the six carbon atoms, and one each from the six hydrogen atoms. Twelve are used in the six C—H σ single bonds, and twelve more in the six C—C σ single bonds. Six electrons remain, along with six p atomic orbitals. It would be logical to use the orbitals in three pairs for making three more covalent bonds. But how are these pairs to be chosen?

Kekulé proposed that the pairs be formed between adjacent carbon atoms around the ring (Figure 10–33). Yet there are two such Kekulé structures that, although indistinguishable in benzene, should produce different isomers in *o*-dichlorobenzene (see Section 7–6). No such isomers exist. Even worse, the C—C bond lengths around the benzene ring should alternate between a single bond length of 1.54 Å and 1.35 Å, the double bond length in ethylene. X-ray analysis shows that all six bonds are equal. Three other structures, with different combinations of three covalent bonds between p orbitals, were proposed by Dewar (Figure 10–33). Each of them by itself is even less satisfactory than a Kekulé structure. It is impossible to draw any one bond structure that explains the benzene molecule. The fault lies in our notion that a bond is something that is formed in a private sort of way between two atoms in a molecule, with the other atoms uninvolved.

In a strict sense, every molecular wave function should include atomic orbitals from all of the atoms of a molecule, as was mentioned in Section 10–6. Usually, all atoms except two make negligible contributions to a given wave function, and we can consider that wave function as an accurate description of the bond between the two atoms. But anomalies such as benzene occur

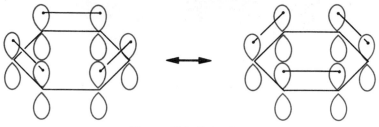

The two Kekulé structures

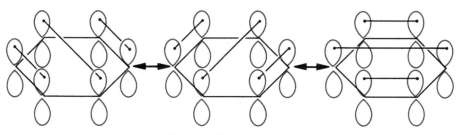

The three Dewar structures

Figure 10–33. The six p orbitals can be combined in pairs to make covalent bonds in several ways.

often enough to remind us of the flaws in our assumption about localized bonds.

Using symmetry arguments beyond the scope of this text, we can construct a set of six full-molecule wave functions and orbitals from the six atomic $2p$ orbitals. Let the six orbitals be labeled z_a, z_b, z_c, z_d, z_e, and z_f, as in Figure 10–32. Let the sign of each z orbital be plus or minus, depending on whether the plus lobe of the p wave function is up or down in Figure 10–32. Then the six full-molecule wave functions are given by

$$\pi_1^b = z_a + z_b + z_c + z_d + z_e + z_f$$
$$\pi_2^b = 2z_a + z_b - z_c - 2z_d - z_e + z_f$$
$$\pi_3^b = z_b + z_c - z_e - z_f$$
$$\pi_1^* = 2z_a - z_b - z_c + 2z_d - z_e - z_f$$
$$\pi_2^* = z_b - z_c + z_e - z_f$$
$$\pi_3^* = z_a - z_b + z_c - z_d + z_e - z_f$$

The squares of these functions are the electron density distributions. These six orbitals are pictured in Figure 10–34. Three are bonding, and three are antibonding. Their energies are shown in Figure 10–35. Note how these orbitals illustrate the rule that, in general, the more nodes an orbital has, the

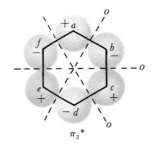

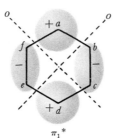

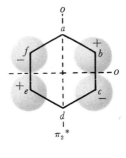

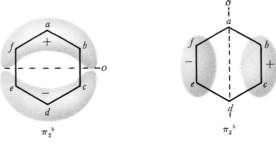

Figure 10–34. The six full-molecule delocalized molecular orbitals obtained from the six 2p orbitals of benzene. The dashed lines indicate nodes of zero electron density, and the plus and minus signs mark the relative signs of the wave functions on either side of a nodal surface. The greater the number of nodal surfaces, the higher the energy of the wave function. All six of these orbitals have a nodal plane in the plane of the paper; for example, the π_1^b orbital has a region of high probability density above the plane of the benzene ring and another region of high density below the plane, in which the wave function itself has the opposite sign. The π_3^* has twelve lobes of high density: six above the plane of the ring as shown here, and another six below the ring with opposite signs for the wave functions.

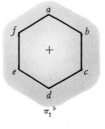

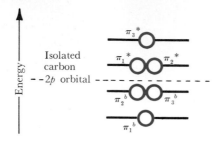

Figure 10–35. Energy-level diagram for the six benzene delocalized orbitals diagramed in Figure 10–34. A generally valid principle in quantum mechanics is that, the more space there is available in which a particle can move, the lower and more closely spaced its energy levels will be. As an extreme example, this is why we notice the quantization of energy of an electron in a hydrogen atom, but do not notice the quantization of energy of a baseball in Yankee Stadium. The mass of the baseball is so great, and the volume in which it can move is so large, that its quantized energy levels are too close together to detect. In addition to the mass difference, the baseball example is a case of extreme delocalization. Even at the molecular level, the more room in which the electrons have to move (the more they are delocalized), the more stable the molecule will be, if other factors are neglected. The delocalization stability of each of the $\pi_1{}^b$, $\pi_2{}^b$, and $\pi_3{}^b$ orbitals is given by its distance below the horizontal dashed line representing the stability of electrons in the isolated 2p orbitals.

higher its energy will be. You can test the validity of this rule with the homo-nuclear and heteronuclear orbitals discussed previously in this chapter, and even with the hydrogen atomic wave functions.

The six unused valence electrons in benzene occupy the three bonding orbitals in Figure 10–35. No one electron pair belongs to any pair of C atoms; thus, the six electrons are said to be *delocalized*. Each carbon–carbon bond consists of one full σ bond and half a π bond. The C—C bond length, 1.390 Å, is intermediate between single and double bond lengths.

Benzene is actually more stable than might be expected for a molecule with six C—C single bonds, six C—H single bonds, and three C—C π bonds. This added stability results because the electrons in the three π bonds are delocalized over all six carbon atoms. Orbital $\pi_1{}^b$ in Figure 10–34 is symmetrical with respect to all six carbon atoms. Orbitals $\pi_2{}^b$ and $\pi_3{}^b$ look unsymmetrical, but the combination of the two is symmetrical. There is nothing special about atoms *a* and *d*; we could have written $\pi_2{}^b$ and $\pi_3{}^b$ so atoms *f* and *c* appeared to be the "axis" of the molecule. If we did not allow the delocalization of electrons in C_6H_6, the bonding would be as in one of the Kekulé or Dewar structures shown in Figure 10–33 or 10–36. Instead, the bonding structure in benzene can be represented best as in the bottom drawing of Figure 10–36. As we shall calculate from experimental data in Section 15–4, the molecule is 40 kcal mole^{-1} more stable than expected from the sum of the bond energies of six C—H, three C—C, and three C=C bonds.

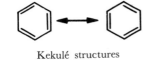

Kekulé structures

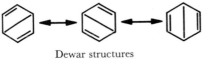

Figure 10–36. The two Kekulé structures for benzene, the three Dewar structures, and a diagrammatic representation of the delocalized electrons in the benzene ring. The Kekulé and Dewar structures sometimes are called resonance structures for benzene. The implication intended by this rather unfortunate terminology is not that the bonds flip back and forth or resonate from one structure to the other, but only that the true structure, which cannot be represented by localized bonds, has something of the character of each of these structures.

Dewar structures

Simple molecular orbital structure

Another way of treating the symmetrical benzene molecule is by using the idea of resonance mentioned previously with Lewis structures. In this treatment, we state that, although the benzene molecule cannot be represented accurately by any one localized-bond model, the real molecule has some of the character of both Kekulé structures and all three Dewar structures in Figure 10–36. We can write the complete wave function of benzene as a linear combination of the wave functions of the two Kekulé structures (K_1 and K_2), and the three Dewar structures (D_1, D_2, and D_3)

$$\psi = uK_1 + vK_2 + wD_1 + xD_2 + yD_3$$

There are ways of calculating the coefficients u through y, and the result is that the best approximation to the real benzene molecule occurs if we assume that each Kekulé structure contributes 39% to the real molecule and each Dewar structure contributes 7%. These five models are called the resonance structures for benzene, and the extra 40 kcal mole^{-1} of stability is called the resonance stabilization energy. Again, beware of the terminology, and *do not* think of the benzene molecule as resonating or flipping back and forth from one resonance structure to another.

10–11   SUMMARY

We now know what a covalent bond is. We have a theory of bonding, the molecular orbital theory, that accounts satisfactorily for bond lengths and bond energies in simple molecules, predicts the presence or absence of unpaired electrons, and even indicates when the molecule should not exist. From the theory we can predict correctly the dipole moment and percent ionic character by an extension of ideas that have been touched upon only qualitatively here. Molecular orbital theory accounts for the tetrahedral geometry of carbon bonding and the shapes of other small molecules such as H_2O, NH_3, and BF_3.

Molecular orbital theory completes the ideas of electron-pair bonding that were initiated by G. N. Lewis. When the Lewis structures are correct, molecular orbital theory interprets them; when they are inadequate, it corrects them. The Lewis structures are now convenient memory devices for recalling the conclusions of molecular orbital theory. This theory offers an explanation of how double and triple bonds are formed, and why a structure with a double bond is constrained in a planar configuration.

For most of the examples we have used what is called the localized molecular orbital assumption. This assumption is usually adequate, but in benzene we were forced to explain the molecular structure in terms of delocalized, full-molecule orbitals.

W. H. Wollaston was quoted at the beginning of Chapter 7 as lamenting, in 1808, that "It is perhaps too much to hope that the geometrical arrangement of primary particles will ever be perfectly known." More than a century and a half later we know many of these arrangements, and, thanks to molecular orbital theory, we even may understand them.

SUGGESTED READING

G. M. Barrow, *The Structure of Molecules*, W. A. Benjamin, Menlo Park, Calif., 1963. An introduction to molecular spectroscopy. One step beyond this chapter. Tells how we find out the structures of molecules from the ways in which they absorb and emit radiation.

E. Cartmell and G. W. A. Fowles, *Valency and Molecular Structure*, Reinhold, New York, 1967.

I. Cohen and T. Bustard, "Atomic Orbitals: Limitations and Variations," *J. Chem. Educ.* **43**, 187 (1966).

A. Companion, *Chemical Bonding*, McGraw-Hill, New York, 1964.

H. B. Gray, *Chemical Bonds*, W. A. Benjamin, Menlo Park, Calif., 1973. Includes an extensive introduction to molecular orbital theory.

H. B. Gray, *Electrons and Chemical Bonding*, W. A. Benjamin, Menlo Park, Calif.,

1965. A somewhat more extended treatment of the material in this chapter, from a slightly different viewpoint.

R. C. Johnson and R. R. Rettew, "Shapes of Atoms," *J. Chem. Educ.* **42**, 145 (1965).

E. A. Ogryzlo and G. B. Porter, "Contour Surfaces for Atomic and Molecular Orbitals," *J. Chem. Educ.* **40**, 256 (1963).

L. Pauling, *The Nature of the Chemical Bond*, Cornell University Press, Ithaca, New York, 1960, 3rd ed.

G. E. Ryschkewitsch, *Chemical Bonding and the Geometry of Molecules*, Reinhold, New York, 1972, 2nd ed.

D. K. Sebera, *Electronic Structure and Chemical Bonding*, Blaisdell, Waltham, Massachusetts, 1964.

A. C. Wahl, "Chemistry by Computer," *Scientific American*, April, 1970.

QUESTIONS

1 What experimental evidence is there that O_2 has a bond order of two, and that N_2 has a bond order of three?

2 How are N_2, O_2, and F_2 represented with Lewis structures?

3 What is meant by the statement that the ammonium ion and methane are isoelectronic? If they are, then why do they not have similar chemical properties?

4 What combination of boron and hydrogen would be isoelectronic with methane and the ammonium ion? Can you think of any good reasons why such a boron compound might or might not exist?

5 Is the addition compound of BF_3 and NH_3 isoelectronic with any organic compound?

6 How are Lewis acids and bases different from Brønsted acids and bases, and from Arrhenius acids and bases?

7 In the water solution of ammonia,

$$NH_3 + H_2O \rightarrow NH_4^+ + OH^-$$

which of the species on the left (or the species that can be derived from them) is the acid and which is the base in the definitions of Arrhenius, Brønsted, and Lewis?

8 What is the difference between oxidation number and formal charge? How is each one calculated for an atom in a molecule?

9 Why does the potential energy curve in Figure 10–2 have a minimum? How do we represent vibration of the H_2 molecule—stretching and compressing of its bond length—in a diagram such as Figure 10–2? At what point in its vibration do its two atoms have their greatest potential energy? Least potential energy? At what point in the vibration do the atoms have the greatest kinetic energy?

Least kinetic energy? What remains constant during the vibration of the molecule?

10 In Figure 10–3(a) are the profiles through the two atomic electron-density clouds, $[1s_a]^2$ and $[1s_b]^2$, and the profile through the bonding molecular orbital formed from $[1s_a + 1s_b]^2$. What would profiles through the two atomic wave functions, $1s_a$ and $1s_b$, look like? Draw the profile through the antibonding molecular wave function, $1s_a - 1s_b$, and through the resulting electron density distribution in the antibonding molecular orbital, $[1s_a - 1s_b]^2$.

11 What do the symmetry symbols for molecular orbitals, σ and π, indicate?

12 Why does molecular orbital theory predict that He_2 should not exist, whereas He_2^+ should, under the right conditions?

13 What is the equivalent of a Lewis covalent bond in molecular orbital theory?

14 What is wrong with the general statement that bonding orbitals have low energies and antibonding orbitals have high energies? What feature of a bonding orbital makes it a bonding orbital?

15 What is the order of increasing energy among the molecular orbitals formed from the $2s$ and $2p$ atomic orbitals in diatomic molecules? What experimental evidence is there that this order is correct?

16 What are homonuclear and heteronuclear diatomic molecules?

17 What do the small positive and negative charges in the lobes of the orbitals in Figure 10–7 signify? What do the small line drawings at the upper right of each orbital drawing represent?

18 Why is the bond energy in Li_2 less than in H_2?

19 Each boron atom has three valence electrons. Why is the B_2 molecule not held together by a triple bond as is N_2?

20 What is the ground-state electronic configuration of the C_2 molecule?

21 Why is O_2 paramagnetic, whereas N_2 is not? Support your argument with the ground-state electronic configurations of the two molecules.

22 What happens to the molecular orbitals when two atomic orbitals with radically different energies on two different atoms are combined? If we combine two atomic orbitals with drastically different energies, and place two electrons in the lower of the two resulting molecular orbitals, how would we describe the electrons? Are they bonding electrons?

23 Why cannot the $2p_x$ and $1s$ atomic orbitals in Figure 10–11 be combined to produce two molecular orbitals?

24 In Figure 10–12, is the σ^b molecular orbital more like the hydrogen $1s$ or the fluorine $2p$ atomic orbital? Which atomic orbital makes more of a contribution to the σ^* molecular orbital?

25 What experimental data give us the relative heights in Figure 10–12 of the hydrogen $1s$ and fluorine $2p$ atomic orbitals?

26 If the hydrogen $1s$ and fluorine $2p$ atomic orbitals were by some process made equal in energy, what effect would this have on the character of the bond in HF?

27 What is a purely ionic bond, in the language of molecular orbital theory and Figure 10–12?

28 How can dipole moments give us an estimate of the ionic character of a bond? How ionic is the HF bond?

29 Why do the orbital drawings for heteronuclear diatomic molecules in Figure 10–14 differ from those for homonuclear diatomic molecules in Figure 10–7?

30 What are meant by localized and delocalized molecular orbitals? What evidence suggests when we can use localized orbitals? When can we not use them?

31 How do we know that all four C—H bonds in methane are equivalent?

32 If energy is required to unpair the two $2s$ electrons in a carbon atom and to create the electronic configuration $2s^1 2p_x^1 2p_y^1 2p_z^1$, then why does carbon not remain in the $2s^2 2p_x^1 2p_y^1$ state and show a combining capacity of two rather than four?

33 What happens to the third p orbital in sp^2 hybridization?

34 Why is sp^2 trigonal hybridization wrong for the ammonia molecule?

35 What is the VSEPR theory, and how can it be used to predict the structures of H_2O and NH_3? Why is the theory less accurate with H_2S than with H_2O?

36 How is the second bond of the C=C double bond in ethylene formed?

37 Why does the presence of the double bond in ethylene keep the molecule planar? What would happen if the two ends of the ethylene molecule were twisted around the C=C bond axis?

38 Why is the sp^3 hybridization model for ethylene less correct than the sp^2 hybridization model?

39 What experimental evidence do we have that a Kekulé structure for bonding in benzene is wrong?

40 What is the relationship between the number of nodes in the wave functions for benzene and their energies?

PROBLEMS

1 For the following species, assign formal charges to each of the atoms and indicate the overall net charge on each:

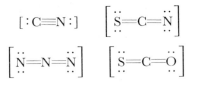

2 Write Lewis structures for the atoms Na, C, Si, Cl, and Kr.

3 Write Lewis structures for the atoms or ions Ca^{2+}, K^+, Ar, Cl^-, and S^{2-}.

4 What are the Lewis structures for the diatomic molecules or ions O_2, CO, Li_2^+, and CN^-?

5 What are the Lewis structures for Cl_2, N_2, NO, and HCl? Which of these molecules would you expect to be paramagnetic, and why?

6 Write Lewis structures for $BaCl_2$, PH_3, NH_4Cl, HOCl (no H—Cl bond), H_2O, H_2O_2, and NO_2^-.

7 Write Lewis structures for $CaCl_2$, SiH_4, CS_2 (no S—S bond), ClO_2 (chlorine dioxide), ClO_2^- (chlorite ion), and N_2O.

8 In diazomethane (H_2CNN), one nitrogen atom is attached directly to the carbon atom, and the second nitrogen atom is attached to the first. Draw Lewis structures for this molecule in which (a) the two N are joined by a triple bond and (b) the middle nitrogen atom forms two double bonds to C and to N. When correctly drawn, each C and N atom should have eight electrons in its valence shell. What is the formal charge on each atom in structures (a) and (b)?

9 For each of the following reactions, indicate which molecule or ion is the Lewis acid and which is the Lewis base:

a) $Ag^+ + 2NH_3 \rightarrow Ag(NH_3)_2^+$

b) $C_2H_3O_2^- + HF \rightarrow$
$$HC_2H_3O_2 + F^-$$

c) $NH_2^- + NH_4^+ \rightarrow 2NH_3$

d) $2(CH_3)_3P + O_2 \rightarrow 2(CH_3)_3PO$

10 Write Lewis structures for BrO_4^-, SiH_4, PCl_4^+, CH_2Cl_2, and BF_4^-.

11 Describe the electronic structure of the diatomic molecule NO by using molecular orbitals. From the molecular orbital diagram, would you expect the molecule to be paramagnetic? Does your answer agree with the predictions that you can make from the Lewis structure? Would you predict the heat of dissociation of NO to be greater than, equal to, or less than that of the ion NO^+?

12 Describe the electronic structure of the diatomic molecule O_2 by using molecular orbitals. Is O_2 paramagnetic, from predictions based on molecular orbital theory, and does this prediction agree with the possible predictions based on the Lewis structure? Which molecule would you expect to have the greater heat of dissociation, O_2 or NO?

13 Using molecular orbitals, describe the electronic structures of the peroxide ion, O_2^{2-}, and the superoxide ion, O_2^-. Are these ions diamagnetic or paramagnetic? How does the strength of the oxygen–oxygen bond in each of these ions compare to that in O_2?

14 Which molecule has the greatest bond energy: O_2, N_2, or F_2?

15 Which of the following molecules are paramagnetic: CO_2, CO, Cl_2, NO, N_2?

16 Which of the following molecules will be expected to have a dipole moment: H_2, O_2, HF, HI, I_2, CH_4? Assuming that the molecules with dipole moments are completely ionic, calculate their dipole moments. (Needed data are in Tables 10–4 and 10–6.)

17 The measured dipole moment of carbon monoxide (CO) is 0.112 debye. What is the partial ionic character of the C—O bond?

18 Carbon dioxide, CO_2, has *no* dipole moment. What does this indicate about its molecular structure? Water, H_2O, *has* a dipole moment. What does this indicate about its structure?

19 Which substance has bonds of greater ionic character, KI or BaO? What is the percent ionic character of each? (Dipole moment data are available in the CRC *Handbook of Chemistry and Physics*.)

20 Compare the percent ionic character of the bonds in HCl, CsCl, and TlCl. Can you interpret your results in terms of the periodic table? (Tl—Cl bond length: 3.2 Å. Measured dipole moment for TlCl molecules in the gas phase: 4.44 debyes; for CsCl, 10.42 debyes.)

21 The molecular orbital description of the B_2 molecule can be written: KK $(\sigma_s{}^b)^2(\sigma_s{}^*)^2(\pi_{x,y}^b)^2$. What does the KK symbol mean? In the same notation, what is the molecular orbital description of F_2?

22 What are the molecular orbital descriptions of Li_2 and Be_2? Which molecule should not exist, and why?

23 Use Table 10–7 to predict the heat of formation of the straight-chain hydrocarbon $C_{10}H_{22}$.

24 What is the geometry of bonding around the central atom in each of the following compounds, and what is the hybridization of atomic orbitals on the central atom: CH_4, BF_3, NF_3, $ICl_4{}^-$, H_2O?

25 What is the geometry of bonding around the central atom in each of the following compounds, and what is the hybridization of atomic orbitals on the central atom: $BrO_3{}^-$, $CHCl_3$, $ClO_4{}^-$, H_2S?

26 Give examples of ions or molecules that have the following structures:

a) $[AB_3]^{2-}$ planar
b) $[AB_3]$ planar
c) $[AB_3]$ pyramidal
d) $[AB_3]^-$ pyramidal
e) $[AB_4]^-$ tetrahedral
f) $[AB_4]^{2-}$ tetrahedral
g) $[AB_2]$ linear
h) $[AB_2]$ bent

27 What is the geometry of bonding around each carbon atom in acetic acid,

What is the hybridization of atomic orbitals around each carbon atom and each oxygen atom? Describe the bonding in terms of molecular orbitals. Which carbon–oxygen bond will be longer?

28 When acetic acid is ionized in solution, the two carbon–oxygen bonds have the same length. Write two resonance structures for the acetate ion that account for this fact. What experimental methods might we use to measure the frequency with which the ion "resonates" between the two structures?

29 How can you explain the equal bond lengths of Problem 28 by using molecular orbital theory? Which explanation do you prefer, molecular orbital theory or resonance structures?

30 Three isomers of dichlorobenzene have measured gas phase dipole moments of 0.00, 1.72, and 2.50 debyes. Match these dipole moments to the structures shown in Figure 7–9.

31 Chemists often speak of the percentage *s* character of a particular bond. For example, each C—H bond of methane is said to be 25% *s* character because the sp^3 hybrid orbital is made up of one *s* and three *p* orbitals. (a) Calculate the percent *s* character for sp^2 and sp hybrid orbitals. (b) Consider the atomic orbitals used by sulfur in H_2S and by nitrogen in NH_3. Calculate the percent *s* character of the orbitals used by S and N to make S—H and N—H bonds. (*Hint:* The cosine of the interorbital angle is equal to the ratio of the *s* character to the *s* character minus one.)

11 COORDINATION COMPOUNDS

In Chapter 7, we introduced some compounds of Pt, Co, and other transition metals that have strange empirical formulas and are often brightly colored. These are *coordination compounds*. Their major distinguishing feature is the presence of two, four, five, six, and sometimes more chemical groups positioned geometrically around the metal ion. These groups can be neutral molecules, cations, or anions. Each coordinating group can be a separate entity, or all the groups can be connected in one long, flexible molecule that wraps itself around the metal. Coordinating groups change significantly the chemical behavior of a metal. The colors of the compounds provide clues about their electronic energy levels.

For instance, every plant depends on the green Mg coordination complex known as chlorophyll. The combination of Mg and its coordinating groups in chlorophyll has electronic properties that the free metal or ion does not have, and can absorb visible light and use the energy for chemical synthesis. Every oxygen-breathing organism requires cytochromes. These are coordination compounds of Fe that are essential to the breakdown and combustion of foods and the storage of the energy released by the breakdown. Most larger organisms need

Table 11–1.   Platinum Complexes, Numbers of Ions Produced, and Complex Structures

Complex	Molar conductivity (ohm^{-1})	Number of Cl$^-$ ions precipitated by Ag$^+$	Total number of ions	Ions produced
$PtCl_4 \cdot 6NH_3$	523	4	5	$Pt(NH_3)_6^{4+}$; $4Cl^-$
$PtCl_4 \cdot 5NH_3$	404	3	4	$Pt(NH_3)_5Cl^{3+}$; $3Cl^-$
$PtCl_4 \cdot 4NH_3$	229	2	3	$Pt(NH_3)_4Cl_2^{2+}$; $2Cl^-$
$PtCl_4 \cdot 3NH_3$	97	1	2	$Pt(NH_3)_3Cl_3^+$; Cl^-
$PtCl_4 \cdot 2NH_3$	0	0	0	$Pt(NH_3)_2Cl_4^0$
$PtCl_4 \cdot NH_3 \cdot KCl$	109	0	2	K^+; $Pt(NH_3)Cl_5^-$
$PtCl_4 \cdot 2KCl$	256	0	3	$2K^+$; $PtCl_6^{2-}$

hemoglobin, another Fe complex in which the coordinating groups enable the iron to bind oxygen molecules without being oxidized. Large areas of biochemistry are really applied transition-metal chemistry. In this chapter we shall look at the structures and properties of some coordination compounds. We shall try to explain their behavior in terms of the molecular orbital theory developed in Chapter 10.

11–1   PROPERTIES OF TRANSITION-METAL COMPLEXES

The transition metals often are encountered in highly colored compounds with complex formulas. Although $PtCl_4$ exists as a simple compound, there are other compounds in which $PtCl_4$ is combined with two to six NH_3 molecules or with KCl (Table 11–1). Why should such apparently independent, neutral compounds associate with other molecules, and why should they do so in varying proportions? Measurements of electrical conductivity of solutions, and the precipitation of Cl$^-$ ions by Ag$^+$, indicate how many ions are present in aqueous solution. As we saw in Chapter 7, this and other evidence lead us to propose the ionic structures listed at the right of the table. These substances that contain ammonia all are coordination compounds, in which the NH_3 molecules are arranged around a central Pt^{4+} ion. The Pt(IV) complexes are octahedrally coordinated. In contrast, complexes of Pt(II) are in a square planar coordination, with a coordination number of four. Complexes of metals with a coordination number of four also may be tetrahedral. The coordination number two also is found, and these compounds are linear.

Color

Color is a distinctive property of coordination compounds of transition metals. The octahedral complexes of cobalt exist in a wide spectrum of colors, which depend on the groups coordinated to it (Table 11–2). Such

Table 11–2. Octahedral Complexes of Co(III), Their Colors, and Estimates of Electronic Transition Energy

Complex[a]	Color	Spectral color absorbed	Approximate wavelength (Å)	Approximate transition energy (wave number, cm⁻¹)
$Co(NH_3)_6^{3+}$	Yellow	Indigo	4300	23,200
$Co(NH_3)_5NCS^{2+}$	Orange	Blue	4700	21,200
$Co(NH_3)_5H_2O^{3+}$	Red	Blue-green	5000	20,000
cis-$Co(en)_2(H_2O)_2^{3+}$	Red	Blue-green	5000	20,000
$Co(NH_3)_5OH^{2+}$	Pink	Blue-green	5000	20,000
$Co(NH_3)_5CO_3^+$	Pink	Blue-green	5000	20,000
$Co(NH_3)_5Cl^{2+}$	Purple	Green	5300	18,900
$Co(EDTA)^-$	Violet	Yellow	5600	17,800
cis-$Co(NH_3)_4Cl_2^+$	Violet	Yellow	5600	17,800
$trans$-$Co(en)_2Br(NCS)^+$	Blue	Orange	6100	16,400
$trans$-$Co(NH_3)_4Cl_2^+$	Green	Red	6800	14,700
$trans$-$Co(en)_2Br_2^+$	Green	Red	6800	14,700

[a] (en) is an abbreviation for ethylenediamine, $NH_2CH_2CH_2NH_2$.

coordinated groups are called *ligands*. In solution, color arises from the association of solvent molecules with the metal as ligands, and not from the metal cation itself. In concentrated sulfuric acid (a potent dehydrating agent) Cu^{2+} is colorless, in water it is aquamarine, and in liquid ammonia it is a deep ultramarine. The colors that Arrhenius ascribed to the metal ions are actually those of hydrates such as $Cu(H_2O)_6^{2+}$ (blue), $Co(H_2O)_6^{2+}$ (pink), and $Ni(H_2O)_6^{2+}$ (green). Metals in high oxidation states have brilliant colors if they absorb energy in the visible spectrum: CrO_4^{2-} is bright yellow and MnO_4^- is an intense purple.

Whenever a certain energy of visible electromagnetic radiation, E, is absorbed by a compound during the excitation of an electron to a higher quantum state, the wavelength of light absorbed can be calculated from the expression

$$E = h\nu = hc\bar{\nu} = hc/\lambda$$

If the energy is given in wave numbers, $\bar{\nu}$, as frequently is done, then the wavelength is simply the reciprocal: $\lambda = 1/\bar{\nu}$. The color that we see in the compound is the *complementary* color to the color absorbed; it is the color that remains in the spectrum after the particular spectral color has been removed. These colors are listed in Table 11–3. If the energy absorbed is so small that it corresponds to wavelengths in the infrared, or so large that it occurs in the ultraviolet (which is usually the case in compounds of representative elements), then the compound will be colorless or white. With transition-metal compounds, interesting things happen during absorption in the visible spectrum region of energy. Often you can learn something about trends in chemical behavior by "eyeball spectroscopy."

Table 11–3. Colors of Compounds, Spectral Colors, Wavelengths, and Energies

Color of compound	Spectral color absorbed	Approximate wavelength (Å)	Energy difference between electronic levels (wave number, cm^{-1})
Colorless	Ultraviolet	< 4000	> 25,000
Lemon yellow	Violet	4100	24,400
Yellow	Indigo	4300	23,200
Orange	Blue	4800	20,800
Red	Blue-green	5000	20,000
Purple	Green	5300	18,900
Violet	Lemon yellow	5600	17,900
Indigo	Yellow	5800	17,300
Blue	Orange	6100	16,400
Blue-green	Red	6800	14,700
Green	Purple-red	7200	13,900
Colorless	Infrared	7200	< 13,900

The colors of transition-metal complexes explain the trick of writing with invisible ink made from $CoCl_2$. If you write with a pale-pink solution of $CoCl_2$, the writing is virtually undetectable on paper. If the paper is heated gently over a candle flame, the message appears in a bright blue. Upon cooling, the writing slowly fades. The pink color is that of the octahedral hydrated cobalt ion: $Co(H_2O)_6{}^{2+}$. Heating drives away the water and leaves a blue chloride complex with tetrahedral geometry. This compound is *hygroscopic;* that is, it absorbs water from the atmosphere and fades to the pale-pink hydrate again.

Challenge. To see if you can predict the color of a biologically important coordination compound, try Problem 11–5 in *Relevant Problems for Chemical Principles* by Butler and Grosser.

Isomers and Geometry

The compound whose empirical formula is $CoCl_3 \cdot 4NH_3$ can be either green or violet. This fact provided transition-metal chemists with convincing evidence that the coordination in this compound is octahedral. Both the green and violet $CoCl_3 \cdot 4NH_3$ dissociate to produce only one Cl^- ion per molecule, so the cation must be $Co(NH_3)_4Cl_2{}^+$, with a coordination number of six. How are the six ligands arranged? We can suggest three possibilities: a flat six-membered ring, a trigonal prism, or an octahedron. In each of these three structures there is more than one way of placing the two Cl^- ions among the six coordination positions. Such structures, which differ only in the arrangement of the same ligands around the central metal, are called

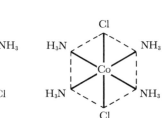

(a)

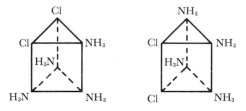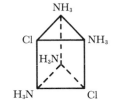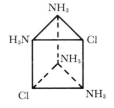

(b)

Figure 11–1. The six coordinating groups around Co in $Co(NH_3)_4Cl_2^+$ can be arranged in three possible symmetrical ways: a flat hexagon (a), a triangular prism (b), or an octahedron (c). Three different arrangements of coordinating groups, or geometrical isomers, can be produced in the hexagon and four in the triangular prism, but only two in the octahedron. Since only two isomers of $Co(NH_3)_4Cl_2^+$ have been found, an octahedral structure is the most probable structure.

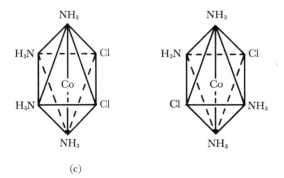

(c)

geometrical isomers. As Figure 11–1 shows, the existence of only two geometrical isomers of $Co(NH_3)_4Cl_2^+$ is convincing evidence for the octahedral structure. Octahedral coordination, with a coordination number of six, is by far the most common structure for such transition-metal compounds.

Fourfold coordination also is found. Is this coordination tetrahedral or square? Again, data on the number of variant forms of a compound with the same empirical formula provide the answer. The compound with the formula $PtCl_2 \cdot 2NH_3$, or $Pt(NH_3)_2Cl_2$, occurs in two forms that presumably are geometrical isomers. Both isomers are a creamy white, but they differ in solubility and chemical properties. As illustrated in Figure 11–2, there cannot be isomers for the tetrahedral structure, whereas the square planar structure has two. Therefore, the compound $Pt(NH_3)_2Cl_2$ must be square

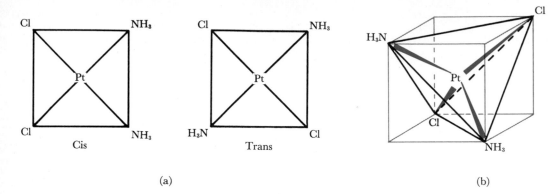

(a)   (b)

Figure 11–2. If the neutral molecule $Pt(NH_3)_2Cl_2$ has square planar coordination, two isomers are possible (a). But if the coordination is tetrahedral, only one can exist (b). Two isomers have been found, so the tetrahedral structure is eliminated. Why is this proof more convincing than that of Figure 11–1?

planar. (As an example of a comparable tetrahedral structure, CH_2Cl_2 has only one form and not two.)

Square planar geometry is characteristic of Pd(II), Pt(II), and Au(III), all of whose cations have eight d electrons, or a d^8 structure (Table 11–4). Tetrahedral coordination is encountered most often in transition-metal compounds in which the coordinating group is O^{2-}, as in CrO_4^{2-} and MnO_4^-. Now coordination structures can be examined directly by x-ray crystallography, and the conclusions about geometrical isomers from other experiments have been confirmed.

Table 11–4. Valence Electronic Configurations of Transition Metals[a]

Configuration:	d^1s^2	d^2s^2	d^3s^2	d^4s^2	d^5s^2	d^6s^2	d^7s^2	d^8s^2	d^9s^2	$d^{10}s^2$
Elements:	Sc	Ti	V	Cr	Mn	Fe	Co	Ni	Cu	Zn
	Y	Zr	Nb	Mo	Tc	Ru	Rh	Pd	Ag	Cd
	La	Hf	Ta	W	Re	Os	Ir	Pt	Au	Hg
Number of valence electrons:										
Neutral atom	3	4	5	6	7	8	9	10	11	12
M^{2+} ion (d electrons)	1	2	3	4	5	6	7	8	9	10
M^{3+} ion (d electrons)	0	1	2	3	4	5	6	7	8	9

[a] All configurations are given as $d^n s^2$, since what is of interest here is the number of electrons in the ion and not the electronic configuration of the neutral atom.

Magnetic Properties

Some transition-metal complexes are diamagnetic, which indicates no unpaired electrons. Many others are paramagnetic and have one or more unpaired electrons. For example, $Co(NH_3)_6^{3+}$ is diamagnetic, whereas CoF_6^{3-} is paramagnetic with four unpaired electrons per ion. The ionic charge is not the governing factor, since $Fe(H_2O)_6^{2+}$ is paramagnetic, with four unpaired electrons, yet $Fe(CN)_6^{4-}$ is diamagnetic. The magnetic properties of several other octahedral complexes are illustrated in Figure 11–3. One of our goals will be to explain this magnetic behavior in terms of electronic arrangement.

Lability and Inertness

A coordination complex that rapidly exchanges its ligands for others is *labile;* a complex that releases its ligands slowly is *inert.* Inertness is not the same as *stability* in the thermodynamic sense. A complex can be unstable, which means that it is not the most favored state according to the principles of thermodynamics discussed in Chapter 15. Given enough time, the complex will change to some other state. Yet if the transition to the most favored state is extremely slow, the unstable complex is inert. As an example of inert yet unstable compounds, H_2 and O_2 can be kept as a mixture for years without an appreciable spontaneous formation of water. However, if a small amount of platinum black (finely divided Pt) is supplied as a catalyst, or if a flame is brought near, the reaction to make the more stable H_2O is sudden, complete, and violent. A mixture of H_2 and O_2 by itself is *unstable,* yet *inert.*

Returning to coordination compounds, we note that $Cu(NH_3)_4SO_4$ can be dissolved in water and the $Cu(NH_3)_4^{2+}$ can be allowed to react with dilute acid to produce NH_4^+ and $Cu(H_2O)_6^{2+}$ as fast as the solutions are mixed. In contrast, $Co(NH_3)_6Cl_3$ can be heated in concentrated sulfuric acid to drive off HCl gas and make $[Co(NH_3)_6^{3+}]_2(SO_4^{2-})_3$ without breaking the bonds between Co and NH_3. The copper complex is labile; the cobalt complex is inert. Tripositive ions with three or six d electrons form especially inert complexes.

Oxidation Number and Structure

In Chapter 3 we observed that ions are most stable when surrounded by ions of opposite charge. The geometry of ionic crystals is determined almost completely by the sizes of the ions and the magnitudes of their charges. Transition-metal ions achieve stability by association with ligands. Since the size of most transition metals permits six ligands of the usual type (H_2O, NH_3, Cl^-) to pack around them, it is not surprising to find six as the most prominent coordination number.

Coordination number six appears to be optimal for ions with oxidation numbers $+2$ and $+3$; these include many transition-metal compounds. An

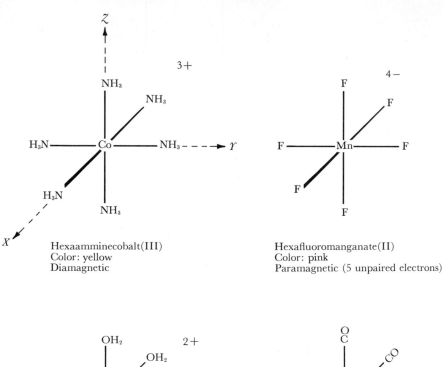

Hexaamminecobalt(III)
Color: yellow
Diamagnetic

Hexafluoromanganate(II)
Color: pink
Paramagnetic (5 unpaired electrons)

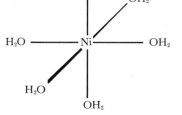

Hexaaquonickel(II)
Color: green
Paramagnetic (2 unpaired electrons)

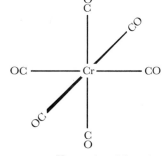

Hexacarbonylchromium(0)
Color: white
Diamagnetic

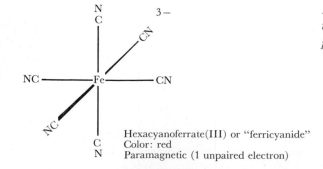

Figure 11–3. Several octahedral complexes, with systematic names, colors, and magnetic properties.

Hexacyanoferrate(III) or "ferricyanide"
Color: red
Paramagnetic (1 unpaired electron)

oxidation number of $+1$ is too low to attract six electron-donor groups to build a complex ion. Most complexes of $+1$ ions have smaller coordination numbers such as two for Ag^+ and Cu^+ in $Ag(NH_3)_2^+$ and $CuCl_2^-$. Stable complexes of rather high coordination number do occur with $+1$ ions and neutral atoms. But in most of these instances, such as $Mn(CN)_6^{5-}$ and $Mo(CO)_6$, these ligands have special π-bonding features that transcend simple electron donation.

Complexes of central ions having oxidation numbers greater than $+3$ are rare. They usually exist only with O^{2-} and F^-. We expect stronger bonding as the oxidation number of the central ion increases. However, if the oxidation number becomes too high, the central ion attracts ligand electrons so strongly that they are pulled completely away from the ligand. Then the complex is not stable, and the metal is reduced to a lower oxidation state. For this reason, Fe^{3+} forms no complex with I^-; instead, it oxidizes I^- to I_2. Since O and F are so electronegative, and since O^{2-} and F^- when bound to a central metal are so difficult to oxidize, they can exist in complexes in which the central ion has an oxidation state higher than the usual $+2$ or $+3$.

Influence of the Number of d Electrons

Much of coordination chemistry can be understood in terms of the number of d electrons on the central metal ion. As we already have mentioned, the $+2$ and $+3$ oxidation states are most common. The number of net d electrons for neutral atoms and for $+2$ and $+3$ cations are listed for reference in Table 11–4; we shall use this table frequently. In addition to this preference for $+2$ and $+3$ states, ions with the d-shell configurations d^0, d^5, and d^{10} are particularly favored.

Noble-gas shell, d^0. The noble-gas configuration with no d electrons is especially stable. The ion Sc^{3+} has this configuration, as does Ti(IV) in TiF_6^{2-}. It is increasingly difficult for ions to attain the d^0 structure from left to right across the periodic table. The reason is that the resulting charge on the central metal ion increases. Stabilization is then possible only by coordination to oxide ions. Therefore, we find VO_4^{3-} instead of V^{5+}, CrO_4^{2-} instead of Cr^{6+}, and MnO_4^- instead of Mn^{7+}.

This series of oxide complexes is a good example of the application of *eyeball spectroscopy*. Photons of the appropriate energy can excite electrons from the ligand oxygen atoms to the empty d orbitals of the metal. This process is called *charge transfer* and is a common origin for color in transition-metal complexes. The higher the oxidation state of the metal, the easier it is for electrons to transfer, and the lower is the energy of the photons required to bring about the transfer. The required energy in VO_4^{3-} occurs with photons in the ultraviolet region. Therefore, the VO_4^{3-} ion is colorless. In CrO_4^{2-}, the absorption of photons is in the violet region, at approximately

23,200 cm^{-1}; thus the chromate ion in solution appears yellow from the frequencies of light that are *not* absorbed (Table 11–3). (In accordance with standard spectroscopic practice, we express energy in wave numbers, cm^{-1}. See Section 8–2.) The Mn^{7+} ion has the highest oxidation state of all, and absorbs green light (around 19,000 cm^{-1}) for the charge-transfer excitation. Therefore, MnO$_4^-$ appears purple. The colors in these charge-transfer complexes are usually quite intense, which indicates strong absorption. Increasing the size of the central ion makes charge transfer more difficult and moves the absorption into the ultraviolet; thus MoO$_4^{2-}$, WO$_4^{2-}$, and ReO$_4^-$ all are colorless.

The greater attraction of a large positive charge on the central ion for the negative charge on the ligands is reflected in the decreasing tendency of ligands in the coordination ion to bind to other cations. In the series VO$_4^{3-}$, CrO$_4^{2-}$, and MnO$_4^-$, the vanadate ion is a fairly strong base and will bind H$^+$ or other cations. The chromate ion is a reasonably strong base also. But the permanganate ion is a weak base; the compound HMnO$_4$ is completely ionized in water. The acid HMnO$_4$ is one of the strongest known (Table 10–2). Reactions of the type

$$2VO_4^{3-} + 2H^+ \rightarrow {}^{2-}O_3V{-}O{-}VO_3^{2-} + H_2O$$

occur easily with the vanadate ion, which forms polyvanadates with many —O— bridges, and with the chromate ion, which forms dichromate, Cr$_2$O$_7^{2-}$, in acid. In contrast, Mn$_2$O$_7$ can be made only in concentrated sulfuric acid, which acts as a powerful dehydrating agent. Once formed, it is so unstable that it is a dangerous explosive.

Filled and half-filled shells. The filled d^{10} structure in Zn^{2+} and Ag$^+$, and the half-filled d^5 structure in Mn^{2+} and Fe^{3+}, make these ions particularly stable, even though the complexes that Mn^{2+}, Fe^{3+}, and Zn^{2+} form are relatively weak and contribute little to stabilizing the metal. This behavior is another example of the stability of filled and half-filled shells that we have seen so often.

Ions with d^3, d^6, or d^8 structures. The prominence of the oxidation number $+3$ for Cr(d^3) and for Co(d^6), plus the remarkable inertness of their complexes in chemical reactions [recall Co(NH$_3$)$_6$Cl$_3$ in hot sulfuric acid] cannot be explained on the basis of the ideas presented so far. Nor can we account for the special tendency of d^8 ions to adopt square planar rather than octahedral or tetrahedral coordination. To explain these structures and the existence of complexes of metals with oxidation number zero, we must examine how d orbitals participate in bonding with the ligands.

Instability of d^4. The ion Cr^{2+} (d^4) is a powerful reducing agent that is oxidized to a d^3 arrangement. Also a d^4 ion, Mn^{3+} is an equally powerful oxidant and is reduced to a d^5 ion. And Co^{5+}, likewise with a d^4 structure, is the only point in the pyramid of Table 7–2 for which no compound has ever been found. Any theory of bonding in coordination complexes will have to interpret this extreme instability of the d^4 configuration.

11–2   NOMENCLATURE FOR COORDINATION COMPOUNDS

Many complex transition-metal salts have common names that were given to them before their chemical identity was known. Some of the names are slightly informative: "potassium ferricyanide" for $K_3Fe(CN)_6$ and "potassium ferrocyanide" for $K_4Fe(CN)_6$. "Luteocobaltic chloride" for $Co(NH_3)_6Cl_3$ and "praseocobaltic chloride" for $trans$-$[Co(NH_3)_4Cl_2]Cl$ are informative only if you know the Latin and Greek for yellow (*luteus*) and green (*praseos*). Luteoiridium chloride, $Ir(NH_3)_6Cl_3$, is not even yellow, and only was given that name because it has the analogous chemical formula to the cobalt salt. And "Reinecke's salt," "Erdmann's salt," and "Zeise's salt" are completely useless names.

The Stock system of nomenclature gradually is replacing these older names. The new system is based on the following rules:

1) In naming the entire complex, the name of the cation is given first and the anion second (just as for sodium chloride), no matter whether the cation or the anion is the complex species.

2) In the complex ion, the name of the ligand or ligands precedes that of the central metal atom. Special ligand names are "aquo" for water, "ammine" for NH_3, and "carbonyl" for CO.

3) Ligand names generally end in "o" if the ligand is negative ("chloro" for Cl^-, "cyano" for CN^-) and "ium" in the rare cases in which the ligand is positive ("hydrazinium" for $NH_2NH_3^+$). The names are unmodified if the ligand is neutral ("methylamine" for CH_3NH_2, "ethylenediamine" for $NH_2CH_2CH_2NH_2$).

4) A Greek prefix (mono, di, tri, tetra, penta, hexa, etc.) indicates the number of each ligand (mono is often omitted for a single ligand of a given type). If the name of the ligand itself contains the terms "mono," "di," and so forth (ethylenediamine, en; diethylenetriamine, dien), then the ligand is enclosed in parentheses and its number is given with the alternative prefixes "bis" and "tris," instead of "di" and "tri." Hence, for example, $Pt(en)_3Br_4$ is tris(ethylenediamine)platinum(IV) bromide.

5) A Roman numeral or a zero in parentheses indicates the oxidation number of the central metal atom.

6) If the complex ion is negative, the name of the metal ends in "ate."

7) If more than one ligand is present in a species, the order of ligands in the name is negative, neutral, and positive.

Some examples of systematic nomenclature are

$Pt(NH_3)_6Cl_4$	Hexaammineplatinum(IV) chloride
$[Pt(NH_3)_5Cl]Cl_3$	Chloropentaammineplatinum(IV) chloride
$[Pt(NH_3)_3Cl_3]Cl$	Trichlorotriammineplatinum(IV) chloride
$Pt(NH_3)_2Cl_4$	Tetrachlorodiammineplatinum(IV)
$KPt(NH_3)Cl_5$	Potassium pentachloromonoammine-platinate(IV)
K_2PtCl_4	Potassium tetrachloroplatinate(II)
K_2CuCl_4	Potassium tetrachlorocuprate(II)
$Fe(CO)_5$	Pentacarbonyliron(0)
$[Ni(H_2O)_6](ClO_4)_2$	Hexaaquonickel(II) perchlorate
$K_4Fe(CN)_6$	Potassium hexacyanoferrate(II)
$K_3Fe(CN)_6$	Potassium hexacyanoferrate(III)
$[Pt(en)_2Cl_2]Br_2$	Dichlorobis(ethylenediamine)platinum(IV) bromide
$[Pt(NH_3)_4](PtCl_4)$	Tetraammineplatinum(II) tetrachloro-platinate(II)

Some common ligands are listed in Table 11–5. All of these ligands are *monofunctional;* that is, each ligand binds to the central ion at only one point. Other ligands are bifunctional, trifunctional, or even hexafunctional (Table 11–6). Three molecules of ethylenediamine, NH_2—CH_2—CH_2—NH_2, can coordinate octahedrally to Pt to produce the cation illustrated in Figure 11–4.

Table 11–5. Common Monofunctional Ligands[a]

Ligand	Name
F^-, Cl^-, Br^-, I^-	Fluoro, chloro, bromo, iodo
$:NO_2^-$ and $:ONO^-$	Nitro and nitrito
$:CN^-$	Cyano
$:SCN^-$ and $:NCS^-$	Thiocyanato and isothiocyanato
$:OH^-$	Hydroxo
$CH_3COO:^-$	Acetato
H_2O	Aquo
NH_3	Ammine
CO	Carbonyl
NO^+	Nitrosyl
py	Pyridine, C_5H_5N

[a] The electron pairs are shown to remind you which atom bonds to the central metal. They ordinarily are omitted.

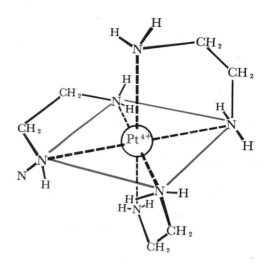

Figure 11–4. The structure of the tris(ethylenediamine)-platinum(IV) ion. Each molecule of ethylenediamine, $NH_2—CH_2—CH_2—NH_2$, coordinates to the platinum ion at two points. Such bifunctional and multifunctional ligands are called chelating groups, and the compounds are called chelates, from the Greek chela, "claw."

The ethylenediaminetetraacetato ion listed in Table 11–6 can wrap itself around a metal ion and coordinate with all six octahedral positions at once (Figure 11–5). Ethylenediaminetetraacetate, or EDTA, is so efficient a scavenger for Ca, Mg, Mo, Fe, Cu, and Zn that it will remove the essential metal atom from an enzyme, and will completely block its enzymatic activity. EDTA also is a useful scavenger in removing traces of metals from distilled and purified water. A molecule or ion that coordinates more than once with a metal ion is called a *chelating group*, and the total complex is called a *chelate*.

Table 11–6.   Common Chelating Groups or Multifunctional Ligands

Symbol	Ligand name	Formula	Bonds
en	Ethylenediamine	$\overset{..}{N}H_2—CH_2—CH_2—\overset{..}{N}H_2$	2
pn	Propylenediamine	$\overset{..}{N}H_2—CH_2—CH—\overset{..}{N}H_2$ $\quad\qquad\qquad\underset{\displaystyle CH_3}{\vert}$	2
dien	Diethylenetriamine	$\overset{..}{N}H_2—CH_2CH_2—\overset{..}{N}H—CH_2CH_2—\overset{..}{N}H_2$	3
trien	Triethylenetetraamine	$\overset{..}{N}H_2—CH_2CH_2—\overset{..}{N}H—CH_2CH_2—\overset{..}{N}H—CH_2CH_2—\overset{..}{N}H_2$	4
EDTA	Ethylenediaminetetraacetato	$^-{:}OOC—CH_2 \qquad\qquad CH_2—COO{:}^-$ $\qquad\quad\vert \qquad\qquad\qquad\quad \vert$ $\qquad {:}N—CH_2CH_2—N{:}$ $\qquad\quad\vert \qquad\qquad\qquad\quad \vert$ $^-{:}OOC—CH_2 \qquad\qquad CH_2—COO{:}^-$	6
ox or C_2O_4	Oxalato	$^-{:}OOC—COO{:}^-$	2

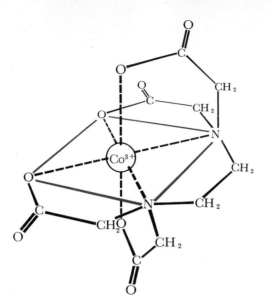

Figure 11–5. One molecule of EDTA, *or ethylenediaminetetraacetate, can completely enclose a metal ion in octahedral coordination.* EDTA's *attraction for metals is so strong that it will remove metals from enzymes and will inhibit their catalytic activity completely.*

Challenge. You can check your knowledge of nomenclature with Problem 11–12 (economics of fuel gas production) and of chelation with Problems 11–18 and 11–19 (principles underlying the treatment of heavy-metal poisoning) in the Butler and Grosser book.

Isomerism

Three types of isomers are found in coordination complexes: structural, geometrical, and stereo or optical isomers. *Structural isomers* have the same overall chemical formula but different ways of connecting component parts. Ethyl alcohol (CH_3CH_2OH) and dimethyl ether (CH_3—O—CH_3) are structural isomers. The material with the formula $Cr(H_2O)_6Cl_3$ exists in three structural isomers:

$[Cr(H_2O)_6{}^{3+}](Cl^-)_3$	Hexaaquochromium(III) chloride
$[Cr(H_2O)_5Cl^{2+}](Cl^-)_2 \cdot H_2O$	Chloropentaaquochromium(III) chloride monohydrate
$[Cr(H_2O)_4Cl_2{}^+](Cl^-) \cdot 2H_2O$	Dichlorotetraaquochromium(III) chloride dihydrate

The first of these is violet, the second is light green, and the third is dark green. Their structures can be demonstrated by precipitation of Cl^- with Ag^+ and by elimination of zero, one, or two waters of hydration by drying over H_2SO_4. The compound $Co(NH_3)_5SO_4Br$ mentioned in Section 7–1 has two structural isomers.

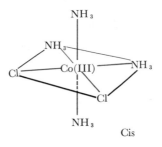
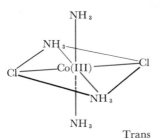

Cis   Trans

(a)

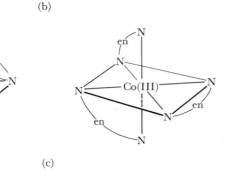
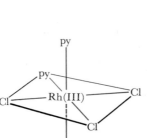

Figure 11–6. Geometrical and optical isomers of octahedral complexes. (a) cis- and trans-dichlorotetraamminecobalt(III) ions; (b) cis- and trans-trichlorotripy-ridinerhenium(III). Both (a) and (b) are pairs of geometrical isomers. (c) The two stereoisomers or optical isomers of the tris(ethylenediamine)cobalt(III) ion. Can you prove to your own satisfaction that, for all three compounds, only two such isomers exist?

Cis   Trans

(b)

(c)

Geometrical isomers differ in the arrangement of groups around the same center, as illustrated in Figure 11–6. The prefix "cis" indicates that two identical groups are adjacent; "trans" means that they are across from one another, or at least not adjacent. In Figure 11–6(a), the two Cl's in the cis isomer are adjacent to one another along one edge of the octahedron, whereas in the trans isomer they are across a diagonal through the octahedron. In Figure 11–6(b), the cis isomer has three Cl's clustered around one face of the octahedron; in the trans form they are arranged in a belt around the octahedron.

Optical or stereo isomers have the same groups connected in the same relative arrangement, but in the reverse sense as your right hand is to your left. Optical isomers arising from the arrangement of groups about a central atom always occur in pairs, one of which is the mirror image of the other. These pairs are called *enantiomers*. An example of two enantiomers is the two

Co(en)$_3^{3+}$ complexes shown in Figure 11–6(c). A central atom around which such isomers can be formed is called an *asymmetric center*. Another example of a pair of enantiomers is L- and D-alanine, shown in Figure 12–12. Many optical isomers can be formed when several asymmetric carbon atoms are connected in a chain. (We shall return to optical isomerism in the next chapter.)

11–3   THEORIES OF BONDING IN COORDINATION COMPLEXES

The maximum number of σ bonds that can be constructed with s and p valence orbitals is four. Thus, four is the highest coordination number commonly encountered in the representative elements in Period 2. These elements do not have filled d orbitals or access to empty d orbitals in the next higher shell. For example, in CH_4 the central carbon atom is "saturated" with four σ bonds. However, with a first-row transition metal as the central atom, there are five d valence orbitals in addition to the four s and p orbitals.

If the central metal made full use of its d, s, and p valence orbitals in σ bonding, a total of *nine* ligands could be attached. However, because of the bulkiness of most ligands it is extremely difficult to achieve a coordination number of nine. With rhenium (Re), a large third-row transition metal atom, and H, a small ligand, the coordination number of nine is found in the complex ReH_9^{2-}, mentioned in Section 1–9. The structure of this interesting complex is illustrated in Figure 11–7.

The bonds in most coordination complexes, however, use fewer than the nine atomic orbitals from the metal. We shall turn now to the theories that have been developed to explain this bonding and the properties of the complexes formed. There have been four stages in the development of transition-metal bonding theory. These are the simple electrostatic theory, the valence bond or localized molecular orbital theory, the crystal field and ligand field theories, and the delocalized molecular orbital theory. Each of these theories

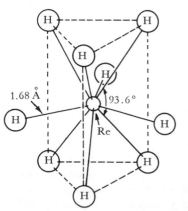

Figure 11–7. The structure of the ReH_9^{2-} ion. There are six H atoms at the corners of a trigonal prism, and three more H atoms around the Re atom on a plane halfway between the triangular end faces of the prism.

is an improvement on its predecessor. Considered together, they are a good case study of how bonding ideas develop, and how the same physical facts can be explained by different and seemingly contradictory assumptions.

We shall devote most of the discussion to octahedral coordination because it is both the most common and the easiest to understand. Keep in mind the following questions, which we shall try to answer when developing the theories.

1) How can we explain the difference in absorption of energy (manifested in the color) by a complex as the nature of the ligands is changed (recall Table 11–2)?

2) How can we explain that a complex such as $Co(NH_3)_6{}^{3+}$ is diamagnetic, yet others such as $CoF_6{}^{3-}$ are paramagnetic and have one or several unpaired electrons?

3) The stability of d^0, d^5, and d^{10} electron arrangements can be explained. But why are d^3 and d^6 so stable (recall Cr^{3+} and Co^{3+})?

4) Why do certain ions with the d^8 configuration, such as Pt(II) and Pd(II), prefer square planar geometry to tetrahedral or octahedral?

Electrostatic Theory

The simple electrostatic theory assumes only that the ligands, with negative charges, approach the positively charged central ion. Ligands and central ion attract one another, but ligands repel one another. The electrostatic repulsion between ligands leads to a prediction that a coordination number of two will be linear, and three ligands will lie at the corners of an equilateral triangle with the central atom at the center of the triangle. Four ligands will be tetrahedral, and six will be octahedral. This electrostatic theory cannot explain the existence of square planar complexes. Also, it cannot explain why complexes form with neutral molecules (CO, H_2O, NH_3) or with positive ions ($NH_2NH_3{}^+$). Finally, the theory does not discuss magnetic properties of complexes or their electronic energy levels as revealed by their colors and spectra.

Valence Bond or Localized Molecular Orbital Theory

One of the first definite advances toward understanding why octahedral geometry occurs was when Pauling showed, in 1931, that a set of six s, p, and d orbitals could be hybridized in a manner similar to the sp^3 and sp^2 hybridization to produce six equivalent orbitals directed to the vertices of an octahedron. The orbitals required are the s, the three p, and the $d_{x^2-y^2}$ and d_{z^2} orbitals lying either just below or just above these s and p orbitals. The two d orbitals are chosen because they have lobes of maximum density pointing in the six axial directions of an octahedron, as do the three p orbitals. The resulting six octahedrally oriented orbitals are called d^2sp^3 or sp^3d^2 hybrid

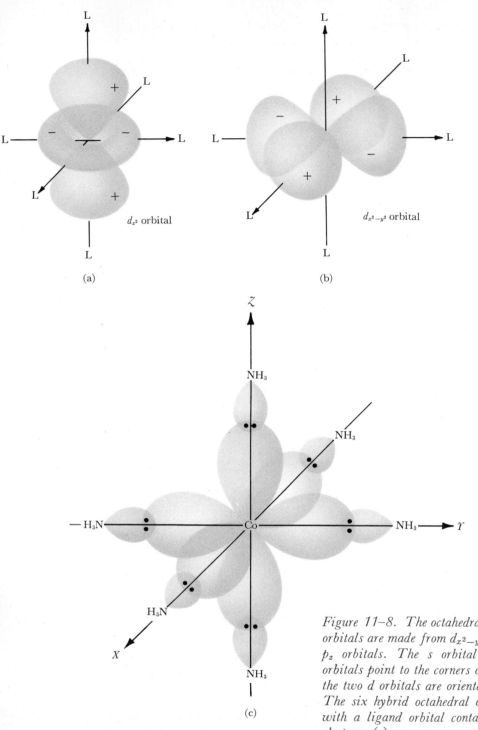

d_{z^2} orbital

(a)

$d_{x^2-y^2}$ orbital

(b)

(c)

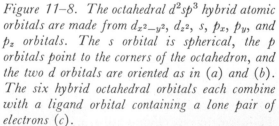

Figure 11–8. The octahedral d^2sp^3 hybrid atomic orbitals are made from $d_{x^2-y^2}$, d_{z^2}, s, p_x, p_y, and p_z orbitals. The s orbital is spherical, the p orbitals point to the corners of the octahedron, and the two d orbitals are oriented as in (a) and (b). The six hybrid octahedral orbitals each combine with a ligand orbital containing a lone pair of electrons (c).

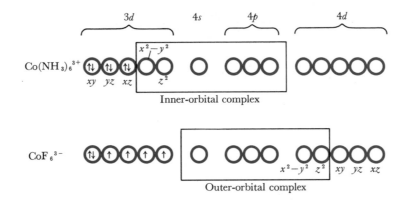

Figure 11–9. The valence bond theory postulates that in inner-orbital complexes of cobalt (low-spin) such as $Co(NH_3)_6^{3+}$, six electrons from the metal are spin-paired in d_{xy}, d_{yz}, and d_{xz} orbitals; the octahedral hybrid orbitals are produced from s, three p, and the two d from the level beneath. In outer-orbital cobalt complexes (high-spin), all five of the underlevel d orbitals are used for electrons from the metal, now not completely paired. The octahedral hybrids use two d orbitals from the same quantum level as s and p. In either case, lone-pair electrons on the ligands fill the bonding orbitals formed between ligand orbitals and the six metal orbitals of the octahedral hybrid.

orbitals, depending on whether the principal quantum number of the d orbitals is one less than or is the same as that of the s and p orbitals.

Each of the hybrid orbitals can be combined with an orbital from a ligand to make a bonding and an antibonding orbital, each with σ symmetry around the metal–ligand bond axis. The lone pair of electrons from each ligand orbital goes into the bonding molecular orbital, and six covalent bonds are produced (Figure 11–8). Similarly, four equivalent hybrid orbitals directed to the corners of a square in the xy plane can be produced from the $d_{x^2-y^2}$, s, p_x, and p_y metal orbitals.

The valence bond theory has not been successful in making quantitative predictions about energies, but at least it gives a rationalization for the magnetic properties of octahedral complexes. Pauling proposed that two types of complexes could be prepared: outer-orbital sp^3d^2 complexes in which the d orbitals lie above the s and p orbitals, and inner-orbital d^2sp^3 complexes in which they lie below (Figure 11–9). In inner-orbital complexes, the number of d orbitals left to hold the d electrons that remain on the metal ion is restricted. Only the d_{xy}, d_{yz}, and d_{xz} are available; the other two are used in octahedral hybridization.

Let us use cobalt as an example of the valence bond explanation of magnetic properties. The neutral cobalt atom has nine electrons beyond the Ar noble-gas shell, and can be represented as

The Co^{3+} ion has six electrons, which by Hund's rule will be distributed among all five $3d$ orbitals:

Now let us assume that six ligands, each with an electron pair, are to form six covalent bonds with hybridized metal orbitals that are octahedrally oriented. If an outer complex is formed with $4s$, $4p$, and $4d$ metal orbitals, the electrons in the $3d$ orbitals are undisturbed (Figure 11–9):

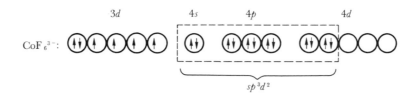

Four electrons will remain unpaired and, by this theory, CoF_6^{3-} should be paramagnetic, as it is observed to be.

In contrast, if an inner complex is formed with $3d$ orbitals in the octahedral hybridization, then only three $3d$ orbitals will be left for the six valence electrons originally present in the Co^{3+} ion:

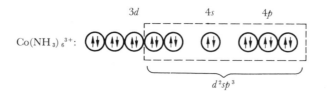

Hence we predict $Co(NH_3)_6^{3+}$ to be diamagnetic, and it is.

In Figure 11–3, the Mn^{2+} ion in hexafluoromanganate(II) has a d^5 structure:

Mn²⁺:

If hexafluoromanganate(II) were an inner-orbital complex, its five electrons would be compressed into three d orbitals and one electron would be unpaired:

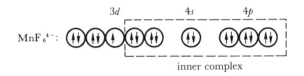

inner complex

Conversely, if it were an outer-orbital complex, all five electrons would be unpaired in the five d orbitals. Both possible complexes would be paramagnetic, but they would differ in the magnitude of the magnetic moment. Experimental data indicate that the complex has five unpaired spins, so it must be an outer-orbital complex. The Fe^{3+} ion also has a d^5 configuration; however, because magnetic data show that hexacyanoferrate(III) has one unpaired electron, it is described in the valence bond theory as an inner-orbital complex. Ligands such as CN^- and CO tend to form inner-orbital complexes, and ligands such as F^-, Cl^-, Br^-, and I^- usually form outer-orbital complexes.

The valence bond theory produces the correct two alternatives for the number of unpaired electrons, but it offers little help in making the choice between them. It does predict that inner-orbital complexes will be relatively inert. The experimental observation that outer-orbital (high-spin) complexes *are* usually more labile than inner-orbital (low-spin) complexes gives us confidence that the valence bond theory is at least a step in the right direction. It was a landmark at the time that it was proposed; however, it has been supplanted by crystal field theory and a more complete molecular orbital theory.

Before you go on.   The participation of d orbitals in the bonding of both non-metal compounds and transition-metal complexes in terms of localized molecular orbital theory is treated in Review 11 of *Programed Reviews of Chemical Principles* by Lassila *et al*.

Crystal Field and Ligand Field Theories

From a localized molecular orbital theory, the pendulum now swings the other way to a purely electrostatic theory that regards the bonding between metal and ligand as ionic. The simple electrostatic theory predicts that octahedral coordination will arise for the same reason that six unit charges, constrained to move on the surface of a sphere, will adopt an octahedral arrangement as the one of lowest energy. This is simply the electron-pair repulsion idea of Section 10–7.

Crystal field theory is more realistic. With this theory we consider what happens to the five metal d orbitals when six negative charges are brought near the metal in an octahedral array along the three principal axes of the

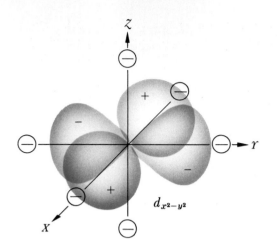

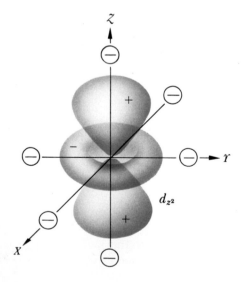

Figure 11–10. By crystal field theory, the six ligands of an octahedral complex may be represented as six negative charges, which point directly at the electron-density lobes of the metal $d_{x^2-y^2}$ and d_{z^2} orbitals. Any electrons in these two d orbitals will be repelled by the negative charges. More energy is required to force electrons into these two d orbitals on the metal than into the d_{xy}, d_{yz}, and d_{xz} metal orbitals, all of which point between the ligands.

d orbitals. The negative charges represent the lone pairs on the ligands. They are considered to remain with the ligands rather than being involved in any type of covalent bonding with the metal. Therefore, crystal field theory assumes purely *ionic* bonding.

The $d_{x^2-y^2}$ and d_{z^2} orbitals are most affected by the negative charges, which represent the ligands. The orbitals point directly at these charges (Figure 11–10). Any electrons in these d orbitals will respond to the electrostatic repulsion from the ligand lone pairs. Electrons in these two d orbitals will have higher energies than those in the other three. In contrast, the d_{xy}, d_{yz}, and d_{xz} orbitals have their lobes of maximum density directed *between* the ligands (see Figure 8–24). Electrons in these orbitals are more stable. The net result of this electrostatic interaction with the ligands is that the five d orbitals are split into two energy levels separated by a *crystal-field splitting energy*, Δ_o, as shown in Figure 11–11. The lower level is called the t_{2g} level, and the upper, the e_g. The names come from group theory, and their origin need not concern us here.

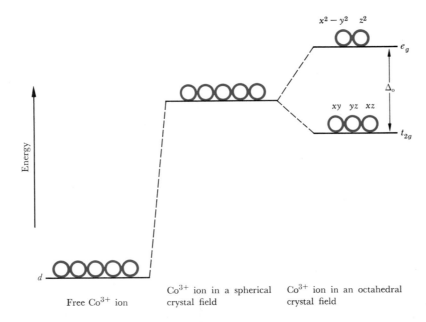

Figure 11–11. Energy-level diagram for the five d orbitals of a metal ion in an octahedral crystal field. On the left is the energy of electrons in the d orbitals of a free ion. In the center is the energy of electrons in the d orbitals if the ion were surrounded by a spherical cloud of negative charges. On the right is the splitting in energies of the d orbitals produced if the negative charges are arranged octahedrally around the metal. The three d orbitals that point between the ligands have lower energies than the two orbitals that point directly at the ligands.

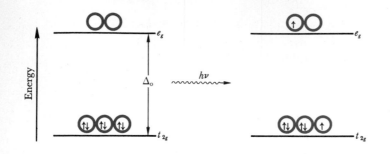

Figure 11–12. When $Co(NH_3)_6^{3+}$ *absorbs a photon of violet light and transmits those frequencies that give it its yellow color, the electronic configuration goes from the one at the left to the one at the right.*

The crystal-field splitting energy, Δ_o, is obtained by measuring the energy absorbed when one electron is promoted from the t_{2g} level to the e_g level (Figure 11–12). This splitting energy is crucial in accounting for magnetic properties. If Δ_o is small, as in CoF_6^{3-}, the six d electrons of Co^{3+} are spread out among all five d orbitals (Figure 11–13). There is a saving of energy if as few electrons as possible are paired. Conversely, if the splitting constant is large enough to overcome the energy of pairing two electrons in the same orbital, the more stable arrangement will be for the three low-lying orbitals of the t_{2g} level to contain one pair of electrons each and for the two upper orbitals to be vacant. This is the situation in $Co(NH_3)_6^{3+}$. Because of the different numbers of unpaired electrons in the two structures, $Co(NH_3)_6^{3+}$ is called a low-spin complex and CoF_6^{3-} a high-spin complex.

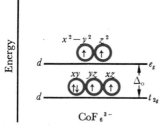

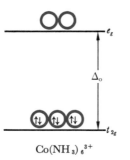

Figure 11–13. Crystal field (and ligand field) theory explanations of high-spin and low-spin complexes. The crystal-field splitting produced by the F^- ion is small, and the energy required to place two electrons in the upper level is less than the energy required to pair them with others. Therefore, the high-spin CoF_6^{3-} complex spreads its electrons among all five orbitals, and has four unpaired spins. The NH_3 group produces such a large crystal-field splitting that it is easier to pair electrons in the bottom three orbitals. The low-spin $Co(NH_3)_6^{3+}$ complex has no unpaired electrons.

Note how the same facts are explained by two quite different theories, the valence bond and crystal field theories. Both theories state that low-spin octahedral complexes arise when only three d orbitals of low energy are available for electrons originally from the central metal ion. High-spin octahedral complexes occur when there are *five* low-lying d orbitals. However, valence bond theory accounts for the presence of three or five such orbitals in terms of the set of six orbitals used in octahedral hybridization. In contrast, crystal field theory invokes a small or a large energy gap between a low-lying set of three d orbitals and a less stable set of two. In the valence bond theory, the operating factor is hybridization of orbitals from the metal, and the bonds to the ligands are entirely covalent. In the crystal field theory, the operating factor is electrostatic repulsion between ligand electron pairs and electrons on the metal ion, and the bonds to the ligands are entirely ionic. The effects are the same, but the explanations are radically different. Which theory is true?

Some chemists dislike the word "true" and prefer circumlocutions such as "successful in accounting for the facts." But unless the chemist also is a mystic who believes in some sort of inner reality beyond that which can be apprehended by the senses, the two sets of terminology are equivalent. No theory can ever be proven to be true in the absolute sense. All we can say is that one theory is "truer" than another because it can account for more observed properties of its subject than another theory. By this criterion, crystal field theory is better than valence bond theory. The common ligands can be ranked according to the magnitude of crystal-field splitting, Δ_o, that they produce, and this order can be justified to a certain extent.

The stronger the electrostatic field created by the ligand, the greater the splitting should be. Small ions with their lone pairs concentrated in one place, as in F^-, should produce a greater effect than larger groups with electrons diffused over a larger volume, as in Cl^-. Beyond this size argument, we can list the ligands in order but cannot explain the order:

$$CO, CN^- > en > NH_3 > -NCS^- > H_2O > OH^-, F^- > Cl^- > Br^- > I^-$$

| strong-field | intermediate-field | weak-field ligands |
| ligands | ligands | |

We write the isothiocyanate ion as $-NCS^-$ to emphasize that the metal–ligand bond is through the N atom in these cobalt(III) complexes discussed in this chapter.

Without spectroscopes or prisms, we can quickly check the order of ligands in this list merely by looking at the colors of complexes with these ligands. The absorption of visible light during the excitation of metal d electrons from t_{2g} to e_g orbitals is the other important source of color in transition-metal complexes in addition to charge-transfer absorption. For metals in $+2$ and $+3$ oxidation states, the charge-transfer absorption is usually in the ultraviolet, and the colors we see are from ligand-field splitting. These colors are not as intense as the charge-transfer absorption colors of CrO_4^{2-} and MnO_4^-. Table 11–2 contains a list of cobalt complexes, their colors, the colors absorbed in the electronic transition of lowest energy, and the approximate wavelengths and energies involved. Replacing even one NH_3 in the complex by —NCS^-, H_2O, OH^-, or Cl^- decreases the energy difference between levels, or the transition energy, in the order listed above. Substitution of Br^- for —NCS^- in the ethylenediamine complex lowers the transition energy by approximately 10% and changes the ion from blue to green. Replacing —NCS^- by a halide, Cl^-, in the presence of five NH_3, also lowers the transition energy by 10% and changes the salt from orange to purple.

Why is this list of relative strengths in energy-level splittings as it is? We cannot say from crystal field theory. An extension and improvement of this theory is *ligand field theory*. In this theory, the ligands are considered as more than simple negatively charged bodies. The orbitals on the ligands are taken into account, both those that contain the electron pairs to be shared with the metal and those that contain lone electron-pairs not directly associated with the metal. By giving the ligands some structure, we can explain more of the order of splitting energies. However, ligand field theory is really a halfway house toward a complete molecular orbital theory. This extended molecular orbital theory contains both crystal field and valence bond theories as extreme cases, much as the framework of the molecular orbital theory in Section 10–4 includes both ionic and covalent bonding.

Molecular Orbital Theory

With ligand field theory we take into account the orbitals on the ligands, and consider the ligands as something more than mere spherical charges. In the delocalized molecular orbital treatment, six ligand orbitals, assumed as a first approximation to have σ symmetry around the metal–ligand bond lines, are combined with six of the nine metal s, p, and d orbitals: $d_{x^2-y^2}$, d_{z^2}, s, p_x, p_y, and p_z. These are the same orbitals that Pauling used to synthesize his six hybrid orbitals. Now we shall combine all of them with the six ligand atomic orbitals to produce six delocalized bonding orbitals and six antibonding orbitals (Figure 11–14). The d_{xy}, d_{yz}, and d_{xz} orbitals, having the wrong symmetry for combining with σ-like ligand orbitals, are *nonbonding*. Electron pairs in these orbitals have no effect on holding ligands and metal together, and are described as metal lone pairs. A similar example of incorrect symmetry that makes an orbital nonbonding is shown in Figure 10–11.

The resulting energy-level diagram appears in Figure 11–14. The six bonding orbitals at the bottom are filled with electron pairs. We can think of them as being the six pairs donated by the ligands and forget about them. The upper four antibonding orbitals are similarly irrelevant; they will be empty except in extreme cases of electronic excitation, which we shall ignore. The nonbonding level and the lowest antibonding level correspond to the two levels, t_{2g} and e_g, produced by crystal-field splitting in Figure 11–13. We shall

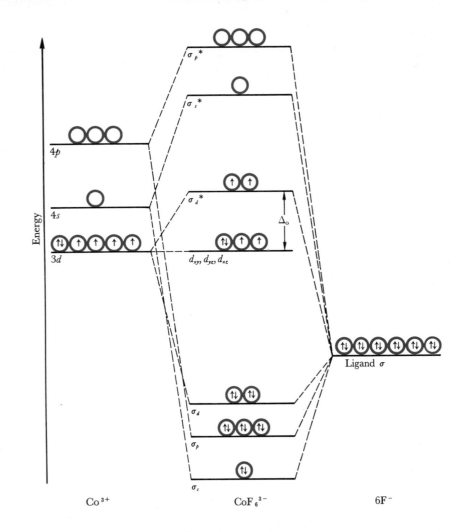

Figure 11–14. In the delocalized molecular orbital treatment of octahedral coordination, the same six metal orbitals that were used in the valence bond theory ($d_{x^2-y^2}$, d_{z^2}, s, p_x, p_y, and p_z) now combine with the six lone-pair-containing ligand orbitals to produce six bonding molecular orbitals (σ_s, σ_p, and σ_d) and six antibonding orbitals (σ_d^*, σ_s^*, and σ_p^*). The d_{xy}, d_{yz}, and d_{xz} metal orbitals are nonbonding. The low-lying six bonding orbitals fill with the electron pairs from the ligand to make six electron-pair bonds between metal and ligand. The d electrons of the metal ion are in the nonbonding and lowest antibonding levels, which are separated by the energy Δ_o. These two levels correspond to those in Figure 11–11, but the explanation of their origin is different.

continue to call them by these names, even in the molecular orbital treatment. But note the difference in the explanation of how this splitting occurred. In crystal field theory, it is the consequence of electrostatic repulsion; in ligand field theory, it is a consequence of the preparation of molecular orbitals. As we saw in Chapter 10 for HF and KCl, the same molecular orbital theory can accommodate everything from purely ionic to purely covalent bonding. The choice between these two theories is accordingly a pseudochoice, a consequence of being committed to two extreme models. In CoF_6^{3-} there is a certain ionic character to the bonding, because as you can see in Figure 11–14, the ligand orbitals are lower than those of the metal and closer in energy to the bonding molecular orbitals. Therefore, the bonding orbitals will have more of the character of the ligand orbitals, and there will be a displacement of negative charge toward the ligands. Thus, the bonds will be partially ionic.

With the molecular orbital theory, we can do a much better job of predicting which ligands will cause large energy differences between the t_{2g} and e_g levels in octahedral coordination, and which will produce small splittings. For this prediction we must look at the interactions of d_{xy}, d_{yz}, and d_{xz} orbitals in the t_{2g} level with atomic orbitals on ligands that have π symmetry around the metal–ligand bond.

The crystal field theory assumes that there are no such ligand orbitals and that each ligand is a featureless sphere of charge. Ligand field theory considers the ligand orbitals that form bonds to the metal ion, and also the two unhybridized p orbitals at right angles to the metal–ligand bond. These unhybridized p orbitals strongly influence the ligand-field splitting energy, Δ_o.

Figure 11–15 depicts four of these chloride p orbitals overlapping one of the three d orbitals in the t_{2g} energy level. If there are electrons in this d orbital, they are repelled by the lone-pair electrons in these p orbitals, and the energy of the t_{2g} level is raised. Any ligand with filled orbitals having such π symmetry around the ligand–metal axis decreases the ligand-field splitting energy, Δ_o. If we retain the crystal field theory terminology, such ligands (OH^-, Cl^-, Br^-, I^-) are called *weak-field ligands*. Fluoride ion is not as efficient at this process because it holds its electrons so tightly. Such an interaction is called a ligand-to-metal(π) or $L \rightarrow M(\pi)$ interaction.

Polyatomic groups that have an unfilled antibonding orbital with π symmetry behave differently. The cyanide ion (Figure 11–16) has a triple bond made from one bonding σ^b orbital and two bonding π^b molecular orbitals. One of these π^b bonding orbitals is shown in Figure 11–16(a). This orbital destabilizes or raises the t_{2g} level by a $L \rightarrow M(\pi)$ process just as in Cl^-. But most of the electron density of the π^b orbital lies between the C and the N, *not* in the direction of the metal atom. It is the antibonding π^* orbital [Figure 11–16(b)] that interacts more with the metal t_{2g} level. Here the effect is the reverse of that in Cl^-. Electrons in the metal t_{2g} orbitals can become partially delocalized and flow into the π^* orbital on the ligand. This delocalization stabilizes the t_{2g} orbital and lowers its energy. Therefore,

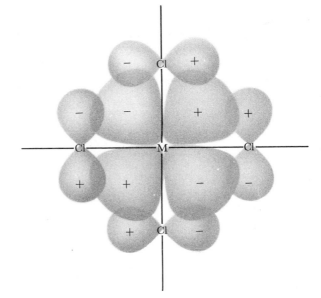

Figure 11–15. The lone-pair electrons in the π orbitals of Cl^- repel electrons in the d_{xy}, d_{yz}, and d_{xz} orbitals of the metal, thereby making the levels less stable. The t_{2g} level in Figure 11–11 rises and the splitting energy, Δ_o, decreases.

the splitting energy, Δ_o, increases. This process is called $M \rightarrow L(\pi)$ bonding or back bonding. Ligands that increase the splitting of the levels in this way (CO, CN^-, NO_2^-) are called *strong-field ligands* in crystal field terminology. Single atoms with many lone pairs of electrons, such as the halide ions, are weak-field ligands because they donate electrons. Bonded groups of atoms such as CO are more likely to be strong-field ligands because their bonding orbitals of π symmetry are concentrated between pairs of atoms and away from the metal, while the empty antibonding molecular orbitals extend closer to the metal.

    The nature of the metal itself also has a large influence on the size of the ligand-field splitting. Metal atoms or ions utilizing $4d$ and $5d$ valence orbitals give rise to much larger splittings than in corresponding complexes involving $3d$-orbital metals. For example, the Δ_o values for $Co(NH_3)_6^{3+}$, $Rh(NH_3)_6^{3+}$, and $Ir(NH_3)_6^{3+}$ are $22,900 \text{ cm}^{-1}$, $34,100 \text{ cm}^{-1}$, and $40,000 \text{ cm}^{-1}$, respectively. Presumably the $4d$ and $5d$ valence orbitals of the ion are more suitable for σ bonding with the ligands than are the $3d$ orbitals, but the reason for this is not well understood. An important consequence of the much larger Δ_o values of $4d$ and $5d$ central metal ions is that *all* second- and third-row metal complexes have low-spin ground states, even complexes such as $RhBr_6^{3-}$, which contain ligands at the weak-field end of the spectrochemical series.

    We have discovered that both the magnetic properties and the colors of transition-metal complexes depend on the nature of the ligand and metal by their effects on the ligand-field splitting energy, Δ_o. Thus, two of our goals listed at the beginning of this section have been realized. We also can explain the unusual stability of d^3 and d^6 configurations in complexes with strong-field

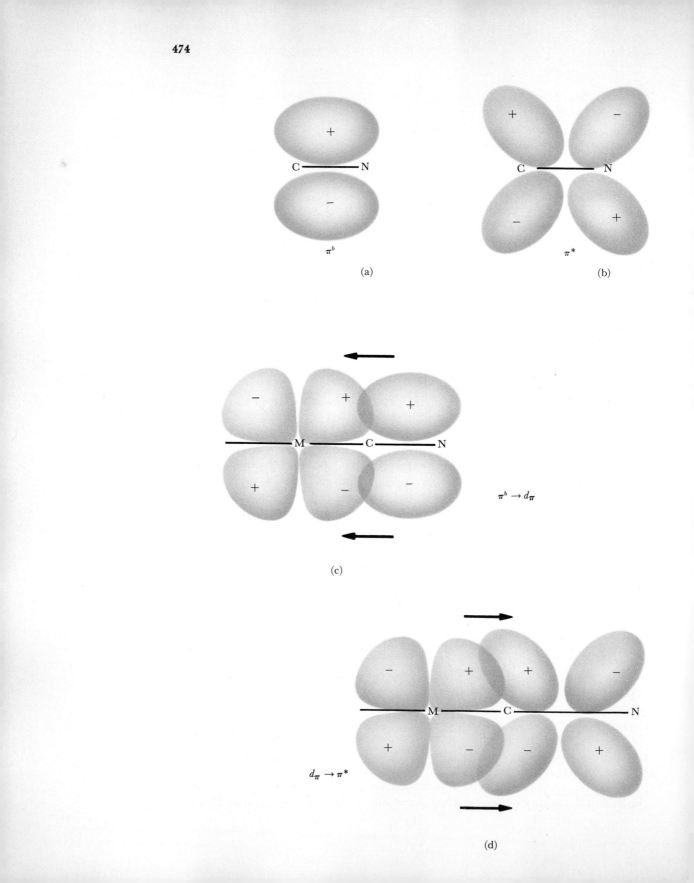

π^b

(a)

π^*

(b)

$\pi^b \to d_\pi$

(c)

$d_\pi \to \pi^*$

(d)

◀ *Figure 11–16. The effect of π bonding in cyano complexes. (a) In the CN⁻ ion, the bonding π^b molecular orbital contains an electron pair, and the antibonding π^* orbital (b) is empty. (c) The metal orbitals of the t_{2g} type are more stable in the presence of simple σ symmetrical ligands because the t_{2g} orbitals do not concentrate their electrons in the directions of the ligands. But if the ligand has filled π orbitals, then these orbitals interact with the metal t_{2g} orbitals and make them less stable. The splitting constant decreases. (d) If the metal has filled t_{2g} orbitals that interact with the empty antibonding π ligand orbitals, then the metal electrons are delocalized, the energy of the orbitals falls, and the splitting energy increases. This last effect predominates in most CN⁻ complexes, and we say that CN⁻ produces a large ligand-field splitting.*

ligands. The d^3 and d^6 arrangements are half-filled and completely filled t_{2g} levels. When the level splitting is large, these arrangements have the same significance in terms of stability that d^5 and d^{10} configurations do when all five d levels have the same energy. The stability of d^5 and d^{10} arrangements is most noticeable in weak-field complexes, when the ligand-field splitting is small.

11–4   TETRAHEDRAL AND SQUARE PLANAR COORDINATION

Energy levels estimated from ligand field theory for ligands of a given strength in different geometrical arrangements around the metal are compared in Figure 11–17. The relative order of energies in tetrahedral coordination is the reverse of octahedral, and it is not difficult to understand why. Ligands in a tetrahedral complex approach the metal from four of the eight corners of a cube [Figure 11–2(b)]. It is precisely the $d_{x^2-y^2}$ and d_{z^2} orbitals that do *not* point to the corners of the cube around the metal atom. As you can verify from Figure 8–24, the density lobes of the d_{xy}, d_{yz}, and d_{xz} orbitals point to the midpoints of the twelve edges of a cube, whereas the other two point to the midpoints of the six faces. The set of three d orbitals, being closer to the tetrahedral ligands, will be less stable, even though the splitting is not as pronounced as for octahedral geometry.

Square-planar splitting is almost as straightforward. Since we usually work with d_{z^2} and $d_{x^2-y^2}$ orbitals, let us take the xy plane as the plane of the complex, and assume that the ligands are at equal distance in the $\pm x$ and $\pm y$ directions. The $d_{x^2-y^2}$ orbital then points directly at the four ligands and is least stable. The d_{z^2} orbital points perpendicularly out of the plane of the ligands and is most stable (Figure 11–17). The other three orbitals have intermediate stability; d_{yz} and d_{xz} are more stable than d_{xy} because they are out of the plane of the ligands.

The octahedral arrangement is intrinsically more stable than the square planar because six bonds are formed instead of four. A typical covalent

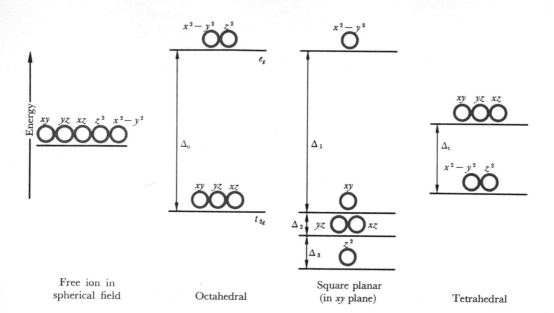

Figure 11–17. *Energy levels for the five d orbitals in the free ion in a spherical field of electrical charge and in the three common coordination geometries, all calculated for the same strength ligand. The relative order of levels is explained in the text. Δ_o, Δ_1, Δ_2, Δ_3, and Δ_t represent the ligand-field splitting energies.*

single bond and a typical ionic bond both have a bond energy of 50 to 100 kcal mole^{-1}. This corresponds to 18,000 to 35,000 cm^{-1} in the units in which splitting energies are given in Table 11–7. An octahedral complex, with two more bonds than either square planar or tetrahedral, has an intrinsic energy advantage of 35,000 to 70,000 cm^{-1}. Although it appears from Figure 11–17 that square planar coordination is preferable for d^1 through d^6, the extra bond energy causes octahedral coordination to predominate. However, the seventh and eighth electrons are forced into the high-energy e_g orbitals in octahedral coordination, whereas the much more stable d_{xy} is available in square planar. This extra stability is decisive for d^8 configurations in which the ligand-field splitting is large: They are found in square planar coordination. The ligand-field splitting is larger at higher atomic numbers. Hence, Pt(II) and Pd(II) regularly have square planar coordination, whereas Ni(II) is usually octahedral. The ninth and tenth electrons tip the balance back in favor of octahedral because of the extra stability gained from the two additional bonds.

Tetrahedral coordination seldom is preferred and is relatively rare. In addition to the smaller number of bonds in comparison with octahedral, the tetrahedral coordination also suffers from the double disadvantage of a less stable lower level and the necessity of commencing the upper level at the third electron rather than the fourth (in the high-spin complexes).

Table 11-7. Ligand-Field Splitting Energies for Representative Metal Complexes

Octahedral complexes	Δ_0, cm^{-1}	Octahedral complexes	Δ_0, cm^{-1}
$Ti(H_2O)_6^{3+}$	20,300	CoF_6^{3-}	13,000
TiF_6^{3-}	17,000	$Co(H_2O)_6^{3+}$	18,200
$V(H_2O)_6^{3+}$	17,850	$Co(NH_3)_6^{3+}$	22,900
$V(H_2O)_6^{2+}$	12,400	$Co(CN)_6^{3-}$	34,500
$Cr(H_2O)_6^{3+}$	17,400	$Co(H_2O)_6^{2+}$	9,300
$Cr(NH_3)_6^{3+}$	21,600	$Ni(H_2O)_6^{2+}$	8,500
$Cr(CN)_6^{3-}$	26,600	$Ni(NH_3)_6^{2+}$	10,800
$Cr(CO)_6$	32,200	$RhCl_6^{3-}$	22,800
$Fe(CN)_6^{3-}$	35,000	$Rh(NH_3)_6^{3+}$	34,100
$Fe(CN)_6^{4-}$	33,800	$RhBr_6^{3-}$	19,000
$Fe(H_2O)_6^{3+}$	13,700	$IrCl_6^{3-}$	27,600
$Fe(H_2O)_6^{2+}$	10,400	$Ir(NH_3)_6^{3+}$	40,000

Tetrahedral complexes	Δ_t, cm^{-1}
VCl_4	9010
$CoCl_4^{2-}$	3300
$CoBr_4^{2-}$	2900
CoI_4^{2-}	2700
$Co(NCS)_4^{2-}$	4700

Square planar complexes	Δ_1, cm^{-1}	Δ_2, cm^{-1}	Δ_3, cm^{-1}	Total Δ, cm^{-1}
$PdCl_4^{2-}$	23,600	3900	7400	34,900
$PtCl_4^{2-}$	29,700	4700	6800	41,200

A selection of measured ligand-field splitting energies for all three coordinations is in Table 11-7. See whether the octahedral data are compatible with the order of ligand splitting strengths given previously in this section. Also, note how close our guess for the splitting of $Co(NH_3)_6^{3+}$, based purely on color (Table 11-2), really was.

11-5 EQUILIBRIA INVOLVING COMPLEX IONS

When we write Co^{2+} to represent an ion in aqueous solution, we understand implicitly that the bare ion is not present, but that water molecules of hydration are coordinated to the metal. Therefore, the chemistry of complex ions in solution is the chemistry of the substitution of one ligand molecule or ion for another in the coordination shell around a metal. Nevertheless, it is customary, for simplicity, to write the formation of the ammine complex of

Table 11–8. Overall Formation Constants for Some Complexes in Aqueous Solution[a] at 298°K

Complex, ML_n	K_f, $[ML_n]/[M][L]^n$	Complex, ML_n	K_f, $[ML_n]/[M][L]^n$
L = NH₃		**L = $H_2NCH_2CH_2NH_2$ (en)**	
$Ag(NH_3)_2^+$	1×10^8	$Mn(en)_3^{2+}$	5×10^5
$Cu(NH_3)_4^{2+}$	1×10^{12}	$Fe(en)_3^{2+}$	4×10^9
$Zn(NH_3)_4^{2+}$	5×10^8	$Co(en)_3^{2+}$	8×10^{13}
$Cd(NH_3)_4^{2+}$	1×10^7	$Ni(en)_3^{2+}$	4×10^{18}
$Ni(NH_3)_6^{2+}$	6×10^8	$Cu(en)_2^{2+}$	1.6×10^{20}
$Co(NH_3)_6^{2+}$	1×10^5	$Zn(en)_3^{2+}$	1.2×10^{13}
[b]L = F⁻		**L = Cl⁻**	
AlF_6^{3-}	7×10^{19}	$MgCl^+$	4.0
SnF_3^-	8×10^9	$CuCl^+$	1.0
SnF_6^{2-}	10^{25}	$CuCl_4^{2-}$	10^{-5}
ZnF^+	5.0	$AgCl_2^-$	1×10^2
FeF^{2+}	3×10^5	$HgCl_4^{2-}$	1.6×10^{16}
MgF^+	65	$TlCl_4^-$	7.5×10^{18}
HgF^+	10	$BiCl_6^{3-}$	4×10^6
CuF^+	10	$SnCl_4^{2-}$	1.1×10^2
		$PbCl_4^{2-}$	4×10^2
		$FeCl^{2+}$	3.0
		$FeCl_4^-$	6×10^{-2}
[c]L = OH⁻		**L = CN⁻**	
$Cr(OH)^{2+}$	1×10^{10}	$Fe(CN)_6^{3-}$	10^{31}
$Fe(OH)^{2+}$	1×10^{11}	$Fe(CN)_6^{4-}$	10^{24}
$Co(OH)^{2+}$	1×10^{12}	$Ni(CN)_4^{2-}$	10^{30}
$Al(OH)^{2+}$	2×10^{28}	$Zn(CN)_4^{2-}$	5×10^{16}
$In(OH)_4^-$	1.5×10^{35}	$Cd(CN)_4^{2-}$	6×10^{18}
$Mn(OH)^+$	3×10^4	$Hg(CN)_4^{2-}$	4×10^{41}
$Fe(OH)^+$	1×10^7	$Ag(CN)_2^-$	10^{21}
$Co(OH)^+$	2.5×10^4		
$Ni(OH)^+$	1×10^5		
$Cu(OH)^+$	1×10^7		
$Zn(OH)^+$	1×10^5		
$Ag(OH)$	1×10^3		
$Zn(OH)_4^{2-}$	5×10^{14}		
$Pb(OH)_3^-$	8×10^{13}		

[a] In a strict sense, values should be accompanied by a more detailed description of solvent media and method of measurement. These values are approximate and are useful only for comparisons of similar species. For $n = 1$ assume that three or five water molecules are also in the complex.

[b] Many stable complexes such as SiF_6^{2-} and AsF_6^- form, but they hydrolyze in water to give oxions or oxides.

[c] Most polypositive metal ions tend to form polynuclear complexes with

bridges in the presence of OH⁻, as in Fe—O—Fe⁴⁺, $Bi_6(OH)_{12}^{6+}$, $Cr_2(OH)_2^{4+}$, and so on, not to mention extremely insoluble hydroxide precipitates.

Co^{2+}, for example, as if it were the addition of NH_3 to dipositive cobalt ions:

$$Co^{2+} + 6NH_3 \rightleftarrows Co(NH_3)_6{}^{2+} \tag{11-1}$$

We can write an equilibrium constant for this reaction:

$$K_f = \frac{[Co(NH_3)_6{}^{2+}]}{[Co^{2+}][NH_3]^6} \tag{11-2}$$

Since the equilibrium concerns the formation of a complex, K_f is known as a formation constant. For the formation of hexaamminecobalt(II), $K_f = 1 \times 10^5$.

There is no difference in principle between the mathematics of formation constant problems and that of dissociation of acids or bases. The parallel would be somewhat more apparent if Equation 11-1 were written as a dissociation of $Co(NH_3)_6{}^{2+}$ rather than an association, and if a dissociation constant that is the inverse of K_f were used. Formation constants, however, are customary.

As soon as NH_3 is added to a solution of Co^{2+}, some of it combines with Co^{2+} and produces some complex ions. At equilibrium after the addition of NH_3, the concentrations of the complex ion, NH_3, and free Co^{2+} (actually hydrated) can be calculated from Equation 11-2.

Example. Enough NH_3 is added to a 0.100-molar solution of Ag^+ to make the initial concentration of NH_3 1 mole liter^{-1}. After equilibrium is restored, what will be the concentrations of Ag^+ and of $Ag(NH_3)_2{}^+$?

Solution. The formation constant for $Ag(NH_3)_2{}^+$ is given in Table 11-8 as $K_f = 1 \times 10^8$. Therefore, the equilibrium constant expression is

$$K_f = \frac{[Ag(NH_3)_2{}^+]}{[Ag^+][NH_3]^2} = 1 \times 10^8$$

Because the formation constant is so large, we can assume that the formation reaction is effectively complete and that the concentration of $Ag(NH_3)_2{}^+$ is equal to the initial concentration of Ag^+. Since this quantity is appreciable, the concentration of NH_3 remaining at equilibrium is the original concentration less the amount reacted with Ag^+:

$[Ag(NH_3)_2{}^+] = 0.100$ mole liter^{-1}

$[NH_3] = 1.000 - 0.200 = 0.800$ mole liter^{-1}

(Two moles of NH_3 react for every mole of $Ag(NH_3)_2{}^+$ produced.) Therefore, the concentration of silver ion left at equilibrium is

$$[Ag^+] = \frac{0.100}{(0.800)^2} \times 1 \times 10^{-8} = 1.6 \times 10^{-9} \text{ mole liter}^{-1}$$

The assumption that the formation reaction is effectively complete is justified by the small Ag^+ concentration.

Exercise. What will be the final concentration of Ni^{2+} hydrated ion if 1.00-molar NH_3 is added to 0.100-molar Ni^{2+}?

(*Answer:* $Ni^{2+} = 4 \times 10^{-8}$ mole liter^{-1}.)

Exercise. What will be the final concentration of Ni^{2+} hydrated ion if 1.00-molar ethylenediamine (en) is added to 0.100-molar Ni^{2+}?

(*Answer:* $Ni^{2+} = 7.5 \times 10^{-20}$ mole liter^{-1}.)

[*Hint:* In the two preceding exercises, remember to use the proper power of the ligand concentration and to calculate correctly the amount of unbound ligand left at equilibrium.]

These two exercises illustrate the considerably greater attraction that a chelating agent has for a metal ion as compared with a related monofunctional ligand. Formation constants for ethylenediamine complexes in Table 11–8 are eight to ten orders of magnitude, or about a billion times, as large as formation constants for NH_3 complexes of the same metal ion. The bonding of ammonia and amine chelates to the metal is similar; in both cases the lone-pair electrons on an ammonia or amine nitrogen atom interact with the metal. The difference in formation constants between NH_3 and ethylenediamine reflects the increased stability when the bonding atoms of ligands are combined in a chelate molecule. However, the cyanide ion, CN^-, (which bonds through the carbon) has an intrinsically stronger attraction for metals than does an amine nitrogen atom. As Table 11–8 shows, the formation constants for cyanide complexes are three to thirteen orders of magnitude greater even than those of the corresponding ethylenediamine complexes!

Because formation constants are usually so large, we ordinarily can assume in complex-ion equilibrium problems that the concentration of the complex is the same as the total concentration of metal ion, as we have in the previous examples. However, for complexes of F^- this approximation is incorrect.

Example. What are the final concentrations of F^-, Hg^{2+}, and HgF^+ if 50 ml of a solution 2.00 molar in F^- are added to 50 ml of a solution 0.200 molar in Hg^{2+}?

Solution. It is best to begin with a table as in Chapter 5:

	F^-	$+ Hg^{2+}$	$\rightleftarrows HgF^+$
Initial conditions:	1.00	0.100	0 mole liter^{-1}
At equilibrium:	$1.00 - x$	$0.100 - x$	x mole liter^{-1}

$$K_f = \frac{x}{(1.00 - x)(0.100 - x)} = 10$$

Solve the equation for x by using the quadratic formula. $x = 0.090$ mole liter^{-1} = [HgF$^+$]. [F$^-$] = 0.910 mole liter^{-1} and [Hg^{2+}] = 0.010 mole liter^{-1}.

11–6 SUMMARY

This chapter has been a brief introduction to a rich area of chemistry, that of transition-metal complexes. Much of the richness (and the confusion) in their chemistry results from the presence of closely spaced energy levels involving d orbitals of the metal. The key to understanding transition-metal chemistry is the explanation of how the ligands perturb these metal energy levels. Valence bond theory and crystal field theory offer partial explanations, but currently the most successful theory is ligand field theory.

The story of these three theories is an illustration of the dictum: "You can always prove a theory wrong, but you can never prove it right." The success of valence bond theory in accounting for the coordination geometry and magnetic properties is no guarantee that the theory is right, or even that this way of looking at the problem is correct. For example, does the splitting of t_{2g} and e_g levels come about because of the formation of molecular orbitals (ligand field theory), electrostatic repulsion (crystal field theory), or the choice of six orbitals for hybridization (valence bond theory)? Or are all three theories incomplete, and will we some day regard ligand field theory with the same skeptical tolerance with which we now view the old valence bond theory?

For the present, ligand field theory works in many ways and accounts for much of the behavior of transition-metal complexes. Using it, we can explain the absorption of light and the observed magnetic properties of ions. It accounts successfully for the effect of the ligand on the splitting of energy levels. It explains why the d^0, d^3, d^5, d^6, and d^{10} electronic configurations are especially favored, and why d^8 leads to square planar geometry. Ligand field theory also can help us to predict the relative rates of reaction of complexes, as we shall explore in Chapter 18.

In Section 11–7, you will find an example of coordination chemistry in living organisms. The complexing molecule, porphyrin, is a chelate with fourfold coordination to the metal ion. As we continue through the coming chapters, we shall return to coordination chemistry again and again for examples of important chemical behavior.

11–7 POSTSCRIPT: COORDINATION COMPLEXES AND LIVING SYSTEMS

Since we first realized that we lived on a planet circling one sun among many, rather than being fixed at the center of creation, we have wondered whether we were a onetime miracle (or accident) or part of a general pattern

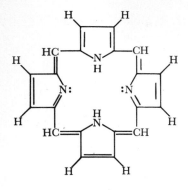

Figure 11–18. The porphin molecule. Porphin molecules with side groups substituted at the eight outermost hydrogen atoms around the ring are called porphyrins. A vertex where several bond lines meet, without a letter symbol, by convention is assumed to be a carbon atom. The four carbon atoms explicitly shown here by "C" could have been left out.

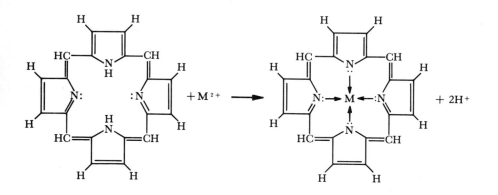

Figure 11–19. A porphyrin molecule can act as a tetradentate chelating group around an ion of a metal such as Mg, Fe, Zn, *or* Cu.

of living things. The astronomer Johannes Kepler (1571–1630) wrote a science fiction novel, *Somnium*, in which he described life on the moon as seen with a new invention, the telescope. He imagined intelligent humanoids and fast-growing plant life that sprouted, matured, and died in the course of one lunar day.

Today we know that any humanoids on the moon or Mars will be immigrants. However, it is possible that we will find the remains of simple life forms or the possible precursors of life forms on Mars, and that these will suggest something about how life evolved on earth. For years, scientists have extracted and analyzed organic matter from meteorites. They have debated whether this organic matter is truly meteoric or only terrestrial contamination, and whether it is of biological origin.

One of the compounds whose presence in meteorite samples is most suggestive of extraterrestrial life is porphin (Figure 11–18), and its deriva-

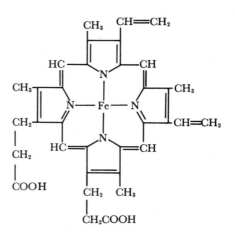

Figure 11–20. The iron–porphyrin complex with the side chains shown here is called a heme group.

tives, the *porphyrins*. The porphyrins are flat molecules that can act as tetradentate chelating groups[1] for metals such as Mg, Fe, Zn, Ni, Co, Cu, and Ag in a square planar complex as in Figure 11–19. The iron complex with the side chains shown in Figure 11–20 is called *heme*. The magnesium complex of porphyrin, with the organic side chain shown in Figure 11–21, is chlorophyll.

These two compounds, chlorophyll and heme, are the key components in the elaborate mechanism by which solar energy is trapped and converted for use by living organisms. We have noticed that a peculiar feature of transition-metal complexes is their closely spaced d levels that permit them to absorb light in the visible part of the spectrum and to appear colored. The porphyrin ring around the Mg^{2+} ion in chlorophyll serves the same function. Chlorophyll in plants can absorb photons of visible light and go to an excited electronic state (Figure 11–22). This energy of excitation can initiate a chain of chemical synthesis that ultimately produces sugars from carbon dioxide and water:

$$6CO_2 + 6H_2O \xrightarrow{h\nu} C_6H_{12}O_6 + 6O_2$$
$$\text{glucose}$$

Most compounds of the representative elements cannot absorb visible light; there are no electronic energy levels close enough together. Neither can Mg^{2+} alone. But the coordination complex of Mg^{2+} plus its square planar chelate has such levels, and chlorophyll is able to trap light and to use its energy in chemical synthesis.

---

[1] The name "chelate" comes from the Greek for "claw"; "tetradentate" literally means "four-toothed." Chelates with twofold, threefold, or fourfold coordination to the metal ion are called bidentate, tridentate, or tetradentate. It may seem illogical to speak of claws with teeth, but lovers of lobster or crab will appreciate the usage.

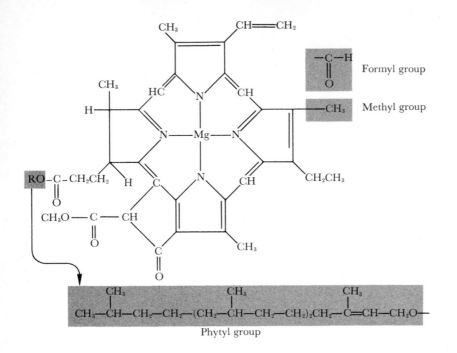

Formyl group

Methyl group

Phytyl group

Figure 11–21. The magnesium-porphyrin derivative shown here is called chlorophyll a, and is the essential molecule in photosynthesis. Chlorophyll b has a formyl group in place of the methyl group.

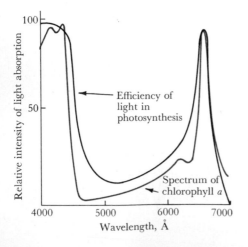

Figure 11–22. Chlorophyll a absorbs visible light except in the region around 5000 Å (green light), and thus appears green.

Scientists now believe that life evolved on earth in the presence of a *reducing* atmosphere, an atmosphere with ammonia, methane, water, and carbon dioxide but *no* free oxygen. Free oxygen would degrade organic compounds faster than they could be synthesized by natural processes (electrical discharge, ultraviolet radiation, heat, or natural radioactivity). In the absence of free oxygen, such organic compounds would accumulate in the oceans for eons until finally a packaged, localized bit of chemicals developed that we would call "living."

Living organisms, once developed, would exist by degrading these naturally occurring organic compounds for their energy. The amount of life on the planet would be limited severely if this were the only source of energy. Fortunately for us, around three billion years ago, the right combination of metal and porphyrin occurred and an entirely new source of energy was tapped—the sun. The first step that lifted life on earth above the humble role of a scavenger of high-energy organic compounds was an application of coordination chemistry.

Unfortunately, photosynthesis (as the chlorophyll photon-trapping process is called) liberates a dangerous byproduct, oxygen. Oxygen was not only useless to these early organisms, it competed with them by oxidizing the naturally occurring organic compounds before they could be oxidized within the metabolism of the organisms. Oxygen was a far more efficient scavenger of high-energy compounds than living matter was. Even worse, the ozone (O_3) screen that slowly developed in the upper atmosphere cut off the supply of ultraviolet radiation from the sun and made the natural synthesis of more organic compounds even slower. From all contemporary points of view, the appearance of free oxygen in the atmosphere was a disaster.

As so often happens, life bypassed the obstacle, absorbed it, and turned a disaster into an advantage. The waste products of the original simple organisms had been compounds such as lactic acid or ethanol. These are not nearly so energetic as sugars, but they can release large amounts of energy if oxidized completely to CO_2 and H_2O. Living organisms evolved that were able to "fix" the poisonous O_2 as H_2O and CO_2, and to gain, in the bargain, the energy of combustion of what were once its waste products. Aerobic metabolism had evolved.

Again, the significant development was an advance in coordination chemistry. The central components in the new machinery for aerobic metabolism, by which the combustion of organic molecules was brought to completion, are the *cytochromes*. These are molecules in which an iron atom is complexed with a porphyrin to make a heme (Figure 11–20), and the heme is surrounded with protein. The iron atom changes from iron(II) to iron(III) and back again as electrons are transferred from one component in the chain to another. The entire aerobic machinery is a carefully interlocked set of oxidation–reduction or redox reactions, in which the overall result is the reverse of the photosynthetic process:

$$6O_2 + C_6H_{12}O_6 \longrightarrow 6CO_2 + 6H_2O$$
glucose

Figure 11–23. Cytochrome c is a globular protein with 104 amino acids in one protein chain and an iron-containing heme group. In this schematic drawing, each amino acid is represented by a numbered sphere, and only key amino acid side chains are shown. The heme group is seen nearly edgewise in a vertical crevice in the molecule. Copyright © 1972 R. E. Dickerson and I. Geis; Scientific American, page 62, April, 1972.

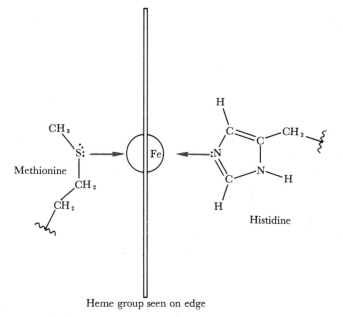

Figure 11–24. The iron atom in cytochrome c is octahedrally coordinated through five bonds to nitrogen atoms and one to a sulfur atom. One nitrogen atom and the sulfur atom come from side groups on the protein chain. The other four nitrogen atoms are from the porphyrin ring of the heme.

Heme group seen on edge

The energy liberated is stored in the organism for use as needed. The entire, elaborate, chlorophyll–cytochrome system can be regarded as a mechanism for converting the energy of solar photons into stored chemical energy in the muscles of living creatures.

Iron atoms usually exhibit octahedral coordination. What happens to the two coordination positions above and below the plane of the porphyrin ring? In cytochrome c, the heme group sits in a crevice in the surface of the protein molecule (Figure 11–23). From each wall of this crevice, one new ligand extends toward the heme: on one side a nitrogen lone electron pair from a *histidine* side chain on the protein, and on the other side a sulfur lone pair from a *methionine* side chain (Figure 11–24). Therefore, the octahedral coordination positions on the iron are directed to five nitrogen atoms and one sulfur atom.

How does the cytochrome c molecule operate? This is not yet known. The structure of the version with iron(III) was only determined in 1969 by x-ray diffraction, and that of the reduced iron in 1971. The ligands in the complex around the iron, and the protein wrapped around the whole structure, both modify the redox chemistry of the iron atom and ensure that oxidation and reduction are coupled to the earlier and later links in the terminal oxidation chain.

There is one more step in the story of metal–porphyrin complexes. Parkinson might add a subclause to his well-known law: Organisms expand to accommodate the food supplies available. With the guarantee of new energy sources, multicelled organisms evolved. At this point arose the problem, not of obtaining foods or oxygen, but of transporting oxygen to the

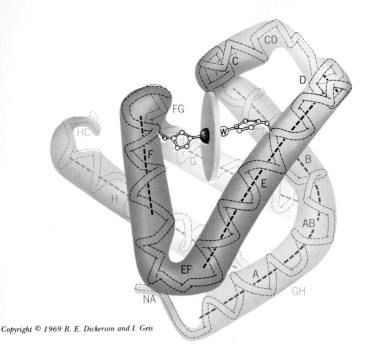

Figure 11–25. The myoglobin molecule is a storage unit for an oxygen molecule in muscle tissue. The heme group is represented by a flat disk, and the iron atom by a ball at the center. The circled W marks the binding site for O_2. The path of the polypeptide chain is shown by double dashed lines.

proper place in the organism. Simple gaseous diffusion through body fluids will work for small organisms but not for large, multicelled creatures. Again, a natural limit was placed on evolution.

For the third time, the way out of the impasse was found with coordination chemistry. Molecules of iron, porphyrin, and protein evolved, in which the iron could *bind* a molecule of oxygen without being oxidized by it. The oxidation of Fe(II) was, in a sense, "aborted" after the first binding step. Oxygen merely was carried along, to be released under the proper conditions of acidity and oxygen scarcity. Two compounds evolved, *hemoglobin*, which carries O_2 in the blood, and *myoglobin*, which receives and stores O_2 in the muscles until it is needed in the cytochrome process.

The myoglobin molecule is depicted in Figure 11–25. As in cytochrome *c*, four of the six octahedral iron positions are taken by heme nitrogen atoms. The fifth position has the nitrogen atom of a histidine. However, the sixth position has *no ligand*. This is the place where the oxygen molecule binds, marked by the circled W. In myoglobin, the iron is in the Fe(II) state. If the iron is oxidized, the molecule is inactivated and a water molecule occupies the oxygen position.

Hemoglobin is a package of four myoglobinlike molecules (Figure 11–26). In the past decade, these two structures have been determined by x-ray crystallography. It has become apparent that the four subunits of hemoglobin shift by 7 Å relative to one another when oxygen binds. Hemoglobin and myoglobin now become a model system for transition-metal chemists to study. Why does binding at the sixth ligand site of the iron complex cause

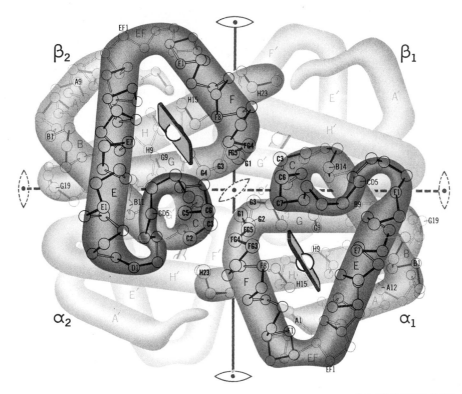

Figure 11–26. The hemoglobin molecule is the carrier of oxygen in the bloodstream. It is built from four subunits, each of which is constructed like a myoglobin molecule. This figure and that of myoglobin are reprinted from R. E. Dickerson and I. Geis, The Structure and Action of Proteins, W. A. Benjamin, Menlo Park, Calif., 1969.

the protein subunits to rearrange? Why does the oxygen molecule fall away from hemoglobin in an acid environment (such as in oxygen-poor muscle tissue)? How is the coordination chemistry of hemoglobin and myoglobin so carefully meshed that myoglobin binds oxygen just as hemoglobin releases it at the tissues?

Heme, or iron porphyrin, also is at the active sites of enzymes such as peroxidase and catalase (Section 12–9). Many other transition metals are essential components in enzyme catalysis; we shall discuss some of them in Chapter 12. With the evolution of myoglobin and hemoglobin, the size limitation was removed from living organisms. Thereafter, all of the multicelled animals that we ordinarily see around us evolved. In the sense that transition metals and double-bonded organic ring systems such as porphyrin are uniquely suited for absorbing visible light, and their combinations have a particularly rich redox chemistry, life is indeed applied coordination chemistry.

SUGGESTED READING

F. Basolo and R. Johnson, *Coordination Chemistry*, W. A. Benjamin, Menlo Park, Calif., 1964. Perhaps the best introduction to coordination chemistry for beginners. Clear and easy to understand.

F. A. Cotton and G. Wilkinson, *Advanced Inorganic Chemistry*, Wiley, New York, 1972, 3rd ed. A good standard reference work.

R. E. Dickerson, "The Structure and History of an Ancient Protein [cytochrome *c*]," *Scientific American*, April, 1972.

H. B. Gray, *Chemical Bonds*, W. A. Benjamin, Menlo Park, Calif., 1973.

H. B. Gray, *Electrons and Chemical Bonding*, W. A. Benjamin, Menlo Park, Calif., 1965.

E. M. Larsen, *Transitional Elements*, W. A. Benjamin, Menlo Park, Calif., 1965. Background information and descriptive chemistry.

L. E. Orgel, *An Introduction to Transition Chemistry*, Methuen, London, 1966, 2nd ed. This book and Basolo and Johnson are undoubtedly the best next steps in coordination chemistry, beyond this chapter.

E. G. Rochow, *Organometallic Chemistry*, Reinhold, New York, 1964.

QUESTIONS

1 How can you account for the series of compounds with the formulas $CrCl_3$, $CrCl_3 \cdot 3NH_3$, $CrCl_3 \cdot 4NH_3$, $CrCl_3 \cdot 5NH_3$, and $CrCl_3 \cdot 6NH_3$? Why would you not expect to find the missing members of the series $CrCl_3 \cdot 2NH_3$ and $CrCl_3 \cdot NH_3$?

2 If you found the compound $CrCl_3 \cdot NaCl \cdot xNH_3$, what would you expect x to be?

3 How many different isomers of this compound, $CrCl_3 \cdot NaCl \cdot xNH_3$, would you expect to find?

4 What assumption about the geometry of bonding around the Cr molecule did you make in answering Question 3?

5 How does the number of isomers of a compound distinguish between the possible geometrical arrangements around the central metal ion? Illustrate with tetrahedral and square planar geometry.

6 What is the difference between paramagnetic and diamagnetic compounds? How are these distinguished from one another by experiment?

7 What is the difference between stability and inertness? Can a chemical system be stable yet not be inert? Can it be inert yet not be stable?

8 Why are complexes with electronic configurations of d^5 or d^{10} on the central metal atom stable? Why are complexes with d^3 and d^6 arrangements stable? Which configurations would you predict to be more important for stability in complexes with ligands of large splitting energies? Of small splitting energies?

9 How would you name the following compounds in a systematic way:

$Ir(NH_3)_3Cl_3$ $Rh(en)_2Cl_2Ir(en)Cl_4$

$Co(NH_3)_6Cl_3$ $Rh(en)Cl_4Ir(en)_2Cl_2$

$Rh(en)_3IrCl_6$ $RhCl_6Ir(en)_3$

10 Sketch each of the four Rh–Ir complexes of Question 9.

11 Sketch each of the following complex ions or molecules:

cis-dichlorotetraammine-
    chromium(III) ion

trans-dichlorotetraammine-
  chromium(III) ion

cis-trichlorotripyridinerhodium(III)

trans-trichlorotripyridinerhodium(III)

Indicate the charge on each complex.

12 What is the difference between structural, geometrical, and stereo isomers? Find examples in Questions 9 and 11 of structural and geometrical isomers.

13 Why do complexes in which the central metal ion has the d^8 electronic configuration exist with square planar geometry?

14 What will be the number of unpaired electrons in $FeCl_6^{3-}$? In $Fe(CN)_6^{3-}$?

15 All octahedral complexes of vanadium(III) have the same number of unpaired electrons, no matter what the nature of the ligand. Why is this so?

16 What is the difference in the way that valence bond theory and crystal field theory explain the magnetic properties of complex ions?

17 How does ligand field theory account for the observed order of ligands in terms of the sizes of their splitting energies?

18 Why, in the crystal field theory, are the five d orbitals on the metal atom divided into two energy levels in the way they are? Where do the corresponding energy levels come from in the molecular orbital theory of complex ion structure?

19 Why are the same groupings of the five d orbitals made in tetrahedral coordination as in octahedral, but with the relative energies of these two groups reversed?

20 What is a chelate? If porphyrin is a tetradentate chelating group, and ethylenediamine is a bidentate chelating group, how would triethylenetetramine, diethylenetriamine, and EDTA be described?

21 What is a heme group? How does it function in hemoglobin and in cytochrome c?

PROBLEMS

1 A student was given 1.00 g of ammonium dichromate for the preparation of a coordination compound. The sample was ignited, thereby producing chromium(III) oxide, water, and nitrogen gas. The chromium(III) oxide was allowed to react at 600°C with carbon tetrachloride to yield chromium(III) chloride and phosgene ($COCl_2$). Upon treatment with excess liquid ammonia, the chromium(III) chloride reacted to produce hexaamminechromium(III) chloride. Calculate the maximum amount of hexaamminechromium(III) chloride that the student could prepare from his 1.00-g sample of ammonium dichromate.

2 When silver nitrate is added to a solution of a substance with the empirical formula $CoCl_3 \cdot 5NH_3$, how many moles of AgCl will be precipitated per mole of cobalt present? Why?

3 Co(III) occurs in octahedral complexes with the general empirical formula $CoCl_m \cdot nNH_3$. What values of n and m are possible? What are the values of n and m for the complex that precipitates 1 mole of AgCl for every mole of Co present?

4 How many ions per mole will you expect to find in solution when a compound with the empirical formula $PtCl_4 \cdot 3NH_3$ is dissolved in water? What about $PtCl_2 \cdot 3NH_3$? Draw diagrams of each of the complex cations.

5 Each of the following is dissolved in water to make a 0.001-molar solution.

Rank the compounds in order of decreasing conductivity of their solutions: K_2PtCl_6, $Co(NH_3)_6Cl_3$, $Cr(NH_3)_4Cl_3$, $Pt(NH_3)_6Cl_4$. Rewrite each compound by using brackets to distinguish the complex ion present in aqueous solution.

6 Give the systematic names of $[Co(NH_3)_4Cl_2]Br$, $K_3[Cr(CN)_6]$, and $Na_2[CoCl_4]$.

7 Write the formulas for each of the following compounds by using brackets to distinguish the complex ion from the other ions: (a) hexaaquonickel(II) perchlorate, (b) trichlorotriammineplatinum(IV) bromide, (c) dichlorotetraammineplatinum(IV) sulfate, (d) potassium monochloropentacyanoferrate(III).

8 Write the formula for each of the following by using brackets to distinguish the complex ion: (a) hydroxopentaaquoaluminum(III) chloride, (b) sodium tricarbonatocobaltate(III), (c) sodium hexacyanoferrate(II), (d) ammonium hexanitrocobaltate(III).

9 How many isomers are there of the compound $[Cr(NH_3)_4Cl_2]Cl$? Sketch them.

10 Sketch all of the geometrical and optical isomers of $PtCl_2I_2(NH_3)_2$.

11 How many geometrical and optical isomers are there of the complex ion $[Co(en)_2Cl_2]^+$? Of these, how many pairs of isomers are there differing only by a mirror reflection? How many isomers have a plane of symmetry and hence do not exist in pairs of optical isomers?

12 Repeat Problem 11 with propylenediamine substituted for ethylenediamine. Ignore optical isomers from the propylene carbon.

13 How many different *structural* isomers are there of a substance with the empirical formula $FeBrCl \cdot 3NH_3 \cdot 2H_2O$? For each different structural isomer, how many different *geometrical* isomers exist? How many of these can be grouped into right-handed and left-handed pairs of *stereoisomers*?

14 The Co^{2+} ion in aqueous solution is octahedrally coordinated and paramagnetic, with three unpaired electrons. Which one or ones of the following statements follow from this observation: (a) $Co(H_2O)_4^{2+}$ is square planar; (b) $Co(H_2O)_4^{2+}$ is tetrahedral; (c) $Co(H_2O)_6^{2+}$ has a Δ_o that is larger than the electron-pairing energy; (d) the d levels are split in energy and filled as follows: $(t_{2g})^5(e_g)^2$; (e) the d levels are split in energy and filled as follows: $(t_{2g})^6(e_g)^1$.

15 The coordination compound potassium hexafluorochromate(III) is paramagnetic. What is the formula for this compound? What is the configuration of the Cr d electrons?

16 How many unpaired electrons are there in Cr^{3+}, Cr^{2+}, Mn^{2+}, Fe^{2+}, Co^{3+}, Co^{2+} in (a) a strong octahedral electrostatic field and (b) in a very weak octahedral field?

17 A low-spin tetrahedral complex has never been reported, although numerous high-spin complexes of this geometry have been prepared. What conclusion may be drawn regarding the magnitude of Δ_t from this fact?

18 Certain platinum complexes have been found to be active antitumor agents. Among these are *cis*-$Pt(NH_3)_2$-Cl_4, *cis*-$Pt(NH_3)_2Cl_2$, and *cis*-$Pt(en)Cl_2$ (none of the trans isomers are effective). Use valence bond theory to account for the diamagnetism of these complexes. Are these inner or outer complexes? What kinds of hybrid orbitals are used in bonding?

19 Construct a molecular-orbital diagram for $[Cr(NH_3)_6]Cl_3$. How many unpaired electrons are present? If six Br^- groups were substituted for the six NH_3 groups to give $[CrBr_6]^{3-}$, would you expect Δ_o to increase or decrease?

20 Diagram the electronic arrangements in $Fe(H_2O)_6{}^{2+}$ and $Fe(CN)_6{}^{4-}$ for both the valence bond and crystal field models. Briefly compare these models.

21 For each of the following, sketch the d-orbital energy levels and the distribution of d electrons among them:

a) $Ni(CN)_4{}^{2-}$ (square planar)
b) $Ti(H_2O)_6{}^{2+}$ (octahedral)
c) $NiCl_4{}^{2-}$ (tetrahedral)
d) $CoF_6{}^{3-}$ (high-spin complex)
e) $Co(NH_3)_6{}^{3+}$ (low-spin complex)

22 Co(III) can occur in the complex ion $Co(NH_3)_6{}^{3+}$. (a) What is the geometry of this ion? In the valence bond theory, what Co orbitals are used in making bonds to the ligands? (b) What is the systematic name for the chloride salt of this ion? (c) Using crystal field theory, draw two possible d-electron configurations for this ion. Assign to them the labels high spin, low spin, paramagnetic, diamagnetic. Which two labels are correct for the ammine complex? (d) $Co(NH_3)_6{}^{3+}$ can be reduced to $Co(NH_3)_6{}^{2+}$ by adding an electron. Draw the preferred d-electron configuration for this reduced ion. Why is it preferred?

23 Pt(II) can occur in the complex ion $PtCl_4{}^{2-}$. (a) What is the geometry of this ion? In the valence bond theory, what Pt orbitals are used in making bonds to the Cl^- ions? (b) What is the systematic name for the sodium salt of this ion? (c) Using crystal field theory, draw the d-electron configuration for this ion. Is the ion paramagnetic or diamagnetic? (d) Pt(II) can be oxidized to Pt(IV). Draw the d-electron configuration for the chloride complex ion of Pt(IV). Explain the difference between this configuration and that of Pt(II). Is the Pt(IV) chloride complex ion paramagnetic or diamagnetic?

24 A solution is prepared that is 0.025 molar in tetraamminecopper(II),

$Cu(NH_3)_4{}^{2+}$. What will be the concentration of Cu^{2+} hydrated copper ion if the ammonia concentration is 0.10, 0.50, 1.00, and 3.00 molar, respectively? What ammonia concentration is needed to keep the Cu^{2+} concentration less than 10^{-15} molar?

25 From the data in Table 11–8, calculate the pH of a 0.10-molar solution of Cr^{3+} ion. *Hint:* Consider the reactions

$$Cr^{3+} + H_2O \rightleftarrows Cr(OH)^{2+} + H^+$$
$$K = ?$$
$$Cr^{3+} + OH^- \rightleftarrows Cr(OH)^{2+}$$
$$K_f = 1 \times 10^{10}$$
$$H^+ + OH^- \rightleftarrows H_2O \qquad K_w = ?$$

26 From the data in Table 11–8, calculate the pH of 0.10-molar solutions of Mn^{2+}, Fe^{2+}, and Ag^+. See Problem 25 if you need help. From the results of these two problems, can you correlate the "acidity" of positive ions with their charge?

27 The ion $Co(NH_3)_6{}^{3+}$ is very stable, with $K_f = 2.3 \times 10^{34}$. If the hydrolysis constant for the ammonium ion, K_b, is 5×10^{-10}, show that the equilibrium in the reaction

$$Co(NH_3)_6{}^{3+} + 6H^+ \rightleftarrows Co^{3+} + 6NH_4{}^+$$

lies far to the right. Then why does $Co(NH_3)_6{}^{3+}$ remain intact in hot concentrated sulfuric acid?

28 What is the concentration of chromate ion, $CrO_4{}^{2-}$, when solid $BaCrO_4$ is placed in contact with water? What is the chromate ion concentration when solid $BaCrO_4$ is placed in contact with a solution of 0.2-molar Ba^{2+}? $BaCrO_4$ can be dissolved in a solution of pyridine (py), producing the complex $Ba(py)_2{}^{2+}$, with a formation constant of 4×10^{12}. If 0.10-molar $BaCrO_4$ is dissolved in a solution with a constant pyridine concentration of 1.0 mole liter^{-1}, what is the concentration of Ba^{2+} ion?

29 What is the solubility of $Cu(OH)_2$ in pure water? In buffer at pH 6? Copper(II) forms a complex with NH_3, $Cu(NH_3)_4{}^{2+}$, with $K_f = 1.0 \times 10^{12}$. What concentration of ammonia must be maintained in a solution to dissolve 0.10 mole of $Cu(OH)_2$ per liter of solution?

30 Calculate the silver ion concentration in a saturated solution of AgCl in water. Silver ions react with an excess of Cl^- as follows:

$$Ag^+ + 2Cl^- \rightleftarrows AgCl_2{}^- \quad K_f = 1 \times 10^2$$

Calculate the concentration of $AgCl_2{}^-$ and show that you were justified in ignoring the complex ion formation in calculating the silver ion concentration at the beginning of the problem.

31 The formation constant for the pyridine complex of silver

$$Ag^+ + 2py \rightleftarrows Ag(py)_2{}^+$$

is $K_f = 1 \times 10^{10}$. If a solution is initially 0.10 molar in $AgNO_3$ and 1.0 molar in pyridine, what are the equilibrium concentrations of silver ion, pyridine, and the complex ion?

32 In 0.10-molar NaCl, the concentration of silver ions cannot exceed 10^{-9} mole $liter^{-1}$ because AgCl is so slightly soluble. What concentration of pyridine must be added to dissolve 0.10 mole of AgCl per liter of solution?

Organic chemistry is so called because it treats of the substances which form the structure of organized beings, and of their products, whether animal or vegetable.
William Gregory (1846)

12 THE SPECIAL ROLE OF CARBON

This chapter is designed to give you a brief look at two large areas of chemistry: organic chemistry and biochemistry. Some of the chapters of this book deal with fundamentals and essential techniques, and should be studied intensively. Others, such as this one, are designed to give you a general impression of an area of chemistry. As you read this chapter, you should try to understand and appreciate rather than memorize. If we may use the traditional (and in some circles, unfashionable) divisions of chemistry, most of the material in the first eleven chapters would be classified as inorganic and physical chemistry, and Chapters 13–18 will be largely so as well. Now, we shall attempt in one chapter to give an overview of two equally large areas of chemistry. Although the array of facts and terms is too large to be absorbed in its entirety, hopefully you will end this chapter with some feeling for what organic and biochemistry are all about.

The dividing line between these two areas is not sharp, and the distinction is primarily historical. The word "organic" was introduced by Berzelius, in 1780, to denote chemicals that came from living sources rather than from nonliving or laboratory reactions. The accepted doctrine then was that a special vital force was necessary for the synthesis of organic compounds. In 1828, Friedrich Wöhler (1800–1882) disproved this vital force idea when he prepared an organic compound, urea, by heating an inorganic substance, ammonium cyanate:

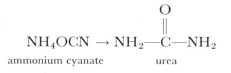

$$NH_4OCN \rightarrow \underset{\text{urea}}{NH_2\!-\!\overset{\displaystyle O}{\overset{\|}{C}}\!-\!NH_2}$$

ammonium cyanate   urea

The synthesis of other organic substances followed rapidly. The original reason for the classification disappeared, but the classification remained useful.

Better practical definitions of organic and biochemistry now can be given: Organic chemistry is the chemistry of the compounds of carbon; biochemistry is the branch of organic chemistry that deals with reactions in living systems. These definitions imply that there is something special about carbon. Why is the chemistry of nitrogen compounds, or of boron compounds, not distinguished as one of the traditional classifications of chemistry of past generations? What is special about the sixth element?

"The chemistry of carbon compounds" is a useful category because of the great profusion of such compounds. Well over a million different organic compounds are known, and there is no limit to the possibilities. In comparison, the total number of inorganic compounds of all the other elements is around 100,000. So many carbon compounds exist because carbon can link with itself to make straight chains and branched chains that no other element can. Some of these compounds are shown in Figure 12–1. Chains made by the repetition of a subunit are called *polymers*, and the repeated unit is called a *monomer*.

Hydrocarbons are polymers of the subunit —CH_2—, with the ends of the polymer terminated by hydrogen atoms. Butane is a tetramer (four subunits), and is a gas used for heating and cooking. Five- to twelve-carbon polymers are gasolines; heptane in Figure 12–1 is one example. Kerosene is a mixture of molecules with 12 to 16 carbon atoms, and lubricating oils and paraffin wax are mixtures of chains with 17 and more carbons. Polyethylene plastic has roughly 1500 —CH_2— monomer units per chain. There are many other organic chains. Neoprene rubber, Teflon, and Dacron in Figure 12–1 are synthetic polymers, and the polypeptide chain at the bottom of Figure 12–1 is the polymer from which all proteins are built.

Because carbon can make as many as four bonds, branched and cross-linked chains can be built. Isobutane (Figure 12–1) is a branched-chain isomer of C_4H_{10}. In Figure 12–2, silk and its synthetic analogue, nylon, are constructed from parallel, covalently bonded chains that are cross-linked into a sheet of hydrogen bonds (Section 14–4). Bakelite and Melamac are hard, inflexible plastics because their monomers are covalently linked in three dimensions.

The other distinguishing feature of carbon is its ability to make double bonds with itself and with other elements, and to do so in the middle of these

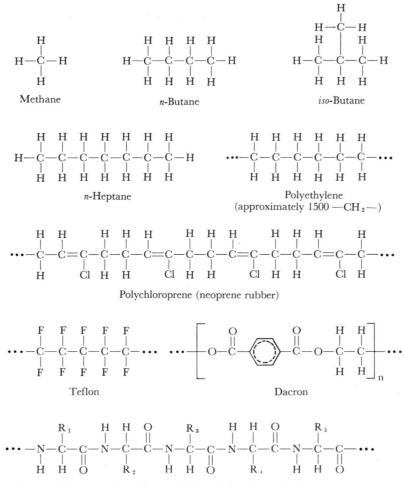

Polypeptides or proteins (silk, wool, hair, collagen, hemoglobin, enzymes, antibodies). R_1, R_2, R_3, and so forth are chemically different side chains.

Figure 12–1. Natural and synthetic chains of carbon atoms. The first two rows are hydrocarbons of increasing chain length from methane, through the commercial heating gases (butane) and gasolines (heptane), to polyethylene plastic. The double bond at every fourth carbon connection in polychloroprene is typical of natural and synthetic rubbers. Dacron shows two kinds of multiple bonding: C=O double bonds of the familiar π^b type, and delocalized benzenelike bonding. Polypeptide chains are cross-linked one to another as in Figure 12–2.

chains. Neoprene rubber (Figure 12–1) has such double bonds between carbon atoms. Dacron has double bonds between C and O, and also has the delocalized multiple bonding that we saw in Chapter 10 for benzene. Figure 12–3 depicts some other examples of double bonds in carbon compounds.

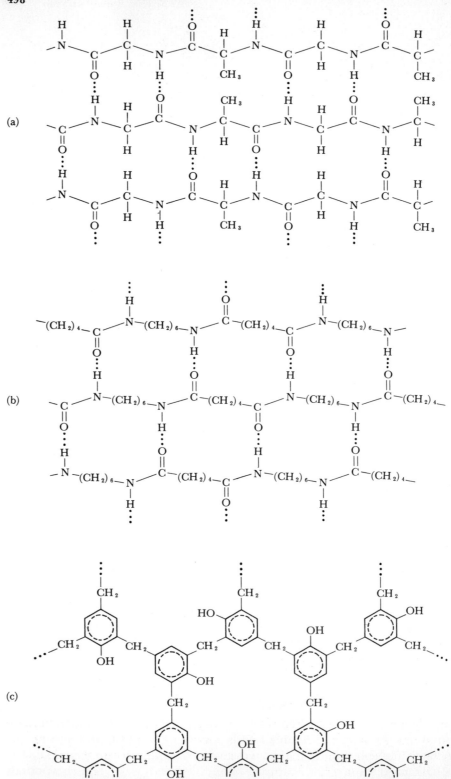

(a)

(b)

(c)

◄ *Figure 12–2. Three varieties of natural and synthetic polymers. (a) Silk, made from polypeptide chains. The chains are cross-linked into sheets by hydrogen bonds. A hydrogen bond is a primarily electrostatic bond between a partially positive hydrogen atom and a small, partially negative atom such as O or F. Its bond energy is around 6 kcal mole^{-1}. (b) Nylon 66 is closely patterned after silk. It was invented, in 1935, by W. H. Carothers at E. I. du Pont de Nemours & Co., Inc. It has hydrogen bonding similar to silk, but at longer intervals down the chains. In both fibers, the fiber axis is horizontal in the figure and parallel to the covalently bonded chains. (c) Bakelite is one of the earliest synthetic plastics, having been invented, in 1909, by L. H. Baekeland, an American chemist who also contributed to the chemistry of photography. Bakelite is one member of a class of phenol–formaldehyde resins that are strong and hard because of their three-dimensional network of covalent bonds.*

Since the double bond often can be converted to a single bond by adding an atom at each end of the bond, such double-bond compounds are called *unsaturated:*

$$CH_2\text{==}CH_2 + H_2 \;\rightarrow\; CH_3\text{---}CH_3$$

      ethylene   ethane

$$CH_2\text{==}CH_2 + HCl \rightarrow CH_3\text{---}CH_2\text{---}Cl$$

      ethylene   ethyl chloride

Compounds with delocalized, benzenelike multiple bonds are called *aromatic* compounds. Dacron (Figure 12–1) and naphthalene, DDT, adenine, and riboflavin (Figure 12–3) all have aromatic components. Adenine and riboflavin also show that carbon can make double bonds to nitrogen, and that nitrogen can participate in a delocalized, aromatic ring. Much of organic chemistry involves the special properties of aromatic ring systems. Aromatic, unsaturated molecules and transition-metal complexes are the two main classes of compounds in which the energy required to excite an electron falls in the visible part of the spectrum. Hence, these compounds are involved in dyes of all descriptions, and in mechanisms for trapping and transferring photon energy.

The four distinguishing features of organic compounds can be summarized as follows:

1) Long-chain polymers with C—C bonds.
2) Branched and cross-linked chains.
3) Double and triple bonds.
4) Delocalized aromatic bonds.

How many of these characteristics are exhibited by the immediate neighbors of carbon—B, N, and Si? What can carbon do that these elements cannot do, and why is this so? What particular combination of electrons and orbitals makes carbon so versatile?

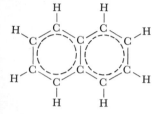

Propylene

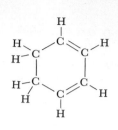

1, 3-Cyclohexadiene

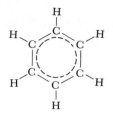

Benzene

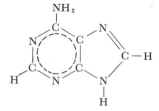

Naphthalene

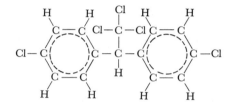

DDT

Adenine

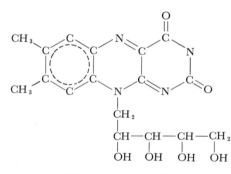

Riboflavin (vitamin B_2)

Figure 12–3. Examples of double bonds and delocalized bonds in organic compounds. Adenine, an essential component of the genetic polymer DNA *(deoxyribonucleic acid) and of the energy-storing molecule* ATP *(adenosine triphosphate), is a pentamer of* HCN. *It has been prepared from* HCN *under conditions simulating those of earth in the early stages of the evolution of life.*

12–1   THE CHEMISTRY OF THE NEIGHBORS OF CARBON

Boron, carbon, and nitrogen all are second-period elements of similar size. They differ in the number of valence electrons that they possess: three electrons in B, four in C, and five in N. Silicon, a third-period element, is like carbon in that it has four valence electrons, but they are one major energy level farther from the nucleus and have a principal quantum number of 3 instead of 2. Below the valence electrons, Si has *ten* inner-orbital electrons, two with principal quantum number 1 and eight with quantum number 2. In contrast, B, C, and N have only *two* electrons below their valence orbitals. All of the differences in chemical properties among B, C, N, and Si that will concern us in this chapter come from these two factors: the number of valence electrons and the number of electrons in completed inner orbitals.

Boron has three valence electrons and four valence orbitals per atom. It commonly uses three orbitals in sp^2 hybridization in compounds such as BF_3. Carbon has four valence electrons and four orbitals. Except when involved in multiple bonds, it uses sp^3 hybrid orbitals. Nitrogen has five electrons and four orbitals. It typically makes three bonds to other atoms in tetrahedral configurations, the fourth sp^3 atomic orbital is occupied by the lone electron pair (Section 10–7). Both carbon and nitrogen can make double and triple bonds involving the π overlap bond discussed in Section 10–9. The bond length for both elements decreases by 13% in a double bond and 22% in a triple bond. The atoms are held more tightly because of the electrons in π^b molecular orbitals derived from overlapping $2p$ atomic orbitals. Conversely, the overlap of these orbitals is too small for significant bonding unless the atoms are closer together. This is the reason why Si and other elements in the third period of the table and beyond cannot form multiple bonds. Silicon has ten inner-orbital electrons instead of two as in C and N. The repulsion between these inner-orbital electrons does not permit two Si atoms to come close enough for p-orbital overlap and the formation of a double bond. Although chemists are actively trying to synthesize compounds with Si=Si and Si=C bonds, none has been prepared yet. With one or two exceptions, double and triple bonds are confined to elements in the second period of the periodic table, with no more than two inner-orbital electrons per atom. Exceptions such as S=O, P=O, and Si=O use overlap between p and d orbitals, as we shall see later in this section with Si.

Boron

The reasons why boron is not a good candidate for a carbonlike chemistry can be understood by looking at the series of boron hydrides. The hydride BH_3 does not exist except as a short-lived decomposition product of higher hydrides. Other hydrides are known: B_2H_6, B_4H_{10}, B_5H_9, B_5H_{11}, B_6H_{10}, B_6H_{12}, B_8H_{12}, B_9H_{15}, and $B_{10}H_{14}$. The simplest boron hydride, B_2H_6, has eight atoms and only 12 valence electrons. If it were to have an ethanelike structure, it would need 14 electrons for the seven covalent bonds. But diborane has only 12 valence electrons: It is an *electron-deficient* compound. Its true structure is

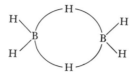

Each B atom has two normal two-center covalent B—H bonds, using a total of eight electrons. The remaining four electrons are used in two *three-center* B—H—B bonds, in which each of the three atoms contributes an orbital to the bonding molecular orbital. This concept of the three-center bond is enough to explain the structures of all of the boron hydrides. It also explains why boron cannot do the things that carbon can do.

For most compounds in which the number of valence electrons is at least as great as the number of valence orbitals, the idea of a two-atom chemical bond is meaningful, and we need consider only two atoms at a time. However, as we learned in the discussion of benzene (Section 10–10), localized molecular orbitals are only an approximation to reality. Sometimes we must construct delocalized molecular orbitals from atomic orbitals contributed by several, or occasionally all, of the atoms in a molecule. In benzene the C—H and the C—C σ bonds can be dealt with individually, but the six p orbitals must be considered together.

To explain the behavior of boron, the smallest unit of bonding that we can consider is sometimes *three atoms*. Three atomic orbitals, one from each atom, can combine to make three molecular orbitals: one bonding, one antibonding, and one nonbonding—the latter with virtually the same energy as the original atomic orbitals (Figure 12–4). We have seen nonbonding orbitals before. In HF (Figures 10–11 and 10–12), the $2p_x$ and $2p_y$ orbitals of F are nonbonding, as are the d_{xy}, d_{yz}, and d_{xz} orbitals of the metal in an octahedral coordination complex (Figure 11–14).

Two electrons in one of these *bonding* three-center molecular orbitals can hold three atoms together. This economy in bonding helps to compensate for the electron deficiency in boron. However, it also forces a cramped geometry on its compounds that makes boron unsuitable as a rival for carbon. Vast molecular networks can be constructed from the straight- and branched-chain carbon hydrides (hydrocarbons), in which the atoms are connected two at a time. In contrast, the boron hydrides, in which the atoms are connected three at a time, build structures whose boron frameworks are fragments of an icosahedron [Figure 12–5(a)]. The hydride B_4H_{10} is a small fragment of the icosahedron [Figure 12–5(b)]. It has six normal two-center bonds between B and H, one two-center B—B bond, and four three-center B—H—B bonds. Each of these bonds requires one electron pair. In this way, 14 atoms are held together by using 26 atomic orbitals but only 22 electrons. The hydride B_9H_{15} is three fourths of a complete icosahedron [Figure

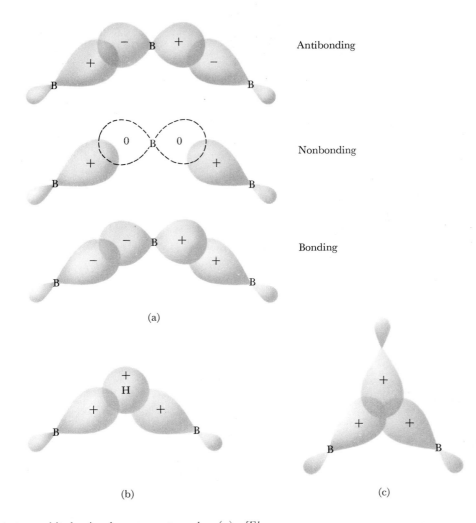

Figure 12–4. Three-center orbitals in boron compounds. (a) Three boron atoms each can donate one orbital (two sp^3 and one p) to make a bonding, a nonbonding, and an antibonding orbital. One electron pair in the bonding orbital holds all three atoms together. This arrangement is called an open three-center bond. (b) The arrangement of atomic orbitals in a bonding orbital for a B—H—B bridge bond. (c) The arrangement of atomic orbitals in a closed three-center bond. Such three-center bonds are found in electron-deficient compounds involving B and Al.

12–5(c)]. In this compound, 24 atoms are held together with 51 atomic orbitals and only 42 bonding electrons. The complete B_{12} icosahedron is found in crystalline boron. The manner in which such three-center bonds are used in the larger boron hydrides is shown for B_5H_{11} in Figure 12–6.

In conclusion, boron is an unsuitable candidate for organic chemistry because of its electron deficiency, which leads to three-center bonding and a

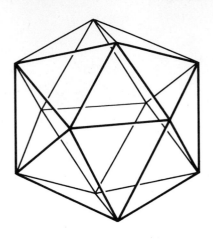

(a)

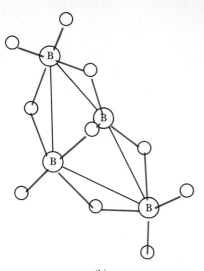

(b)

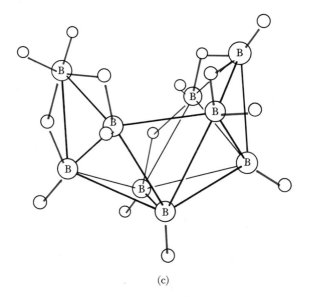

(c)

Figure 12–5. (a) The icosahedron is the boron framework for almost all of the boron hydrides. An icosahedron has twelve vertices and twenty equilateral triangular faces. (b) Tetraborane-10, B_4H_{10}, has its four boron atoms outlining two faces of the icosahedron. Bonds are marked in color. Six of the hydrogen atoms make normal two-center covalent bonds to boron; the others participate in four B—H—B bridges. The two central boron atoms are joined by a conventional two-center bond. (c) Enneaborane-15, B_9H_{15}, has a framework that is derived from the icosahedron by removing any three adjacent boron vertices that do not form an equilateral triangle. Ten hydrogen atoms make two-center covalent bonds to the boron atoms; the other five hydrogen atoms participate in B—H—B bridges.

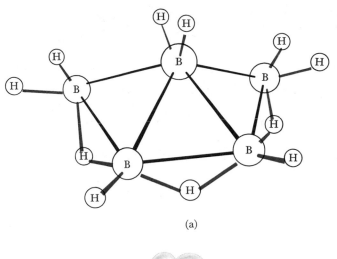

(a)

Figure 12–6. Structure and bonding orbitals in pentaborane-11, B_5H_{11}. Each of the boron atoms has sp^3 hybridization except the central one, which has three sp^2 and one unhybridized p orbitals. The closed three-center bond uses two sp^3 orbitals and one sp^2 orbital. The open three-center bond involving the central B atom uses two sp^3 and one p, as in Figure 12–4(a). The entire molecule uses 31 atomic orbitals but only 26 electrons.

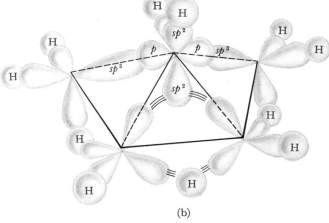

(b)

tendency for boron structures to close in upon themselves. Even worse, the geometrical arrangement produced makes it impossible for p orbitals to lie parallel on adjacent atoms and to form π bonds. In terms of the desirable properties in carbon, boron is a good try, but not good enough.

Nitrogen

Nitrogen, like carbon, can make double and triple bonds to itself and to other first- and second-period atoms. But nitrogen suffers from the opposite defect from boron; it has too many electrons. Repulsions between lone electron pairs on neighboring nitrogen atoms make the N—N single bond energy only 38 kcal mole^{-1}, in comparison with 83 kcal mole^{-1} for a C—C bond. In the C—N bond, in which one of these repelling lone electron pairs is absent, the bond energy increases to 70 kcal mole^{-1}.

Some compounds with chains of linked nitrogen atoms exist:

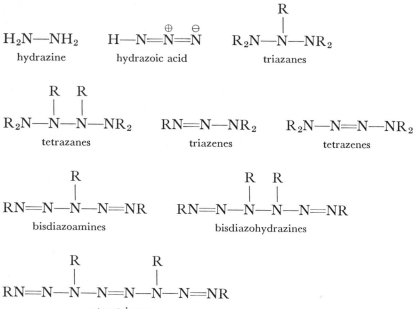

$$H_2N—NH_2$$
hydrazine

$$H—\overset{\oplus}{N}=\overset{\ominus}{N}=N$$
hydrazoic acid

$$R_2N—\overset{\overset{\displaystyle R}{|}}{N}—NR_2$$
triazanes

$$R_2N—\overset{\overset{\displaystyle R}{|}}{N}—\overset{\overset{\displaystyle R}{|}}{N}—NR_2$$
tetrazanes

$$RN=N—NR_2$$
triazenes

$$R_2N—N=N—NR_2$$
tetrazenes

$$RN=N—\overset{\overset{\displaystyle R}{|}}{N}—N=NR$$
bisdiazoamines

$$RN=N—\overset{\overset{\displaystyle R}{|}}{N}—\overset{\overset{\displaystyle R}{|}}{N}—N=NR$$
bisdiazohydrazines

$$RN=N—\overset{\overset{\displaystyle R}{|}}{N}—N=N—\overset{\overset{\displaystyle R}{|}}{N}—N=NR$$
octazotrienes

Hydrazine is used as a rocket fuel. Hydrazoic acid is extremely explosive and toxic. It sometimes is used in detonators for explosives. The higher *hydronitrogens*, as these compounds are called by analogy with the hydrocarbons, seldom can be prepared in the simplest forms, with hydrogen atoms replacing the R's shown in the preceding structures. Those sufficiently stable even to exist have phenyl groups (benzene rings) for R, or methyl or ethyl groups (CH_3— or CH_3CH_2—). They all are extremely unstable, and most are explosively so. They decompose rapidly under all conditions. Or as one scientist has said, "They stand on the edge of existence."

An important factor in the instability of nitrogen chains is the unusual stability of the triple bond in the N≡N molecule. The N_2 triple bond, whose bond energy is 225 kcal mole^{-1}, is *six times* as strong as the N—N single bond, whereas the C≡C triple bond in acetylene is only 2.3 times the strength of the C—C single bond. A long nitrogen chain is far less stable than the system after the chain breaks into a series of N_2 molecules.

Nitrogen participates in chains and rings with carbon and forms double bonds like carbon does. Diazomethane,

$$H_2C=\overset{\oplus}{N}=\overset{\ominus}{N}$$

is one of the most versatile and useful reagents in organic chemistry, despite the fact that it is highly toxic, dangerously explosive, and cannot be stored without decomposition. Two or more adjacent nitrogen atoms in such structures are rarely stable.

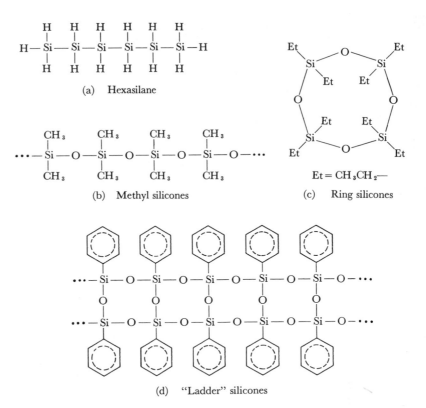

(a) Hexasilane

(b) Methyl silicones

(c) Ring silicones

Et = CH_3CH_2-

(d) "Ladder" silicones

Figure 12–7. Silicon can exist in two types of polymers: the reactive silanes, in which Si atoms are bonded directly, and the inert siloxanes or silicones, in which each connection is through a bridging oxygen atom. The silicones are chemically inert, heat resistant, electrically nonconducting oils and rubbers used as lubricants, insulators, and protective coatings. Three of the four Si bonds are to bridging oxygen atoms in the ladder silicones, which are rubbery or plastic materials. When all four Si bonds are involved in oxygen bridges, the silicate minerals result.

Silicon

The critical difference between Si and C is the greater number of inner-orbital electrons in Si, and the consequent inability to bring two silicon atoms close enough together for double and triple bonds. Silicon forms *silanes* analogous to the alkane hydrocarbons to be discussed in Section 12–2. Silanes have the general formula Si_nH_{2n+2}. The longest of these chains that has been prepared is only hexasilane (Figure 12–7). These silanes, like the hydronitrogens, all are dangerously reactive. The smallest silanes are stable in a vacuum, but all are spontaneously inflammable in air, and all react explosively with halogens. They are powerful reducing agents.

Table 12–1.   Relative Bonding Abilities of Carbon and Its Neighbors

Element, R	B	C	N	Si
Valence electrons	3	4	5	4
Usual coordination	Threefold (sp^2) (or fourfold with 3-center bonds)	Fourfold (sp^3)	Fourfold (sp^3) (including lone pair)	Fourfold (sp^3)
Single bond energies (kcal mole^{-1})				
R—R		83.1	38.4	42.2
R—C		(83.1)	69.7	69.3
R—N		69.7	(38.4)	–
R—O		84.0	~55	88.2
R—H		98.8	93.4	70.4
Electron-to-orbital ratio and bonding behavior	Electron-deficient; 3-center bonds (in multiple-B compounds)	Electron match; 2-center bonds	Electron surfeit; 2-center bonds. Lone-pair repulsions	Electron match; 2-center bonds
Linkage of like atoms	Icosahedral shells	Extensive chains	Chains, limited extent	Chains, limited extent
Double and triple bonds, π bonding	π bonding impossible with 3-center bonds in icosahedral framework	Good π overlap in double and triple bonds. Entire bonding capacity *cannot* be satisfied with one other like atom. Builds networks	Good π overlap in double and triple bonds. Entire bonding capacity can be satisfied with one other like atom. Builds N_2 molecules	Double and triple bonds impossible. Builds networks, principally using the stable Si—O bond rather than the less stable Si—Si bond

The silanes are so unstable and susceptible to oxidation because the Si—O bond is so much more stable than the Si—Si bond: 88 kcal mole^{-1} versus 42 kcal mole^{-1}. In contrast, with carbon the C—O and C—C bond energies are almost the same: 84 kcal mole^{-1} and 83 kcal mole^{-1} (Table 12–1). Hydrocarbons are oxidized much less easily than are the silanes. Although the reaction

$$\text{H—Si—Si—Si—Si—H} + 6\tfrac{1}{2}O_2 \rightarrow 4SiO_2 + 5H_2O$$

is explosively spontaneous, the analogous reaction with butane

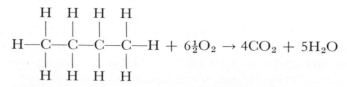

$$\text{H—C—C—C—C—H} + 6\tfrac{1}{2}O_2 \rightarrow 4CO_2 + 5H_2O$$

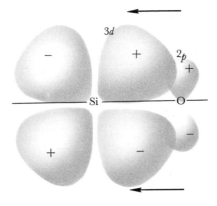

Figure 12–8. The strength of the Si—O *bond is due to an unusual partial double-bond character. One of the filled 2p lone-pair orbitals of oxygen shares its electrons with an empty 3d orbital of* Si *that has similar energy. For this reason, the* Si—O *bond energy is 88 kcal* mole^{-1}; *whereas the comparable silicon bond with* C, *which lacks the lone electron pairs, is only 69 kcal* mole^{-1}.

must be ignited by heat, and continues under ordinary conditions only because the heat released by the reaction keeps the reactants at a high temperature.

Part of this difference in oxidation of C and Si compounds arises because the Si—Si bond is weaker than the C—C bond. This is to be expected from the greater size of Si. The bonding electrons are farther from each nucleus, and the bond is not as strong. The same effect gives Si a lower ionization energy than C and makes it less electronegative (Figures 9–3 and 9–7). But an even more important factor in the difference between the behavior of C and Si is the anomalously high strength of the Si—O bond. In carbon, the empty 3d orbitals have a much higher energy than the filled lone-pair 2p orbitals of oxygen. There is no interaction between them. However, in silicon the added nuclear charge lowers the energy of the empty 3d atomic orbitals closer to the energy of the oxygen 2p orbitals. Oxygen then can share part of its lone-pair electrons with Si (Figure 12–8) in a back bonding similar to the L → M(π) and M → L(π) sharing in coordination complexes discussed in Section 11–3. Since the d_{xy}-type orbitals of Si extend farther toward O than a Si p orbital of a π bond, Si and O need not come as close as if they formed a $p\pi$–$p\pi$ double bond. The result of this sharing of oxygen lone electron pairs is that, although the Si—Si bond is 41 kcal mole^{-1} weaker than the C—C bond, the Si—O bond is 4 kcal mole^{-1} *stronger* than the C—O bond.

These results suggest that compounds in which Si atoms are linked by bridging oxygen atoms might be stable. This is so, and these compounds are the silicones, which we already have encountered in Section 9–5. As shown in Figure 12–7, silicones can exist as straight chains, rings, or as "ladder" compounds with two parallel linked chains. The silicones, as we have mentioned previously, are extremely inert compounds. The silanes are much *more* reactive than the hydrocarbons; the silicones are much *less* reactive.

Comparison of B, N, and Si

Each of the neighboring elements of carbon is unable to do the things that make carbon so important: to build long, stable chains with branching, cross-linking, and double bonds, and rings with delocalized electrons. The relative behavior of these elements is summarized in Table 12–1. Boron is forced into an unfavorable geometry by its deficiency of electrons and cannot overlap p orbitals to make double bonds. Although N occasionally can replace C in carbon rings and chains, and can form double bonds as easily

Table 12–2.   Some Common Saturated Hydrocarbons

Formula	Common name	Systematic name
CH_4	Methane	Methane
CH_3—CH_3	Ethane	Ethane
CH_3—CH_2—CH_3	Propane	Propane
CH_3—CH_2—CH_2—CH_3	*n*-Butane	Butane
CH_3—$\underset{\underset{CH_3}{\mid}}{CH}$—$CH_3$	*iso*-Butane	Methylpropane
CH_3—CH_2—CH_2—CH_2—CH_3	*n*-Pentane	Pentane
CH_3—CH_2—$\underset{\underset{CH_3}{\mid}}{CH}$—$CH_3$	*iso*-Pentane	Methylbutane
CH_3—$\overset{\overset{CH_3}{\mid}}{\underset{\underset{CH_3}{\mid}}{C}}$—$CH_3$	*neo*-Pentane	Dimethylpropane
CH_3—$\overset{\overset{CH_3}{\mid}}{\underset{\underset{CH_3}{\mid}}{C}}$—$CH_2$—$\overset{\overset{CH_3}{\mid}}{CH}$—$CH_3$	*iso*-Octane	2,2,4-Trimethylpentane
Cyclohexane structure	Cyclohexane	Cyclohexane

as carbon can, long chains of nitrogen atoms are unstable. Silicon is hampered by the weakness of its Si—Si bond in comparison with the Si—O bond and by its inability to make double bonds.

Carbon, then, is the fortunate combination of a small atom that has as many valence electrons as valence orbitals, and a bond to itself that is as strong as a bond to oxygen. Science fiction writers have long speculated on totally alien extraterrestrial life based on nonaqueous chemistry and an element other than carbon. Silicon has been the favorite element, and Mars has been the favorite homeland for rock-metabolizing, silicone-putty-fleshed monsters. But the more we learn about what carbon compounds do in terrestrial living creatures, the less easy it is to imagine silicon compounds performing even remotely similar roles. Carbon *is* special, and its properties can be duplicated by no other element.

12–2 PARAFFIN HYDROCARBONS OR ALKANES

Compounds that contain only carbon and hydrogen are called hydrocarbons, and those in which all carbon atoms form four single bonds are called saturated hydrocarbons, paraffins, or *alkanes*. The word "paraffin" originates from the Greek word for "little reactivity," and the chemical properties of these paraffins are in marked contrast to those of the silanes and hydronitrogens.

The general chemical formula for the alkanes is C_nH_{2n+2}. Alkanes exhibit a regular increase in melting point and boiling point with increasing molecular weight. Methane, ethane, propane, and butane are gases; pentane through $C_{20}H_{42}$ are liquids; and $C_{21}H_{44}$ and heavier compounds are waxy solids.

Several examples of the alkanes are given in Table 12–2. The first four have common names; those with five through 19 carbon atoms commonly are described by a Greek prefix indicating the total number of carbon atoms, and the standard suffix "ane." Above 19 carbon atoms the chemical formula usually is employed as the name.

If there are four or more carbon atoms, more than one way of connecting the carbons is possible. Consequently, isomers can exist. The five isomers of hexane have the following carbon skeletons and systematic names:

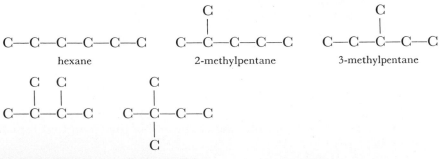

The old labels of "normal" for straight chain, "iso" for a branched chain, and "neo" for a third isomer rapidly become confusing, and the systematic nomenclature of the right column in Table 12–2 must be used. With the systematic nomenclature, the compound is given the name corresponding to the longest carbon chain that can be traced through the molecule. The molecule is stretched along this longest chain, and the carbon atoms are counted by beginning with the end that has the nearest branch point. The side chains then are identified, and located by giving the number of the carbon to which they are attached on the main chain. Hydrocarbon side chains are named by analogy with the hydrocarbons: CH_3—, methyl; CH_3CH_2—, ethyl; $CH_3CH_2CH_2$—, propyl; and

$$CH_3—CH—$$
$$|$$
$$CH_3$$

isopropyl

Thus, neopentane in systematic nomenclature is dimethylpropane, and not trimethylethane or even tetramethylmethane, because the longest continuous carbon chain has three carbon atoms as in propane.

Exercise. Draw the carbon skeletons and give the systematic names for the five isomers of heptane.

(*Answer:* You will find one heptane, two substituted hexanes, four pentanes, and one butane. The latter compound is 2,2,3-trimethylbutane.)

Example. What is the systematic name of the following compound:

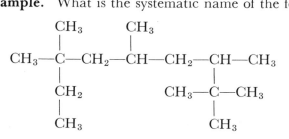

Solution. It has been written to suggest the name 2,4-dimethyl-2-ethyl-6-isopropylheptane. But a chain longer than seven C atoms can be found, and the proper name should be 2,2,3,5,7,7-hexamethylnonane.

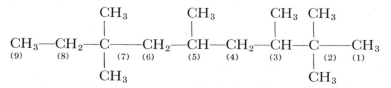

It is customary to begin the numbering at the end of the chain that is nearest the first branch point.

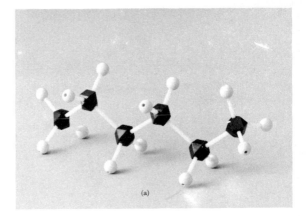

Figure 12–9. n-Hexane, C_6H_{14}, *and cyclohexane,* C_6H_{12}. *(a) The straight-chain hydrocarbon n-hexane. (b) The boat configuration of cyclohexane. The two top hydrogen atoms at the prow and stern of the boat are too close; this configuration is less stable than (c), the chair form. This same type of ring occurs in hexose sugars, which also adopt the chair form.*

Hydrocarbons can form rings as well as chains. The smallest is the three-carbon ring of cyclopropane:

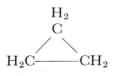

As you might imagine, this ring is highly strained. The optimum bond angle is 109° (the tetrahedral angle), but the angles in this three-membered ring are 60°. Cyclobutane and cyclopentane are less strained, and six-membered rings with the cyclohexane structure are extremely common. Cyclohexane can have two different structures, called the boat and the chair forms (Figure 12–9). The boat form is less stable because of the close approach of two hydrogen atoms across the top of the ring. Sugars and other substances whose molecules have a cyclohexanelike ring almost always occur in the chair form.

Reactions of Alkanes

As an example of the chemical unreactivity of the alkanes, the compound *n*-hexane is not attacked by boiling HNO_3, concentrated H_2SO_4, the strong oxidizing agent $KMnO_4$, or molten $NaOH$. The inertness of the alkanes makes them useful as lubricating oils, plastic films, and solid plastics for tubing and containers. Polyethylene is a familiar example. Virtually the only chemical reactions of the alkanes are combustion, dehydrogenation, and halogenation.

Combustion makes the alkanes useful as fuels:

$$CH_3\!-\!CH_2\!-\!CH_2\!-\!CH_3 + 6\tfrac{1}{2}O_2 \rightarrow 4CO_2 + 5H_2O$$

butane gas

Propane and butane gas, gasolines, and kerosenes all are alkanes whose value lies in their combustibility.

Dehydrogenation is the removal of two atoms of hydrogen and the creation of a double bond. This process usually occurs at high temperatures and in the presence of a catalyst such as Cr_2O_3:

$$H_3C\!-\!CH_3 \xrightarrow[\text{Cr}_2\text{O}_3 \text{ catalyst}]{500°C} H_2C\!\!=\!\!CH_2 + H_2$$

These dehydrogenated products are called *alkenes* or *olefins*. We shall discuss them further in Section 12–4.

Halogenation is the reaction of a hydrocarbon with F_2, Cl_2, or Br_2 (I_2 is too inert under ordinary conditions) and the replacement of one or more H atoms by halogen atoms:

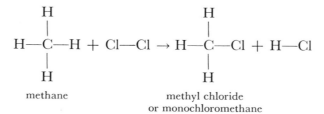

methane   methyl chloride
or monochloromethane

These halogenated hydrocarbons are the gateway to a great many other chemical reactions.

12–3   DERIVATIVES OF HYDROCARBONS: FUNCTIONAL GROUPS

In a *chlorination* reaction, one or more hydrogen atoms can be replaced by Cl, and many isomers are possible. Some examples, with their systematic names, are

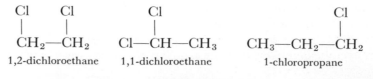

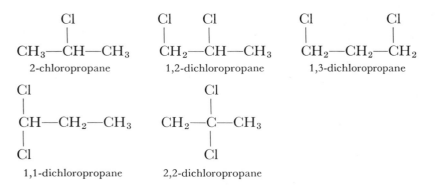

Exercise. How many different isomers are there of trichloropropane, and what are they?

(*Answer:* Five. 1,2,3; 1,2,2; 1,1,3; 1,1,2; 1,1,1.)

These chlorinated hydrocarbons are the starting materials for the preparation of many classes of compounds that cannot be prepared directly from the hydrocarbons. Their chemical reactivity lies in the C—Cl bond, and the rest of the molecule acts as a unit in many reactions. Therefore, it is convenient to think of the hydrocarbon part of the molecule as a *radical* attached to a *functional group*. Ethyl chloride, CH_3CH_2—Cl, behaves chemically like the combination of an ethyl radical, CH_3CH_2— or C_2H_5—, and a chloride group, —Cl. Many *replacement reactions* can occur, given the proper temperatures and catalysts:

$$C_2H_5{-}Cl + H_2O \rightarrow C_2H_5{-}OH + HCl$$
ethyl chloride   ethyl alcohol

$$C_2H_5{-}Cl + H_2S \rightarrow C_2H_5{-}SH + HCl$$
                              ethyl mercaptan

$$C_2H_5{-}Cl + NH_3 \rightarrow C_2H_5{-}NH_2 + HCl$$
                              ethylamine

$$C_2H_5{-}Cl + AgCN \rightarrow C_2H_5{-}CN + AgCl$$
                              ethyl cyanide

In subsequent reactions of these products, the ethyl group usually remains intact, while chemical activity takes place at the bond between the ethyl radical and the functional group.

Several common functional groups are listed in Table 12–3 and are shown in three-dimensional skeletal models in Figures 12–10 and 12–11. The *alcohols* are good solvents for organic materials, and the lower molecular weight alcohols are soluble in water. Methanol, or "wood alcohol," is a toxic alcohol that produces blindness and death when ingested. It attacks the nervous system by dissolving fatty material at nerve endings. The less toxic ethanol, or "grain alcohol," is the end product of energy extraction in

Table 12–3.   Hydrocarbon Derivatives and Functional Groups

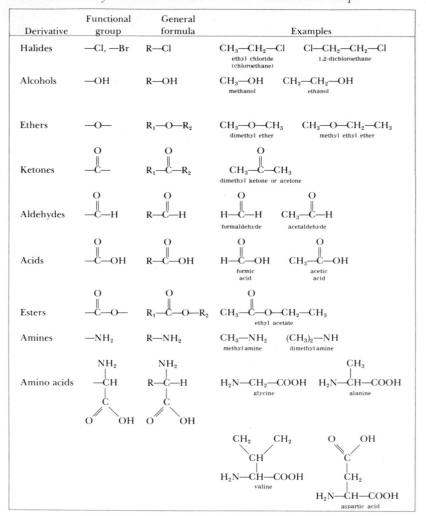

Derivative	Functional group	General formula	Examples

(Table contents)

Halides — —Cl, —Br — R—Cl — CH₃—CH₂—Cl (ethyl chloride / chloroethane)   Cl—CH₂—CH₂—Cl (1,2-dichloroethane)

Alcohols — —OH — R—OH — CH₃—OH (methanol)   CH₃—CH₂—OH (ethanol)

Ethers — —O— — R₁—O—R₂ — CH₃—O—CH₃ (dimethyl ether)   CH₃—O—CH₂—CH₃ (methyl ethyl ether)

Ketones — R₁—C(=O)—R₂ — CH₃—C(=O)—CH₃ (dimethyl ketone or acetone)

Aldehydes — R—C(=O)—H — H—C(=O)—H (formaldehyde)   CH₃—C(=O)—H (acetaldehyde)

Acids — R—C(=O)—OH — H—C(=O)—OH (formic acid)   CH₃—C(=O)—OH (acetic acid)

Esters — R₁—C(=O)—O—R₂ — CH₃—C(=O)—O—CH₂—CH₃ (ethyl acetate)

Amines — —NH₂ — R—NH₂ — CH₃—NH₂ (methyl amine)   (CH₃)₂—NH (dimethyl amine)

Amino acids — H₂N—CH₂—COOH (glycine)   H₂N—CH(CH₃)—COOH (alanine)   valine, aspartic acid

anaerobic (nonoxygen-using) organisms such as yeasts:

$$C_6H_{12}O_6 \xrightarrow[\text{enzymes}]{\text{yeast}} 2C_2H_5OH + 2CO_2$$

Methanol and ethanol are employed in vast quantities both as solvents and as raw materials for chemical syntheses. Methanol is synthesized commercially from carbon dioxide and hydrogen:

$$CO_2 + 3H_2 \rightarrow CH_3OH + H_2O$$

and ethanol is produced from ethylene:

$$CH_2{=}CH_2 + H_2O \rightarrow CH_3CH_2OH$$

Figure 12–10. Examples of hydrocarbon derivatives, showing typical functional groups. (a) Methyl alcohol, with the —OH group, (b) acetaldehyde (named as a derivative of acetic acid, CH_3COOH), with the —CHO aldehyde group, (c) dimethyl ether, with the —O— ether bridge, and (d) dimethyl ketone, or acetone, with the ketone linkage,

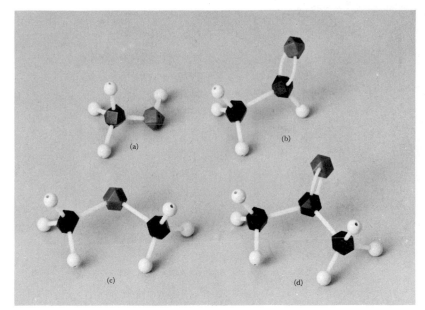

Figure 12–11. Organic acids and bases, and their derivatives. (a) Acetic acid, shown with its carboxyl group ionized. (b) Methyl acetate, with the characteristic

ester linkage. (c) Methylamine, with the amine —NH_2 group. The model shows the amine in its ionic form, —NH_3^+.

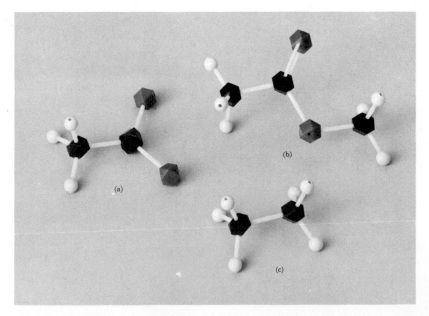

Table 12–4. Names of Hydrocarbon Derivatives[a]

R	Alcohols R—OH	Aldehydes R—CHO	Acids R—COOH	Esters CH_3—COO—R	Esters R—COO—CH_3
H—	Water	Formaldehyde (methanal)	Formic	Acetic acid	Methyl formate
CH_3—	Methyl (methanol)	Acetaldehyde (ethanal)	Acetic (ethanoic)	Methyl acetate	Methyl acetate (methyl ethanoate)
C_2H_5—	Ethyl (ethanol)	Propionaldehyde (propanal)	Propionic (propanoic)	Ethyl acetate	Methyl propionate (methyl propanoate)
C_3H_7—	Propyl (propanol)	Butyraldehyde (butanal)	Butyric (butanoic)	Propyl acetate	Methyl butyrate (methyl butanoate)
C_4H_9—	Butyl (butanol)	Valeraldehyde (pentanal)	Valeric (pentanoic)	Butyl acetate	Methyl valerate (methyl pentanoate)
C_5H_{11}—	Pentyl (pentanol)	Caproaldehyde (hexanal)	Caproic (hexanoic)	Pentyl acetate	Methyl caproate (methyl hexanoate)
C_6H_{13}—	Hexyl (hexanol)	Heptaldehyde (heptanal)	Heptanoic (heptanoic)	Hexyl acetate	Methyl heptylate (methyl heptanoate)
C_7H_{15}—	Heptyl (heptanol)	Octaldehyde (octanal)	Caprylic (octanoic)	Heptyl acetate	Methyl caprylate (methyl octanoate)
C_8H_{17}—	Octyl (octanol)	Pelargonic aldehyde (nonanal)	Pelargonic (nonanoic)	Octyl acetate	Methyl pelargonate (methyl nonanoate)
$C_{11}H_{23}$—	Undecyl (undecanol)	Lauric aldehyde (dodecanal)	Lauric	Undecyl acetate	Methyl laurate
$C_{15}H_{31}$—	Pentadecyl (pentadecanol)	Palmitic aldehyde (hexadecanal)	Palmitic	Pentadecyl acetate	Methyl palmitate
$C_{17}H_{35}$—	Heptadecyl (heptadecanol)	Stearic aldehyde (octadecanal)	Stearic	Heptadecyl acetate	Methyl stearate
$CH_3(CH_2)_7CH{=}CH(CH_2)_7$— (*cis*-isomer)			Oleic		Methyl oleate

[a] The problem of names in organic chemistry is formidable. There are two parallel systems: the common names, and systematic names agreed upon by the International Union of Pure and Applied Chemistry (IUPAC). Common names are generally shorter and more convenient, but are only labels. From the systematic name you usually can determine most of the molecule's structure. Systematic names are given in parentheses in this table.

Common names are based on two series: those of the alkanes and those of the acids. The alkane series begins with arbitrary names but quickly shifts to the Greek prefixes indicating the number of carbon atoms: methyl, ethyl, propyl, butyl, pentyl, hexyl, and so forth. Unfortunately, the acid series retains its nonnumerical names, which usually reflect the source of the material.

Note that the numerical prefixes for acids are one place out of step with the alcohols because the carbon atom of the carboxyl group is included in the counting. Thus, C_5H_{11}COOH is *hexa*noic acid and not pentanoic.

Aldehydes and the carbon-linked part of esters use the acid nomenclature. Alcohols, ethers, ketones, amines, and the oxygen-linked part of esters use the alkane nomenclature.

You should know the names through C_4, and should understand the principles of systematic nomenclature beyond this point.

(For the names of some alcohols and other hydrocarbon derivatives, see Table 12–4.)

Ethers are relatively volatile compounds obtained when alcohols condense in the presence of sulfuric acid to eliminate water:

$$CH_3CH_2—O\underset{\text{ethyl alcohol}}{\underline{|—H}} + \underset{\text{ethyl alcohol}}{\underline{H—O|}}—CH_2CH_3$$

$$\xrightarrow{H_2SO_4} \underset{\text{diethyl ether}}{CH_3CH_2—O—CH_2CH_3} + H_2O$$

Diethyl ether is the familiar ether used as an anesthetic. Ethers are valuable as solvents for waxes, fats, and other water-insoluble organic substances.

Aldehydes and *ketones* are the first step in the oxidation of alcohols:

$$\underset{\text{ethanol}}{CH_3CH_2OH} + \tfrac{1}{2}O_2 \rightarrow \underset{\text{acetaldehyde}}{CH_3—\overset{\overset{\textstyle O}{\|}}{C}—H} + H_2O$$

This reaction occurs at moderately high temperatures in the presence of a catalyst such as finely divided silver, or a mixture of powdered iron and molybdenum oxide. The second step in oxidation leads to a *carboxylic acid*, an acid with the carboxyl group,

$$—\overset{\overset{\textstyle O}{\|}}{C}—OH$$

For example,

$$\underset{\text{acetaldehyde}}{CH_3—\overset{\overset{\textstyle O}{\|}}{C}—H} + \tfrac{1}{2}O_2 \rightarrow \underset{\text{acetic acid}}{CH_3—\overset{\overset{\textstyle O}{\|}}{C}—OH}$$

Aldehydes and ketones are used as solvents and as raw materials for chemical syntheses. Formaldehyde,

$$H—\overset{\overset{\textstyle O}{\|}}{C}—H$$

is the starting point for phenyl–formaldehyde resins such as Bakelite. Acetone,

$$CH_3—\overset{\overset{\textstyle O}{\|}}{C}—CH_3$$

is one of the most common laboratory solvents.

The carboxylic acids are relatively weak acids; they dissociate to only a limited extent in aqueous solution. When the carboxyl group does dissociate, the negative charge is spread over both oxygen atoms. The three p orbitals on the two oxygen atoms and the carbon atom connecting them are combined into one delocalized molecular orbital:

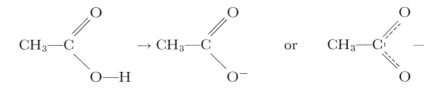

Both carbon–oxygen bonds in the ionized carboxyl group have the *same length*. The negative charge is spread over all three atoms. (The middle structure in the preceding equation can be considered as one of the two resonance structures contributing to the true carboxyl ion structure. What would the other resonance structure look like?) With metal hydroxides and carbonates, the carboxylic acids react as any other acid would to make salts:

$$C_2H_5COOH + NaOH \rightarrow C_2H_5COONa + H_2O$$

propionic acid   sodium propionate

Sodium propionate is dissociated in aqueous solution and is obtained as a salt only on drying.

Formic acid, $HCOOH$, is the main irritant in insect stings. Acetic acid, CH_3COOH, is the acid in vinegar. The acids from butyric (C_4) to heptanoic (C_7) have acrid odors that are encountered in rancid butter and strong cheese.

Esters are made by allowing acids and alcohols to react:

acetic acid   *n*-butanol   butyl acetate

Esters are not ionized and are volatile liquids with pleasing, fruity odors. Butyl acetate gives bananas their odor and therefore is called banana oil. Ethyl butyrate, $C_3H_7COOC_2H_5$, has the odor of pineapples, and octyl acetate, $CH_3COOC_8H_{17}$, the odor of oranges. Oils such as linseed, cottonseed, and olive oil, and fats such as butter, lard, and tallow, are esters of the trihydroxyl alcohol glycerol,

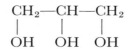

with large molecular weight acids such as palmitic, $C_{15}H_{31}COOH$; stearic, $C_{17}H_{35}COOH$; and oleic, $C_{17}H_{33}COOH$.

Soluble soaps are the alkali metal salts of these fatty acids, obtained by treating animal fats with alkali metal hydroxides, especially NaOH:

$$(C_{17}H_{35}COO)_3C_3H_5 + 3NaOH \rightarrow 3C_{17}H_{35}COONa + C_3H_5(OH)_3$$

glyceryl stearate   sodium stearate   glycerol
(from animal fat)   (a soap)

In aqueous solution, a soap molecule has a hydrocarbon end and a charged end. Soaps "lift" dirt into solution by surrounding a small amount of grease with many molecules; all their hydrocarbon tails point in toward the grease and their carboxyl groups point out. The soap molecules thus "package" the grease in droplets or *micelles* that can be taken up into the solution and washed away.

The most common organic bases are called *amines* and can be thought of as derivatives of ammonia:

$$CH_3—NH_2 \qquad CH_3CH_2—NH_2 \qquad CH_3—NH—CH_3$$

methylamine   ethylamine   dimethylamine

$$\begin{array}{c} CH_3 \\ | \\ CH_3—N—CH_3 \end{array} \qquad CH_3—NH—C_2H_5$$

trimethylamine   methylethylamine

$$H_2N—CH_2—CH_2—NH_2 \qquad HO—NH_2$$

ethylenediamine   hydroxylamine

They are called primary, secondary, or tertiary amines, depending on how many of the hydrogen atoms of NH_3 are replaced by organic radicals. These organic bases are about as strong as ammonia, and add a proton to produce the ionic form:

$$CH_3CH_2—NH_2 + H^+ \rightarrow CH_3CH_2—NH_3{}^+$$

Methylamine is shown in Figure 12–11(c) in its ionic form.

The amines as a group have fishy odors and are generally toxic. Triethylamine in moderate concentrations has a choking odor of rotting fish. In toxic concentrations, the olfactory receptors are saturated, and only the ammonia smell is sensed.

The *amino acids* are an important combination of carboxylic acid and amine in one molecule. They have the general formula

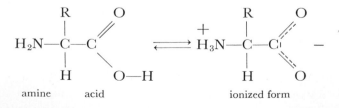

amine   acid   ionized form

Figure 12–12. The two optical isomers of an amino acid, showing the side chain branching in opposite directions from the central carbon atom, which is called the α carbon. (a) L-Alanine, and (b) D-alanine. Both the carboxyl group and the amine group are shown in their ionized form

$$CH_3$$
$$|$$
$$^+H_3N-CH-COO^-$$

The proteins of all living organisms are built from only the L-amino acids. Their mirror images, the D-amino acids, are found in small amounts in bacterial cell walls and in antibiotics produced by some micro-organisms. One of the problems in explaining the evolution of life is accounting for this asymmetry of the components of living organisms. Any carbon atom that has four different atoms bonded to it will have two different configurations possible, which are mirror images of one another. Such a carbon atom is called an asymmetric carbon.

in which —R indicates the side group that gives each amino acid its identity and chemical properties (Figure 12–12). The carbon from which the side group branches off is called the α carbon. As Figure 12–12 shows, the α carbon is an asymmetric carbon, and there are two optical isomers or enantio-morphs of an amino acid.

In solution, both the amine end and the carboxyl end of an amino acid are ionized, and the charged molecule is known as a *zwitterion*. All proteins

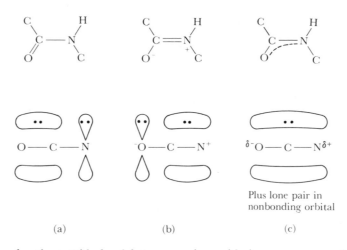

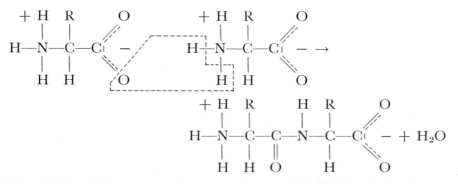

Figure 12–13. A structure for the peptide bond between amino acids in proteins can be drawn in which a double bond connects C and O, and the peptide bond itself is a single C—N bond (a). Another structure can be drawn with a single bond from C to O and a double peptide bond (b). This structure places charges on O and N, and is therefore less favorable. The true situation can be represented by combining p atomic orbitals from O, C, and N to make bonding, nonbonding, and antibonding delocalized molecular orbitals. The delocalized bonding orbital extends over all three atoms (c), and is therefore more stable. The nonbonding orbital is not shown. The extra stability of the delocalized electrons more than compensates for the slight charge separation at O and N. The partial double bond character of the C—N peptide bond prevents rotation around the bond and keeps the peptide unit planar.

are built from polymers of amino acids in which water is removed and a *peptide bond* is formed:

Once the peptide bond is made, electrons in the C=O double bond become delocalized and the peptide C—N bond acquires a partial double bond character. The peptide unit (Figure 12–13) is thus forced to remain planar.

This unit is the cornerstone of all protein structure and is one of the most important examples of delocalization of π bonding in chemical systems.

Before you go on. If you need help to learn the names and structures of simple organic compounds, study Review 12 of *Programed Reviews of Chemical Principles* by Lassila *et al.* Section 12–1 deals with hydrocarbons, and Section 12–2 is on functional groups.

12–4   UNSATURATED HYDROCARBONS

The third reaction of hydrocarbons mentioned in Section 12–2, in addition to oxidation and halogenation, is *dehydrogenation:*

$$CH_3\text{—}CH_2\text{—}CH_3 \xrightarrow{\text{heat}} CH_2\text{=}CH\text{—}CH_3 + H_2$$

propane   propylene

In the cracking process for petroleum, heat and catalysts break long-chain hydrocarbons into saturated hydrocarbons in the gasoline range, and unsaturated alkenes such as propylene, ethylene, and butadiene. Double bonds also can be produced by removing HCl from alkyl halides with KOH in alcohol, or by removing H_2O from alcohols with acid:

$$CH_3\text{—}CH_2\text{—}Cl + KOH \xrightarrow{\text{alcohol}} CH_2\text{=}CH_2 + KCl + H_2O$$

ethyl chloride   ethylene

$$CH_3\text{—}CH_2\text{—}OH \xrightarrow{\text{acid}} CH_2\text{=}CH_2 + H_2O$$

Triple bonds also can be formed, as in acetylene, $HC\equiv CH$, but these are not as important or as widespread as double bonds. By analogy with the alkanes, compounds with double bonds are called *alkenes*, and those with triple bonds are *alkynes* in the systematic International Union of Pure and Applied Chemistry (IUPAC) nomenclature. The systematic names of ethane, ethylene, and acetylene are ethane, ethene, and ethyne. Perhaps in a generation, ethylene and acetylene will go the way of carburetted hydrogen and olefiant gas, but at present the nonsystematic nomenclature is used more often for these common chemicals.

Because of the double bond, there is restricted rotation around the central bond, and *geometrical isomers* result. Thus, $CH_3CH\text{=}CHCH_3$, 2-butene, can exist as two isomers:

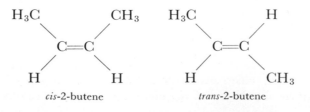

cis-2-butene   trans-2-butene

Because of the double bond, the two central C atoms, and the C and H attached directly to them, lie in a plane. As with the isomers of coordination complexes, the prefix "cis" indicates adjacent positioning of similar groups, and "trans" means "across" or, at least, not adjacent. The *trans*-2-butene molecule is slightly more stable than the cis form because its bulky methyl groups are farther apart. We shall find that *steric hindrance*, or the bumping of bulky groups, plays a significant role in determining the structures of organic and biological molecules.

In the longer paraffins, dehydrogenation leads to a mixture of several products with the double bond in different places. The straight-chain isomer of butane, *n*-butane, can lead to two *structural isomers* of butene with one double bond, and two isomers of butadiene with two double bonds:

$CH_2\!\!=\!\!CH\!-\!CH_2\!-\!CH_3$ 1-butene

$CH_3\!-\!CH\!\!=\!\!CH\!-\!CH_3$ 2-butene

$CH_2\!\!=\!\!CH\!-\!CH\!\!=\!\!CH_2$ 1,3-butadiene

$CH_2\!\!=\!\!C\!\!=\!\!CH\!-\!CH_3$ 1,2-butadiene

The numbers 1, 2, and 3 locate the positions of the double bonds.

Addition reactions can occur at the double bonds with H_2, HCl, or Cl_2. For example,

$$\underset{\text{1-butene}}{CH_2\!\!=\!\!CH\!-\!CH_2\!-\!CH_3} + Cl_2 \rightarrow \underset{\text{1,2-dichlorobutane}}{\overset{\displaystyle Cl \quad\; Cl}{\overset{\displaystyle |\quad\;\; |}{CH_2\!-\!CH\!-\!CH_2\!-\!CH_3}}}$$

The corresponding 1,3-butadiene addition reaction is peculiar; the addition takes place at the extreme ends of the two double bonds, in what appears to be a simultaneous (concerted) process. One double bond disappears in the reaction, and the other moves to the center of the molecule:

$$\underset{\text{1,3-butadiene}}{CH_2\!\!=\!\!CH\!-\!CH\!\!=\!\!CH_2} + Cl_2 \rightarrow \underset{\text{1,4-dichloro-2-butene}}{\overset{\displaystyle Cl \qquad\qquad\quad Cl}{\overset{\displaystyle |\qquad\qquad\quad\; |}{CH_2\!-\!CH\!\!=\!\!CH\!-\!CH_2}}}$$

This unusual behavior occurs because the double bonds in the 1,3-butadiene molecule are delocalized. Such an alternating arrangement of double and single bonds ($-C\!\!=\!\!C\!-\!C\!\!=\!\!C-$) is called a *conjugated* system. When such conjugated double bonds occur in flat, closed rings with all atoms in a plane, we call the compounds aromatic.

Challenge. To see whether you can recognize and name hydrocarbons and their derivatives used in refrigerants and cough drops, try Problems 12–1 and 12–3 in *Relevant Problems for Chemical Principles* by Butler and Grosser.

12–5   AROMATIC COMPOUNDS

Aromatic compounds are ring compounds with delocalized electrons. The simplest of these compounds is benzene, C_6H_6. The delocalized electrons give aromatic compounds the special properties that differentiate them from *aliphatic* compounds such as we have been examining so far. The benzene ring commonly is written as one of the Kekulé structures,

although a better representation of delocalization is

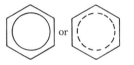

The delocalization can extend over more than one adjacent ring as in naphthalene:

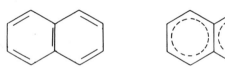

and anthracene:

Coronene has seven adjoining rings:

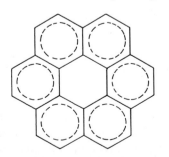

The ultimate limit of this process is graphite, with its sheets of hexagonal rings and delocalization over the entire sheet (Figure 6–10). Because of these delocalized electrons, graphite is a good conductor of electricity, whereas diamond is not. In a sense, graphite is a "two-dimensional metal" whose electron mobility is restricted to the individual stacked sheets.

Benzene is surprisingly unreactive in comparison to alkenes such as butene. In its lack of reactivity it is more like the saturated alkanes. It does not undergo addition reactions at a double bond; if it did so, it would reduce the extent of the delocalization of electrons (Section 10–10). Because of this delocalization, benzene is 40 kcal mole^{-1} more stable than would have been expected of a compound with three single and three double bonds (Section 15–4). In general, the larger the region in a molecule over which delocalization occurs, the more stable the molecule is.

Instead of addition, the typical reaction of aromatic rings is *substitution*:

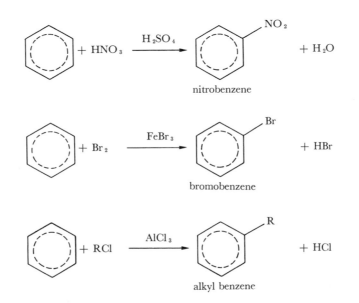

In the first of these reactions:

nitrobenzene

bromobenzene

alkyl benzene

(In writing structures of this sort, it is common practice to omit the H's bonded to the ring carbons; each apex of a hexagonal ring represents a C to which one H is attached.) In the first of these reactions, sulfuric acid aids the reaction by converting HNO_3 to NO_2^+, the species that attacks the benzene ring. It also acts as a dehydrating agent to remove the water formed as a product. The compounds $FeBr_3$ and $AlCl_3$ are *catalysts*. To see why they are necessary we must look at the mechanism of the reaction. Aromatic rings are particularly susceptible to attack by *electrophilic groups*, or Lewis acids, which have a strong affinity for electron pairs. In the bromination reaction, Br_2 is not electrophilic; in the absence of the $FeBr_3$ catalyst, no reaction takes place within a reasonable time. However, $FeBr_3$ itself has an attraction for another Br^- ion with its electron pair and will tear a Br_2 molecule into Br^-

and Br^+ ions:

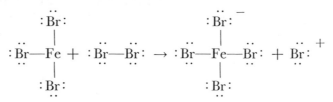

The electrophilic Br^+ then attacks the aromatic ring and attracts an electron pair to make a C—Br bond. This intermediate compound is unstable and dissociates either by ejecting the Br^+ to make the starting material again, or by ejecting a hydrogen ion to make the end product, bromobenzene:

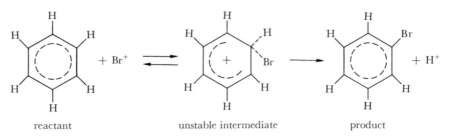

reactant   unstable intermediate   product

The liberated H^+ reacts with the $FeBr_4^-$ to produce HBr and the original $FeBr_3$. The NO_2^+ in the nitrobenzene reaction also is an electrophilic group, and that reaction proceeds by a similar mechanism.

Two important aspects of chemical reactions are illustrated by this mechanism: the lack of completeness of most reactions, and the use of catalysts. Not every molecule of unstable intermediate that decomposes produces bromobenzene; many molecules break down to yield the reactant again. The result of most syntheses is a mixture in which the desired end product is one component (hopefully a major one) of a range of possible products. One of the challenges of chemical synthesis is to devise procedures and synthetic pathways that maximize the yield of the desired product. Often the long way around is better than an obvious one-step synthesis because the more involved synthesis produces essentially a single product.

A *catalyst* is a substance that accelerates a chemical reaction by providing an easier pathway, without itself being used up in the reaction (Section 18–5). This does not mean that it is uninvolved. The $FeBr_3$ plays an important part in the stepwise mechanism previously outlined. But at the end of the reaction the $FeBr_3$ is regenerated in its original form. This is the general and defining behavior of a catalyst. A mixture of H_2 and O_2 can remain for years at room temperature without appreciable reaction, but the introduction of a small amount of platinum black causes an instant explosion. It has the same effect on butane gas or alcohol vapor and O_2. (Cigarette lighters with platinum black instead of a wheel and flint once were manufactured, but they soon ceased to operate because of the poisoning of the catalytic surface by impurities in the butane gas.) Platinum black acts as a catalyst by aiding the dissociation of diatomic gas molecules adsorbed on its surface. These dissociated atoms (e.g., H or O) are much more reactive than the original molecules in the gas phase. A catalyst does not affect the *overall* energy of a reaction or enter irreversibly into the reaction. It only provides an easier mechanism or pathway that makes the reaction go more rapidly.

Many catalysts, but not all, are surface-acting agents like platinum black. The substances catalyzed, called the *substrates*, bind to the surface of the catalyst. If chemical groups on the surface of the catalyst weaken a bond in a substrate, the cleavage of the substrate becomes easier. This is what happens with platinum catalysis. The finely divided black powder of Pt is a more efficient catalyst than a block of Pt only because it has much more surface exposed.

Example.   One sample of Pt is cast into a sphere with 1-cm^3 volume, and a second sample of the same amount of Pt is precipitated from solution as tiny spheres of 1000-Å diameter. What is the ratio of the total surface area in the two samples?

Solution.   The radius of the large sphere is determined by

$$V = \frac{4}{3} \pi R^3 = 1 \text{ cm}^3$$

from which $R = 0.621$ cm. Therefore, the surface area of the large sphere is

$$A = 4\pi R^2 = 4.85 \text{ cm}^2$$

The volume of a small sphere is

$$v = \frac{4}{3} \pi (10^{-5} \text{ cm})^3 = 4.18 \times 10^{-15} \text{ cm}^3$$

The number of such small spheres, N, is the ratio of volumes:

$$N = \frac{V}{v} = \frac{1 \text{ cm}^3}{4.18 \times 10^{-15} \text{ cm}^3} = 2.39 \times 10^{14}$$

The area of one small sphere is

$$a = 4\pi (10^{-5} \text{ cm})^2 = 1.256 \times 10^{-9} \text{ cm}^2$$

Hence, the total area of all small spheres is

$$a_T = 2.39 \times 10^{14} \times 1.256 \times 10^{-9} \text{ cm}^2 = 301,000 \text{ cm}^2$$

The ratio of areas in the finely divided and single-sphere samples of Pt is

$$\frac{a_T}{A} = \frac{301,000 \text{ cm}^2}{4.85 \text{ cm}^2} = 62,000$$

The advantages of platinum black over a sphere of solid Pt in surface catalysis are obvious.

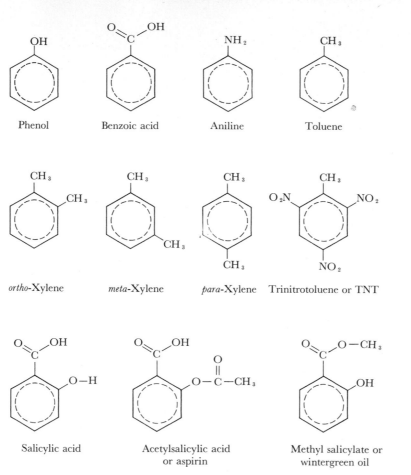

Figure 12–14. *Some representative derivatives of benzene. Salicylic acid can form an ester in two ways: by using either its acid group in methyl salicylate or its hydroxyl group in aspirin. Unlike alcohols, phenols are acids, although they are usually far weaker acids than the carboxylic acids. Aromatic amines such as aniline are weaker bases than aliphatic amines. Ortho-, meta-, and para- (frequently abbreviated o-, m-, and p-) denote relative positions of groups attached to the benzene ring, as for the three isomers of xylene.*

The compound $AlCl_3$ plays a catalytic role in the alkylation reaction to produce benzene derivatives with alkyl side chains. An important class of biological catalysts are the protein molecules called *enzymes*. These molecules have regions on their surface, called *active sites*, where catalysis occurs. Transition metals frequently are bound to the enzymes at their active sites, and are essential participants in catalysis. We shall look at an example of enzymatic catalysis in Section 12–11.

Several of the derivatives of benzene are shown in Figure 12–14. Phenol is weakly acidic, unlike the alcohols of which it appears to be an aromatic analogue. This ability of phenol and its derivatives to lose the hydroxyl proton arises because electrons of the oxygen become partially involved in the delocalization. The bond from ring to oxygen attains a partial double bond character, and hydrogen, robbed of some of its bonding electron pair, dissociates easily. However, the acidity of the phenols is generally less than that of the carboxylic acids.

For the same reason, aniline is a weaker base than ammonia or the aliphatic amines. The nitrogen lone electron pair that would have attracted a proton is partially involved with the aromatic ring and is less able to attract the proton and to ionize the molecule.

12–6 AROMATIC COMPOUNDS AND THE ABSORPTION OF LIGHT

Aromatic ring compounds with delocalized electrons, like transition-metal complexes with d orbitals, frequently have energy levels close enough together to absorb visible light. Hence, these two classes of compounds often are brightly colored. When a photon of energy is absorbed, one electron in a π^b bonding orbital (Figure 10–35) is promoted to the lowest π^* antibonding molecular orbital. Therefore, this absorption is called a $\pi \rightarrow \pi^*$ transition. In benzene and in naphthalene the levels are too far apart for the absorption to be in the visible spectrum, and these compounds are colorless. But if two nitro groups are added to make 1,3-dinitronaphthalene, the electronic-level spacing falls below 25,000 cm^{-1} and the compound appears pale yellow. (Table 11–3 with its spectral and complementary colors will be helpful throughout this section.) This phenomenon happens because the delocalized electron system has been enlarged to include the two nitro groups, and the energy levels (and the spacings between them) have fallen accordingly. The effect is continued in Martius Yellow, a common dye for wool and silk. An added hydroxyl group enlarges the conjugated system even more, and the energy of a $\pi \rightarrow \pi^*$ transition decreases. The color of the compound changes to yellow-orange. The light actually absorbed by the three compounds is ultraviolet for naphthalene, violet for 1,3-dinitronaphthalene, and blue for Martius Yellow.

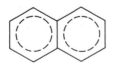

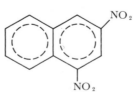

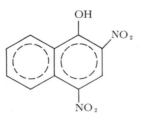

naphthalene   1,3-dinitronaphthalene   Martius Yellow, or
                                                     2,4-dinitro-1-naphthol

Many substances such as phenolphthalein, methyl orange, and litmus have different colors in acidic or basic solutions; hence, they are useful *acid–base indicators* (Figure 5–7). *p*-Nitrophenol is a poor indicator because its color change is not vivid, but it is a simple molecule for showing what happens when an indicator changes color. Since phenols are weak acids, the reaction in solution is

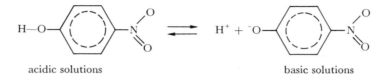

acidic solutions   basic solutions

p-Nitrophenol in basic solutions is a deep yellow. Its maximum absorption occurs at a wavelength of 4000 Å. In the basic form, the oxygen atom and the nitro group can combine with the aromatic ring to make one large delocalized system:

Such an extended delocalized structure is called a quinone structure, after the yellow benzoquinone:

In the acidic form of *p*-nitrophenol, the negative charge on the oxygen atom is lacking. It is not as easy to involve the oxygen lone electron-pairs in delocalization; therefore, the energy level of the first excited electronic state is not lowered as much. Absorption occurs with a maximum just inside the ultraviolet, at 3200 Å, and the compound appears a pale greenish-yellow. Phenolphthalein, which is colorless in acid and pink in base, is a more complicated molecule that works by exactly the same principle.

A particularly good way of expanding a delocalized system is illustrated by the azo dyes, which have two aromatic rings bridged by the —N=N— group. Methyl orange, another acid–base indicator (Figure 5–6), is an azo dye:

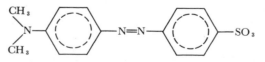

It is red in acid and yellow in base. (In which conditions are its electronic energy levels more widely spaced? Can you figure out why?)

An extremely important example of delocalization and energy absorption is chlorophyll, which was discussed in Section 11–7. The aromatic ring surrounding the Mg^{2+} ion is an extended delocalized system derived from porphyrin (Figure 11–21). The electronic energy levels are such that one absorption occurs in the violet, at 4300 Å, and a second in the red, at 6900 Å (Figure 11–22). When light is absorbed by chlorophyll molecules, the energy excites an electron to a higher energy level, thereby enabling it to reduce the Fe^{3+} ions in *ferredoxin*, which is a protein of molecular weight 13,000 that has two iron atoms coordinated to sulfur. The reoxidation of ferredoxin supplies the energy to drive other reactions that eventually lead to the splitting of water, the reduction of carbon dioxide, and ultimately the synthesis of glucose, $C_6H_{12}O_6$.

12–7   CARBOHYDRATES

The sugar glucose, produced in the leaves of green plants, is a *carbohydrate*. The name "carbohydrate" comes from an early misconception about the structures of these compounds. The formula for glucose, $C_6H_{12}O_6$, can be written as $(C \cdot H_2O)_6$. Substances whose formulas could be represented by equal amounts of carbon and water were called carbohydrates.

The glucose molecule is polymerized in chains of thousands of monomer units in plants to make cellulose, and in a slightly different way to make starch. A close relative of glucose, *N*-acetylglucosamine (NAG), is polymerized to form chitin, the material from which the shells of insects are made. NAG and a close variant, *N*-acetylmuramic acid (NAM), are copolymerized in alternating sequence in chains that make up part of the walls of bacterial cells. Glucose is decomposed in a stepwise fashion to produce the energy that a living organism requires. Excess glucose is carried in the bloodstream to the liver and is converted into the animal starch glycogen, which is reconverted to glucose when needed. Glucose, cellulose, starch, and glycogen all are carbohydrates.

Carbohydrates, in the form of starch, are the primary sources of energy from foods. To obtain this energy, we either eat the grains in which the starch is stored, or feed the grains to animals and let them synthesize meat protein before we eat *them*. In either case, the energy that we obtain ultimately originates from starch, the polymerized product of photosynthesis. We encounter cellulose fibers in cotton and linen, and in the artificial products cellulose acetate and rayon. The shelter over our heads probably is cellulose in the form of wood. This book is a processed cellulose called paper. Even our money, having ceased to be made from noble metals, is well on its way to becoming notarized cellulose. In this section we shall look very briefly at what carbohydrates are and how they are used.

The most fundamental unit of a carbohydrate is a *monosaccharide*, or simple sugar. Such sugars can have three, four, five, or six carbon atoms, in which case they are called *trioses*, *tetroses*, *pentoses*, or *hexoses*. We shall look only at

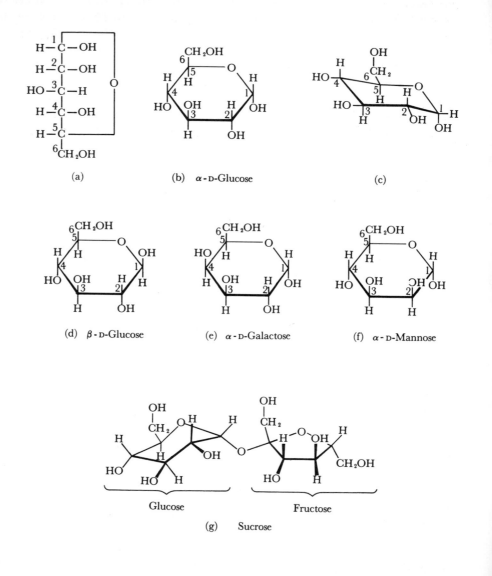

Figure 12–15. (a) α-D-Glucose in the Fischer representation, (b) in the flat hexagon diagram, and (c) in a form that most closely represents its actual shape. (d) β-D-Glucose, which differs from the α form only at carbon 1. (e) α-D-Galactose, which differs from glucose at carbon 4. β-D-Galactose is produced by exchanging —H and —OH at carbon 1. (f) α-D-Mannose, which differs from α-D-glucose only at carbon 2. (g) Sucrose, a dimer of α-D-glucose and β-D-fructose.

hexoses, and especially at the most common one, D-glucose. The structure of D-glucose is depicted in Figure 12–15(a) through (c). Figure 12–15(a) shows the numbering of the six carbon atoms, and the Fischer convention of writing formulas to indicate the structure around an asymmetric carbon.

An asymmetric carbon atom is one that is bonded to four different groups, as are carbon atoms 1 through 5 in glucose. As we saw for the α carbon of an amino acid, each such asymmetric carbon atom has two different arrangements of the four groups, which are related by a mirror reflection. With five asymmetric carbon atoms, and two different configurations around each, there are a total of $2^5 = 32$ different isomers of the hexose sugars.

By the Fischer convention, the bonds to the right and left in Figure 12–15(a) lead from the central atom to atoms that lie above the plane of the page. Bonds extending up or down from the central atom go to atoms below the plane of the diagram. A change in configuration at any asymmetric carbon atom in the hexose is produced by exchanging the —H and —OH groups right for left in the Fischer diagram. This asymmetry is easier to see in the flat hexagon representation of the same molecule in Figure 12–15(b). The actual shape of the molecule, with its tetrahedral geometry at the carbon atoms, is depicted more accurately by Figure 12–15(c). Glucose has the chair conformation, which we first saw with cyclohexane, rather than the boat form.

In the 32 isomers of hexose that arise from the 32 possible interchanges of arrangement at carbon atoms 1 through 5, the positions of —H and —OH at carbon atom 1 are indicated by the prefixes α and β. In α-hexoses the hydroxyl group points down as in Figure 12–15(b) or (c); in β-hexoses it points up as in Figure 12–15(d). A complete mirror reversal of a hexose at all five asymmetric carbon atoms simultaneously produces an L-hexose from a D-hexose. Therefore, for each type of hexose there are four variants: α-D, α-L, β-D, and β-L. There must be $32/4 = 8$ different types of hexose. However, only three of these occur naturally: glucose, galactose, and mannose. These three sugars differ at carbon atoms 2 and 4, and are compared in Figures 12–15(d), (e), and (f). Galactose occurs in the milk sugar lactose, and mannose is a plant product (named for the Biblical *manna*). However, the most common hexose by far is glucose.

Of the hexose sugars with a five-membered ring, the most common is *fructose*. Fructose occurs naturally in honey and fruit (hence its name), and is combined with glucose as the common table sugar *sucrose* [Figure 12–15(g)].

Challenge. Isomerization in organic compounds is crucial to the sex life of butterflies, as well as in malaria treatment, LSD, antibiotics, and rat control. To see how, try Problems 12–6, 12–11, and 12–13 in the Butler and Grosser book.

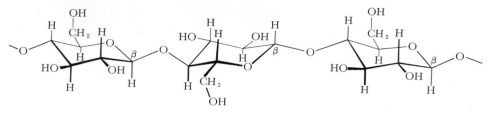

(a)   Cellulose

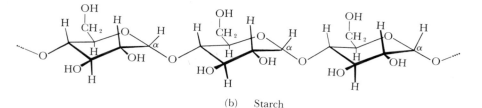

(b)   Starch

Figure 12–16. (a) Cellulose, a polymer of β-D-glucose. (b) Starch, a polymer of α-D-glucose.

Polysaccharides

Cellulose is the structural fiber in trees and plants. It is used as wood, cotton, and linen, and, in a modified form, in paper. Cellulose is a polymer of β-D-glucose, with a typical chain length of about 3000 monomer units. The connection from one β-glucose to another, shown in Figure 12–16(a), is called a β-glucoside link.

The hydroxyl groups of glucose can form esters; the treatment of cellulose with acetic anhydride, acetic acid, and a small amount of sulfuric acid produces the derivative cellulose acetate. The chains are broken to a length of 200–300 monomers, and an average of two acetate groups attach to each monomer. Cellulose acetate is the material for photographic film backing: it also is dissolved in acetone and extruded through fine holes in a metal cup to form threads of rayon.

Cellulose is not a source of food. With the exception of termites, and ruminants such as cows, which carry cellulose-digesting microorganisms in their stomachs, animals are incapable of breaking the β-glucosidic bond. The cleavage is an enzymatically catalyzed process, and we lack the enzymes. In 1967, a process was developed for degrading cellulose to produce an artificial flour that, although usable in baking like starch flour, had no nutritive value. It was touted briefly as a dieting aid, but rapidly sank into obscurity. (*Life* magazine referred to it as "non-food," and suggested that its inventor be paid in non-money.) However, it has been suggested quite seriously that if man could in some way learn to live in a symbiotic relationship with cellulose-digesting microorganisms in his intestines as ruminants do, his food problems would be resolved for many centuries.[1]

Starch is also a polymer of glucose, but with the α linkage of Figure 12–16(b). Starch is the standard storage medium for glucose to be used as a food supply in plants, and is our chief source of trapped solar energy. It is stored in plant stalks, leaves, roots, seeds, and grains. All organisms possess the enzymes necessary to digest starch. The first step in fermentation, whether it takes place in the stomach or in the brewer's vat, is the breakdown of starch to glucose. A piece of bread held in your mouth eventually will taste sweet because the enzymes in the saliva can digest bread starch to sugar.

Polymers of hexose derivatives are the structural materials in insect shells (chitin) and bacterial cell walls. In insect chitin, a hexose derivative called *N*-acetylglucosamine is polymerized without cross-linking. One layer in the walls of bacteria is a polymer of hexose derivatives, cross-linked with short chains of four amino acids for strength. We and all other higher organisms have evolved an enzyme, lysozyme, to protect us by lysing or dissolving this polysaccharide wall structure in invading bacteria. Lysozyme is found in most external secretions such as sweat and tears. One of the few places where D-amino acids exist in nature is in the walls of certain bacteria. One view is that they have been placed there by the bacteria simply to keep them out of the way, but it also has been speculated that they might have evolved in the structure of the wall as a defensive measure against attack by enzymes (not lysozymes, which attack the β-glucoside link) that operate most effectively against the common L-amino acids.

---

[1] This is one of those superficially attractive suggestions with possibly disastrous social consequences. It is an illustration of the unpleasant fact that a blind application of science and technology to isolated problems often creates more problems than it solves. What would be the most likely consequences if cellulose suddenly became an apparently unlimited source of cheap food? We can list a few of the results.

1) A rapid diminution of interest in the crisis of overpopulation, and an upsurge in the total population of the planet.

2) Great changes in standards and modes of living in the face of the population increases made possible by "unlimited" cheap food.

3) Deforestation of large areas of the world where people are starving. This deforestation would lead to flooding, erosion of topsoil, crop failures of conventional foods, and probable starvation again.

The introduction of edible cellulose, *by itself*, would be as short-sighted and disastrous a step as finding an efficient way of diffusing the smog of Los Angeles over the farms of Iowa. It would not avoid the eventual crisis; it merely would delay its arrival. Our planet is so carefully structured that almost any major technological or scientific change introduced without thinking is likely to lead to trouble. Who could have convinced Henry Ford that his mass transportation machines would eventually become a curse on the landscape?

12–8 ENERGY AND METABOLISM IN LIVING SYSTEMS

To mountaineers who take their hobby seriously, a particularly challenging operation is a "dynamic traverse." This is a traverse across a difficult piece of terrain where, at each instant, the climber is in an unstable situation, and where he is prevented from a disastrous fall only by his momentum. In a sense, every living organism is engaged continually in a dynamic traverse. One of the most important generalizations in science, the second law of thermodynamics, states that in any process taking place in a closed system (i.e., the object studied plus its entire environment with which it exchanges matter or energy), the disorder of the system as a whole *increases*. We shall look more closely at this principle in Chapter 15. A living organism is an intricate chemical machine, evolved to a high level of complexity. It is faced constantly with the dilemma that every chemical reaction that takes place inside it increases the disorder and reduces its complexity. A constant supply of energy is needed from outside sources, not only to provide the power to do physical work, but also to keep down the level of disorder, or *entropy*, within. If this supply of energy fails, then death and the breakdown of the chemical machine is only a matter of time: a day for the fast-living shrew, a few weeks for man. Just as momentum saves the climber from falling, so a constant influx of energy keeps the living machine from collapsing. The degradation of high-energy fuels and the extraction of their energy is called *metabolism*. In this section we shall trace briefly the outlines of the metabolism that is common to all oxygen-using living organisms. Since glucose is so important a metabolite, we shall use it as an illustration.

The Combustion of Glucose

If 180.16 g of glucose, or 1 mole, are burned, 673 kcal of heat are liberated. Such an uncontrolled combustion is wasteful; only a small fraction, if any, of the energy stored in glucose is put to good use. It is more efficient to feed glucose to horses and use them to pull a load than it is to burn glucose and operate a locomotive with it. This is because in the horse's metabolism glucose is broken down in a series of small steps. The energy released at each step is stored in the chemical bonds of a special molecule, adenosine triphosphate (ATP), and is available for use in other chemical reactions that make muscles do work. Combustion in the horse is controlled and efficient; combustion in the locomotive is uncontrolled and wasteful.

We could study chemical energy by looking at heats of reaction or changes in enthalpy. However, it is more informative to look at these heats after correcting for the change in order or disorder produced by the reaction. This corrected energy is the total energy that is free and available to do chemical work. Therefore, it is called the *free energy*, G, and the change in free energy in a reaction is represented by ΔG. We shall see in Chapter 15 how to obtain free energy changes from the changes in enthalpy and the increase or decrease in disorder in a process. For the moment, we want only to use free energy as a bookkeeping device for following energy in metabolic processes. The standard free energy change in the combustion of glucose,

$$C_6H_{12}O_6 + 6O_2 \rightarrow 6H_2O + 6CO_2$$

is $\Delta G^0 = -686$ kcal mole^{-1}. (For comparison purposes, the standard enthalpy change or the heat of reaction is $\Delta H^0 = -673$ kcal mole^{-1}.) In the horse, 44% of the liberated free energy is saved by using it to synthesize 38 molecules of ATP for every molecule of decomposed glucose. It is this orderly breakdown process that we shall examine.

The Three-Step Process in Metabolic Oxidation

There are three parts to the combustion process in living, oxygen-using organisms. In the first part, all foods, no matter what their chemical nature, are degraded to pyruvic acid,

$$\begin{array}{cc} O & O \\ \| & \| \\ CH_3\text{---}C\text{---}C\text{---}OH \end{array}$$

Not much energy is obtained from this process. Its main purpose is to reduce everything to a common set of chemicals and prepare for the real energy-producing steps. In the second part, called the citric acid cycle, pyruvate is oxidized to CO_2, and the hydrogen atoms from pyruvate reduce two important carrier molecules that we shall examine shortly, nicotinamide adenine dinucleotide (NAD) and flavin adenine dinucleotide (FAD). Again, only a small amount of free energy is stored in ATP in this cycle. Its principal purpose is to break up the 273 kcal mole^{-1} of free energy in pyruvate into four smaller and more easily handled packages, of around 50 kcal mole^{-1}, in the form of four moles of reduced carrier molecule. The third part of the process, the respiratory chain, accepts these reduced carrier molecules. It reoxidizes them, uses the hydrogen atoms obtained during the oxidation to reduce O_2 to water, and uses the free energy obtained to synthesize ATP.

We can see two objectives to this three-part machinery: to reduce the thousands of different possible foods to a common set of chemical reactions as rapidly as possible, and to break inconveniently large packages of free energy into several smaller ones that can be handled by the machinery for synthesizing ATP. Let us now look more closely at each of these three parts.

Step 1: Glycolysis

The first part of this combustion process does not require oxygen. It is common to all living organisms, and is known as *anaerobic fermentation*. If O_2 is present, the end product is pyruvic acid, as we have stated. But in other organisms that do not use oxygen, or in some oxygen-using microorganisms deprived of oxygen, other compounds are produced. Yeast cells produce ethanol under anaerobic conditions, certain types of bacteria produce acetone, and human muscle cells produce lactic acid:

$$C_6H_{12}O_6 \rightarrow 2CH_3CH(OH)COOH \qquad \Delta G^0 = -47.4 \text{ kcal}$$

$\qquad$ glucose $\qquad\qquad\qquad$ lactic acid

It is this accumulation of lactic acid in our muscles that produces muscle cramps during sudden exertion when the oxygen supply in the muscles is exhausted. As more oxygen is brought to the muscles, the lactic acid is reconverted to pyruvate, the normal product:

$$2CH_3CH(OH)COOH + O_2 \rightarrow$$

$\qquad$ lactic acid

$$2CH_3COCOOH + 2H_2O \qquad \Delta G^0 = -92.7 \text{ kcal}$$

$\qquad\qquad$ pyruvic acid

This first part of metabolism is known as *glycolysis*. It occurs in eleven chemical steps, in which glucose is degraded to fructose and then to three-carbon glyceraldehyde derivatives. Only in the last step or two does the process branch into separate pathways to produce pyruvic acid, or lactic acid, or ethanol, or acetone. Each step of the breakdown is controlled by a catalyst (Section 18–5), an enzyme with a molecular weight ranging from 30,000 to 500,000.

Glycolysis probably is a chemical fossil from the time before oxygen existed in the atmosphere, when one-celled organisms lived by degrading naturally occurring organic compounds, as was suggested in Section 11–7. When organisms increased in size and complexity and in their energy requirements, and when oxygen appeared in our atmosphere, a more complex and much more energetic biochemical process, known as the *citric acid cycle* (Step 2), began to evolve. Before we explore this process, we must look at the universal method of storing chemical energy in every kind of living organism.

Energy Storage and Carrier Molecules

The structure of the key molecule in the energy storage process, adenosine triphosphate (ATP), is illustrated in Figure 12–17. It is built from adenine (Figure 12–3), ribose (a five-carbon sugar), and three linked phosphate groups. The terminal phosphate group in ATP can be hydrolyzed, or split off with the addition of OH^- and H^+ from water, to yield phosphate and adenosine diphosphate (ADP). ADP can be decomposed even further to produce another phosphate group and adenosine monophosphate (AMP). Finally, the last phosphate group can be removed to make adenosine. The first two cleavages liberate 8 kcal mole^{-1} of free energy each, whereas the third cleavage liberates only 2 kcal mole^{-1}. It is this substance, and more particularly the first phosphate bond (farthest left in the figure), that is the principal means of energy storage in any living cell. Every time a molecule of glucose is degraded biochemically to two molecules of pyruvate, eight

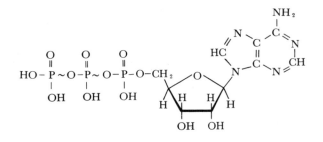

Figure 12–17. The structure of adenosine triphosphate (ATP). *The bonds marked by wavy lines in the phosphate groups liberate an unusually large amount of energy when they are cleaved by hydrolysis, and are the means by which chemical energy is stored in the* ATP *molecule.*

molecules of ATP form from 8ADP:

$$\text{glucose} + 8\text{ADP} + 8 \text{ phosphate} \rightarrow 2\text{CH}_3\text{COCOOH} + 8\text{ATP}$$

This results in the storage of $8 \times 8 = 64 \text{ kcal mole}^{-1}$ of free energy. The enzymes that control all of the steps of the breakdown ensure that the energy released at a step of this process is used to synthesize an ATP molecule rather than being wasted as heat.

There are two other carriers that we should examine before going on to the citric acid cycle. One is nicotinamide adenine dinucleotide, whose structure is shown in Figure 12–18. It resembles ATP in having an adenine group, ribose, and phosphate. However, the essential part is a nicotine ring that can be reduced and oxidized. This molecule is a redox carrier. When a metabolite is oxidized at one step in the citric acid cycle, the oxidized form of nicotinamide adenine dinucleotide, NAD^+, can be reduced to NADH and H^+. The other important carrier is FAD or flavin adenine dinucleotide, which is reduced to FADH_2. Both of these carriers feed into the last production line of the energy storage factory, the *terminal oxidation chain*. This is a four-step pathway, involving the cytochrome enzymes, in which the reduced electron carriers, NADH and FADH_2, are reoxidized. In this process, oxygen is reduced to water, and the energy released is stored as ATP. Each time a reduced carrier transporting a pair of electrons passes down the cytochrome oxidation chain and loses its relatively high energy, the energy is conserved by the synthesis of several molecules of ATP.

Many of the essential vitamins are the semifinished components for energy carriers such as these. Small amounts of these vitamins must be supplied from outside sources because we have lost the ability to synthesize them ourselves. For example, the chemical group that is oxidized and reduced in FAD is a flavin. Some organisms can synthesize flavins; perhaps we or our ancestor species also could synthesize them at one time, but we cannot do so now. To keep the energy-transfer system operating and to replace

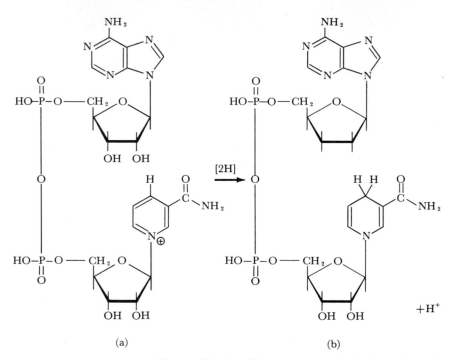

Figure 12–18. The structure of nicotinamide adenine dinucleotide (a) oxidized, NAD⁺, and (b) reduced, NADH. Note the similarity between the top half of the molecule as drawn here and ATP. Reduced NADH is a carrier molecule that can pass on its stored energy in chemical syntheses in a similar way as reduced ferredoxin does in photosynthesis.

the FAD molecules as they are gradually worn out, we need small amounts of *riboflavin*, or vitamin B_2, supplied from external sources (Figure 12–3).

Step 2: Citric Acid Cycle

Now we are ready to look at the second part of the three-part energy-retrieval process of metabolism. The main role of this part, the citric acid cycle, is to convert the more than 250 kcal mole^{-1} of free energy contained in pyruvic acid molecules into four packages of 50 kcal mole^{-1}, which are in the form of reduced NADH and $FADH_2$. The "primer" step before the cycle begins is the combination of pyruvic acid with a molecule called reduced coenzyme A (CoA—SH) to form acetyl coenzyme A. Acetyl coenzyme A is the raw material for the citric acid cycle (Figure 12–19):

$$CH_3COCOOH + NAD + CoA\text{—}SH \rightarrow$$
pyruvic acid

$$CH_3CO\text{—}S\text{—}CoA + CO_2 + NADH_2$$
acetyl coenzyme A

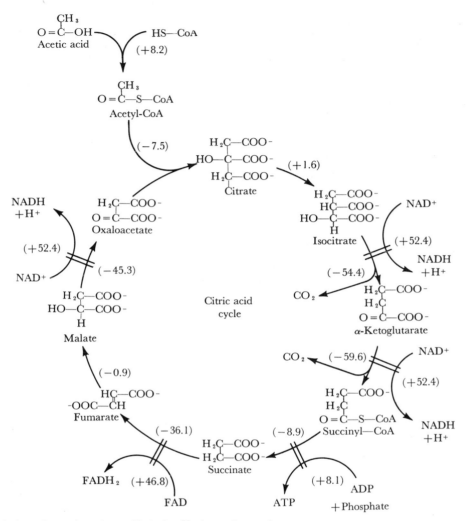

Figure 12–19. The citric acid cycle, also called the Krebs cycle or the tricarboxylic acid cycle. The numbers in parentheses are the standard free energies for the reactions shown. A double bar indicates that the oxidation in the cycle and the reduction of a carrier molecule are made to occur together by an enzyme. The reduced carrier molecules feed into the terminal oxidation chain, where they are oxidized again, and O_2 is reduced to H_2O.

In the cycle, the two-carbon acetate group is combined first with four-carbon oxaloacetate to make the six-carbon citrate ion. Then citrate is degraded in seven steps to release two of its carbon atoms as CO_2 and to restore the oxaloacetate again. Each of the steps in the citric acid cycle is either an oxidation (isocitrate to α-ketoglutarate, malate to oxaloacetate) or a rearrangement in preparation for the next oxidation (citrate to isocitrate). In five of the oxidation steps, the liberated energy reduces a carrier molecule: NAD^+, FAD, or ADP.

Step 3: Terminal Oxidation Chain

Each NADH molecule, no matter what its origin, funnels into the third part of the metabolic process, the terminal oxidation chain, and produces three molecules of ATP. Also, each $FADH_2$ arrives midway through the same chain and produces two ATP. You can count the score in Figure 12–19 and see that each turn of the citric acid cycle, as it destroys one mole of acetate, ultimately leads to the synthesis of 12 moles of ATP. The overall reaction is

$$CH_3COOH + 2O_2 \rightarrow 2CO_2 + 2H_2O \qquad \Delta G^0 = -211 \text{ kcal}$$
$$12 \text{ ADP} + 12 \text{ phosphate} \rightarrow 12 \text{ ATP} \qquad \Delta G^0 = 12 \times 8 = +96 \text{ kcal}$$

In the total degradation of one mole of glucose—which involves glycolysis to two molecules of pyruvate, its conversion to two moles of acetate, and then two turns of the citric acid cycle—the balance sheet looks like this (for simplicity, the conversion of pyruvate to acetyl CoA has been included in glycolysis):

Degradation
$$C_6H_{12}O_6 + 2O_2 \rightarrow 2CH_3COOH + 2H_2O + 2CO_2$$
$$\Delta G^0 = -264 \text{ kcal}$$
$$2CH_3COOH + 4O_2 \rightarrow 4CO_2 + 4H_2O \qquad \Delta G^0 = -422 \text{ kcal}$$
$$\overline{C_6H_{12}O_6 + 6O_2 \rightarrow 6CO_2 + 6H_2O} \qquad \overline{\Delta G^0 = -686 \text{ kcal}}$$

Synthesis
$$14 \text{ ADP} + 14 \text{ phosphate} \rightarrow 14 \text{ ATP} \qquad \Delta G^0 = 14 \times 8 \text{ kcal}$$
$$= +112 \text{ kcal}$$
$$24 \text{ ADP} + 24 \text{ phosphate} \rightarrow 24 \text{ ATP} \qquad \Delta G^0 = 24 \times 8 \text{ kcal}$$
$$= +192 \text{ kcal}$$
$$\overline{38 \text{ ADP} + 38 \text{ phosphate} \rightarrow 38 \text{ ATP}} \qquad \overline{\Delta G^0 = +304 \text{ kcal}}$$

The overall efficiency of energy conversion from glucose as food to ATP stored in the muscles is $304/686 = 0.44$.

This outline should not mislead you into thinking that the entire process of energy conversion is known. The outlines of the glycolytic pathway, the citric acid cycle, and the terminal oxidation chain are known, and many of the details of their reactions are understood. Yet the *mechanisms* by which they occur are not as well known. What is the molecular machinery by which the energy of degradation becomes involved in connecting a phosphate group to ADP during the process known as *oxidative phosphorylation?* We do not know, but there is intensive research being carried out to answer the question. What is it that reduces one molecule of NAD^+ every time one molecule of malate is oxidized to oxaloacetate? We are just beginning to see a possible explanation. The *what* of metabolic chemistry is reasonably clear; the *why* or the *how* of the processes is still unknown. The controlling agents in all of these metabolic reactions are enzymes. It is these molecules that we want to consider next.

12–9 ENZYMES AND PROTEINS

A protein is a folded polymer of amino acids. Such a polymer is shown at the bottom of Figure 12–1, and a model of a single amino acid appears in Figure 12–12. Enzymes are one class of proteins, and perhaps the most glamorous class. They are approximately globular molecules with molecular weights from 10,000 to several million and diameters of 20 Å and more. Other globular protein molecules such as myoglobin and hemoglobin are carriers and storage units for molecular oxygen (Section 11–7). The cytochromes are oxidation–reduction proteins that serve as intermediate links in the terminal oxidation chain mentioned in the preceding section. The gamma globulins are antibody molecules with a molecular weight of 150,000. They attach themselves to viruses, bacteria, or other foreign bodies, and precipitate them from body fluids to protect their host. All of these proteins are *globular proteins*.

The other large class of proteins is the *fibrous proteins*. These are mainly structural materials. *Keratin*, found in skin, hair, wool, nails, and beaks, is a fibrous protein. *Collagen* in tendons, the underlayers of skin, and the cornea of the eye is another type of fibrous protein, as are *silks* and many kinds of insect fibers. Proteins in combination with carbohydrates (polysaccharides) and with lipids (long-chain fats and fatty acids) are the structural materials of all living organisms.

The chief distinction between a protein chain and a chain of polyethylene or Dacron is that not all side chains in a protein are alike. In fibrous proteins, it is the repetition of the sequence of side groups that gives a particular fibrous protein—silk or hair or collagen—its special mechanical properties. Globular proteins are even more intricate. These molecules typically have 100 to 500 amino acids polymerized in one long chain, and the complete sequence of side groups is the same in every molecule of the same globular protein. The side groups can be hydrocarbonlike, acidic, basic, or neutral but polar. Both the folding of the protein chain to make a compact globular molecule and the chemical behavior of the molecule once it is folded are determined by the kind and sequence of amino acid side groups.

Only 20 different kinds of amino acid side groups ordinarily are found in living organisms. These side groups are shown in Table 12–5. Some of them are hydrocarbonlike, such as Val, Leu, Ile, and Phe. These *hydrophobic* groups are more stable if they can be removed from an aqueous environment. A protein chain in aqueous solution tends to fold into a molecule with these groups *inside*. Other side groups are charged, such as the acids Asp and Glu, and the bases Lys and Arg. Others, although uncharged, are compatible

Table 12–5.   Side Groups of the Twenty Common Amino Acids

Symbol	Name	Side group[a]
Acidic side groups		
Asp	Aspartic acid	$-CH_2COOH$
Glu	Glutamic acid	$-CH_2CH_2COOH$
Basic side groups		
Lys	Lysine	$-CH_2CH_2CH_2CH_2NH_2$
Arg	Arginine	$-CH_2CH_2CH_2NH-C(NH)NH_2$
His	Histidine	$-CH_2-C\underset{\displaystyle CH-N}{\overset{\displaystyle NH}{\diagup\diagdown}}CH$
Uncharged but polar side groups		
Asn	Asparagine	$-CH_2-CO-NH_2$
Gln	Glutamine	$-CH_2CH_2-CO-NH_2$
Ser	Serine	$-CH_2OH$
Thr	Threonine	$-CH(OH)-CH_3$
Gly	Glycine	$-H$
Sulfur-containing side groups		
Cys	Cysteine	$-CH_2SH$
Met	Methionine	$-CH_2CH_2-S-CH_3$
Aliphatic side groups		
Ala	Alanine	$-CH_3$
Val	Valine	$-CH-CH_3$ with CH_3 below
Leu	Leucine	$-CH_2-CH-CH_3$ with CH_3 below
Ile	Isoleucine	$-CH-CH_2-CH_3$ with CH_3 below
Pro	Proline (entire amino acid shown)	$\begin{array}{c} CH_2 \\ CH_2 \quad CH_2 \\ {}^+H_2N-CH-COO^- \end{array}$

(continued)

Table 12–5 continued

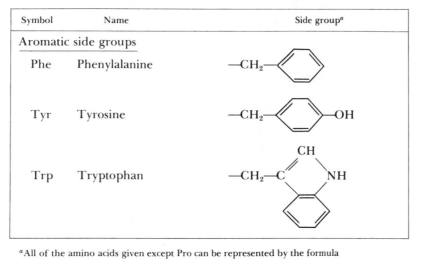

Symbol	Name	Side group[a]
Aromatic side groups		
Phe	Phenylalanine	
Tyr	Tyrosine	
Trp	Tryptophan	

[a]All of the amino acids given except Pro can be represented by the formula

in which the side groups, —R, are listed above. The entire molecule is shown for Pro.

with an aqueous environment, such as Asn, Gln, and Ser. One of the most important factors in determining how a protein chain will fold into a globular molecule is the stability that results if hydrophobic groups are buried on the inside of the molecule and charged groups are outside. Although either optical isomer shown in Figure 12–12 might seem equally probable, all of the amino acids in proteins are L-amino acids [Figure 12–12(a)].

A protein chain is particularly stable when coiled into a right-handed α helix (Figure 12–20). In this structure, the side groups point away from the axis of the helix, and the C=O groups of one turn of the helix are bonded to the H—N groups of the next higher turn by a hydrogen bond.

A hydrogen bond (Section 14–4) is a bond formed between an especially electronegative atom, such as F or O, and a hydrogen atom with a slight local excess of partial positive charge. Such a bond is mainly electrostatic and depends on the two atoms being able to approach closely. For this reason, O and F can make such bonds, but the larger Cl ordinarily cannot. Hydrogen bonds are present in ice, where they hold the water molecules in an open framework that collapses when the ice melts. This is the reason why ice is less dense than water, and why ice floats rather than sinks to the bottom of a pond in winter. Hydrogen bonds are quite common in proteins because of the presence of the carbonyl oxygen and amine hydrogen on the polypeptide chain. As Figure 12–13 shows, the partial double bond character of the C—N peptide bond not only keeps the linkage planar, it also makes the oxygen

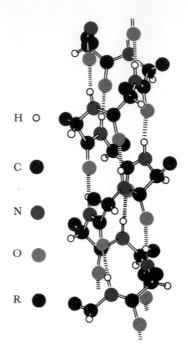

H O

C

N

O

R

Figure 12–20. The α helix, a type of protein chain folding found in both fibrous and globular proteins. The α helix was proposed first by L. Pauling and R. B. Corey from model-building experiments based on the bond lengths and bond angles determined in x-ray analyses of individual amino acids and polymers of two and three amino acids. The structure since has been discovered in hair and wool, in skin keratin, and in globular proteins such as myoglobin and hemoglobin.

atom slightly negative and the nitrogen atom (and eventually its hydrogen atom) slightly positive. These are favorable conditions for hydrogen bonding.

In hair, wool, and other keratins, α helices are twisted into threads, strands, and cables to make the fibers that we can see and manipulate. In silk, the chains are stretched full length rather than in an α helix, and are cross-connected by hydrogen bonds into the sheets illustrated in Figure 12–2(a). In the globular proteins, the strands can be neither fully extended nor fully α-helical; there must be some bending back and forth to keep the molecule compact. In myoglobin (Figure 11–25), the 153 amino acids in the protein chain are coiled into eight lengths of α helix (lettered A through H), which then are folded back and forth to make a compact molecule. Helices E and F form a pocket into which the heme group fits, and the oxygen molecule binds to the iron of this heme. Hemoglobin is constructed along similar lines and has four such myoglobinlike units (Figure 11–26). The very small protein cytochrome *c* (104 amino acids) has no room for α helices. Its protein chain is wound around its heme like a cocoon. In larger enzymes such as chymotrypsin (241 amino acids) and carboxypeptidase (307 amino acids), there are regions in the center of the molecule where the protein chain is in several parallel strands held together by hydrogen bonds very much as in silk [Figure 12–2(a)].

The purpose of an enzyme is to provide a surface to which its substrates (the molecules acted upon) can bind, and to facilitate the formation or rupture of bonds in these molecules. The site on the surface of the enzyme where these activities take place is called the *active site*. The enzyme has two functions: *recognition* and *catalysis*. If it bound every molecule that came along, it would spend only a small proportion of its time catalyzing the reaction that it was supposed to catalyze. Conversely, even if it bound the right molecules, it would be useless if it could not assist in the making or breaking of the proper bonds. Enzymes recognize their proper substrates by having properly positioned amino acid side chains at their active site that can interact with the substrate molecule by way of charge interaction, hydrogen bonding, or the attraction of hydrophobic groups. This selection of molecules that an enzyme will or will not bind is called its *specificity*.

Once bound, the substrate is subject to attack by groups on the enzyme. Many enzymes involved in bond-breaking reactions use metals such as Zn, Mg, Mn, or Fe. Sometimes one part of the substrate will coordinate to the metal; in other cases, the metal draws electrons from the substrate and weakens a bond. Both roles are illustrated in the catalytic action of carboxypeptidase, discussed in Section 12–11.

Only recently have we known the molecular structures of proteins. The first x-ray analysis of a protein, that of myoglobin, was completed in 1959. That of the first enzyme, lysozyme, was accomplished in 1964. Research on larger enzymes, electron carriers, and antibodies is progressing rapidly. We now know the detailed molecular framework of more than 25 proteins. Biochemistry, in these areas, merges imperceptibly with its sister field of molecular biology.

12–10 SUMMARY

We have come a long way in this chapter from speculating about the relative chemistry of B, C, N, and Si. Carbon undoubtedly has a special role, conferred upon it by its adequate supply of electrons for orbitals, absence of repulsive lone electron pairs, and ability to make double and triple bonds. The simple alkanes, or single-bonded compounds of hydrogen and carbon, have illustrated the diversity of compounds that carbon can build because of its ability to make stable chains. The alkyl halides are the bridges from the relatively unreactive alkanes to the wealth of hydrocarbon derivatives: alcohols, ethers, aldehydes, ketones, esters, acids, amines, amino acids, and others that we will not examine here. The ability of carbon to make double and triple bonds has been seen in the alkenes and alkynes; this subject introduced us to the special type of multiply bonded substances known as aromatic compounds.

The aromatic compounds are special because a certain number of their electrons are delocalized, or not confined to the region between two bonded atoms. In general, the greater the delocalization, the more stable the molecule will be. Benzene is 40 kcal mole^{-1} more stable than a calculation of the bond energies of a Kekulé structure would lead one to predict. Moreover, the first unfilled orbitals in such delocalized-electron molecules are lower and closer to the filled orbitals than in other molecules. The energy required to excite an electron often falls in the visible part of the spectrum, thereby making these compounds colored. The ability to absorb visible light is crucial in a large delocalized ring molecule known as chlorophyll, because the energy of the light absorbed is used to synthesize glucose. Without photosynthesis, life on this planet would have remained a rare and irrelevant scavenger, breaking down once again those organic molecules synthesized naturally by electric discharge or ultraviolet radiation.

The molecule that is synthesized as an energy-storing agent in photosynthesis is glucose, a carbohydrate. Besides being a storehouse for chemical energy, carbohydrates are important structural materials in plants: wood, cotton, woody stalk tissue in softer plants. Glucose is polymerized into *cellulose*, which is the basis of the structural carbohydrates, and cannot be redigested, and *starch*, which is stored in seeds, grains, and roots, and is destined to be degraded later for its glucose.

When needed as an energy source, the glucose is not burned in one step. Instead, it is degraded in a series of more than 25 separate steps. During many of these steps the energy released is saved by synthesizing molecules of ATP. The end products of glucose degradation are carbon dioxide and water, themselves the raw materials for photosynthesis again. The entire photosynthetic and metabolic process is an elaborate chemical machine for converting the energy of photons from the sun into the energy of the phosphate bonds in ATP. All of the hundreds of chemical substances involved are, in a sense, only middlemen.

The controllers of all of this chemistry are the biological catalysts known as enzymes. These are proteins—long-chain polymers of amino acids—whose molecular properties are determined by the sequence of 20 different amino acids along the protein chain. The enzyme molecules are not extended chains; they are compact, folded arrangements of chains in which the right amino acid side groups are brought together to bind the molecules on which the enzyme acts and then to facilitate a chemical reaction.

At the end of the preceding chapter we made the half-facetious remark that life is only applied transition-metal chemistry. This may be so, in the sense it was intended, but the application of transition-metal chemistry is to compounds of carbon.

12–11 POSTSCRIPT: THE CATALYTIC ACTION OF CARBOXYPEPTIDASE

The molecular structures of a few enzyme molecules now are known, and we can begin to propose mechanisms for their catalytic activity. One of the clearest examples is the enzyme *carboxypeptidase A*, whose structure was determined from x-ray diffraction experiments, in 1967, by W. N. Lipscomb and his group at Harvard.

Carboxypeptidase is an enzyme of molecular weight 34,600, with a single polypeptide chain of 307 amino acids. It is secreted by the pancreas

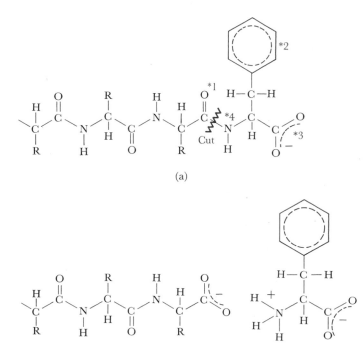

Figure 12–21. The cleavage of the carboxyl terminal amino acid from a polypeptide chain by carboxypeptidase A. (a) The chain prior to cleavage showing the cleavage point by a wavy line, the carbonyl oxygen that becomes a ligand to the Zn (*1), the hydrophobic group that is recognized by the pocket on the enzyme (*2), the carboxyl end of the chain that is attracted by an arginine on the protein (*3), and the nitrogen atom that receives a proton from a tyrosine side group (*4). (b) The chain after cleavage. The nitrogen atom is shown in its zwitterion, charged form. The third proton comes from H^+ *ions in solution.*

to help digest foods in the intestine. Its particular function is to cut away the last amino acid residue from the carboxyl end of a protein chain. There are two variants of carboxypeptidase, A and B. Carboxypeptidase A is most efficient when the side chain on the amino acid to be removed is large and hydrophobic. Therefore, the *substrate* of carboxypeptidase A is a polypeptide chain (Figure 12–21), its *specificity* is for chains with bulky hydrophobic groups at the carboxyl end, and its *catalytic activity* is the cleavage of the final peptide bond in the chain [Figure 12–23(b)]. How does the enzyme operate?

The enzyme is a football-shaped object, 45 × 45 × 55 Å in diameter, with a depression at one end and a groove running around the side of the molecule from this depression. The depression is the active site, and the groove is the place where the polypeptide chain that is to be cut, binds. The carboxyl end of the chain fits into the active site.

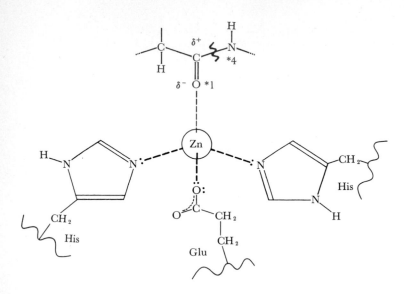

Figure 12–22. The coordination of the Zn(II) *ion at the active site of carboxypeptidase A. The three lower ligands are side chains of amino acids in the protein itself. The upper ligand, in color, is the carbonyl oxygen (*1) next to the bond to be cut in the substrate. The* Zn *ion polarizes the double bond and makes the* O *slightly negative and the* C *slightly positive.*

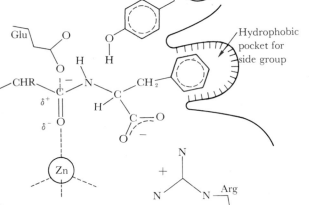

(a)

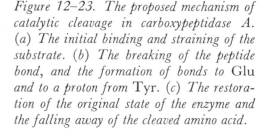

Figure 12–23. The proposed mechanism of catalytic cleavage in carboxypeptidase A. (a) The initial binding and straining of the substrate. (b) The breaking of the peptide bond, and the formation of bonds to Glu *and to a proton from* Tyr. *(c) The restoration of the original state of the enzyme and the falling away of the cleaved amino acid.*

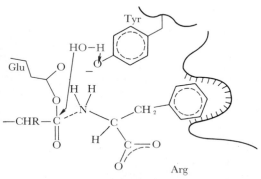

(b)

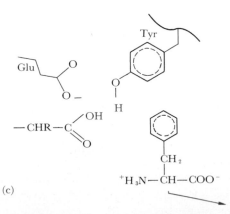

(c)

In the middle of the active site depression is a Zn(II) ion, which is the key to the catalytic activity of the enzyme. It is tetrahedrally coordinated and has three of its four ligands coming from N or O on amino acid side chains on the protein (Figure 12–22). These ligands keep the Zn ion in place. The fourth coordination site is open until the polypeptide substrate binds to the enzyme. When it does, the oxygen of the C=O group next to the bond to be cut becomes the fourth Zn(II) ligand. The Zn ion attracts the lone-pair electrons of the oxygen atom and polarizes the C=O double bond. The O atom acquires a slight negative charge, and the C atom, a slight positive charge (Figure 12–22). This polarization of charge is stabilized by the negative charge on a glutamic acid side chain from the protein that swings down next to the slightly positive C atom [Figure 12–23(a)].

The negatively charged carboxyl end of the chain to be cut (*3 in Figure 12–21) is held in place by the positive charge on a basic side chain of the protein, arginine. The bulky hydrophobic side chain of this terminal amino acid (*2 in Figure 12–21) fits neatly into a pit or pocket at one end of the active site depression, which is lined with hydrophobic groups such as valine, leucine, and isoleucine. This hydrophobic pocket confers the specificity upon the enzyme. A polypeptide chain with a positively or negatively charged side chain would not be able to fit it into the pocket as easily, and would be less likely to be bound to the enzyme and cleaved. Hence the specificity of carboxypeptidase A for chains that end with hydrophobic groups.

The polypeptide chain is recognized, bound, and strained. Now the cleavage begins. As the chain binds to the enzyme, a tyrosine molecule with an aromatic —OH group swings around a distance of 14 Å to bring its —OH group next to the N of the bond to be cut [Figure 12–23(a)]. The proton from this —OH group draws electrons from the C—N bond, and the polarization of the C=O group also draws electrons away from the bond. The bond is subjected to an electrophilic attack in the same way that benzene was in Section 12–5. The bond weakens and gives way at the same time that the tyrosine proton attaches to the nitrogen atom and the carbon atom makes a bond to the glutamic group [Figure 12–23(b)]. The last amino acid in the polypeptide chain now is cleaved successfully and falls away.

However, the remainder of the chain still is attached covalently to the glutamic acid side chain of the protein, and the tyrosine side chain is ionized. Both defects are remedied when a water molecule dissociates, restoring Tyr with the proton and cleaving the bond to Glu with the hydroxyl ion [Figure 12–23(c)]. Tyr and Glu both are restored to their original condition, and the cut end of the polypeptide chain is capped and ready to fall away from the enzyme. The Glu, Tyr, and Arg groups swing back to their free-enzyme positions, and the enzyme is ready for another polypeptide chain.

Figure 12–24 is an accurate drawing of the active site of carboxypeptidase A, with a substrate in place. The original from which this drawing was prepared was drawn by a computer-controlled Calcomp plotter from the coordinates of the atoms. The structure of the enzyme was examined by

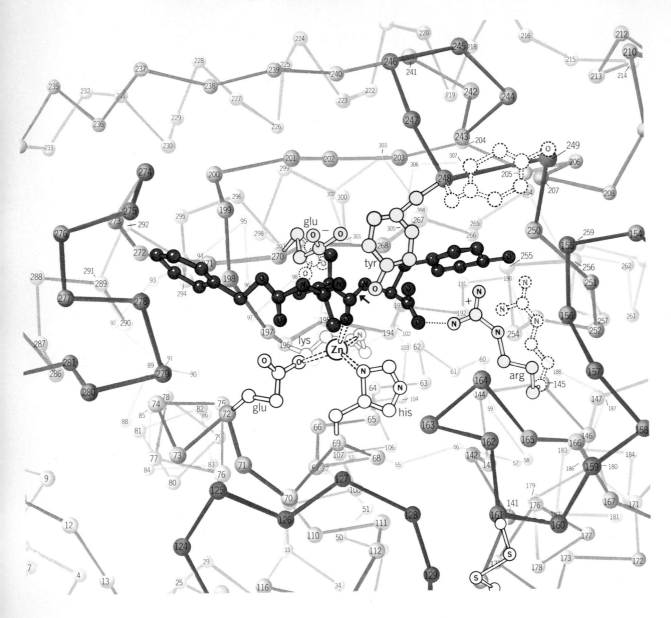

Inhibitor

Zn-liganding side chains

Moving side chains *without* inhibitor

Same, *with* bound inhibitor

◀ *Figure 12–24. A view into the active site of carboxypeptidase A, with the substrate bound in place. The three protein side chains that interact with the substrate, Arg, Tyr, and Glu, are shown in their positions without substrate bound (dotted) and with substrate (solid). Note the tetrahedral coordination of the Zn atom. Since this drawing was made, a reexamination of the x-ray results has shown that the Lys side chain coordinating to the Zn should be a His as in Figure 12–22. For clarity, the protein chain of the enzyme is represented only by its α-carbon atoms. The atomic positions from which this painting was prepared were supplied from the x-ray analysis by Professor W. N. Lipscomb of Harvard, and the rough drafts from which the painting was made were drawn by an IBM computer and plotter. The final painting is by I. Geis, and is reproduced from R. E. Dickerson and I. Geis, The Structure and Action of Proteins, W. A. Benjamin, Menlo Park, Calif., 1969.*

x-ray diffraction, both with and without a nonreactive analogue of the substrate in place, to be sure about any motion of side chains during binding of the substrate. In Figure 12–24, the side chains that move are shown as they are before and after binding of the substrate. You can see the pocket into which the substrate side chain fits. For the sake of clarity, all amino acids except those involved in catalysis or binding are represented only by numbered spheres at the positions of their α-carbon atoms.

When enzyme chemists of only a generation ago spoke of the "mechanism" of enzyme catalysis, they did so figuratively, even though they had a literal meaning in mind. Now we can diagram a mechanism that is every bit as literal and mechanical as the mechanism of a combination lock. What we now can do for carboxypeptidase and two or three other enzymes, we should be able to do in a few years for numerous kinds of enzymes, including those involved in the glucose metabolism that we were examining earlier. Even now, knowing what transition metals and side chains on enzymes do in carboxypeptidase, we can make intelligent guesses as to how related enzymes work. Much research remains to be done, but now much can be anticipated.

SUGGESTED READING

N. L. Allinger and J. Allinger, *Structures of Organic Molecules*, Prentice-Hall, Englewood Cliffs, N. J., 1965.

"The Biosphere," *Scientific American*, September, 1970. An issue devoted to a common theme: the chemistry of our planet and life.

R. Breslow, "The Nature of Aromatic Molecules," *Scientific American*, August, 1972.

R. E. Dickerson and I. Geis, *The Structure and Action of Proteins*, W. A. Benjamin, Menlo Park, Calif., 1969. A discussion of the principles of protein folding and structure, and of the molecular basis for the chemical behavior of enzymes and other proteins. Similar level and approach to Section 12–11.

E. Frieden, "The Chemical Elements of Life," *Scientific American*, July, 1972.

W. Herz, *The Shape of Carbon Compounds*, W. A. Benjamin, Menlo Park, Calif., 1963. An elementary introductory monograph on organic chemistry, written as a supplement to a freshman course. Not as structure-oriented as its title suggests. Good introduction to organic mechanisms.

F. R. Jevons, *The Biochemical Approach to Life*, Basic Books, New York, 1968, 2nd ed. An unusually readable introduction to some of the central ideas of biochemistry.

R. C. Johnson, *Introductory Descriptive Chemistry*, W. A. Benjamin, Menlo Park, Calif., 1966. Elementary introduction

with useful information on boron and nitrogen compounds.

J. Lambert, "The Shapes of Organic Molecules," *Scientific American*, January, 1970.

R. J. Light, *A Brief Introduction to Biochemistry*, W. A. Benjamin, Menlo Park, Calif., 1968. Broader coverage of topics than Jevons, but at about the same introductory level.

Structure and Function of Proteins at the Three-Dimensional Level, Cold Spring Harbor Symposia on Quantitative Biology, Vol. 36, 1972. The state of protein structure knowledge in mid-1971. Red-green stereo drawings of protein molecules.

M. Yudkin and R. Offord, *A Guidebook to Biochemistry*, Cambridge University Press, Cambridge, 1971, 3rd ed. Covers more ground than this chapter, but elementary and readable. Similar to Jevons but more up to date.

QUESTIONS

1 Why cannot Si form double bonds by using its p orbitals?

2 Why does B not form double bonds by using its p orbitals?

3 Why is the N—N bond weaker than the C—C bond?

4 Why is the Si—Si bond weaker than the C—C bond?

5 Why is the Si—O bond stronger than the C—O bond?

6 Draw a diagram of the atomic orbitals used in diborane, B_2H_6, and show how they are combined into molecular orbitals. How many atomic orbitals are employed, and how many valence electrons are used to make the bonds?

7 How many molecular orbitals result when three atomic orbitals are combined as in a three-center bond? What bonding character do the molecular orbitals have? How many of them are filled with electron pairs in boron hydrides?

8 In addition to the weakness of the N—N single bond, what other factor makes the long-chain nitrogen compounds extremely unstable?

9 What is the difference in structures of silanes and silicones? What is the difference in chemical reactivity? How do you explain this difference?

10 What are alkanes? How does their chemical reactivity compare with that of silanes, silicones, and hydronitrogens?

11 What is the smallest alkane for which isomers exist?

12 Why are the halogen-substituted alkane derivatives so important in chemical synthesis?

13 What is an organic radical? Why is the terminology useful?

14 How does the chemical behavior of the hydrogen atom in the —OH group differ in the compounds CH_3OH and CH_3—CO—OH? What general class of organic compounds does each of these belong to, and what is each compound's name?

15 What is the difference between the chemical structure of an ester and an ether? Which can be thought of as a derivative of an organic acid?

16 If the wrong microorganisms get into a new batch of wine, or if it is mistreated in any of several ways, the ethanol turns to acetic acid, and vinegar results. Is this conversion an oxidation or reduction? Write a balanced equation for the reaction.

17 How are aldehydes and ketones related in structure to acids? How might an aldehyde be formed from an alcohol? Write a reaction for such a process; start from ethanol. What aldehyde would result? Is this an oxidation or a reduction?

18 Where is formic acid found naturally? Butyric acid? (Its name is a clue.)

19 What are soaps? How do they function?

20 What are amines? Are they acidic, basic, or neutral? Write a balanced equation showing what happens when an amine ionizes in aqueous solution.

21 What is an amino acid? What is its zwitterion form? (You may want to look up the term *"zwitter"* in a German–English dictionary to see why the ion is so named.)

22 What is a peptide bond? Why is it planar? What atoms in the peptide lie in one plane? What effect does this have on charge distribution at the O and N atoms?

23 What is the difference between alkanes, alkenes, and alkynes?

24 What is the difference in structure between 1-butene and 2-butene, and how do they differ from butane? How many isomers are there of each of these two butenes?

25 What alkene would you chlorinate to obtain each of the following dichloro compounds?

a) 1,2-dichlorobutane

b) 2,3-dichlorobutane

c) 1,4-dichloro-2-butene

How might you then obtain the following?

d) 1,2,3,4-tetrachlorobutane

e) 1,4-dichlorobutane

26 Which form of solid carbon is the logical limit of the series: benzene, naphthalene, anthracene, . . . coronene, . . . ? In what sense is that form the limit of trends in the given series of compounds?

27 Why does benzene not undergo addition reactions like the butenes?

28 Is a Lewis acid electrophilic or nucleophilic? What does the term "nucleophilic" mean?

29 How does an electrophilic group aid in addition reactions to benzene?

30 What is an aromatic compound? Are aromatic hydroxyl compounds such as phenol, C_6H_5—OH, more or less acidic than aliphatic hydroxyl compounds such as methanol, CH_3—OH? Why this difference in behavior?

31 What is a $\pi \rightarrow \pi^*$ transition in an aromatic compound? Illustrate with benzene.

32 What effect on the electronic energy levels of a molecule does a greater degree of delocalization of electrons have? How is this effect used in acid–base indicators?

33 What is the difference between a carbohydrate and a hydrocarbon?

34 What is the difference between the molecular structures of starch and cellulose? Why can we digest one but not the other? What is the monomer unit of each?

35 What is an asymmetric carbon atom? How many such asymmetric carbon atoms are present in glucose? In glycine? Do any of the compounds in Figure 12–3 have asymmetric carbon atoms? How many?

36 Which form of cyclohexane is more stable, the chair or boat form? Why? Which form is adopted in glucose?

37 What is the difference between *ortho-*, *meta-*, and *para-*dichlorobenzene?

38 Without worrying about the exact details, what sort of differences distinguish the three sugars glucose, galactose, and mannose?

39 Where does the starch in plants come from?

40 If the carbon atoms in CO_2 in the upper atmosphere were dated each year like vintage wine, then a clever homicide detective could always establish the year of death of a victim, if not the exact time of death. Explain why. (Breathing in CO_2 is not the answer.)

41 How much heat is obtained by burning one mole of glucose? Why is it more efficient for living organisms to decompose glucose slowly in a series of steps? What happens to the chemical energy formerly present in the glucose molecules? What are the chemical end products of glucose degradation?

42 What is the three-step machinery by which glucose is degraded in living organisms? What do we mean when we say

that the first step, in many organisms, is anaerobic? Under what conditions is this first step carried out anaerobically in man, what alteration in the end product of this first step occurs, and what are the physiological distress symptoms that result?

43 What is the principal purpose of the citric acid cycle, within the larger framework of the complete machinery for degrading compounds such as glucose? How does the citric acid cycle transfer chemical energy to the third and last step of the process?

44 What are ATP, ADP, NAD^+, and FAD? What are they good for?

45 What are vitamins? What connection do many vitamins have with the substances in Question 44?

46 Can you think of any reasons why it might be an advantage for us *not* to be able to synthesize every chemical compound we need? (This is really a question in molecular genetics, not chemistry.)

47 What does the terminal oxidation chain do? What part do the cytochromes play in it?

48 The overall result of the three-step glucose degradation process is that glucose and oxygen are converted to carbon dioxide and water. At which of the three steps is carbon dioxide evolved? At which of the three steps is water produced? Why are they both not produced in the same step of the process?

49 In which of the three steps is the greatest amount of ADP converted to ATP, the compound that stores chemical energy?

50 Which of the three steps is probably the oldest, in terms of the evolution of life? Why can we draw such a conclusion?

51 What is a protein? What is an enzyme? Give some examples of proteins that are not enzymes.

52 What are the two main structural classes of proteins? To which class do the following belong: hemoglobin, keratin, silk, carboxypeptidase A?

53 What is the fundamental difference between long-chain molecules such as are found in Teflon, Dacron, and polyethylene in one group, and myoglobin, hemoglobin, and carboxypeptidase in another? (The difference is more fundamental than the nature of the backbone structures of the chains.)

54 What is the difference in chemical behavior between the side chains of the amino acids lysine, leucine, and serine?

55 What is a hydrophobic side chain? What influence does it have on the structure of a protein molecule?

56 How are hydrogen bonds involved in protein structure? How does a side effect of the planarity of a peptide group favor hydrogen bonding in proteins?

57 What is the purpose of a catalyst in chemical reactions?

58 How do enzymes exert a catalytic influence on a chemical reaction? What is the substrate in an enzymatic reaction?

59 What do we mean by the statement that an enzyme has two roles, recognition and catalysis?

60 In Chapter 11, we remarked that "Life is really applied transition-metal chemistry." In what way is this exaggeration supported in enzyme catalysis?

PROBLEMS

1 a) Calculate the heat of formation of

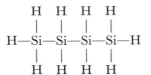

The Si—Si bond energy is 42.0 kcal mole^{-1}, the Si—H bond energy is 74.0 kcal mole^{-1}, and the H—H bond energy is 103.0 kcal mole^{-1}. The heat of formation of Si(g) is 88.0 kcal mole^{-1}.

b) Calculate the heat liberated in the reaction

$$
\begin{array}{cccc}
H & H & H & H \\
| & | & | & | \\
H-Si-Si-Si-Si-H(g) + 6\tfrac{1}{2}O_2(g) \rightarrow \\
| & | & | & | \\
H & H & H & H
\end{array}
$$
$$4SiO_2(s) + 5H_2O(l)$$

The heats of formation for $SiO_2(s)$ and $H_2O(l)$ are -205.4 kcal mole^{-1} and -68.3 kcal mole^{-1}, respectively.

2 Solid hyponitrous acid ($H_2N_2O_2$) explodes at the slightest provocation. Draw an electron-dot structure for the compound.

3 There is evidence that the methonium ion, $CH_5{}^+$, forms when CH_4 is in contact with a very strong acid. Present a possible model to explain the bonding in the electron-deficient methonium ion.

Problems 4–9. What is the systematic name of each of the following compounds?

4 $CH_3-CH-CH-CH_3$
$$\quad\quad\quad | \quad\ |$$
$$\quad\quad\quad CH_3\ CH_3$$

5 $CH_3-CH_2-CH(CH_3)_2$

6 $CH_2{=}CH-CH{=}CH_2$

7
$$\quad\quad\quad\quad\quad CH_2-CH_3$$
$$\quad\quad\quad\quad\quad |$$
$$Cl-(CH_2)_2-CH-CH-C{\equiv}CH$$
$$\quad\quad\quad\quad\quad |$$
$$\quad\quad\quad\quad\quad CH_3$$

8

9

$$CH_3-\overset{\overset{\displaystyle O}{\|}}{C}-CH_2-CH_2-CH_3$$

Problems 10–15. Name the following compounds in any correct manner you can, systematic names or common names.

10 HCOOH   **11** CH_3-OH

12

13

$$CH_3-\overset{\overset{\displaystyle O}{\|}}{C}-H \qquad CH_3-\overset{\overset{\displaystyle O}{\|}}{C}-O-H$$

14

$$CH_3-\overset{\overset{\displaystyle O}{\|}}{C}-CH_3$$

15

$$CH_3-\overset{\overset{\displaystyle O}{\|}}{C}-O-CH_3$$

Problems 16–31. Draw the chemical formula for each compound.

16 *o*-Nitrobromobenzene

17 *m*-Nitrophenol   **18** *n*-Butylamine

19 Methylethylketone   **20** Urea

21 Hydrazine   **22** Methyl mercaptan

23 Ethyl cyanide

24 Sodium propionate

25 Sodium stearate

26 Ethylenediamine

27 Pyruvic acid   **28** Glycine   **29** Valine

30 Glutamic acid   **31** Glucose

32 Name the *functional group* that most likely accounts for each of the following phenomena:

a) A compound reacts with an alcohol to produce an ester.

b) A substance has a fruity odor.

c) A gas has a fishy odor.

d) A compound is produced, along with the salt of an acid, upon hydrolysis of an ester.

e) A compound, when mixed with a solution of Br_2, causes the Br_2 to lose its color.

f) A compound reacts with an amine to form a peptide linkage.

33 L-Ascorbic acid (vitamin C), whether synthetic or of natural origin, has the molecular formula $C_6H_8O_6$. The molecule has the following structural features:

1) It contains a 5-membered ring.

2) There is a double bond between carbon atoms 2 and 3 of the ring.

3) Carbon atoms 1 and 4 are connected by an oxygen bridge.

4) An oxygen atom is double-bonded to carbon atom 1.

5) Carbon atoms 2, 3, 5, and 6 are bound to OH groups.

Draw the structure of vitamin C.

34 How many asymmetric carbon atoms are present in

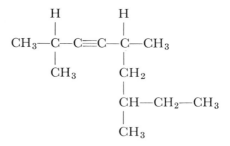

35 How many isomers are there of dichloromethane? Of C_2H_4BrCl?

36 How many isomers are there of tartaric acid,

$$HOOC-\underset{\underset{\displaystyle OH}{|}}{CH}-\underset{\underset{\displaystyle OH}{|}}{CH}-COOH$$

37 1,3-Dichloropentane has only one form, whereas 1,3-dichlorocyclopentane has two geometrical isomers. Why? Illustrate with drawings of the molecules.

38 Calculate the number of grams of glucose that must be degraded to make 100 moles of ATP from ADP and phosphate.

Even if we resolve all matter into one kind, that kind will need explaining. And so on for ever and ever deeper into the pit at whose bottom truth lies, without ever reaching it. For the pit is bottomless.
O. Heaviside

13 NUCLEAR CHEMISTRY

While chemists were assimilating and using the results of the quantum revolution of the 1920's, physicists were continuing their study of the nucleus. Rutherford transmuted nitrogen to oxygen, in 1919, by bombarding it with alpha particles. In 1930, J. D. Cockroft and E. T. S. Walton developed an electrostatic accelerator to produce a high-velocity beam of protons for use in nuclear bombardment experiments. In the same year, E. O. Lawrence invented the cyclotron to accelerate other nuclei for the same purpose. Groups of physicists in Rome and Berlin, Copenhagen and Columbia, Berkeley and Chicago began to study nuclear reactions and to try to prepare transuranium elements. Out of this activity of the 1930's, carried on mostly by physicists, came nuclear fission and the frightening destructiveness of atomic weapons. Most post-World War II international relations have been shaped by the existence of atomic weapons.

From the wartime acceleration of research in nuclear reactions has come the knowledge and the availability of radioactive isotopes that have led to so many chemical applications. Isotopes, radioactive and stable, have enabled us to find answers to chemical problems that are solvable in no other way. Radioisotopes also have given us a means of accurately dating past events, both historical and geological. And they have revealed comparative ages for the earth and the moon that have disproved some old theories about the origin of the moon.

13–1   THE NUCLEUS

The nucleus is comprised of protons and neutrons, both of which are called *nucleons*. It has a positive charge equal to the number of protons it contains, and this number, Z, is the *atomic number*. A neutral atom has an equal number of electrons around the nucleus. Since these outer electrons give each atom its chemical properties, all neutral atoms with the same number of electrons and protons are classified as the same element. Hence, the atomic number identifies an element. The total number of protons and neutrons in a nucleus is its *mass number*, A.

Atoms with the same number of protons but different numbers of neutrons are called *isotopes*. In the representation of an isotope, the atomic number, Z, is written as a left subscript to the symbol of the element, and the mass number, A, is written as a left superscript. Thus, the mercury isotope with 80 protons and 116 neutrons is written $^{196}_{80}Hg$ ($80 + 116 = 196$). The mass of the nucleus, in atomic mass units (amu), is nearly equal to its mass number, A. One amu is defined as exactly one twelfth the mass of one atom of carbon in the $^{12}_{6}C$ isotope. One amu is 1.66053×10^{-24} g. Other mass conversions are given in Appendix 1.

Elements found in nature usually are mixtures of several isotopes. For example, hydrogen has three isotopes, $^{1}_{1}H$, $^{2}_{1}H$, and $^{3}_{1}H$, of which the first two occur in nature in the proportions indicated in Table 13–1. The nuclei of these three isotopes contain a proton, a proton and a neutron, and a proton and two neutrons, respectively. Mercury has isotopes that contain from 189 to 206 neutrons and protons, or from 109 to 126 neutrons. The seven isotopes listed in Table 13–1 occur naturally in the relative amounts listed. The observed atomic weight, 200.59, is the weighted average of the atomic weights of the individual naturally occurring isotopes.

Size and Shape

Rutherford made the first measurements of the size of a nucleus during his α-particle scattering experiments. Better measurements can be made with neutron scattering because neutrons are not deflected by electrostatic repulsion. Numerous neutron-scattering experiments have shown that the radius of a nucleus is proportional to the cube root of the number of nucleons within it:

$$r = 1.33 \times 10^{-13} A^{1/3} \text{ cm}$$

Table 13–1.   Natural Isotopes of Hydrogen and Mercury[a]

Isotope	Mass (amu)	Percent natural abundance	Name
$^{1}_{1}H$	1.0078	99.98	Hydrogen
$^{2}_{1}H$ or D	2.0141	0.02	Deuterium
$^{3}_{1}H$ or T	—	—	Tritium
Natural mixture 1.0080			
$^{196}_{80}Hg$	195.9658	0.15	Mercury-196
$^{198}_{80}Hg$	197.9668	10.02	Mercury-198
$^{199}_{80}Hg$	198.9683	16.84	Mercury-199
$^{200}_{80}Hg$	199.9683	23.13	Mercury-200
$^{201}_{80}Hg$	200.9703	13.22	Mercury-201
$^{202}_{80}Hg$	201.9706	29.80	Mercury-202
$^{204}_{80}Hg$	203.9735	6.85	Mercury-204
Natural mixture 200.59			

[a] Values from the *Handbook of Chemistry and Physics,* Chemical Rubber Co., Cleveland, Ohio, 1971–72, 52nd ed.

The isotope $^{1}_{1}H$ has a radius of 1.33×10^{-13} cm, and $^{238}_{92}U$ has a radius of 8.25×10^{-13} cm. The atomic radii, including electron clouds, are twenty thousand times the size of the nucleus.

If a collection of charges is not perfectly spherical, then it can have what is called an electric quadrupole moment even though it does not have a dipole moment. These quadrupole moments can be measured, although they will not concern us here. Such measurements have revealed that many nuclei are spherical; and most of those that are not spherical are elongated like a football, in which the ratio of longest to shortest diameter never exceeds 1.2.

Binding Energy

The mass *number*, *A*, and the mass of a nucleus in amu are not the same, partly because the mass of a proton or a neutron is not exactly 1 amu. The mass of a proton (Appendix 1) is 1.0073 amu, and the mass of a neutron is 1.0087 amu. There is another reason as well: An atom of a stable isotope weighs *less* than the sum of masses of electrons, protons, and neutrons from which it is built.

Example.   What is the total mass of the particles of which an atom of $^{200}_{80}Hg$ is composed?

Solution.   The mass of a neutron is 1.008665 amu, and that of a proton *and* the electron in the electron cloud which neutralizes the proton's charge is 1.007763 amu. (Masses of neutral atoms usually are tabulated, rather than masses of nuclei.) The given isotope of mercury has 80 protons and 200 —

80 = 120 neutrons. The total mass of the component particles then is

$$80 \times 1.007763 \text{ amu} = 80.62104 \text{ amu of protons and electrons}$$

$$120 \times 1.008665 \text{ amu} = \underline{121.03980} \text{ amu of neutrons}$$

$$\text{Total} = 201.66084 \text{ amu of component parts of } {}^{200}_{80}\text{Hg}$$

Yet the observed atomic mass of ${}^{200}_{80}\text{Hg}$ is only 199.9683 amu. What has happened to the other 1.6925 amu of matter?

The missing mass has been converted to energy that is given off, energy which is needed before the nucleus breaks apart again. This "binding energy" of the nucleus can be calculated from Einstein's famous relationship between mass and energy:

$$E = mc^2$$

$$\text{(ergs)} \quad \text{(g)(cm}^2 \text{ sec}^{-2})$$

Here c is the speed of light, 2.998×10^{10} cm sec^{-1}. The energy into which 1.00000 amu of matter can be converted is

$$E = 1.0000 \text{ amu} \times \frac{1.66053 \times 10^{-24} \text{ g}}{1 \text{ amu}} \times (2.998 \times 10^{10})^2 \text{ cm}^2 \text{ sec}^{-2}$$

$$E = 1.492 \times 10^{-3} \text{ erg} \times \frac{6.2420 \times 10^{11} \text{ eV}}{1 \text{ erg}} = 9.31 \times 10^8 \text{ eV}$$

(The conversions from amu to g, and from ergs to eV, are given in Appendix 1, as is the conversion, 1 erg = 1 g cm^2 sec^{-2}.) It is useful to remember that 1 amu of mass is equivalent to 931 MeV (million electron volts) of energy.

In making ${}^{200}_{80}\text{Hg}$ from electrons, protons, and neutrons, the loss in mass *per nucleon* is 1.6925 amu/200 nucleons = 0.00846 amu nucleon^{-1}. This mass corresponds to a binding energy of 0.00846 amu nucleon$^{-1} \times$ 931 MeV amu^{-1} = 7.88 MeV nucleon^{-1}. The binding energy per nucleon is plotted against mass number, A, for the entire range of elements in Figure 13–1. In the first few elements, the binding energy per nucleon is low and is approximately proportional to the number of nucleons. Beyond oxygen, the binding energy per nucleon is almost constant at 8 MeV per nucleon. This fact means that forces between nucleons are quite short-range, and that each nucleon essentially interacts only with its immediate neighbors. If nuclei were held together by longer-range forces so every nucleon in the nucleus interacted with every other one, the *total* binding energy would increase with the square of the number of nucleons and not with the first power, as it does. The initial increase of binding energy per nucleon with number of nucleons is reasonable when we realize that in such small nuclei each nucleon does not have its full complement of neighbors. Therefore, each new nucleon increases the binding of a preexisting nucleon by increasing its packing coordination number. Protons and neutrons are not simply closely packed in the nucleus like marbles

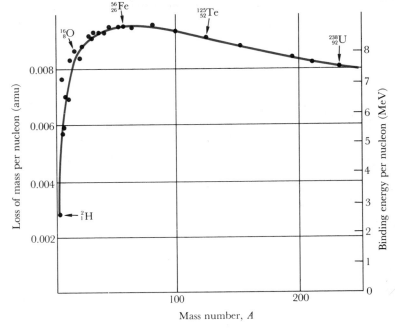

Figure 13–1. The loss of mass and the binding energy per nucleon for the formation of nuclei from electrons, protons, and neutrons. Beyond oxygen, the total binding energy is nearly proportional to the number of nucleons in the nucleus; that is, the binding energy per nucleon is approximately constant.

in a box. (We soon shall see evidence for structure within the nucleus.) But with regard to binding energy, they behave as if they were.

Why have we waited until now to introduce the idea that energy changes in a reaction are accompanied by mass changes? If the total binding energy of a mercury nucleus is equivalent to 1.6925 amu of matter, why have we not made comparable calculations for the bond energies of molecules? The answer lies in the enormously greater energies involved in nuclear reactions. As an example, let us calculate the amount of mass that is equivalent to the entire bond energy of a Cl_2 molecule.

Example. The enthalpy of the following reaction is 58.02 kcal mole^{-1} of Cl_2:

$$Cl_2(g) \rightarrow 2Cl(g)$$

What is the mass equivalent of this energy, in amu molecule^{-1}?

Solution

$$\Delta H = 58.02 \text{ kcal mole}^{-1} \times \frac{1 \text{ eV molecule}^{-1}}{23.06 \text{ kcal mole}^{-1}} = 2.52 \text{ eV molecule}^{-1}$$

$$\Delta m = 2.52 \text{ eV} \times \frac{1 \text{ amu}}{9.31 \times 10^8 \text{ eV}} = 2.71 \times 10^{-9} \text{ amu}$$

The total energy involved in the dissociation of a molecule of Cl_2 amounts to a mass of only five millionths of an electron mass. Chemical reactions usually involve energies of a few electron volts, whereas nuclear energies are in the million electron volt range. One MeV per molecule equals 23.06 *million* kcal mole^{-1}, which is quite outside the range of most chemical reactions. Therefore, in chemical reactions we can work with the separate principles of conservation of mass and of energy. Interconversions are not detectable. In contrast, for nuclear reactions the interconversion of mass and energy is normal; here we must use the more general principle of the conservation of mass–energy. In any nuclear reaction, the total of *mass and energy* of the reacting species and their surroundings does not change during the course of the reaction.

13–2 NUCLEAR DECAY

Many nuclei do not decay. These are called the *stable* isotopes. Others break down spontaneously to new elements, and they are the *radioactive* isotopes. The radiation that Becquerel observed in 1896 (Chapter 8), was α particles (^4_2He nuclei) from the decay of uranium:

$$^{238}_{92}U \rightarrow {}^{234}_{90}Th + {}^4_2He$$

One single isotope of polonium, found in uranium ore by the Curies in 1898, decays in three different ways:

$$^{207}_{84}Po \rightarrow {}^{203}_{82}Pb + {}^4_2He \qquad (\alpha \text{ emission})$$
$$^{207}_{84}Po \rightarrow {}^{207}_{83}Bi + {}^{\ 0}_{+1}e \qquad (\beta^+ \text{ emission})$$
$$^{207}_{84}Po \rightarrow {}^{207}_{83}Bi \qquad (\text{electron capture})$$

In the first reaction, an α particle is emitted and polonium changes to lead. In the second and third reactions, one proton in the nucleus changes to a neutron. This is accomplished in the second reaction by emitting a *positron* (β^+), a particle with the mass of an electron but with a unit positive charge:

$$^{\ 1}_{+1}p \rightarrow {}^1_0n + {}^{\ 0}_{+1}e$$

In the third reaction, the change occurs by capturing one of the electrons from the orbitals around the nucleus:

$$^{\ 1}_{+1}p + {}^{\ 0}_{-1}e \rightarrow {}^1_0n$$

(The two nuclear reactions of a proton just written do not occur in the simple manner implied here. The overall result is as shown, but the actual mechanism is more complicated. Such a simplification is analogous to the way we write most chemical reactions.) In either of these last two reactions, the atomic number decreases by one, but the atomic mass remains the same. Radium-228 decays by another mechanism, with the emission of an electron:

$$^{228}_{88}Ra \rightarrow {}^{228}_{89}Ac + {}^{\ 0}_{-1}e \qquad (\beta^- \text{ emission})$$

These examples illustrate the four types of decay: β^- emission, electron capture, β^+ emission, and α-particle emission.

β^- or Electron Emission

In spontaneous β^- decay, one of the neutrons in a nucleus becomes a proton and an electron that is emitted from the nucleus. Energy accompanies the stream of emitted electrons from a sample undergoing decay, and the calculation of this energy is relatively straightforward. Consider the decay of carbon-14 to nitrogen-14:

$$^{14}_{6}C \rightarrow {}^{14}_{7}N + {}^{0}_{-1}e$$

During this reaction, a carbon nucleus and six electrons are converted into a nitrogen ion having six electrons (one too few), and one electron as a β^- particle. Thus, the masses of reactants and products are equal to the masses of neutral atoms of C and N:

Reactants: mass of $^{14}_{6}C$ atom: 14.003242 amu

Products: mass of $^{14}_{7}N$ atom: 14.003074 amu

Loss of mass in reaction: 0.000168 amu

The energy equivalent to this loss of mass is

$$E = 0.000168 \text{ amu} \times 931 \text{ MeV amu}^{-1} = 0.156 \text{ MeV}$$

Therefore, the emitted electron has an energy of 0.156 MeV. In β^- decay, the atomic number always increases by one unit.

Orbital Electron Capture, EC

In orbital electron capture, one electron from the cloud surrounding the nucleus is captured by the nucleus and combines with a proton to form a neutron. An example is the decay of beryllium-7 to lithium-7:

$$^{7}_{4}Be \xrightarrow{\text{EC}} {}^{7}_{3}Li$$

In this reaction, a nucleus and four orbital electrons are changed to a nucleus that is heavier by one electron, and three orbital electrons. Again, the masses of reactants and products are the same as the ones for the neutral atoms

Reactants: mass of $^{7}_{4}Be$ atom: 7.0169 amu

Products: mass of $^{7}_{3}Li$ atom: 7.0160 amu

Loss of mass in reaction: 0.0009 amu or 0.84 MeV

In electron capture, the atomic number always decreases by one.

β^+ or Positron Emission

Carbon-11 decays by the emission of positrons:

$$^{11}_{6}C \rightarrow {}^{11}_{5}B + {}^{0}_{+1}e$$

The energy calculation contains a trap. Carbon and its six electrons are converted to boron with six electrons (one too many) *plus* the mass of the positron. In the mass balance, a neutral carbon atom is converted to a neutral boron atom and *two* electron masses:

Products: one $^{11}_{5}$B atom: 11.009305 amu

two electrons: $\underline{0.001098}$ amu

11.010403 amu

Reactants: one $^{11}_{6}$C atom: $\underline{11.011443}$ amu

Loss of mass in reaction: 0.001040 amu or 0.968 MeV

In positron emission, as in electron capture, the atomic number decreases by one.

α-Particle Emission

For the heavier elements, above $A = 200$, a common mode of decay is the emission of a helium nucleus or an α particle:

$$^{232}_{90}\text{Th} \rightarrow {}^{228}_{88}\text{Ra} + {}^{4}_{2}\text{He}$$

In α-particle emission, the atomic number decreases by two, and the mass by four.

γ Emission During α Decay

High-energy electromagnetic radiation also can be emitted during a nuclear decay. In the uranium-238 reaction,

$$^{238}_{92}\text{U} \rightarrow {}^{234}_{90}\text{Th} + {}^{4}_{2}\text{He}$$

α particles of two different energies are liberated, 4.18 MeV and 4.13 MeV. Electromagnetic radiation accompanies these with an energy equal to the difference in α-particle energies, 0.05 MeV. Radiation of such energy occurs at the extreme right of the electromagnetic spectrum, as shown in Figure 8–5; this is called γ radiation.

Example. What is the wavelength of 0.05-MeV γ radiation?

Solution

$$E = 0.05 \times 10^6 \text{ eV} \times \frac{1.602 \times 10^{-12} \text{ erg}}{1 \text{ eV}} = 0.08 \times 10^{-6} \text{ erg}$$

$$E = h\nu = hc/\lambda \qquad (\lambda = hc/E)$$

$$\lambda = \frac{6.626 \times 10^{-27} \text{ erg sec} \times 3.00 \times 10^{10} \text{ cm sec}^{-1}}{0.08 \times 10^{-6} \text{ erg}}$$

$$\lambda = 0.25 \times 10^{-8} \text{ cm} = 0.25 \text{ Å}$$

Figure 13–2. A nuclear energy-level diagram showing the ground state and an excited state for $^{234}_{90}$Th. This diagram explains the occurrence of α particles with 4.18 and 4.13 MeV energy, and γ radiation with 0.05 MeV energy, in the decay of $^{238}_{92}$U. More complicated patterns of γ radiation during decay reveal even more about nuclear energy levels.

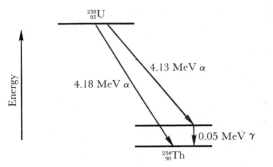

Radiation in this wavelength region is called either x ray or γ ray, depending on the source of the radiation. That from nuclear decay traditionally is called γ radiation, and that obtained by bombarding a metal anode with a beam of electrons is called x radiation. Both are electromagnetic radiation.

The 4.13-MeV α particles are emitted when ^{238}U is converted to ^{234}Th in an *excited* nuclear state. When ^{234}Th drops to its *ground* state, 0.05-MeV γ radiation is liberated. The two energy levels for ^{234}Th are shown in Figure 13–2. Thus, a form of "nuclear spectroscopy" is possible, analogous to atomic spectroscopy with the hydrogen atom. In the future, such studies may reveal the substructure of the nucleus.

Stability and Half-Life

The number of nuclei that decay in a given time in a sample of material is always proportional to the amount of material left to decay. Nuclear decay is an independent, nucleus-by-nucleus process. The probability that a given nucleus will decay in a given time is constant and independent of the surroundings of the nucleus. Therefore, the total number of nuclei that decay in a given time is proportional to the total number of nuclei present. If the number of nuclei is n, and if the change in this number is Δn during a time Δt, then the decay is represented by

$$\text{Nuclei decaying per unit of time} = -\frac{\Delta n}{\Delta t} = kn \qquad (13\text{–}1)$$

This is known as a *first-order* rate equation since the rate depends on the first power of a concentration term, n. Using calculus, we can convert this rate equation into an expression relating the number of nuclei, n, that remain at time t to the number originally present, n_0, at time $t = 0$. This integrated expression is

$$n = n_0 e^{-kt} \qquad (13\text{–}2)$$

This expression is plotted in Figure 13–3 for the decay of $^{14}_{6}$C, starting with $n_0 = 1$ g of carbon-14. Although this is a plot of Equation 13–2, you can verify that it is compatible with Equation 13–1 by noting that the slope of

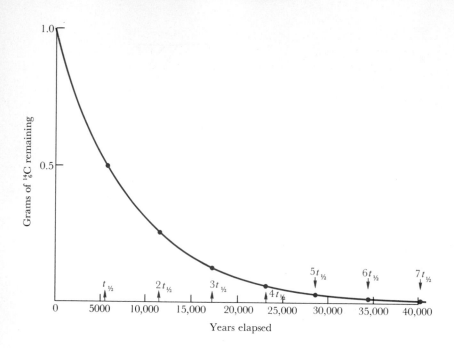

Figure 13–3. The nuclear decay curve for carbon-14: $^{14}_{6}C \rightarrow$ $^{14}_{7}N + ^{0}_{-1}e$. Every 5570 years, the amount of carbon remaining is halved. This half-life of 5570 years is given the symbol $t_{1/2}$. The regular decay of carbon-14 is the basis for dating carbon-containing objects that ceased to be alive within the last 20,000 years.

the curve, which is approximately equal to $\Delta n / \Delta t$, is proportional to the amount of carbon left at every point on the curve.

An important property of first-order decay is that the time required for any amount of material to decay to half that amount is constant and independent of the amount of material present. If this time is defined as the *half-life*, $t_{1/2}$, then

$$\frac{n_0}{2} = n_0 e^{-kt_{1/2}}$$

$$2 = e^{+kt_{1/2}}$$

$$t_{1/2} = \frac{\ln 2}{k} = \frac{0.693}{k}$$

If either the half-life, $t_{1/2}$, or the first-order rate constant, k, is known, the other can be calculated. If the half-life of a radioactive substance is 10 days, the same time (10 days) is required for 1 g of material to decay to a half gram, for 16 g to decay to 8 g, for 10 tons to decay to 5 tons, or for 2 mg to decay to 1 mg. This may seem paradoxical, until you recall that 10 tons of radioactive material contain many more atoms than 2 mg, so it is natural that many more should decay in the 10-day period if *each atom* has precisely the same probability of decaying within 10 days' time.

The half-lives for the reactions that we have examined vary enormously. The half-life for the reaction

$$^{238}_{92}\text{U} \rightarrow ^{234}_{90}\text{Th} + ^{4}_{2}\text{He}$$

is 4.51 billion years. The half-life for polonium decay, according to the reaction

$$^{207}_{84}\text{Po} \rightarrow ^{203}_{82}\text{Pb} + ^{4}_{2}\text{He}$$

is 5.7 hours. The decay of carbon-14 in Figure 13–3 has a half-life of 5570 years, and the decay of astatine to bismuth,

$$^{216}_{85}\text{At} \rightarrow ^{212}_{83}\text{Bi} + ^{4}_{2}\text{He}$$

has a half-life of only 3×10^{-4} sec.

13–3 STABILITY SERIES

In Figure 13–4 are plotted the stable isotopes for all elements (in black) and the radioactive isotopes for elements below atomic numbers of 35 and above 75 (in color). Notice that in stable isotopes beyond H and He the number of protons is never greater than the number of neutrons, and that the most stable isotopes have an excess of neutrons. The neutrons "dilute" the positive charges of protons and help to stabilize the nucleus against charge repulsion.

Note also the zig-zag appearance of the band of stable isotopes, and the predominance of isotopes with even numbers of protons, or neutrons, or both. This is suggestive of some type of pairing and of substructure within the nucleus. This preference for even numbers of each type of nucleon is even more apparent in Table 13–2.

Isotopes to the right of and below the region of greatest stability in Figure 13–4 can reach this region by losing electrons. β^- Emission is the rule when such nuclei decay. Isotopes to the left and above the stable region can decay to stable isotopes by electron capture or positron emission. Above the region of approximately $Z = 80$, α-particle emission predominates. β^- Emission moves the isotope diagonally up to the left by one square; either electron capture or positron emission proceed in the reverse direction, down and to the right by one square. α-Particle emission takes the nucleus down and to the left by *two* squares, almost along the line of greatest stability. This mode of decomposition is used by atoms beyond the end of the stable region in Figure 13–4.

Figure 13–4 shows only the *existence* of stable (nonradioactive) isotopes, and not their degree of nuclear stability or their abundance. Nuclei are particularly stable and abundant when they have either Z or n (the number of neutrons) equal to 2, 8, 20, 28, 50, 82, or 126. These have been called *magic*

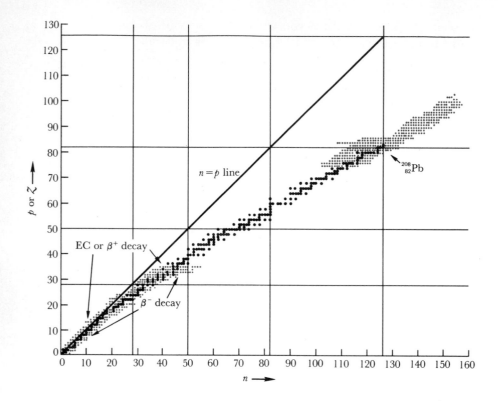

Figure 13–4. A plot of stable isotopes (black dots) and radioactive isotopes (colored dots) as a function of their number of protons, p or Z, and number of neutrons, n. Radioactive isotopes from Z = 35 to 75 are omitted for clarity. The band of stable isotopes is bordered on both sides by radioactive isotopes. On this plot, radioisotopes lying above the band of stable isotopes decay to stable isotopes by electron capture (EC) or positron emission (β⁺). Radioisotopes below the stable band decay by electron (β⁻) emission. The stable band ends at $^{209}_{83}$Bi, but there are naturally occurring isotopes with quite long half-lives beyond this point. For example, $^{232}_{90}$Th has a half-life of 13.9 billion years, and $^{238}_{92}$U has a half-life of 4.51 billion years.

Table 13–2.   Occurrence of Stable Nuclei with Even and Odd Numbers of Protons and Neutrons

		Neutrons	
		Even	Odd
Protons	Even	166	53
	Odd	57	8

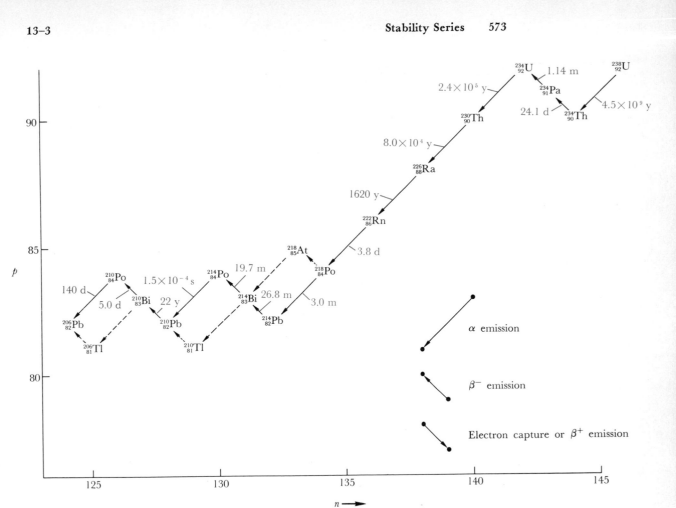

Figure 13–5. The natural radioactivity series beginning with $^{238}_{92}U$ and ending with the stable isotope $^{206}_{82}Pb$. Note that the overall effect has been to come backward along the stable isotope band by a series of α-particle and β⁻ emissions. y = years, d = days, m = minutes, and s = seconds. Dashed lines are alternative decay schemes.

numbers. Although they convey information about the shell structure of the nucleus, we do not yet have a theory that will explain them. They are comparable to the set of magic numbers of 2, 10, 18, 36, 54, and 86 for the atomic numbers of the particularly stable noble gases. There must be an explanation in terms of nuclear shell structure, and these nuclear quantum shells must exist independently for protons and neutrons. A magic number of either protons or neutrons bestows stability upon the nucleus; atoms such as $^{208}_{82}Pb$ with magic numbers of each are exceptionally stable. Such nuclei are also nearly spherical, as determined from their quadrupole moments. The shell theories of the nucleus that have been proposed have led to some useful predictions, but we are presently at the same point with nuclear structure that Bohr and Sommerfeld were with electronic structures of atoms. Someone in the next generation of scientists is needed to give us the equivalent of the Schrödinger equation and wave mechanics for nuclei.

Natural Radioactive Series

The heavy, unstable elements at the upper right end of the stability curve in Figure 13–4 decay in a set of four pathways. One of these, starting with $^{238}_{92}U$ and ending with the stable $^{206}_{82}Pb$, is illustrated in Figure 13–5. This enlarged fragment of Figure 13–4 shows more clearly the changes produced on the plot by α emission, β^- emission, and electron capture or β^+ emission. A second such series begins with $^{232}_{90}Th$ and ends, after 10 steps, at $^{208}_{82}Pb$. A third begins at $^{235}_{92}U$ and ends at $^{207}_{82}Pb$, and a fourth begins at the artificial $^{241}_{94}Pu$ and ends at $^{209}_{83}Bi$.

13–4 NUCLEAR REACTIONS

In 1919, Rutherford carried out the first transmutation of one element into another by bombardment. He used α particles to change nitrogen into oxygen:

$$^{14}_{7}N + {}^{4}_{2}He \rightarrow {}^{17}_{8}O + {}^{1}_{1}H$$

He demonstrated the occurrence of this reaction by detecting the protons emitted. In this reaction, the α particle fuses with the nitrogen nucleus to form an unstable and excited intermediate, $^{18}_{9}F$, which then decomposes to oxygen and the proton. In nuclear reactions such as Rutherford's it is difficult to force a charged particle close enough to the nucleus to react. One of the chief goals of the development of particle accelerators such as the linear accelerator and the cyclotron has been to produce beams of positively charged nuclei having enough energy to make them react with target nuclei.

Neutron beams need not be as energetic, because the neutrons themselves are not electrostatically repelled by the target nuclei. For example, neutron beams from atomic reactors are used to prepare tritium, $^{3}_{1}H$, for medical or for chemical tracer work:

$$^{10}_{5}B + {}^{1}_{0}n \rightarrow {}^{3}_{1}H + 2\,{}^{4}_{2}He$$
$$^{6}_{3}Li + {}^{1}_{0}n \rightarrow {}^{3}_{1}H + {}^{4}_{2}He$$

Radioactive cobalt-60, which is used in cancer therapy, is prepared from the stable isotope by neutron bombardment:

$$^{59}_{27}Co + {}^{1}_{0}n \rightarrow {}^{60}_{27}Co$$

Artificial Elements

One of the most interesting applications of high-energy accelerators has been the preparation of new transuranium elements. Elements 93 through 105 have been prepared by the following bombardment reactions:

$$^{238}_{92}U \; + \; ^{2}_{1}H \; \rightarrow \; ^{238}_{93}Np \; + \; 2\,^{1}_{0}n \qquad \text{(Neptunium)}$$

$$^{238}_{92}U \; + \; ^{4}_{2}He \; \rightarrow \; ^{239}_{94}Pu \; + \; 3\,^{1}_{0}n \qquad \text{(Plutonium)}$$

$$^{239}_{94}Pu \; + \; ^{4}_{2}He \; \rightarrow \; ^{240}_{95}Am \; + \; ^{1}_{1}p \; + \; 2\,^{1}_{0}n \qquad \text{(Americium)}$$

$$^{239}_{94}Pu \; + \; ^{4}_{2}He \; \rightarrow \; ^{242}_{96}Cm \; + \; ^{1}_{0}n \qquad \text{(Curium)}$$

$$^{244}_{96}Cm \; + \; ^{4}_{2}He \; \rightarrow \; ^{245}_{97}Bk \; + \; ^{1}_{1}p \; + \; 2\,^{1}_{0}n \qquad \text{(Berkelium)}$$

$$^{238}_{92}U \; + \; ^{12}_{6}C \; \rightarrow \; ^{246}_{98}Cf \; + \; 4\,^{1}_{0}n \qquad \text{(Californium)}$$

$$^{238}_{92}U \; + \; ^{14}_{7}N \; \rightarrow \; ^{247}_{99}Es \; + \; 5\,^{1}_{0}n \qquad \text{(Einsteinium)}$$

$$^{238}_{92}U \; + \; ^{16}_{8}O \; \rightarrow \; ^{249}_{100}Fm \; + \; 5\,^{1}_{0}n \qquad \text{(Fermium)}$$

$$^{253}_{99}Es \; + \; ^{4}_{2}He \; \rightarrow \; ^{256}_{101}Md \; + \; ^{1}_{0}n \qquad \text{(Mendelevium)}$$

$$^{246}_{96}Cm \; + \; ^{13}_{6}C \; \rightarrow \; ^{254}_{102}No \; + \; 5\,^{1}_{0}n \qquad \text{(Nobelium)}$$

$$^{252}_{98}Cf \; + \; ^{10}_{5}B \; \rightarrow \; ^{257}_{103}Lr \; + \; 5\,^{1}_{0}n \qquad \text{(Lawrencium)}$$

$$^{249}_{98}Cf \; + \; ^{12}_{6}C \; \rightarrow \; ^{257}_{104}XX \; + \; 4\,^{1}_{0}n \qquad \text{(Unnamed)}$$

$$^{249}_{98}Cf \; + \; ^{15}_{7}N \; \rightarrow \; ^{260}_{105}XX \; + \; 4\,^{1}_{0}n \qquad \text{(Unnamed)}$$

As you can surmise from the names given these new elements, many of them have been produced at the Lawrence Radiation Laboratory at the University of California, Berkeley. Perhaps now that Mendeleev has been honored, it would be appropriate to use the name *Newlandium* for element 105, since 105 is evenly divisible into octaves. (See Newlands' law of octaves, page 216.)

What lies ahead for the synthesis of transuranium elements? Will there be more radioactive and extremely short-lived species such as 97 through 105? It now appears as if there is a chance of reaching a new zone of stability that might even include some nonradioactive elements. Calculations with nuclear-shell models have led to the expectation that element $^{298}114$, with 114 protons and 184 neutrons (both magic numbers in the nuclear shell theory) would be on an island of stability in a sea of instability. Figure 13–4 is replotted with a third dimension in Figure 13–6, whose vertical axis is a measure of nuclear stability. If some means can be found to reach the elements in the neighborhood of $^{298}114$, we may have a set of relatively long-lived species. Some attempts have been made at Berkeley, and the following reactions are possibilities:

$$^{248}_{96}Cm \; + \; ^{40}_{18}Ar \; \rightarrow \; ^{284}114 \; + \; 4\,^{1}_{0}n$$

$$^{244}_{94}Pu \; + \; ^{48}_{20}Ca \; \rightarrow \; ^{288}114 \; + \; 4\,^{1}_{0}n$$

The first reaction has been unsuccessful, probably because its product nuclei are neutron-deficient and hence unstable; they lie just to the left of the island of stability in Figure 13–6. The second reaction is more promising, but could

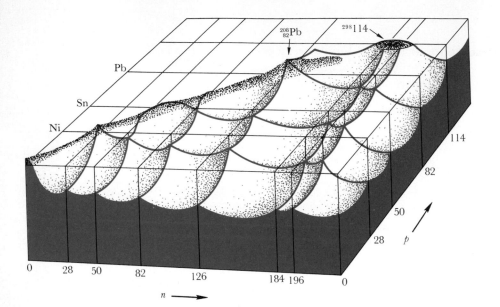

Figure 13–6. The relative stabilities of isotopes as a function of the number of protons, p, and of neutrons, n. The axes in the horizontal plane are the ones of Figure 13–4. The vertical axis indicates the relative stability of the nuclei, with stable nuclei rising above an imaginary "sea level" in a long peninsula. Nuclei that have a magic number of either protons or neutrons (28, 50, 82, 126) are less susceptible to breakdown than their neighbors. These correspond to ridges on the ocean floor. The doubly magic $^{208}_{82}$Pb nucleus is shown as a mountain, and the suspected region of stability around ^{298}114 as a hill above the waterline. [Redrawn from G. T. Seaborg and J. L. Bloom, "The Synthetic Elements: IV," Copyright © April, 1969 by Scientific American. Also see S. G. Thompson and C. F. Tsang, "Superheavy Elements," Science 178, 1047 (1972).]

not be tried until recently because of the unavailability of suitable heavy-ion accelerators.

With the realization that it may be possible to jump to a new region of stability, it is of interest to extend the periodic table further. Figure 13–7 extends the table through the end of the presently partially complete seventh period and a new eighth period. In this period, for the first time, there appear *g* orbitals, the 5*g*. There may be some initial uncertainties about the order of filling the 5*g*, 6*f*, and 7*d*. However, recent calculations at Los Alamos have indicated that after the first one or two electrons, the remainder fill the 5*g* orbitals in an orderly manner. These elements might be called the *hypotransition metals*, from "hypo," meaning "deeply buried" or "below."

Fission

One of the most famous (or infamous) of nuclear reactions is

$$^{235}_{92}U + {}^1_0n \rightarrow {}^{139}_{56}Ba + {}^{94}_{36}Kr + 3\,{}^1_0n$$

This is the reaction of the ^{235}U atomic bomb. This reaction was produced first in the late 1930's by Enrico Fermi and his colleagues in Rome, and by Otto Hahn, Lise Meitner, and Fritz Strassman at the Kaiser Wilhelm Institute in Berlin. Both groups were trying to produce transuranium elements by the methods used later by the Berkeley group and others. No one expected *fission* (or fragmentation) into two approximately equal pieces to occur. In 1938, Hahn identified one of the decomposition products as barium, thereby indicating that splitting or fission was taking place. He passed the news privately by letter to Lise Meitner, who had been forced to leave Nazi Germany and was working in Scandinavia. Otto Frisch, her cousin, discovered that tremendous quantities of energy were emitted during the reaction, and he realized that fission might have military applications. At the time (early 1939), Frisch was working in Niels Bohr's laboratory in Copenhagen. On a visit to the United States in January of 1939, Bohr related this information about nuclear fission to physicists in the United States. Fermi had fled the Fascists in Italy and was at Columbia University. He verified Bohr's news, as did others at Berkeley.

The Hungarian physicist Leo Szilard, one of many European scientists who sought political asylum in England and the U.S. in the late 1930's, acted as a spokesman for many concerned physicists. Foreseeing the significance of fission, he persuaded Albert Einstein to write a letter to President Roosevelt in August, 1939. In the now-famous letter, Einstein outlined the military possibilities of nuclear fission and mentioned the grave possibilities that the Nazis might develop it as a weapon in the war that was brewing in Europe. Roosevelt responded by establishing the "Manhattan Project" and supporting the enormous research effort that led, in 1945, to the first bomb test at Trinity Flats, New Mexico, and the two bombs dropped on Hiroshima and Nagasaki. The atomic bomb, more than any other development, made scientists aware that they had to be concerned with the political and social impact of their discoveries, and led to the increased involvement of scientists in public affairs that has typified the post-World War II era.

The uranium fission reaction is a potential chain reaction because it produces three neutrons for every one used to initiate a fission of a nucleus. Naturally occurring uranium has only a small amount of ^{235}U dispersed among the more abundant ^{238}U. However, if ^{235}U is purified, and enough of it is brought together, each fission of a nucleus will liberate neutrons that will cause the fission of *more* than one nucleus. Thus, a branched-chain explosion will occur. If the piece of ^{235}U is too small, the losses of neutrons to the surroundings will prevent a chain reaction. The mass at which losses to the surroundings become smaller than the rate of production of neutrons is the *critical mass* of uranium.

In a nuclear reactor, the reaction takes place rapidly enough to produce usable heat, but not rapidly enough to become a branched-chain explosion. The control of such a reactor is achieved by means of cadmium rods extended through the atomic pile to absorb neutrons. When the rods are pushed into the pile, so many neutrons are absorbed by the cadmium rods that fission occurs at a very low level. As the rods are withdrawn, fewer neutrons are absorbed by the cadmium and more fission occurs.

Fusion

One of the reactions by which the sun produces energy is the fusion of hydrogen nuclei to form helium:

$$_1^2H + _1^3H \rightarrow _2^4He + _0^1n + 17.6 \text{ MeV}$$

To overcome electrostatic repulsion between hydrogen nuclei, the collision energy must be about 0.02 MeV. This energy can be reached in accelerators

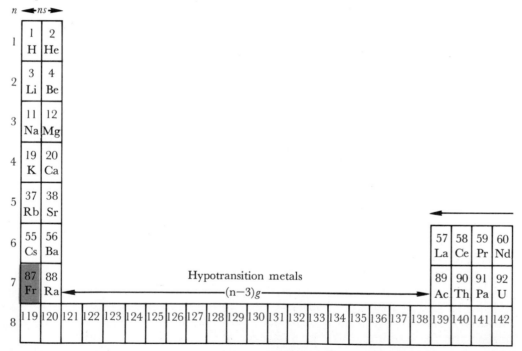

Figure 13–7. The hyperlong form of the periodic table, extended to include the eighth period and the insertion of the 5g orbitals to create a series of hypotransition metals. Artificial elements are printed in color. (Francium and astatine do exist in nature as short-lived intermediates in nuclear decay processes, but it is estimated that there is never more than an ounce of either element in the earth's crust at any given moment.) Elements with colored atomic numbers

on a small scale, but the energy obtained is far less than that needed to operate the accelerators. If we could discover some way to make these reactions work on a large scale and in a controlled manner, large quantities of power would be available. Unfortunately, military applications again are farther advanced than peaceful ones. Temperatures of more than 200 million degrees are required to start the fusion process; these temperatures are attained in hydrogen bombs by using an atom bomb as a kind of match. It is much more difficult to carry out fusion in a controlled manner, and the successful fusion reactor has not yet been designed. At the necessary temperatures for fusion, there are no solids and the normal concepts of containers for reactants are irrelevant. The most promising approach is to utilize magnetic lines of force to keep the hot ionized atoms constrained and away from the walls of the reactor. However, so far no approach has succeeded.

Challenge. Radioactive strontium-90 "fall out" and the maximum permissible dose in human adults are dealt with in Problems 13–10 through 13–12 in *Relevant Problems for Chemical Principles* by Butler and Grosser.

have not yet been prepared. The new artificial transuranium elements from 95 to 105 have completed the second inner-transition series and are beginning a new fourth transition-metal series. The island of stability that Seaborg and his co-workers at Berkeley hope to find lies below lead in the new seventh-period representative elements. The g orbitals being filled in the first hypotransition metals would be so deeply buried that the elements would be extremely difficult to characterize and separate.

13–5   APPLICATIONS OF NUCLEAR CHEMISTRY AND ISOTOPES

Isotopes are useful in studying nonisotopic and nonradioactive chemical reactions because they are a tag by which the pieces of a reactant can be identified in the products of a reaction. The most common nonradioactive isotopes that are used as chemical markers are ^2H, ^{15}N, and ^{18}O. Radioactive isotopes have the additional advantage that they can be detected and their concentrations measured by counting their radioactivity, rather than by chemical analysis. Some radioactive tracer nuclei are ^3H, ^{14}C, ^{32}P, and ^{35}S.

Chemical Markers

As an example of the usefulness of chemical tags in determining the mechanism of a reaction, consider the cleavage of ATP to make ADP and phosphate:

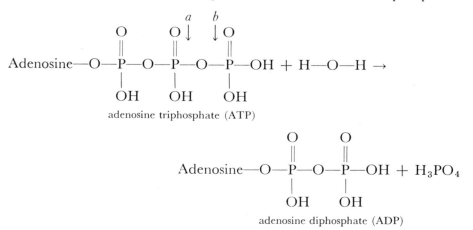

Is the bond marked a or b cleaved by hydrolysis? Without isotopes there is no way to find out. But if the reaction is run in water enriched with ^{18}O, the question is answered by seeing where the ^{18}O occurs in the products. If cleavage is at bond a, the hydroxide from water will bond to ADP, and ADP will contain the ^{18}O. If cleavage occurs at b, the ^{18}O will appear in the phosphate. This experiment has been carried out, and cleavage is at b.

Melvin Calvin and his colleagues at the University of California, Berkeley, determined the molecular mechanism of photosynthesis with isotope tracers by using $^{14}CO_2$ as the starting material.

Radiometric Analysis

Radiometric analysis often is faster and more accurate than conventional chemical analysis. In the analysis for small quantities of Zn(II), the Zn is precipitated by an excess of $(NH_4)_2HPO_4$, in which the P is the ^{32}P radioisotope. The insoluble $Zn(NH_4)PO_4$ precipitate is washed, and its radioactivity measured. From the known radioactivity of pure ^{32}P, we can calculate the concentration of the phosphate precipitate and hence of Zn. This method is faster than conventional weighing analysis. No weighings are required; and the product need not be pure, so long as all of the radioactive $(NH_4)_2HPO_4$ reagent is washed away.

Isotope-Dilution Methods

Isotope dilution is used when it is difficult to separate all of the substance to be analyzed from a complex mixture. By this method, a small amount of the component to be analyzed in the mixture is added *to* the mixture. However, the added material has 100% (or at least a known percent) radioactive species for some atom in the compound. Radioactivities are calculated as *specific activities*, or radioactive disintegrations per second per gram of substance. The added substance is thoroughly mixed. Then the component to be analyzed is isolated by a method that yields, not quantitative separation, but a small amount of extremely pure compound. The dilution of the specific activity of the added compound by the nonradioactive version of the same compound in the mixture leads to the amount of the compound in the initial mixture. For example, if the specific activity of the purified sample is the *same* as that of the added compound, then there was no such compound in the mixture; we are detecting only what we put into the mixture. If the specific activity is half that of the added compound, the compound was present in the mixture in an amount equal to the amount that we added. If the specific radioactivity has fallen to a tenth of its initial value, there must have been nine times as much compound in the mixture as was added. In general, if

W_m = weight of a compound in a mixture

W_a = weight of a radioactive version of the same compound added to the mixture

A_i = specific activity or radioactive decompositions per second per gram of the added compound

A_f = specific activity of the final purified sample

then the weight of the compound in the mixture can be calculated from the expression

$$\frac{W_m}{W_a} = \frac{A_i}{A_f} - 1$$

Example. At the end of a synthesis of benzoic acid (C_6H_5COOH), 10 mg of pure benzoic acid with a specific activity for ^{14}C of 1600 counts min^{-1} mg^{-1} are added to the products of reaction and stirred well. A sample of 40 mg of pure benzoic acid is extracted and shows a specific activity for ^{14}C decay of 190 counts min^{-1} mg^{-1}. How much benzoic acid was produced in the reaction?

Solution. From the expression given, with $W_a = 10$ mg, $A_i = 1600$ cpm mg^{-1}, and $A_f = 190$ cpm mg^{-1}

$$W_m = 10 \text{ mg} \left(\frac{1600}{190} - 1\right) = 74.3 \text{ mg benzoic acid}$$

Exercise. A 1.0-ml sample of an aqueous solution containing 2×10^6 counts sec^{-1} of tritium is injected into the bloodstream of an animal. After time is allowed for complete circulatory mixing, a 1.0-ml sample of blood is withdrawn and found to have an activity of 1.5×10^4 counts sec^{-1}. Calculate the blood volume of the animal.

(*Answer:* 133 ml. This is both more accurate and less harmful than the old method of draining the beast dry.)

Radiocarbon Dating

Reactions in the upper atmosphere involving cosmic neutrons maintain a constant supply of $^{14}_{6}C$ in the atmosphere according to the reaction

$$^{14}_{7}N + ^{1}_{0}n \rightarrow ^{14}_{6}C + ^{1}_{1}H$$

So long as an organism is alive, it maintains a constant ratio of carbon-14 to the stable carbon-12. Carbon is being lost continually by a living organism in the form of CO_2 and organic waste products, and new carbon is ingested in foods. In plants, the intake of carbon from the atmosphere is direct, by means of photosynthesis. Animals that eat plants, or eat the protein from other animals that have eaten plants, also maintain a steady intake of carbon from the photosynthetic process. Since the turnover of carbon atoms in the food chain is so rapid compared with the half-life of decay of ^{14}C, the isotope composition of carbon in a living organism is the *same* as that in the atmosphere around it. But as soon as the organism dies, it ceases to equilibrate its carbon isotopes with the atmosphere. Carbon-14 decay by the reaction

$$^{14}_{6}C \rightarrow ^{14}_{7}N + ^{0}_{-1}e$$

is no longer compensated by more carbon of the atmospheric ratio. Thus, the proportion of carbon-14 in the dead organism decreases. The ratio of ^{14}C to ^{12}C can be used to reveal the age of a sample of a carbon-containing substance (or more precisely, at what time in the past it ceased to be alive).

Living organisms have mixtures of carbon-14 and carbon-12 that produce 15.3 ± 0.1 disintegrations of carbon-14 atoms per minute per gram of carbon. The number of ^{14}C atoms in a sample is proportional to the disintegration rate: $n \propto d$. From the disintegration equation, Equation 13–2, we can write

$$kt = \ln\left(\frac{n_0}{n}\right) = \ln\left(\frac{d_0}{d}\right)$$

in which d is the disintegration rate in counts per minute (cpm) per gram

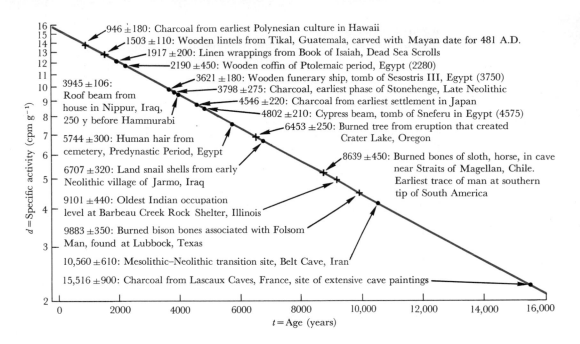

Figure 13–8. Plot of carbon-14 decay rate (in counts per minute per gram of carbon from the sample) against the age of the sample in years. This plot is prepared from the equation $t = 18,600 \log_{10} (15.3/d)$. It is what is called a "semilog plot." Although the vertical axis is marked in units of d, it is really measured in equal units of $\log_{10} d$. Such historically datable points as the Ptolemaic period and the period of Sneferu in Egypt permit us to check the whole concept of radiocarbon dating. The agreement is excellent.

of carbon. Since the rate constant, k, is related to the half-life, $t_{1/2}$, by $t_{1/2} = 0.693/k$, we can write

$$t = \frac{1}{k} \ln \left(\frac{d_0}{d}\right) = \frac{t_{1/2}}{0.693} \ln \left(\frac{d_0}{d}\right) = t_{1/2} \times \frac{2.303 \log_{10} (d_0/d)}{0.693}$$

$$t = t_{1/2} \times 3.33 \log_{10} \left(\frac{d_0}{d}\right) \tag{13–3}$$

This equation is true in general; however, for the special case of $^{14}_{6}C$, which has a half-life of 5570 years and an initial decay rate of 15.3 cpm g^{-1}, it can be written

$$t = 18,600 \log_{10} \left(\frac{15.3}{d}\right) \text{ years}$$

The results of several analyses are shown in Figure 13–8. Carbon-14 dating has been of the utmost importance in straightening out the chronology of prehistoric cultures of Europe and the Middle East. The last ice age ended around 10,000 years ago, and the resulting climatic change in the Middle East led to the invention of farming, the domestication of animals, and the beginning of settled village life—in short, the Neolithic Revolution. It is fortunate for archaeologists that a common element such as carbon has an isotope with a convenient half-life of 5570 years, suitable for dating events in the past 10,000 years. For some of the results, and some of the difficulties in applying carbon-14 dating to archaeology and prehistory, see Renfrew, and Protsch and Berger in the Suggested Reading.

The Age of the Earth and the Moon

The ^{238}U radioactive series shown in Figure 13–5 can be used to date the rocks of the earth by measuring the ratio of ^{238}U to the stable end product, ^{206}Pb (assuming that all of this isotope of lead came from a decomposition of ^{238}U). Since the half-life of ^{238}U, 4.5 billion years, is 20,000 times that of the next longest half-life, we can assume that the time required is for a decay of ^{238}U to ^{234}Th, and that the rest of the chain is relatively instantaneous. Then Equation 13–3 applies, and $t_{1/2} = 4.5 \times 10^9$ years.

Example. In a sample of uranium ore, there are 0.277 g of ^{206}Pb for every 1.667 g ^{238}U. How old is the ore?

Solution. If the lead is entirely a decay product of ^{238}U, then to produce 0.277 g lead, the following amount of ^{238}U had to decay:

$$0.277 \text{ g} \times \frac{238}{206} = 0.320 \text{ g uranium}$$

Hence, the initial amount of uranium was $0.320 \text{ g} + 1.667 \text{ g} = 1.987$ g. Applying Equation 13–3, and realizing that the ratio of disintegrations is the same as the ratio of grams, we find that

$$t = 4.5 \times 10^9 \times 3.33 \log_{10} \frac{1.987}{1.667} \text{ years}$$

$$t = 1.13 \times 10^9 \text{ years}$$

We can correct for our simplifying assumption that all of the ^{206}Pb came from ^{238}U by looking at the ratio of different lead isotopes in materials containing and lacking ^{238}U. The oldest rocks yet found, granite from the west coast of Greenland, have an age of slightly more than 3.7×10^9 years. We know from dating stony meteorites that the earth was formed by accretion of rocky matter and dust approximately 4.6 billion years ago. (Independent checks of this figure are obtained by rubidium–strontium and potassium–argon dating.) The first billion years of our planet's history have been erased by the continual processes of weathering, erosion and rebuilding, that are typical of the earth.

The moon tells us a different story. Radioisotope dating of rock and soil samples brought back from the moon by the Apollo 11–17 expeditions has transformed our ideas of the moon's origin and history. Before the Apollo program, astronomers argued about whether the lunar craters all had arisen from meteoritic impact, or whether volcanic activity had helped to shape the surface. Two rival theories were proposed for the origin of the moon: that it grew by accretion of dust and rock in the same way and at the same time that the earth did, or that it split off from the earth much later, perhaps leaving a scar that we now see as the Pacific basin.

Apollo 11 and 12 landed in two "maria" or flat lowland regions of the moon in 1969: the Sea of Tranquility and the Ocean of Storms. In both regions the astronauts found a terrain made up of two kinds of material: crystalline rocks of volcanic origin resembling basalt, and breccia, a conglomerate rock produced by shattering and reforming of fragments and dust over the eons. The volcanic basalt samples brought back to earth were dated by three isotope-ratio methods: potassium–argon, rubidium–strontium, and uranium–thorium–lead. All gave the same remarkably great age: between 3.6 billion and 4.2 billion years.[1] It became apparent that the maria or "seas" on the moon are the result of outfloodings of lava from the interior, which took place in the first billion years of the moon's 4.6-billion-year existence. The cratering in the maria arose from meteorite impact after the lava hardened.

There is even more to be learned from isotope dating. The lava on the floor of the Ocean of Storms is younger by 300 million years than that on the Sea of Tranquility. This is not surprising, since the Ocean of Storms is less heavily cratered by meteorites and therefore appears younger. But the difference in age is not as large as the difference in cratering suggests. To reconcile cratering and age, we must conclude that meteorite collisions were more frequent during the moon's first billion years, and gradually have diminished. This makes very good sense in terms of a picture of the moon initially nucleating out along with the earth and the other planets from a mass of dust and debris around the sun. The rate of meteoritic collision would have been very high during the accretion period, and then would have fallen slowly as most of the solid matter in the vicinity was swept up into the moon or earth.

In early 1971, Apollo 14 landed near Fra Mauro, on a blanket of material that was thrown out during the meteoritic collision that produced the Mare Imbrium. Later that year, Apollo 15 visited the floor of the Mare Imbrium, near Hadley Rille. Isotope dating by Rb–Sr, K–Ar, and, ^{40}Ar–^{39}Ar methods again produced a surprise. Basalt samples were found at both sites, but the floor of the Mare Imbrium near Hadley Rille proved to be younger than the debris thrown out by impact: 3.3 billion versus 3.8 billion years. This only could mean that subsequent lava flow occurred years after the mare was formed. Not only did meteorite craters pockmark lava beds,

---

[1] The variation is in the rock samples, not the dating methods.

lava flows occurred later in the bottoms of large craters. Both processes transpired continuously to shape the moon.

The "Genesis rock" that Scott and Irwin were so excited about during the Apollo 15 moonwalk was anorthosite, a calcium aluminum silicate of a type that is found in the very oldest terrestrial rocks, crystallized from a molten magma. The astronauts had been trained to look for this type of rock as possible samples of the original lunar crust, and their Genesis rock proved to be 4.2 billion years old. The Apollo 16 and 17 missions to the lunar highlands found mainly this same ancient anorthosite, not covered over by subsequent lava flow as the maria had been. These minerals were compressed into breccias at least 200 meters deep, which are the lunar records of a surface plowed up, crushed, remelted and compacted by 4.2 billion years of meteoritic collisions.

The Antiquity of Life on Earth

Terrestrial geological strata in which fossils of invertebrates first appear in profusion are classified in the Cambrian era; the beginning of this era has been dated by radioisotopic methods to 600 million years ago. Yet almost every present-day phylum of invertebrates is already represented in Cambrian fossils. Where are the fossils of their predecessors during the period of differentiation of phyla?

These pre-Cambrian fossils are more difficult to find; for the animals were soft-bodied, and the rocks into which their sediment hardened has undergone metamorphoses that tended to destroy the fossil record. Fossil plant remains have been unearthed in the Nonesuch shale deposits of northern Michigan, and rubidium–strontium methods date this deposit at about 1.05 billion years. Fossil plants resembling blue-green algae, and others resembling no known species, have been discovered in the Gunflint iron formations of Ontario; these have been dated by potassium–argon methods as about 2 billion years old. The oldest fossil remains of what are believed to be living organisms are fossil bacteria in the Fig Tree deposits in the Transvaal, South Africa. These have been dated by the rubidium–strontium method as 3.1 billion years old. While the floors of the great lunar maria were being blasted out by meteorites and covered by lava flows, the first traces of life were beginning to appear in the oceans of our planet.

13–6   SUMMARY

We are approximately at the same position with nuclei and nuclear structure now that we were with atoms and atomic structure in 1925. We can measure, describe, and classify nuclei, but we do not have a general theory to explain them. Nuclei are made of protons and neutrons that pack together and interact most strongly only with their immediate neighbors in the nucleus. In some ways (as binding energy) they act like packed droplets of uniform particles, and in other ways (preference for even nucleons and magic numbers) they act as if they had shell structures similar to electronic shell structures. Energy-level diagrams can be drawn for nuclei from their γ-radiation spectrum in nuclear transformations. There are ground states and excited states for nuclei, as well as for electrons around the nucleus.

Nuclei decay spontaneously by electron capture, electron or positron emission, and α-particle emission. This decay is a first-order rate process. The half-life, or time for the amount of material to decrease to half its initial value, is independent of the amount of material present.

Nuclear reactions can occur when target nuclei are bombarded with other nuclei that are accelerated to sufficient speed to overcome electrostatic repulsion between the positively charged nuclei. Neutrons interact more easily because they have no charge. One important use for nuclear reactions is the preparation of isotopes for chemistry, industry, and medicine. Another is the synthesis of new transuranium elements. Elements to $Z = 105$ have been prepared, and there is hope that those around 114 might be more stable than those prepared recently.

Nuclear fission, having had its beginnings in war, is now being adapted as a practical source of peacetime power. Fusion has yet to be made practical as a power source, but remains promising.

Isotopes, radioactive or not, are of great utility in chemistry as a means of tagging reactants and following them through the reaction to products. Radioactive isotopes are especially useful in chemical analyses.

In 1650, Archbishop Ussher made one of the first serious calculations of the age of the earth. On the basis of Biblical genealogies, he placed Creation Day at 4004 B.C.; this gave the earth an age of 5654 years. More recent values based on nuclear genealogies set the figure closer to 4.5 billion years. For those who like coincidences, Ussher's original age of the earth is almost exactly one half-life of carbon-14. The first living organism of which we have any fossil remains are fossil bacteria with an age of approximately 3.1 billion years. Somewhere in the first 1.4 billion years of this planet's existence, chemical evolution developed to the point at which bacterialike organisms existed. From these organisms, during the following 3.1 billion years evolved the vast diversity of living organisms that we now find around (and including) us.

13–7 POSTSCRIPT: THE ETHICAL DILEMMA OF SCIENTISTS

The anguish of physicists over the atomic bomb illustrates sharply the ethical dilemma that often faces scientists now: To what extent is a scientist responsible for the uses that are made of his discoveries? The atomic bomb was built, in spite of the profound misgivings of many who worked on the project, as a result of the fear that the Nazis might develop it first. Einstein said after the war: "If I had known that the Germans would not succeed in constructing the atom bomb, I would never have lifted a finger." A few physicists declined to work on the atom bomb project. Others plunged into the work wholeheartedly, believing that it was necessary for the defense of their country. A larger group, which included many of the leaders of the effort, participated only because they considered the atomic bomb to be the lesser of the two evils.

The issues were clearer in World War II than in these days of muddled motives and doubtful goals. If ever a war was, World War II was a "Just War" in the Medieval sense. The Nazi regime openly espoused racial superiority, conquest, and the subjugation of the conquered populations as matters of State policy. The racial extermination programs, which were revealed in their full horror after the fall of Germany, were at least rumored, if not always believed. The Nazi regime made no pretense of "liberating" its conquered regions, and was usually vehemently opposed by underground movements of citizens in occupied areas.

Was it morally right for physicists to develop a weapon as terrible as the atom bomb for use against such a regime? The question can have two different answers: one if it is known that the Nazis will develop the same weapon, another if it is known that they will not. But what if you cannot know their actions? To say automatically that the end never justifies the means evades the issues. Frequently we are faced, not with a choice between good and evil, but with a choice between two evils of varying degree.

The same two men, Szilard and Einstein, whose 1939 letter to President Roosevelt initiated the atom bomb project, again wrote to him in April, 1945, urging that the bomb not be used in actual warfare, and warning of the dangers of a nuclear arms race. This second letter was found unopened on Roosevelt's desk after his sudden death on April 12, 1945.

The bomb was never to be used against the Nazi regime. Germany surrendered on May 7, 1945. The full attention of the allies then was devoted to ending the war in the Pacific. The physicists were faced with another dilemma. If the atom bomb was developed in response to a threat of a similar bomb from the enemy, and if this threat had now vanished, were we morally able to use the bomb against a different enemy that lacked this weapon and was in the last stages of defeat? The battle for Okinawa in May, 1945, had been unexpectedly bitter. Some 12,500 American soldiers had been killed, and three times that number wounded; 110,000 Japanese had died in battle. If the battle for the homeland of Japan would be equally bitter, was it not better to end the war quickly with atomic weapons?

The Franck Report, prepared for the University of Chicago by Nobel Laureate James Franck, Szilard, Glen T. Seaborg, and other scientists, urged that the bomb not be used against cities in Japan. It proposed a public demonstration in some barren site to be attended by observers from the United Nations. The report was forwarded to the Secretary of War on June 11, and referred to a panel of government scientists. J. Robert Oppenheimer, the director of the atom bomb project at Los Alamos, New Mexico, recalled the reaction of the panel to the Franck Report:

> We said that we didn't think that being scientists especially qualified us as to
> how to answer this question of how the bombs should be used or not; opinion
> was divided among us as it would be among other people if they knew about it.
> We thought the two overriding considerations were the saving of lives in the
> war and the effect of our actions on the stability, on our strength, and the
> stability of the postwar world. We did say that we did not think that exploding
> one of these things as a firecracker over a desert was likely to be very impressive.

The first atom bomb was detonated on July 16, 1945, near Alamogordo,
New Mexico. The second of three bombs built obliterated the Japanese city
of Hiroshima on August 6, and the last one destroyed Nagasaki on August 9.
On August 11, Japan surrendered. About 114,000 people, mostly civilians,
died in the two blasts.

Was it right to destroy two cities and to kill 114,000 people in order to
prevent the killing of possibly a greater number of soldiers and civilians during
a protracted and bloody invasion of the islands of Japan? Was the total
destruction of two cities preferable to the piecemeal destruction of more of
the country in a ground war? Was the prediction that the Japanese would
fight to the very end even correct? Had all the avenues of negotiations for
surrender been investigated, and would a public demonstration of the atom
bomb have produced surrender? Such questions were formerly the province
of politics, philosophy, and morality. With the coming of the atom bomb the
questions became the concern of the scientist as well, because the scientist
created the conditions from which these questions arose.

After the war, the misgivings of scientists about the child of their efforts
have sometimes been exemplified in the story of one man, J. Robert Oppenheimer. Yet virtually every physicist connected with the project faced the
same ethical questions about his actions as Oppenheimer did. Many left
physics; some, such as Szilard and E. Rabinowich, have made valuable contributions to molecular biology. Rabinowich and others founded an influential
magazine, the *Bulletin of the Atomic Scientists*, which constantly calls attention
to the dangers created by the improper or thoughtless use of science. Szilard
also became known as a writer of articles and essays with the same urgent
message: A scientist must always think about the consequences of his work,
and must be willing to exert himself to see that science is not used for the
destruction of mankind.

The issues are not as clearly delineated now as in 1945, but the dangers
are still with us. There are more ways to destroy a world than by nuclear
holocaust, although this threat is certainly still present. We can permit the
misapplication of science in such a way that the planet is made unfit for
habitation, and also can sit inactively while underprivileged segments of the
world's population explode in despair and involve us all in war. Neutrality
is a luxury that is becoming increasingly costly to maintain. In times of rising
pressures, the refusal to take a stand becomes a definite stand. Oppenheimer

was right; scientists have no special monopoly on ethical decision making when technology is involved. But nonscientists must know enough about science and technology to help make such decisions intelligently.

SUGGESTED READING

G. Choppin, *Nuclei and Radioactivity*, *Elements of Nuclear Chemistry*, W. A. Benjamin, Menlo Park, Calif., 1964. A more extensive introduction to nuclear chemistry than this chapter, but at the same level.

R. Jungk, *Brighter than a Thousand Suns*, Harcourt, Brace & World, New York, 1958. An exceedingly well-written narrative of the rise of the nuclear age, from quantum mechanics in the 1920's through the development of the atom bomb and the rise of the cold war. Raises some ethical issues that are relevant now.

D. H. Kenyon and G. Steinman, *Biochemical Predestination*, McGraw-Hill, New York, 1969. Ignore the title; the book is good. An introduction to what we know about the chemical evolution of life on the primitive earth. Chapter 2, on radio-isotope dating and the antiquity of terrestrial life, is particularly relevant.

W. F. Libby, *Radiocarbon Dating*, University of Chicago Press, Chicago, 1955, 2nd ed. Interesting monograph with a discussion of the methods of dating objects with ^{14}C, and a summary of the findings in Western Hemisphere and Near Eastern archaeology.

R. Protsch and R. Berger, "Earliest Dates for Domestication of Animals," *Science* **179,** 235 (1973).

C. Renfrew, "Carbon-14 and the Prehistory of Europe," *Scientific American*, October, 1971.

G. T. Seaborg, *Man-Made Transuranium Elements*, Prentice-Hall, Englewood Cliffs, N. J., 1963.

S. C. Thompson and C. F. Tsang, "Superheavy Elements," *Science* **178,** 1047 (1972).

QUESTIONS

1 How are isotopes distinguished when writing chemical symbols?

2 How does the size of a nucleus compare with that of the entire atom?

3 What is the binding energy of a nucleus? How is it calculated, and what does it signify? Where does the energy come from?

4 What is the difference between β^+ emission and electron capture if both result in the decrease of atomic number by one?

5 How can the gamma radiation accompanying α emission give us information about the nucleus?

6 What is the half-life of a radioactive element?

7 Why does it take as long for 10 g of a radioactive substance to decay to 5 g as for 1 g to decay to $\frac{1}{2}$ g? If in the first reaction, 10 g decay to 5 g in a year, why won't 7 g decay to 2 g (also a loss of 5 g) in the same time?

8 What are the magic numbers in nuclear physics, and what do they imply about the nucleus? Are there analogous numbers in atomic structure? What do they represent?

9 How does fission differ from decay?

10 What is a nuclear chain reaction?

11 What is the difference between nuclear fission and fusion?

12 Why is it useful to use radioactive isotopes, or at least unusual isotopes, in studying chemical reactions? What advantages do ^{32}P and ^{35}S have that ^{15}N and ^{18}O do not have?

13 How does radiometric analysis work?

14 With isotope-dilution methods, you can determine how much of a chemical substance is present in a sample *without* doing a true quantitative separation and analysis. How can you do this?

15 What experimental measurements are made in the radiocarbon dating process? How are these converted to an age for the sample?

16 A scientist interested in the age of a Greek temple investigated the ^{14}C content of wood from the roof beams and $CaCO_3$ from the marble walls, and found radically different results. Why? Which result would be a better means of dating the age of the temple?

17 What inferences might you draw about the history of the moon from the information that the moon contains a greater amount of elements with high melting and boiling points than does the earth?

PROBLEMS

The masses of isotopes needed for many of these problems are tabulated in the Chemical Rubber Company *Handbook of Chemistry and Physics* and other standard physics reference books.

1 Determine which naturally occurring isotope has a nuclear radius that is most nearly three times the radius of the hydrogen nucleus. Which isotope has a radius most nearly four times that of the hydrogen nucleus?

2 (a) The heat of combustion of $CH_4(g)$ is -118.3 kcal mole^{-1}. Calculate the mass equivalent of this energy in grams. (b) When ^{14}C decays to $^{14}_7N + ^{0}_{-1}e$, a mass loss of 0.000168 amu occurs. How many moles of $CH_4(g)$ would have to be burned to produce the same amount of energy as would be produced from the decay of one mole of ^{14}C?

3 When two ^1_1H nuclei and two neutrons combine to form ^4_2He, the mass of the product helium nuclei is not the same as the sum of the masses of the reactant particles. Calculate the energy, in calories per mole of helium atoms, that is equivalent to the change in mass during the reaction. If the energy per helium atom were released as a single photon,

what would be its wavelength? How does this wavelength compare with the approximate radius of the helium nucleus?

4 Write equations for each of the following nuclear processes: (a) positron emission by $^{120}_{51}Sb$, (b) electron emission by $^{35}_{16}S$, (c) α-particle emission by $^{226}_{88}Ra$, (d) electron capture by ^7_4Be.

5 Complete the following nuclear equations and supply symbols or values for X or x:

a) $^x_{88}Ra \rightarrow {}^4_2X + {}^{222}_xX$

b) $^{14}_xC \rightarrow {}^x_xN + {}^0_{-1}e$

c) $^x_xNe \rightarrow {}^{19}_xF + {}^0_{+1}e$

d) $^{73}_xAs + {}^0_{-1}e \rightarrow {}^x_{32}X$

e) $^{176}Lu \rightarrow {}^x_xX + {}^x_{-1}e$

f) $^{235}_xU + {}^1_0n \rightarrow x^1_0n + {}^{94}_{36}X + {}^{139}_xX$

g) $^{59}_xCo + {}^2_1X \rightarrow {}^{60}_xCo + {}^x_xX$

h) $^{19}_9X + {}^1_xH \rightarrow {}^{16}_8X + {}^4_xX$

i) $^{16}_8X + {}^2_1X \rightarrow {}^{14}_7X + {}^4_xX$

j) $^{26}_xMg + {}^1_0X \rightarrow {}^4_xHe + {}^x_xX$

6 Calculate the binding energy per nucleon for the following nuclear species: (a) $^{12}_6C$ ($m = 12.0000$ amu), (b) $^{37}_{17}Cl$ ($m = 36.96590$ amu), (c) $^{208}_{82}Pb$ ($m = 207.9766$ amu), (d) $^{32}_{16}S$ ($m = 31.97207$ amu), (e) $^{16}_8O$ ($m = 15.99491$ amu).

7 The masses of $^{22}_{11}$Na and $^{22}_{10}$Ne atoms are 21.994435 amu and 21.991385 amu, respectively. In terms of energy, is it possible for ^{22}Na to decay to ^{22}Ne by positron emission?

8 The only stable isotope of fluorine is ^{19}F. What radioactivity would you expect from the isotopes ^{17}F, ^{18}F, and ^{21}F?

9 When an electron and a positron meet, they are annihilated and two photons of equal energy result. What are the wavelengths of these photons?

10 If 1 g of ^{99}Mo decays by beta emission to $\frac{1}{8}$ g in 200 h, what is the half-life of ^{99}Mo?

11 The number of alpha particles emitted per second by 1 g of radium is 3.608×10^{10}. Determine the decay constant and the half-life of radium.

12 The alpha particles emitted by radium have energies of 4.795 MeV and 4.611 MeV. What is the wavelength of the gamma radiation accompanying the decay?

13 The half-lives of uranium-235 and uranium-238 are 7.1×10^8 years and 4.5×10^9 years, respectively. Presently, uranium is 99.28% ^{238}U and 0.72% ^{235}U. Calculate the percent natural abundance of each uranium isotope when the earth was formed 4.5 billion years ago.

14 The decay of $^{234}_{92}$U ultimately leads to $^{206}_{82}$Pb. The process proceeds via the sequence of α and β^- decay steps: α, α, α, α, α, β^-, α, β^-, β^-, β^-, α. Write the symbol for each of the isotopes produced in the decay process.

15 The human body contains about 18% carbon by weight. Of that carbon, 1.56×10^{-10}% is ^{14}C. Calculate the mass of ^{14}C present in a 150-pound body. Calculate the number of disintegrations occurring per minute in a body of this weight. (The half-life of ^{14}C is 5570 years.)

16 Radioactive Na with a half-life of 14.8 h is injected into an animal in a tracer experiment. How many days will it take for the radioactivity to fall to 0.10 of its original intensity?

17 One atom of ^{235}U evolves about 200 MeV when it undergoes fission. How does this heat of fission compare on a weight basis with the heat evolved in the combustion of 1 g of carbon? How many tons of coke (carbon) will give as much heat on combustion as 1 lb (454 g) of ^{235}U evolves on fission?

18 Ten grams of a protein are digested and decomposed into amino acids. To this mixture is added 100 mg of pure deuterium-substituted alanine, H_2N—$CH(CD_3)$—$COOH$. After thorough mixing, some of the alanine is separated from the mixture and purified by crystallization. This crystalline alanine contains 1.03% deuterium by weight. How many grams of alanine were originally present in the protein digest?

19 A man weighing 70.8 kg was injected with 5.09 ml of water containing tritium $(9 \times 10^9$ cpm). After 3 h, the tritiated water had equilibrated with the body water of the patient. A 1-ml sample of plasma water then showed an activity of 1.8×10^5 cpm. Estimate the weight percent of water in the human body.

20 A sample of wood from an Egyptian mummy case gives 9.4 cpm per gram of carbon of ^{14}C disintegrations. How old is the mummy case?

21 In the skeleton of a fish caught in the Pacific Ocean in 1960, the ^{14}C disintegration rate is 17.2 cpm per gram of carbon. How can this be? What does this imply for Pacific archaeology?

22 In a sample of uraninite ore from the Black Hills of South Dakota, the weight of Pb is 22.8% the weight of the U present. Estimate a minimum age for the earth from this information.

Imagine two hundred brilliant violin players playing the same piece with perfectly tuned instruments, but commencing at different places selected at random. The effect would not be pleasing, and even the finest ear could not recognize what was being played. Such music is made for us by the molecules of gases, liquids, and ordinary solids. . . . A crystal, on the other hand, corresponds with the orchestra led by a vigorous conductor when all eyes intently follow his nod, and all hands follow the exact beat. . . . To me, the music of physical law sounds forth in no other department in such full and rich accord as in crystal physics.
W. Voigt

14 BONDING IN SOLIDS AND LIQUIDS

Now that we know how bonding occurs between small numbers of atoms, we can look at bonding in solids and liquids. A simple but quite useful theory of the electrical properties of solids considers the entire solid body as one large molecule and uses delocalized molecular orbitals that extend over the entire solid. This is the band theory of metals and insulators.

How are compounds that definitely exist as molecules held together in the solid? Why are not Br_2, I_2, and all organic substances gases at room temperature? What is the force that keeps the hydrocarbon molecules of gasoline in the liquid state? Why is sugar crystalline if no covalent or ionic bonds hold one molecule to another? Molecular solids are comprehensible as soon as we recognize the contributions of the weak forces known as van der Waals attraction and hydrogen bonding.

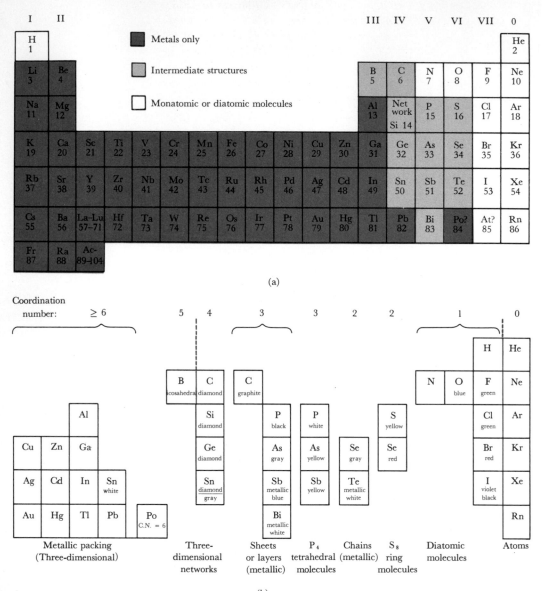

Figure 14–1. Solid structures of the elements. (a) General trends in bonding in solid elements. (b) The types of bonding in the variable zone of (a). The structures of the nonmetals are determined by their coordination numbers, which are eight less the group number except when multiple bonds are used in graphite, N_2, and O_2. See Table 14–1.

Table 14–1.   The Correlation between Coordination Number and Structure in Elemental Solids

Bonding coordination number	Type of solid structure
0	Atomic solids, low melting and boiling points
1	Diatomic molecular solids, low melting and boiling points
2	Rings or chains. Solids with packed ring molecules are less metallic than those with packed chains
3	P_4 tetrahedra or sheets. Solids with packed tetrahedral molecules are less metallic than those with packed sheets
4	Three-dimensional nonmetallic networks
5	B sheet curved in on itself in a B_{12} icosahedron
6 or more	Packed metallic solids

The four types of chemical bonding—covalent bonding, ionic attraction, hydrogen bonding, and van der Waals attraction, in approximate order of decreasing strength—are sufficient to explain atomic and molecular interactions in solids, liquids, and gases. Each of these bonding types contributes a different kind of stability to the atoms it binds. In what follows, we shall examine the ways in which they differ.

Chapter 12 was devoted entirely to the special role of carbon as the framework material for all of our planet's organic and living matter. Its neighbor, silicon, plays an equally important role as the framework material for the planet itself. Most of the rocks, earth, sand, and mineral matter that we encounter are some kind of silicate mineral. If the carbon compounds are the pigments with which this planet is colored, the silicates are the canvas. We shall look briefly at some of these silicates to discover how their molecular structure accounts for their physical and chemical properties.

14–1   BONDING FORCES IN SOLIDS

We have looked at ions in solids in Chapter 3, at bonding in nonmetallic elements in Chapter 4, and at the theory behind such bonding in Chapter 11. The information that we have calculated on the solid structures of elements is summarized in Figure 14–1 and Table 14–1.

Solids that are built by weak attractive interactions between individual molecules are called *molecular solids*. Examples of molecular solids include iodine crystals, composed of discrete I_2 molecules, and paraffin wax, composed of long-chain alkane molecules. At very low temperatures the noble gases exist as molecular solids that are held together by weak interatomic forces. For example, argon freezes at $-189°C$ to make the close-packed

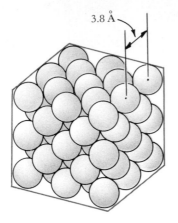

3.8 Å

Figure 14–2. The structure of solid argon. Each sphere is an individual atom of Ar, *in cubic close packing with* 3.8 Å *between atomic centers.*

structure shown in Figure 14–2. Examples of nonpolar molecules that crystallize at low temperatures to give molecular solids include Br_2, which freezes at $-7°C$ to build the structure shown in Figure 14–3. Methane, CH_4, freezes at $-183°C$ to form the close-packed crystal shown in Figure 14–4.

Nonmetallic network solids consist of infinite arrays of bonded atoms; no discrete molecules can be distinguished. Thus any given piece of a network solid may be considered a giant, covalently bonded molecule. Network solids generally are poor conductors of heat and electricity. Strong covalent bonds among neighboring atoms throughout the structure give these solids strength and high melting temperatures. Some of the hardest substances known are nonmetallic network solids.

Diamond, the hardest allotrope of carbon, has the network structure shown in Figure 14–5(a). Diamond sublimes (volatilizes directly to a gas), rather than melts, at 3500°C and above. Graphite, a softer allotrope of carbon,

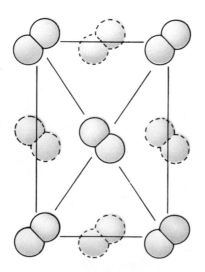

Figure 14–3. The structure of crystalline bromine made of Br_2 *molecules. The solid outlines indicate one layer of packed molecules, and the dashed outline indicates a layer beneath. The molecules have been shrunk for clarity in this drawing; they are actually in close contact within a layer, and the layers are packed against one another.*

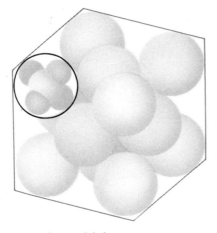

Figure 14–4. The structure of solid methane. Each large sphere represents a methane molecule, CH_4, as indicated at the upper left. The methane molecules are arranged in cubic close packing.

has the layered structure shown in Figure 14–5(b). Quartz, SiO_2, which melts at 1610°C, has a structure similar to that shown in Figure 14–6.

One feature that distinguishes network solids from metals is the lower coordination number of atoms in network structures. In the preceding examples the coordination number of C, in diamond, and Si is four, and that of O in quartz is two. In Section 14–3, we will see that a localized molecular orbital picture of bonding satisfactorily accounts for the properties of diamond.

Figure 14–5. Crystalline carbon. (a) Diamond structure. The coordination number of carbon in diamond is 4. Each atom is surrounded tetrahedrally by four equidistant atoms. The C—C bond distance is 1.54 Å. (b) Graphite structure. This is the more stable structure of carbon. Strong carbon–carbon bonding occurs within a layer, weaker bonding between layers.

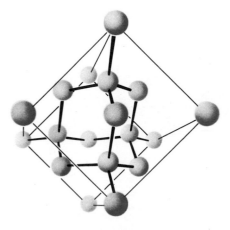

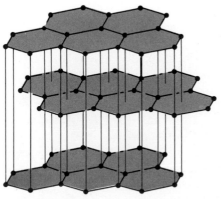

(a)   (b)

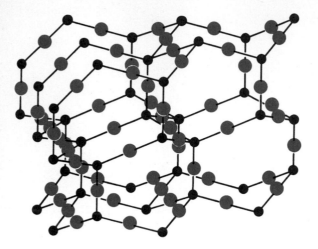

Figure 14–6. The three-dimensional network of silicate tetrahedra in tridymite, one crystalline form of silicon dioxide $(SiO_2)_n$. Quartz has a similar structure, also in which all O atoms in SiO_4^{4-} are shared with other Si atoms. The feldspar minerals have varying amounts of Al substituted for Si, up to 50%.

● Silicon atom—each attached to 4 oxygen atoms

● Oxygen atom—each attached to 2 silicon atoms

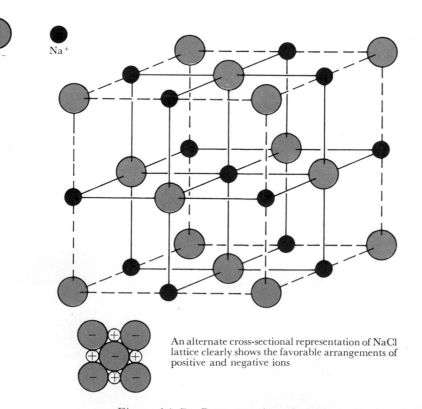

Cl^- Na^+

An alternate cross-sectional representation of NaCl lattice clearly shows the favorable arrangements of positive and negative ions

Figure 14–7. Representation of the ionic NaCl structure. The bottom figure is a representation of a cross section of the NaCl structure.

Metallic solids also consist of infinite arrays of bonded atoms, but in contrast to nonmetals each atom in a metal has a high coordination number: sometimes four or six, but more often eight or twelve. We have discussed the most common packing schemes, ccp, hcp, and bcc, in Section 3–7. The band theory of delocalized molecular orbitals will be developed in Section 14–3 to explain the fact that metals generally are good conductors of electricity.

In the periodic table shown in Figure 14–1 the elemental solids are classified as metallic, network nonmetallic, or molecular. The majority of metals crystallize in close-packed structures in which each atom has a high coordination number. Included as metals are elements such as tin and bismuth, which crystallize in structures with relatively low atomic coordination numbers but which still have strong metallic properties. The light colored area of the periodic table includes elements that have borderline properties. Although germanium crystallizes in a diamondlike structure in which the coordination number of each Ge atom is only four, certain of its properties resemble those of metals. This similarity to metals indicates that the valence electrons in germanium are not held as tightly as would be expected in a true nonmetallic network solid. Arsenic, antimony, and selenium exist as either molecular or metallic solids, although the so-called metallic structures have relatively low atomic coordination numbers. We know that tellurium crystallizes in a metallic structure, and it seems reasonable to predict that it also may exist as a molecular solid. From its position in the periodic table we predict intermediate properties for astatine, which has not been studied in detail.

Ionic solids consist of infinite arrays of positive and negative ions that are held together by electrostatic forces. These forces are the same as those that hold a molecule of NaCl together in the vapor phase. In solid NaCl the Na^+ and Cl^- ions are arranged to maximize the electrostatic attraction, as shown in Figure 14–7. The coordination number of each Na^+ ion is six, and each Cl^- ion similarly is surrounded by six Na^+ ions. Because ionic bonds are very strong, much energy is required to break down the structure in solid-to-liquid or liquid-to-gas transitions. Thus ionic compounds have high melting and boiling temperatures.

The preceding discussion has distinguished four types of solids—molecular, nonmetallic network, metallic, and ionic. Of these types, by far the weakest bonding is found in molecular solids, in which only *intermolecular* forces hold the crystal together. In the next section we will examine in more detail the nature of these intermolecular forces.

14–2 MOLECULAR SOLIDS AND THE VAN DER WAALS BOND

In N_2, O_2, Cl_2, P_4, and S_8, all the valence orbitals are either used in bonding or are occupied with nonbonding electrons. Thus, any *intermolecular* bonding that holds molecules together in the solid must be weak in comparison with

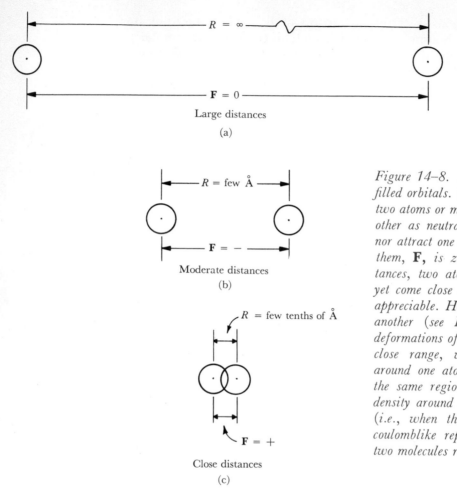

Large distances

(a)

Moderate distances

(b)

Close distances

(c)

*Figure 14–8. Repulsion of electrons in filled orbitals. (a) At very large distances, two atoms or molecules behave toward each other as neutral species and neither repel nor attract one another. The force between them, **F**, is zero. (b) At moderate distances, two atoms or molecules have not yet come close enough for repulsion to be appreciable. However, they do attract one another (see Figure 14–9) because of deformations of their charge clouds. (c) At close range, when the electron density around one atom or molecule is large in the same region of space as the electron density around the other atom or molecule (i.e., when the filled orbitals overlap), coulomblike repulsion dominates and the two molecules repel one another.*

the strength of the *intramolecular* bonding in the molecule. The weak forces that contribute to intermolecular bonding are called van der Waals forces.

Van der Waals forces

There are two principal van der Waals forces. The most important force at short range is the repulsion between electrons in the filled orbitals of atoms on neighboring molecules. This electron-pair repulsion is illustrated in Figure 14–8. The analytical expression commonly used to describe the energy resulting from this interaction is

$$\text{van der Waals repulsion energy} = be^{-aR} \qquad (14-1)$$

in which b and a are constants for two interacting atoms. Notice that this repulsion term is very small at large values of the interatomic distance, R.

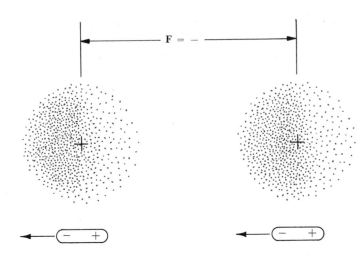

Instantaneous polarization of an atom leaves, for a moment, more electron density on the left than on the right, thus creating an "instantaneous dipole"

This "instantaneous dipole" can polarize another atom by attracting more electron density to the left, thereby creating an "induced dipole"

*Figure 14–9. Schematic illustration of the instantaneous dipole–induced dipole interaction that gives rise to a weak attraction. For the brief instant that this figure describes, there is an attractive force, **F**, between the instantaneous dipole and the induced dipole. The effect is reciprocal; each atom induces a polarization in the other.*

The second force is the attraction that results when electrons in the occupied orbitals of the interacting atoms synchronize their motion to avoid each other as much as possible. For example, as shown in Figure 14–9, electrons in orbitals of atoms belonging to interacting molecules can synchronize their motion to produce an instantaneous dipole-induced dipole attraction. If at any instant the left atom in Figure 14–9 had more of its electron density at the left, as shown, then the atom would be a tiny dipole with a negative left side and a positive right side. This positive side would attract electrons on the right atom in the figure and would change this atom into a dipole with similar orientation. Therefore, these two atoms would attract each other because the positive end of the left atom and the negative end of the right atom are close. Similarly, fluctuation in electron density of the right atom will induce a temporary dipole, or asymmetry of electron density, in the left atom. The electron densities are fluctuating continually, yet the net effect is an extremely small but important attraction between atoms. The energy resulting from this attractive force is known as the *London energy*, after Fritz London, who derived the quantum mechanical theory for this attraction in 1930. The London energy varies inversely with the sixth

Table 14–2.   Van der Waals Energy Parameters

Interaction pair	a (au)$^{-1}$ ^{a}	b, kcal mole^{-1}	d, kcal mole^{-1} (au)^6
He—He	2.10	4.1×10^3	1.5×10^3
He—Ne	2.27	20.7×10^3	2.9×10^3
He—Ar	2.01	30.0×10^3	9.7×10^3
He—Kr	1.85	16.4×10^3	13.7×10^3
He—Xe	1.83	26.6×10^3	21.3×10^3
Ne—Ne	2.44	104.8×10^3	5.7×10^3
Ne—Ar	2.18	151.8×10^3	19.2×10^3
Ne—Kr	2.02	82.8×10^3	26.7×10^3
Ne—Xe	2.00	134.2×10^3	41.5×10^3
Ar—Ar	1.95	219.5×10^3	64.6×10^3
Ar—Kr	1.76	119.8×10^3	90.1×10^3
Ar—Xe	1.74	194.4×10^3	139.3×10^3
Kr—Kr	1.61	65.2×10^3	125.4×10^3
Kr—Xe	1.58	106.0×10^3	194.4×10^3
Xe—Xe	1.55	171.8×10^3	301.1×10^3

^{a} 1 au = 1 atomic unit = 0.529 Å. The value of R in Equation 14–3 must be expressed in atomic units as well.

power of the separation between atoms:

$$\text{London energy} = -\frac{d}{R^6} \tag{14–2}$$

in which d is a constant and R is the distance between atoms. This "inverse sixth" attractive energy decreases rapidly with increasing R, but not nearly as rapidly as the van der Waals repulsion energy. Thus at longer distances the London attraction is more important than the van der Waals repulsion, consequently a small net attraction results.

The total potential energy of van der Waals interactions is the sum of the attractive energy of Equation 14–2 and the repulsive energy of Equation 14–1:

$$PE = be^{-aR} - \frac{d}{R^6} \tag{14–3}$$

The total van der Waals potential energy can be compared quantitatively with ordinary covalent bond energies by examining systems for which the curves of potential energy versus interatomic distance, R, are known accurately. We can calculate values for the constants a, b, and d from experimental data on the deviation of real gases from ideal gas behavior. Some of these values for interactions of noble gases are listed in Table 14–2.

The potential energy curve for van der Waals interactions between helium atoms is illustrated in Figure 14–10. At separations of more than

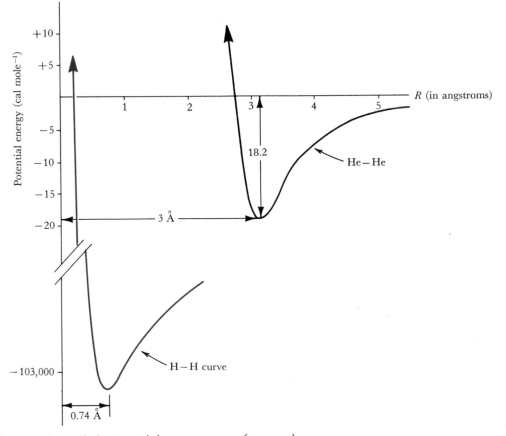

Figure 14–10. A comparison of the potential energy curves for van der Waals attraction between two He atoms (black curve) and covalent bonding between two H atoms (colored curve). Note that the energy scale is in cal mole^{-1}, rather than kcal mole^{-1}. The covalent bond is over 5000 times as stable as the van der Waals bond.

3.5 Å, the second term in Equation 14–3 predominates. As the atoms move closer together they attract each other more, and the energy of the system decreases. However, at distances closer than 3 Å the strong electron-pair repulsion overwhelms the London attraction, and the potential energy curve in Figure 14–10 rises. A balance between attraction and repulsion exists at 3-Å separation, and the He---He "molecule" is 18.2 cal mole^{-1} more stable than two isolated atoms.

Figure 14–10 also shows the marked contrast between van der Waals attraction and covalent bonding. In the H_2 molecule strong electron–proton attractions in the bonding molecular orbital cause the potential energy to decrease as the H atoms approach one another, and it is proton–proton

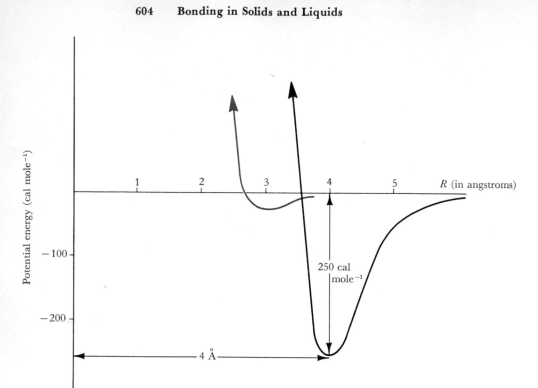

Figure 14–11. A comparison of the potential energy curves for van der Waals attraction between two atoms of Ar (black curve) and two atoms of He (colored curve). The larger Ar atoms are held more tightly, although the bond energy is still one four-hundredth that of a H—H bond.

repulsion that makes the energy increase sharply if the atoms are pushed too closely together. This proton–proton repulsion operates at smaller distances than the electronic repulsion between the two He atoms. The H—H bond length in the H_2 molecule is 0.74 Å, whereas the equilibrium distance of van der Waals-bonded He atoms is 3 Å. Moreover, a covalent bond is much stronger than a weak van der Waals interaction. Only 18.2 cal mole^{-1} is required to separate helium atoms at their equilibrium distance, but 103,000 cal mole^{-1} is needed to break the covalent bond in H_2.

Molecular solids, in which only van der Waals intermolecular bonding exists, generally melt at low temperatures. This is because relatively little energy of thermal motion is needed to overcome the energy of van der Waals bonding. The liquid and solid phases of helium, which result from weak van der Waals "bonds," exist only at temperatures below 4.6°K. Even at temperatures near absolute zero, solid helium can be produced only at high pressures (29.6 atm at 1.76°K).

Exercise. Assuming ideal gas behavior for helium (which is wrong but not drastically so), use Equation 2–24 to calculate the kinetic energy of a mole

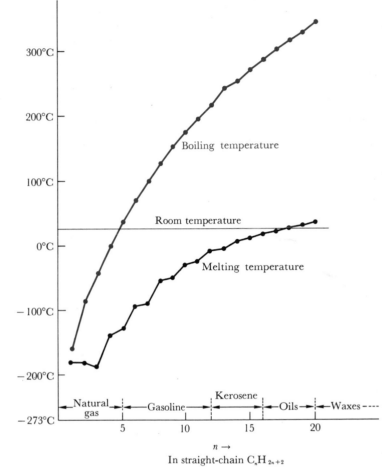

Figure 14–12. Melting temperatures and boiling temperatures of the straight-chain hydrocarbons as a function of the length of the carbon chain. More energy is required to separate two molecules of eicosane (20 carbon atoms) than ethane (2 carbon atoms) because of the more numerous van der Waals interactions between the two larger molecules.

of He gas at $10°K$, at $5°K$, and at $1°K$. At what temperature does this kinetic energy become equal to the energy of van der Waals bonding in solid He? If we allow for the crudity of our ideal-gas assumption, this temperature should be the melting point of He.

(*Answer:* 29.7 cal mole^{-1}; 14.9 cal mole^{-1}; 2.97 cal mole^{-1}; $6.1°K$.)

Van der Waals bonds in molecular solids generally are stronger with increasing size of the atoms and molecules involved. For example, as the atomic number of the noble gas elements increases, the strength of the van der Waals bonding increases also, as is shown by the Ar–Ar potential energy curve in Figure 14–11. The attraction in the larger atoms is stronger, presumably because the outer electrons are held more loosely, and larger *instantaneous dipoles* and *induced dipoles* are possible. Because of this stronger van der Waals bonding, solid argon melts at $-184°C$, or $89°K$, which is considerably higher than the melting temperature of solid helium. The

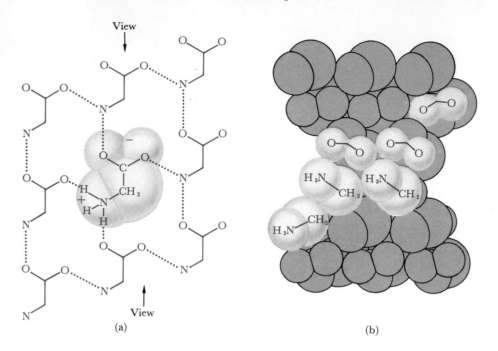

Figure 14–13. The bonding in solid glycine, ^+H_3N—CH_2—COO^-.
*(a) Molecules in a layer are tightly packed and are held together by
van der Waals attraction and by hydrogen bonds (dotted). (b) The layers
are stacked on top of one another and held together by van der Waals
attractions. With this perspective the layers are on edge, and in a horizontal position. The view of the layers in (b) is marked by arrows in (a).*

relationship between van der Waals forces and molecular size also is shown
by hydrocarbons. The melting and boiling temperatures of the straight-chain
alkanes, with formulas C_nH_{2n+2}, are depicted in Figure 14–12 for $n = 1$
through 20. A large molecule such as eicosane ($C_{20}H_{42}$) has a large surface
area of contact with its neighbors, in comparison with a small molecule such
as ethane. The energy per mole needed to set eicosane molecules in continual
motion past one another, that is, to liquefy solid eicosane, is accordingly
greater than the energy required to liquefy solid ethane. The difference
between eicosane and ethane is even greater when the two liquids are vaporized. For vaporization, the energy demanded is not just that needed to slip
one molecule past another. In this process, the energy required is that
sufficient to separate neighboring molecules and bring them into a gas phase.
Therefore, the boiling temperatures of substances increase faster with molecular weight than do the melting temperatures.

The bonding in another molecular solid, glycine, is shown in Figure
14–13. Glycine molecules are held in layers by van der Waals forces and
hydrogen bonds, a second common bond-type in molecular solids. The

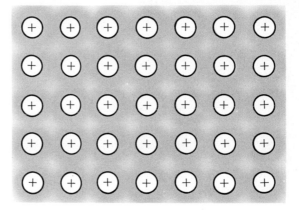

Figure 14–14. Cross section of a crystal lattice of a metal with the sea of electrons. Each positive lattice site represents the nucleus and filled non-valence electron shells of a metal atom. The shaded area surrounding the positive metal ions indicates schematically the mobile sea of electrons. The variation of density in the shading qualitatively represents the greater probability that an electron will be near a positive ion.

glycine layers are held together by van der Waals attractions to build crystals. Some molecules have *permanent* dipoles that enhance van der Waals attraction; such dipoles are especially prominent in compounds with N—H and O—H bonds. These dipole molecules already have been seen in Section 10–4, and will be discussed again in Section 14–4.

14–3 METALS, INSULATORS, AND SEMICONDUCTORS

Most of the elements are metals. Metals are characterized by always having many more valence atomic orbitals than there are valence electrons to fill them. In a sense there is a logical progression from carbon and the other nonmetals, through boron, to the metals. Carbon and the other nonmetals have at least one electron per valence orbital and can construct localized electron-pair bonds. Boron has three electrons for four orbitals. It has to use delocalized three-center orbitals to hold the framework of the molecule together. Sodium has only one valence electron for its $3s$ and $3p$ orbitals. Hence, it is impossible for Na to build the sort of covalently bonded network that carbon does.

Metallic crystals behave as though their valence electrons were relatively free to move through the crystal lattice structure. The electrons form a sea of negative charges that hold the atoms tightly together in a metallic solid. Figure 14–14 illustrates a cross section of such a metal structure. The crystal lattice sites are the positively charged ions remaining when outer electrons are stripped away, thereby leaving the nucleus and the filled electron shells. Since metals generally have high melting points and high densities, especially in comparison with molecular solids, the "electron sea" must bind the positive ions strongly in the crystal lattice.

The simple electron-sea model for metallic bonding is also consistent with two other commonly observed properties of metals: malleability and ductility. A malleable material can be hammered easily into sheets; a ductile material can be drawn into thin wires. For metals to be shaped and drawn

Shift of metallic crystal lattice along a plane
with no resultant strong repulsive forces

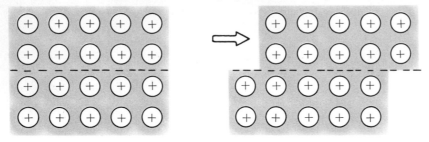

Shift of ionic crystal lattice along a plane
with resultant strong repulsive forces and lattice distortion

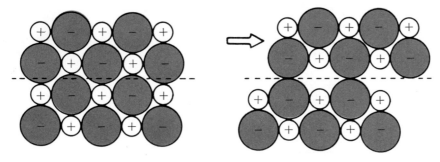

Figure 14–15. The relative softness of most metals in comparison to ionic solids such as NaCl, *and the ductility and malleability of metals, can be explained by the relative ease with which one layer of metal atoms can be slipped over another.*

without fracturing, the atoms in the crystal structure must be displaced easily in planes with respect to each other. This displacement does not result in the development of strong repulsive forces in metals because the mobile sea of electrons provides a constant buffer or shield between the positive ions. This situation is in direct contrast to ionic crystals, in which the binding forces are due almost entirely to electrostatic attractions between oppositely charged ions. In an ionic lattice, valence electrons are bound firmly at ionic lattice sites. Displacement of layers of ions in such a crystal brings ions of like charge together and causes strong repulsions that can lead to the rupture of the crystal (Figure 14–15).

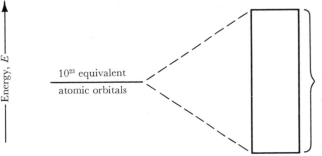

Figure 14–16. Two atomic orbitals can combine to produce two molecular orbitals, as in the H_2 molecule. Six atomic p orbitals can combine to produce six delocalized molecular orbitals in benzene. In a similar way, 10^{23} atomic orbitals in a metal can combine to produce 10^{23} metallic orbitals that are so closely spaced that they can be treated as a continuous band of energy. The simple band theory of metals can explain many of their properties.

Electronic Bands in Metals

The delocalized molecular orbital theory provides a more detailed (and more informative) model for metallic bonding. In this model, the entire block of metal is considered as a giant molecule, and delocalized molecular orbitals are obtained that cover the entire metal. All the atomic orbitals of a particular type in the crystal interact to form a set of delocalized orbitals. For a particular crystal, assume that the number of valence orbitals is of the order of 10^{23}. Figure 14–16 depicts the combination of approximately 10^{23} equivalent atomic orbitals in a crystal to form 10^{23} delocalized orbitals. All of these orbitals cannot have the same energy if they are delocalized. However, instead of producing an antibonding and a bonding molecular orbital, as a diatomic molecule does, they produce a *band* of closely spaced energy levels.

Figure 14–17 illustrates the three bands of energy levels formed by the $1s$, $2s$, and $2p$ orbitals of the simplest metallic element, lithium. The $1s$ molecular orbitals are filled completely since the $1s$ atomic orbitals in isolated lithium atoms are filled. These electrons make no contribution to bonding. They are part of the positive ion cores and can be eliminated from the discussion.

Atomic lithium has one valence electron in a $2s$ orbital. If there are 10^{23} atoms in a lithium crystal, the 10^{23} $2s$ orbitals interact to form a band of 10^{23} delocalized orbitals. As usual, each of these orbitals can accommodate two electrons, so the capacity of the band is 2×10^{23} electrons. The lithium metal has enough electrons to fill only the lower half of the $2s$ band, as illustrated in Figure 14–17.

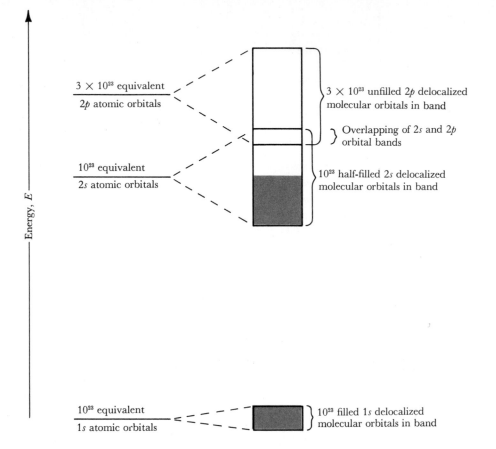

Figure 14–17. Delocalized metallic orbital bands in lithium. The original 2s and 2p atomic orbitals are so close in energy that the molecular or metallic orbital bands overlap. Lithium has one electron for every 2s atomic orbital, and hence half as many electrons as can be accommodated in the 2s atomic orbitals or in the delocalized molecular orbital band. There are unfilled energy states an infinitesimal distance above the highest energy filled state, so an infinitesimal energy is required to excite an electron and send it moving through the metal. Thus lithium is a conductor.

The presence of a partially filled band of delocalized orbitals accounts for bonding and electrical conduction in metals. Electrons in the lower filled orbitals move throughout the lattice in a random fashion such that their motion results in no net separation of electrons and positive ions in the metal. For a metal to conduct an electric current, single electrons must be excited to unfilled crystal orbitals so their motion in one direction is not exactly canceled by electrons moving in the opposite direction. Concerted electron move-

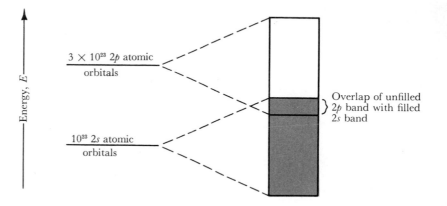

Figure 14–18. Metallic band-filling diagram for beryllium. A Be *atom has enough electrons (two) to fill its 2s atomic orbital, so* Be *metal has enough electrons to fill its 2s delocalized molecular orbital band. If the 2s and 2p bands did not overlap,* Be *would be an insulator because an appreciable amount of energy would be required to make electrons flow in the solid. But with the band overlap shown here, an infinitesimal amount of energy excites electrons to the 2p orbitals and electrons flow.*

ment occurs only upon the application of an electric potential difference between two regions of a metal. Electrons are excited to the unfilled delocalized orbitals, which are part of the same band and just slightly higher in energy. Therefore, we can expect a metal to have small resistance to electrical conduction. However, conduction is limited by the frequent collisions of electrons with the positive ions, which have kinetic energy and thus "rattle around" in a random manner about their lattice sites. As the temperature rises, lattice vibration of the positive ions increases, and collisions with the conduction electrons are more frequent. Therefore, electrical conductivity in metals decreases as the temperature increases.

Beryllium is a more complicated example. An isolated beryllium atom has exactly enough electrons to fill its 2s orbital. Accordingly, beryllium metal has enough electrons to fill its 2s delocalized band. If the 2p band did not overlap the 2s band (Figure 14–18), beryllium would be an insulator because an energy equal to the gap between bands would be required before electrons could move through the solid. However, the two bands do overlap, and beryllium has unoccupied orbitals that are only infinitesimally higher in energy than the most energetic filled orbitals. Beryllium is a metallic conductor.

Insulators

Nonmetallic network materials such as boron or carbon are *insulators;* that is, they do not conduct electric currents. Figure 9–6 shows the abrupt loss of

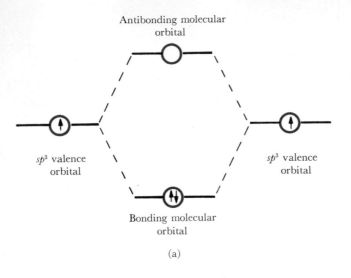

(a)

Figure 14–19. Bonding in diamond crystals. (a) Localized orbital energy levels in diamond crystals. Each pair of neighboring localized atomic sp^3 orbitals produces a bonding orbital and an antibonding orbital. (b) Schematic representation of the overlap of the four sp^3 hybrid orbitals of a carbon atom with similar orbitals from four other carbon atoms.

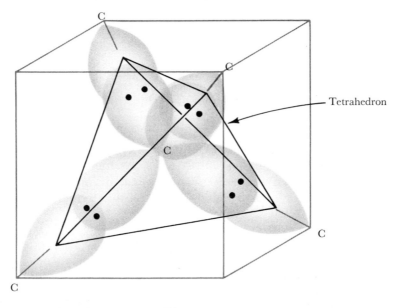

(b)

conductivity from Be to B. One way to visualize the difference between nonmetallic insulators and metals is to use the approximation of localized orbitals for insulators. We can use localized bonds to describe insulators quite accurately because their coordination numbers are relatively low. Because of the low coordination there usually are enough electrons in the valence orbitals to form three or four simple covalent bonds between each atom and its nearest neighbors. The construction of these bonds resembles the formation of localized bonds in a polyatomic molecule.

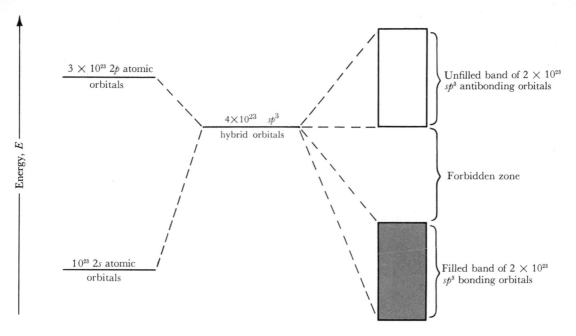

Figure 14–20. Delocalized molecular orbital bands in an insulator, formed from equivalent sp^3 localized hybrid orbitals. Note the relatively large zone between the filled band of sp^3 bonding orbitals and the unfilled band of sp^3 antibonding orbitals.

For diamond we begin to construct the bonding model by assigning each carbon atom four localized tetrahedral sp^3 hybrid orbitals. One such orbital from each of two neighboring carbon atoms combine to make one bonding and one antibonding molecular orbital (Figure 14–19). The four valence electrons in each carbon atom are sufficient to fill these bonding orbitals. Thus all electrons in diamond are used for bonding, thereby leaving none to move freely to conduct electricity.

To construct the band model of delocalized orbitals for an insulating network solid such as diamond, we will proceed as follows. Assume 10^{23} carbon atoms. When the 4×10^{23} localized sp^3 orbitals interact with each other, two bands of delocalized orbitals are formed, one from the 2×10^{23} bonding orbitals of Figure 14–19 and one from the 2×10^{23} antibonding orbitals. These are depicted in Figure 14–20, in which the atomic orbitals are drawn at the left to remind you that these orbitals originally came from the $2s$ and $2p$ atomic orbitals. The important fact in this diagram is that the band filled with electrons does not overlap with the next higher energy band, which has completely unfilled orbitals. There is a forbidden energy zone or gap between what is called the *valence band* below and the *conduction band* above. There are 4×10^{23} valence electrons per 10^{23} carbon atoms, enough to fill completely the 2×10^{23} orbitals in the valence band.

For an insulator to conduct, energy is required that is sufficient to excite electrons in the filled band across this forbidden energy zone into the unfilled molecular orbitals. This energy is the activation energy of the conduction process. Only high temperatures or extremely strong electrical fields will provide enough energy to an appreciable number of electrons for conduction to occur. In diamond the gap between the top of the valence band and the bottom of the conduction band is 5.2 eV, or 120 kcal mole^{-1}.

Semiconductors

The border line between metallic and nonmetallic network structures of elements in the periodic table is not sharp (Figure 14–1). This is shown by the fact that several elemental solids have properties that are intermediate between conductors and insulators. Silicon, germanium, and α-gray tin all have the diamond structure. However, the forbidden energy gap between filled and empty bands for these solids is much smaller than for carbon. Rather than 120 kcal mole^{-1} for carbon, the gap for silicon is only 25 kcal mole^{-1}. For germanium it is 14 kcal mole^{-1}, and for α-gray tin it is 1.8 kcal mole^{-1}. The metalloids silicon and germanium are called *semiconductors*. Figure 14–21 shows the band diagram for a semiconductor, with a small forbidden energy zone.

A semiconductor can carry a current if the relatively small energy required to excite electrons from the lower filled valence band to the upper empty conduction band is provided. Since the number of excited electrons increases as the temperature increases, the conductance of the semiconductor increases with temperature. This behavior is exactly the opposite of that of metals.

Conduction in materials such as silicon and germanium can be enhanced by adding small amounts of certain impurities. Although there is a forbidden energy gap in silicon, it can be narrowed effectively if impurities such as boron or phosphorus are added to silicon crystals. Small amounts of boron or phosphorus (a few parts per million) can be incorporated into the silicon structure when the crystal is grown. Phosphorus has five valence electrons and thus has an extra free electron even after four electrons have been used in the covalent bonds of the silicon structure. This fifth electron can be moved away from a phosphorus atom by an electric field; hence we say phosphorus is an electron donor. Only 0.25 kcal mole^{-1} is required to free the donated electrons, thereby making a conductor out of silicon to which a small amount of phosphorus has been added. The opposite effect occurs if boron instead of phosphorus is added to silicon. Atomic boron has one too few electrons for complete covalent bonding. Thus for each boron atom in the silicon crystal there is a single vacancy in a bonding orbital. It is possible to excite the valence electrons of silicon into these vacant orbitals in the boron atoms, thereby causing the electrons to move through the crystal. To accomplish this conduction an electron from a silicon neighbor drops into the empty boron orbital. Then an electron that is two atoms away can fill the silicon

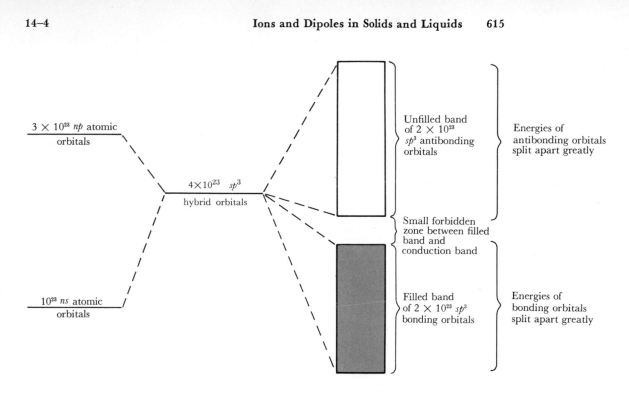

Figure 14–21. Bands of delocalized orbitals in semiconductors formed from equivalent sp³ localized hybrid orbitals. The forbidden zone between filled and empty bands is smaller than in an insulator.

atom's newly created vacancy. The result is a cascade effect, whereby an electron from each of a row of atoms moves one place to the neighboring atom. Physicists prefer to describe this phenomenon as a hole moving in the opposite direction. No matter which description is used, it is a fact that less energy is required to make a material such as silicon conduct if the crystal contains small amounts of either an electron donor such as phosphorus or an electron acceptor such as boron.

14–4   IONS AND DIPOLES IN SOLIDS AND LIQUIDS

In solid NaCl, each positive ion is surrounded by six negative ions in octahedral coordination, and each negative ion is surrounded by six positive ions (Figure 14–7). Electrostatic attractions between these oppositely charged ions keep the crystal together. How, then, can water dissolve NaCl? Table salt is soluble in water, although it is not soluble in gasoline. Water can decompose the solid lattice and separate the oppositely charged ions because the stability lost in separating the ions is recovered during the formation of hydrated ions (Figure 14–22). Each Na^+ ion again has an octahedron of negative charges around it, but instead of being Cl^-, they are the negatively charged oxygen atoms of the water molecules. Water is a polar solvent, and

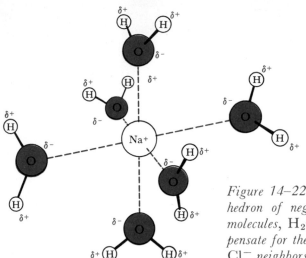

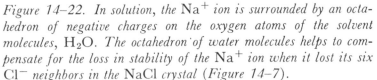

Figure 14–22. In solution, the Na^+ *ion is surrounded by an octahedron of negative charges on the oxygen atoms of the solvent molecules,* H_2O*. The octahedron of water molecules helps to compensate for the loss in stability of the* Na^+ *ion when it lost its six* Cl^- *neighbors in the* $NaCl$ *crystal (Figure 14–7).*

each of its molecules is a small dipole. The Cl^- ions also are hydrated, but it is the positively charged hydrogen atoms on the water molecules that approach the Cl^- ions. A nonpolar solvent such as gasoline, whose molecules are hydrocarbons, cannot form such ion–dipole bonds with Na^+ and Cl^-. The ions are much more stable when associated with one another than when separated and dispersed in gasoline as a solvent. As a consequence, NaCl (and other salts as well) is insoluble in gasoline.

Polar Interactions: Hydrogen Bonds

Polar molecules have a small separation of charge and hence are dipoles. In H_2O, the oxygen atom is slightly negative, and the hydrogen atoms are positive. We discussed molecules with dipole moments, in connection with the ionic character of a bond, in Section 10–4. In polar molecules, different parts of the structure have different charges, but the molecules do not separate into ions. Polar molecules are stabilized in the solid by the interaction of oppositely charged ends of the molecules (Figure 14–23). This is called *dipole–dipole interaction.* Such solids are dissolved by polar solvents for the same reason that ionic solids are: The stability lost when the crystal is disrupted is regained by interactions between polar molecules from the crystal and polar solvent molecules (Figure 14–24). Ice is soluble in liquid ammonia but not in benzene; this is because liquid ammonia is a polar solvent, whereas benzene is nonpolar.

A particularly important kind of polar interaction is the *hydrogen bond.* This is the bond, primarily electrostatic, between a positively charged hydrogen atom and a small, electronegative atom, usually F or O. Ice has such bonds (Figure 14–25). Each oxygen atom is tetrahedrally coordinated to four

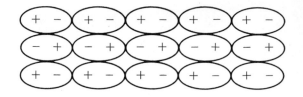

Figure 14–23. Diagrammatic representation of the packing of polar molecules into a crystalline solid. Packing occurs in a way such that partial charges of opposite sign can be in close proximity.

Energy is required to break up a solid molecular lattice

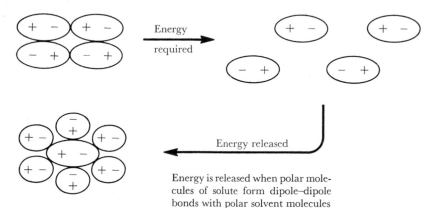

Energy is released when polar molecules of solute form dipole–dipole bonds with polar solvent molecules

Figure 14–24. When a crystalline solid with polar molecules dissolves, stability is lost when oppositely charged ends of neighboring molecules are removed. This loss is compensated by the stability produced by hydrating the polar molecules in solution. A solvent that cannot provide such stabilization cannot dissolve the solid.

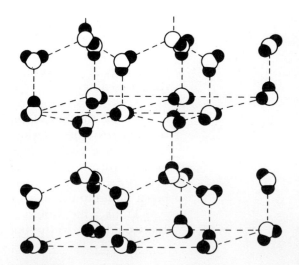

Figure 14–25. In crystalline ice, each H_2O molecule is hydrogen bonded to two others by means of its own hydrogen atoms, and bonded to two more H_2O molecules by means of their hydrogen atoms. The coordination is tetrahedral, and the lattice is similar but not the same as that of diamond. The hydrogen bonding gives ice its low density compared to water, and gives water its high freezing and boiling temperatures compared to H_2S.

other oxygen atoms in a structure that resembles diamond, but is not quite the same. Each oxygen atom is bound to its four neighbor oxygen atoms by hydrogen bonds; one such bond is illustrated by the dotted line:

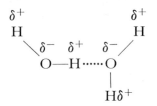

In two of these hydrogen bonds, the central oxygen atom supplies the hydrogen atoms; in the other two bonds, the hydrogens come from neighboring water molecules. Such bonds are relatively weak compared with covalent bonds. A typical covalent bond energy is about 100 kcal mole^{-1}, whereas a hydrogen bond between H and O is approximately 5 kcal mole^{-1}. But hydrogen bonds are important for the same reason that van der Waals bonds are: They may be weak but there are many of them.

Hydrogen bonding in water is responsible for many of its most important properties. Because of hydrogen bonds in the solid and the liquid, both the melting point and boiling point of water are unexpectedly high when compared with those of H_2S, H_2Se, and H_2Te, which are hydrogen compounds of elements in the same group of the periodic table. Solid and liquid ammonia and HF show the same effect, and for the same reason (Figure 14–26). Hydrogen bonding in ammonia is less pronounced than in water for two reasons: N is less electronegative than O, and NH_3 has only one lone pair of electrons to attract the H from a neighboring molecule. Hydrogen fluoride is less well hydrogen-bonded than H_2O in spite of the greater electronegativity of F and the presence of three lone electron pairs. This is because HF has only one hydrogen atom to donate in making such bonds.

Since hydrogen bonding builds an open network structure in ice (Figure 14–25), ice is less dense than water at the melting point. Upon melting, part of this open-cage structure collapses, and the liquid is more compact than the solid. The measured heat of fusion of ice is only 1.4 kcal mole^{-1}, whereas the energy of its hydrogen bonds is 5 kcal mole^{-1}. This information indicates that only about 28% of the hydrogen bonds of ice are broken when it melts. Water is not composed of isolated, unbonded molecules of H_2O; it has regions or clusters of hydrogen-bonded molecules. Part of the hydrogen-bonded structure of the solid persists in the liquid. As the temperature is raised, these clusters break up, and the volume continues to shrink. If the temperature is raised still higher, the expected thermal expansion dominates over the shrinkage caused by the collapse of the cage structures. Liquid water has a minimum molar volume, or a maximum density, at 4°C.

Think what would happen if water were not hydrogen bonded as it is. Ice would sink to the bottom of its liquid as do most solids. The bottoms of

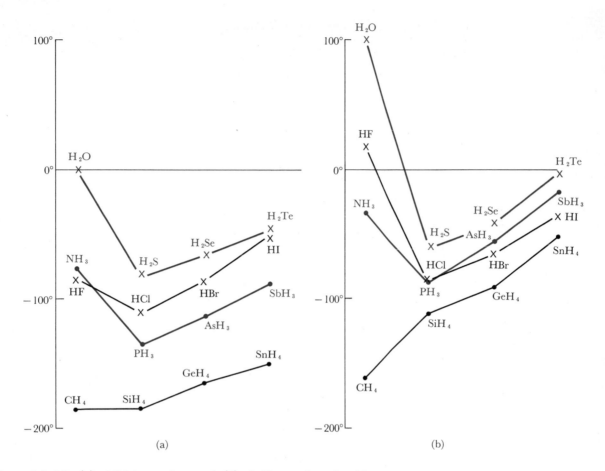

Figure 14–26. (a) Melting points and (b) boiling points for binary hydrogen compounds of some representative elements. In general, melting points increase with molecular weight within a group. The anomalous compounds, HF, H₂O, and NH₃, all have hydrogen bonds between molecules in both solid and liquid states.

lakes and oceans would have deposits of ice for the entire year; the ice would freeze a little higher in the winter and melt back to a lower level in the summer. The warm water, being lighter, would remain on top. There would be no mixing by convection as we now have. Oceans would not be the relatively uniform-temperature heat reservoirs that they are now, but would manifest radical temperature changes with depth.

Because the hydrogen-bonded clusters are broken slowly as heat is added and the temperature rises, water has a higher specific heat than any other common liquid except ammonia. (The specific heat of a substance is the amount of heat required to raise the temperature of one gram of the substance by 1°C.) Water also has an unusually high heat of fusion and heat of vaporization. All three of these properties mean that water acts as a large thermostat that confines the temperature on the planet within moderate limits. Ice absorbs a large amount of heat when it melts, and water can absorb more heat per unit of temperature rise than almost any other substance. Correspondingly, as it cools, water gives off more heat to its surroundings than do other substances. Coastal regions never experience the extremes of heat and cold that are typical of continental regions like the American Great Plains and the Steppes of Central Asia and Siberia. It is unlikely that life could evolve and develop to a high level on planets where the extremes of temperature were not moderated by a high specific heat liquid such as H_2O.

Hydrogen bonds are even more important to life than the water structure suggests. They are one of the principal means of holding protein molecules together, as we saw in Section 12–9. Without such bonds between carbonyl oxygen atoms and amine hydrogen atoms, no polypeptide chain would fold properly to make its protein.

Challenge. If you want to work out the hydrogen-bonded structures involved in a glass-etching fluid and in the DNA molecule, try Problems 14–5 and 14–13 in *Relevant Problems for Chemical Principles* by Butler and Grosser.

14–5 THE FRAMEWORK OF THE PLANET: SILICATE MINERALS

From the microscopic world of hydrogen-bonded protein molecules, we turn to the nucleus of the planet earth. The core of our planet is believed to be mainly iron and nickel, with a radius of approximately 2200 miles. This core gives earth its magnetic field, which the moon and our neighbor planets, Mars and Venus, apparently lack. The earth's core is at a high pressure and temperature, and is probably fluid. An old theory of the origin of our planet presumed that it formed when hot gases collected and cooled. By this theory, the core is a relic of the first hot period; it has not solidified because of the insulating effect of the outer layers.

The current view is that the earth grew by cold accretion of solid debris and dust. After a certain critical mass was reached, heat from the interior could not be lost to the surroundings as rapidly as it was generated by natural radioactivity and pressure, and the center of the planet liquefied. This could occur only in a planet above a critical size, which presumably explains the absence of a molten core and a strong magnetic field in Mars and the moon. This phenomenon is similar to that of critical mass in uranium fission. Below a certain mass of ^{235}U the loss of neutrons to the outside is too great to sustain a chain reaction, so nuclear explosion occurs only in pieces of ^{235}U above this critical size.

For 1800 miles above the core of the earth extends the *mantle*, a layer probably composed of a dense silicate material similar to basalt. The upper 20 miles under the continents, or as little as three miles under the ocean beds, is the *crust*, the only part about which we have any real chemical knowledge.

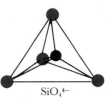

$SiO_4{}^{4-}$

● Silicon

● Oxygen

Figure 14–27. The $SiO_4{}^{4-}$ tetrahedron, which is the building block of most silicate minerals. The Si atom (black) is covalently bonded to four oxygen atoms at the corners of a tetrahedron (color). The black lines between oxygen atoms are included only to give form to the tetrahedron.

On top of this crust—itself only 1.5% of the volume of the planet—is spread the thin layer containing virtually all the matter with which we are concerned.

The crust is 48% oxygen by weight, in the form of the various silicate minerals. It is also 26% silicon, 8% aluminum, 5% iron, and 2% to 5% calcium, sodium, potassium, and magnesium. Remarkably enough, these silicate minerals are more than 90% oxygen by volume. Most of the crust is some form of the mixture of silicate minerals known as *granite*.

The basic building block in silicates is the orthosilicate ion, $SiO_4{}^{4-}$, shown in Figure 14–27. Each silicon atom is covalently bonded to four oxygen atoms at the corners of a tetrahedron. The $SiO_4{}^{4-}$ anion occurs in simple minerals such as zircon ($ZrSiO_4$), garnet, and topaz. Two tetrahedra can share a corner oxygen atom to form a discrete $Si_2O_7{}^{6-}$ anion, or three tetrahedra can form a ring, shown in Figure 14–28. Benitoite, $BaTiSi_3O_9$, is the best known example of this uncommon kind of silicate. Beryl, $Be_3Al_2Si_6$-O_{18}, a common source of beryllium, has anions composed of rings of six tetrahedra with six shared oxygen atoms.

Chain Structures

All of the silicates mentioned so far are made from discrete anions. A second class is made of endless strands or chains of linked tetrahedra. Some minerals

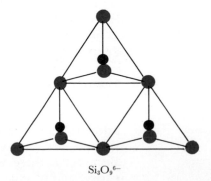

$Si_3O_9{}^{6-}$

Figure 14–28. A ring of three tetrahedra, with three oxygen atoms shared between pairs of tetrahedra, has the formula $Si_3O_9{}^{6-}$. This structure occurs as the anion in soft, crumbly rocks such as benitoite, $BaTiSi_3O_9$.

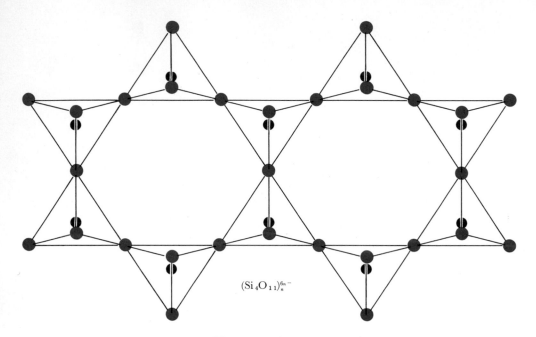

$(Si_4O_{11})_n^{6n-}$

Figure 14–29. Long double-stranded chains of silicate tetrahedra are in fibrous minerals such as asbestos.

have single silicate strands with the formula $(SiO_3)_n^{2n-}$. A form of asbestos has the double-stranded structure shown in Figure 14–29. The double-stranded chains are held together by electrostatic forces between themselves and the Na^+, Fe^{2+}, and Fe^{3+} cations packed around them. The chains can be pulled apart with much less effort than is required to snap the covalent bonds within a chain. Therefore, asbestos has a stringy, fibrous texture. Aluminum ions can replace as many as one quarter of the silicon ions in the tetrahedra. However, each replacement requires one more positive charge from another cation (such as K^+) to balance the charge on the silicate oxygen atoms. The physical properties of the silicate minerals are influenced strongly by how many Al^{3+} ions replace Si^{4+} ions, and by how many extra cations therefore are needed to balance the charge.

Sheet Structures

Continuous broadening of double-stranded silicate chains produces planar sheets of silicate structures (Figure 14–30). Talc, or soapstone, has this structure, in which none of the Si^{4+} is replaced with Al^{3+}. Therefore, no additional cations between the sheets are required to balance charges. The silicate sheets in talc are held together primarily by van der Waals forces. Because of these weak forces the layers slide past one another relatively easily, and produce the slippery feel that is characteristic of talcum powder.

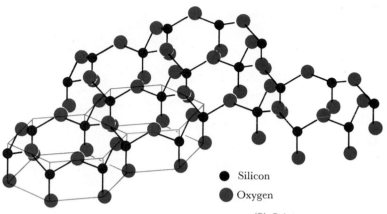

Figure 14–30. In talc, mica, and the clay minerals, silicate tetrahedra each share three of their corner oxygen atoms to make endless sheets. All of the unshared oxygen atoms point down in this drawing on the same side of the sheet.

Mica resembles talc, but one quarter of the Si^{4+} in the tetrahedra is replaced by Al^{3+}. Thus an additional positive charge is required for each replacement to balance charges. Mica has the layer structure shown in Figure 14–31. The layers of cations (Al^{3+} serves as a cation between layers as well as a substitute in the silicate tetrahedra) hold the silicate sheets together electrostatically with much greater strength than in talc. Thus mica is not slippery to the touch and is not a good lubricant. However, it cleaves easily, thereby splitting into sheets parallel to the silicate layers. Little effort is required to flake off a chip of mica, but much more strength is needed to bend the flake across the middle and break it.

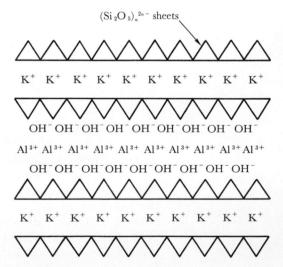

Figure 14–31. In mica [muscovite, $K_2Al_4Si_6Al_2O_{20}$-$(OH)_4$], anionic sheets of silicate tetrahedra, as in Figure 14–30, alternate with layers of potassium ions and aluminum ions sandwiched between hydroxide ions. This layer structure gives mica its flaky cleavage properties.

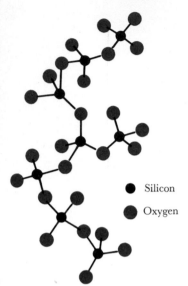

● Silicon

● Oxygen

Figure 14–32. Glasses are amorphous, disordered chains of silicate tetrahedra that are fused with metal oxides or carbonates such as Na_2CO_3 or $CaCO_3$.

The clay minerals are silicates with sheet structures such as in mica. These layer structures have enormous "inner surfaces" and often can absorb large amounts of water and other substances between the silicate layers. This is why clay soils are such useful growth media for plants. This property also is why clays are used as beds for metal catalysts. The common catalyst platinum black is finely divided platinum metal obtained by precipitation from solution. The catalytic activity of platinum black is enhanced by the large amount of exposed metal surface (see Section 12–5). The same effect can be achieved by precipitating a metal to be used as a catalyst (Pt, Ni, or Co) onto clays. The metal atoms coat the interior walls of the silicate sheets, and the clay structure prohibits the metal from consolidating into a useless mass. J. D. Bernal has suggested that the first catalyzed reactions in the early stages of the evolution of life, before biological catalysts (enzymes) existed, may have occurred on the surfaces of clay minerals.

Three-Dimensional Networks

The three-dimensional silicate networks, in which all four oxygen atoms of SiO_4^{4-} are shared with other Si^{4+}, are typified by quartz, $(SiO_2)_n$, (Figure 14–6). In quartz, all of the tetrahedral structures have Si^{4+} ions, but in other network minerals, up to half the Si^{4+} can be replaced with Al^{3+}. These minerals include the feldspars, with a typical empirical formula $KAlSi_3O_8$. Feldspars are nearly as hard as quartz. Basalt, which may be the material of the mantle of the earth, is a compact mineral related to feldspar. Granite, the chief component of the earth's crust, is a mixture of crystallites of mica, feldspar, and quartz.

Glasses are amorphous, disordered, noncrystalline aggregates with linked silicate chains of the sort depicted in Figure 14–32. Common soda-lime

Table 14–3. Types of Bonding in Solids

	Molecular	Nonmetallic network	Metallic	Ionic
Structural unit:	Molecule	Atom	Atom	Ion
Principal bonding between units:	Weak van der Waals and, in polar molecules, a stronger dipole–dipole bond	Strong covalent bonds	Delocalized electron sea through a system of metallic positive ion centers	Strong ionic bonding (electrostatic)
Properties:	Soft Low melting point Insulator	Hard High melting point Insulator or semiconductor	Wide range of hardness Wide range of melting points Conductor	Hard High melting point Insulator
Usually occurs in:	Nonmetals at right of periodic table and compounds predominantly composed of nonmetals	Nonmetals in center of periodic table	Metals in left half of periodic table	Compounds of metals and nonmetals
Examples:	O_2, C_6H_6, H_2N—CH_2—$COOH$	Diamond, Si, ZnS, SiO_2	Na, Zn, Au, brass, bronze	KI, Na_2CO_3, LiH

glass is made with sand (SiO_2), limestone ($CaCO_3$), and sodium carbonate (Na_2CO_3) or sodium sulfate (Na_2SO_4), which are melted together and allowed to cool. Other glasses with special properties are made by using other metal carbonates and oxides. Pyrex glass has boron as well as silicon and some aluminum in its silicate framework. Glasses are not true solids, but are extremely viscous liquids. If you examine the panes of glass in a very old New England home, you can sometimes see that the bottom of the pane is slightly thicker than the top because of two centuries of slow, viscous flow of the glass.

14–6   SUMMARY

The types of bonding in solids that we have seen are summarized in Table 14–3. Ionic or electrostatic bonds and electron-pair covalent bonds both have bond energies of approximately 100 kcal mole^{-1}. Metallic bonds are variable, but are of comparable strength. Hydrogen bonds are much weaker: The bond between O and H is about 5 kcal mole^{-1}. Van der Waals attractions are weaker yet: from a few tenths of a kcal to half a kcal mole^{-1}. Hydrogen bonds and van der Waals attractions assume a greater importance than their strength would suggest because of the large number of such bonds that can form.

The first four types of bonding, ionic, covalent, metallic, and hydrogen bonding, can be interpreted in terms of molecular orbitals. Covalent bonding with localized molecular orbitals was the subject of Chapter 10. In Section 10–5 we considered partial ionic bonding in a molecule such as HF as the limit of extreme polarity of a covalent bond. Both metals and covalent network solids can be interpreted by using delocalized molecular orbitals, in which the "molecule" extends over the entire piece of matter under study. This band theory accounts for many of the observed properties of conductors, semiconductors, and insulators. Hydrogen bonding can be regarded as an ionic bond between a positive hydrogen atom and an electronegative atom of sufficiently small radius that the proton can approach closely. Oxygen and fluorine can participate in such hydrogen bonds, as can nitrogen and carbon to a lesser degree, but chlorine usually is too large. Hydrogen bonds are responsible for many of the familiar properties of water and ice, and are essential in holding protein molecules together correctly.

The silicates have framework structures of a diversity that is suggestive of, if not comparable to, the branched-chain compounds of carbon. The basic subunit, a SiO_4^{4-} tetrahedron, can be organized into rings, chains, sheets, and three-dimensional networks. Aluminum can replace some silicon, but other cations must be added to balance the charge, thereby increasing the electrostatic contribution to holding the solid together. The silicates illustrate four of the five types of bonding that have been discussed in this chapter: covalent bonding between Si and O in the tetrahedra, van der Waals forces between silicate sheets in talc, ionic attractions between charged sheets and chains, and hydrogen bonds between water molecules and the silicate oxygen atoms in clays. If we include Ni catalysts prepared on a clay support, the fifth type of bonding (metallic) is represented as well.

SUGGESTED READING

A. H. Cottrell, "The Nature of Metals," *Scientific American*, September, 1967.

W. A. Deer, R. A. Howie, and J. Zussman, *Rock Forming Minerals*, Vols. 1–5, Wiley, New York, 1962.

T. L. Hill, *Matter and Equilibrium*, W. A. Benjamin, Menlo Park, Calif., 1966. Good treatments of states of matter, ideal and nonideal gases, intermolecular forces, solid structures, and liquids.

W. J. Moore, *Seven Solid States*, W. A. Benjamin, Menlo Park, Calif., 1967. Seven solids: NaCl, gold, silicon, steel, nickel oxide, ruby, and anthracene, used as a framework on which to develop a great amount of the theory of solids. Goes beyond this chapter, but contains a lot that is relevant to it.

N. Mott, "The Solid State," *Scientific American*, September, 1967.

H. Reiss, "Chemical Properties of Materials," *Scientific American*, September, 1967.

A. F. Wells, *Structural Inorganic Chemistry*, Oxford University Press, New York, 1962, 3rd ed.

QUESTIONS

1 What types of forces hold molecules together in crystals and liquids?

2 What effect do hydrogen bonds have on the boiling temperatures of liquids?

3 Why are nonmetallic network solids usually quite hard?

4 What physical effect is responsible for the attraction in van der Waals interactions? What is responsible for the repulsion in such interactions? Compare the origin of attraction and repulsion in van der Waals interactions with that in ionic and covalent bonds.

5 How do we determine an experimental value for the van der Waals radius of hydrogen?

6 If van der Waals bonds are extremely weak, why are they discussed at all?

7 In the delocalized molecular orbital theory of metals, in what sense do we say that the entire piece of metal is a large molecule?

8 Why would beryllium be an insulator if the $2s$ and $2p$ molecular-orbital bands did not overlap?

9 What is the structural difference between metals, semiconductors, and insulators?

10 What effect do small amounts of boron or phosphorus have on the conducting properties of silicon?

11 How do hydrogen bonds participate in the structure of ice? What effects do they have on its properties?

12 How do we know that some hydrogen bonding in water persists in the liquid phase?

13 Provide a structural explanation for the fact that quartz is hard, asbestos fibrous and stringy, and mica platelike.

14 Why are clays useful in industrial catalysis?

15 Explain the trend in the melting temperatures of the following tetrahedral molecules: CF_4, $90°K$; CCl_4, $250°K$; CBr_4, $350°K$; CI_4, $440°K$.

16 Where does the "band" of energy levels come from in the delocalized molecular theory of electronic structure of metals?

17 How are hydrogen bonds involved in protein structures? Can you find examples of hydrogen bonding in Figures 12–2, 12–20, and 12–24?

18 How do glasses differ from quartz?

19 What type of solid will BF_3 and NF_3 molecules build? What kinds of intermolecular interactions are likely to be important in each case? Which compound should have the higher melting temperature?

PROBLEMS

1 Draw curves of the ways in which the repulsion and attraction terms of the van der Waals interaction (Equations 14–1 and 14–2) vary with distance R between atomic centers. Add these two curves in an approximate way and satisfy yourself that a potential curve such as Figure 14–10 is the result.

2 It requires 5.2 eV, or 120 kcal mole^{-1}, to excite electrons in a diamond crystal from the valence band to the conduction band. What frequency of light is needed to bring about this excitation? What wavelength? What wave number? What part of the electromagnetic spectrum does this correspond to?

3 Using data given in this chapter, repeat Problem 2 for the semiconductors silicon and germanium.

4 Construct the potential energy curve for the Kr–Kr van der Waals interaction. How strong is the Kr–Kr van der Waals bond? Estimate the Kr–Kr bond distance in solid krypton.

5 The molecule RbBr is held together primarily by an ionic bond. The distance between Rb^+ and Br^- in the molecule is 2.945 Å. The closed electron shells of Rb^+ and Br^- both have the configuration of the noble gas Kr. From the energy curve constructed for Problem 4, estimate the van der Waals energy between Rb^+ and Br^-, assuming that the energy is the same as for a pair of Kr atoms separated by a distance of 2.945 Å. Is the repulsive part or the attractive part of the interaction dominant? How important is the van der Waals energy compared to the overall bond energy of 90 kcal mole^{-1} in RbBr? Examine the Kr–Kr van der Waals energy for distances of 2 Å and 1 Å and then explain what prevents Rb^+ and Br^- ions from approaching each other too closely in an ionic solid.

PART FOUR
CHEMICAL DYNAMICS

The next four chapters deal with topics that can be called chemical dynamics. Names are always difficult to deal with for they inevitably distort the thing they represent. They place boundaries where no boundaries should exist, and they imply attributes that the object named may not possess. Still, one of man's most primitive emotions is the feeling that if he can name a thing, he somehow gains control over it or understands it. Names are useful levers with which to move ideas, so long as they are not confused with the ideas themselves.

Chemistry as a profession—as opposed to the occasional hobby of the well-to-do and curious—was practically unheard of before Lavoisier. The metallurgists and alchemists were primarily technicians and secondarily systematic investigators. Those who were interested in fundamental principles of the behavior of matter were more likely to call themselves natural philosophers, rather than chemists.

The division between *inorganic* and *organic* chemistry, which began in Lavoisier's time, is therefore as old as the profession itself. The weakness of this division became apparent as soon as Wöhler synthesized urea, in 1828. Yet the division was retained because it was convenient. Organic and inorganic chemists in those early days usually worked with different substances and thought about them in different ways. In the latter half of the last century, *analytical* chemistry arose in response to a practical need, and *physical* chemistry developed as the study of the theory of chemical reactions. *Biochemistry* evolved from the work of Pasteur and others on enzymes and metabolism. Other subdivisions of chemistry have been used: nuclear chemistry, geochemistry, electrochemistry, agricultural chemistry, petrochemistry, and more. These have never been accorded the fundamental importance given to the first five divisions: inorganic, organic, analytical, physical, and biochemistry.

As we learn more about chemical behavior, these old categories are being challenged. The organic chemist adopts quantum mechanics to explain reactions and studies organic lasers and semiconductors. The inorganic chemist finds that transition-metal complexes lead him to enzymatic catalysis, the biochemist turns to the thermodynamics of metabolic processes, and the physical chemist applies nuclear magnetic resonance to the stacking of side chains in DNA. Where does organic chemistry end and biochemistry begin? What is analytical chemistry if not the intelligent application of physical, inorganic, and organic chemistry to a certain type of problem? What organic research chemist can avoid using physical methods of analysis and structure determination? To what does physical chemistry apply its theories if not to the subjects of the other branches of chemistry? Are these old labels now containers, or barriers?

A new classification has been gaining favor in the past few years; now chemistry is divided into structural chemistry, chemical dynamics, and chemical synthesis. So far, most of what we have dealt with in this book could be called structural chemistry. We have been concerned with the structures of atoms and molecules, and how these structures explain their physical and chemical properties. The distinction is not clearcut. Our reason for wanting to know structures is to help us to explain reactions. The new classification has all of the inherent arbitrariness and weaknesses of the old ones. Proficient chemists must be a blend of all three categories. A man who determines the structures of compounds but cares nothing for their reactions is a useful person, but he is not a chemist. A chemical dynamicist with no interest in structure is setting himself a hopeless task. A synthetic chemist with no knowledge of structure or dynamics will never synthesize anything, except by accident. The new categories also have another weakness that the old ones avoided. The new categories purport to describe how a chemist thinks, whereas the old ones were content to describe what he thinks about. It is always dangerous to presume another man's patterns of thought.

Although the categories of structure, dynamics, and synthesis have scant value as a professional classification of working chemists, they are valuable for organizing chemistry in the learning process. Within any of the areas of chemistry, a man can focus his attention primarily on the structural, the dynamic, or the synthetic aspects of his particular problem, but he must integrate all three if he is to succeed.

In these last four chapters, we shall look at the dynamic aspects of chemistry. The two central ideas will be those of *equilibrium* and *reaction rates*. We already have discussed equilibria from a purely descriptive viewpoint in Chapter 5. Now we will give these same ideas a foundation in thermodynamics. We shall see how to predict and detect equilibrium, and how to use equilibrium information in nonequilibrium situations. We shall examine how fast a reaction approaches equilibrium and how this rate is measured. We shall develop theories to account for this rate of reaction in terms of the molecular mechanisms by which chemical reactions occur. In all of this material, the emphasis will be upon *change*. We cannot understand change without knowing the structure of the things that change; but once we are thoroughly grounded in the architecture of chemical substances, the proper application of that knowledge is to the understanding of chemical transformations. This is the area of chemical dynamics.

Heat and cold are nature's two hands by which she chiefly worketh.
Francis Bacon (1627)

15 ENERGY AND ENTROPY IN CHEMICAL SYSTEMS

An old motto from the time of World War II (and probably earlier) is, "The difficult we do at once; the impossible takes a little longer." In this chapter we shall discover what is *possible* in chemical reactions. This does not mean that everything that is possible by the laws of thermodynamics will take place in a short time. When the chemical thermodynamicist says that a reaction is spontaneous, he makes no predictions whatsoever about elapsed time; he only says that, given *enough* time, the reaction can happen. To the thermodynamicist, the explosion produced by dropping sodium in water and the weathering away of the entire North American continent are both spontaneous processes.

To the chemist, it is important to know whether a reaction is spontaneous in the thermodynamic sense. If it is slow but spontaneous, then some means may be found to hasten the process, such as catalysis. If the reaction is not spontaneous, such a search is doomed at the start: Another means must be devised to force the desired reaction to occur.

By what criterion does a chemist say that a reaction is spontaneous? In Chapter 5 we discussed the ideas of spontaneity and equilibrium, but we took the numerical values of equilibrium constants on faith. Now we shall see how these constants can be related to other measurable properties of a reaction. Most spontaneous reactions release heat. Is this a valid generalization for all reactions? Why do some reactions go to completion so thoroughly that essentially no reactants are left, whereas others appear to halt when a mixture of reactants and products is present? Can we predict in advance that a given reaction will behave in either of these two ways? What effect does the amount of a reactant or a product have on the spontaneity of a reaction?

These are some of the questions that we will answer in the course of this chapter. However, you should not forget that thermodynamics only describes what *can* happen (or better, what is not forbidden). Making it happen, and making it happen in a reasonable time, is the task of the research chemist.

15-1 WORK, HEAT, AND CALORIC

One of Lavoisier's great contributions to chemistry was to undercut the phlogiston theory, as we have seen in Chapter 1. He demonstrated that combustion was a combination with oxygen and not a loss of phlogiston. He was less perceptive in his ideas of where the heat came from, which is so prominent a feature of combustion. Lavoisier coined the term "caloric," in 1789, for what he regarded as the "imponderable matter of heat." Heat was considered to be a fluid, probably weightless, that surrounded the atoms of substances and could be drained away in reactions that produced heat.

Dalton conceived of each atom as existing in an "atmosphere" of heat. In 1808, he wrote:

> The most probable opinion concerning the nature of caloric is that of its being an elastic fluid of great subtility, the particles of which repel one another, but are attracted by all other bodies.

According to this generally accepted idea, a gas is heated when it is compressed because the particles of caloric, repelling one another as they do, are squeezed out of the gas. Heat of friction develops when the frictional motion strips caloric away from its atoms. The caloric theory of heat was accepted by most scientists for the first half of the nineteenth century.

The Cannons of Bavaria

In 1798, Benjamin Thompson (Count Rumford) conducted some experiments with friction that, if they had been appreciated fully, would have done away with caloric as Lavoisier did away with phlogiston. Thompson was superintending the boring of cannons at the military arsenal in Munich. The process involved cast metal cannon blanks and drill bits that were turned by horses. Thompson was impressed by the considerable heat evolved during the drilling. He tried boring the cannons under water and determined that the same length of time always was required to bring a given amount of water to the boiling point. He also observed that the generation of heat apparently could be continued indefinitely. He interpreted what he saw correctly; the work provided by the horses was being converted into heat. Thompson wrote:

> It is hardly necessary to add that anything which any insulated body or system of bodies can continually be furnished without limitation cannot possibly be a material substance, and it appears to me to be extremely difficult if not quite impossible to form any distinct ideas of anything capable of being excited and communicated in the manner the *Heat* was excited and communicated in these experiments except it be *Motion*.

His experiments failed to convince others. Those who believed in the caloric theory were ready with the explanation that the friction of the drill bit rubbed caloric away from the metal atoms and brought it to the surface. They failed to appreciate the significance of Count Rumford's ability to continue to produce heat indefinitely. According to the caloric theory, after the supply of caloric had been rubbed away from the metal, further boring should not produce heat. Unfortunately, scientists were not accustomed to thinking of heat in quantitative terms, just as they were not accustomed to thinking of matter in quantitative terms before Lavoisier's proposal. Rumford's work had little impact.

Blood, Sweat, and Gears

The men who finally convinced scientists that heat and work were equivalent, and that both were forms of energy, were Julius Mayer (1814–1878) and Hermann von Helmholtz (1821–1894), both German physicians, and James Joule (1818–1889), the son of an English brewer. In 1840, Mayer signed onto a ship bound for Java as a ship's doctor. He noted that the blood from the veins of the Javanese and from his own ship's crew was a brighter red than that he had seen from patients in Germany. He interpreted this correctly as indicating that more oxygen remained in the veins of inhabitants from the tropics than of people from cold climates, because less combustion of foods was required to maintain a constant body temperature in the tropics. This train of thought led him to the further conclusion that the heat of combustion of foods was used both to maintain body temperature and to carry out the work done by an individual. Heat could be converted into work, and both were forms of the same thing, energy. On his return to Germany he tried to calculate the conversion factor between heat and work by using stirring devices for water and expanding gases into chambers. The experiments were difficult to perform accurately, because the temperature increases were fractions of a degree. Nevertheless, he obtained an approximate value for the mechanical equivalent of heat and submitted an account of his work to the *Annalen der Physik*. The *Annalen der Physik* rejected his paper as unfit for publication. He reworked it and submitted it to the *Annalen der Chemie und Pharmacie* instead. It was published in 1842, and aroused no comment whatever.

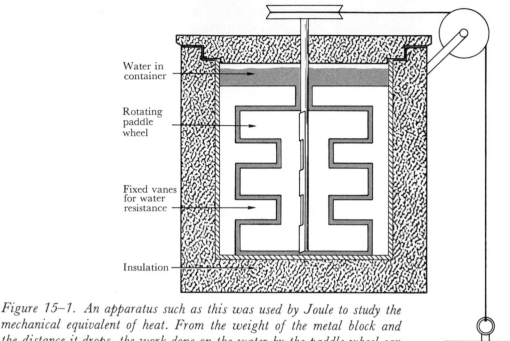

Figure 15–1. An apparatus such as this was used by Joule to study the mechanical equivalent of heat. From the weight of the metal block and the distance it drops, the work done on the water by the paddle wheel can be calculated. The increase in temperature of the water can be measured with a sensitive thermometer. Because the material being heated includes the fixed vanes and paddle wheel as well as the water, the apparatus must be calibrated by introducing known amounts of heat and observing the temperature increase.

Like Newlands, Mayer had expected controversy, but encountered only indifference.

At the same time, Joule, in England, was doing virtually the same experiments and meeting the same indifference and disbelief. Joule was a student of Dalton and the son of a Lancashire brewer. At the age of 19, he began building electric motors and generators with the intention of converting the brewery from steam power to electricity. These attempts were abortive, but Joule became interested in the relationship among the work of cranking the dynamo, the electricity generated, and the heat produced by electricity. Later he dropped the electricity from the sequence and studied the heat produced by mechanically stirring water with paddles driven by a falling weight (Figure 15–1). Like Mayer, Joule found the experiments difficult because of the small temperature changes produced. In spite of this, he obtained a conversion factor that, expressed in metric units, is 42.4 kg cm cal^{-1}, within 1% of the currently accepted value of 42.686 kg cm cal^{-1}. That is, a 1-kg weight, falling through a distance of 42.686 cm, can do enough work (by turning a stirring paddle, for example) to add 1 cal of heat to the water. If the experiment is performed with an insulated 1-liter container of water, then, since the specific heat of water is 1 cal deg^{-1} g^{-1}, the temperature will increase only by one thousandth of a degree. It was a remarkable achievement to come so close to the best modern value with Joule's home-made and home-calibrated thermometers.

In 1843, Joule submitted his results to the British Association. It was received with disbelief and general silence. A year later a paper on the subject was rejected by the Royal Society. In 1845, Joule again presented his ideas on the equivalence of work and heat to the British Association. He suggested that the water at the bottom of Niagara Falls should be 0.2°F warmer than the water at the top because of the energy gained in the fall. He also proposed the idea of an absolute zero of temperature, based on the thermal expansion of gases, at −480°F (−266°C). No one listened. He tried again in 1847, and in Joule's own words, written in 1885:

> The communication would have passed without comment if a young man had not risen in the section, and by his intelligent observations created a lively interest in the new theory. The young man was William Thomson, who had two years previously passed the University of Cambridge with the highest honour, and is now probably the foremost scientific authority of the age.

Thomson, later to become Lord Kelvin, was 26 years old at the time. Neither he nor Faraday, who was also at the meeting, were convinced by Joule's case, which depended on temperature increases of hundredths of a degree; but at long last, Joule had forced his peers to discuss his ideas. Thomson wrote later that, two weeks after the 1847 meeting, he was walking from Chamonix to begin a tour of Mont Blanc when

> ...whom should I meet walking up but Joule, with a long thermometer in his hand, and a carriage with a lady in it not far off. He told me that he had been married since we parted at Oxford! and he was going to try for elevation of temperatures in waterfalls.

In 1849, a paper by Joule entitled "On the Mechanical Equivalent of Heat" was communicated to the Royal Society by Faraday, and it appeared in their *Philosophical Transactions* the next year.

Mayer suffered the same pangs that Newlands did; he saw what he regarded as his own ideas being acclaimed by others but attributed to Joule. Mayer's despondency led to a suicide attempt in 1850 and commitment to a mental asylum for two years thereafter. He continued to receive little credit or attention until late in his life, when John Tyndall, in England, and Rudolf Clausius and Hermann Helmholtz, in Germany, made a concerted effort to secure proper recognition for Mayer.

The man who finally convinced scientists of the validity of the equivalence of heat and work was Helmholtz. In 1847, he submitted a paper to the *Annalen der Physik* that outlined the principle of the *conservation of energy* and the equivalence of heat and work in more general terms than either

Mayer or Joule had done. The paper was rejected. Helmholtz presented the paper at a meeting in Berlin and had it published privately.

Helmholtz' analysis of heat, work, and energy convinced Faraday and Thomson. Joule's experiments gradually began to be accepted. Ultimately, the German physicist Rudolf Clausius (1822–1888) stated, in 1850, the first law of thermodynamics as it usually is given today: *In any process, energy can be changed from one form to another (including heat and work), but it is never created or destroyed.* Helmholtz' conservation of energy joined Lavoisier's conservation of mass as one of the great generalizations of science.

15–2   THE FIRST LAW OF THERMODYNAMICS

Thermodynamicists talk continually about thermodynamic *systems* and their *surroundings;* so will we. We shall look at the work that a system does on its surroundings, or the work that the surroundings do on the system. We shall note the loss or gain of heat of a system to or from its surroundings. What is a thermodynamic system?

A *thermodynamic system* is any part of the universe that we want to focus attention upon, and its surroundings are that part of the universe with which it can exchange energy, heat, or work. A suitable system could be a balloon full of gas, or a flask with reacting chemicals, or a locomotive engine, or just the cylinders and pistons of the engine. If we are looking at the energy balance on our planet, then the earth itself would be a thermodynamic system, and the sun would be part of its surroundings. We also could consider the earth and sun together as a good approximation to a closed system. A *closed system* is one that does not exchange energy, heat, or work with its surroundings. So far as thermodynamics is concerned, it has no surroundings. The word "system" is a pointing finger on an old-fashioned signboard; it calls attention to whatever region of matter that we want to examine.

It is often easiest to think of an ideal gas in some type of enclosure as a typical thermodynamic system. Many of the thermodynamic properties common to all systems are comprehended most readily in such a simple system. When we heat a gas, it expands unless it is constrained. As it expands, it pushes against the pressure of the atmosphere and therefore does work against this pressure. We say that heat, q, has been added to the gas from its surroundings, and that the gas has done work on the surroundings. If we add heat to the gas but constrain it so it cannot expand, the temperature and pressure increase, as given by the ideal gas law,

$$PV = nRT \tag{15-1}$$

Heat again has been added to the gas, but no work has been done by the gas. If the gas is initially at a high pressure, we can allow it to expand without heating it. In this case, the gas does work against its environment or surroundings, without having heat added to it. However, the gas at the conclusion of the expansion is cooler than it was initially.

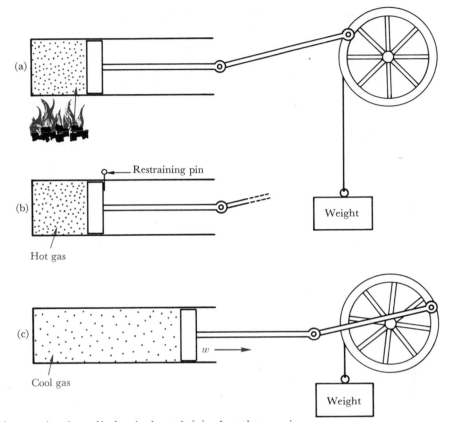

Figure 15–2. If the gas in the cylinder is heated (a), but the gas is prevented from expanding (b), its temperature increases. Conversely, if the gas is allowed to expand (c), it can do mechanical work, and its final temperature will not be so high. In the first example, heat is converted into internal energy; in the second example, it is converted into work.

A variation of this experiment is shown in Figure 15–2. In this experiment, the way in which the expanding gas can be made to do work is somewhat more obvious. Heat, q, can be added to the gas with or without its doing work, and work, w, can be obtained from the gas if heat is added and sometimes even if it is not.

How is work measured in an expanding gas? Work is defined in physics as the product of the force against which motion takes place times the distance moved. Endless motion of an object produces no work if there is no resisting force to the motion. Moreover, no matter how large the resisting force to the motion of an object might be, no work is done unless the object moves against that force. For an infinitesimal movement, ds, against a force, F, the infinitesimal amount of work done is $dw = F\,ds$. If the object moves through a finite distance, Δs, against a constant force, F, the work done is $w = F\,\Delta s$.

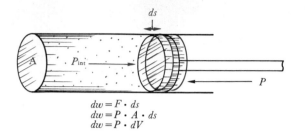

$$dw = F \cdot ds$$
$$dw = P \cdot A \cdot ds$$
$$dw = P \cdot dV$$

Figure 15–3. The work done by the gas in the cylinder when it moves the piston an infinitesimal distance, ds, against an opposing pressure, P, is dw = Fds = PAds = PdV. (A is the area over which the pressure is exerted.) The pressure in the work expression is the external pressure against which the motion takes place, not the internal pressure of the expanding gas. However, for an expansion to occur, P_{int} must be greater than the external pressure.

Let us suppose that the gas is enclosed in a cylinder with a piston (Figure 15–3) and that the pressure inside the cylinder, P_{int}, is greater than the constant atmospheric pressure outside, P. As the gas expands and moves the piston an infinitesimal distance, ds, the force against the piston from the outside remains constant and equal to the product of the pressure, P, and the area, A, over which this pressure is exerted. The work done, as shown in the figure, is the product of the volume increase times the external pressure against which the expansion takes place: $dw = P\, dV$. For an expansion such as this one, in which the resisting pressure remains constant, the work done in a measurable volume change of ΔV is $w = P\, \Delta V$. These relationships, although derived here only for gas expanding in a cylinder, are generally true in gas expansions. This kind of work commonly is called "expansion work," or "PV work." Other kinds of work are possible. We can do gravitational work by lifting a weight to a position where it has greater potential energy and can fall to its original position. We can do electrical work by moving charged ions or other objects in an electrical potential field. We can do magnetic work by pulling a compass needle away from the direction to which it points if undisturbed. All of these types of work are included in the generalization known as the first law of thermodynamics.

In a thermodynamic system, heat can go in or out, and work can be done on or by the system. The first law states that, in all of these processes, *energy in the system is neither created nor destroyed.* The energy of the system is not necessarily constant; it can rise or fall, depending on what we do to, or with, the system. But the *change* in energy of the system is equal to the *net* heat added to the system less the *net* work done by the system on its surroundings:

$$\Delta E = q - w \tag{15–2}$$

A Different View of the First Law

Another way of looking at the first law of thermodynamics is more meaningful for chemists. In this view we think of Equation 15–2 as no more than a definition of a bookkeeping function known as the *internal energy*, *E*. Recall from the discussion of Figure 15–2 that we can heat a gas and do work with it. We also can reverse the process. We can do work on a gas by compressing it, and drain away the heat produced. We can heat a gas without letting it do work, and the temperature increases. Conversely, we can let a gas at high pressure expand and do work without being heated, but we will find that the gas cools in the process. With the right conditions, *q* and *w* can be manipulated independently. We can keep track of what is happening better if we define change in the internal energy, ΔE, as the *difference* between the heat added and the work done, as in Equation 15–2. If heat is added and an exactly equivalent amount of work is done, the internal energy of the system is unchanged. If we heat the gas but constrain it so it cannot expand and do work, the internal energy increases by an amount equal to the heat added. Finally, if we use the gas to do work without adding heat, the internal energy decreases by an amount equal to the work done. Our common-sense observations in this paragraph about when the gas heats and cools suggest that internal energy and temperature should be related.

Thus far, we have done nothing remarkable. With this viewpoint, Equation 15–2 is not the first law; it is the definition of a bookkeeping device or a fudge factor. The first law is the statement that this new bookkeeping function is a state function.

State Functions

State functions are extremely important in thermodynamics, especially to chemists. A *state function* is a property whose value is determined completely by the state of a system at a given instant, and is not dependent on the past history of the system.

Let us illustrate what this means with an imaginary Cold War scenario. Suppose that a river flows from West Germany into East Germany at Einstadt and reemerges to West Germany many miles further at Ausdorf. West German observers are not allowed behind the Iron Curtain. The East German authorities are suspected of building an atomic power station by the banks of the river and of using the river water as a coolant and working medium for their steam turbines. Can the presence of the reactor be detected by the West German observers?

Let us suppose that the heat from the reactor converts river water to steam, which then drives steam turbines and generates electricity. The river water is now the thermodynamic system. The water is heated and vaporized by the heat supplied by the atomic pile and then is cooled as it expands in the turbine and does work to turn the turbine rotor. The East German power authorities suspect that their Western colleagues are investigating, and they take care to cool the water before dumping it back so it is at the same temperature as it was at the intake channel.

Finally, suppose that May Day is a holiday at the reactor, and that on May 1 of every year the engineers shut down the reactor and close the intake channel at the river. Can the West German observers detect the presence of the power station by taking measurements on the river at Einstadt and Ausdorf?

Unfortunately for the peace of mind of Western Intelligence, they cannot. What physical measurements might be made on the thermodynamic system (the river)? They could measure the temperature of the water, its density, viscosity, molar volume, electrical conductivity, ^{16}O-to-^{18}O isotope ratio, melting and boiling points, chemical purity, and many other properties. All of these properties are state functions. The temperature of the water depends on its present state and not on its past history. (That is, the present temperature may be a consequence of what happened in the past, but we do not need to know that history to measure the temperature.) The change in temperature of the water as it flows from Einstadt to Ausdorf can be determined by measuring the temperature at the two towns:

$$\Delta T = T_{\text{Aus}} - T_{\text{Ein}} = T_2 - T_1$$

Similarly, the change in density, viscosity, or any other state function is obtained by taking the difference between the density, or viscosity, or other property, at Einstadt and at Ausdorf. We do not need to know what happened to the river in East Germany.

All of this may seem trivial until you realize that we cannot do the same thing with heat or work. There is no such thing as a "heat content" that we can measure at Einstadt and Ausdorf to indicate how much heat has been added to the water during its route through East Germany. Similarly, there is no property known as "work content" that can be measured at the two towns to determine how much work the East Germans obtained from the water. So long as the power authorities are careful to make the temperature of the exit water from the reactor match the temperature of the intake water, there will be no difference in the water at Ausdorf when the reactor is running, or on May Day when it is not. The existence or non-existence of the power station will be a mystery to the Western observers, at least from their measurements on the river water.

If we observe that the change in water temperature at Einstadt and Ausdorf is the same on May Day as on any other day, then we can say that whatever amount of heat was added to the water in the reactor must have been balanced precisely by the work done in the turbines (or by the post-cooling before the water was dumped back into the river). Conversely, every bit of work obtained in the turbines must have been compensated by heat from the reactor or else the water would have been cooler at Ausdorf. No information about the water at the two towns will suggest the amount of heat, q, added to the water or the amount of work done, w. Neither heat nor work are state functions, *but their difference is*. If the quantity $q - w$ is not kept constant, differences will appear in measurable properties of the water at Ausdorf. The most obvious such property is temperature, but molar volume, density, electrical conductivity, and other properties will change as well. Turning this statement around, if we specify the state of the water at Einstadt and Ausdorf, then we have specified the quantity, $q - w$, during the transit to East Germany, even though we have not specified q and w individually. Their difference is the *change in a state function, E*.

This internal energy, E, is the same function as the mean molar kinetic energy, E_k, that we encountered in Section 2–6. For an ideal gas, the internal energy is directly proportional to temperature:

$$E = \tfrac{3}{2}PV = \tfrac{3}{2}RT \qquad \text{(for 1 mole of ideal monatomic gas)} \qquad (2\text{--}24)$$

For nonideal gases, E will be approximately proportional to temperature; and for substances in general, an increase or decrease in temperature accompanies a rise or fall in internal energy.

State functions are useful to chemists precisely because they do not depend on the history of a chemical system. Energy is a state function. So are pressure, temperature, volume, and all of the other quantities that we commonly think of as properties of a substance. The very term "property" suggests something that a substance has, independent of any factors other than its present condition. We never speak of the work that a substance has, and should not speak of the heat that it possesses. If the final state of a system is identified by a subscript "2," and the initial state is given the subscript "1," then Equation 15–2 becomes the first law of thermodynamics if we expand it to

$$\Delta E = E_2 - E_1 = q - w \qquad (15\text{--}3)$$

The state-function properties are implicit in the middle term.

15–3 THE FIRST LAW AND CHEMICAL REACTIONS

When chemists first began a systematic study of heats of reaction, they discovered that a particularly convenient type of reaction was one constrained to a fixed volume in a *bomb calorimeter* (Figure 4–2). This is a sturdy steel container with a tight lid, immersed in a water bath and provided with electrical leads to detonate the reaction inside. The heat evolved in such a reaction at constant volume (hopefully) is measured by the increase in temperature of the water bath.

If the chemical system inside the metal container cannot change in volume, it cannot do PV work. If no other types of work are involved, the heat liberated by the reaction is equal to the decrease in internal energy:

$$\Delta E = q_V \qquad \text{(at constant volume)} \qquad (15\text{--}4)$$

In the absence of any work effects, the gain or loss of heat by the contents of the container is a direct measure of the increase or decrease of internal energy of the reacting substances. If the reaction releases heat, it is called an *exothermic* reaction; if it absorbs heat, it is *endothermic*.

Most reactions take place at constant pressure rather than constant volume, and it would be useful to have a thermodynamic function that behaves at constant pressure like E does at constant volume, that is, a measure of the heat of the reaction under those conditions. Such a function is the *enthalpy*, H, defined by

$$H \equiv E + PV \tag{15-5}$$

At constant pressure, the change in enthalpy of a system is

$$\Delta H = \Delta E + P\,\Delta V$$

From the first law, we can write this as

$$\Delta H = q - w + P\,\Delta V$$

If we rule out electrical, gravitational, magnetic, and all kinds of work other than PV work, the last two terms cancel in the preceding equation. We then have the statement that the heat of a reaction at constant pressure is equal to the change in enthalpy of the system:

$$\Delta H = q_P \tag{15-6}$$

(The subscript, P, indicates that the heat change takes place at constant pressure.) At constant pressure, enthalpy *increases* in an endothermic reaction as heat flows into the system, and enthalpy *decreases* in an exothermic process as heat flows out of the system.

All of the functions on the right of the equivalence sign in Equation 15-5 are state functions, so H is a state function as well. The change in enthalpy of a system depends on the enthalpy of the system before and after a process and not at all on the path by which the system went from the initial state to the final state; thus

$$\Delta H = H_2 - H_1 \tag{15-6a}$$

Equation 15-6a is the most important single consequence of the first law of thermodynamics for chemistry. It states that the heat of a reaction carried out at constant pressure is a state function. The heat of reaction is the difference between the enthalpy of the products and the enthalpy of the reactants and is the same whether the actual reaction occurs in one step, or in half a dozen intermediate steps. In our example from Section 4-5 on the synthesis of diamond, the heat of preparation of diamond from methane is the same, regardless of whether diamond is made directly from methane, or the methane is oxidized to carbon dioxide, then the carbon dioxide is used to make diamond:

$$CH_4(g) + 2O_2(g) \rightarrow CO_2(g) + 2H_2O(l) \qquad \Delta H_{298} = -212.8 \text{ kcal}$$
$$\underline{CO_2(g) \rightarrow O_2(g) \ + C(dia) \qquad\qquad \Delta H_{298} = \ +94.5 \text{ kcal}}$$
$$CH_4(g) + O_2(g) \ \rightarrow C(dia) \ + 2H_2O(l) \qquad \Delta H_{298} = -118.3 \text{ kcal}$$

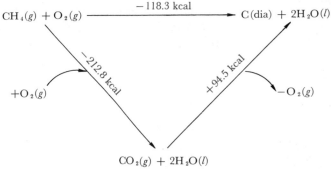

Figure 15–4. The heat of preparation of diamond from methane is the same whether the reaction occurs in one step or whether CO_2 is made from methane, and then diamond from CO_2. Such a statement can be made only because the heat of a reaction at constant pressure and temperature is equal to the change in enthalpy, H, and enthalpy is a state function.

Because enthalpy, H, is a state function, heats of reaction are additive in the same way that the reactions to which they pertain are. This statement is Hess' law.

This independence of enthalpy change of the path of a reaction can be diagramed for the diamond synthesis reactions by a cycle as in Figure 15–4. The first law states that either way around the cycle (the one- or two-step path) leads to the same ΔH and heat of reaction. The enthalpy changes also can be represented on an energy-level diagram (Figure 15–5). Note that the absolute numerical value of the enthalpy is not defined, but only the changes in going from one state of reactants or products to another. In the past, every time that we drew an energy-level diagram we unconsciously were

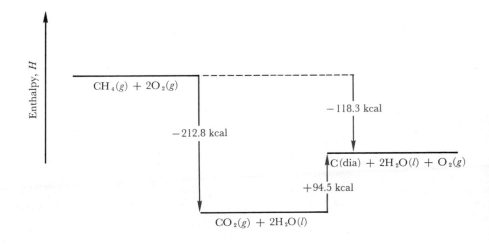

Figure 15–5. For the same reasons that the reactions of Figure 15–4 can be drawn as a cycle, the enthalpies can be represented in an energy-level diagram such as this one. The enthalpy change in going from one state to another depends only on the levels of the two states in this diagram and not on the manner of going from one state to the other.

using the state function properties of energy. Neither heat nor work can be represented on such a diagram (except for the special cases in which one or the other of these is equal to a state function).

All of the discussion of Section 4–5 is valid because H is a state function. It is unnecessary to tabulate the heats of all reactions; we need to list only those from which all other reactions can be obtained by a proper combination of reactions. The reactions chosen are the reactions for the formation of compounds from their elements in standard states. The *standard state* of a gas at a chosen temperature is 1 atm partial pressure; that of a liquid or solid is the pure liquid or solid at 1 atm external pressure. The chosen temperature usually is 25°C for most thermodynamic tabulations. Standard heats of formation for many substances are listed in Appendix 2.

Example. What is the standard heat of the reaction for the formation of anhydrous crystalline copper sulfate from its elements in their standard states?

Solution. The reaction is

$$Cu(s) + S(s) + 2O_2(g) \rightarrow CuSO_4(s)$$

This is the reaction for which the heat of formation is tabulated in Appendix 2. For this reaction, $\Delta H^0_{298} = -184.00$ kcal mole^{-1}.

Example. The standard heat of formation of gaseous B_5H_9 is given in Appendix 2 as $+15.0$ kcal mole^{-1}. Of what reaction is this the heat of reaction?

Solution. The reaction is that of synthesis of B_5H_9 from elemental solid boron and hydrogen gas at 1 atm and 298°K:

$$5B(s) + 4\tfrac{1}{2}H_2(g) \rightarrow B_5H_9(g) \qquad \Delta H^0_{298} = +15.0 \text{ kcal}$$

When a heat term is written after a reaction, as above, the units are not kcal mole^{-1} but kcal per stoichiometric unit of reaction as written. The heat of the reaction above is 15.0 kcal per mole of B_5H_9, but only 3.0 kcal per mole of boron used, or 15.0 kcal for every $4\tfrac{1}{2}$ moles of hydrogen gas used. Heats of formation always are tabulated per mole of the compound formed.

Example. B_5H_9 ignites spontaneously in air with a green flash to produce B_2O_3 and water. What is the heat of the reaction under standard conditions?

Solution. The reaction in unbalanced form is

$$B_5H_9(g) + O_2(g) \rightarrow B_2O_3(s) + H_2O(l)$$

Two moles of B_5H_9 are needed for every 5 moles of B_2O_3 to account for the boron. Hence, the 18 hydrogen atoms must appear as nine water molecules. Balancing the oxygen yields

$$2B_5H_9(g) + 12O_2(g) \rightarrow 5B_2O_3(s) + 9H_2O(l)$$

The tabulated standard heats of formation of reactants and products are

Substance	ΔH^0_{298}
$B_5H_9(g)$	$+15.0$ kcal mole^{-1}
$O_2(g)$	0.0 kcal mole^{-1}
$B_2O_3(s)$	-302.0 kcal mole^{-1}
$H_2O(l)$	-68.3 kcal mole^{-1}

The three reactions of formation that, when added, produce the desired reaction are

	ΔH^0_{298}
$2B_5H_9(g) \rightarrow 10B(s) + 9H_2(g)$	$-2 \times (+15.0\,\text{kcal}) = -30.0\,\text{kcal}$
$10B(s) + 7\frac{1}{2}O_2(g) \rightarrow 5B_2O_3(s)$	$+5 \times (-302.0\,\text{kcal}) = -1510.0\,\text{kcal}$
$9H_2(g) + 4\frac{1}{2}O_2(g) \rightarrow 9H_2O(l)$	$+9 \times (-68.3\,\text{kcal}) = -614.7\,\text{kcal}$
$2B_5H_9(g) + 12O_2(g) \rightarrow 5B_2O_3(s) + 9H_2O(l)$	$\Delta H^0_{298} = -2154.7\,\text{kcal}$

Boron hydrides once were considered as rocket fuels because of their extremely high heats of combustion.

We can take a shortcut with these tables of enthalpies or heats of formation by pretending that the numbers are absolute enthalpies of the compounds, rather than enthalpies of formation from elements. The result is the same, since reactants and products must be composed of the same number and same kind of atoms. Then for the reaction

$$2B_5H_9(g) + 12O_2(g) \rightarrow 5B_2O_3(s) + 9H_2O(l)$$

the enthalpy of reaction is nine times the standard enthalpy of liquid water, plus five times the standard enthalpy of solid B_2O_3, less two times the standard enthalpy of gaseous B_5H_9. The standard enthalpy of elemental O_2 is zero.

Exercise. B_5H_9 can be prepared from diborane, B_2H_6, which reacts at the proper temperature to give B_5H_9 and H_2. Is the reaction exothermic or endothermic? What is the heat of reaction per mole of diborane consumed?

(*Answer:* Exothermic; $\Delta H^0_{298} = -1.5$ kcal mole^{-1} of B_2H_6.)

Before you go on. You can find a review of energy diagrams and heats of formation, including their definitions and method of addition, in Section 4–5 of *Programed Reviews of Chemical Principles* by Lassila *et al.*

15-4 BOND ENERGIES

With the localized bond model for methane, CH_4, we say that the molecule is held together by four equivalent C—H single bonds. If this idea is valid, the heat of decomposition of methane to isolated carbon and hydrogen atoms should be four times the *bond energy* of a C—H bond. (Although we shall work consistently with enthalpies, we shall adopt the common but loose terminology and refer to our results as bond energies rather than bond enthalpies. The difference is small and is within the limits of accuracy of the bond-energy approach itself.)

The heat of formation reaction for methane is

$$C(gr) + 2H_2(g) \rightarrow CH_4(g) \qquad \Delta H^0_{298} = -17.889 \text{ kcal}$$

But for bond energies we need the decomposition to atomic gaseous carbon and hydrogen, not solid graphite and diatomic H_2. The atomization reactions are

$$C(gr) \rightarrow C(g) \qquad \Delta H^0_{298} = +171.698 \text{ kcal}$$
$$\tfrac{1}{2}H_2(g) \rightarrow H(g) \qquad \Delta H^0_{298} = +52.089 \text{ kcal}$$

These are the standard heats of formation of the gaseous atoms from the elements in their standard states, and are tabulated in Appendix 2 along with the other heats of formation.

The desired reaction for decomposing methane into isolated atoms can be constructed from the preceding reactions:

	ΔH^0_{298}		
$CH_4(g) \rightarrow C(gr) + 2H_2(g)$	$-1 \times (-17.889 \text{ kcal})$	$=$	$+17.889 \text{ kcal}$
$C(gr) \rightarrow C(g)$	$+1 \times (+171.698 \text{ kcal})$	$=$	$+171.698 \text{ kcal}$
$2H_2(g) \rightarrow 4H(g)$	$+4 \times (+52.089 \text{ kcal})$	$=$	$+208.356 \text{ kcal}$
$CH_4(g) \rightarrow C(g) + 4H(g)$		$\Delta H^0_{298} =$	$+397.943 \text{ kcal}$

If this is the heat needed to break four C—H bonds, the bond energy (strictly speaking, the bond enthalpy) of one C—H bond is one fourth this figure. The bond energy of a C—H bond in methane is 99.5 kcal mole^{-1} of bonds.

Bond Energy of a C—C Single Bond

From the heat of formation of ethane, C_2H_6, we can obtain a value for the bond energy of a carbon single bond. From the tables in Appendix 2,

	ΔH^0_{298}		
$C_2H_6(g) \rightarrow 2C(gr) + 3H_2(g)$	$-1 \times (-20.24 \text{ kcal})$	$=$	$+20.24 \text{ kcal}$
$2C(gr) \rightarrow 2C(g)$	$+2 \times (+171.70 \text{ kcal})$	$=$	$+343.40 \text{ kcal}$
$3H_2(g) \rightarrow 6H(g)$	$+6 \times (+52.09 \text{ kcal})$	$=$	$+312.54 \text{ kcal}$
$C_2H_6(g) \rightarrow 2C(g) + 6H(g)$		$\Delta H^0_{298} =$	$+676.18 \text{ kcal}$

In the localized bond model of ethane, the molecule has six C—H bonds and one C—C bond. If the value of 99.5 kcal mole^{-1} is accepted for the C—H bonds in methane, the six C—H bonds in ethane must account for 597.0

kcal mole^{-1}. The remaining 79.18 kcal must be the bond energy of a mole of C—C single bonds.

We can test the validity of this entire approach by calculating the expected heat of formation of propane, C_3H_8, from graphite and hydrogen gas. In the localized bond model, propane has two C—C bonds and eight C—H bonds. The heat of formation is calculated as follows:

	ΔH^0_{298}
$3C(g) \; + \; 8H(g) \; \rightarrow \; C_3H_8(g)$	$2 \times (-79.2 \text{ kcal}) + 8 \times (-99.5 \text{ kcal})$
	$= -954.4 \text{ kcal}$
$3C(gr) \rightarrow 3C(g)$	$3 \times (+171.7 \text{ kcal}) \qquad = \quad 515.1 \text{ kcal}$
$4H_2(g) \rightarrow 8H(g)$	$8 \times (+52.1 \text{ kcal}) \qquad = \quad 416.8 \text{ kcal}$
$3C(gr) + 4H_2(g) \rightarrow C_3H_8(g)$	$\Delta H^0_{298} = -954.4 + 931.9 = \quad -22.5 \text{ kcal}$

The observed heat of formation of propane is -24.8 kcal mole^{-1}, which gives you some idea of the degree of accuracy of bond-energy calculations. We are unfortunately in the difficult position of wanting the small difference between two large numbers. Errors and approximations in the data and in the assumption of localized bonds contribute to an error of 2.3 kcal mole^{-1}.

Tabulation of Bond Energies

We now can proceed to calculate the bond energies of bonds of all types.

Exercise. From the data for ethylene in Appendix 2, calculate the bond energy of a C=C double bond.

(*Answer:* 141.3 kcal mole^{-1}.)

Exercise. From the data for water in Appendix 2, calculate the bond energy of an O—H bond.

(*Answer:* 110.6 kcal mole^{-1}. Note that you need the heat of atomization of oxygen and that you must use the heat of formation of water vapor, not liquid water.)

The most useful bond energies are obtained not from the heats of formation of individual compounds like methane or ethane, but by averaging the values obtained from entire classes of compounds, such as the hydrocarbons for C—H and C—C bond energies. These adjusted best values for several types of bonds are given in Table 15–1. Note that the adjusted value for a C—H bond differs by 0.7 kcal mole^{-1} from the value obtained from methane alone. Errors of 1 or 2 kcal mole^{-1} are considered acceptable in bond-energy calculations.

Table 15–1. Approximate Bond Energies[a] at 298°K (kcal mole^{-1})[b,c]

	C	N	O	F	Si	P	S	Cl	Br	I
H—	98.8	93.4	110.6	134.6	70.4	76.4	81.1	103.2	87.5	71.4
C—	83.1	69.7	84.0	105.4	69.3		62.0	78.5	65.9	57.4
C=	147	147	174				114			
C≡	194	213								
N—	69.7	38.4		64.5				47.7		
N=	147	100								
N≡	213	226								
O—	84.0		33.2	44.2	88.2			48.5		
O=	174									

[a] This is an example of loose but convenient terminology. These are actually bond *enthalpies* at 298°K, in the sense that they were obtained from enthalpies of formation.

[b] From L. Pauling, *The Nature of the Chemical Bond*, Cornell University Press, Ithaca, N.Y., 1960, 3rd ed. See also T. L. Cottrell, *The Strengths of Chemical Bonds*, Butterworths, London, 1958, 2nd ed.

[c] Heats of atomization of elements in their standard states can be found in Appendix 2.

The Heat of Formation of Benzene

The bond-energy method is a conspicuous failure in predicting the heat of formation of benzene. This failure suggests a great deal about the benzene molecule. Let us assume that benzene has one of the Kekulé structures:

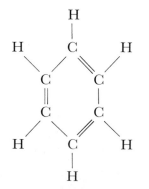

From this model, benzene contains six C—H single bonds, three C—C single bonds, and three C=C double bonds. Hence per mole of benzene the total bond energy is (using values from Table 15–1)

Six C—H bonds:	6 × 98.8 kcal =	592.8 kcal
Three C—C bonds:	3 × 83.1 kcal =	249.3 kcal
Three C=C bonds:	3 × 147 kcal =	441 kcal
	Total bond energy =	1283 kcal

The heat of formation reaction then is constructed:

	ΔH^0_{298}
$6C(g) \; + \; 6H(g) \; \rightarrow \; C_6H_6(g)$	$-(\text{Bond energy}) = -1283 \text{ kcal}$
$6C(gr) \rightarrow 6C(g)$	$6 \times (171.7 \text{ kcal}) = +1030 \text{ kcal}$
$3H_2(g) \rightarrow 6H(g)$	$6 \times (52.1 \text{ kcal}) \; = \; +313 \text{ kcal}$
$6C(gr) + 3H_2(g) \rightarrow C_6H_6(g)$	$\Delta H^0_{298} = \; +60 \text{ kcal}$

There is a flaw somewhere, because the standard heat of formation of gaseous benzene, as measured in the laboratory, is not 60 kcal mole^{-1} of benzene, but only 19.82 kcal mole^{-1}. The benzene molecule is more stable by 40 kcal mole^{-1} than predicted for a molecule with the Kekulé structure.

The flaw lies in the assumption of the Kekulé structure. We saw first in Section 7–6 that the Kekulé structure did not agree with the observed number of isomers of dichlorobenzene. In Section 10–10 we discovered that the Kekulé structure failed to explain the six equal bond lengths between carbon atoms in the benzene ring, but that a delocalized molecular orbital theory could account satisfactorily for bonding. In Section 12–5, we looked at the large class of aromatic compounds—compounds with just such delocalized electrons. We mentioned that delocalization makes the molecule more stable by lowering the energy of the delocalized electrons. Now we have a way to calculate this stabilization from measurements of heats of formation of aromatic compounds.

Exercise. Adding six hydrogen atoms to benzene produces cyclohexane, C_6H_{12}. From bond energies calculate the standard heat of formation of gaseous cyclohexane, and compare it with the measured value of -29.43 kcal mole^{-1}. On the basis of your calculation, would you predict any delocalization stabilization in cyclohexane?

(*Answer:* -29.4 kcal mole^{-1}; no.)

Exercise. Calculate the standard heat of formation of carbon dioxide, $O{=}C{=}O$. Assume the presence of two double $C{=}O$ bonds. Compare your value with the measured value in Appendix 2. Do your figures predict delocalization in CO_2?

(*Answer:* $\Delta H^0_{298} = -58$ kcal mole^{-1}. Yes, stabilization energy of 36 kcal mole^{-1}.)

15–5   SPONTANEITY, REVERSIBILITY, AND EQUILIBRIUM

If we place equivalent amounts of hydrogen and oxygen gases in a container and apply a flame or a Pt catalyst, there will be a violent explosion. The H_2 and O_2 will disappear, and water vapor will form in its place. Similarly, a mixture of H_2 and Cl_2, if triggered by light, will explode and produce HCl gas. In contrast, a mixture of H_2 and N_2 gas will react much less violently, and the final product will be a mixture of H_2, N_2, and NH_3 gases.

The water and HCl reactions are good illustrations of highly spontaneous processes. As we saw first in Chapter 5, a *spontaneous process* is one that has enough impetus to proceed on its own without further input from the rest of the universe.[1] We shall learn in Section 15–8 how that impetus is measured. For the moment we can say that the reaction to produce HCl from H_2 and Cl_2,

$$H_2 + Cl_2 \rightarrow 2HCl$$

has a far greater tendency to occur than the reverse dissociation of HCl,

$$2HCl \rightarrow H_2 + Cl_2$$

The reaction synthesizing ammonia from H_2 and N_2 also has a greater initial tendency to occur than the decomposition reaction. Reaction 15–7 has more of a drive than Reaction 15–8:

$$3H_2 + N_2 \rightarrow 2NH_3 \tag{15-7}$$

$$2NH_3 \rightarrow 3H_2 + N_2 \tag{15-8}$$

But as more NH_3 accumulates, and as less N_2 and H_2 are left, Reaction 15–7 becomes slower and Reaction 15–8 accelerates. Ammonia decomposes more rapidly as more of it is present to decompose. At some concentration of H_2, N_2, and NH_3, Reactions 15–7 and 15–8 proceed at the same rate. Ammonia is produced exactly as fast as it is broken down. Although synthesis and decomposition still occur at the molecular level, we see no net change in the composition of the gas mixture. The gas gives the appearance of having ceased to change. This condition of balance between two opposing reactions is called *chemical equilibrium*.

A reaction that is at equilibrium is a *reversible reaction*. To understand what this means, let us examine our equilibrium mixture of H_2, N_2, and NH_3. An increase in pressure favors Reaction 15–7 over 15–8 because 15–7 leads to a smaller number of moles (or molecules) of gas and relieves the stress on the system caused by the pressure increase. Similarly, a decrease in pressure favors the decomposition of ammonia to produce more moles of gas. Both of these are applications of *Le Chatelier's principle: When a system at equilibrium is subjected to a stress of any kind, the system shifts toward a new equilibrium condition in such a way as to relieve that stress.* The synthesis of ammonia from its elements is an exothermic process: The ΔH_{298}^0 of Reaction 15–7 is -22.08

---

[1] In a strict sense, spontaneity has nothing to do with time. A thermodynamically spontaneous reaction is one that will occur on its own, even if it requires virtually forever to do so. The role of a catalyst is to bring about in a short time that which would occur anyway, but only over a longer time interval. Thermodynamics answers the question, "Will it occur, eventually?" To answer the question, "How soon will it occur?," we must turn to kinetics (Chapter 18).

kcal per stoichiometric unit, as written, or -11.04 kcal mole^{-1} of ammonia. If the temperature of the container is raised, Reaction 15–7 will be hindered and Reaction 15–8 will be favored because 15–8 absorbs heat and partially counteracts the temperature increase. If more ammonia is added to the container from an outside source, Reaction 15–7 will be hindered again and 15–8 will be favored because it relieves the stress of the added ammonia. Le Chatelier's principle is useful in predicting qualitatively what an equilibrium system will do when acted upon by an outside influence.

If the N_2–H_2–NH_3 system is truly at equilibrium, the changes in pressure, temperature, or concentration of one component required to alter the relative rates of Reactions 15–7 and 15–8 are infinitesimally small. Just as the lightest weight can tip a balance in mechanical equilibrium, so the smallest change can affect a system in chemical equilibrium. This is why the term "reversible" is applied to such situations. A fingertip touch cannot halt a falling boulder, and an infinitesimal change in pressure, temperature, concentration, or any other variable cannot halt the explosion of H_2 and Cl_2 or the less spectacular reaction of N_2 and H_2 before equilibrium is reached. Such chemical systems are not at equilibrium; their processes are *irreversible*.

In summary, an equilibrium process is reversible, and a nonequilibrium or spontaneous process is irreversible. We shall want to know how to calculate the equilibrium conditions for a system of chemical substances, both because it often is useful to know the relative amounts of reactants and products at equilibrium, and because the distance of a given chemical situation from equilibrium is a measure of how strong the drive is in a direction toward equilibrium.

Challenge. If you want to apply the ideas of heat and energy problems to nutrition, meteorology, and the design of a steam engine-powered automobile, try Problems 15–1, 15–2, 15–5, and 15–7 in *Relevant Problems for Chemical Principles* by Butler and Grosser.

15–6   HEAT, ENERGY, AND MOLECULAR MOTION

When an object is heated, its molecular motion increases. Heat is not a fluid that can be forced away from the atoms by friction. Instead, it is an expression of the *state of motion* of the molecules and atoms, which are made to move more rapidly by the mechanical forces of friction. These conclusions were suggested by experiments that demonstrated the mechanical equivalence of work and heat, and were made palatable by the kinetic theory of gases and its extension to the molecular theory of liquids and solids.

In previous chapters, we have talked about two types of energy that an object could have: kinetic energy and potential energy. Kinetic energy is possessed by a moving body and is represented by $E_k = \frac{1}{2}mv^2$. Potential energy is possessed by a body because of where that body is. If a mass can perform work by moving from Point A to Point B in space, we say that the body has a greater gravitational potential energy at A than at B. If we want, we can talk about a gravitational potential field in which the body moves, but we are only rephrasing the observation, not explaining it. The idea of a gravitational field *comes from* the observation that work can be done when the body moves from one place to another. Similarly, if a positive or negative charge can be made to do work as it moves from Point A to Point B, we say that the charge has a greater electrostatic potential energy at A than at B. Again, we can describe (not explain) the observation by talking about an electrical field.

Now we have a third kind of energy to deal with: the energy possessed by a body because its atoms and molecules are in a state of motion, even though the body might be stationary. This molecular motion is heat, and it is measured by the temperature of the object. The temperature scale is based on the expansion behavior of an ideal gas, as we noted in Chapter 2. Heat is measured in the same units as work and energy. The amount of heat required for a mole of a substance to experience a temperature increase of 1°C is called the *heat capacity* of the substance and is measured in cal deg^{-1} $mole^{-1}$.

If this third kind of energy did not exist, the first law of thermodynamics would be

$$\Delta E = E_2 - E_1 = -w$$

and would read as follows: The change in the internal energy of a system balances any work that the system does on its surroundings. Or, since $-w$ represents work done on the system by its surroundings, the equation also would state that the increase in internal energy of the system equals the work done on it from the outside. This work could be used to accelerate objects in the system and give them greater kinetic energy, or it could be used to lift them and give them greater potential energy.

The full statement of the first law,

$$\Delta E = E_2 - E_1 = q - w$$

is as follows: The change in the internal energy of a system is the sum of the work done on it by the environment and the increase in random motion given its molecules by the environment. This increase in molecular motion is described as a flow of heat.

It is always possible to change work into heat. Friction often is used as an example because it is so simple. A block, moving as a unit with a large velocity, but with its molecules in relatively slow random motion, comes to a stop on a surface because of friction. After it stops, it has no velocity of motion as a unit. However, its molecules, and the molecules of the surface on which it had been sliding, are moving with greater individual speed than before. If the object is partly a gas, this motion may be straight-line motion throughout the container. If it is a solid, the motion will be vibration of atoms and molecules around average positions in the crystal. In either case, large-scale motion has been converted to microscopic motion.

This process is not completely reversible. The example of a skidding automobile in Section 2–6 is an illustration of this fact. Generally, we cannot take random molecular motion and convert it to coordinated motion of the entire object with 100% efficiency. The expression of our inability to do this is the *second law of thermodynamics*. Two slightly different versions of the second law were proposed in the middle of the nineteenth century. One version, by William Thomson, says: *One cannot convert a quantity of heat completely into work without wasting some of this heat at a lower temperature*. The other version, by Rudolf Clausius, says: *One cannot transfer heat from a cold object to a hot object without using work to make the transfer*. Both statements are *summaries of experience*, and are "statements of impotence." They are statements about the *limitations* on what we can do in the real world. Either can be shown to follow if the other is assumed first.

15–7 ENTROPY AND DISORDER

Either form of the second law of thermodynamics leads, with the help of some calculus that we shall not reproduce here, to a new state function, S. This new state function is called the *entropy* of the system. We shall not use the most general expression for entropy, but one special case is useful. If an amount of heat, q, is added to a system *in a reversible manner* at a temperature T, then the entropy of the system increases by

$$\Delta S = \frac{q}{T} \tag{15–9}$$

If the heat is added irreversibly, the entropy increases by more than q/T:

$$\Delta S > \frac{q_{irr}}{T} \tag{15–9a}$$

The quantity q/T is a lower limit to the entropy increase that is applicable only when the heat transfer is *reversible*, that is, when the object being heated is in thermal equilibrium with the object donating the heat. Heat flows from one body to another because they are *not* in equilibrium, and it would take an infinite time for reversible heat flow to take place. Truly reversible processes are idealizations for real, irreversible processes. What we should say is that in any real (irreversible) process, the entropy increase will be greater than q/T; but the more slowly and carefully we carry out the heat transfer, the less ΔS will exceed q/T.

The entropy as derived from the second law has no obvious molecular interpretation. But Ludwig Boltzmann (1844–1906), an Austrian physicist, showed in 1877 that entropy had a fundamental molecular significance. The entropy of a system of chemical substances is a measure of the *disorder* of the system. Boltzmann's equation relating entropy to disorder (it is carved on his tombstone in Vienna) is

$$S = k \ln W \tag{15–10}$$

In a more convenient form for computation, the equation is

$$S = 2.303k \log_{10} W \tag{15-10a}$$

The quantity W is the number of equivalent ways of producing an arrangement of a system, as we shall illustrate first with a deck of cards as our system. The constant k is called Boltzmann's constant and is the gas constant, R, divided by Avogadro's number. It is the gas constant per molecule rather than per mole. In its most useful form, $k = 1.38062 \times 10^{-16}$ erg deg^{-1} molecule^{-1}.

Poker Entropy

To see what Boltzmann's formula means, let us look at a deck of 52 cards instead of a box containing approximately 10^{23} molecules. Rather than examining measurable arrangements of molecules, let us look at the traditionally accepted poker hands. If you are dealt five cards, one after the other, there are 52 possibilities for the first card. For the second card there are only 51 possibilities, since one card already has been dealt. For the third, there are 50 possibilities, for the fourth, 49, and for the last card, 48. The total number of ways in which you can be dealt a five-card hand is the five-number multiple: $52 \cdot 51 \cdot 50 \cdot 49 \cdot 48$.

This result is true if you keep the cards in front of you in the order in which they were dealt. However, in poker the order of receiving the cards is irrelevant. Any scrambling of these five cards leaves the poker hand unchanged. Therefore, we must divide by the number of ways of arranging or scrambling five objects. To find this number, suppose that you say to a friend, "Draw these five cards, one at a time, and keep track of the order in which you drew them." How many different ways can he do this? He has five choices for his first draw, four for his second, three for his third, a choice between two for his fourth, and only one choice for his last card. The one arrangement that he chooses is only one among $5 \cdot 4 \cdot 3 \cdot 2 \cdot 1$ possibilities. This number is written as 5!, and is called "five factorial." The total number of ways of dealing a poker hand, in which the order of cards is irrelevant, is therefore

$$W_T = \frac{52 \cdot 51 \cdot 50 \cdot 49 \cdot 48}{5 \cdot 4 \cdot 3 \cdot 2 \cdot 1} = 2{,}598{,}960 \text{ different hands}$$

The highest hand in poker is a royal flush, consisting of ace, king, queen, jack, and 10 of one suit. There are only four royal flushes among the 2,598,960 possible poker hands. The straight flush is a sequence of five cards in one suit; they do not necessarily have to be the top five. There are 40 of these, or 36 if the royal flush is excluded as a separate kind of hand. The probability of four of a kind is interesting in its construction. The first card can be any one of the 52. But as soon as this card is received, the value of the four is fixed. The second card may be only one of the three cards in

Table 15–2. Poker Entropy

Hand	W	$S = R \ln W$[a]
Royal flush (AKQ J10, one suit)	4	2.76
Straight flush (sequence, one suit)	36	7.13
Four of a kind (XXXXo)	624	12.81
Full house (XXYYY)	3,744	16.40
Flush (all same suit)	5,108	17.02
Straight (5-card sequence)	10,200	18.40
Three of a kind (XXXoo)	54,912	21.72
Two pair (XXYYo)	123,552	23.30
One pair (XXooo)	1,098,240	27.60
Bust hand	1,302,540	28.10
Total combinations of 52 cards	2,598,960	29.35

[a] To keep the numbers simple, we have used R rather than k; thus the entropy is for a mole of poker hands.

other suits with the same face value. The third card similarly can be only one of the two cards left with the same face value, and the fourth card is predetermined if the resulting hand is to consist of four cards of different suits but the same face value. The fifth card can be anything at all; since four cards have been dealt, there are 48 possibilities for the fifth card. Therefore, the number of ways of obtaining four of a kind is $52 \cdot 3 \cdot 2 \cdot 1 \cdot 48$. Again we have overcounted. The number of ways of shuffling the four cards whose order mattered in the dealing scheme given above (the four cards of the same face value) is $4 \cdot 3 \cdot 2 \cdot 1$. If we correct for the irrelevance of the order of the four cards to the value of a poker hand, the number of different ways of drawing four of a kind is

$$W = \frac{52 \cdot 3 \cdot 2 \cdot 1 \cdot 48}{4 \cdot 3 \cdot 2 \cdot 1} = 624$$

A table of the 10 poker hands, including the worthless "bust" category that includes all remaining hands not counted in the first nine, is given in Table 15–2. For each of these hands, the Boltzmann entropy has been calculated. Both W and S are plotted in Figure 15–6. The fundamentally important feature of the entropy definition is its logarithmic dependence on the number of ways of producing an arrangement, not the constant before the logarithm. In this example, we have used the constant R instead of Boltzmann's constant, k.

Note several advantages of using entropies, S, rather than numbers of possibilities, W.

1) Entropies, being logarithmic, vary more slowly than W. Although W varies from 4 to 2.5 million, S varies only between 3 and 30. This is the same reason why exponential, power-of-ten notation is used for large numbers.

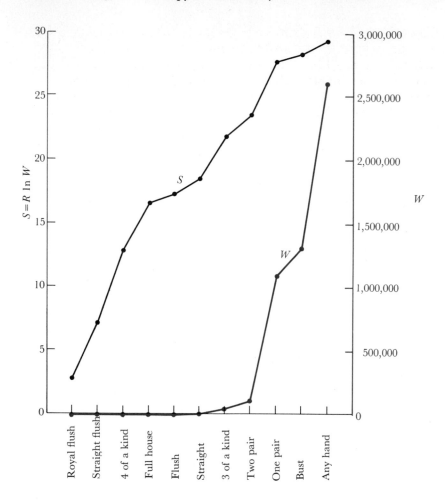

Figure 15–6. A plot of the number of different ways of making each of the accepted hands in poker, W, and the entropy of a "mole" of each hand, S, which is proportional to the logarithm of W. The W plot is less informative than the S plot because for over half the hands the curve is flattened against the horizontal axis and is unreadable.

2) Entropies are additive, whereas numbers of possibilities, W, are multiplicative. If one box of gas can be in five different states, while another can be in six different states, then there are $5 \cdot 6 = 30$ different combinations of the two samples of gas. However, the entropy of the two boxes is the *sum* of the entropy of the individual boxes:

$$S = k \ln (5 \cdot 6) = k \ln 5 + k \ln 6$$

Entropy is an additive property of a system, like its energy, enthalpy, and volume.

3) If two states have the same *difference* in entropy, ΔS, as some other two
   states have, the two pairs of states have the same relative probability.
   For example, in our poker hands, the difference in entropy between four
   of a kind and a straight flush is $12.81 - 7.13 = 5.68$. The entropy difference between a straight and four of a kind is $18.40 - 12.81 = 5.59$,
   approximately the same as the first difference. So the ratios of probabilities should be approximately the same in the two cases. They are;
   four of a kind is $624/36 = 17.3$ times as likely as a straight flush, and
   a simple straight is $10,200/624 = 16.4$ times as likely as four of a kind.

A hidden assumption in all of the foregoing discussion is that the dealer
does not cheat, and that any one combination of five cards is as likely as any
other. Your chances of receiving a full house in such a fair game depend on
how many of the 2,598,960 possible deals of equal probability are classed as
a full house. The answer, from Table 15–2, is 3744 deals. The more different
deals that are classified as a given poker hand, the more likely that hand will
be. An intelligent poker player bases his betting strategy, in part, on his
knowledge of the relative probabilities of receiving the different hands, since
the ranking of hands by the rules of the game follows this list of probabilities.

Real Substances

In thermodynamics and statistical mechanics, we replace the 52 cards by a
collection of molecules. The "hands" are the different experimentally detectable states of the molecules such as, "all molecules in a crystal of the
substance," or "all molecules moving freely through the available volume as
a gas," or, "half of the molecules in a liquid and the other half in a gas above
the liquid." The "deals" are the individual microscopic arrangements that
produce the large-scale, measurable states. These deals are called *microstates*,
and the measurable states are called *macrostates*. The value of W is the number
of microstates that result in the same observable macrostate. If we assume that
the cosmic dealer does not cheat (meaning that there are no hidden physical
laws connecting microstates that we have omitted and that would change
the answer), then we assume that one microstate is just as likely as another. If
this is true, the probability of finding a given experimental situation is proportional to W.

The probability of finding a given experimental situation decreases as
we add more provisos to the description of the state. If we ask, "What is the
number of ways or the relative probability of being dealt any poker hand?,"
the answer is 2,598,960 ways. But if we now add the restriction that they all
be in one of the four suits, the number drops to 5108; and if we add that they
must be in a sequence of five adjacent numbers or values as well, it drops to
$36 + 4 = 40$. Similarly, if we ask what the probability will be of finding a
paperweight with its molecules moving in random vibration about fixed
equilibrium positions, we find a high probability. If we ask what the probability is that *every* molecule in the paperweight will be moving in the same

direction at a given instant so the paperweight suddenly rockets through the wall, we find that the probability is extremely remote; it is so small that ordinarily we never consider that occurrence as possible.

The specifications for a crystalline solid are more restrictive than those for a gas. In the solid we have a molecule in each of many determined places. In the gas, we know only that a certain number of molecules are present in a certain volume, but we specify nothing about their locations other than that they are spread uniformly throughout the container. The solid is like a royal flush, and the gas is like a pair. Therefore, the entropy per mole of a solid will be less than that of a gas. When we use the word "disorder," we use it in this sense: The more microscopically possible states there are that produce the same macrostate, the more disordered that macrostate is and the more likely we are to encounter such a state in nature.

Third-Law Entropies

This is the view of entropy that Boltzmann contributed. In Appendix 2, along with standard heats of formation, there are tabulated standard entropies, S_{298}^0, of substances. These values are *not* obtained from Boltzmann's $S = k \ln W$ expression. They are the results of thermal measurements of heat capacities, heats of fusion, and heats of vaporization. You will learn in advanced courses how to calculate values for S from such thermal data. This S_{298}^0 sometimes is called the "third-law entropy" because the connection between thermal entropy and Boltzmann entropy was not complete without the assumption of the *third law of thermodynamics: The entropy of a perfect crystal at absolute zero is zero.*

The beauty of third-law entropies is that, although they were not derived from Boltzmann's statistical interpretation, they agree with it so well. In Figure 15–7 are plotted the molar entropies for pure representative elements in different physical states. All metallic solids have entropies below 20 cal deg^{-1} $mole^{-1}$, monatomic gases lie between 30 and 45 cal deg^{-1} $mole^{-1}$, and diatomic and polyatomic gases have higher entropies yet. Some of the trends are shown in Table 15–3. Gases have higher entropies than liquids, and liquids have higher entropies than solids. Soft substances have higher entropies than hard substances; the bonds in lead are not as rigid as those in diamond, and the "specifications" for solid lead are not as stringent as for solid diamond. When a substance has two crystalline forms, one metallic and one network nonmetallic, the entropy of the network nonmetallic form is smaller. (See C and Sn in Table 15–3E.) The entropy of solids and liquids increases when they are dissolved in a solvent such as water; the entropy of gases decreases when dissolved in a solvent. The extent of disorder in a solution is more than in a solid or pure liquid, but less than in a gas. Entropy increases with chemical complexity, whether this complexity comes from the number of different ions in a lattice, the number of different molecules

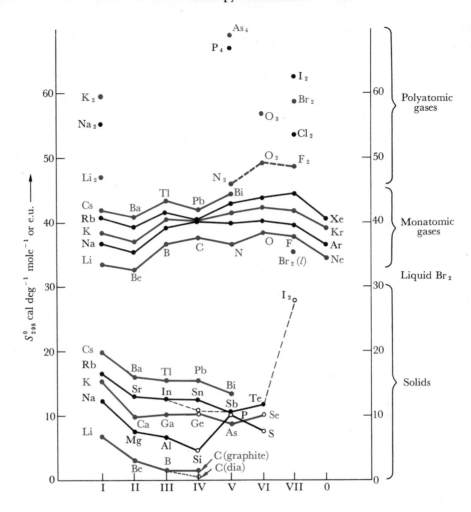

Figure 15–7. Third-law entropies in cal deg^{-1} mole^{-1} for various elements as solids, liquids, and monatomic and polyatomic gases. Polyatomic gases have higher entropies than monatomic gases because the mass of the molecular unit is greater. All monatomic gases have approximately the same entropy per mole, with a gradual increase in entropy with mass. Solids with stronger bonds have lower entropies. Filled circles in solids indicate metallic structures; open circles, nonmetallic structures. The two structures for carbon are graphite (filled) and diamond (open). The two structures for tin are metallic white tin (filled) and gray tin with the diamond structure (open). The entropy of the molecular solid of I_2 molecules is similar to those of crystals of other polyatomic small molecules such as ICN (30.8), glycine (26.1), oxalic acid (28.7), and urea (25.0).

Table 15–3. Entropy and Physical Properties[a]

A. Entropy increases with mass

$F_2(g)$	$Cl_2(g)$	$Br_2(g)$	$I_2(g)$		$O(g)$	$O_2(g)$	$O_3(g)$		$P_4(g)$	$As_4(g)$
48.6	53.3	58.6	62.3		38.5	49.0	56.8		66.9	69.0

B. Entropy increases with vaporization or sublimation to a gas

$I_2(s)$	$I_2(g)$		$Br_2(l)$	$Br_2(g)$		$H_2O(l)$	$H_2O(g)$		$CH_3OH(l)$	$CH_3OH(g)$
27.9	62.3		36.4	58.6		16.7	45.1		30.3	56.8

C. Entropy increases when a solid or liquid is dissolved in water

$HCOOH(l)$	$HCOOH(aq)$		$CH_3OH(l)$	$CH_3OH(aq)$		$NaCl(s)$	$Na^+(aq) + Cl^-(aq)$
30.82	39.1		30.3	31.6		17.3	14.4 + 13.2

D. Entropy decreases when a gas is dissolved in water

$HCOOH(g)$	$HCOOH(aq)$		$CH_3OH(g)$	$CH_3OH(aq)$		$HCl(g)$	$H^+(aq) + Cl^-(aq)$
60.0	39.1		56.8	31.6		44.6	0 + 13.2

E. Entropy is lower in network solids than in metallic solids

$C(gr)$	$C(dia)$		Sn(white, metallic)		Sn(gray, diamond)
1.36	0.58		12.3		10.7

F. Entropy decreases with softness and weak bonds between atoms

$C(dia)$	$W(s)$	$SiO_2(s)$	$Pb(s)$	$Hg(l)$	$Hg(g)$
0.58	8.0	10.0	15.5	18.5	41.8

G. Entropy increases with chemical complexity

$Mg(s)$	$NaCl(s)$	$MgCl_2(s)$	$AlCl_3(s)$		$CuSO_4(s)$	$CuSO_4 \cdot H_2O(s)$	$CuSO_4 \cdot 3H_2O(s)$	$CuSO_4 \cdot 5H_2O(s)$
7.8	17.3	21.4	40		27.1	35.8	53.8	73.0

$CH_4(g)$	$C_2H_6(g)$	$C_3H_8(g)$	$n\text{-}C_4H_{10}(g)$	$iso\text{-}C_4H_{10}(g)$
44.0	54.9	64.5	74.1	70.4

[a] All entropies are in cal deg^{-1} mole^{-1}.

crystallized with a salt (such as water of hydration), or the number of atoms in a molecule (Table 15–3G). Entropy also increases with mass, which is not so easy to understand without quantum mechanics. But, as we have said several times, the spacing between energy levels available to a substance decreases as the mass of the object increases. (This is one of the reasons why we see quantization in electrons and not in baseballs.) A heavier object has more quantum levels available at the same temperature, and hence has more ways of existing in the observed situation. Its entropy is accordingly greater.

With these general trends in mind, let us see what effect entropy has on chemical reactions.

15–8 FREE ENERGY AND SPONTANEITY IN CHEMICAL REACTIONS

What determines whether a chemical reaction is spontaneous or not? What measurable or calculable properties for the system of H_2, Cl_2, and HCl indicate that the reaction of H_2 and Cl_2 is explosively spontaneous under conditions for which the decomposition of HCl to H_2 and Cl_2 is scarcely

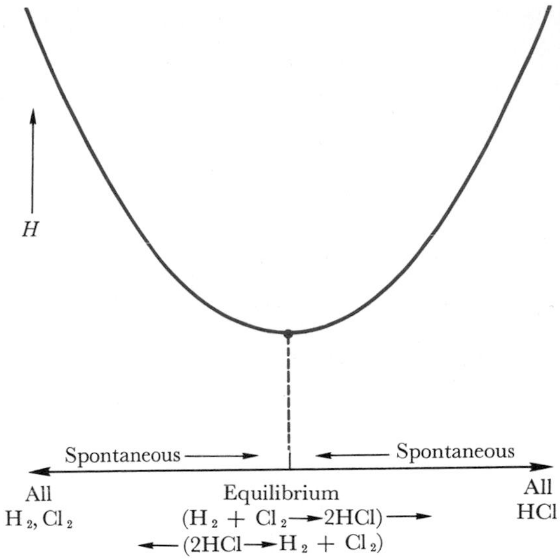

Figure 15–8. If the principle of Berthelot and Thomsen were correct and all spontaneous reactions liberated heat, then enthalpy, H, would be a chemical potential function that is minimized at equilibrium. This is not so; we can find spontaneous processes that absorb heat. The most obvious is the evaporation of a liquid.

detectable at all? Marcellin Berthelot and Julius Thomsen,[2] French and Danish thermodynamicists, proposed the wrong answer, in 1878, in the form of their *Principle of Berthelot and Thomsen:* Every chemical change accomplished without the intervention of an external energy tends toward the production of the body or the system of bodies that sets free the most heat. In other words, all spontaneous reactions are exothermic.

If the principle of Berthelot and Thomsen were correct, and if the enthalpy of a system of reacting chemicals *decreased* during any spontaneous process, equilibrium would occur at the minimum of enthalpy since any spontaneous process is moving toward equilibrium. The plot of enthalpy, *H*, against the extent of reaction on the horizontal axis would look like Figure 15–8, which is drawn for a typical reaction.

---

[2] This Berthelot is not the Berthollet you already have encountered, nor is this Thomsen any of the Thomsons or Thompsons of past reference. For clarity, herewith is a brief glossary of these gentlemen.

Claude Berthollet (1748–1822). French chemist, opponent of definite proportions for compounds.

Marcellin Berthelot (1827–1907). French thermodynamicist.

Benjamin Thompson (1753–1814). American adventurer and spy, Bavarian munitions maker, founder of Royal Institution, London.

William Thomson (1824–1907). British thermodynamicist, later Lord Kelvin.

Julius Thomsen (1826–1909). Danish thermodynamicist.

J. J. Thomson (1856–1940). British physicist, discoverer of electron, Nobel Prize in 1906.

G. P. Thomson (1892–   ). British physicist, Nobel Prize in 1937 for electron diffraction. Son of J. J. Thomson.

Unfortunately for the principle, we can find exceptions: reactions that are spontaneous but which absorb heat in the process. One of these is the vaporization of water or any other substance at a partial pressure less than its vapor pressure. When a pan of water evaporates, heat is absorbed:

$$H_2O(l) \rightarrow H_2O(g) \qquad \Delta H^0_{298} = +10.52 \text{ kcal mole}^{-1}$$

If the principle of Berthelot and Thomsen were true, all gases would condense spontaneously to liquids, and all liquids would solidify to solids since by doing so they would give off enthalpy.

Similarly, the solution of ammonium chloride crystals in water is both spontaneous and endothermic:

$$NH_4Cl(s) + H_2O \rightarrow NH_4(aq)^+ + Cl(aq)^- \qquad \Delta H^0_{298} = +3.62 \text{ kcal}$$

Adding water to solid NH_4Cl makes a beaker cold enough to freeze pure water on the outside of the beaker. Yet we do not see dilute solutions of ammonium chloride separate spontaneously into crystals and pure water simply because heat would be released in the process.

As a last example, dinitrogen pentoxide is an unstable solid that reacts, sometimes explosively, to produce NO_2 and O_2:

$$N_2O_5(s) \rightarrow 2NO_2(g) + \tfrac{1}{2}O_2(g) \qquad \Delta H^0_{298} = +26.18 \text{ kcal}$$

However, a large quantity of heat is absorbed in this decomposition.

There are many other examples of spontaneous processes that absorb heat. We cannot find the point of equilibrium by minimizing H. Enthalpy is not a measure of the tendency of a reaction to proceed spontaneously.

The three preceding reactions occur despite their requirement of heat because their products are much more disordered than their reactants. Water vapor is more diosrdered and has a higher entropy than liquid water. Hydrated $NH_4{}^+$ and Cl^- ions have a higher entropy than crystalline NH_4Cl. Gaseous NO_2 and O_2 are more disordered and have a higher entropy than solid N_2O_5. A chemical system seeks not only the state of lowest energy or enthalpy, but also the state of greatest disorder, probability, or entropy. A new state function can be defined, the *free energy, G:*

$$G \equiv H - TS \tag{15-11}$$

We can show very simply that, for reactions at constant pressure and temperature, *any reaction is spontaneous whose free energy decreases.* If we look at the total free energy, G, of a collection of compounds in a beaker at constant temperature, the change in total free energy brought about by chemical reaction is related to the changes in enthalpy and entropy by

$$\Delta G = \Delta H - T \Delta S \tag{15-12}$$

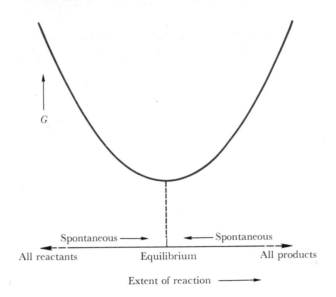

Figure 15–9. The true chemical potential function under conditions of constant pressure and temperature is the free energy, G. In all spontaneous reactions, the free energy decreases, and at equilibrium the free energy change of the reaction in either direction is zero.

But $H = E + PV$, and at constant pressure

$$\Delta H = \Delta E + P\,\Delta V$$

and by the first law of thermodynamics

$$\Delta E = q - w$$

Let us assume that no electrical or other work is done, aside from PV work as the reactants and products expand and contract. Then $w = P\,\Delta V$. And if we substitute into the preceding equations in reverse order,

$$\Delta E = q - P\,\Delta V$$
$$\Delta H = q - P\,\Delta V + P\,\Delta V = q$$

This last equation states that the enthalpy change equals the heat of reaction at constant pressure, which we already have encountered. Substituting q for ΔH into Equation 15–12 yields

$$\Delta G = q - T\,\Delta S \qquad (15\text{–}13)$$

For a reversible reaction, we mentioned previously (Equation 15–9) that $\Delta S = q/T$. Therefore, $q = T\,\Delta S$, and

$$\Delta G = 0 \qquad \text{(for a reversible reaction carried out at constant pressure and temperature)} \qquad (15\text{–}14)$$

What happens in an irreversible reaction? Does the free energy increase, decrease, or do both, depending on the circumstances?

We can answer this question by using Equation 15–9a. In any real, irreversible reaction, the entropy change is *greater* than q/T: $\Delta S > q/T$, or $T \Delta S > q$. Therefore, by Equation 15–13, $\Delta G < 0$ for such a reaction.

Both results can be summarized: *In any spontaneous reaction at constant pressure and temperature, the free energy, G, always decreases. When the reaction system reaches equilibrium, G is at a minimum, and ΔG equals zero.* This behavior of G is represented in Figure 15–9.

It is difficult to overestimate the importance of these results to chemists. Free energy is now the touchstone by which we can determine in advance whether a given reaction will proceed spontaneously, will remain at equilibrium, or will occur spontaneously in the reverse direction.

One last item of thermodynamic ingenuity needs to be mentioned. Since G, like the functions H, T, and S of which it is composed, is a state function, it does not matter if the pressure and temperature change during a reaction, so long as they are brought back to the starting pressure and temperature at the conclusion of the reaction. The preceding comments about G and equilibrium, although derived for unchanging pressure and unchanging temperature, apply equally well to a high-temperature explosion, provided that the reaction vessel and its contents are returned to 298°K and 1 atm at the conclusion of the reaction.[3]

Challenge. You now should be able to use the concepts of free energy to solve problems in polymer degradation and plant growth; try Problems 15–21 and 15–22 in *Relevant Problems for Chemical Principles* by Butler and Grosser.

Free Energy Changes When External Work is Done

In a reversible process in which no external work is involved, we have just shown that the free energy of a reacting system does not change: $\Delta G = 0$. What happens if, during the reversible process, the system does electrical, magnetic, or gravitational work on its surroundings? We shall need the answer to this question in Chapter 17, where we treat electrochemical cells.

If PV work is not the only kind of work involved, then

$$w = P \Delta V + w_{ext} \tag{15–15}$$

in which w_{ext} represents all other kinds of work. Hence,

$$\Delta E = q - P \Delta V - w_{ext}$$

---

[3] In this chapter we deal with reactions at 298°K only. Free energies are sensitive to temperature changes, and we will see examples in Section 16–4.

and the enthalpy and free energy changes are derived as in the preceding pages:

$$\Delta H = q - w_{ext}$$
$$\Delta G = q - T\,\Delta S - w_{ext} = -w_{ext}$$

This final result is the object of our derivation:

$$\Delta G = -w_{ext} \tag{15–16}$$

When a chemical system does work on its surroundings in a reversible manner, the decrease in free energy of the system exactly balances the work done *other than* pressure–volume work. In an electrochemical cell, the work done by the cell is a measure of the decrease of free energy within the cell. Conversely, if a potential is applied across the terminals of an electrolysis cell of the type discussed in Section 3–1, the electrical work done on the cell (measured by methods that we shall examine in Chapter 17) is identical to the increase in free energy of the chemicals in the cell. When water is dissociated electrolytically by passing a current through it, the electrical work required is stored as the increase in free energy of hydrogen and oxygen gas from the free energy of liquid water:

$$H_2O(l) \rightarrow H_2(g) + \tfrac{1}{2}O_2(g) \qquad \Delta G^0 = +56.69 \text{ kcal}$$

This free energy can be recovered as heat when hydrogen and oxygen gases are burned. Alternatively, if the proper apparatus is used, the free energy can be converted to work again. (A fuel cell such as the one for generating electricity in lunar space capsules uses this $H_2 + \tfrac{1}{2}O_2$ reaction. If the gases are simply burned, part of the free energy is converted to heat and, by the second law, is not recoverable as work. The trick in efficient utilization of energy is to avoid turning it into heat at any step in the process. This is the secret of the efficiency of metabolic processes, as we note in Section 12–8, and of fuel cells, in Section 17–9.)

Calculations with Standard Free Energies

Standard free energies of formation of compounds from elements in their standard states are tabulated in Appendix 2. The standard state for a gas, pure liquid, or pure solid is the same as with enthalpies: gas at 1 atm partial pressure, pure liquid, pure solid—usually at 298°K. The standard state for a solute in solution is a concentration of 1 mole per liter of solution, or a 1-molar solution. The standard state of a solution component for tabulating enthalpy was not this 1-molar solution, but was a sufficiently dilute solution that adding more solvent had no additional heat effect. However, since enthalpy does not change very much with concentration (unlike free energy, as we shall see in Section 15–9), we also can use the tabulated enthalpy values as if they were for 1-molar solutions.

Let us return to our chemical examples and interpret them in terms of free energy. The explosive reaction of H_2 with Cl_2 (Section 15–5) has the following molar free energies, enthalpies, and entropies of reaction:

$$H_2(g) + Cl_2(g) \rightarrow 2HCl(g)$$

ΔH^0(kcal mole^{-1})	0.0	0.0	−22.06
ΔG^0(kcal mole^{-1})	0.0	0.0	−22.77
ΔS^0(cal deg^{-1} mole^{-1})	31.21	53.29	44.62

The reaction liberates 22.06 kcal of heat *per mole* of HCl produced, and the free energy decreases by more than this amount: 22.77 kcal mole^{-1} HCl. Where does this extra impetus for reaction come from?

For the reaction as written, which produces 2 moles of HCl, $\Delta H^0 = -44.12$ kcal and $\Delta G^0 = -45.54$ kcal. The entropy of the reaction is $\Delta S^0 = 2 \times (44.62) - 53.29 - 31.21 = +4.74$ e.u. (The units of cal deg^{-1} often are called *entropy units*, e.u.) Thus, $T \Delta S^0$ is $298° \times 4.74$ e.u. $= 1413$ cal or 1.413 kcal. Equation 15–12 is verified:

$$\Delta G = \Delta H - T \Delta S$$
$$-45.54 = -44.12 - 1.41$$
$$-45.54 \simeq -45.53$$

The discrepancy is within the limits of error of the data. Two moles of HCl are slightly more disordered and have a slightly higher entropy than 1 mole each of H_2 and Cl_2 gases. Most of the drive behind the reaction, as expressed by the free energy, comes from the liberation of heat, but 3% of it originates because the products have a higher entropy than the reactants.

A standard free energy of -45 kcal indicates a tremendous impetus toward reaction—the impetus that often accompanies an explosion. Let us look at a gentler reaction, that of making ammonia from hydrogen and nitrogen.

Example. What are the changes in standard free energy, enthalpy, and entropy for the reaction $3H_2(g) + N_2(g) \rightarrow 2NH_3(g)$? Do the heat and the disorder factors promote or oppose this reaction in the direction written?

Solution. From the data in Appendix 2

$$\Delta G^0 = 2(-3.98) - 3(0.0) - (0.0) = -7.96 \text{ kcal}$$
$$\Delta H^0 = 2(-11.04) - 3(0.0) - (0.0) = -22.08 \text{ kcal}$$
$$\Delta S^0 = 2(46.01) - 3(31.21) - (45.77) = -47.38 \text{ e.u.}$$

and at 298°K, $T \Delta S^0 = -14.12$ kcal.

Checking Equation 15–12, we find that

$$\Delta G \quad = \Delta H \quad\quad - T\,\Delta S$$
$$-7.96 = -22.08 + 14.12$$

The reaction is favored by the liberation of 22.08 kcal of heat, but opposed by 14.12 kcal because the products are so much more ordered and have 47.38 e.u. lower entropy than the reactants. Another way of considering the reaction is to say that, of the 22.08 kcal of heat liberated, 14.12 kcal were required to pay for the creation of an ordered system, and only 7.96 kcal remain to drive the reaction.

The drive toward the synthesis of ammonia is much less than the drive toward the synthesis of HCl, primarily because of the entropy factor. Is it ever possible for the entropy term to overwhelm the heat term and send the reaction in the opposite direction to that indicated by enthalpy alone? Yes, it is, and these are precisely the instances when the principle of Berthelot and Thomsen fails.

Exercise. Calculate the free energy, enthalpy, and entropy changes for the vaporization of liquid water. Check Equation 15–12 with your results. Which term is responsible for the evaporation taking place?

(*Answer:* This is a trick question. The *standard* values are $\Delta G^0 = +2.06$ kcal, $\Delta H^0 = +10.52$ kcal, $\Delta S^0 = 28.39$ e.u., $T\,\Delta S^0 = +8.46$ kcal. It still looks as if water should not evaporate. But all that this means is that if water vapor at a partial pressure of 1 atm is present at 298°K, it will condense spontaneously. If the partial pressure of the water is only a few millimeters, it will evaporate spontaneously instead. The dependence of ΔG on concentration is the topic of the next section.)

Exercise. Calculate the free energy, enthalpy, and entropy changes for the reaction $NH_4Cl(s) \rightarrow NH_4(aq)^+ + Cl(aq)^-$. (The solvent, water, is assumed to be present in equal concentrations before and after reaction, and is omitted.) Do the heat and entropy terms promote or oppose the solution of ammonium chloride?

(*Answer:* $\Delta G^0 = -1.62$ kcal; $\Delta H^0 = +3.62$ kcal; $\Delta S^0 = +17.6$ e.u.; $T\,\Delta S^0 = +5.2$ kcal. The enthalpy change opposes the reaction by 3.62 kcal, but the entropy change favors it by 5.2 kcal. The net drive in free energy is 1.62 kcal.)

Example. Calculate the free energy, enthalpy, and entropy changes for the decomposition of N_2O_5,

$$N_2O_5(s) \rightarrow 2NO_2(g) + \tfrac{1}{2}O_2(g)$$

Solution

$$\Delta G^0 = 2(+12.39) + \tfrac{1}{2}(0.0) \quad - (+32.0) = -7.22 \text{ kcal}$$
$$\Delta H^0 = 2(+8.09) \ + \tfrac{1}{2}(0.0) \quad - (-10.0) = +26.18 \text{ kcal}$$
$$\Delta S^0 = 2(+57.47) + \tfrac{1}{2}(49.00) - (+27.1) = +112.3 \text{ e.u.}$$

$$T \Delta S^0 = \frac{298° \times (112.3 \text{ cal deg}^{-1})}{1000 \text{ cal kcal}^{-1}} = +33.5 \text{ kcal}$$

$$\Delta G^0 \quad = \Delta H^0 \quad - T \Delta S^0$$
$$-7.22 = +26.18 - 33.5$$
$$-7.22 \simeq -7.32$$

The disadvantage of having to absorb about 26 kcal of heat on decomposition is more than compensated by the greater entropy of the gaseous products, and the reaction proceeds with a standard driving force of about 7 kcal of free energy.

In summary, the drive of a reaction, carried out at constant pressure and temperature, is measured by its free energy change. If the free energy change is negative, the reaction is spontaneous; if the free energy change is positive, the reaction is spontaneous in the reverse direction; if the free energy change is zero, reactants and products are at equilibrium. The free energy change has two components: $\Delta G = \Delta H - T \Delta S$. A large decrease in enthalpy, meaning the emission of heat, favors a reaction. But there is a second factor as well. A large increase in entropy when reactants form products also favors reaction. The entropy term at normal temperatures is generally small, so ΔG and ΔH have the same sign. In such cases, spontaneous reactions *are* exothermic. Yet there are other instances in which the entropy and enthalpy terms work against one another, and even in which the entropy term dominates. This is especially true in reactions in which solids or liquids change to gases or solutions as products.

Thus far, we have used only standard concentrations, meaning 1 atm partial pressure for gases, pure liquids, and pure solids for condensed phases, and 1-molar solutions for solutes. How does free energy change with changes in concentration?

15–9 FREE ENERGY AND CONCENTRATION

To this point, we have managed to avoid calculus. We shall need it briefly to derive the equation for the dependence of free energy on concentration, but will not use it thereafter. Recall the basic definition of free energy,

$$G = H - TS = E + PV - TS$$

Instead of looking at small but finite changes (ΔG, ΔE, ΔS) at constant temperature, as in Equation 15–12, let us examine infinitesimally small changes (dG, dE, and dS) under the most general experimental conditions. Then the equation for the basic definition of free energy becomes

$$dG = dE + P\,dV + V\,dP - T\,dS - S\,dT$$

The first law of thermodynamics, expressed in the form of infinitesimally small changes, is

$$dE = dq - dw$$

and the free energy expression is

$$dG = dq - dw + P\,dV + V\,dP - T\,dS - S\,dT$$

We can simplify this expression considerably. If the reaction takes place at constant temperature, then $S\,dT = 0$ since dT, the change in temperature, is zero. If the reaction is reversible, $dq = T\,dS$, and if only PV or expansion work is permitted, $dw = P\,dV$. All terms on the right except one cancel, and

$$dG = V\,dP$$

For 1 mole of an ideal gas, $V = RT/P$, and

$$dG = RT\frac{dP}{P} = RT\,d\ln P$$

Our last use of calculus is the integration (summing all the infinitesimal changes) of this equation:

$$G_2 = G_1 + RT\ln\frac{P_2}{P_1} \tag{15–17}$$

This equation means that if we know the molar free energy of an ideal gas at partial pressure P_1 to be G_1, the molar free energy at some other partial pressure P_2 is G_2. Although we derived this for a reversible change of conditions from P_1 to P_2, once we have it we can use it for irreversible changes as well, since by the state function properties of G it is irrelevant how we go from State 1 to State 2.

Now let us make State 1 our chosen standard state of 1 atm partial pressure and make State 2 any state at all. The more general form of Equation 15–17 then is

$$G = G^0 + RT\ln\left(\frac{P}{P^0}\right)$$

in which $P^0 = 1$ atm. For the free energy of the compound in the standard state, G^0, we can use instead the free energy of formation from Appendix 2, ΔG^0 at 298°K. Because, in any chemical reaction, matter is neither created nor destroyed, and both reactants and products must be made from the *same type and quantity of elements* in their standard states.

We now can calculate how the free energy of ammonia depends on its partial pressure in a mixture of gases (or more precisely, the free energy of formation of ammonia from its elements in their standard states). Because

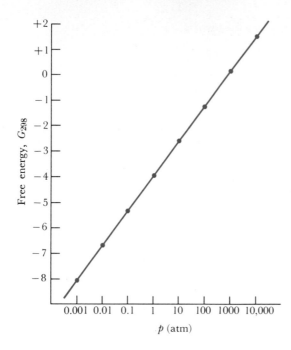

Figure 15–10. *The free energy of a gas depends on its partial pressure according to the expression $G = G^0 + RT \ln (p/p^0)$. This plot is of the free energy of ammonia, as given by $G_{NH_3} = -3.976 + 1.364 \log_{10} (p_{NH_3}/1 \text{ atm})$. Pressure, p, is plotted on logarithmic scale.*

the standard free energy of ammonia at 298°K is -3.976 kcal mole^{-1},

$$G_{NH_3} = -3.976 + RT \ln \left(\frac{p_{NH_3}}{1 \text{ atm}}\right)$$

$$G_{NH_3} = -3.976 + 2.303 \, RT \log_{10} \left(\frac{p_{NH_3}}{1 \text{ atm}}\right)$$

Since $R = 1.987$ cal deg^{-1} mole^{-1} and $T = 298$°K, $2.303 \, RT = 1364$ cal mole^{-1} or 1.364 kcal mole^{-1} (this is a handy number to remember). Thus

$$G_{NH_3} = -3.976 + 1.364 \log_{10} \left(\frac{p_{NH_3}}{1 \text{ atm}}\right)$$

The molar free energy of ammonia is plotted against partial pressure in Figure 15–10. Note that the plot is a straight line because pressure was plotted on a logarithmic scale. The ratio of partial pressure to partial pressure in the standard state usually is abbreviated as the *activity, a*:

$$a = \frac{p}{p^0}$$

Since the standard state is 1 atm, the activity is numerically equal to the partial pressure of the gas, but activity has no units. This makes activity easier to use; we do not find ourselves wondering how, in the expression "$\ln p$," we can take the logarithm of a number with units attached to it, and then having to remember that the proper form of the logarithm is "$\ln (p/1 \text{ atm})$."

In summary, for ammonia we can write the general expression for the relationship between free energy and pressure as

$$G_{NH_3} = G^0_{NH_3} + RT \ln a_{NH_3}$$

We can do the same for all the reactants and products, j, in a chemical process:

$$G_j = G^0_j + RT \ln a_j = G^0_j + 2.303\, RT \log_{10} a_j \qquad (15\text{–}18)$$

Let us apply this to the synthesis of ammonia and see what we can learn. In writing the equation as

$$N_2 + 3H_2 \rightleftarrows 2NH_3$$

we have changed from the single arrow to a double arrow because we now are considering both the forward and the reverse reactions. For each of the reactants and products, we can write

$$G_{N_2} = G^0_{N_2} + RT \ln a_{N_2}$$
$$G_{H_2} = G^0_{H_2} + RT \ln a_{H_2}$$
$$G_{NH_3} = G^0_{NH_3} + RT \ln a_{NH_3}$$

The total free energy of the chemical reaction is

$$\Delta G = 2G_{NH_3} - G_{N_2} - 3G_{H_2}$$
$$\Delta G = 2G^0_{NH_3} - G^0_{N_2} - 3G^0_{H_2} + 2RT \ln a_{NH_3} - RT \ln a_{N_2} - 3RT \ln a_{H_2}$$
$$(15\text{–}19)$$

The first three terms on the right are the *standard* free energy of reaction that we already have used, and are combined as ΔG^0. Then if we take the coefficients in front of the RT terms inside the logarithms as exponents, we can write

$$\Delta G = \Delta G^0 + RT \ln a^2_{NH_3} - RT \ln a_{N_2} - RT \ln a^3_{H_2}$$

Finally,

$$\Delta G = \Delta G^0 + RT \ln \left(\frac{a^2_{NH_3}}{a_{N_2} a^3_{H_2}} \right) \qquad (15\text{–}20)$$

Take a close look at the ratio in parentheses. It is the ratio of activities of products to reactants, and in fact is simply the reaction quotient, Q, of Chapter 5, with concentrations expressed in terms of activities, rather than moles per liter:

$$Q \equiv \left(\frac{a^2_{NH_3}}{a_{N_2} a^3_{H_2}} \right) \qquad (15\text{–}21)$$

This is a more general treatment than we saw previously, and will lead to a more general concept of equilibrium. This reaction quotient, Q, can be calculated for any given set of experimental conditions from the partial pressures of reactants and products.

The true free energy change in the ammonia reaction is the combination of the standard free energy change (for which $Q = 1$ and $\ln Q = 0$) and the reaction quotient term describing the actual experimental conditions. With numbers inserted, Equation 15–20 becomes

$$\Delta G = -7.96 + 1.364 \log_{10} Q \qquad \text{(units of kcal)} \qquad (15\text{–}22)$$

Note that since $\Delta G^0 = 2G^0_{\text{NH}_3} - G^0_{\text{N}_2} - 3G^0_{\text{H}_2} = 2\,\Delta G^0_{\text{NH}_3}$, the free energy value in Equation 15–22 is twice the value in Appendix 2 for the free energy of formation of 1 mole of NH_3 gas.

Table 15–4 lists the results of applying Equation 15–22 to eleven different sets of starting conditions for ammonia synthesis. These are not successive points in the same reaction from the same starting conditions; they are each separate conditions. If the concentrations of reactants and products were as shown in each experiment, what would be the free energy change in the ammonia synthesis reaction? These are the ΔG values at the right of the table. We shall defer until Chapter 16 the question of how to follow a given reaction from start to finish.

In Experiment a, nitrogen and hydrogen gases are at 1 atm partial pressure, and there is no ammonia present. Hence, the drive for the production of ammonia is infinitely strong. But as soon as the ammonia concentration is even 10^{-3} atm, the free energy change has increased from $-\infty$ to -16.136, and the drive to produce more ammonia has slackened (Experiment b). If we were to set up Experiment c, with N_2 and H_2 at 1 atm partial pressure and NH_3 at 0.1 atm, we would find that the free energy is -10.680. The more products and the less reactants there are, the smaller is the drive to produce more products. At uniform concentrations of 1 atm, the standard free energy results (Experiment d). In Experiment e, the excess of ammonia makes the drive less than in the standard state. In Experiments f, g, and h, the drive toward products is stopped entirely. This is the *equilibrium state*. At equilibrium

$$\Delta G = 0 \qquad \text{and} \qquad \Delta G^0 = -RT \ln Q$$

As we saw in Chapter 5, the reaction quotient at equilibrium is the *equilibrium constant*, K_{eq}. We can calculate the value of this equilibrium constant from the standard free energy of the reaction:

$$K_{\text{eq}} = \frac{a^2_{\text{NH}_3}}{a_{\text{N}_2} a^3_{\text{H}_2}} = e^{-(\Delta G^0/RT)} = 10^{-[(-7.952)/2.303\,RT]} = 6.8 \times 10^5$$

If the reaction quotient is less than the equilibrium constant, products will be formed spontaneously. Whenever conditions are such that the reaction quotient is greater than the equilibrium constant, K_{eq}, the reaction will be spontaneous in the reverse direction (Experiments i, j, and k in Table 15–4).

Table 15–4.   Free Energy of Synthesis of Ammonia at 298°K

$$N_2(g) + 3H_2(g) \rightleftarrows 2NH_3(g)$$

$$\Delta G = -7.952 + 1.364 \log_{10} Q$$

$$Q = \frac{a^2_{NH_3}}{a_{N_2}a^3_{H_2}}$$

Experiment	a_{N_2}	a_{H_2}	a_{NH_3}	Q	$1.364 \log_{10} Q$	ΔG
					(kcal)	(kcal)
a	1	1	0	0	$-\infty$	$-\infty$
b	1	1	0.001	10^{-6}	-8.184	-16.136
c	1	1	0.1	10^{-2}	-2.728	-10.680
d	1	1	1	1	0	-7.952
e	1	1	100	10^4	$+5.456$	-2.496
f	1	1	825	6.8×10^5	$+7.952$	0
g	1.47	0.01	1	6.8×10^5	$+7.952$	0
h	0.01	0.1	2.61	6.8×10^5	$+7.952$	0
i	0.01	0.1	26.1	6.8×10^7	$+10.700$	$+2.748$
j	0.01	0.01	100	10^{12}	$+16.368$	$+8.416$
k	0	1	1	∞	$+\infty$	$+\infty$

When the reaction quotient equals the equilibrium constant, $Q = K_{eq}$, the forward and reverse reactions proceed at the same rate, and the reacting system of chemicals is at equilibrium.

General Expressions

So far, we have stated the free energy derivations in terms of the ammonia reaction. Let us now generalize to a reaction in which r moles of compound R combine with s moles of S to produce t moles of T and u moles of U:

$$r R + s S = t T + u U \tag{15–23}$$

The free energy change for the reaction, ΔG, is

$$\Delta G = t G_T + u G_U - r G_R - s G_S \tag{15–24}$$

Each free energy can be expressed in terms of activity:

$$
\begin{aligned}
G_R &= G_R^0 + RT \ln a_R \\
G_S &= G_S^0 + RT \ln a_S \\
G_T &= G_T^0 + RT \ln a_T \\
G_U &= G_U^0 + RT \ln a_U
\end{aligned}
\tag{15–25}
$$

These can be substituted into Equation 15–24,

$$\Delta G = \Delta G^0 + RT \ln Q \tag{15–26}$$

in which

$$\Delta G^0 = t\, G_T^0 + u\, G_U^0 - r\, G_R^0 - s\, G_S^0 \tag{15-27}$$

and

$$Q = \frac{a_T^t a_U^u}{a_R^r a_S^s} \tag{15-28}$$

Equation 15–26 gives the free energy change for the reaction under any conditions. For the special case of equilibrium, $\Delta G = 0$, and

$$Q = K_{eq} \quad \text{(the equilibrium constant)} \tag{15-29}$$

$$-\Delta G^0 = RT \ln K_{eq} \tag{15-30}$$

$$K_{eq} = e^{[-\Delta G^0/RT]} \tag{15-31}$$

Notice that the form of the reaction quotient, Q, and the equilibrium constant, K_{eq}, depends only on the overall stoichiometry of the reaction, and not on any particular reaction mechanism. One need not know in molecular detail how a reaction takes place to write the equilibrium-constant expression. We first mentioned this labor-saving fact in Chapter 5; now we have given it a thermodynamic proof.

Example. Calculate the equilibrium constant at 298°K for the reaction $H_2(g) + Cl_2(g) = 2HCl(g)$, and compare it with the equilibrium constant for the ammonia synthesis.

Solution. We discussed in a previous example that the free energy of this reaction is -22.769 kcal mole^{-1} of HCl, or -45.538 kcal for the reaction as written to produce 2 moles of HCl. Thus

$$K_{eq} = e^{[-(-45.538)/RT]} = 10^{+(45.538/1.364)} = 10^{33.4} = 2.5 \times 10^{33}$$

The K_{eq} for the ammonia synthesis is only 6.8×10^5. The HCl reaction is different; equilibrium lies far on the side of much HCl and few reactants. The expression for the amounts of H_2, Cl_2, and HCl at equilibrium is

$$\frac{a_{HCl}^2}{a_{H_2} a_{Cl_2}} = K_{eq} = 2.5 \times 10^{33}$$

If pure HCl gas is enclosed in a container at 1 atm, enough HCl can dissociate spontaneously to yield partial pressures of H_2 and Cl_2 that are approximately 2×10^{-17} atm, hardly a significant amount of either reactant. In Chapter 16 we shall see how to make such statements for reactions that are less completely skewed in one direction than the HCl reaction is.

15–10 COLLIGATIVE PROPERTIES

When a liquid and a vapor are in equilibrium, the molar free energy of the substance must be the same in each phase; otherwise there would be a spontaneous evaporation of more liquid or condensation of more vapor. We can treat the evaporation–condensation process as a chemical reaction:

$$H_2O(l) \rightleftarrows H_2O(g)$$

At 298°K, the free energy change for this reaction is

$$\Delta G = \Delta G^0 + RT \ln \frac{a(g)}{a(l)} \tag{15–32}$$

The standard free energy of vaporization, from Appendix 2, is

$$\Delta G^0 = (-54.635 \text{ kcal}) - (-56.690 \text{ kcal}) = +2.055 \text{ kcal mole}^{-1}$$

This value indicates that if water vapor at 1 atm partial pressure is at 298°K, part of it will condense spontaneously to liquid. We now can calculate how low the partial pressure of water vapor must be before it is in equilibrium with liquid water. Since the standard state of a liquid is the pure liquid, the activity of the liquid water, $a(l)$, is unity. We can write Equation 15–32 as

$$\Delta G = +2.055 + 1.364 \log_{10} a(g)$$

At equilibrium, $\Delta G = 0$, and

$$a(g) = 10^{-(2.055/1.364)} = 10^{-1.506} = 0.0321$$

The partial pressure of water vapor at equilibrium is therefore 0.0321 atm, or 24.4 torr. The vapor pressure of any liquid is the partial pressure of its vapor that is in equilibrium with it at the specified temperature. The vapor pressure of a liquid increases with temperature. The normal boiling point is the temperature at which the vapor pressure, or the equilibrium partial pressure, becomes equal to 1 atm. At 100°C, liquid water is in equilibrium with water vapor at 1 atm, and the phenomenon of boiling occurs.

The Four Colligative Properties

Four phenomena, important to early chemists because of the information that they gave about the number of particles of solute in a solution, are classed together as the *colligative properties*. These phenomena are:

1) The lowering of vapor pressure of a solvent by a solute.
2) The raising of the boiling point of a solvent by a solute.
3) The lowering of the freezing point of a solvent by a solute.
4) Osmotic pressure.

All four phenomena depend on *how many* particles of solute are present in solution, but not on what these solute particles are (so long as they are not volatile and appear only in the liquid phase). The equations that we shall derive are idealizations of real behavior, like the ideal gas law. In a strict sense, they are valid only in the limit of infinitely dilute solutions, but they are useful for real, dilute solutions as well. These colligative properties were valuable to Arrhenius because he could show that the effects in solutions of electrolytes indicated the presence of more particles than there were molecules of solute. This was strong evidence that the solute molecules were breaking up (into ions). These colligative properties currently are used most often to determine molecular weights of unknown materials. We shall see how this is done.

Vapor pressure lowering. When a pure liquid, B, is in equilibrium with its vapor, the free energy of liquid and gaseous B must be the same. Evaporation and condensation are occurring at the same rate. If a small amount of nonvolatile solute, A, is added to the liquid, the free energy or escaping tendency of B in the solution is lowered. However, the reverse tendency, the condensation of vapor to liquid, is unaffected. At a constant temperature, the frequency with which a molecule in the solution approaches the surface with enough kinetic energy to escape to the gas phase is the same in both pure B and in solution (Figure 15–11). Not all molecules approaching with the right energy can escape from the solution because a certain fraction of the molecules will be solute, A, instead of B. If 1% of the molecules in solution are A, then the vapor pressure of B will be only 99% as high as it was in pure B. If we define the mole fraction, X, of B as the number of moles of B divided by the total number of moles,

$$X_B \equiv \frac{n_B}{n_A + n_B}$$

then the vapor pressure of B in a solution of concentration X_B is

$$P_B = X_B P_B^0$$

The term P_B^0 is the vapor pressure of pure B, which can be calculated from the free energy of vaporization as we have done previously in this section. The change in vapor pressure is

$$\Delta P = P_B - P_B^0 = (1 - X_B)P_B^0 = X_A P_B^0 \tag{15–33}$$

The change in vapor pressure is proportional to the amount of A present, but not to the particular nature of the A molecules.

Boiling point elevation. Look again at Figure 15–11, which shows liquid and vapor at equilibrium in (a). If we dilute the liquid by adding molecules of a nonvolatile substance A (in color), the escaping tendency of B from the liquid to the gas, expressed in terms of free energy, is less:

$$G_B(l) < G_B(g)$$

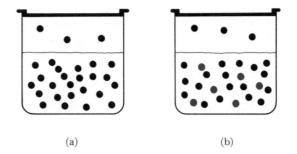

(a)   (b)

Figure 15-11. (a) Equilibrium exists between a liquid and its vapor when the molar free energy or escaping tendency is the same in both liquid and vapor. (b) Adding solute (colored dots) decreases the escaping tendency of the solvent and causes the system to shift toward more condensation of gas. Le Chatelier's principle can be invoked here: When solvent is diluted by solute, the equilibrium position shifts in a direction that will add more solvent.

If the pure liquid had been at its boiling point, the solution no longer will be so. We must raise the temperature until the escaping tendency of the B molecules that remain in solution has compensated for the smaller proportion of B molecules present. We cannot derive the expression for how high the temperature must be raised before boiling takes place again because we do not have an expression for how free energy depends on temperature. This expression requires calculus for its derivation and is beyond the limitations of this text. However, the principle that we invoke is that the temperature must be raised until the molar free energy of B is again the same in liquid and gas:

$$G_B(l) = G_B(g) \qquad \text{(at an elevated temperature)}$$

The result of this derivation is the statement that the elevation in boiling point, ΔT_b, is proportional to the *molal* concentration of solute, A:

$$\Delta T_b = k_b m_A \tag{15-34}$$

The *molality* of a solution, m, as defined in Section 4-4, is the number of moles of solute per 1000 g of solvent. The molal boiling point elevation constant, k_b, depends only on properties of the solvent, B:

$$k_b = \frac{RT_b^2 M_B}{1000 \, \Delta H_v^0}$$

in which T_b is the normal boiling point of B, M_B is its molecular weight, and ΔH_v^0 is the molar heat of vaporization of pure B. The change in boiling point of the solvent depends only on how many particles of solute there are and not on what kind of particles they might be. Equation 15-34 holds only for dilute

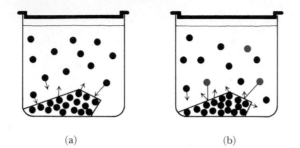

(a)   (b)

Figure 15–12. Pure liquid and solid in equilibrium (a). When solute (colored dots) is added to solvent (b), the equilibrium position shifts in favor of the dissolution of more pure solid. Equilibrium is reestablished only by lowering the temperature. This is the phenomenon of freezing point lowering.

solutions, because at several steps in its derivation, we assume that X_A is much less than X_B, and that X_B is close to 1.00.

Freezing point lowering. In Figure 15–12(a), pure liquid B is in equilibrium with solid B, and the free energies of B in the two phases are the same:

$$G_B(s) = G_B(l)$$

If solute, A, is added, then not every molecule that strikes the surface of the solid can stick, and the escaping tendency of B will be lowered. The free energy of B in the solution will be less than that in the solid, and the solid will begin to dissolve. This is why $CaCl_2$ or $NaCl$ melts ice when it is sprinkled on slippery roads in winter. The reaction

$$B(s) \rightarrow B(l)$$

has a negative free energy.

To restore equilibrium between phases, we must lower the temperature until the escaping tendency of the fewer B molecules in the solution is equal to that of the larger number of B molecules in the pure liquid. At the new freezing point, there are fewer molecules of B in the solution to strike the solid, but they are moving more slowly and have a greater probability of being captured by the solid phase.

As before, we will not derive the expression for freezing point depression, but it is

$$\Delta T_f = -k_f m_A \tag{15–35}$$

The molal freezing point depression constant, k_f, is given by the equation

$$k_f = \frac{RT_f^2 M_B}{1000 \, \Delta H_f^0}$$

in which T_f is the normal freezing point of the solvent, B, M_B is its molecular

Table 15–5. Freezing Point and Boiling Point Constants for Common Solvents[a]

Solvent	T_f, °K	k_f[b]	T_b, °K	k_b[b]	M, g
Water (H_2O)	273.16	1.86	373.0	+0.52	18.0
Carbon tetrachloride (CCl_4)	250.5	~30.00	350.0	5.03	154.0
Chloroform ($CHCl_3$)	209.6	4.70	334.4	3.63	119.5
Benzene (C_6H_6)	278.6	5.12	353.3	2.53	78.0
Carbon disulfide (CS_2)	164.2	3.83	319.4	2.34	76.0
Ether ($C_4H_{10}O$)	156.9	1.79	307.8	2.02	74.0
Camphor ($C_{10}H_{16}O$)	453.0	40.0			152.2

[a] T_f is the normal freezing point; k_f, the molal freezing point lowering constant; T_b, the normal boiling point; k_b, the molal boiling point elevation constant; M, the molecular weight of the substance.

[b] In units of degrees per mole per kilogram.

weight, and ΔH_f^0 is its molar heat of fusion. Again, the magnitude of the freezing point lowering, like that of boiling point elevation, depends on the number of particles of A, and not on their chemical character.

Table 15–5 gives the molal freezing point and boiling point constants for several common solvents. Let us now try some problems involving freezing point and boiling point changes and see how we use the changes to calculate molecular weights.

Molecular Weight Determination

Example. A solution is made up with 1 g of acetamide (CH_3CONH_2) in 100 ml of water. What is its freezing point depression?

Solution. The molecular weight of acetamide is 59.07. A weight of 1 g in 100 ml of water is equivalent to 10 g in 1 kg of water. A weight of 10 g of acetamide is $10/59.07 = 0.169$ mole of acetamide. The molality of the solution is therefore 0.169. The freezing point depression is 1.86 times this number, or 0.32°C.

Example. A saturated solution of glutamic acid in water has 1.500 g glutamic acid per 100 g of water. The freezing point of this solution is −0.189°C. What is the molecular weight of glutamic acid?

Solution. From the freezing point depression formula, Equation 15–35,

$$\Delta T_f = -k_f m_A$$
$$-0.189 = -1.86 m_A$$
$$m_A = 0.189/1.86 = 0.102 \text{ mole kg}^{-1}$$

The solution was prepared from 1.500 g per 100 g water, or from 15.00 g kg^{-1} of water. So 0.102 mole of glutamic acid is 15.00 g, and the molecular weight is

$$M_A = \frac{15.00 \text{ g}}{0.102 \text{ mole}} = 147 \text{ g mole}^{-1}$$

Exercise. A compound with a molecular weight of 329.26 is dissolved in water, with 300 mg (300 × 10^{-3} g) of compound in 10 ml of water. What will be the freezing point of the solution?

(*Answer:* Expected freezing point is −0.169°C.)

Exercise. In the experiment of the preceding exercise, the observed freezing point was −0.676°C. What can you say about what happens when the compound dissolves? [The compound is $K_3Fe(CN)_6$.]

(*Answer:* The compound dissolves to produce four ions per molecular weight unit.)

Exercise. An amount of 200 mg of the protein cytochrome *c* is dissolved in 10 ml of water. The molecular weight of cytochrome *c* is 12,400. What will be the freezing point depression of the solution?

(*Answer:* The freezing point depression is 0.003°C.)

As the cytochrome *c* example illustrates, freezing and boiling point methods are useless for finding molecular weights of large macromolecules. (Why?) The fourth and last colligative property is especially useful for such large molecules.

Osmotic pressure. Many membranes have pores large enough to let some molecules pass through, but too small to allow others to pass. These are called semipermeable membranes. Some will allow water through but not salt ions. Others, with larger pores, will allow water, salts, and small molecules through, but not macromolecules with molecular weights of several thousand. These membranes are used widely for separation purposes, but also can be used in finding molecular weights of large molecules.

In Figure 15–13(a), suppose that the beaker has pure water, whereas the thistle tube with the membrane across its bottom has a water solution of some dissolved substance, A. Further suppose that A cannot pass through the membrane. The rate of flow of water molecules into the thistle tube from the beaker solution is unimpaired, but the rate of flow from the thistle tube is lessened. The molar free energy of the water in the thistle tube will be lessened by the presence of solute A particles just as it was with the other three colligative properties. More water will flow in than out, and solution will rise in the tube, as shown in Figure 15–13(b).

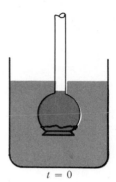

Figure 15–13. When a solute is dissolved in a solvent, and the solution is separated from a reservoir of pure solvent by a membrane that allows solvent molecules but not solute particles through it, the solvent flows into the solution chamber until an extra pressure known as the osmotic pressure builds up to oppose it. (a) Solution inside thistle tube (gray), pure solvent outside (colored). (b) After a time, the solution has risen in the tube. (c) Equilibrium, with the solution in the tube stationary. The osmotic pressure can be calculated from the height of the solution in the tube and the density of the solution.

$t = 0$

(a)

As the pressure inside the thistle tube increases (this pressure increase can be measured by the height of the column of solution in the tube), the free energy of water molecules in the thistle tube increases. A point eventually is reached where the increase in free energy of water because of the increased pressure exactly balances the decrease in free energy of water because of the presence of solute particles, A. At this *osmotic pressure*, π, equilibrium between water molecules on both sides of the membrane is restored, and the flow in both directions is the same.[4]

The derivation of the expression for osmotic pressure leads to the expression

$$\pi = c_A RT$$

The osmotic pressure required to restore equilibrium is proportional to the molarity, c_A (not molality, m_A) of the solution inside the thistle tube.

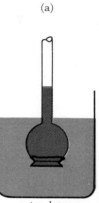

$t = $ later

(b)

Example. An amount of 200 mg of cytochrome c is dissolved in enough water to make a final volume of 10 ml. The molecular weight of cytochrome c is 12,400. If the solution is placed inside the thistle tube of Figure 15–13, and the bottom is sealed by a membrane that will allow water molecules through but not molecules of cytochrome c, how high will the solution rise in the upper tube when osmotic equilibrium is restored?

Solution. The 200 mg in 10 ml is equivalent to 20 g in 1000 ml solution. And 20 g are $20/12{,}400 = 1.61 \times 10^{-3}$ mole; thus, the solution is 1.61×10^{-3} molar. The osmotic pressure (at 298°K) is

$$\pi = 1.61 \times 10^{-3} \text{ mole liter}^{-1} \times 0.08205 \text{ liter atm deg}^{-1} \text{ mole}^{-1} \times 298°K$$

$$\pi = 0.0394 \text{ atm}$$

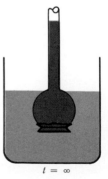

$t = \infty$

(c)

---

[4] In a strict sense, the pressure measured by the height of the column of solution is equal to the osmotic pressure, π, only if the diameter of the column is so small that the dilution of solution within the thistle tube from solvent flowing through the membrane is negligible. The osmotic pressure is really that pressure which, if applied to the solution, would *prevent* a net flow of solvent through the membrane into the solution.

A pressure of 1 atm will support a column of mercury 760 mm or 0.760 meter high. Since mercury has a density 13.645 times that of water, 1 atm will support a column of water 0.760×13.645 meters or 10.4 meters high. The amount of 0.0394 atm thus represents a column of water 0.410 meter or 41.0 cm high.

Measuring a 41.0-cm high column of water can be done with much greater accuracy than measuring a 0.003° temperature drop. Osmotic pressure measurements are definitely preferred for obtaining molecular weights of large molecules.

Before you go on. If you do not understand completely the calculations involving colligative properties, you may want to work the problems on molality and mole fraction, vapor pressure lowering, boiling point elevation, freezing point depression, and osmotic pressure in Review 13 of *Programed Reviews of Chemical Principles* by Lassila *et al.*

15–11 SUMMARY

This chapter probably has more important new ideas packed into it than any other chapter so far. From this chapter you should not obtain facts, but new concepts and ways of thinking about chemical facts. The main ideas that you should retain from this chapter are:

1) Heat and work are alternative forms of energy, and in any chemical reaction, energy is conserved. The change in energy of the reacting system exactly balances the gain or loss of energy from or to the surroundings.

2) Energy and enthalpy are state functions. The change in either quantity during a reaction is therefore dependent only on the starting and ending point, and not on how the reaction is accomplished. Heats of reaction are accordingly additive in the same way that the reactions themselves are.

3) The enthalpy of a molecule can be approximated by the sum of the bond energies of localized bonds between pairs of atoms. The localized bond idea thus gains experimental support. In some classes of molecules, the aromatic ring systems in particular, this localized bond idea is inadequate and cannot be used. Delocalization of electrons in a molecule makes the molecule more stable.

4) Entropy, another state function, is a measure of the disorder of a system. It can be calculated from the number of different microscopic ways of building the same observable situation. Third-law entropies, obtained from purely thermal measurements, correlate well with what we expect for different substances from the statistical explanation of entropy.

5) The spontaneity of a reaction at constant pressure and temperature is measured by its change in free energy per stoichiometric unit of reaction.

For a reaction in which no work other than pressure–volume work is done, if ΔG is large and negative, the reaction is spontaneous. If ΔG is large and positive, the reaction is spontaneous in the reverse direction. If ΔG is zero, the reaction is at equilibrium. Expressed another way, the free energy is the chemical potential function that we minimize to find the point of chemical equilibrium.

6) If electrical work, or some form of work other than pressure–volume work, is involved in a chemical process accomplished in a reversible manner, then the free energy change during the reaction is not zero, as in Point 5. Instead, the free energy of the reacting system decreases by an amount equal to the work done on the surroundings by the reacting system: $\Delta G = -w_{ext}$.

7) The free energy of a reaction is the net result of two effects, heat and disorder: $\Delta G = \Delta H - T \Delta S$. A reaction is favored if it releases heat (ΔH is large and negative) and if the products are more disordered than the reactants ($-T \Delta S$ is large and negative, or ΔS is large and positive). ΔH is usually, but not always, the dominant term on the right side of the equation.

8) The free energy of a gas varies with its partial pressure by the relationship $G_2 = G_1 + RT \ln (p_2/p_1)$. The activity of the gas, a, is the ratio of its pressure to that in a standard state of 1 atm pressure. Thus the free energy at any pressure is given by $G = G^0 + RT \ln a$.

9) The free energy change of a gas reaction varies with the partial pressures of its components according to the expression $\Delta G = \Delta G^0 + RT \ln Q$. For the special case of equilibrium, the free energy of reaction is zero: $\Delta G = 0$. The reaction quotient, Q, is then the equilibrium constant, K_{eq}. The equilibrium constant and standard free energy of reaction are related by

$$\Delta G^0 = -RT \ln K_{eq} \qquad \text{or} \qquad K_{eq} = e^{-(\Delta G^0/RT)}$$

10) For any given set of experimental conditions, if the reaction quotient is smaller than the equilibrium constant, the reaction is spontaneous in the forward direction. If the reaction quotient is greater than K_{eq}, the reverse reaction is spontaneous. At equilibrium, $Q = K_{eq}$.

11) Equilibrium between different phases is attained when each chemical substance has the same free energy in every phase in which it appears. Phase changes can be treated with the same mathematical machinery as chemical reactions. The vapor pressure of a liquid is the partial pressure of its gas for which the free energy of the gas is the same as that of the liquid.

12) When a solute is added to a pure liquid, the molar free energy of the liquid is decreased. This decrease causes more pure liquid to enter the solution either from the gas phase, the pure solid, or through a membrane from a reservoir of pure liquid. The influx of new liquid by con-

densation of the gas can be halted by raising the temperature, the influx by solution of solid can be halted by lowering the temperature, and the influx by diffusion through a porous membrane can be halted by increasing the pressure. When balance has been restored and the free energies of the solvent are again equal in all phases, we observe the boiling point elevation, freezing point depression, or osmotic pressure phenomena.

The next chapter, on free energy and equilibrium, will draw heavily on the ideas in Points 8, 9, and 10. These and Points 5 and 7 will be used in Chapter 18 when we interpret rates of reaction in terms of the rates of breakdown of activated complexes (transition states). Point 6 is especially useful in Chapter 17 (electrochemistry).

15–12 POSTSCRIPT: COUNT RUMFORD VERSUS THE WORLD

In *Order and Chaos*, S. W. Angrist and L. G. Hepler remark:

> It is well established that people with widely differing backgrounds and positions have made significant contributions to the theory of heat and energy. Consider the following list of contributors to the science and practice of heat: (1) a spy for the British government in the employ of General Gage, who was British Commandant in Boston at the time of the American Revolution, (2) the Secretary of the Province of Georgia in the British Foreign Office in 1779, (3) The Undersecretary of State for the Northern Department in the British Foreign Office in 1780, (4) a lieutenant colonel in the King's American Dragoons, (5) a Knight in the court of George III, (6) a British spy in the court of the Elector of Bavaria, (7) the founder of the Munich Military Workhouse, (8) the designer of Munich's English Gardens, (9) a lieutenant general in the service of the Elector of Bavaria, (10) a member of the Polish Order of St. Stanislaus with the rank of White Eagle, (11) a Count of the Holy Roman Empire, (12) the founder of the Royal Institution, (13) a foreign associate of the French Academy of Sciences, and (14) Lavoisier's widow's second husband.

As the authors point out, all of these people are really only one individual, Benjamin Thompson, later Count Rumford.

To this list they could have added: the inventor of the combustion calorimeter, the comparative photometer with the International Standard Candle, the kitchen range or cookstove, the double boiler, the baking oven, the portable stove and army field kitchen, the drip coffee maker, the modern steam heating system, the smoke shelf and damper system now used in all fireplaces, an improved oil study lamp of unprecedented illumination, a naval signaling system used by Great Britain, and an improved ballistic pendulum for measuring the force of gunpowder; the discoverer of convection currents in gases and liquids, the maximum density of water at 4°C, and the superior absorption and emission of radiation by black instead of polished objects; one of the earliest investigators of the tensile strength of fibers and the insulating properties of cloth; the founder of one of the earliest public schools, and of the first international scientific medal and prize (still awarded); and the intended first head of West Point (declined by prearrangement for political reasons). The list is still incomplete. Thompson was a practical genius and inventor in the same league with Thomas Edison. He revolutionized nutrition in Europe, in the late 1700's, in the same way that, a century later, Edison revolutionized life with the practical use of electricity. He was certainly a more prolific inventor than Franklin, and probably a better scientist. Why, then, is he virtually unknown except to historians of science and students of thermodynamics?

The reason lies largely in the personality of the man. Thompson was ambitious and utterly without scruples or principles. He toadied to his superiors, was caustic and treacherous to his peers, and tyrannical to his subordinates. No one could work with him, and he made a host of enemies wherever he went. He was, in short, an intolerable genius.

Thompson was born in Woburn, Massachusetts, in 1753. He was a member of a large farm family. He appears to have been a compulsive organizer and student. Notebooks from his youth give a daily schedule of subjects to be studied ("Munday—Anatomy, Tewsday—Anatomy, Wednesday—Institutes of Physick, Thursday—Surgery, Fryday—Chimistry with the Materia Medica, Saturday—Physick $\frac{1}{2}$ and Surgery $\frac{1}{2}$") as well as hourly schedules for each day. He was first apprenticed to a drygoods dealer, then to a local doctor. Neither apprenticeship was satisfactory, and he became a schoolteacher in Concord, New Hampshire (originally called Rumford, N.H.). Here Thompson made his first—and in many respects typical—step upward in life. In 1772, he married the young widow of a wealthy New Hampshire landowner.

Thompson's natural autocratic leanings and the connections of his wife and her late husband led him to become a favorite of the British Royal Governor of New Hampshire. He became an informer and spy for the British, who needed information on caches of arms and supplies that the Colonial Militia and Minuteman groups were secreting about the New England countryside. He was suspected of being an informer, and the New Hampshire Committee of Public Safety called him before them to answer charges that he was "unfriendly to the cause of freedom." Nothing could be proved against him. But a week before Christmas, 1774, Thompson learned that a group of "patriots" was coming for him that evening with tar and feathers. He left his wife, baby daughter, and elderly father-in-law to face the mob alone and rode for Boston. He never came back.

Thompson continued to spy for the British in Massachusetts, and had another brush with the Committee of Public Safety in that state. Again, he was too clever for anything to be proved against him, but thereafter he was watched carefully and lost his advantage as a spy. When the British army was forced out of Boston in March, 1776, Thompson went with them. He soon arrived in London, where he found employment, first as an expert on the Revolutionary War (the equivalent of our "Kremlinologists"), then in several governmental posts. After seven years, during which time he made several important inventions, was suspected of slipping British naval intelligence to the French in the La Motte case, and made innumerable personal enemies, he felt obliged to seek employment elsewhere. He soon appeared in Munich as a colonel and military advisor to the Elector Karl Theodore of Bavaria. (He sent several military intelligence reports on the state of the army of his new employer back to England in cipher.)

The Bavarian army was in wretched shape. It had no discipline, training, decent equipment, supply procedures, or morale, and was ridden with graft, corruption, and inefficiency. Thompson was given the responsibility for whipping it into a decent fighting force. His situation in Munich was similar to Lavoisier's in the *Ferme générale*, or Tax Farm. The businessmen of the *Ferme générale* contracted with the French Crown to deliver a certain amount in tax revenues to the treasury each year. Any taxes that they collected over and above this amount, they could keep. Colonel Thompson was given a fixed sum of money each year to run the Bavarian army. If the operation of the army became more effective and Thompson simultaneously found ways to cut expenses, the money saved was his own. It paid both Lavoisier and Thompson to carry out their duties in the most efficient possible manner. Thompson's experiments with clothing, nutrition, cannon boring and the Munich military workhouses were all part of his plan to make the Bavarian army efficient. When conservative manufacturers refused to weave cloth and construct equipment to his specifications, he used the army to round up the thousands of street beggars of Munich in one night's sweep and set up the military workhouses as his factories. He gave each worker room and board, and set up free schools for their children (until they were old enough to work). He built the famous Munich "English Gardens" as demonstration gardens for his innovations in agriculture and nutrition. His "Rumford soup," developed for the workhouses, was an attempt to provide a complete food for the lowest possible cost. He introduced potatoes to Bavaria, although he had to smuggle the first ones into his kitchens by stealth because the Bavarians considered them unfit to eat. He propagandized coffee as a stimulating substitute for alcohol and invented the drip percolator to make it popular. Soldiers in European armies of the time found food where they could get it, and cooked it themselves over open fires in camp. Thompson first designed a collapsible one-man field stove and then conceived the idea of a traveling field kitchen to cook for the army. His cannon-boring experiments, virtually the only achievement for which he is still remembered, were only an incident in a colorful career in Munich.

From a colonel in the Bavarian army, Thompson rose to be minister of war, minister of police, major general, chamberlain of the Bavarian Court, and state councillor. He held all of these offices simultaneously and was the second most powerful man in Bavaria, after the elector himself. His ultimate title was that of count of the Holy Roman Empire. Thompson chose as his title the original name of Concord, New Hampshire, and, after 1792, insisted on being addressed as "Count Rumford" rather than as Benjamin Thompson. The choice of "Rumford" may have been a belated acknowledgment of his

Figure 15–14. Count Rumford supervising a public lecture at his Royal Institution in London, in 1802. Rumford is the hook-nosed figure smiling benignly at the upper right. The lecturer is Thomas Young, a Professor of Natural Philosophy at the Royal Institution, and his assistant with the bellows and an evil leer is the young Humphrey Davy. The "victim" of the demonstration is Sir John Hippisley, manager of the Royal Institution. Davy worked extensively with the physiological effects of various gases. He had almost killed himself inhaling methane two years before, and caused a sensation at a lecture in 1801 by giving laughing gas (nitrous oxide) to volunteers from the audience. James Gillray, the artist, was the Herblock or Mauldin of his era and was famous for his devastating political cartoons. He considered these Royal Institution lectures a sham because, although intended as an education for working people, they had become the fashionable entertainment of the wealthy, as caricatured here. Davy and Michael Faraday continued a tradition of public lectures which has been maintained to the present day. One of us [R.E.D.] gave a Royal Institution lecture in 1970 in what was recognizably the same lecture hall as shown here in 1802. (Photograph of the original etching courtesy The Fisher Collection.)

wife and child, whom he deserted 18 years earlier, or it might have derived from his pretentions in Europe that he had come from a wealthy landowning family in the Colonies.

By 1795, his intensive work had begun to damage his health, and his many enemies in the Court of Bavaria were becoming too powerful. He left Munich and returned in triumph to London. He was given almost overwhelming adulation, by both governmental figures and the general public, as a great philanthropist, philosopher, and benefactor. No matter what his difficulties in getting along with people and his personal defects might have been, Rumford's improvements in housing, lighting, clothing, and nutrition made a real difference to the average citizen of Europe of the time. The Royal Institution of Great Britain, now a respected research laboratory, was initially created as a showcase for Thompson's inventions and innovations. He brought in a young country boy by the name of Humphry Davy (the Sir Humphry Davy of the Dalton postscript to Chapter 2) to assist in giving public demonstrations and lectures (Figure 15–14). Typically, Thompson conceived of the Royal Institution as a place where the uninformed would come to ask Count Rumford how their lives should be run. The fact that he was so often right made little difference to those who were put off by his arrogance. Within two years, he was forced out of active control of the Royal Institution, although the Institution went on to be the brilliant showcase and laboratory for Humphry Davy, Michael Faraday, and a continuing procession of noted scientists.

At this point in his career, Rumford was only 49 years old. We must leave him, except to remark that he began a new career in France by marrying the widow of Lavoisier. His first wife had conveniently died by this time. The new marriage was notoriously stormy and lasted only two years. But by the end of this time, he was as firmly established in French affairs as he had been in those of Munich and London.

Rumford died suddenly in 1814. He managed his death as efficiently as he had his life. In a curious will, he left all his possessions to Harvard University, which still looks after his grave in Auteil, France. Upon his death, he sank into obscurity as rapidly as he had risen from it. He was not remembered, as were Lavoisier, or Dalton, or Franklin. People who tried to live and work with him found it so difficult to give him credit for his real achievements that when his life was over they simply forgot him as fast as possible. He had proclaimed so often in life what a great man he was, that people were content to let the issue rest after his death.

In a funeral eulogy before the French Academy, the naturalist Baron Cuvier summarized the flaws in this remarkable man:

> He considered the Chinese government as the nearest to perfection, because in delivering up the people to the absolute power of men of knowledge alone, and in raising each of these in the hierarchy according to the degree of his knowledge, it made in some measure so many millions of hands the passive organs of the will of a few good heads. An empire such as he conceived would not have been

more difficult for him to manage than his barracks and poorhouses. . . . The world requires a little more freedom and is so constituted that a certain height of perfection often appears to it a defect, when the person does not take as much pains to conceal his knowledge as he has taken to acquire it.

SUGGESTED READING

S. W. Angrist and L. G. Hepler, *Order and Chaos: Laws of Energy and Entropy*, Basic Books, New York, 1967. The best nonmathematical introduction to the meaning of thermodynamics. Extensive discussion of applications of thermodynamics in the everyday life of a non-scientist. Well written and highly recommended.

H. A. Bent, *The Second Law: An Introduction to Classical and Statistical Thermodynamics*, Oxford University Press, New York, 1965. A difficult book to learn thermodynamics from because of its highly unorthodox approach. A delight to read after you have been through the subject once. Sound, well written, with an eye for the unusual sidelight to the topics at hand. Many good problems.

S. C. Brown, *Count Rumford, Physicist Extraordinary*, Anchor, New York, 1962. An entertaining biography of one of the greatest practical geniuses and black-guards in the annals of science.

R. E. Dickerson, *Molecular Thermodynamics*, W. A. Benjamin, Menlo Park, Calif., 1969. Somewhat more advanced than Mahan, Nash, or Waser, but with

a strong emphasis on the statistical interpretation of entropy.

I. M. Klotz and R. M. Rosenberg, *Introduction to Chemical Thermodynamics*, W. A. Benjamin, Menlo Park, Calif., 1972, 2nd ed.

B. H. Mahan, *Elementary Chemical Thermodynamics*, W. A. Benjamin, Menlo Park, Calif., 1963. A good introduction to classical thermodynamics (meaning without the statistical interpretation of entropy). Moderate use of calculus, which is mostly explained as it is introduced.

L. Nash, *ChemThermo: A Statistical Approach to Classical Chemical Thermodynamics*, Addison-Wesley, Reading, Mass., 1972. Similar to Dickerson in being a statistical rather than a heat-engine approach to thermodynamics.

L. Nash, *Introduction to Chemical Thermodynamics*, Addison-Wesley, Reading, Mass., 1963.

J. Waser, *Basic Chemical Thermodynamics*, W. A. Benjamin, Menlo Park, Calif., 1966. Both this and Nash's book are at a similar level to Mahan and this chapter. Highly recommended.

QUESTIONS

1 What is the difference between a spontaneous process and a rapid process?

2 Now that we have a thermodynamic definition of spontaneity, what is the difference between a stable complex and

an inert complex, as discussed in Chapter 11?

3 When we define the molar energy of a monatomic ideal gas as the sum of the kinetic energies of the individual molecules, or as Avogadro's number times the average molecular kinetic energy, as in Chapter 2, we are unconsciously saying that E is a state function. Why is this so?

4 Interpret the skidding automobile example of Section 2–6 in terms of the first and second laws of thermodynamics.

5 Electric current does not flow spontaneously through Faraday's electrolysis cells of Section 3–1; work, in the form of an applied electrical potential, is required to make it flow. Where does this work go in such a cell, and can it ever be recovered?

6 In the freezing point experiments of Section 3–5, why does the freezing point of a liquid *decrease* instead of increase like the boiling point when a nonvolatile solute is added? What did the observation that the freezing point depression in these experiments was greater than would be predicted from the molar concentration of solute indicate about the solute?

7 Why does the first law of thermodynamics permit the addition of heats of reaction along with the reactions themselves, as is done in Section 4–6?

8 Suppose that you take two identical clock springs, leave one slack, wind the other tightly and tie it with catgut, and then dissolve each in a beaker of acid. What happens to the work that you exerted to wind the second spring?

9 In Figure 6–7 we studied the stability of the hydrated Ca^{2+} ion by considering a process in which metallic Ca is sublimed to vapor, the vapor is ionized, and the ions are dissolved and hydrated. Why are the conclusions from this study valid when real Ca^{2+} ions in solution are not formed in this way?

10 What is a thermodynamic system, and what is the difference between a closed and an open system? Is a human being a closed or an open thermodynamic system? Can you think of a way in

which a human being might be converted to the other type of system, and what would be the result?

11 What is "PV work"? What other kinds of work are there? How much PV work can be done at constant pressure? At constant volume?

12 In the expression, $\Delta E = q - w$, why do q and w have opposite signs? What is meant by each symbol?

13 What is the difference in meaning of the two symbols: ΔE and dE?

14 Suppose that you decide to travel from San Francisco to Denver. One function that is obviously not a state function for this process is time. Name three other functions that are not state functions (other than the obvious q and w). Name two other functions that are state functions (other than the obvious ones from this chapter: P, T, V, E, H, S, and G).

15 What would happen to the petroleum industry if work were a state function?

16 Why can we use the standard enthalpies of formation tabulated in Appendix 2 as if they were the actual enthalpies of the compounds formed?

17 Why, when actual free energies of compounds are called for as in Equation 15–17 and the subsequent treatment, can we use instead the free energies of formation of these compounds from the elements?

18 How can you measure the experimental bond energy of C—H bonds in methane?

19 Suppose that you want to calculate an experimental C—O single bond energy in methanol, CH_3OH, but that you have no information other than what is contained in Appendix 2. How would you go about it?

20 For what types of compounds are bond-energy tables useful, and for what types of compounds do they lead to erroneous results?

21 Under what conditions is $q = T\,\Delta S$? Under what conditions is the equality incorrect? Which is then greater, q or $T\,\Delta S$?

22 In the expression $S = k \ln W$, what is W?

23 Why is the entropy of a solid less than that of a gas of the same substance?

24 Does an aqueous solution of Ca^{2+} ions have a larger entropy before or after hydration of the ions? Why, then, are the ions hydrated?

25 What is wrong with this statement: "In a spontaneous chemical process, the system goes to a state of lower energy"? What is the corresponding correct statement?

26 What happens to the free energy of a system in a reversible process when only PV work is done? What happens when other kinds of work, such as electrical work, are done as well?

27 What happens to the free energy of a system in an irreversible process when only PV work is done? When other kinds of work are done also?

28 How does the free energy of a gas depend on its pressure? How does the enthalpy of a gas depend on its pressure?

29 In terms of thermodynamics, what is meant by the activity of a substance? How are the activity and the partial pressure of an ideal gas in a mixture related?

30 What is the relationship between the reaction quotient and the equilibrium constant for a reaction?

31 How does the free energy of a collection of gases vary with the composition of the gas mixture?

32 How can you tell from the numerical value of the reaction quotient whether a given reaction is at equilibrium, will tend to go forward spontaneously, or will go spontaneously in reverse? (Assume that you know the standard free energy of the reaction.)

33 Why is the vapor pressure of a solvent lowered by adding to it a nonvolatile solute?

34 Why are freezing point and boiling point changes useless for determining the molecular weights of proteins?

35 What causes the phenomenon of osmotic pressure?

PROBLEMS

1 When a bottle containing copper shot is shaken, the temperature rises. Explain this in terms of the first law of thermodynamics.

2 What is the enthalpy change, ΔH^0, for the reaction

$$CH_4(g) \rightarrow C(gr) + 2H_2(g)$$

If the reaction is performed at 20 torr and 1500°C, then ΔH is -123 kcal mole^{-1} of methane. What is the energy change (ΔE) for the reaction under these conditions?

3 The thermodynamic system in Problem 2 can be considered to be the mixture of the two gases and the solid graphite. Is work done on the system by its surroundings during the reaction or on the surroundings by the system?

4 Using bond energies and heats of formation of free atoms, estimate the heat of formation of ethanol vapor at 25°C. How does this compare with the measured value?

5 From bond energies (Table 15–1) and atomic heats of formation (Appendix 2), calculate the standard heat of formation of $C_2H_6(g)$, $CH_3SH(g)$, and $HCOOH(g)$. Compare your answers with the measured values in Appendix 2.

6 Estimate the heat of the formation of 1 mole of water vapor from 1 mole of hydrogen gas and $\frac{1}{2}$ mole of oxygen gas by using bond energies from Table 15–1. How does your estimate compare with the measured value in Appendix 2?

7 The heat of combustion of gaseous isoprene, $CH_2{=}CH{-}C(CH_3){=}CH_2$, to $CO_2(g)$ and $H_2O(l)$ is -761.5 kcal mole^{-1}. Calculate the heat of formation and, by comparison with a bond-energy calculation, estimate the resonance energy of isoprene. Can you draw possible resonance structures?

8 Which state in each of the following pairs of states has the higher entropy: (a) A mole of liquid water or a mole of water vapor at 1 atm and 373°K? (b) A mole of Dry Ice or a mole of CO_2 vapor at 1 atm and 195°K? (c) Five dimes on a table-top showing four heads and one tail, or showing three heads and two tails? (d) One mole of liquid H_2O in a beaker and 1 mole of liquid D_2O in another beaker, or both liquids mixed together in the first beaker? (D is an alternative symbol for ^2_1H.)

9 Will the entropy change in each of the following processes be positive or negative? Will the disorder in each process increase or decrease?

a) 1 mole of solid methanol →
                1 mole of gaseous methanol.
b) 1 mole of solid methanol →
                1 mole of liquid methanol.
c) $\frac{1}{2}$ mole of gaseous O_2 + 2 moles of solid Na → 1 mole of solid Na_2O.
d) 1 mole of solid XeO_4 →
                1 mole of gaseous Xe
                + 2 moles of gaseous O_2.

Rank these four processes in order of increasing ΔS.

10 In 1884, Frederick Trouton discovered that for many liquids the heat of vaporization is directly proportional to the normal boiling point, or that the ratio of heat of vaporization to boiling point is a constant:

$$\frac{\Delta H_{vap}}{T_b} = 21 \text{ cal deg}^{-1} \text{ mole}^{-1}$$

We now would explain Trouton's rule by saying that the molar entropy of vaporization of many liquids is approximately the same. But the molar entropy of vaporization of liquid HF is significantly higher, 26 cal deg^{-1} mole^{-1}. Why is this so? (The molar entropy of HF gas is not sufficiently different from that of other gases to account for the difference.)

11 The reaction

$$2H_2(g) + O_2(g) \rightarrow 2H_2O(g)$$

proceeds spontaneously even though there is an increase in order within the system. How can this be?

12 Calculate the standard free energy change, ΔG^0, for the reaction

$$Fe_2O_3(s) + 3C(gr) \rightarrow \\ 2Fe(s) + 3CO(g)$$

Is this reaction spontaneous at 25°C? Calculate the standard enthalpy, ΔH^0, and entropy, ΔS^0, of the reaction at 25°C, and show that $\Delta G^0 = \Delta H^0 - T\Delta S^0$. Do the enthalpy change and entropy change each work for or against spontaneity for the reaction? Which factor predominates?

13 Calculate ΔG^0, ΔH^0, and ΔS^0 at 25°C for the reaction

$$2Ag(s) + Hg_2Cl_2(s) \rightarrow \\ 2AgCl(s) + 2Hg(l)$$

Show that $\Delta G^0 = \Delta H^0 - T\Delta S^0$. Is the reaction endothermic or exothermic? Is the reaction spontaneous or not? Do the enthalpy change and entropy change each work for or against spontaneity for the reaction? Which factor predominates? Explain the entropy effect on physical grounds.

14 What is the standard free energy change at 25°C for the reaction

$$Cl_2(g) + I_2(s) \rightleftarrows 2ICl(g)$$

Will the reaction as written be spontaneous? What is the equilibrium constant, K_{eq}, for this reaction?

15 (a) For the reaction

$$2C(gr) + H_2(g) \rightarrow HC\equiv CH(g)$$

ΔG^0 is 50.0 kcal mole^{-1}. Do you think HC≡CH (acetylene) can be prepared by this reaction? Do you think it could be prepared by any reaction? (b) Calculate the free energy change, ΔG^0, for

$$2CH_4(g) + \tfrac{3}{2}O_2(g) \rightarrow$$
$$HC\equiv CH(g) + 3H_2O(g)$$

Comment on the spontaneity of the reaction.

16 Calculate the change in free energy for the following reaction:

$$C(diamond) \rightarrow C(graphite)$$

Why isn't diamond converted spontaneously into graphite?

17 If 79.7 cal of heat are required to melt 1 g of ice at 0°C, what is the molar heat of fusion of ice at this temperature? What is the entropy change when 1 g of ice melts at 0°C? What is the free energy change for the process?

18 At 25°C, ammonia reacts with HCl to produce ammonium chloride:

$$NH_3(g) + HCl(g) \rightarrow NH_4Cl(s)$$

What is the standard free energy change for the reaction? What is the standard enthalpy change? Use these two quantities to calculate the standard entropy change, and compare your answer with the one obtained directly from the tables in Appendix 2. Give a physical explanation for the sign of the entropy term.

19 One mole of benzene is vaporized at its boiling point under a constant pressure of 1 atm. The heat of vaporization measured in a calorimeter at constant pressure is 7300 cal mole^{-1}. The boiling point of benzene at 1 atm is 80°C. Calculate ΔH^0, ΔG^0, and ΔS^0 for this process. Calculate ΔE^0 for the process; assume that benzene vapor is an ideal gas.

20 Compare your answer to Problem 19 to the standard heat of vaporization obtained from Appendix 2. Why are they different? Can you explain the difference by using Le Chatelier's principle?

21 Step I: A reaction, $A \rightarrow B$, was carried out at 298°K in such a way that no useful work was done. In this process, 10,000 cal mole^{-1} of heat were evolved. Step II: The same reaction was carried out in such a way that the maximum amount of useful work was done. In this process, 400 cal mole^{-1} of heat were evolved. For Steps I and II calculate q, w, ΔE^0, ΔH^0, ΔS^0, and ΔG^0.

22 The change in free energy associated with the reaction

$$CH_3OH(l) + \tfrac{3}{2}O_2(g) \rightarrow$$
$$CO_2(g) + 2H_2O(l)$$

at 298°K is -167.9 kcal mole^{-1}. The free energy of formation for $CO_2(g)$ is -94.3 kcal mole^{-1} and for $H_2O(l)$ it is -56.7 kcal mole^{-1}. Calculate the free energy of formation of $CH_3OH(l)$ from these data. What effect would an increase in temperature have on the spontaneity of the reaction?

23 a) For the evaporation–condensation process

$$H_2O(l) \rightleftarrows H_2O(g)$$

calculate ΔH^0, ΔG^0, and ΔS^0 from the following data at 298°K:

	ΔH_{298}^0	ΔG_{298}^0
$H_2O(l)$	-68.317	-56.690
$H_2O(g)$	-57.798	-54.635

b) Recalling that $\Delta H^0 - T\Delta S^0 = -2.30\, RT \log [a(g)]/[a(l)]$, derive an expression for the vapor pressure of water as a function of temperature.

c) Calculate the equilibrium vapor pressure of water at 50°C and at 100°C. (The quantities ΔH and ΔS are essentially constant in this temperature range.)

24 The vapor pressure of pure benzene (C_6H_6) at 20°C is 75 torr, whereas that for toluene (C_7H_8) is 22 torr. Calculate the vapor pressure of a solution consisting of 10 g of benzene and 10 g of toluene. Calculate the mole fraction of each component in the vapor.

25 Show that there are 54,912 ways of being dealt three cards of a kind in a poker game.

26 What is the standard free energy for the synthesis of ammonia at 25°C? The reaction is

$$3H_2(g) + N_2(g) \rightleftarrows 2NH_3(g)$$

Calculate the equilibrium constant, K_{eq}, and write the expression for K_{eq} in terms of the quantities of reactants and products present at equilibrium.

27 Which of the following substances would you expect to give a 0.1-molal aqueous solution with the lowest freezing point: HNO_3, $NaCl$, glucose, $CuSO_4$, $BaCl_2$?

28 Which will have a greater effect on the colligative properties of an aqueous solution, 20 g of $NaCl$ or 10 g of $MgCl_2$? Assume complete solubility in each case.

29 Calculate the freezing point of a solution made by dissolving 1 g of $NaCl$ in 10 g of water. Repeat your calculation by using 1 g of $CaCl_2$ instead. Which of these salts is more effective as an antifreeze, on a weight basis?

30 If 35.5 g of solid chlorine (Cl_2) are dissolved in 32 g of liquid methane (CH_4) at the boiling point of methane, by how much will the vapor pressure of methane be lowered?

31 A solution is prepared by mixing 20 g of a nonvolatile solute, having a molecular weight of 100 g, with 500 g of solvent, having a molecular weight of 75. The boiling point of the solvent rises from 84.00°C to 85.00°C. Calculate the boiling point elevation constant for this solvent.

32 How many grams of methanol must be added to 10.0 kg of water to lower the freezing point of the solution to 263°K? What is the normal boiling point of this solution?

33 A solution is prepared by dissolving 0.40 g of an unknown hydrocarbon in 25.0 g of acetic acid. The freezing point of the solution falls from 16.60°C for pure acetic acid to 16.15°C. The molal freezing point depression constant for acetic acid is 3.60 deg mole^{-1}. What is the molecular weight of the hydrocarbon? Analysis of this compound shows that it contains 93.75% carbon by weight and 6.25% hydrogen. What is its molecular formula?

34 Benzoic acid is 68.9% carbon, 26.2% oxygen, and 4.96% hydrogen. One gram of the acid in 20 g of water freezes at 272.38°K, whereas 1 g in 20 g of benzene freezes at 277.56°K. What is the apparent molecular formula for benzoic acid in each solvent? Can you explain your results?

35 An important hormone that controls the rate of metabolism in the body, thyroxine, can be isolated from the thyroid gland. If 0.455 g of thyroxine is dissolved in 10.0 g of benzene, the freezing point of the solution is 5.144°C. Pure benzene freezes at 5.444°C. What is the molecular weight of thyroxine?

16 FREE ENERGY AND EQUILIBRIUM

In this chapter we shall provide a theoretical basis for some ideas about spontaneity and equilibrium, which were presented only as observed facts in Chapter 5. When we first encountered equilibrium constants, they were purely experimental numbers, which were convenient for predicting how a chemical reaction would behave after enough time had elapsed. There was no way to calculate them from other measurable properties of the chemical substances. Nor could we prove that the equilibrium constant, K_{eq}, which varies with temperature but is independent of concentrations, *should* exist for a reaction. We tried to rationalize K_{eq} by mass-action arguments, with the warning that the arguments held only for the simplest, one-step reactions. All that we really could say was that from observation this was how nature behaves.

The thermodynamics discussed in the preceding chapter placed K_{eq} on a much firmer footing. We now can prove that a K_{eq} exists for every reaction from the equation

$$\Delta G^0 = -RT \ln K_{eq}$$

or

$$K_{eq} = e^{-(\Delta G^0/RT)}$$

in which ΔG^0 is the free energy of the reaction at a specified temperature when all reactants and products are in their standard states. Since this determines the concentrations (1 molar for solutions, 1 atm partial pressure for gases, etc.), the standard free energy, ΔG^0, is independent of the actual concentrations in any real experiment. Therefore, K_{eq} also is independent of concentrations. However, since the standard free energy is defined for a given temperature, and will have a different numerical value at different temperatures, the equilibrium constant also will be a function of temperature.

In this chapter we will look more closely at the relationships between ΔG, K_{eq}, and T. In the process we shall review the most important properties of equilibrium and the equilibrium constant, which were introduced empirically in Chapter 5; but now we will be able to give more satisfactory explanations of these properties.

16-1   THE PROPERTIES OF EQUILIBRIUM

Several fundamental features of equilibrium should be kept in mind as you study this chapter.

1) Equilibrium is a dynamic rather than a static process. Consider the reaction for the dissociation of phosphorus pentachloride vapor that starts with pure PCl_5:

> Dissociation $\rightarrow$
>
> $PCl_5(g) \rightleftarrows PCl_3(g) + Cl_2(g)$ $\qquad\qquad\qquad$ (16-1)
> $\leftarrow$ Recombination

If the reactants and products are confined in a closed vessel, eventually no more chlorine gas will appear to form. When the reaction is monitored by measuring the total pressure with a manometer, we discover that the pressure increases more slowly with time and eventually reaches a constant, maximum value. Dissociation of PCl_5 has not stopped; rather, dissociation and recombination are occurring at equal rates so there is no *net* change. You can prove this by performing an experiment at equilibrium conditions, using chlorine gas composed of the pure ^{37}Cl isotope instead of the natural 3/1 mixture of $^{35}Cl/^{37}Cl$. Even though equilibrium is maintained and no reaction seems to be occurring, you soon will find that the ^{37}Cl isotope is present both in PCl_5 and in PCl_3. Some of the "labeled" Cl_2 has combined with PCl_3 that has the normal Cl isotope ratio to make ^{37}Cl-rich PCl_5; then some of this PCl_5 has dissociated to produce PCl_3 that contains ^{37}Cl. The reversible nature of equilibrium is symbolized in chemical equations by using double horizontal arrows pointing in both directions, or by using an "equals" sign rather than a single arrow to the right.

2) There is a spontaneous tendency toward equilibrium. This statement implies nothing about the *rate* at which a reaction may reach equilibrium; it deals only with the *drive toward* equilibrium. This chemical drive is

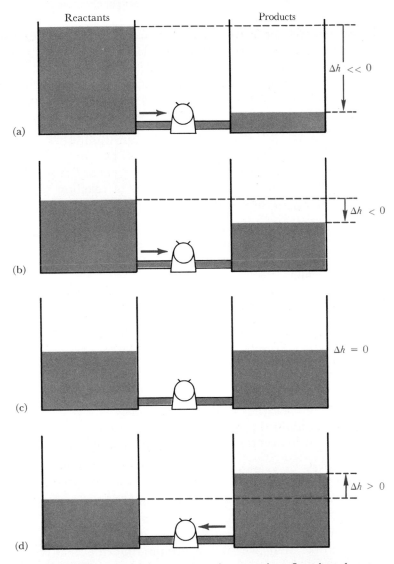

Reactants   Products

$\Delta h \ll 0$

(a)

$\Delta h < 0$

(b)

$\Delta h = 0$

(c)

$\Delta h > 0$

(d)

Figure 16–1. Hydrostatic analogy to the free energy of a reaction. Imagine that two water reservoirs are connected by a pipe that leads through a turbine for generating electricity. The amount of work that each cubic meter of water can do as it passes through the turbine depends on the hydrostatic head, or the distance from reactant water level to product water level, Δh. (a) At the start, a maximum amount of work is obtained per unit of water because Δh is large and negative. But every unit of water run through the turbine decreases the difference in water level and decreases the amount of work that the next unit of water will do. (b) At this stage in the process, each cubic meter of water yields less than a third as much work as at the start. (c) When the water levels in both reservoirs are equal, no more work can be obtained. The system is at equilibrium, and $\Delta h = 0$. (d) This last state will never be attained from the first state spontaneously, because Δh can never spontaneously increase in the positive direction. If the experiment is begun as in (d), water will flow spontaneously in the reverse direction.

measured as the free energy of the reaction. The *free energy change* represents the amount of energy that is available either to do work, or to serve as the driving force in a chemical reaction. A hydrostatic analogy to free energy of reaction is shown in Figure 16–1, in which the drop in water level between reservoir and outflow tank plays the role of the free energy change, ΔG.

3) The driving force toward equilibrium diminishes as equilibrium is approached. Thus the appearance of products decreases the forward impetus of the reaction, and the free energy change, ΔG, becomes less negative. Each successive increment of reaction gives rise to a smaller free energy change and hence can do less work.

4) Equilibrium is reached when ΔG becomes zero. The hydrostatic analogy of chemical equilibrium is shown in Figure 16–1(c). Chemical equilibrium is a balance between two effects, heat and entropy or disorder. At equilibrium

$$\Delta G = 0 = \Delta H - T \, \Delta S$$
$$\Delta H = T \, \Delta S$$

Reaction in one direction is favored if it releases heat and leads to a lower enthalpy. Reaction in the other direction is favored if it leads to less ordered substances with higher entropy. At equilibrium these two effects balance each other.

5) The equilibrium position is the same at constant temperature, no matter from which direction it is approached. Thus in the dissociation of N_2O_4,

$$N_2O_4(g) \rightleftarrows 2NO_2(g) \tag{16–2}$$

the *relative* concentrations of N_2O_4 and NO_2 are identical at equilibrium (assuming constant pressure and temperature) whether the initial material was almost pure N_2O_4 or almost pure NO_2. It is important to remember that we are assuming that the relative mole numbers of reactants and products are chosen properly from the stoichiometry of the equilibrium equation. For example, in the dissociation of PCl_5 (Equation 16–1) the equilibrium point will be the same if you begin with pure PCl_5 or with *equimolar amounts* of PCl_3 and Cl_2, since PCl_3 and Cl_2 combine in equimolar amounts. However, if you begin with twice the number of moles of PCl_3 as Cl_2, the equilibrium will differ from the equilibrium that will be achieved by beginning with pure PCl_5.

6) Even without taking the proper stoichiometry into account, equilibrium for a reaction always is characterized by an equilibrium constant, K_{eq}, expressed in terms of the activities, a, of reactants and products:

$$K_{eq} = \frac{a_{PCl_3} a_{Cl_2}}{a_{PCl_5}} \quad \text{(for Reaction 16–1)} \tag{16–3}$$

Any amounts of PCl_5, PCl_3, and Cl_2 that satisfy the equilibrium-constant expression will lead to an equilibrium situation. For nonideal systems, the activity of a substance is proportional to its concentration and sometimes is called its "effective" concentration.

7) The equilibrium constant for a chemical reaction can be calculated from the standard free energy change of reaction, ΔG^0:

$$\Delta G^0 = -RT \ln K_{eq} \qquad \text{or} \qquad K_{eq} = e^{-(\Delta G^0/RT)} \qquad (16\text{–}4)$$

Since, in the standard free energy change for reaction, ΔG^0, we assume that the partial pressures of all components are 1 atm, the standard free energy change is not a function of pressure. Any pressure variations in reactants or products are expressed by the logarithmic term in the general free energy equation

$$\Delta G = \Delta G^0 + RT \ln Q \qquad (16\text{–}5)$$

Because ΔG^0 is independent of pressure, the equilibrium constant, K_{eq}, also is independent of pressure. No matter how much we compress a mixture of PCl_5, PCl_3, and Cl_2 gases, the *ratio* of their activities at equilibrium (Equation 16–3) will not change. When the gases are compressed, the equilibrium conditions will shift so that the equilibrium constant is the *same* after the shift as before.

     The standard free energy change *does* vary with temperature. In all of our free energy calculations, we have assumed that the temperature is 298°K and often have omitted the subscript 298 from the term ΔG^0_{298}. But the free energy of a gas reaction with all gases at 1 atm partial pressure, for example, generally will not be the same at 1000°K as at 298°K. Hence, the equilibrium constant, K_{eq}, also will vary with temperature. In what follows we shall see some examples of this behavior.

Stoichiometry and the Equilibrium Constant

The expression for the equilibrium constant depends on how the equation for the reaction is written. For example, the reaction for the production of ammonia can be written in terms of producing one mole of ammonia:

1) $\quad \dfrac{1}{2} N_2(g) + \dfrac{3}{2} H_2(g) \rightleftarrows NH_3(g) \qquad K_1 = \dfrac{a_{NH_3}}{a_{N_2}^{1/2} a_{H_2}^{3/2}}$

or of using one mole of nitrogen or one mole of hydrogen:

2) $\quad N_2(g) + 3H_2(g) \rightleftarrows 2NH_3(g) \qquad K_2 = \dfrac{a_{NH_3}^{2}}{a_{N_2} a_{H_2}^{3}}$

3) $\quad \dfrac{1}{3} N_2(g) + H_2(g) \rightleftarrows \dfrac{2}{3} NH_3(g) \qquad K_3 = \dfrac{a_{NH_3}^{2/3}}{a_{N_2}^{1/3} a_{H_2}}$

In each case the exponent of the activity, a, in the equilibrium-constant expression is the coefficient of that substance in the chemical equation. Reaction 2 is twice Reaction 1 and three times Reaction 3; thus equilibrium constant K_2 is the square of K_1 and the cube of K_3. In general, multiplication of the reaction equation by any number, n, raises the corresponding equilibrium constant to the nth power.

The reverse of Reaction 2 is

$$4)\quad 2NH_3(g) \rightleftarrows N_2(g) + 3H_2(g) \qquad \text{and} \qquad K_4 = \frac{a_{N_2}a_{H_2}^3}{a_{NH_3}^2} = \frac{1}{K_2}$$

Thus the equilibrium constant for the reverse reaction is the inverse of the equilibrium constant for the forward reaction.

If we add two reactions to give a third reaction, the equilibrium constant of the third reaction is the *product* of the equilibrium constants of the first two:

$$1)\quad C(s) + O_2(g) \qquad\qquad \rightleftarrows CO_2(g) \qquad\qquad K_1 = \frac{a_{CO_2}}{a_C a_{O_2}}$$

$$2)\quad H_2(g) + CO_2(g) \qquad\qquad \rightleftarrows H_2O(g) + CO(g) \qquad K_2 = \frac{a_{H_2O}a_{CO}}{a_{H_2}a_{CO_2}}$$

$$3)\quad H_2(g) + C(s) + O_2(g) \rightleftarrows H_2O(g) + CO(g) \qquad K_3 = \frac{a_{H_2O}a_{CO}}{a_{H_2}a_C a_{O_2}}$$

$$= K_1 K_2$$

If $(1) + (2) = (3)$, then $K_1 K_2 = K_3$.

Standard States and Activities

We can extend the concept of activity, and therefore of free energy changes and equilibrium constants, to solids, liquids, and components of solutions, if we define the activity of any substance as the ratio of the concentration of that substance to its concentration in a chosen standard state. When calculating equilibrium constants, this standard state obviously must be the same as the standard state for which the thermodynamic data are tabulated if we are to calculate K_{eq} from such data. The standard states used for calculating the free energy values in Appendix 2 are listed in Table 16–1.

For gas reactions the activities of reactants and products are unitless quantities that are numerically equal to the partial pressures in atmospheres. It is common to use equilibrium constants calculated from the partial pressures of the reactants and products. For example,

$$K_p = \frac{p_{PCl_3}p_{Cl_2}}{p_{PCl_5}}$$

We shall use this terminology occasionally since it is so common. However,

Table 16–1.   Standard States and Activities in Free Energy Tabulations and Equilibrium Constant Calculations

Substance	Standard state	Activity, a
Gas, pure or in a mixture	1 atm partial pressure of gas in question	Numerically equal to the partial pressure: $a_j = |p_j|$
Pure liquid or solid	Pure liquid or solid	Unit activity: $a_j = 1$
Solvent in a dilute solution	Pure solvent	Activity nearly equal to 1: $a_j \simeq 1$
Solute	1 molar solution of the solute	Numerically equal to the molar concentration of solute: $a_j = |c_j|$

you should remember that, in the strict sense, these "partial pressures" are really the activities of the individual gases, and are the ratios of partial pressures under experimental conditions to the partial pressures in the standard state of 1 atm. When activities are used the equilibrium constant, K_{eq}, is a unitless number, as it must be if $\ln K_{eq}$ is to have any meaning.

A pure solid or liquid component in a reaction acts like an infinite reservoir of material, and the amount of solid or liquid does not affect the equilibrium as long as some solid or liquid is present. In the decomposition of limestone to lime and carbon dioxide,

$$CaCO_3(s) \rightleftarrows CaO(s) + CO_2(g)$$

the free energy of the reaction depends only on the partial pressure of the carbon dioxide above the solids, and not on the amount of limestone or lime that is present. The choice of the standard state for solids as the pure solid itself makes the activities unity and eliminates them from the equilibrium constant and reaction quotient:

$$\Delta G = \Delta G^0 + RT \ln \left(\frac{1 \cdot a_{CO_2}}{1} \right) = \Delta G^0 + RT \ln a_{CO_2}$$

$$K_{eq} = e^{-(\Delta G^0 / RT)} = a_{CO_2}$$

At equilibrium, the partial pressure of CO_2 in a container with CaO and $CaCO_3$ is constant for a given temperature, exactly as the equations predict. In all of our applications a solvent in a *dilute* solution can be considered an inexhaustible source of pure solvent material. Thus the activity of a solvent also is unity and does not enter into the equilibrium constant.

Challenge. To see how equilibrium is involved in storing energy in the body, try Problem 16–6 in Butler and Grosser.

16–2   REACTIONS INVOLVING GASES

Let us apply the previous conclusions to reactions in which all reactants and products are gases. We shall do this in a series of examples, each of which will illustrate a new idea.

Experimental Measurement of Equilibrium Constants

Ammonia, nitrogen, and hydrogen are at equilibrium in a steel tank at 298°K. An analysis of the contents of the tank shows the following partial pressures of the three gases:

$$p_{N_2} = 0.080 \text{ atm}$$
$$p_{H_2} = 0.050 \text{ atm}$$
$$p_{NH_3} = 2.60 \text{ atm}$$

What is the equilibrium constant, K_{eq}, for the reaction

$$N_2(g) + 3H_2(g) \rightleftarrows 2NH_3(g) \tag{16–6}$$

The equilibrium constant is calculated to be

$$K_{eq} = \frac{a_{NH_3}^2}{a_{N_2} a_{H_2}^3} = \frac{(2.60)^2}{(0.080)(0.050)^3} = 6.8 \times 10^5$$

Exercise.   The preparation of sulfur trioxide from sulfur dioxide is an important step in making sulfuric acid. A mixture of SO_2 and O_2 gases is passed slowly through a tube containing a platinum catalyst heated to 1000°K. The outflow gases are analyzed and the following partial pressures are found:

$$p_{SO_2} = 0.559 \text{ atm}$$
$$p_{O_2} = 0.101 \text{ atm}$$
$$p_{SO_3} = 0.331 \text{ atm}$$

What is the equilibrium constant, K_{eq}, for the reaction

$$2SO_2(g) + O_2(g) \rightleftarrows 2SO_3(g) \tag{16–7}$$

(*Answer:* $K_{eq} = 3.47$.)

Calculation of Equilibrium Constants

Now we can calculate equilibrium constants from thermodynamic data, which we could not do in Chapter 5. Given either the equilibrium constant or the standard free energy of a reaction, we can calculate the other quantity.

Example.   Calculate the equilibrium constant for Reaction 16–6 from standard free energies of formation.

Solution

$$\Delta G^0 = 2(-3.976) - (0.0) - 3(0.0) = -7.952 \, \text{kcal}$$
$$K_{eq} = e^{-(\Delta G^0/RT)} = 10^{-(-7.952/1.364)} = 10^{+5.830} = 6.8 \times 10^5$$

You should not attach any importance to the fact that this is the same value that we obtained previously from partial pressures. Equilibrium constants rarely are known to better than 5% accuracy.

Example. Calculate the dissociation constant for acetic acid from the free energy tables in Appendix 2.

Solution. The reaction is

$$CH_3COOH(aq) \rightarrow H^+(aq) + CH_3COO^-(aq)$$

and the free energy of dissociation is

$$\Delta G^0 = 0.0 + (-89.02) - (-95.51) = +6.49 \, \text{kcal}$$

The dissociation constant is found from

$$K_a = e^{-(\Delta G^0/RT)} = 10^{-(6.49/1.364)} = 10^{-4.75}$$
$$pK_a = 4.75 \quad \text{and} \quad K_a = 1.76 \times 10^{-5}$$

You can check this result against the table of dissociation constants in Chapter 5. Notice that if the dissociation reaction had been written

$$CH_3COOH(aq) + H_2O \rightarrow CH_3COO^-(aq) + H_3O^+(aq)$$

the result using data in Appendix 2 would have been the same.

Exercise. From the equilibrium constant that you calculated previously for the sulfur trioxide reaction, find the standard free energy of Reaction 16-7 at 1000°K.

(*Answer:* $\Delta G^0_{1000} = -2.47$ kcal, or -1.24 kcal mole^{-1} of SO_3.)

The Partial Pressure of One Component

We can calculate the partial pressure of one component of a system at equilibrium if we know K_{eq} and the partial pressures of the other components of the system.

Example. In another experiment with the ammonia reaction in a steel tank, the partial pressures of ammonia and hydrogen at 298°K were found to be

$$p_{NH_3} = 1.53 \, \text{atm}$$
$$p_{H_2} = 0.50 \, \text{atm}$$

No nitrogen could be detected. What must the partial pressure of N_2 have been if the contents of the tank were at equilibrium?

Solution.   Let y be the unknown partial pressure of nitrogen. Then

$$K_{eq} = \frac{(1.53)^2}{y(0.50)^3} = 6.7 \times 10^5$$

$$y = \frac{(1.53)^2}{0.125 \times 6.7 \times 10^5} = 2.8 \times 10^{-5} \, \text{atm}$$

Because we have calculated the value of the equilibrium constant for this reaction from other experiments with more nearly equal amounts of reactants and products, it is possible to calculate how much nitrogen is present even if we cannot measure it.

Exercise.   In the SO_3 equilibrium at 1000°K (Equation 16–7), what would the partial pressure of oxygen gas have to be to have equal amounts of SO_2 and SO_3?

(*Answer:* $p_{O_2} = 0.288$ atm.)

Alteration of Stoichiometry

What is the equilibrium constant for the reaction

$$2NH_3(g) \rightleftharpoons N_2(g) + 3H_2(g) \tag{16–8}$$

This reaction is the reverse of Reaction 16–6, so the equilibrium constant is the reciprocal of the forward reaction:

$$K_{eq} = \frac{a_{N_2} a_{H_2}^3}{a_{NH_3}^2} = \frac{1}{6.8 \times 10^5} = 1.5 \times 10^{-6}$$

Exercise.   What is the equilibrium constant for the reaction

$$NH_3(g) = \tfrac{1}{2}N_2(g) + \tfrac{3}{2}H_2(g) \tag{16–9}$$

(*Answer:* $K_{eq} = 1.2 \times 10^{-3}$.)

Extent of Reaction

Often it is useful to be able to calculate the extent of reaction at equilibrium, expressed as the fraction or percentage of the pure starting material that has reacted. For example, we can express the progress of the decomposition of

Table 16–2.   *Dissociation of Ammonia at* $298°K$

	$2NH_3(g) \rightleftarrows N_2(g) + 3H_2(g)$			Total
Start:	$2n$	0	0	$2n$
Equilibrium:	$2n(1-\alpha)$	$n\alpha$	$3n\alpha$	$2n(1+\alpha)$
Mole fraction:	$\dfrac{(1-\alpha)}{1+\alpha}$	$\dfrac{\alpha}{2(1+\alpha)}$	$\dfrac{3\alpha}{2(1+\alpha)}$	1
Partial pressure:	$\dfrac{(1-\alpha)P}{1+\alpha}$	$\dfrac{\alpha P}{2(1+\alpha)}$	$\dfrac{3\alpha P}{2(1+\alpha)}$	P

$$K_{eq} = \frac{a_{N_2}a_{H_2}^3}{a_{NH_3}^2} = \frac{\alpha P}{2(1+\alpha)} \cdot \frac{27\alpha^3 P^3}{8(1+\alpha)^3} \cdot \frac{(1+\alpha)^2}{(1-\alpha)^2 P^2} = \frac{27}{16}\frac{\alpha^4 P^2}{(1-\alpha^2)^2}$$

$$\Delta G^0 = (0.0) + 3(0.0) - 2(-3.976) = +7.952 \text{ kcal}$$

$$K_{eq} = 10^{-(7.952/1.364)} = 10^{-5.830} = 1.5 \times 10^{-6}$$

$$K_{eq} = \frac{27}{16}\frac{\alpha^4 P^2}{(1-\alpha^2)^2} = 1.5 \times 10^{-6}$$

$$\frac{\alpha^2 P}{1-\alpha^2} = 0.94 \times 10^{-3}$$

ammonia into nitrogen and hydrogen in terms of a *percent dissociation* of pure ammonia. Pure ammonia at a total pressure of 1 atm will dissociate spontaneously (to a small extent) to H_2 and N_2. If the pressure is kept constant at 1 atm and the temperature is 298°K, what fraction of the ammonia is dissociated?

The reaction is shown in Equation 16–8 and at the top of Table 16–2. Since the equilibrium-constant expression is in terms of partial pressures, our first task is to express concentrations in partial pressures also. We then can set up the equilibrium-constant expression, use the known value of the equilibrium constant, and solve for the percent dissociation.

It is helpful to make a table such as Table 16–2. Assume that pure NH_3 was present at the beginning of the experiment and that no N_2 or H_2 was present. This description need not correspond to any real experimental situation; it is merely a framework for calculating concentrations at equilibrium. Since two moles of ammonia are involved in the reaction as written, assume that $2n$ moles of ammonia are present at the start. Now assume that the system comes to equilibrium at some later time when a fraction, α, of the ammonia has dissociated. Then, of the original $2n$ moles of ammonia, $2n\alpha$ moles will have reacted and $2n(1-\alpha)$ moles will be left.

For every two moles of ammonia that decompose, one mole of nitrogen and three moles of hydrogen are formed. At equilibrium, the $2n\alpha$ moles of dissociating ammonia produce $n\alpha$ moles of nitrogen and $3n\alpha$ moles of hydrogen, as shown on the line of Table 16–2 marked "Equilibrium." The total number of moles of the three types of molecules has increased as a result of the dissociation. There now are $2n(1 - \alpha) + n\alpha + 3n\alpha = 2n(1 + \alpha)$ moles of gas, whereas prior to dissociation there were only $2n$ moles. The total number of moles is recorded at the right of the table.

The mole fraction of each component in the reaction mixture is found by dividing the number of moles of each component by the total number of moles. At this stage, the advantage of using *fractional* dissociation, α, is obvious, because the number of moles, n, drops out of the problem. In the third line of the table, headed "Mole fraction," the mole fraction of each component at equilibrium is given in terms of the degree of dissociation of ammonia.

In the next line of the table, each mole fraction is multiplied by the total pressure to yield the partial pressure of that component. These partial pressures then are substituted into the equilibrium-constant expression, assuming the activities of the three gases to be numerically equal to their partial pressures. The result, after simplifying as much as possible, is an expression for the equilibrium constant in terms of the degree of dissociation:

$$K_{eq} = \frac{27\alpha^4 P^2}{16(1 - \alpha^2)^2}$$

From the free energy change for the reaction, $+7.952$ kcal, we can use Equation 16–4 to calculate the numerical value of the equilibrium constant to be

$$K_{eq} = 1.5 \times 10^{-6}$$

Using this value, we have an expression,

$$\frac{27\alpha^4 P^2}{16(1 - \alpha^2)^2} = 1.5 \times 10^{-6}$$

which after combining numerical constants and taking the square root of both sides of the equation becomes

$$\frac{\alpha^2 P}{1 - \alpha^2} = 0.94 \times 10^{-3}$$

This expression can be solved exactly for α^2:

$$\alpha^2 = \frac{0.00094}{P + 0.00094}$$

Since $P = 1$ atm, we can neglect the 0.00094 term in the denominator and solve for $\alpha = 0.0307$. The answer to our original question is that at 298°K and 1 atm pressure, ammonia is approximately 3% dissociated.

Exercise. A molecule of N_2O_4 gas dissociates spontaneously into two molecules of NO_2. Derive an expression for the degree of dissociation, α, as a

Table 16–3.   Dissociation of N_2O_4 at 298°K

	$N_2O_4(g)$	$\rightleftarrows$	$2NO_2(g)$	Total
Start:	n		0	n
Equilibrium:	$n(1 - \alpha)$		$2n\alpha$	$n(1 + \alpha)$
Mole fraction:	$\dfrac{1 - \alpha}{1 + \alpha}$		$\dfrac{2\alpha}{1 + \alpha}$	1
Partial pressure:	$\dfrac{1 - \alpha}{1 + \alpha} \cdot P$		$\dfrac{2\alpha}{1 + \alpha} \cdot P$	P

$$K_{eq} = \frac{a_{NO_2}^2}{a_{N_2O_4}} = \frac{4\alpha^2 P^2}{(1 + \alpha)^2} \cdot \frac{(1 + \alpha)}{(1 - \alpha)P} = \frac{4\alpha^2 P}{1 - \alpha^2}$$

$$\Delta G^0 = 2(12.390 \text{ kcal}) - (23.491 \text{ kcal}) = +1.289 \text{ kcal}$$

$$K_{eq} = e^{-(1.289/RT)} = 10^{-(1.289/1.364)} = 10^{-0.945} = 0.114$$

$$\frac{4\alpha^2 P}{1 - \alpha^2} = 0.114 \qquad \text{or} \qquad \frac{\alpha^2 P}{1 - \alpha^2} = 0.0285$$

function of the total pressure on the system at 298°K. What fraction of the N_2O_4 will have dissociated at 1 atm pressure?

(*Answer*: The derivation is shown in Table 16–3, to which you can refer if you have trouble. The expression for K_{eq} in the table again can be solved exactly for α^2:

$$\alpha^2 = \frac{K_{eq}}{4P + K_{eq}} = \frac{1}{(4P/K_{eq}) + 1} \tag{16–10}$$

and at $P = 1$ atm and $K_{eq} = 0.114$, $\alpha = 0.167$. At 1 atm and 298°K, one N_2O_4 molecule in six is dissociated. Note that in this problem, K_{eq} was large enough that it could not be neglected in the denominator in comparison with $4P$ when $P = 1$ atm.)

What will be the effect on the dissociation of N_2O_4 when we increase the pressure? Since reassociation decreases the total number of moles, Le Chatelier's principle predicts that an increase in pressure will favor reassociation. In Figure 16–2(a) the degree of dissociation, α, obtained from Equation 16–10, is plotted against the total pressure. You can see that Le Chatelier's principle is borne out. Above 1000 atm pressure the gas mixture is nearly all N_2O_4; below 0.001 atm, it is almost entirely NO_2. In Figure 16–2(a), the pressure was plotted as a logarithmic function, that is, in increasing powers of 10. The change in α with changing pressure is displayed clearly. In contrast, in Figure 16–2(b), which has pressure plotted on a linear scale, the curve flattens over half the α range on the vertical axis. For any α above 0.5, Figure 16–2(b) gives no clear description of what is happening. Any other choice of a linear plotting range for pressure (0 to 150 atm, 0 to 1.5 atm, or 0 to 0.15 atm) would have similar drawbacks. We shall find that logarithmic plots are used widely in equilibrium problems.

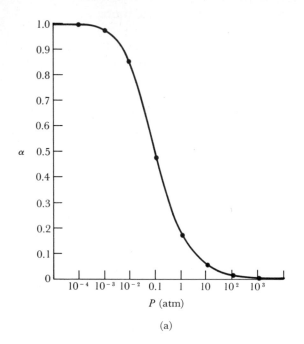

(a)

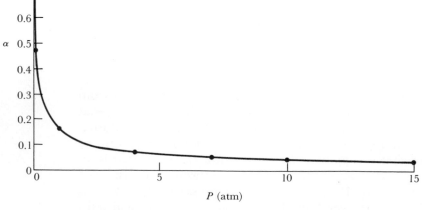

Figure 16–2. Plots of the degree of dissociation of N_2O_4, α, as a function of pressure at 298°K. The function plotted is Equation 16–10. (a) Plot of α against $\log_{10} P$ showing the full detail across the entire range of α. (b) A less informative plot of α against P that obscures the behavior of the curve above $\alpha = 0.5$.

(b)

These two examples, the ammonia reaction and the dissociation of N_2O_4, produced the same mathematical expression for the degree of dissociation. With ammonia, there was 3% dissociation; with N_2O_4, there was 16.7% dissociation. These relative values could have been predicted from the standard free energy changes for the reactions: $+3.976$ kcal mole^{-1} for NH_3 and $+1.289$ kcal mole^{-1} for N_2O_4. The standard free energy change for dissociation is higher for NH_3 than for N_2O_4; thus the drive toward dissociation is less and the equilibrium point is closer to the side of no dissociation.

Before you go on. Equilibrium calculations for homogeneous reactions are treated in terms of concentrations in moles per liter and of partial pressures in Section 5–1 of *Programed Reviews of Chemical Principles* by Lassila *et al.*

16–3 LE CHATELIER'S PRINCIPLE

Le Chatelier's principle states that, when a stress is applied to a system at equilibrium, the equilibrium conditions shift in such a way as to relieve the stress. In the two dissociation reactions we have examined so far (those of N_2O_4 and NH_3) the pressure dependence has appeared in the numerator of the equilibrium-constant expression. Since K_{eq} is independent of pressure, α must decrease as P increases if the quantity on the right side of the equilibrium-constant expression is to remain unchanged. Le Chatelier's principle predicts that at higher pressures the equilibrium conditions will shift in the direction that produces the smaller number of moles of gas, which is in agreement with the equilibrium-constant derivations.

What would this principle predict about equilibrium at higher pressures for the water-gas reaction,

$$H_2(g) + CO_2(g) \rightleftarrows H_2O(g) + CO(g) \tag{16–11}$$

If you set up a table such as Table 16–2 for this reaction, and let α be the fraction of H_2 that has reacted, you will find that the equilibrium constant is

$$K_{eq} = \frac{a_{H_2O}a_{CO}}{a_{H_2}a_{CO_2}} = \frac{\alpha^2}{(1-\alpha)^2} \tag{16–12}$$

and that pressure does not appear in the equilibrium-constant expression. This is because the same number of moles of gas are found on each side of the equation, and all of the pressure terms cancel. For the same reason, Le Chatelier's principle predicts that the water-gas equilibrium will be insensitive to changes in pressure.

The Effect of Temperature

What does Le Chatelier's principle predict about the effect of temperature on the equilibrium constant? The temperature of a reacting system is raised by adding heat. The stress of the added heat can be relieved if the equilibrium

Table 16–4. Variation of K_{eq} with Temperature for the Reaction
$$2SO_3(g) \rightleftarrows 2SO_2(g) + O_2(g)$$

Temperature (°K)	ΔG^0 (kcal per 2 moles SO_3)	K_{eq}	
298	33.46	2.82	$\times 10^{-25}$
400	28.85	1.78	$\times 10^{-16}$
500	24.25	2.51	$\times 10^{-11}$
600	19.63	1.94	$\times 10^{-8}$
700	15.20	1.82	$\times 10^{-5}$
800	10.60	1.29	$\times 10^{-3}$
900	6.63	0.0248	
1000	2.65	0.264	
1100	−1.40	1.89	
1200	−5.50	10.0	
1300	−9.60	40.8	
1400	−13.63	132	

$$K_{eq} = \frac{a^2_{SO_2} a_{O_2}}{a^2_{SO_3}}$$

conditions shift in the direction that absorbs heat. If a reaction is endothermic, its equilibrium constant will increase with temperature; if it is exothermic, its equilibrium constant will decrease.

The SO_3 dissociation,

$$2SO_3(g) \rightleftarrows 2SO_2(g) + O_2(g)$$

is highly endothermic. At 298°K,

$$\Delta G^0 = 2(-71.79) - 2(-88.52) = +33.46 \text{ kcal per 2 moles of } SO_3$$
$$\Delta H^0 = 2(-70.76) - 2(-94.45) = +47.38 \text{ kcal per 2 moles of } SO_3$$
$$\Delta S^0 = 2(59.40) + (49.00) - 2(61.24) = +45.32 \text{ cal deg}^{-1} \text{ per}$$
$$\text{2 moles of } SO_3 \text{ dissociated}$$

Since so much heat is absorbed in the dissociation, this dissociation is strongly favored by higher temperatures. Thus, a difference in 1100°K results in a change in K_{eq} of 27 orders of magnitude (Table 16–4).

16–4 THE ANATOMY OF A REACTION

Le Chatelier's principle gives an indication of how a reaction will proceed; however, it does not explain why, except in the most intuitive way. Why does the position of equilibrium change with temperature? Why does the driving impetus of the SO_3 dissociation reaction increase so sharply with temperature? To answer these questions we must look at the behavior of the free energy, enthalpy, and entropy of reaction as the temperature changes.

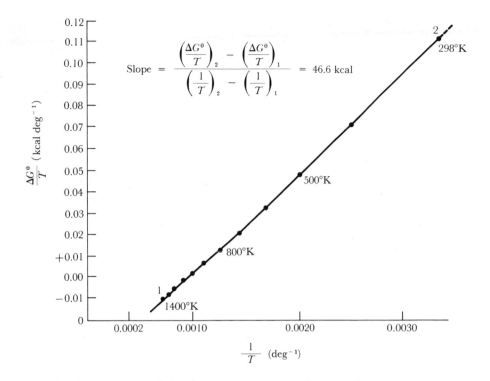

Figure 16–3. A Gibbs–Helmholtz plot for the dissociation of SO_3 to SO_2 and O_2. The slope at any point of the curve (which here appears to be nearly a straight line) gives the heat of the reaction at that point. The enthalpy of the SO_3 dissociation does not vary significantly with temperature from 298°K to 1400°K. The ΔG^0 data are from Table 16–4.

At a given temperature the standard free energy, enthalpy, and entropy of reaction are related by the expression

$$\Delta G^0 = \Delta H^0 - T\,\Delta S^0 \tag{16–13}$$

The standard free energy changes for the SO_3 dissociation reaction at various temperatures, as calculated from experimental dissociation constants, are listed in Table 16–4. As the temperature is raised, the standard free energy change for the reaction becomes more negative, the equilibrium constant becomes larger, and the reaction proceeds farther to the right before equilibrium is attained. The heats and entropies of reaction are hidden in these data as well. From Equation 16–13 and the use of calculus, we can derive a relationship known as the Gibbs–Helmholtz equation. This equation need not concern us except for its prediction that, if we plot $\Delta G^0/T$ against $1/T$, the slope of the curve at any point is ΔH^0 at that temperature.

A plot for the data in Table 16–4 appears in Figure 16–3. The curve is nearly a straight line, which indicates that the standard heat of the SO_3

dissociation reaction does not change very much between 298°K and 1400°K. The average slope over this entire temperature range yields an average enthalpy of reaction of +46.6 kcal, whereas the measured value at one extreme end of the range, at 298°K, is +47.38 kcal. To a close approximation, we can consider the heat of the reaction to be constant for all temperatures.

The heat of reaction, ΔH^0, and the free energy change, ΔG^0, both are plotted against temperature in Figure 16–4. The difference between them, $\Delta H^0 - \Delta G^0$, is $T \Delta S^0$. If both ΔH^0 and ΔG^0 are approximated by straight lines, then $T \Delta S^0$ is proportional to T; thus ΔS^0 also is approximately independent of temperature. If we extrapolate the ΔH^0 and ΔG^0 lines to absolute zero, we see that they intersect. At 0°K, $T \Delta S^0 = 0$, and $\Delta H^0 = \Delta G^0$.

Now we can explain what happens in this reaction as the temperature changes. When two molecules of SO_3 dissociate,

$$2SO_3(g) \rightleftarrows 2SO_2(g) + O_2(g) \tag{16-14}$$

two S—O bonds are broken and one O—O bond is made. The enthalpy required to accomplish this, 47 kcal for two moles of SO_3, is so large that the minor fluctuations arising with temperature can be ignored, and ΔH^0 is approximately constant. This enthalpy factor by itself would keep all of the SO_3 undissociated. Berthelot and Thomsen would have been satisfied with this conclusion (Section 15–8).

However, there is a second factor to consider, entropy. Two molecules of SO_2 and one molecule of O_2 are more disordered than two molecules of SO_3. The relative molar entropies are exactly what you might predict for the three molecules based on the complexity of each molecule. From Appendix 2,

Molecule:	O_2	SO_2	SO_3
Molar entropy, S^0: (cal deg^{-1} mole^{-1})	49.00	59.40	61.24

However, the entropies per mole are numerically so similar for the three kinds of molecules that the deciding factor is the change in the number of molecules during the reaction. Two molecules of reactant (SO_3) yield three molecules of product ($2SO_2$, O_2), and the overall entropy change is

$$\Delta S^0 = 2(59.40) + (49.00) - 2(61.24) = +45.32 \text{ cal deg}^{-1}$$

Entropy, like enthalpy, is not very sensitive to temperature. The disorder produced per unit of reaction at 298°K is approximately the same as that produced at 1400°K. Nevertheless, the effect of this entropy change is more marked at high temperatures because the entropy term is multiplied by absolute temperature, T. The higher the temperature, the more of an effect a given increase in disorder has on the reaction. The impetus toward reaction, or the standard free energy, is the combination of the heat effect and the entropy effect. As you can see from Figure 16–4, enthalpy continues to oppose

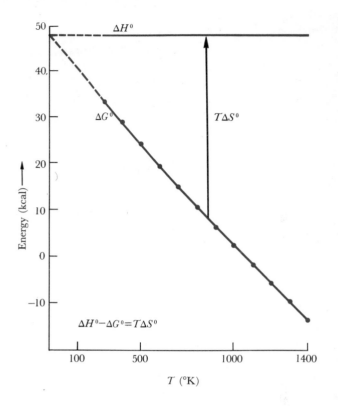

Figure 16–4. Enthalpy, entropy, and free energy for the SO_3 *dissociation as a function of temperature. In this reaction, enthalpy and entropy are approximately independent of temperature. The free energy decreases as temperature increases because of the T factor in* $\Delta G^0 = \Delta H^0 - T \Delta S^0$.

the dissociation of SO_3 to approximately the same extent at all temperatures. But entropy provides a steadily increasing driving force in favor of dissociation. Thus, at 298°K, the dissociation constant is only 2.82×10^{-25}, whereas at 1400°K it is 132.

Exercise. What would be the value of the dissociation constant, K_{eq}, if the reactants and products in the SO_3 dissociation were of equal entropy? This also would be the dissociation constant at absolute zero if the dashed-line extrapolations in Figure 16–4 were valid. (They nearly are.)

(*Answer:* $K_{eq} = 6 \times 10^{-35}$.)

Thus far our conclusions have been derived from a special case, the dissociation of SO_3. Yet it generally is true that at low temperatures the enthalpy or heat is more important in determining which way a chemical reaction will go, and at high temperatures the entropy or disorder is more important.

16–5 EQUILIBRIA WITH CONDENSED PHASES PRESENT

In Section 16–1 we stated that the equilibrium constant, K_{eq}, could be used with solids and liquids as well as with gases, if the standard states of the condensed phases were defined properly. A pure solid or pure liquid is considered to be an infinite reservoir of material as long as some of the solid or liquid is present. The equilibrium is not affected by the amount of solid or liquid.

Why can we say this? For example, doesn't the vapor pressure of water above the liquid water confuse the issue? In the reaction

$$H_2(g) + CO_2(g) \rightleftarrows H_2O(l) + CO(g) \tag{16-15}$$

shouldn't we use the vapor pressure of water for a_{H_2O} rather than set $a_{H_2O} = 1$?

In fact, we *are* using the vapor pressure of water for a_{H_2O}, although this is disguised somewhat. For the reaction between gases only:

$$H_2(g) + CO_2(g) \rightleftarrows H_2O(g) + CO(g)$$
$$\Delta G_1^0 = \Delta G_{H_2O(g)}^0 + \Delta G_{CO(g)}^0 - \Delta G_{H_2(g)}^0 - \Delta G_{CO_2(g)}^0 \tag{16-16}$$
$$K_1 = e^{-(\Delta G_1^0/RT)} = \frac{a_{H_2O(g)} a_{CO(g)}}{a_{H_2(g)} a_{CO_2(g)}}$$

If the water vapor is in equilibrium with liquid water, then the value of $a_{H_2O(g)}$ is restricted to the vapor pressure of water at that temperature. As we saw in Section 15–10, the vapor pressure is related to the molar free energy of vaporization, ΔG_v^0:

$$H_2O(l) \rightleftarrows H_2O(g)$$
$$\Delta G_v^0 = \Delta G_{H_2O(g)}^0 - \Delta G_{H_2O(l)}^0 = -RT \ln p_{H_2O(g)}$$

The activity of water vapor then is numerically equal to its vapor pressure in atmospheres:

$$a_{H_2O(g)} = p_{H_2O(g)} = e^{-(\Delta G_v^0/RT)}$$

If liquid water is present so that $a_{H_2O(g)}$ is fixed, we can bring this activity to the other side of the equation and incorporate it into the equilibrium constant:

$$K_2 = \frac{K_1}{a_{H_2O(g)}} = \frac{a_{CO(g)}}{a_{H_2(g)} a_{CO_2(g)}} = e^{-(\Delta G_1^0/RT)} e^{+(\Delta G_v^0/RT)} = e^{-[(\Delta G_1^0 - \Delta G_v^0)/RT]}$$

The free energies in the exponent, from which K_2 is calculated, are

$$\Delta G_2^0 = \Delta G_1^0 - \Delta G_v^0$$

If you calculated these values, you would find that ΔG_2^0 is simply the standard free energy of Reaction 16–15, with *liquid* water instead of water vapor. Thus, we can think of what we have done as incorporating the activity of H_2O into the equilibrium constant because that activity is fixed by an equilibrium between vapor and a condensed phase. We also could think of it as calculating the equilibrium constant from data on free energies of formation by using the *condensed* phase rather than the gas phase and by setting the activity of the condensed phase equal to one. The two procedures give identical results. In the language of mathematicians, we would say that every new equation relating the variables of the problem (such as an equilibrium equation between liquid and vapor of one component) reduces the number of independent variables by one.

Example. Will the dissociation equilibrium for H_2S gas,

$$H_2S \rightleftarrows H_2 + S$$

be affected by pressure at 1000°K? At 298°K?

Solution. At 1000°K, both products are gases. Two moles of gas are produced for every mole of H_2S dissociated. (In reality, this figure is less than two, depending on how much S_8 is present in the vapor, but there are always more moles of products than reactants.) Le Chatelier's principle predicts that an increase in pressure at 1000°K will inhibit dissociation of H_2S. In contrast, at 298°K most of the sulfur is solid. The vapor pressure above the solid sulfur will not be influenced by how much H_2S has dissociated, because as more H_2S dissociates, more sulfur vapor condenses on the solid. Similarly, as the reaction goes to the left, more solid sulfur sublimes, hence the vapor pressure of sulfur remains fixed. At 298°K the number of moles of gas is the same before and after dissociation of H_2S, so Le Chatelier's principle predicts that changes in pressure have no effect on the equilibrium.

16–6 SUMMARY

We now know how to deal with equilibria involving pure gases and mixtures of gases, and pure solids and liquids. We can calculate equilibrium constants either from experimental data on concentrations at equilibrium or from thermodynamic data on standard free energies of formation. We can express the fraction of one component that has reacted, α, in terms of the equilibrium constant, and we can calculate how much of a reactant will have reacted under given conditions of temperature and total pressure.

Le Chatelier's principle summarizes the behavior of equilibrium systems. It states that whenever stress is applied to a system at equilibrium, the equilibrium conditions shift in such a way as to relieve the stress. A reaction that absorbs heat (an endothermic reaction) will be favored by an increase in temperature. A reaction that leads to a decrease in the total number of moles of gas will be favored by an increase in total pressure. Any reaction will be hindered by the addition of more of its products and favored by their removal.

The driving force in any chemical reaction, ΔG, is a combination of the tendency of a chemical system to go to a state of lowest enthalpy, and simultaneously to proceed to a state of least order or greatest entropy. At low temperatures the enthalpy term (ΔH) in the expression

$$\Delta G = \Delta H - T \Delta S$$

dominates, and at high temperatures the entropy term ($T \Delta S$) becomes more important. In the SO_3 example, enthalpy worked against the dissociation, and entropy promoted it. Therefore, the tendency toward dissociation increased with temperature. In other reactions, the opposite may be true. That is, if the products have lower entropy than the reactants, higher temperatures will favor reactants rather than products.

SUGGESTED READING

A. J. Bard, *Chemical Equilibrium*, Harper and Row, New York, 1966.

R. E. Dickerson, *Molecular Thermodynamics*, W. A. Benjamin, Menlo Park, Calif., 1969. Chapter 5, "Thermodynamics of Phase Changes and Chemical Reactions," has a more detailed treatment of thermodynamics and equilibrium.

W. Kauzmann, *Thermodynamics and Statistics, with Applications to Gases*, W. A. Benjamin, Menlo Park, Calif., 1967. Chapter 5 has a good treatment of thermodynamics and equilibrium constants.

I. M. Klotz and R. M. Rosenberg, *Introduction to Chemical Thermodynamics*, W. A. Benjamin, Menlo Park, Calif., 1972, 2nd ed.

B. H. Mahan, *Elementary Chemical Thermodynamics*, W. A. Benjamin, Menlo Park, Calif., 1964. Chapter 3, "The Second Law," has a simpler treatment of thermodynamics and equilibrium than Dickerson, but no applications to real problems.

L. Nash, *ChemThermo: A Statistical Approach to Classical Chemical Thermodynamics*, Addison-Wesley, Reading, Mass., 1972.

M. J. Sienko, *Chemistry Problems*, W. A. Benjamin, Menlo Park, Calif., 1972, 2nd ed. Contains several hundred problems on equilibrium, with answers and often with explanations of the method of solutions. Uses constant-volume problems with concentrations in moles per liter rather than constant-pressure problems with concentrations in partial pressures, even for gas reactions. Emphasis is on practical problem solving, rather than on thermodynamic interpretation of equilibrium.

QUESTIONS

1 How is the equilibrium constant for a reaction related to thermodynamic quantities? How does this relationship prove that the equilibrium constant is independent of concentrations of reactants and products?

2 What can you say about the standard free energy change for a reaction if the reaction is spontaneous when its reactants and products are in their standard states? What can you say about the equilibrium constant for the above reaction? (*Hint:* What is the reaction quotient, Q, for reactants and products in standard states?)

3 If the standard free energy change is negative, is a reaction spontaneous under all conditions?

4 If the heat that can be obtained by burning gasoline is more or less constant and independent of burning conditions, why does the useful work that can be obtained per gram of fuel decrease steadily as equilibrium is approached? (Look again at the hydrostatic analogue in Figure 16–1.) Why is this of little concern in an automobile engine?

5 In Chapter 5, we used concentrations of reactants and products in partial pressures (gases) or moles per liter (solutes). In this chapter, we use activities instead. What is the relationship between the two usages, and why are activities useful?

6 How does the use of activities in equilibrium expressions make the treatment of condensed phases simpler?

7 Why are Tables 16–2, and 16–3, which help in setting up the equilibrium-constant expressions, more complex than tables with a similar goal in Section 5–3? Does this give you a clue as to why we moved as rapidly as possible from gases to ionic solutions in Chapter 5?

8 If hydrogen and iodine react at 300°C, the free energy change for the reaction depends on the concentrations (partial pressures) of H_2, I_2, and HI. In contrast, if the reaction is run at room temperature, only the concentrations of H_2 and HI have any effect on the free energy change for the reaction. Why is this true?

9 What effect will an increase in pressure have on the PCl_5 reaction of Equation 16–1?

10 What effect will an increase in temperature have on the PCl_5 reaction in Equation 16–1?

11 What do you think would happen to the equilibrium of Equation 16–1 if you introduced (a) nitrogen gas at constant pressure, (b) nitrogen gas at constant volume, or (c) chlorine gas at constant pressure?

12 How do bond breaking and creation of disorder influence the equilibrium conditions for a chemical reaction? If bond energies were the only important factor, what would be the equilibrium constant for the dissociation of hydrogen gas molecules into atoms? If entropy were the only operating factor, what would be the equilibrium constant for the hydrogen dissociation? Using the above answers, and the relationship between G, H, and S, explain why dissociation of hydrogen gas should be more pronounced at higher temperatures.

13 How can the vapor pressure of a liquid be calculated from the thermodynamic data in Appendix 2? What equilibrium is involved, and how do our conventions about condensed phases assist in such calculations?

PROBLEMS

1 A system consisting of nitrogen, hydrogen, and ammonia—all gases—is allowed to come to equilibrium at a total pressure of 5 atm. The partial pressures of the three gases were measured to be $p_{N_2} = 1$ atm, $p_{H_2} = 2$ atm, and $p_{NH_3} = 2$ atm. What is K_{eq} for the reaction

$$N_2(g) + 3H_2(g) \rightleftarrows 2NH_3(g)$$

What is K_{eq} for the reaction

$$NH_3(g) \rightleftarrows \tfrac{1}{2}N_2(g) + 1\tfrac{1}{2}H_2(g)$$

2 Steam reacts with iron at 500°C to produce hydrogen gas and Fe_3O_4. Write the equilibrium constant for the reaction in terms of quantities (activities) of reactants and products present.

3 In an experiment at an elevated temperature, a mixture of SO_2, O_2, and SO_3 is allowed to equilibrate by the reaction

$$SO_2(g) + \tfrac{1}{2}O_2(g) \rightleftarrows SO_3(g)$$

The partial pressures are then measured as $p_{SO_2} = 40$ torr, $p_{O_2} = 20$ torr, and $p_{SO_3} = 800$ torr. Evaluate the equilibrium constant, K_{eq}, for the reaction as written.

4 For the reaction of nitric acid with hydrogen sulfide at 25°C,

$$2H^+(aq) + 2NO_3^-(aq) + 3H_2S(aq) \rightleftarrows$$
$$2NO(g) + 4H_2O(l) + 3S(s)$$

K_{eq} is 1×10^{81}. Given that the dissociation constant of H_2S is 1×10^{-19} and the solubility product of CdS is 8×10^{-27}, calculate the equilibrium constant of the reaction

$$3CdS(s) + 8H^+(aq) + 2NO_3^-(aq) \rightleftarrows$$
$$2NO(g) + 4H_2O(l)$$
$$+ 3Cd^{2+}(aq) + 3S(s)$$

Do you think CdS will dissolve in aqueous nitric acid? Calculate ΔG for the reaction.

5 Phosphonium chloride, PH_4Cl, which is unstable even at low temperatures, dissociates into PH_3 and HCl. If one mole of PH_4Cl partially decomposes in a 5-liter flask, determine the equilibrium constant for the reaction

$$PH_4Cl(s) \rightleftarrows PH_3(g) + HCl(g)$$

if the partial pressure of PH_3 is 5.0 atm.

6 The density of ice at 273°K is 0.917 g cm^{-3}, and the density of liquid water is 1.00 g cm^{-3} at the same temperature. What will be the effect of a pressure increase on the freezing point of water?

7 Is the reaction

$$2NO(g) + O_2(g) \rightleftarrows 2NO_2(g)$$

endothermic or exothermic when run under standard conditions? If the temperature increases, will K_{eq} increase, decrease, or remain unchanged? If the pressure is increased, will K_{eq} increase, decrease, or remain unchanged? If a catalyst is added, will K_{eq} increase, decrease, or remain unchanged?

8 At 25°C and 20 atm, the reaction

$$N_2(g) + 3H_2(g) \rightleftarrows 2NH_3(g)$$

has a ΔH of -22.1 kcal. If the temperature is raised to 300°C while the pressure is held at 20 atm, will more or less ammonia be present at equilibrium? If the pressure is increased to 30 atm while the temperature remains at 25°C, will more or less ammonia be present compared with the initial conditions? If half the ammonia is removed and the system allowed to come to equilibrium again, will the amount of nitrogen gas present increase or decrease? What will be the effect on the original equilibrium mixture if a catalyst for ammonia synthesis is added?

9 When 1 mole of gaseous HI is sealed in a 1-liter flask at 225°C, it decomposes to form 0.182 mole each of hydrogen and iodine:

$$2HI(g) \rightleftarrows H_2(g) + I_2(g)$$

What is the value of the equilibrium constant K_{eq} at 225°C? Calculate the standard free energy change, ΔG^0, for the reaction at 225°C. From the data in Appendix 2, calculate the standard free energy change and the equilibrium constant at 25°C. Why the difference in values between 25°C and 225°C?

10 At 298°K, what is the standard enthalpy change of the reaction

$$2NO_2(g) \rightleftarrows N_2O_4(g)$$

Calculate the equilibrium constant for the reaction. How is the extent of conversion of NO_2 to N_2O_4 affected by increasing the temperature? By increasing the pressure? By increasing the volume at constant temperature? Solid Na_2O reacts with NO_2 to produce $NaNO_3$, but will not react with N_2O_4. What is the effect on the extent of conversion of NO_2 to N_2O_4 if Na_2O is added? What is the effect on K_{eq}?

11 At a certain temperature, T, and a total pressure of 2.00 atm, phosphorus pentachloride is 75.0% dissociated:

$$PCl_5(g) \rightleftarrows PCl_3(g) + Cl_2(g)$$

What is K_{eq} at this temperature? What is K_{eq} at 298°K? Do you think that temperature T is greater or less than 298°K? On what evidence do you base your decision?

12 From data in Appendix 2, calculate the standard free energy of the reaction

$$CH_3COOH(aq) \rightleftarrows$$
$$H^+(aq) + CH_3COO^-(aq)$$

Use this value to calculate K_{eq} for the reaction. Compare your value with that in Table 5–4.

13 Use the data in Appendix 2 to calculate the dissociation constant for formic acid in aqueous solution. Compare your result with the value in Table 5–4. What is the pH of a 0.075-molar solution of formic acid?

14 Metacresol (which we shall represent by HCre) is a weak organic acid with $K_{eq} = 1.0 \times 10^{-10}$. Write the equilibrium-constant expression for this process. What is the concentration of the Cre^- ion in a 1.00-molar solution of metacresol in water? What is the pH of this solution? What is the standard free energy change of the metacresol dissociation reaction in aqueous solution?

15 Hydrazine is a weak base that dissociates in water according to the equation

$$N_2H_4 + H_2O \rightleftarrows N_2H_5^+ + OH^-$$

The equilibrium constant for this dissociation at 25°C is 2.0×10^{-6} and the standard free energy of undissociated hydrazine in aqueous solution is 30.56 kcal mole^{-1}. Recalling that the hydrated H^+ ion is assigned (by convention) a standard free energy of 0.00 kcal mole^{-1}, calculate the standard free energy of the hydrazinium ion.

16 The solubility of potassium chloride in water is 347 g liter^{-1} at 20°C, and 802 g liter^{-1} at 100°C. Calculate K_{sp} for KCl at each temperature. Using a Gibbs–Helmholtz plot such as Figure 16–3, calculate the heat of solution of KCl. Is the dissolving process exothermic or endothermic?

17 At 25°C, ΔH^0 is 22.13 kcal for the reaction

$$PCl_5(g) \rightleftarrows PCl_3(g) + Cl_2(g)$$

An equilibrium mixture of $PCl_5(g)$, $PCl_3(g)$, and $Cl_2(g)$ is subjected to various operations (a–e). For each operation determine whether the position of equilibrium shifts to the right, shifts to the left, or remains the same. Also determine whether the equilibrium constant increases, decreases, or remains the same.

a) A catalyst is added.

b) The volume of the container is decreased.

c) $Cl_2(g)$ is added.

d) The temperature is increased.

e) An inert ideal gas is added to the container.

18 Derive an expression for the degree of dissociation, α, of $PCl_5(g)$ as a function of the total pressure of the system. What fraction of $PCl_5(g)$ will have dissociated at 1 atm and 25°C?

19 If 0.50 g of N_2O_4 are allowed to evaporate into a 2-liter flask at 25°C, the reaction $N_2O_4(g) \rightleftarrows 2NO_2(g)$ takes place. Calculate the partial pressure exerted by N_2O_4. K_{eq} for the reaction at 25°C is 0.114.

17 OXIDATION–REDUCTION EQUILIBRIA AND ELECTROCHEMISTRY

The current market prices for ancient Greek coins are not dictated entirely by the value of the gold, silver, or copper in them, or by the relative scarcity of the individual issues. A gold stater of Croesus or a silver teradrachm of Alexander, 2500 years after they were struck, can be miniature art masterpieces in a physical condition that belies their age. In contrast, a typical bronze or copper coin of the same period will be corroded, pitted, and ugly, especially if it has been near moisture or damp earth for a long time. Why does copper corrode, whereas silver and gold do not? The answer lies in the different affinities of the three metals for their electrons.

Under certain circumstances, a person with both gold inlays and silver amalgam fillings in his teeth is in trouble. Every time he bites and brings the two metals into contact, he receives a shock, which makes eating unpleasant. Why the shock? Again, it is a matter of the different affinities that substances have for electrons, and a flow of electrons from where they are not particularly wanted to where they are wanted.

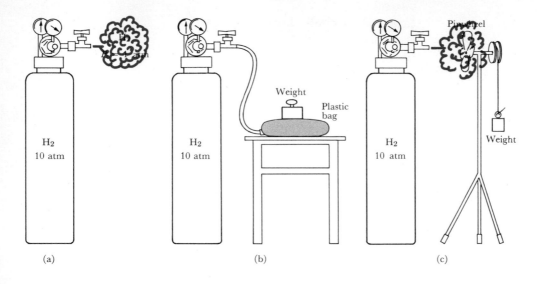

Figure 17–1. Nonproductive and productive use of the free energy stored in hydrogen gas at 10 atm pressure. (a) Expansion to 1 atm with loss of energy. (b) Use of part of the available free energy to lift a weight by inflating a plastic bag. (c) Use of part of the free energy to lift a weight by turning a pinwheel or windmill.

In this chapter we will discuss ways in which chemical reactions can be divided into two physically separated parts; one that gives up electrons easily, and one that accepts them easily. If we can catch these electrons as they flow "downhill" (to use a gravitational analogy), we may be able to use this flow to do outside work. This is the principle of the *galvanic* cell. Moreover, if we can find ways to push the electrons "uphill" from regions where they are wanted to regions where they are not, then we either can store the energy that this requires, for later use, or we can make chemical reactions take place that normally are not spontaneous. This is the principle of *electrolytic* cells.

As in the preceding chapter, we will be interested in the concepts of spontaneity and equilibrium. Only this time, we will have added a new dimension to spontaneity, that of electron flow.

17–1 HARNESSING SPONTANEOUS REACTIONS

If we open a tank of hydrogen gas at ten atmospheres pressure and allow the gas to escape, it will do so spontaneously. No useful work is obtained in this experiment. However, if we connect the outlet tube to a plastic bag or a piston we can lift weights, or if we place a windmill in the path of the outrushing gas we can convert the stored energy of the gas into mechanical work (Figure 17–1). The overall reaction is

$$H_2(g, 10 \text{ atm}) \rightarrow H_2(g, 1 \text{ atm})$$

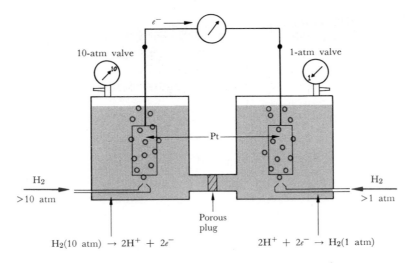

$$H_2(10 \text{ atm}) \rightarrow 2H^+ + 2e^- \qquad 2H^+ + 2e^- \rightarrow H_2(1 \text{ atm})$$

Figure 17–2. A hydrogen pressure cell for converting the free energy of expansion of H_2 at 10 atm into useful work. Two platinum electrodes are immersed in pure water (hydrogen ion concentration of 10^{-7} mole liter^{-1}) in two tanks that are connected by a channel with a porous plug, which permits ion flow but enables a pressure difference to be maintained. Hydrogen gas is bubbled over each electrode, and bleed-off valves and regulators maintain the H_2 pressure at 10 atm in the left tank and 1 atm in the right. In operation, the following reactions occur spontaneously:

Left tank: $H_2(g, 10 \text{ atm}) \rightarrow 2H^+ + 2e^-$
Right tank: $\underline{2H^+ + 2e^- \rightarrow H_2(g, 1 \text{ atm})}$

Overall: $H_2(g, 10 \text{ atm}) \rightarrow H_2(g, 1 \text{ atm})$

Electrons produced in the left-tank reaction flow through the external circuit to the right electrode, where they are used to react with hydrogen ions. (The dial with the arrow in the external electrical circuit symbolizes any current-measuring or work-producing device.) Hydroxide ions diffuse slowly through the porous plug from right to left to maintain electrical neutrality, and combine with the protons produced in the left tank.

and from Chapter 15, the maximum free energy obtainable per mole of H_2 is

$$\Delta G = \Delta G^0 + RT \ln (p_2/p_1)$$
$$\Delta G = 0 + RT \ln (1/10) = -RT \ln (10) = -1.364 \text{ kcal mole}^{-1}$$

(The standard free energy change for the reaction $H_2(g) \rightarrow H_2(g)$ obviously is zero.) This free energy decrease of 1.364 kcal mole^{-1} represents the *maximum* work obtainable when the process is carried out reversibly, which is an impractical upper limit that would take forever. For any real, spontaneous, finite-time process, somewhat less than 1.364 kcal mole^{-1} of useful work will be available.

There is another way in which we can harness the free energy of the escaping gas, the so-called pressure cell (Figure 17–2). Although it may seem to be a somewhat artificial device, it will lead us directly to other kinds of chemical cells. The pressure cell shown in Figure 17–2 carries out the same reaction in two steps. In the left tank, hydrogen gas at 10 atm is "taken apart" into protons and electrons; in the right tank, hydrogen at 1 atm is "reassembled" from electrons and protons. However, the overall reaction still is

$$H_2(g, 10 \text{ atm}) \rightarrow H_2(g, 1 \text{ atm})$$

and the overall free energy yield likewise must be unchanged.

Since an excess of electrons is produced at the left electrode, and reaction at the right electrode cannot proceed without electrons, a flow of electrons from left to right occurs if the two terminals are connected with a wire. (This is analogous to opening the valve on a tank of compressed gas: high-pressure gas is converted to low-pressure gas without doing any useful work.) This tendency for electrons to flow, or this electron "pressure," is measured by a voltage difference between the two terminals. If we want to prevent electrons from flowing, we must supply an opposing voltage equal to the voltage developed by the pressure cell. Alternatively, we can use this electron "pressure" to carry out some useful task (the pressure-cell analogue of lifting weights with a windmill). In terms of the driving force for doing work, the pressure of compressed hydrogen gas has been converted into a voltage difference between an electron-rich terminal and an electron-poor terminal. For this particular cell the voltage is quite small: 0.0296 volt (V) or 29.6 millivolts (mV). Thus it is not a very useful type of cell.

The pressure cell separates the overall reaction into two parts: oxidation and reduction. The H_2 molecule is oxidized to H^+ in the left tank, with a loss of electrons, and H^+ is reduced to H_2 in the right tank, with a gain of electrons. The two terminals or electrodes are named according to whether electrons flow into them or out of them as observed from outside the system. Electrons flow away from the right electrode into the solution[1] during reduction; therefore this electrode is called the *cathode*. (*Cata-* means "away from," as in "catapult.") Conversely, during oxidation electrons flow from the solution into the left electrode, which thus is called the *anode*. (*Ana-* means "back." For those of us for whom Greek is not a second tongue, it is easier to remember that "anode" and "oxidation" both begin with a vowel, and "cathode" and "reduction" begin with a consonant.)

Concentration Cells

The pressure cell is a kind of concentration cell; it can generate an external flow of electrons because the concentrations of H_2 gas in the two electrode

---

[1] In the sense that either a negative ion is formed from a neutral substance, or a positive ion combines with the electrons and is neutralized.

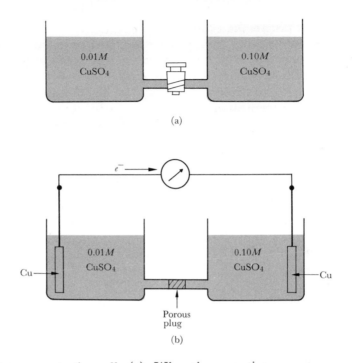

Figure 17–3. Copper sulfate concentration cell. (a) When the connecting stopcock is opened, the concentrated and dilute solutions mix without yielding any useful work. (b) Concentration cell for converting the free energy of mixing into useful work. Two copper electrodes are immersed in the 0.01-molar and 0.10-molar $CuSO_4$ solutions, which are connected by a porous plug that permits slow ion flow without bulk mixing of the solutions. The following reactions occur spontaneously:

Left compartment:　$Cu(s) \rightarrow Cu^{2+}(0.01M) + 2e^-$
Right compartment:　$Cu^{2+}(0.10M) + 2e^- \rightarrow Cu(s)$
Overall reaction:　$Cu^{2+}(0.10M) \rightarrow Cu^{2+}(0.01M)$

Electrons produced in the left compartment flow through the external circuit to the right, where they react with copper ions. The solution in the left compartment gradually becomes more concentrated, and that in the right compartment, more dilute. When the concentrations become equal, no further electron flow occurs.

compartments are different. We can make a similar concentration cell using Cu and $CuSO_4$. If two solutions of different concentrations of copper sulfate are placed in contact, they will mix spontaneously [Figure 17–3(a)]. We can harness this spontaneous reaction by setting up a cell such as that shown in Figure 17–3(b). In the left compartment with a dilute solution, the copper electrode will be eroded slowly as copper is oxidized to form more Cu^{2+} ions. Hence, the left electrode is the anode, and it accumulates an excess of electrons. In the right compartment, with a high Cu^{2+} concentration, some of these copper ions will be reduced and Cu will deposit on the copper cathode. If the two electrodes are connected, electrons will flow from left to right, and sulfate ions will diffuse through the porous plug to maintain electrical neutrality. The result is that the dilute left compartment becomes more concentrated in $CuSO_4$, and the concentrated right compartment becomes more dilute, just as in a free-mixing process. When the two compartments reach equal concentration, electron flow stops.

The overall reaction

$$Cu^{2+}(0.10M) \rightarrow Cu^{2+}(0.01M)$$

has a free energy change that can be calculated from the expression

$$\Delta G = \Delta G^0 + RT \ln (c_2/c_1)$$

in which c_2 and c_1 are final and initial concentrations in moles liter^{-1}. For this example,

$$\Delta G = 0 + RT \ln (0.01/0.10) = -1.364 \text{ kcal mole}^{-1}$$

The free energy change is the same as for the H_2 pressure cell because the ratio of concentrations is the same. As the reaction proceeds and the concentrations become more nearly equal, the free energy per mole of reaction decreases. (A given mass of water dropped over a two-foot high dam can do less work than the same mass of water dropped over a twenty-foot dam. See Figure 16–1.) The initial voltage or potential difference between electrodes is 29.6 mV as in the H_2 cell, and gradually falls to zero as the Cu^{2+} concentrations in the two compartments become equal and the cell runs down.

17–2 ELECTROCHEMICAL CELLS

The two reactions in an electrochemical cell need not be the reverse of one another to produce a useful cell. All that is required is two substances with widely different tendencies to gain or lose electrons. This difference in affinity for electrons then can be harnessed to accomplish useful work.

Zinc and Copper: the Daniell (Gravity) Cell

If a piece of slightly impure zinc is placed in a solution of copper sulfate, it slowly will be pitted and eroded away. At the same time, copper will be deposited on the zinc surface as a spongy brown coating, and the characteristic blue color of the copper sulfate solution gradually will fade. The zinc spontaneously replaces the copper ions in solution by the reaction

$$Zn(s) + Cu^{2+} \rightarrow Zn^{2+} + Cu(s)$$

because copper ions in solution have a greater affinity for electrons than do zinc ions.

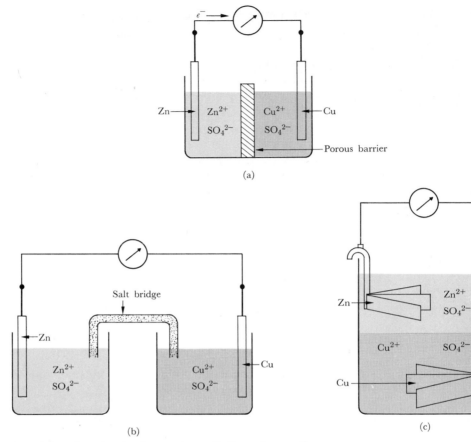

Figure 17–4. Three versions of a simple zinc–copper cell. In each case the oxidation reaction at the anode at the left is

$$Zn(s) \rightarrow Zn^{2+} + 2e^-$$

and the reduction at the cathode at the right is

$$Cu^{2+} + 2e^- \rightarrow Cu(s)$$

The zinc anode is eaten away as copper deposits on the cathode. If $ZnSO_4$ *and* $CuSO_4$ *are present in 1-molar concentrations, this cell develops a potential or emf of* $+1.10$ V. *(a) Two solutions separated by a porous barrier. (b) Solutions separated by a salt bridge. (c) Solutions separated by gravity in a Daniell cell, taking advantage of different solution densities.*

Useful work can be obtained if we separate these two substances in a simple cell, as shown in Figure 17–4(a). Zinc is oxidized spontaneously at the anode (left), and copper ions are reduced to the metal, which deposits on the cathode. Electrons flow through the external circuit from anode to cathode with a potential difference of 1.10 V if the solutions each are one molar. Anions diffuse left through the porous barrier to maintain electrical neutrality.

With a little reflection it should be apparent that neither the $ZnSO_4$ nor the copper rod is essential. Copper metal will deposit at the cathode on any other good conductor, such as a platinum wire, and zinc sulfate solution, in the anode compartment, can be replaced by any other conducting salt that does not react with the zinc anode, such as sodium chloride. The porous barrier has a relatively high resistance to ion diffusion, and hence sets up a relatively high electrical resistance, which cuts down the current that can be drawn from the cell. A better method is to use a salt bridge, which is a glass U-tube containing an electrolyte such as KNO_3 mixed with agar or gelatin to hold it in place [Figure 17–4(b)].

The best setup for a cell that will not be moved is to let gravity separate the solutions, with no internal barrier at all [Figure 17–4(c)]. In this cell a dilute solution of zinc sulfate is layered carefully over a concentrated, more dense copper sulfate solution. In the absence of motion or vibration the cell works quite well. Its internal resistance of almost zero makes it possible for large currents to be drawn from it. This Daniell cell once was used widely as a stationary power source in telegraph offices and for home appliances such as doorbells.

The Hydrogen Electrode

Other combinations of metals can be used in a cell similar to that shown in Figure 17–4(b). If the metals are nickel and copper, nickel is oxidized at the anode, Cu^{2+} ions are reduced at the cathode, and the cell has a voltage or electromotive force (emf) of 0.57 V. If zinc and nickel are used, zinc is oxidized and Ni^{2+} ions are reduced, with an emf of 0.53 V (providing that the ions are at one-molar concentration). Notice that the cell emf's are additive in the same way that the reactions are:

$$
\begin{array}{ll}
Ni + Cu^{2+} \rightarrow Ni^{2+} + Cu & \mathscr{E}^0 = +0.57 \text{ V} \\
\underline{Zn + Ni^{2+} \rightarrow Zn^{2+} + Ni} & \underline{\mathscr{E}^0 = +0.53 \text{ V}} \\
Zn + Cu^{2+} \rightarrow Zn^{2+} + Cu & \mathscr{E}^0 = +1.10 \text{ V}
\end{array}
$$

The sum of these two cell reactions is the Zn–Cu reaction of the Daniell cell, and the sum of the two cell potentials gives the potential of the Daniell cell. (We will draw more conclusions from these observations later.) The symbol $\mathscr{E}^0$, with the zero superscript, indicates a *standard* potential with all reacting ions at one-molar[2] concentrations and a temperature of 298°K. The positive sign of the emf means that the cell equation as written is spontaneous from left to right.

---

[2] Because solutions do not behave ideally when concentrated, the *activity*, or "effective concentration" really should be used instead of molarity. Standard potentials actually are defined at unit activity, rather than unit molarity. But because our goal is an understanding of basic principles rather than laboratory technique, we will ignore the difference between activity and molarity in this chapter.

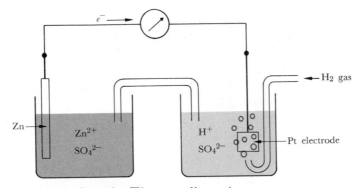

Figure 17–5. Cell with a hydrogen electrode. The two cell reactions are

Anode: $Zn(s) \rightarrow Zn^{2+} + 2e^-$
Cathode: $2H^+ + 2e^- \rightarrow H_2(g)$
_____
Overall: $Zn(s) + 2H^+ \rightarrow Zn^{2+} + H_2(g)$

The same overall reaction could be obtained less productively by immersing a strip of zinc in sulfuric acid. The metal would be eaten away (as at the anode in this cell) and bubbles of hydrogen gas would be given off (as at the cathode).

An electrode reaction need not involve a metal: metals are merely particularly easy substances to shape and machine. Figure 17–5 shows a cell in which the cathode reaction is the liberation of hydrogen gas:

$$2H^+ + 2e^- \rightarrow H_2(g)$$

For a cell with the overall reaction

$$Zn(s) + 2H^+ \rightarrow Zn^{2+} + H_2(g)$$

the standard emf is $+0.76$ V. If the Zn anode is replaced by Cu, and $ZnSO_4$ by $CuSO_4$, electrons will flow in the other direction, from right to left, because copper ions have more of an affinity for electrons than hydrogen ions do. Copper ions will be reduced spontaneously at the left electrode, which therefore will become the cathode:

Cathode: $Cu^{2+} + 2e^- \rightarrow Cu(s)$

Hydrogen gas will be oxidized at the anode on the right to form hydrogen ions:

Anode: $H_2(g) \rightarrow 2H^+ + 2e^-$

The overall spontaneous reaction is

$$Cu^{2+} + H_2(g) \rightarrow Cu(s) + 2H^+$$

and the standard potential of this cell is $\mathscr{E}^0 = +0.34$ V. Once again, notice the additivity of cell potentials. These two hydrogen-electrode reactions can be added to produce the Daniell-cell reaction, and the sum of their emf's

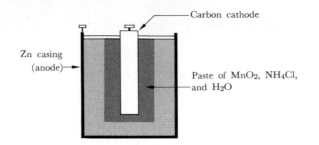

Figure 17–6. Dry cell. The individual electrode reactions are

Anode (zinc casing): $Zn(s) \rightarrow Zn^{2+} + 2e^-$

Cathode (carbon rod): $2MnO_2(s) + 8NH_4^+ + 2e^-$
$$\rightarrow 2Mn^{3+} + 4H_2O + 8NH_3$$

Overall: $Zn(s) + 2MnO_2(s) + 8NH_4^+$
$$\rightarrow Zn^{2+} + 2Mn^{3+} + 8NH_3 + 4H_2O$$

All chemical components are either solid or in the form of a paste, and the cell as a whole is sealed (hence the cell's name). Therefore, it is a highly useful cell for flashlights, radios, and other portable units.

is the emf of the Daniell cell:

$$\begin{array}{ll} Zn(s) + 2H^+ \rightarrow Zn^{2+} + H_2(g) & \mathscr{E}^0 = +0.76 \text{ V} \\ Cu^{2+} + H_2(g) \rightarrow Cu(s) + 2H^+ & \mathscr{E}^0 = +0.34 \text{ V} \\ \hline Zn(s) + Cu^{2+} \rightarrow Zn^{2+} + Cu(s) & \mathscr{E}^0 = +1.10 \text{ V} \end{array}$$

The Dry Cell

Any cell that involves liquids will be difficult or impossible to use in a situation involving motion. (Try to imagine a flashlight powered by a Daniell cell!) The dry cell, shown in Figure 17–6, is particularly convenient because all of its components are either solids or moist pastes, which are sealed tightly from the environment. The anode is the zinc casing of the dry cell itself. Around the carbon-rod cathode is a paste composed of MnO_2, NH_4Cl, and H_2O. At the anode zinc is oxidized to Zn^{2+} ions, and at the cathode MnO_2 is reduced to a mixture of several compounds of Mn in its $+3$ oxidation state. If the cell is used rapidly, ammonia, which is produced by the cathode reaction, can decrease the cell current by forming an insulating layer of gas around the carbon rod. With slower usage, zinc ions from the anode diffuse toward the cathode and combine with the ammonia to form complex ions such as $Zn(NH_3)_4^{2+}$.

Reversible Cells: the Lead Storage Battery

Most of the cells mentioned so far are reversible; that is, if a voltage greater than the cell voltage is applied from the outside, the cell reactions can be reversed, and electrical energy can be stored in the cell for withdrawal later.

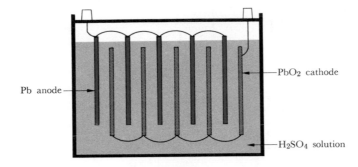

Figure 17–7. Lead storage battery. The electrode reactions are

Anode: $Pb(s) + SO_4{}^{2-} \rightarrow PbSO_4(s) + 2e^-$
Cathode: $PbO_2(s) + 4H^+ + SO_4{}^{2-} + 2e^- \rightarrow PbSO_4(s) + 2H_2O$

This battery is reversible (chargeable) since the product of the cell reaction, lead sulfate at both electrodes, adheres to the plates rather than diffusing or falling away. One cell of a lead storage battery as shown here delivers approximately 2 V, and 6- or 12-V batteries have three or six cells connected in series.

Thus in Figure 17–3(b), if electrons are driven from right to left by an outside voltage, the $CuSO_4$ in the left compartment is made even more dilute, and the $CuSO_4$ in the right compartment is made more concentrated. In Figure 17–4(c), an external voltage can cause additional Zn to deposit on the left terminal of the Daniell cell, and more copper from the right terminal to go into solution. Thus the free energy of the cell is brought to an even higher state, and at some later time this extra free energy can be released as useful work. Unfortunately, the commercial dry cell cannot be recharged in this way. The Zn^{2+} produced at the anode diffuses away, and the ammonia gas from the cathode reaction complexes with it. There is no convenient mechanism for breaking this complex ion and sending each component back to its original position.

What is needed for rechargeability is a cell in which the electrode products stay in place at the electrodes, ready to be reconverted as the cell is charged. An example of such a cell is the lead storage battery, shown in Figure 17–7. The anode is a spongy lead screen, and when the lead is oxidized to lead sulfate, it remains in place in the screen. Similarly, when lead oxide in the cathode is reduced, also to lead sulfate, the reduction products stay in place. The most noticeable change in the cell as it discharges is a dilution of the initially strong sulfuric acid solution. Therefore, the state of charge of the battery can be measured by a floating hygrometer. This instrument records the density of the battery fluid, and hence the strength of the sulfuric acid solution. The emf of this cell is 2 V, but several cells can be connected in series to produce 6-V and 12-V storage batteries.

When the cell has run down, it can be recharged by applying an external voltage in excess of the normal emf of 2 V per cell. The reactions shown in

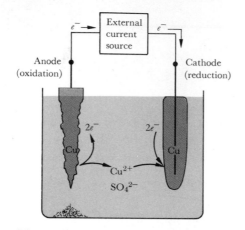

Figure 17–8. Electrolytic purification of crude copper. Impure copper is oxidized at the anode, and pure copper deposites at the cathode. Impurities accumulate below the anode as "anode slime." The rare metals recovered from the anode slime, such as gold, silver, and platinum, are often valuable enough to pay for the cost of the purification process.

the caption of Figure 17–7 are reversed, and the lead sulfate is converted to lead and to lead oxide. If the lead sulfate fell to the bottom of the tank as the cell discharged, this reverse reaction would be impossible. But it does not; it remains in place on the grid, ready for reconversion. Therefore, the lead storage battery is a convenient device for storing electrical energy in the form of chemical free energy.

Electrolytic Cells

The cell shown in Figure 17–8 can be used to produce high-purity copper, if an external battery or other current source is used to drive the cell in a direction that it would not take spontaneously. An ingot of impure "blister" copper, which is prepared from the reduction of copper oxide (CuO) by coke (C), is suspended in a copper sulfate solution along with a "core" wire of very pure copper. Current is passed through the cell in the direction that makes the ingot the anode, and the pure wire the cathode. The ingot is eaten away, and copper ions plate out as very pure metal on the cathode, with the impurities settling to the bottom of the tank below the anode.

Other kinds of electrolytic cells can be used to plate gold or silver over base metals in jewelry, or to make accurate copies of engraving plates for printing. Our U.S. paper money is printed in plates of twelve notes. The engraver makes only one master engraving on steel, which is hardened and then copied by the electroplating process shown in Figure 17–9. The electro-deposited copy, called an "alto" since the engraved lines are now in relief, then is electrocopied to produce a "basso" whose intaglio lines are an exact

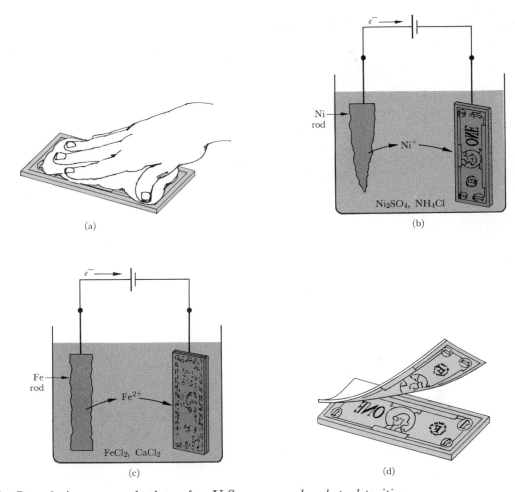

Figure 17–9. Reproducing engraved plates for U.S. currency by electrodeposition. (a) The engraved steel master plate is rubbed thoroughly with wet graphite powder and washed clean. The minute graphite coating aids electrodeposition and will make it possible to separate the master plate from the plated copy at the end. (b) The master is plated with a 0.025-mm layer of nickel during a ten-hour period. (The vertical tall and short lines at the top are the conventional sign for a battery, with the short line being the electron source—the battery anode.) (c) The nickel layer is backed with a 0.1-mm layer of iron for strength, during a fourteen-hour electrodeposition. (d) The electrodeposited reverse image is peeled away and soldered to a steel plate. Repeating this process, starting with the reverse plate just produced, leads to a copy of the master that can be recopied or used for printing banknotes.

copy of the grooves on the engraved master plate. These basso copies then are assembled into a twelve-note plate that can be used either directly for printing or for making one-piece printing plates by repeating the electro-copying process.

Cells such as these, in which an external current source is used, are called *electrolytic* cells; those such as we discussed previously, which use internal chemical reactions to produce an electric current, are called *galvanic* cells. In both types of cells, the terminal at which oxidation occurs is the anode, and the cathode is the site of reduction.

17–3 CELL EMF AND FREE ENERGY

When a charge, q, moves spontaneously through a potential drop of $\mathscr{E}$ volts, the external electrical work that can be done on its surroundings is $w_{ext} = q\mathscr{E}$. Therefore, from Equation 15–16, the free energy change of the system containing the moving electrical charge is

$$\Delta G = -w_{ext} = -q\mathscr{E} \qquad\qquad 17\text{–}1$$

Since the electron has a negative charge, it will move spontaneously from a region of low potential to one of high potential. For one electron moving through a potential increase of $\mathscr{E}$, the free energy change is

$$\Delta G = -e\mathscr{E}$$

For a mole of electrons,

$$\Delta G = -Ne\mathscr{E} = -\mathscr{F}\mathscr{E}$$

in which N is Avogadro's number and $\mathscr{F} = Ne$ is the charge on one mole of electrons, 96,500 coulombs or one faraday. In a reaction involving n electrons per molecule of reaction, or n faradays per mole, the free energy change is

$$\Delta G = -n\mathscr{F}\mathscr{E} \qquad\qquad (17\text{–}2)$$

If a galvanic cell is based on a reaction that yields ΔG kcal mole^{-1} of free energy, and if n moles of electrons are transferred through the external circuit per mole of reaction, then the potential difference between terminals, $\mathscr{E} = \mathscr{E}_{cathode} - \mathscr{E}_{anode}$ is given by

$$\mathscr{E}_{cathode} - \mathscr{E}_{anode} = \mathscr{E} = -\Delta G/n\mathscr{F}$$

Rather than using Faraday's constant as 96,500 coulombs mole^{-1}, multiplying by volts to obtain joules, and converting from joules to kilocalories every time a problem is worked, it is easier to remember that 1 electron volt molecule^{-1} equals 23.06 kcal mole^{-1}:

$$\Delta G_{(kcal)} = -23.06n\mathscr{E}$$

The pressure cell discussed in Section 17–1, for which the two-electron reaction had a standard free energy of -1.364 kcal mole^{-1}, therefore has a standard cell potential of

$$\mathscr{E}^0 = -\frac{-1.364 \text{ kcal mole}^{-1}}{2 \times 23.06 \text{ kcal mole}^{-1}} = +0.0296 \text{ V}$$

The Daniell cell reaction,

$$Zn(s) + Cu^{2+} \rightarrow Zn^{2+} + Cu(s)$$

has a standard cell potential of $+1.10$ V. Therefore, the standard free energy of this reaction must be

$$\Delta G = -2(23.06)(+1.10) = -50.73 \text{ kcal mole}^{-1}$$

In this expression $n = 2$ because two electrons are transferred per ion. You can calculate the standard free energies of other cells that we have discussed so far.

Exercise.　Calculate the standard free energies of the Ni—Cu, Zn—Cu, and Zn—Ni cells discussed in Section 17–2. Demonstrate that the free energies are additive in the same way that the reactions and the potentials are.

(*Answer:*

Ni—Cu cell: $\mathscr{E}^0 = +0.57$ V; $\Delta G = -26.3$ kcal mole^{-1}
Zn—Ni cell: $\mathscr{E}^0 = +0.53$ V; $\Delta G = -24.4$ kcal mole^{-1}
Zn—Cu cell: $\mathscr{E}^0 = +1.10$ V; $\Delta G = -50.7$ kcal mole^{-1})

As you can see, for a cell to deliver even one volt of potential difference between terminals requires a cell reaction that has quite a large free energy change. Since the free energies must be additive (by the first law of thermodynamics), it should not be surprising that emf's are additive also, as was discussed in Section 17–2.

This additivity of emf's is represented in Figure 17–10. Recall from the discussion of free energy in Chapter 15 that we do not need to tabulate the free energy change for every possible reaction. Having once tabulated the free energy change for a certain kind of reaction, namely, the formation of each compound from elements in their standard states, we then can calculate the free energy change for any reaction involving these compounds, because of the additivity of free energies. Similarly, we now do not need to tabulate the potential of every conceivable cell, or every conceivable combination of anode and cathode reactions. Instead, we only need to tabulate voltages of cells in which all electrode reactions are paired with one standard electrode. This amounts to choosing an arbitrary zero in Figure 17–10. We can divide a cell reaction into two half-reactions, one at the anode and the other at the

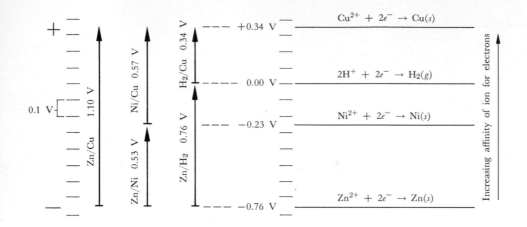

Figure 17–10. Since the potentials for the various cells involving Zn, Cu, Ni, *and* H$_2$ *that we discussed previously are additive, they can be represented as additive distances along a vertical axis measured in volts. The choice of a zero potential along this axis is arbitrary, but once made, all voltages can be specified relative to this zero point. In this figure, the hydrogen electrode has been selected as the reference zero voltage. This is the standard convention in tables, although* Ni, Zn, *or any other electrode, could have been chosen.*

cathode, and can assign to each half-reaction the voltage that would be observed in a cell if that half-reaction were paired with

$$H_2(g) \rightarrow 2H^+ + 2e^-$$

This is the basis for the scale of reduction potentials in Tables 17–1 and 17–2.

Table 17–1.   Standard Reduction Potentials in Acid Solution[a] at 298°K

Half-reaction (couple)	$\mathscr{E}^0$ (volts)
$F_2 + 2e^- \rightarrow 2F^-$	2.87
$Ag^{2+} + e^- \rightarrow Ag^+$	1.99
$H_2O_2 + 2H^+ + 2e^- \rightarrow 2H_2O$	1.78
$MnO_4^- + 4H^+ + 3e^- \rightarrow MnO_2 + 2H_2O$	1.68
$PbO_2 + 4H^+ + SO_4^{2-} + 2e^- \rightarrow PbSO_4 + 2H_2O$	1.69
$MnO_4^- + 8H^+ + 5e^- \rightarrow Mn^{2+} + 4H_2O$	1.49
$PbO_2 + 4H^+ + 2e^- \rightarrow Pb^{2+} + 2H_2O$	1.46
$Cl_2 + 2e^- \rightarrow 2Cl^-$	1.36
$Cr_2O_7^{2-} + 14H^+ + 6e^- \rightarrow 2Cr^{3+} + 7H_2O$	1.33
$MnO_2 + 4H^+ + 2e^- \rightarrow Mn^{2+} + 2H_2O$	1.21

Table 17–1.   *(Continued)*

Half-reaction (couple)	$\mathscr{E}^0$ (volts)
$O_2 + 4H^+ + 4e^- \rightarrow 2H_2O$	1.23
$Br_2(l) + 2e^- \rightarrow 2Br^-$	1.06
$AuCl_4^- + 3e^- \rightarrow Au + 4Cl^-$	0.99
$NO_3^- + 4H^+ + 3e^- \rightarrow NO + 2H_2O$	0.96
$2Hg^{2+} + 2e^- \rightarrow Hg_2^{2+}$	0.90
$Ag^+ + e^- \rightarrow Ag$	0.80
$Hg_2^{2+} + 2e^- \rightarrow 2Hg$	0.80
$Fe^{3+} + e^- \rightarrow Fe^{2+}$	0.77
$O_2 + 2H^+ + 2e^- \rightarrow H_2O_2$	0.68
$MnO_4^- + e^- \rightarrow MnO_4^{2-}$	0.56
$I_2 + 2e^- \rightarrow 2I^-$	0.54
$Cu^+ + e^- \rightarrow Cu$	0.52
$Cu^{2+} + 2e^- \rightarrow Cu$	0.34
$Hg_2Cl_2 + 2e^- \rightarrow 2Hg + 2Cl^-$	0.27
$AgCl + e^- \rightarrow Ag + Cl^-$	0.22
$SO_4^{2-} + 4H^+ + 2e^- \rightarrow H_2SO_3 + H_2O$	0.20
$Cu^{2+} + e^- \rightarrow Cu^+$	0.16
$2H^+ + 2e^- \rightarrow H_2$	0.00
$Pb^{2+} + 2e^- \rightarrow Pb$	-0.13
$Sn^{2+} + 2e^- \rightarrow Sn$	-0.14
$Ni^{2+} + 2e^- \rightarrow Ni$	-0.23
$PbSO_4 + 2e^- \rightarrow Pb + SO_4^{2-}$	-0.36
$Cd^{2+} + 2e^- \rightarrow Cd$	-0.40
$Cr^{3+} + e^- \rightarrow Cr^{2+}$	-0.41
$Fe^{2+} + 2e^- \rightarrow Fe$	-0.41
$Zn^{2+} + 2e^- \rightarrow Zn$	-0.76
$Mn^{2+} + 2e^- \rightarrow Mn$	-1.03
$Al^{3+} + 3e^- \rightarrow Al$	-1.66
$H_2 + 2e^- \rightarrow 2H^-$	-2.23
$Mg^{2+} + 2e^- \rightarrow Mg$	-2.37
$La^{3+} + 3e^- \rightarrow La$	-2.37
$Na^+ + e^- \rightarrow Na$	-2.71
$Ca^{2+} + 2e^- \rightarrow Ca$	-2.76
$Ba^{2+} + 2e^- \rightarrow Ba$	-2.90
$K^+ + e^- \rightarrow K$	-2.92
$Li^+ + e^- \rightarrow Li$	-3.05

[a] Values from the *Handbook of Chemistry and Physics,* Chemical Rubber Co., Cleveland, Ohio, 1971–72, 52nd ed.

Table 17–2.   Standard Reduction Potentials in Basic Solution at 298°K

Half-reaction (couple)[a]	$\mathscr{E}^0$ (volts)
$HO_2^- + H_2O + 2e^- \rightarrow 3OH^-$	0.87
$MnO_4^- + 2H_2O + 3e^- \rightarrow MnO_2 + 4OH^-$	0.59
$O_2 + 4e^- + 2H_2O \rightarrow 4OH^-$	0.40
$Co(NH_3)_6^{3+} + e^- \rightarrow Co(NH_3)_6^{2+}$	0.10
$HgO + H_2O + 2e^- \rightarrow Hg + 2OH^-$	0.10
$MnO_2 + H_2O + 2e^- \rightarrow Mn(OH)_2 + 2OH^-$	−0.05
$O_2 + H_2O + 2e^- \rightarrow HO_2^- + OH^-$	−0.08
$Cu(NH_3)_2^+ + e^- \rightarrow Cu + 2NH_3$	−0.12
$Ag(CN)_2^- + e^- \rightarrow Ag + 2CN^-$	−0.31
$Hg(CN)_4^{2-} + 2e^- \rightarrow Hg + 4CN^-$	−0.37
$S + 2e^- \rightarrow S^{2-}$	−0.51
$Pb(OH)_3^- + 2e^- \rightarrow Pb + 3OH^-$	−0.54
$Fe(OH)_3 + e^- \rightarrow Fe(OH)_2 + OH^-$	−0.56
$Cd(OH)_2 + 2e^- \rightarrow Cd + 2OH^-$	−0.81
$SO_4^{2-} + H_2O + 2e^- \rightarrow SO_3^{2-} + 2OH^-$	−0.92
$Zn(NH_3)_4^{2+} + 2e^- \rightarrow Zn + 4NH_3$	−1.03
$Zn(OH)_4^{2-} + 2e^- \rightarrow Zn + 4OH^-$	−1.22
$Mn(OH)_2 + 2e^- \rightarrow Mn + 2OH^-$	−1.47
$Mg(OH)_2 + 2e^- \rightarrow Mg + 2OH^-$	−2.67
$Ca(OH)_2 + 2e^- \rightarrow Ca + 2OH^-$	−3.02

[a] *Note:* Couples involving ions not affected by changing pH (such as $Na^+|Na$) have the same potential in acid or base.

17–4   HALF-REACTIONS AND REDUCTION POTENTIALS

The reduction potentials in Table 17–1 are the potentials that would be observed if the reduction equation as written were paired with the oxidation of hydrogen:

$$H_2(g) \rightarrow 2H^+ + 2e^-$$

A positive sign indicates that the cell reaction will go spontaneously in the direction indicated. A negative sign means that the reverse reaction will occur; that is, the particular substance will be oxidized and protons will be reduced to hydrogen gas. The more positive the reduction potential, the greater the tendency for the substance to accept electrons and become reduced. A large negative reduction potential indicates a strong favoring of the oxidized state. (These are *standard* values, which means a 1-molar concentration for all reacting ions and 1 atm partial pressure for gases at 298°K.)

Example.   If chlorine gas at 1 atm is bubbled over one platinum electrode in a solution of hydrochloric acid, and hydrogen gas at 1 atm is bubbled over a similar electrode, what will be the overall reaction and what will be the emf of the resulting cell?

Solution. The standard reduction potential for the reaction

$$Cl_2(g) + 2e^- \rightarrow 2Cl^-$$

is $\mathscr{E}^0 = +1.36$ V. This means that when combined with the hydrogen electrode, the chlorine electrode is the cathode (where reduction occurs) and the hydrogen electrode is the anode. The overall reaction

$$Cl_2(g) + H_2(g) \rightarrow 2Cl^- + 2H^+$$

is spontaneous from left to right and has a standard free energy change of

$$\Delta G^0 = -n\mathscr{F}\mathscr{E}^0 = -2(23.06)(+1.36) = -62.7 \text{ kcal mole}^{-1} Cl_2 \text{ or } H_2$$

Verify this free energy value from the data in Appendix 2.

Example. What is the spontaneous reaction when a cadmium electrode and a hydrogen electrode are paired in acid solution? What is the cell potential?

Solution. The standard reduction potential for the half-reaction

$$Cd^{2+} + 2e^- \rightarrow Cd(s)$$

is -0.40 V. Thus, the overall reaction

$$Cd^{2+} + H_2(g) \rightarrow Cd(s) + 2H^+$$

is spontaneous *in reverse*, from right to left. Cadmium will be oxidized spontaneously to ions at the anode, and hydrogen ions will be reduced to hydrogen gas at the cathode. The cell potential is $+0.40$ V.

Example. Will a piece of cadmium, dropped into a 1-molar acid solution, produce bubbles of hydrogen gas? What about a piece of silver?

Solution. From the previous example, the reaction

$$Cd(s) + 2H^+ \rightarrow Cd^{2+} + H_2(g)$$

is spontaneous under standard conditions, so bubbles of hydrogen gas will escape as the cadmium metal is etched away by the acid solution. In contrast, silver has a positive reduction potential and a stronger tendency to remain reduced than hydrogen has. Thus silver, when dropped into a weak acid solution, will remain unaffected. The "nobility" of the noble metals gold, silver, and platinum, is primarily a consequence of their large, positive reduction potentials.

Cell Potentials from Reduction Potentials of Half-Reactions

To find the potential of a cell in which a given reaction is occurring, one first must break up the reaction into its two half-reactions. Choose one of these half-reactions to be a reduction reaction at the cathode, and the other half-reaction to be an oxidation reaction at the anode. Write the second reaction in reverse, as an oxidation. Then find the standard reduction potentials for these two half-reactions and reverse the sign of $\mathscr{E}^0$ for the reaction that has been taken as the oxidation. Add the two half-reactions as a check to be sure that you obtain the original overall reaction, and add the two half-reaction potentials at the same time. After finishing, if you have a positive overall potential, the reaction as written is spontaneous. If the overall potential is negative, you made the wrong assumption about the anode and the reverse reaction is spontaneous.

Example. Find the emf of a cell made up of a Zn electrode in $ZnSO_4$ and a Cu electrode in $CuSO_4$. Which is the anode and which is the cathode?

Solution. This is just the Daniell cell, and you know the answer already; the Zn electrode will be the anode. But assume that you did not know this, and guessed (wrongly) that the copper electrode was the anode. The two half-reactions are

$$Zn^{2+} + 2e^- \rightarrow Zn(s) \qquad \mathscr{E}^0 = -0.76 \text{ V}$$
$$Cu^{2+} + 2e^- \rightarrow Cu(s) \qquad \mathscr{E}^0 = +0.34 \text{ V}$$

It is obvious that by reversing the Zn reaction and changing the sign of its emf we will obtain a positive overall emf. Nevertheless, let us assume that the copper electrode is the anode. The two reactions to be added then are

$$Zn^{2+} + 2e^- \quad \rightarrow Zn(s) \qquad\qquad \mathscr{E}^0 = -0.76 \text{ V}$$
$$\underline{Cu(s) \qquad\quad \rightarrow Cu^{2+} + 2e^- \qquad \mathscr{E}^0 = -0.34 \text{ V}}$$
$$Zn^{2+} + Cu(s) \rightarrow Zn(s) + Cu^{2+} \qquad \mathscr{E}^0 = -1.10 \text{ V}$$

The negative potential tells us that the reverse reaction is spontaneous, and that the cell thus set up will have an emf of $+1.10$ V. This method is fool-proof against errors in the original assumption about anode and cathode.

Example. What is the spontaneous direction in a cell made up of a ferrous–ferric ion electrode and an iodine–iodide ion electrode? What will be the cell emf?

Solution. The half-reactions are

$$Fe^{3+} + e^- \quad \rightarrow Fe^{2+} \qquad \mathscr{E}^0 = +0.77 \text{ V}$$
$$I_2 \quad + 2e^- \rightarrow 2I^- \qquad \mathscr{E}^0 = +0.54 \text{ V}$$

It is obvious that the way to obtain a positive overall voltage is to subtract the second half-reaction from the first, thereby reversing the sign of the 0.54-V term. The spontaneous overall reaction is

$$2Fe^{3+} + 2I^- \rightarrow 2Fe^{2+} + I_2 \qquad \mathscr{E}^0 = +0.23 \text{ V}$$

We had to multiply the first equation by two before subtracting because it involved only one electron, whereas the iodine equation involved two. If we were subtracting free energies we would multiply the standard free energy change for the iron half-reaction by two before subtracting. Should we multiply the potential, $+0.77$ V, by two also? No, because the number of electrons already has been taken into account by n in the expressions:

$$\Delta G^0 = -n\mathscr{F}\mathscr{E}^0 \qquad \text{and} \qquad \mathscr{E}^0 = -\Delta G^0/n\mathscr{F}$$

Half-cell potentials are already on a "per electron" basis and can be combined directly. To use our hydrostatic analogy again, these potentials are electron "pressures" and not energy quantities. The pressure of water behind a dam of specified height does not depend upon whether we take the water from the bottom in one- or two-gallon batches, but the work or energy obtained per batch does.

The factor of two, for two electrons, will be taken into account as soon as we calculate free energy changes. For the iron half-reaction,

$$2Fe^{3+} + 2e^- \rightarrow 2Fe^{2+}$$
$$\Delta G^0 = -2\mathscr{F}\mathscr{E}^0 = -2(23.06)(+0.77) = -35.5 \text{ kcal}$$

For the iodine half-reaction,

$$I_2(s) + 2e^- \rightarrow 2I^-$$
$$\Delta G^0 = -2\mathscr{F}\mathscr{E}^0 = -2(23.06)(+0.54) = -24.9 \text{ kcal}$$

Subtracting the second half-reaction from the first, and the second free energy change from the first, we get

$$\Delta G^0 = -35.5 \text{ kcal} + 24.9 \text{ kcal} = -10.6 \text{ kcal}$$

As a check, verify that the value of this free energy change leads to the overall cell potential, using $\Delta G^0 = -n\mathscr{F}\mathscr{E}^0$.

Any half-reaction with a higher positive reduction potential will dominate over a half-reaction with a lower reduction potential, and send the latter in the reverse direction if the two half-reactions are coupled in a cell. Thus the two half-reactions for the lead storage battery are

$$PbO_2 + 4H^+ + SO_4^{2-} + 2e^- \rightarrow PbSO_4 + 2H_2O \qquad \mathscr{E}^0 = +1.69 \text{ V}$$
$$PbSO_4 + 2e^- \rightarrow Pb + SO_4^{2-} \qquad \mathscr{E}^0 = -0.36 \text{ V}$$

Since the first reaction has the higher reduction potential, it will take place at the cathode, and the reverse of the second reaction will occur at the anode.

The overall cell potential will be

$$\mathscr{E}^0 = (+1.69 \text{ V}) - (-0.36 \text{ V}) = +2.05 \text{ V}$$

Ions of any metal in the reduction potential table will become reduced in the presence of a metal lower in the table. Thus silver from a silver nitrate solution will precipitate in the presence of zinc, iron, cadmium, or even copper or mercury. In the presence of Ag^+ will iron be the ferrous ion, Fe^{2+}, or the ferric ion, Fe^{3+}?

The half-reactions from Table 17–1 are

1) $Ag^+ + e^- \rightarrow Ag$ $\qquad \mathscr{E}^0 = +0.80 \text{ V}$
$$\Delta G^0 = -1(23.06)(+0.80) = -18.45 \text{ kcal mole}^{-1}$$

2) $Fe^{3+} + e^- \rightarrow Fe^{2+}$ $\qquad \mathscr{E}^0 = +0.77 \text{ V}$
$$\Delta G^0 = -1(23.06)(+0.77) = -17.76 \text{ kcal mole}^{-1}$$

3) $Fe^{2+} + 2e^- \rightarrow Fe(s)$ $\quad \mathscr{E}^0 = -0.41 \text{ V}$
$$\Delta G^0 = -2(23.06)(-0.41) = +18.90 \text{ kcal mole}^{-1}$$

But why is no reaction listed for the reduction of Fe^{3+} to metallic iron? The reaction we want,

$$Fe^{3+} + 3e^- \rightarrow Fe(s)$$

can be obtained by adding half-reactions (2) and (3). Can we find the potential by adding the potentials of these two reactions:

$$\mathscr{E}^0 = +0.77 - 0.41 = +0.36 \text{ V}$$

The answer is: *no*, we cannot. The emf of the reduction of Fe^{3+} to $Fe(s)$ is not $+0.36$ V. It is legitimate to subtract one half-cell *potential* from another when the corresponding half-cell *reactions* are subtracted to form a correct overall cell reaction with proper balancing of electron gain and loss. It is not legitimate to add the potentials of a one-electron half-reaction and a two-electron half-reaction to obtain the potential of the resulting three-electron half-reaction.

It is always safe to work with free energies, which is why we bothered to calculate free energies for the three reactions above. In calculating free energies from cell potentials, explicit counts of electrons involved are made via the factor, n, in the expression $\Delta G^0 = -n\mathscr{F}\mathscr{E}^0$. Since the desired half-cell reaction is the sum of half-reactions (2) and (3), the overall free energy is the sum of the two free energies:

$$Fe^{3+} + 3e^- \rightarrow Fe(s) \quad \Delta G^0 = -17.76 + 18.90 = +1.14 \text{ kcal mole}^{-1}$$

The standard emf now can be found by dividing by $-3\mathscr{F}$, since this is a three-electron reaction:

$$\mathscr{E}^0 = \frac{+1.14}{-3(23.06)} = -0.016 \text{ V}$$

But this is not a realistic half-reaction. If both Fe^{3+} and metallic iron were present together, they would combine spontaneously to produce Fe^{2+}, as the free energies indicate:

$$Fe(s) \rightarrow Fe^{2+} + 2e^- \qquad \Delta G^0 = -18.90 \text{ kcal}$$
$$\underline{2Fe^{3+} + 2e^- \rightarrow 2Fe^{2+} \qquad \Delta G^0 = 2(-17.76) \text{ kcal}}$$
$$Fe(s) + 2Fe^{3+} \rightarrow 3Fe^{2+} \qquad \Delta G^0 = -54.42 \text{ kcal}$$

This is a highly spontaneous reaction with a strongly negative standard free energy change. Since a Fe—Fe^{3+} half-cell is physically unrealistic, its potential is not tabulated.

We now have answered the original question. When silver plates out from an Ag^+ solution in the presence of metallic iron, the iron goes into solution as Fe^{2+}. If some Fe^{3+} were present, it immediately would combine with metallic iron by the preceding reactions to make more Fe^{2+}. Thus, the two electrochemical reactions involved are

$$Ag^+ + e^- \rightarrow Ag \qquad \mathscr{E}^0 = +0.80 \text{ V}; \qquad \Delta G^0 = -18.45 \text{ kcal}$$
$$Fe^{2+} + 2e^- \rightarrow Fe(s) \qquad \mathscr{E}^0 = -0.41 \text{ V}; \qquad \Delta G^0 = +18.90 \text{ kcal}$$

Handling of electrode potentials often can be confusing. A safe guideline is: *When in doubt, work with free energies.*

Shorthand Notation for Electrochemical Cells

In the usual abbreviated notation for a cell, the reactants and products are listed from left to right in the form

Anode|Anode solution||Cathode solution|Cathode

A single vertical line indicates a change of phase: solid, liquid, gas, or solution. A double vertical line indicates a porous barrier or salt bridge between two solutions. Thus, some of the cells we have discussed previously would be written

H_2 pressure cell:	$Pt|H_2(g, 10 \text{ atm})|H_2O||H_2O|H_2(g, 1 \text{ atm})|Pt$
Cu concentration cell:	$Cu|Cu^{2+}(0.01M)||Cu^{2+}(0.10M)|Cu$
Daniell cell: (in which x and y are the molarities of the ionic solutions)	$Zn|Zn^{2+}(xM)||Cu^{2+}(yM)|Cu$
Lead storage battery:	$Pb|H_2SO_4(aq)|PbO_2$

Challenge. To see how cell potentials can be applied to the design of a new electrochemical cell for automobile propulsion, try Problems 17–14 through 17–17 in Butler and Grosser.

17–5   THE EFFECT OF CONCENTRATION ON CELL VOLTAGE: THE NERNST EQUATION

The reduction potentials that we have been working with so far all have been standard potentials, that is, with the concentrations of all solutes at 1 mole liter^{-1} and all gases at 1 atm partial pressure,[3] and at 298°K. Does the emf of a cell change with concentration? It does, for the same reasons that the free energy of the cell reaction changes. We saw special examples of this at the beginning of this chapter in connection with concentration cells, and now we will derive a more general relationship.

The relationship between free energy and concentration was given by Equation 15–26:

$$\Delta G = \Delta G^0 + RT \ln Q$$

in which the reaction quotient, Q, is the ratio of products to reactants, each raised to its proper stoichiometric power. We can convert this into a cell-voltage equation, using $\Delta G = -n\mathscr{F}\mathscr{E}$, to yield

$$\mathscr{E} = \mathscr{E}^0 - \frac{RT}{n\mathscr{F}} \ln Q$$

This is called the Nernst equation after Walther Nernst, who first proposed it in 1881. $\mathscr{E}^0$ is the standard cell potential, Q is the ratio of concentrations under the conditions of a given experiment, and $\mathscr{E}$ is the cell voltage measured under these same conditions. For the copper concentration cell discussed at the beginning of this chapter,

$$\Delta G = 0 + RT \ln (c_2/c_1) = 1.364 \log_{10} (c_2/c_1) \qquad \text{at } 298°K$$

$$\mathscr{E} = 0 - \frac{RT}{2\mathscr{F}} \ln (c_2/c_1) = -\frac{1.364}{2\mathscr{F}} \log_{10} (c_2/c_1)$$

The quantity $RT/\mathscr{F}$ at 298°K is encountered so often that it should be evaluated first:

$$RT/\mathscr{F} = \frac{8.32 \text{ joules deg}^{-1} \times 298°K}{96{,}500 \text{ coulombs}} = 0.0257 \text{ joule coulomb}^{-1}$$

$$RT/\mathscr{F} = 0.0257 \text{ V}$$

$$2.303 \, RT/\mathscr{F} = 0.0592 \text{ V}$$

The general expression then is

$$\mathscr{E} = \mathscr{E}^0 - \frac{0.0592}{n} \log_{10} Q$$

---

[3] To be absolutely correct, with all substances at unit activity.

and the emf of the copper concentration cell is

$$\mathscr{E} = -0.0296 \log_{10}(c_2/c_1)$$

As we calculated previously, for a concentration ratio of $1:10$ the emf is $+0.0296$ V or 29.6 mV. This value is true for *any* two-electron concentration cell with a $1:10$ concentration ratio, no matter what the chemical reaction actually is. Thus, the same voltages were calculated previously for the hydrogen cell and for the copper concentration cell. For a ratio of $1:5$ the emf of the cell is

$$\mathscr{E} = +29.6 \log_{10}(5) = +20.7 \text{ mV}$$

Example.   What are the emf and free energy change for the cell

$$\text{Cd}|\text{Cd}^{2+}(0.0500M)||\text{Cl}^-(0.100M)|\text{Cl}_2(1 \text{ atm})|\text{Pt}$$

Solution.   The half-reactions are

$$\text{Cd}^{2+} + 2e^- \rightarrow \text{Cd}(s) \qquad \mathscr{E}^0 = -0.40 \text{ V}$$
$$\text{Cl}_2(g) + 2e^- \rightarrow 2\text{Cl}^- \qquad \mathscr{E}^0 = +1.36 \text{ V}$$

A positive overall voltage will be obtained by taking the $\text{Cd}|\text{Cd}^{2+}$ electrode to be the anode, where oxidation occurs, and subtracting it from the Cl_2 reaction to yield

$$\text{Cl}_2(g) + \text{Cd}(s) \rightarrow 2\text{Cl}^- + \text{Cd}^{2+} \qquad \mathscr{E}^0 = +1.36 - (-0.40)$$
$$= +1.76 \text{ V}$$

The cell voltage at other than standard conditions is found from the expression

$$\mathscr{E} = +1.76 - \frac{0.0592}{2} \log_{10} \frac{[\text{Cl}^-]^2[\text{Cd}^{2+}]}{1 \cdot 1}$$

Chlorine gas at 1 atm and solid Cd both are in their standard states, thereby giving unit activities in the denominator. Hence,

$$\mathscr{E} = +1.76 - 0.0296 \log_{10}[(0.100)^2(0.0500)]$$
$$\mathscr{E} = +1.76 - 0.0296 \log_{10}(0.000500)$$
$$\mathscr{E} = +1.76 + 0.0296(3.301) = +1.86 \text{ V}$$
$$\Delta G = -2\mathscr{FE} = -46.12(1.86) = -85.8 \text{ kcal}$$

Single-Electrode Potentials

It is often more convenient to deal with the concentration dependence of each half-reaction separately, and then to combine the results. The Nernst equation can be broken up by imagining that each half-reaction, oxidation and reduction, is coupled with the standard hydrogen reaction,

$$2\text{H}^+(1.0M) + 2e^- \rightarrow \text{H}_2(g, 1 \text{ atm}) \qquad \mathscr{E}^0 = 0.000 \text{ V}$$

in which both H^+ and H_2 have unit activities. Thus for the $Zn^{2+}|Zn(s)$ half-reaction,

$$\mathscr{E} = -0.76 - 0.0296 \log_{10} \frac{1}{[Zn^{2+}]}$$

since the activity of solid zinc in the numerator is one. The emf of the hydrogen electrode, $H^+|H_2(g)$, depends upon pH. If the pressure of the hydrogen gas is maintained at 1 atm then this dependence is

$$\mathscr{E} = 0.000 - 0.0592 \text{ pH}$$

(Prove to yourself that this is true.) For the half-reaction $Fe^{3+}|Fe^{2+}$,

$$\mathscr{E} = +0.77 - 0.0592 \log_{10} \frac{[Fe^{2+}]}{[Fe^{3+}]}$$

And for the reaction

$$MnO_4^- + 8H^+ + 5e^- \rightarrow Mn^{2+} + 4H_2O$$

$$\mathscr{E} = +1.49 - 0.0118 \log_{10} \frac{[Mn^{2+}]}{[MnO_4^-][H^+]^8}$$

$$= +1.49 - 0.0118 \log_{10} \frac{[Mn^{2+}]}{[MnO_4^-]} - 0.0947 \text{ pH}$$

(Note the division of 0.0592 by 5 to yield 0.0118, and the remultiplication by 8 at the right.) Therefore, any factors that affect ion concentrations will affect electrode potentials.

These individual half-reaction equations, once written properly with respect to concentrations of ions, then can be combined to form the Nernst equation for the cell as a whole.

Example. Write a balanced equation and calculate K_{eq} for the reaction involving the $Fe^{3+}|Fe^{2+}$ and $H^+, MnO_4^-|Mn^{2+}$ couples.

Solution

a) $5e^- + 8H^+ + MnO_4^- \rightarrow Mn^{2+} + 4H_2O$

$$\mathscr{E} = 1.49 - \frac{0.0592}{5} \log_{10} \frac{[Mn^{2+}]}{[MnO_4^-][H^+]^8}$$

b) (reversed) $Fe^{2+} \rightarrow Fe^{3+} + e^-$

$$\mathscr{E} = -0.77 - \frac{0.0592}{1} \log_{10} \frac{[Fe^{3+}]}{[Fe^{2+}]}$$

To provide enough electrons for a balanced equation with the manganese reaction, we must multiply the iron reaction by 5:

c) $5Fe^{2+} \rightarrow 5Fe^{3+} + 5e^-$

$$\mathscr{E} = -0.77 - \frac{0.0592}{5} \log_{10} \frac{[Fe^{3+}]^5}{[Fe^{2+}]^5}$$

Notice how the coefficients in the Fe equations have been treated in the denominator of the 0.0592 term and in the exponents within the logarithm. Recall that

$$\frac{1}{n} \log a^n = \log a$$

Now add (a) and (c):

d) $8H^+ + MnO_4^- + 5Fe^{2+} \rightarrow Mn^{2+} + 4H_2O + 5Fe^{3+}$

$$\mathscr{E} = 0.72 - \frac{0.0592}{5} \log_{10} \frac{[Mn^{2+}][Fe^{3+}]^5}{[H^+]^8[MnO_4^-][Fe^{2+}]^5}$$

At equilibrium, $\mathscr{E} = 0$, and

$$\mathscr{E}^0_{cell} = 0.72 \text{ V} = \frac{0.0592}{5} \log_{10} K_{eq}$$

$$K_{eq} = 5 \times 10^{62} = \frac{[Mn^{2+}][Fe^{3+}]^5}{[H^+]^8[MnO_4^-][Fe^{2+}]^5}$$

To reemphasize the treatment of stoichiometry, we should note that in considering the half-reaction

$$Fe^{3+} + e^- \rightarrow Fe^{2+}$$

the magnitude of the potential is independent of the number of times we use it in the balanced equation:

$$\mathscr{E} = \mathscr{E}^0 - 0.0592 \log_{10} \frac{[Fe^{2+}]}{[Fe^{3+}]} = \mathscr{E}^0 - \frac{0.0592}{5} \log_{10} \frac{[Fe^{2+}]^5}{[Fe^{3+}]^5}$$

Range of K_{eq} for Oxidation-Reduction Reactions

If we examine the reduction potentials in Tables 17–1 and 17–2, we find that they range from about $+3$ to -3 volts. A difference of six volts in half-reaction potentials corresponds to an equilibrium constant of 10^{100n}, in which n is the number of electrons transferred in a redox process:

$$\mathscr{E}^0 = \frac{0.0592}{n} \log_{10} K_{eq} = 6.0$$

$$\log_{10} K_{eq} \simeq 100n$$

$$K_{eq} \simeq 10^{100n}$$

This is an immensely large equilibrium constant in comparison with the maximum K_{eq} value of about 10^{14} encountered in proton-transfer reactions in aqueous solution. This large range of equilibrium-constant values within a redox potential range of six volts means that the chance of two half-cell potentials being close enough to establish equilibrium with significant quantities of both reactants and products present is small. An equilibrium constant as large as 10^{20} would require only a redox potential difference of 0.59 V for a two-electron reaction. Redox reactions tend to be all-or-nothing processes, with either reactants or products present in significant amounts, but not both. For such reactions, cell emf measurements provide a convenient, practical way of determining equilibrium constants.

17–6   SOLUBILITY EQUILIBRIA AND POTENTIALS

One of the difficulties in measuring solubility products directly is that for slightly soluble salts, such as AgCl, the concentrations at equilibrium are too low to measure accurately. However, the half-reaction

$$Ag^+ + e^- \rightarrow Ag$$

for which

$$\mathscr{E} = 0.80 - 0.0592 \log_{10} \frac{1}{[Ag^+]} = 0.80 + 0.0592 \log_{10} [Ag^+]$$

can be used to measure solubility. For example, if the silver ion concentration was as low as 10^{-30} molar, the silver electrode potential still would be -1.0 V, which is clearly within the measurable range.

Example.   Calculate the K_{sp} for AgCl at 298°K from appropriate couples in Table 17–1.

Solution.   Begin with the two half-reactions:

$$AgCl + e^- \rightarrow Ag + Cl^- \qquad \mathscr{E}^0 = 0.22 \text{ V (reduction)}$$
$$Ag \rightarrow Ag^+ + e^- \qquad \mathscr{E}^0 = -0.80 \text{ V (oxidation)}$$

Add the reactions to obtain the equation for the solution of AgCl:

$$AgCl \rightarrow Ag^+ + Cl^- \qquad \mathscr{E}^0 = -0.58 \text{ V}$$

$\mathscr{E}^0$ is negative since the reaction goes spontaneously to the left for 1-molar Ag^+ and Cl^-. For the net reaction,

$$\Delta G^0 = -RT \ln K_{sp} = -n\mathscr{F}\mathscr{E}^0$$

($\mathscr{E}$ in the Nernst equation is zero at equilibrium.)

$$\ln K_{sp} = -\frac{1\mathscr{F}}{RT}(0.58)$$

$$\log_{10} K_{sp} = -\frac{0.58}{0.0592} = -9.8$$

$$K_{sp} = 1.6 \times 10^{-10}$$

Example. Calculate K_{sp} for Hg_2Cl_2 at 298°K from data in Table 17–1.

Solution. The solubility equilibrium is

$$Hg_2Cl_2(s) \rightarrow Hg_2^{2+} + 2Cl^-$$
$$K_{sp} = [Hg_2^{2+}][Cl^-]^2$$

The half-reactions for this equilibrium are

$$Hg_2Cl_2 + 2e^- \rightarrow 2Hg + 2Cl^- \qquad \mathscr{E}^0 = +0.27 \text{ V}$$
$$Hg_2^{2+} + 2e^- \rightarrow 2Hg \qquad \mathscr{E}^0 = +0.80 \text{ V}$$

Adding the first half-reaction to the reverse of the second, we obtain

$$Hg_2Cl_2 \rightarrow Hg_2^{2+} + 2Cl^- \qquad \mathscr{E}^0 = 0.27 - 0.80 = -0.53 \text{ V}$$

The solubility-product expression is

$$0 = -0.53 - \frac{0.0592}{2} \log_{10} K_{sp}$$

$$\log_{10} K_{sp} = -17.9$$
$$K_{sp} = 1.3 \times 10^{-18}$$

An interesting consequence of the effect of solubility on electrode potentials can be pointed out by considering the reaction between silver ion and iodide ion. Table 17–1 "predicts" that silver ion should oxidize iodide ion according to the reaction

$$2Ag^+ + 2I^- \rightarrow 2Ag + I_2$$

If a cell is made of the two half-reactions working separately, silver plates on the cathode and iodine forms at the anode, as predicted. However, if the two ions are mixed directly, the only reaction observed is the formation of insoluble silver iodide:

$$Ag^+ + I^- \rightarrow AgI(s)$$

From the Nernst equation, we can show that since K_{sp} of AgI is about 10^{-16}, the potential of the $Ag^+|Ag$ couple is decreased by 16×0.0592 or nearly 1.0 V in a 1.0-molar I^- solution, whereas the $I_2|I^-$ couple has its potential increased by nearly the same amount in a 1.0-molar Ag^+ solution. Since the $I_2|I^-$ couple is only about 0.25 V below $Ag^+|Ag$, these two changes are more than enough to reverse the position of the couples in the table.

Complex-ion Formation and Reduction Potentials

Consider the equilibrium between silver ion and cyanide ion:

$$Ag^+ + 2CN^- \rightarrow Ag(CN)_2{}^-$$

The equilibrium expression is

$$K_{eq} = \frac{[Ag(CN)_2{}^-]}{[Ag^+][CN^-]^2}$$

or

$$[Ag^+] = \frac{[Ag(CN)_2{}^-]}{K_{eq}[CN^-]^2}$$

For a silver electrode immersed in a solution containing 0.01-molar silver ion with an excess of cyanide ion, virtually all of the silver ion is complexed. The potential is

$$\mathscr{E} = 0.80 + 0.0592 \log_{10} \frac{[Ag(CN)_2{}^-]}{K_{eq}[CN^-]^2}$$

$$\mathscr{E} = 0.80 - 0.1184 - 0.0592 \log_{10} K_{eq} - 0.1184 \log_{10} [CN^-]$$

Such relationships are used in the study of equilibria involving complex ions.

17–7 REDOX CHEMISTRY GONE ASTRAY: CORROSION

Corrosion of metals is a redox process. For example, iron can be oxidized either by molecular oxygen or by acid, if sufficient moisture is present for the chemical reactions to proceed at an appreciable rate.

Oxidation:	$Fe(s) \rightarrow Fe^{2+} + 2e^-$	$\mathscr{E}^0 = +0.41$ V
Reduction:	$\begin{cases} \frac{1}{2}O_2(g) + H_2O(l) + 2e^- \rightarrow 2OH^- \\ 2H^+ + 2e^- \rightarrow H_2(g) \end{cases}$	$\begin{aligned} \mathscr{E}^0 &= +0.40 \text{ V} \\ \mathscr{E}^0 &= 0.00 \text{ V} \end{aligned}$

When iron rusts, metallic iron is oxidized to the $+2$ state, where it deposits as flakes of FeO or other iron oxides. Aluminum corrodes even more vigorously:

$$Al(s) \rightarrow Al^{3+} + 3e^- \qquad \mathscr{E}^0 = +1.66 \text{ V}$$

On the reduction-potential scale, aluminum is more susceptible to oxidation than iron. Yet we think of aluminum as relatively inert to corrosion, whereas the rusting of iron and steel is a serious and expensive problem. Why?

We find out why only when we look at the crystal structures of aluminum, iron, and their oxides. The unit-cell or packing distances in aluminum and its oxide are very similar to one another, thus the aluminum oxide formed at the surface of the metal can adhere tightly to the uncorroded aluminum beneath it. The oxidized surface provides a protective layer that prevents

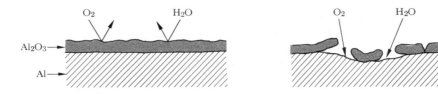

Figure 17–11. A layer of aluminum oxide, once formed, adheres to the surface of the aluminum metal and protects it against further corrosion (left). Unfortunately, iron oxide does not adhere as well to an iron surface. It continually flakes away as rust and exposes a clean surface to further attack by oxygen and moisture. Iron and steel objects must have their surfaces protected by some artificial method.

oxygen from getting to the metal beneath. "Anodized" aluminum kitchenware has had a particularly tough oxide layer applied to it by placing the aluminum object in a situation where corrosion is especially favored—by making it the anode in an electrochemical reaction.

In contrast, the packing dimensions of metallic iron and FeO are not particularly close; thus there is no tendency for an iron oxide layer to adhere to metallic iron. The curse of rust is not that it forms, but that it constantly flakes off and exposes fresh iron surface for attack (Figure 17–11). One way to prevent rusting is to keep moisture and oxygen away from the surface of an iron object by giving it an artificial coating such as paint. A good paint adheres better than FeO does, but it still is not permanent.

Another, and more effective, method is to make electrochemistry work for you. Just as aluminum can be made to form an oxide film by making it the anode in an electrochemical cell, so iron can be *prevented* from oxidizing by making it the cathode. One way to do this is to coat it or plate it with a more reactive metal, yet one which itself forms a protective oxide coating. Aluminum would be a possible candidate. If iron and aluminum are in contact, the iron will behave as the cathode and the aluminum as the anode, as their reduction potentials indicate:

$$Fe^{2+} + 2e^- \rightarrow Fe(s) \qquad \mathscr{E}^0 = -0.41 \text{ V}$$
$$Al^{3+} + 3e^- \rightarrow Al(s) \qquad \mathscr{E}^0 = -1.66 \text{ V}$$

Aluminum will prevent the iron from becoming oxidized, and its own oxide will protect aluminum from continual destructive corrosion.

But if you are going to aluminum-plate iron, you might as well make the objects out of aluminum to begin with, thereby having the advantage of light weight. Unfortunately, aluminum is expensive. An earlier, and cheaper, alternative has been to "galvanize" iron; that is, give it a thin coat of zinc. You can see from Table 17–1 that the principle is the same, although the reduction potentials of zinc and iron are closer. A galvanized steel bucket is

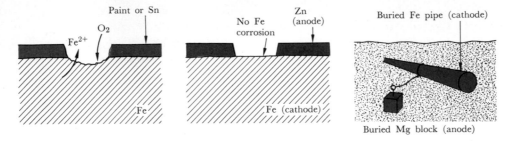

Figure 17–12. One protection for an iron object is an air-tight coating of paint or of another metal such as tin. This works as long as the coating is intact, but when a pit or scratch develops, corrosion begins. A zinc coating provides additional electrochemical protection because iron has a higher reduction potential, and tends to remain reduced while zinc is oxidized. Similarly, magnesium can protect iron pipe by corroding in place of it. In these applications zinc and magnesium are called sacrificial metals.

corrosion-free, not merely because zinc shields iron as paint would, but because zinc electrochemically prevents iron from being reduced. Scratch a galvanized pail and the scratched pail will not corrode; in principle the iron object does not even have to be completely covered.

Tin is another story. A "tin can" is tin-plated iron. You can see from Table 17–1 that Sn lies above Fe in reduction potential, thus the Sn^{2+} ion has a greater tendency to become reduced to the metal than does Fe^{2+}. The tin coating actually *encourages* oxidation, and hence corrosion, of the iron. A tin can is protected from corrosion only as long as the entire tin surface is intact. Scratch a tin can and it most certainly will rust. The tin plating functions only as a particularly tough and adhesive "superpaint." This is perhaps fortunate for our environment. Tin cans eventually self-destruct, whereas aluminum cans do not.

The same electrochemical trickery can be used to keep iron pipes from corroding in moist ground, or steel ship hulls from corroding in salt water. Magnesium rods driven into the ground periodically alongside an iron pipeline and connected electrically to it will make the pipe cathodic and prevent it from corroding. The magnesium itself will corrode, but it is easier and cheaper to replace magnesium rods than to dig up and replace the pipeline (Figure 17–12). Blocks of magnesium attached to a ship's hull in sea water perform the same function. The magnesium is called a "sacrificial" metal; its corrosion is acceptable as an alternative to corrosion of the iron object.

17–8 SUMMARY

Some atoms and ions attract electrons more strongly than others. When we allow electrons to flow from the less attracting ions or atoms to more attracting ones, a more stable situation develops and energy is released. If no special precautions are taken, this energy is dissipated as heat, or as an increase in disorder (entropy). But if the electron-releasing and electron-accepting half-reactions can be separated physically, then the flow of electrons from one place to another can be harnessed to do electrical work. This is the principle of all electrochemical cells.

The electron-losing substance is *oxidized*, and the electrode where this occurs is the *anode*. The electron-accepting substance is *reduced* at the *cathode*. The "pressure" that these electrons exert, measured between anode and cathode, is the cell voltage or electromotive force (emf). A positive cell voltage means that the cell reaction will proceed spontaneously, with electron flow from the anode to the cathode. A negative cell voltage means that the reverse reaction is spontaneous. The cell voltage is related to the free energy of the cell reaction by the expression

$$\Delta G = -n\mathscr{F}\mathscr{E}$$

An overall cell reaction can be separated into two half-reactions that represent the processes at the anode and cathode. Each of these half-reactions can be assigned its own half-cell potential, defined as that potential which would be observed if the half-reaction were paired with the hydrogen half-cell,

$$2H^+ + 2e^- \rightarrow H_2(g)$$

(This amounts to defining the voltage of the hydrogen half-reaction as $\mathscr{E}^0 = 0.000\ldots$ V.) By convention, half-cell reactions are written as *reductions*, and the corresponding *reduction potential* measures the relative tendency for that reduction to take place. (Some older books use oxidations and oxidation potentials. The oxidation potential is the negative of the reduction potential.) A high positive reduction potential indicates a strong tendency toward reduction, and a low negative potential indicates a strong tendency toward the oxidized state. When one half-reaction is subtracted from another to build a complete cell reaction, the corresponding reduction potentials are subtracted. Although one half-cell reaction may have to be multiplied by a stoichiometric constant to make the total number of electrons cancel in the overall reaction, the corresponding reduction potential is not multiplied by the constant. Reduction potentials are effective "electron pressures," and already are on a one-electron basis. Stoichiometry is taken care of by the quantity n in the expression

$$\Delta G = -n\mathscr{F}\mathscr{E}$$

If the concentrations of all ionic species are 1 molar, and all gases are at 1 atm partial pressure, this is the *standard state* for the emf designated $\mathscr{E}^0$ (analogous to the standard free energy, ΔG^0, with a superscript zero). If the reactants and products are not all in their standard concentrations, then the cell voltage is given by the Nernst equation,

$$\mathscr{E} = \mathscr{E}^0 - \frac{RT}{n\mathscr{F}} \ln Q$$

which is the electrochemical analogue of the free energy versus concentration equation discussed in Chapters 15 and 16. Just as an overall cell reaction can be divided into two half-reactions, Nernst equations can be written separately for each half-cell process.

One remarkable feature of redox reactions, in contrast to most other chemical reactions, is that they occur over such a wide range of equilibrium-constant values. For a two-electron reaction, a cell voltage of six volts corresponds to an equilibrium constant of $K_{eq} = 10^{200}$! This means that only rarely will two half-reactions have so similar half-cell potentials that the equilibrium constant for the overall reaction will be of moderate size. Most redox reactions either go to completion (effectively) or do not go at all. However, electrochemical methods can be used to study equilibrium, solubility product, and complex-ion formation in circumstances where one or another component at equilibrium is present in quantities far too small to be detected by standard analytical methods.

Using what we know about electrochemistry, we can design and build cells and batteries that deliver electrical power in small amounts in convenient places, and can use electrical power to bring about desirable chemical reactions. Electroplating and the refining of aluminum are examples. We also can use electrochemical principles to halt corrosion of susceptible metals that have low reduction potentials. What we cannot do yet is produce a cheap, lightweight storage battery with a high-energy density, or an electrochemical fuel cell (see postscript) that will operate with commonly available fuels.

17–9 POSTSCRIPT: THERE'S NO FUEL LIKE AN OLD FUEL—OR IS THERE?

For thousands of years, man has known from experience that when you want energy, you burn something. The first fuel was wood, and then later came coal and some of the oils and gases that seep out of swampy places. The first Industrial Revolution in England was based on an energy source—coal—that had been commonplace since the beginning of recorded history. Man also could burn fats and tallow from animals to get energy, but only in the last two hundred years did he appreciate that this was what the animals were doing with them as well. The energy for the operation of almost all living creatures is obtained by combustion; that is, by the burning of some energy-rich compound in the presence of oxygen. On this particular planet, which thus far has had a sufficient oxygen supply, this is a reasonable thing to do.

There are two main problems in extracting useful energy from chemical substances: finding the substances with the most energy, and finding ways to get the maximum energy out in a usable form. Both of these problems become important when the supply of possible fuels is short, when their use harms the environment with waste products, when the supply of oxygen is not unlimited, or when we have to transport the fuels far from their natural source, as in space exploration.

Table 17–3.   Energy Obtained Per Gram from Typical Fuels During Combustion with Oxygen Gas

Fuel	Physical state	Molecular weight (grams)	ΔH^0_{298} of combustion (kcal mole^{-1})	ΔH^0_{298} per gram
H_2	g	2	$-$ 68.3	-34.2
C_8H_{18} (octane, a typical gasoline)	l	114	-1303	-11.4
$C_{17}H_{35}COOH$ (stearic acid, typical for fats)	s	285	-2712	$-$ 9.5
$H_2NCH(CH_3)COOH$ (alanine, typical for protein)	s	89	$-$ 388	$-$ 4.4
$C_6H_{12}O_6$ (glucose, typical for carbohydrates)	s	180	$-$ 673	$-$ 3.7

The most thoroughly tested fuel systems on Earth are those of living organisms. The tests have been going on for three billion years or more, and the penalty for a bad experiment is extinction. If we look at Table 17–3, we see that living organisms have not done too badly. The most efficient combustion energy source, in terms of kilocalories of heat per gram of fuel, is hydrogen gas. Floating gasbag organisms have never evolved. Aside from the logistics problem, natural sources of hydrogen gas are not that abundant. Hydrocarbons such as gasoline yield one third as much energy per gram, and have the advantage of being in the more compact, liquid phase. Animals store most of their energy as fats, which are esters of long, gasolinelike fatty acids such as stearic acid. As Table 17–3 shows, these fats are very nearly as good energy-storage compounds as gasoline on a weight basis. The esterification helps to make them solids instead of liquids, an additional practical advantage for a creature that must move about.

Proteins (represented by the data for alanine) and carbohydrates (represented by the data for glucose) are only half as efficient in storing energy. But for plants, which do not move about, the energy yield *per gram* of fuel is really not very important. A redwood tree does not need to be weight-conscious. It happens that the biochemistry required to store energy in fatty acid molecules and fats, and to retrieve it again when needed, is complicated. In contrast, carbohydrate synthesis and breakdown are simpler and faster. Plants give up the unimportant energy-per-gram advantage of fats, for the significant advantage of easier carbohydrate biochemistry. In green plants, energy is stored as starches rather than fats. Even animals use the fast-access storage advantage of carbohydrates. We store a limited amount of energy as glycogen, a starchlike molecule, in our blood stream as a buffer against sudden energy needs.

When we leave our natal planet and begin exploring space, the energy-per-gram problem of all animals becomes even more serious. Gasoline then becomes too energy-poor, and we turn to direct combustion of hydrogen for the first stage of the Saturn rockets. We also look for newer and better ways of generating electrical energy to operate the spacecraft systems. One approach is not to attempt to carry energy from our home planet on the mini-planet of a spacecraft, but to generate electricity directly from the energy of the sun by means of solar batteries. As yet, we do not have the technology to make this a source of large power supplies. But if we have to use terrestrial fuel, then it should at least be the best fuel on a weight basis. The answer is the generation of electrical power from a fuel cell that uses the combustion of hydrogen as its energy source:

$$H_2(g) + \tfrac{1}{2}O_2(g) \rightarrow H_2O(l) \qquad \Delta H^0_{298} = -68.3 \text{ kcal}$$
$$\Delta G^0_{298} = -56.7 \text{ kcal}$$

About the worst thing that we could do with hydrogen in our spacecraft would be to burn it like we do most fuels on Earth, and use the heat to generate electricity. At a typical flame temperature of approximately 1500°K, if we used a heat engine with a condenser at 300°K, from the second law of thermodynamics, we only could hope for a maximal energy conversion efficiency of

$$\text{efficiency} = \frac{1500°K - 300°K}{1500°K} = 0.80 \quad \text{or} \quad 80\%$$

This is the maximum that is thermodynamically possible. We still would be faced with the much lower efficiency of any real heat engine, and the losses thereafter in converting heat into electricity via some sort of a generator. We would be lucky to convert as much as 25% of the energy available in the reaction into electricity. And nobody wants a furnace, boiler, and power plant operating within the close confines of a space capsule.

There is a better way. Why convert chemical energy into heat energy and then into electrical energy, if the heat step can be omitted? Heat is like Hungarian Forints: hard to obtain without losses, and even harder to exchange for another currency once you have them. A *fuel cell*, which makes a direct chemical–electrical conversion, is shown in Figure 17–13. The overall reaction is combustion of hydrogen. As with any other electrical cell, the half-reactions have been separated at two electrodes, and the electrons made to flow through an external circuit from anode to cathode. This cell is no different in principle than any of those shown in Figures 17–2 through 17–7. The innovation is that new reactants (H_2 and O_2) are fed constantly into the cell and the product (H_2O) is drawn off so that the voltage of the cell remains constant and the power output is uninterrupted. If we continually drained away zinc ions and replenished copper ions in the Daniell cell (Figure 17–4) then it, too, would be a fuel cell. The great advantage of all fuel cells is that they convert chemical free energy directly into electrical energy without the intermediate conversion through heat energy. The thermodynamic limitations of a heat engine are thereby avoided.

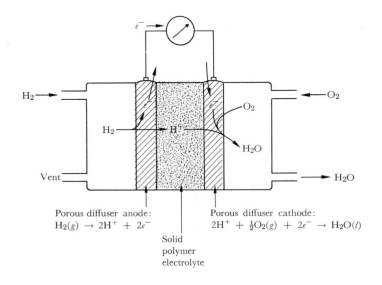

Figure 17–13. A hydrogen–oxygen fuel cell. Hydrogen gas introduced at the left dissociates in the porous, conducting barrier at the left side. Hydrogen ions migrate to the right through the electrolyte (for spacecraft applications, a solid polymer), and the electrons flow through the external circuit. At the right diffuser, these electrons, hydrogen ions from the electrolyte, and molecular oxygen combine to produce water. In a gravity field, the product water will run off into a storage tank; in free space, it can be drawn away from the diffuser cathode by a wick. In either event, the resulting water is available for human use.

The standard emf of the hydrogen fuel cell is calculated to be

$$\mathcal{E}^0 = \frac{-\Delta G^0}{n\mathcal{F}} = \frac{-(-56.69)}{2(23.06)} = +1.23 \text{ V}$$

and the dependence of the emf on gas pressure is given by the Nernst equation:

$$\mathcal{E} = +1.23 - 0.0296 \log_{10}\left(\frac{1}{p_{H_2}p_{O_2}^{1/2}}\right)$$

Methods that are tolerable on board a spacecraft, operating on a NASA budget, do not necessarily work economically on Earth. Why do we persist with antiquated coal-burning power stations, which pollute our skies with smoke and our rivers with warm water? Why not react coal, or at least readily available methane gas, directly in a fuel cell? The problems are all practical, and not theoretical. There is no known reversible electrode suitable for the methane half-reaction,

$$CH_4(g) + 2H_2O(l) \rightarrow CO_2(g) + 8H^+ + 8e^-$$

If there were, we immediately could set up a methane fuel cell similar to Figure 17–13, and a revolution in energy production would be at hand. Many hundreds of man-years have been spent in seeking suitable reversible electrodes for hydrocarbon fuels or coal, but so far with only limited success.

So, in spite of all our knowledge about electrochemical cells, the one key answer that could transform our use of fuels eludes us. The old fuels are good, but what we do with them in the process

$$\text{Bond energy} \rightarrow \text{Heat} \rightarrow \text{Electrical energy} \rightarrow \text{Other uses}$$

is sloppy, inefficient, and polluting. We cannot even escape this thermodynamic trap with our glorious new atomic energy. For after developing this new source of energy from nuclear reactions, what do we do with it as the first step in generating electricity? We use it to boil water! James Watt and Ernest Rutherford, wherever they are, must be laughing.

SUGGESTED READING

A. J. Bard, *Chemical Equilibrium*, Harper and Row, New York, 1966.

L. Hepler, "Electrochemistry," in *Chemical Principles*, Chapter 16, Blaisdell, New York, 1964.

W. M. Latimer, *Oxidation Potentials*, Prentice-Hall, Englewood Cliffs, N. J., 1952, 2nd ed.

C. A. VanderWerf, *Oxidation–Reduction*, Reinhold, New York, 1961.

"Selected Values of Chemical Thermodynamic Properties," *National Bureau of Standards Circular 500*. Department of Commerce, Washington, D. C., 1952.

QUESTIONS

1 Which of the two electrodes, anode or cathode, is associated in all electrochemical cells with oxidation? Which with reduction?

2 What is the driving force that creates a potential in a concentration cell?

3 What would happen if you replaced the semipermeable barrier in a concentration cell by a totally impermeable one? If you removed it altogether?

4 How is cell potential related to concentrations in a concentration cell?

5 How is a Daniell cell constructed, and what is the source of its electrical potential? Why is it unsuitable for powering flashlights or electric cars?

6 Why is a dry cell more suitable for the above applications? What chemical reactions take place in a dry cell? Which reaction produces electrons, and how are they used up? Which reaction takes place at the anode, and which at the cathode?

7 How is cell potential related to the free energy of the cell reaction? What is meant by the "standard potential"?

8 Are the free energies of half-cell reactions always additive? Are the half-cell potentials always additive? Outline the conditions under which each of the preceding statements is true, and is not true.

9 Does a high positive reduction potential for a half-cell reaction indicate a strong tendency for the redox couple to reduce other substances?

10 Given two half-cell reactions with different half-cell potentials, how does one obtain the overall cell voltage? If the resulting cell voltage is positive, what does this indicate about the cell reaction?

11 The two manganese half-reactions in Table 17–1 with standard reduction potentials of $+1.68$ V and $+1.21$ V can be added to produce a third reaction, listed in that table with a potential of $+1.49$ V. Explain why the reduction potential of this third reaction is not the sum of the other two; that is, not $+1.68$ V $+$ 1.21 V $= +2.89$ V. Account for the observed value of $+1.49$ V, in terms of the tabulated potentials of the first two reactions.

12 What is the Nernst equation, and how does it relate cell potentials to concentrations?

13 Why is the first term of the Nernst equation for a concentration cell zero?

14 Why do concentration cells with the same ratio of concentrations always have the same cell potential, no matter what the nature of the chemical substances involved in the reaction?

15 How can oxidation–reduction measurements yield the solubility of a substance when the dissolved substance is not present in large enough amounts to be detected by ordinary analytical methods?

16 How is corrosion of iron impeded by a layer of paint? A layer of zinc metal? A layer of tin metal? Why are these methods unnecessary with aluminum? What is meant by a "sacrificial metal" in preventing corrosion?

PROBLEMS

Data necessary for these problems can be found in Tables 17–1 and 17–2, and in the *Handbook of Chemistry and Physics* or similar references.

1 Determine the amount of useful work done when a mole of zinc powder is allowed to react with a 1.00-molar solution of $Cu(NO_3)_2$ in a constant-temperature calorimeter. If the reaction were carried out reversibly, how much useful work could be accomplished? ΔH^0_{298} for the reaction is -51.40 kcal. Calculate the heat liberated when the reaction is carried out reversibly.

2 A standard $Cl_2|Cl^-$ half-cell has been coupled to a $Cl_2(1.00\,atm)|Cl^-(0.010M)$ half-cell. What is the cell voltage? Determine ΔG_{298} for the reaction.

3 Assuming unit activities for all substances, determine which of the following reactions will be spontaneous:

a) $Zn + Mg^{2+} \rightarrow Zn^{2+} + Mg$
b) $Fe + Cl_2 \rightarrow Fe^{2+} + 2Cl^-$
c) $4Ag + O_2 + 4H^+ \rightarrow 4Ag^+ + 2H_2O$
d) $2AgCl \rightarrow 2Ag + Cl_2$

4 What are the potentials for the following cells or half-cells:

a) $\mathscr{E}^0$ for the cell
$$Zn(s)|Zn^{2+}\|Cu^{2+}|Cu(s)$$
b) $\mathscr{E}$ for the half-cell
$$Zn(s)|Zn^{2+}(0.0010M)$$
c) $\mathscr{E}$ for the half-cell
$$Cu^{2+}(10^{-36}M)|Cu(s)$$

5 What are the standard potentials, $\mathscr{E}^0$, for the half-cells

a) $S^{2-}|CuS(s)|Cu(s)$
b) $NH_3(aq), Zn(NH_3)_4{}^{2+}|Zn(s)$

6 Consider the cell
$$Ag(s)|Ag^+(1.0M)\|Cu^{2+}(1.0M)|Cu(s)$$

(a) Write the chemical reaction that takes place in this cell. In which direction will the reaction proceed spontaneously? (b) What is $\mathcal{E}^0$ for the cell? (c) Do electrons flow from Ag to Cu in the external circuit, or the other way?

7 The following two reactions have the $\mathcal{E}^0$ values given:

$$2Ag + Pt^{2+} \rightarrow 2Ag^+ + Pt$$
$$\mathcal{E}^0 = +0.40 \text{ V}$$
$$2Ag + F_2 \rightarrow 2Ag^+ + 2F^-$$
$$\mathcal{E}^0 = +2.07 \text{ V}$$

If the potential for the reaction $Pt \rightarrow Pt^{2+} + 2e^-$ is assigned a value of zero, calculate the potentials for the half-reactions

a) $Ag \rightarrow Ag^+ + e^-$
b) $F^- \rightarrow \frac{1}{2}F_2 + e^-$

8 Consider the following cell:

$$Ni|Ni^{2+}(0.010M)\|Sn^{2+}(1.0M)|Sn$$

(a) Predict the direction in which spontaneous reactions will occur. (b) Which metal, Ni or Sn, will be the cathode and which the anode? (c) What is $\mathcal{E}^0$ for the cell? (d) What will $\mathcal{E}$ be for the cell with the specified concentrations at 25°C?

9 Use the line notation of the previous problems to represent a cell that uses the following half-reactions:

$$PbO_2 + 4H^+ + 2e^- \rightarrow Pb^{2+} + 2H_2O$$
$$PbSO_4 + 2e^- \rightarrow Pb + SO_4^{2-}$$

a) Which is the reaction at the cathode of the cell? Which way do electrons flow in an external circuit?
b) What is $\mathcal{E}^0$ for this cell?

10 Show that hydrogen peroxide is thermodynamically unstable and should disproportionate to water and oxygen.

11 From the data in Tables 17–1 and 17–2: (a) Will Fe reduce Fe^{3+} to Fe^{2+} (assume unit activities)? (b) Calculate the equilibrium constant, K_{eq}, for the

reaction at 25°C,

$$Fe + 2Fe^{3+} \rightarrow 3Fe^{2+}$$

12 Find the missing standard reduction potentials for the following half-reactions from Table 17–1:

Half-reaction	$\mathcal{E}^0$ (V)
$MnO_4^- + 8H^+ + 5e^- \rightarrow$ $Mn^{2+} + 4H_2O$	+1.49
$Au^{3+} + 3e^- \rightarrow Au(s)$	+1.42
$Cl_2 + 2e^- \rightarrow 2Cl^-$	?
$AuCl_4^- + 3e^- \rightarrow$ $Au(s) + 4Cl^-$	?
$4H^+ + NO_3^- + 3e^- \rightarrow$ $NO + 2H_2O$	?

If we assume that all reactants and products are at unit activity: (a) Which substance in the half-reactions above is the best oxidizing agent? Which is the best reducing agent? (b) Will permanganate oxidize metallic gold? (c) Will metallic gold reduce nitric acid? (d) Will nitric acid oxidize metallic gold in the presence of Cl^- ion? (e) Will metallic gold reduce pure Cl_2 gas in the presence of water? (f) Will chlorine oxidize metallic gold if Cl^- ion is present? (g) Will permanganate oxidize chloride ion?

13 Consider the cell

$$Sn|SnCl_2(0.10M)\|AgCl(s)|Ag$$

(a) Will electrons flow spontaneously from Sn to Ag, or in the reverse direction? (b) What is the standard potential, $\mathcal{E}^0$, for the cell? (c) What will the cell potential, $\mathcal{E}$, be at 25°C?

14 For an electrochemical cell in which the spontaneous reaction is

$$3Cu^{2+} + 2Al \rightarrow 2Al^{3+} + 3Cu$$

what will be the qualitative effect on the cell potential if we add ethylenediamine, a ligand that coordinates strongly with Cu^{2+} but not with Al^{3+}?

15 Find the standard reduction potentials for the following half-reactions:

$$MnO_4^- + 8H^+ + 5e^- \rightarrow Mn^{2+} + 4H_2O$$
$$Al^{3+} \quad + 3e^- \rightarrow Al$$
$$Cl_2 \quad + 2e^- \rightarrow 2Cl^-$$
$$Mg^{2+} \quad + 2e^- \rightarrow Mg$$

(a) Which is the strongest reducing agent? Which is the strongest oxidizing agent? (b) Write the overall reaction for a successful cell made from the Mg and Cl_2 couples. (c) Write the line notation for this cell. (d) Which is the anode of the cell, and which is the cathode? (e) What is $\mathscr{E}^0$ for the cell? (f) What is the equilibrium-constant expression for this cell reaction? Calculate the numerical value of the equilibrium constant at 25°C.

16 Find the standard reduction potentials for the following half-reactions:

$$SO_4^{2-} + 4H^+ + 2e^- \rightarrow H_2SO_3 + H_2O$$
$$Ag^+ + e^- \rightarrow Ag$$

(a) Write the balanced overall reaction for a successful cell made from these two couples. (b) Write the line notation for the cell. (c) What is $\mathscr{E}^0$ for the cell? (d) What is the equilibrium constant for the cell reaction at 25°C? (e) Calculate the ratio of activities of products and reactants, Q, that will produce a cell voltage of 0.51 V.

17 Find the standard reduction potentials for the following half-reactions:

$$Hg_2^{2+} + 2e^- \rightarrow 2Hg$$
$$Cu^{2+} \quad + 2e^- \rightarrow Cu$$

(a) Write the overall reaction for a successful cell made from these two half-reactions. (b) Write the line notation for the cell. Which material, Hg or Cu, is the anode? (c) What is $\mathscr{E}^0$ for the cell? (d) What is the equilibrium constant for the cell? (e) What is the voltage of the cell when $[Hg_2^{2+}] = 0.10$ molar and $[Cu^{2+}] = 0.010$ molar?

18 A galvanic cell consists of a rod of copper immersed in a 10.0-molar solution of $CuSO_4$ and a rod of iron immersed in a 0.10-molar solution of $FeSO_4$. Using the Nernst equation and the reduction potentials for

$$Fe^{2+} + 2e^- \rightarrow Fe$$
$$Cu^{2+} + 2e^- \rightarrow Cu$$

calculate the voltage for the cell as described.

19 What voltage will be generated by a cell that consists of a rod of iron immersed in a 1.00-molar solution of $FeSO_4$ and a rod of manganese immersed in a 0.10-molar solution of $MnSO_4$?

20 A copper–zinc battery is set up under standard conditions with all species at unit activity. Initially, the voltage developed by this cell is 1.10 V. As the battery is used, the concentration of the cupric ion gradually decreases, and that of the zinc ion increases. According to Le Chatelier's principle, should the voltage of the cell increase or decrease? What is the ratio, Q, of the concentrations of zinc and copper ions when the cell voltage is 1.00 V?

21 Two copper electrodes are placed in two copper sulfate solutions of equal concentration and connected to form a concentration cell. What is the cell voltage? One of the solutions is diluted until the concentration of copper ions is one fifth its original value. What is the cell voltage after dilution?

22 The following cell:

$$Ag|Ag^+(0.10M)\|Ag^+(1.0M)|Ag$$

is a concentration cell and is capable of electrical work. (a) What is the cell potential? (b) Which side of the cell is the cathode, and which is the anode? (c) What is ΔG in calories for the spontaneous cell reaction?

23 Consider the following perpetual motion device built around a concentration cell. (a) Two copper electrodes are placed in copper sulfate solutions of equal concentration and connected to form a concentration cell. Initially there is no voltage in the cell. Assume that the two electrodes each contain more copper than is present in either solution. (b) Solution A is diluted until its Cu^{2+} concentration is cut in half, at which point the cell has a potential, $\mathscr{E}$. The cell is run and useful work is done on the surroundings until the concentrations in the two solutions are equalized, at which time the cell voltage has fallen to zero again. (c) Solution B is diluted until its Cu^{2+} concentration is halved, at which time the cell has the same potential, $\mathscr{E}$, as before, but in the opposite direction. Again the cell is run, and work is done until the concentrations in solutions A and B are the same. (d) Steps (b) and (c) are repeated, diluting first one solution and then the other by halving its Cu^{2+} concentration after equilibrium has been attained in the previous step. Since neither concentration ever falls to zero by the halving process, we can maintain this process as long as we like and take an infinite amount of work out of the cell. The operation of the cell actually helps us, for it *raises* the concentration of the solution that we had just diluted. What is wrong with this analysis?

24 Consider the cell

$$Zn|Zn^{2+}(0.0010M)\|Cu^{2+}(0.0010M)|Cu$$

for which $\mathscr{E}^0 = +1.10$ V. Does the cell voltage, $\mathscr{E}$, increase, decrease, or remain unchanged when each of the following changes is made? (a) Excess 1.0-molar ammonia is added to the cathode compartment. (b) Excess 1.0-molar ammonia is added to the anode compartment. (c) Excess 1.0-molar ammonia is added to both compartments at the same time. (d) H_2S gas is bubbled into the Zn^{2+} solution. (e) H_2S gas is bubbled into the Zn^{2+} solution at the same time that excess 1.0-molar ammonia is added to the other solution.

25 Using half-reactions, show that Ag^+ and I^- spontaneously form $AgI(s)$ when mixed directly at unit initial activity. Show that K_{sp} for AgI is 10^{-16}.

26 Consider the galvanic cell

$$Zn|Zn^{2+}\|Cu^{2+}|Cu$$

Calculate the ratio of $[Zn^{2+}]$ to $[Cu^{2+}]$ when the voltage of the cell has dropped to 1.05 V, 1.00 V, and 0.90 V. Notice that a small drop in voltage parallels a large change in concentration. Therefore, a battery that registers 1.00 V is quite "run down."

27 A silver electrode is immersed in a 1.00-molar solution of $AgNO_3$. This half-cell is connected to a hydrogen half-cell in which the hydrogen pressure is 1.00 atm and the H^+ concentration is unknown. The voltage of the cell is 0.78 V. Calculate the pH of the solution.

28 A standard hydrogen half-cell is coupled to a standard silver half-cell. Sodium bromide is added to the silver half-cell, causing precipitation of AgBr, until a concentration of $1.00M$ Br^- is reached. The voltage of the cell at this point is 0.072 V. Calculate the K_{sp} for silver bromide.

29 Chrome-plated automobile trim contains an iron core coated by a thick layer of nickel that is coated by a layer of chromium. Arrange the metals in order of ease of oxidation. What is the purpose of the chromium layer? Of the nickel layer?

Chemical phenomena must be treated as if they were problems in mechanics.
Lothar Meyer (1868)

18 RATES AND MECHANISMS OF CHEMICAL REACTIONS

If chemicals did not react, chemists would find other things to do.[1] The goal of every chemist, no matter what types of chemical compounds he works with, is to understand how and why these chemicals react and change. Yet this is the most difficult task of all. It is not enough to know the structures of all of the reactants and products, although such knowledge is a vital starting point. We also must know how these molecules approach one another and with what energies and with what orientations they interact. Both the concepts of energy and of entropy are important in understanding chemical reactions. In this chapter we shall look at some of the problems that face us because we cannot examine individual molecular events. We shall see how to explain complicated experimental expressions for the rate of reaction in terms of a mechanism of reaction, or a sequence of simple reactions, that lead to the overall chemical change. We shall examine two theories for predicting the rates of such simple reactions and compare their success or lack of it. We shall look at the two factors that often make reactions slow—energy and entropy—and see how catalysts can overcome these factors and accelerate chemical changes. Although

---

[1] If chemicals did not react, neither chemists nor anyone else would be doing anything at all.

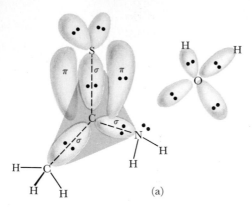

(a)

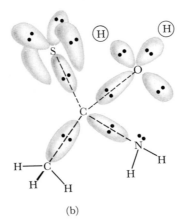

(b)

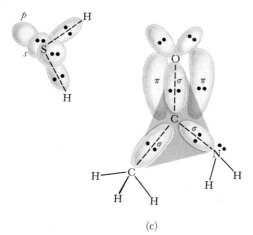

(c)

Figure 18–1. A mechanism for the reaction of thioaceta-mide, CH_3—CS—NH_2, *with water to make acetamide,* CH_3—CO—NH_2, *and* H_2S. *(a) The thioacetamide molecule has all four heavy atoms in one plane, with S, C, and N in an equilateral triangle around a central C. The central C has* sp^2 *hybridization and makes a σ single bond to C and N and a σ, π double bond to S. Molecular orbitals and lone-pair orbitals are in color. The orbitals of the double bond are distorted toward S to represent its greater electronegativity. Orbitals that play no part in the reaction (C—H, N—H, N lone-pair) are not drawn. (b) Intermediate or transition state, with partial bonds from C to both S and O. The former O—H bonding electron pairs are becoming lone pairs. (c) Products of the reaction: acetamide and* H_2S. *The tetrahedral geometry of the transition state has reverted to trigonal planar geometry as the sulfur atom leaves. Bonding around the S atom is shown with unhybridized s and p orbitals, in contrast to the* sp^3 *hybridization in water (a). This is probably close to the truth, since the bond angle in* H_2S *is 92° in contrast to the 105° of* H_2O.

we cannot present a complete theory of chemical reaction (no one can do this yet), we will outline the foundations on which this theory will someday be constructed.

18–1 WHAT HAPPENS WHEN MOLECULES REACT?

Let us suppose that, by some means, we can watch what happens when two molecules react. As an example, consider the reaction of a molecule of thioacetamide, CH_3—CS—NH_2, with water to yield acetamide, CH_3—CO—NH_2, and H_2S (Figure 18–1).

In the original thioacetamide molecule, the central carbon atom is bound to C and to N by σ bonds, and by a σ, π double bond to S [Figure 18–1(a)]. Since S and N both are more electronegative than C, the electron pairs of their bonds to C are slightly displaced toward S and N. These two atoms bear a small negative charge, and the central C has a small positive charge. All four atoms lie in a plane.

The most favorable direction of approach of a water molecule is perpendicularly from either side of the plane of the four heavy atoms. The most favorable orientation of the incoming water molecule is as shown in Figure 18–1(a). Here a lone electron pair from the water is attracted to the positive charge on the central C. As the water molecule approaches this C atom, the lone-pair electrons are drawn to it and begin to form a partial bond. This partial bond formation has two effects: It weakens the bond between C and S by repelling the electrons even more toward S, and it simultaneously weakens the O—H bonds in water by pulling electrons from these bonds toward O as the O lone-pair electrons are attracted toward C. This intermediate state appears in Figure 18–1(b). The central carbon atom now has two single bonds to C and N, and two partial bonds to S and O.

This intermediate state is not stable. If the water molecule falls away again and the situation reverts to that of Figure 18–1(a) (and there is no reason why this could not happen), then we see no net reaction. The water molecule rebounds from a collision with thioacetamide and goes its separate way. It also could happen that the *sulfur* atom falls away, as in Figure 18–1(c). In this eventuality, the two protons released by O as it makes a double bond with C are attracted by the sulfur atom with four electron pairs, and a molecule of H_2S results. The reaction

$$CH_3—CS—NH_2 + H_2O \rightarrow CH_3—CO—NH_2 + H_2S$$

is complete. The reverse reaction also can occur; a molecule of H_2S can collide with one of acetamide and produce water and thioacetamide. We would be less likely to see such an event if we could watch reactions at the molecular level, simply because there are very few H_2S molecules compared to water molecules.

What factors might affect the reaction of thioacetamide? One factor is certainly the geometry of approach of the water molecule. If the water molecule approached *in the plane* of the thioacetamide molecule, it would find its entry blocked by sulfur lone electron pairs and hydrogen atoms (to a greater extent than is apparent in the skeletal drawings of Figure 18–1). Moreover, if the water molecule approached with a *hydrogen atom*, instead of a lone electron pair, pointed at the central C, it would not be as attracted to the thioacetamide molecule and would be more likely to rebound without reaction. If we could watch every collision, we might see that only one collision in ten, or in a hundred, had both molecules properly oriented for reaction.

A second factor is the energy of the two molecules. In the simplest theories this is expressed only as the relative *speed* of the two molecules upon collision. If the relative speed of the two molecules is small upon impact, the intermediate state will be more likely to revert to the starting molecules. A slowly moving water molecule can bounce harmlessly off the thioacetamide. In contrast, a water molecule that slams forcefully against the thioacetamide has more of a chance of driving away the sulfur atom, thereby producing acetamide and H_2S. We might find that we could plot a curve of the probability of reaction as a function of the velocity of approach of the two molecules along a line connecting their centers.

Unfortunately, the kinds of observations we have been describing for a reaction are an unattainable dream. We must try to find out what is happening during a reaction in a more indirect way. Frequently the most that we can say about a proposed mechanism of reaction is that it is not incompatible with the data. There is always the lingering possibility that some other mechanism of reaction might explain the same data just as well. A classic example of this ambiguity is the reaction of H_2 and I_2. In 1893, Max Bodenstein, in Germany, studied the reaction

$$H_2 + I_2 \rightarrow 2HI$$

This was the first comprehensive kinetic study of a reaction occurring in the gas phase. From that time until 1967, virtually every kinetics text and treatise used this reaction as the ideal example of a two-body collision mechanism. One gas molecule of H_2 collides with a molecule of I_2, they reshuffle atoms, and two molecules of HI are the result. But in 1967, J. H. Sullivan showed that this reaction does not take place by a two-body collision at all, but by a complicated chain reaction. We shall see later why the data measured before 1967 could be explained with equal ease with the two-body or a three-body model.

Not only are we unable to watch individual molecules, we cannot choose the orientation of the molecules upon collision. The best we can do is to estimate the probability of the molecules' being suitably oriented and then modify our calculations of rates of reaction by a suitable factor. Such a correction sometimes is used and is called a *steric factor*.

In a gas reaction or a reaction in solution, we cannot even choose the velocity of approach of the reacting molecules. The molecules in a sample of gas will have a distribution of velocities, as shown in Figure 2–11. We can shift the distribution of velocities by varying the temperature of the gas.

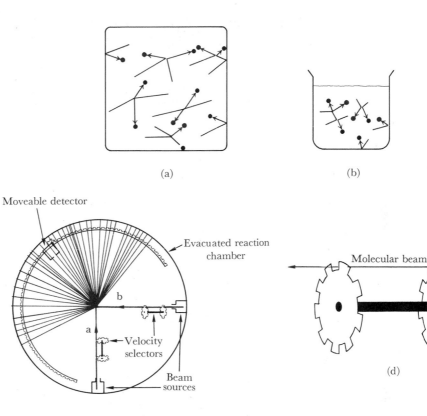

Figure 18–2. The two main classes of experiments in chemical kinetics. In bulk reactions in the gas (a) or liquid phase (b), the orientation of reacting molecules is uncontrolled, and a distribution of molecular velocities exists. In a crossed molecular-beam experiment (c), the orientations are still uncontrolled, but only molecules or ions with certain velocities are used. A typical velocity selecting device (d) is a pair of wheels with suitably spaced sectors cut out so only molecules that take a specified time to travel the distance from one wheel to the next can pass through the open sectors on both wheels. The beam sources typically are ovens that emit a stream of gas molecules, and electrostatic fields that accelerate ions.

As illustrated in Figure 2–11, in nitrogen gas the fraction of all the molecules having a velocity greater than some value such as 1000 m sec^{-1} increases as the temperature increases. At 273°K, only 0.44% of the N_2 molecules have velocities of 1000 m sec^{-1} or greater; at 1273°K, 35% have this velocity or greater; at 2273°K, this fraction increases to 55%. However, nothing that we can do to the system will give us *one specific* velocity.

We can remove the velocity distribution for certain reactions by using the method of crossed molecular beams (Figure 18–2). Instead of reactions occurring between molecules dispersed in a solution or a gas, beams of molecules or ions are passed through one another in an evacuated chamber with negligible amounts of other molecules present. The molecules in the crossed beams react with one another and are scattered from the beams. The products of the reaction, and the unreacted initial molecules, can be observed as a function of angle of scattering by using a moveable detector mounted inside the chamber. This arrangement has the great advantage that the velocity selectors can limit the beam to molecules with velocities in a chosen small range. A knowledge of the products of the reaction as a function of angle of deflection or scatter provides much more information about the process of reaction. The orientation problem remains in a molecular-beam experiment, but one can imagine experiments in which this factor is controlled as well. Intense magnetic or electric fields placed just before the beams intersect might give the majority of the molecules in the beam one preferred orientation in space if the molecules had magnetic moments or dipole moments.

Some of the reactions that have been studied with crossed molecular beams are

$$K + HBr \quad \rightarrow KBr \quad + H$$
$$K + CH_3I \quad \rightarrow CH_3 \quad + KI$$
$$K + C_2H_5I \rightarrow C_2H_5 + KI$$

The reactants, beams of K atoms, HBr, CH_3I, and C_2H_5I molecules, are emitted from heated ovens within the evacuated chamber. The detector is a heated wire filament called a *surface ionization detector*, which is sensitive to alkali metals or compounds of alkali metals.

The disadvantage of molecular-beam experiments is that not all chemical reactions are suitable for study with molecular beams in evacuated chambers. Molecular-beam methods remain a special tool for making complete studies of certain special reactions. The majority of chemical reactions must be studied by bulk methods: gas mixtures, solutions, and (less frequently) solids.

18–2   MEASUREMENT OF REACTION RATES

The rate of reaction usually is followed in bulk methods by watching the disappearance of a reactant or the appearance of a product in a given time. If the chemical reaction is

$$A + 2B \rightarrow 3C$$

then the rate of appearance of product C in a time interval Δt is

$$\frac{\Delta[C]}{\Delta t} \tag{18–1}$$

in which the concentration of C, [C], usually is expressed in moles liter^{-1}. This is the average rate of appearance of C during the time interval Δt.

The limit of this average rate as the time interval becomes smaller is called the *rate of appearance of C at time t*. It is the slope of the curve of [C] versus t at time t. This instantaneous slope or rate is written $d[C]/dt$. Since one molecule of A disappears for every three of C that are produced, and two molecules of B disappear during the same process, the rates of disappearance and appearance of chemical species are related by the expression

$$-\frac{d[A]}{dt} = -\frac{1}{2}\frac{d[B]}{dt} = +\frac{1}{3}\frac{d[C]}{dt}$$

The rate of a chemical reaction will depend on the concentrations of the reactants, although not always in the way that might be expected from the overall chemical equation. For the reaction of hydrogen gas with gaseous iodine to produce HI,

$$H_2 + I_2 \rightarrow 2HI$$

the relationship between rates is

$$\frac{d[HI]}{dt} = -2\frac{d[H_2]}{dt} = -2\frac{d[I_2]}{dt}$$

and as you might intuitively expect, the rate equation is

$$\frac{d[HI]}{dt} = k[H_2][I_2] \tag{18–2}$$

The rate of reaction is proportional both to the concentration of H_2 and of I_2, and dependent on the first power of each concentration. This does not mean that the reaction proceeds by a collision of one H_2 molecule with one I_2; since 1967 we have had evidence that it does not. We must distinguish clearly between the order of a reaction and the molecularity of the reaction.

The *order* of a reaction is the sum of all the exponents of the concentration terms in the rate equation. The HI reaction is first order in each of the reactant concentrations and is second order overall. Order is a purely experimental parameter and describes what is observed about the rate equation rather than implying anything about the mechanism of reaction.

The *molecularity* of a simple one-step reaction is the number of individual molecules that interact in the reaction. Molecularity requires a knowledge of the reaction mechanism. A reaction such as that of hydrogen and iodine actually may take place as a series of half a dozen individual reactions for which we could specify the molecularity of each. The concept of the molecularity of an overall reaction that occurs in a series of steps has no meaning. Most simple one-step reactions are unimolecular (spontaneous decay) or bimolecular (collision). True trimolecular reactions are rare, as three-body collisions are unlikely. Tetramolecular and higher reactions are virtually unheard of. Reactions that from their stoichiometry appear to be trimolecular or higher, after careful study, usually are seen to be the sum of a series of simple unimolecular and bimolecular steps. One of the challenges of chemical kinetics is to determine the true set of reactions in such a case.

The reaction of hydrogen gas with bromine is in complete contrast to that with iodine. The overall reaction is similar:

$$I_2 + Br_2 \rightarrow 2HBr$$

but the experimental dependence of the rate of production of HBr upon concentrations of reactants and products is utterly different from Equation 18–2:

$$\frac{d[HBr]}{dt} = \frac{k[H_2][Br_2]^{1/2}}{1 + k'([HBr]/[Br_2])} \tag{18-3}$$

This expression has two experimental rate constants, k and k'. We cannot talk about the molecularity of the reaction, because the overall process is the result of an elaborate chain of reactions that we shall come back to later. Even the order is a puzzle. At the start of a reaction of H_2 with Br_2, when little HBr is present, the second term in the denominator can be neglected. Then the reaction is effectively $1\frac{1}{2}$ order: first order in H_2 and one half order in Br_2. As the product, HBr, accumulates, it slows down the rate of production of more HBr. Therefore, HBr is called an *inhibitor* of the reaction.

The formation of HCl is even more complicated. The production of HCl is accelerated by light of intensity I and is inhibited by the presence of oxygen gas, even at low oxygen concentrations. The difficulty of purifying the H_2 and Cl_2 gases and eliminating all traces of O_2, led, for many years, to erroneous conclusions about the kinetics of this reaction. The best experimental rate equation for the appearance of HCl is

$$\frac{d[HCl]}{dt} = \frac{k_1 I[H_2][Cl_2]}{k_2[Cl_2] + [O_2]([H_2] + k_3[Cl_2])} \tag{18-4}$$

Notice that, in the limit of the complete absence of oxygen gas, the rate is proportional to the concentration of H_2 gas and not dependent upon the concentration of Cl_2 gas at all! (The second term in the denominator of Equation 18–4 is zero, and the remaining Cl_2 concentrations in the numerator and denominator cancel.) The reaction is additionally complicated by side reactions that take place on the surfaces of the reacting vessels. The results obtained sometimes depend on the size and shape of the reaction container. All of this is a far cry from the simplicity of the HI system. There are side reactions in the HI system, too, but they are not important below 800°K.

Following the Course of a Reaction

How do we measure concentrations of reactants and products during a reaction to find rate equations such as we have been examining? If the total number of moles of gas changes during a gas reaction, the course of the reaction can be measured from the change in pressure at constant volume or the change in volume at constant pressure. These are examples of *physical* measurements that can be performed on the system while it is reacting. They have the advantage of not disturbing the reacting system, and usually they are rapid. With automatic recording devices, we can monitor a physical quantity continuously during the reaction.

Other physical measurements often used in kinetic studies include optical methods such as the rotation of light by a solution (useful if reactants and products have different abilities to rotate polarized light), changes in refractive index of a solution, color, and absorption spectra. Common electrical methods include the electrical conductivity of a solution (especially useful when ions are being produced or consumed), electrical potential in a cell, and mass spectrometry. Thermal conductivity, viscosity of a polymerizing solution, heats of reaction, and freezing points also have been employed. The disadvantage of all such methods is that they are indirect. The property observed must be calibrated in terms of concentrations of reactants and products. The calibration is subject to systematic errors, especially if side reactions are occurring.

Chemical methods are more straightforward and yield concentrations directly. With such methods, a small sample is extracted from the reacting mixture, and the reaction is halted by dilution or cooling long enough to measure concentrations. The serious disadvantages are that we are removing a part of the reacting system and thereby gradually changing it. Moreover, if the reaction cannot be stopped in the sample removed for analysis, then the analysis is that much less accurate. In the gas-phase reactions between H_2 and Cl_2, Br_2 or I_2, there is no change in the number of moles of gas before and after reaction, so pressure- or volume-change methods cannot be used. To study these reactions, samples are taken, and the gas mixtures are analyzed chemically for their compositions.

A First-Order Rate Equation and the Decay of ^{14}C

In a first-order process, the rate of disappearance of reactant is proportional to the amount of reactant present. Each reactant molecule has the same probability of breakdown in a given time interval, and the total rate of breakdown simply depends on how many molecules are present. We first saw this rate expression in Section 13–2. The expression is

$$\frac{dn}{dt} = -kn \tag{13–1}$$

with n being the total number of molecules present. This rate expression can be integrated to yield the concentration as a function of time:

$$n = n_0 e^{-kt} \tag{13–2}$$

Table 18–1.   Decomposition of N_2O_5 in CCl_4 Solution at $45°C^a$

Time, t (sec)	$[N_2O_5]$ (mole liter^{-1})	$\Delta[N_2O_5]$	Δt	$-\dfrac{\Delta[N_2O_5]}{\Delta t}$ (mole liter^{-1} sec^{-1})	$k^b = -\dfrac{1}{[N_2O_5]}\dfrac{\Delta[N_2O_5]}{\Delta t}$
				$N_2O_5 \to 2NO_2 + \frac{1}{2}O_2(g)$	
0	2.33				
		-0.25	184	1.36×10^{-3}	6.2×10^{-4}
184	2.08				
		-0.17	135	1.26×10^{-3}	6.3×10^{-4}
319	1.91				
		-0.24	207	1.16×10^{-3}	6.5×10^{-4}
526	1.67				
		-0.32	341	0.94×10^{-3}	6.2×10^{-4}
867	1.35				
		-0.24	331	0.72×10^{-3}	5.9×10^{-4}
1198	1.11				
		-0.39	679	0.57×10^{-3}	6.2×10^{-4}
1877	0.72				

^{a}From H. Eyring and F. Daniels, *J. Am. Chem. Soc.* **52**, 1472 (1930).
$^b[N_2O_5]$ = average of concentrations at beginning and end of this time interval. For the first entry, $[N_2O_5] = (2.33 + 2.08)/2$.

This rate equation was used in the example of dating with carbon-14 in Section 13–5, where the expressions in terms of the concentration of ^{14}C are

$$\frac{d[^{14}C]}{dt} = -k[^{14}C] \tag{18–5}$$

$$[^{14}C] = [^{14}C]_0 e^{-kt} \tag{18–6}$$

The integrated Equation 18–6 is plotted in Figure 13–3. Taking the logarithm of both sides of Equation 18–6 yields

$$\ln[^{14}C] = \ln[^{14}C]_0 - kt \tag{18–7}$$

This is the equation of Figure 13–8. The plot is a straight line, with a negative slope equal to the rate constant, k. The slope of the plot of Equation 18–6 at any time t, as shown in Figure 13–3, is proportional to the concentration of ^{14}C remaining at that time. This, in words, is what Equation 18–5 means.

Decomposition of N_2O_5

In Section 15–8, we encountered the decomposition of solid N_2O_5 as an example of a reaction that is spontaneous yet strongly endothermic. Now we shall look at the decomposition of N_2O_5 dissolved in carbon tetrachloride as an example of a first-order chemical reaction. Solid N_2O_5 and one product,

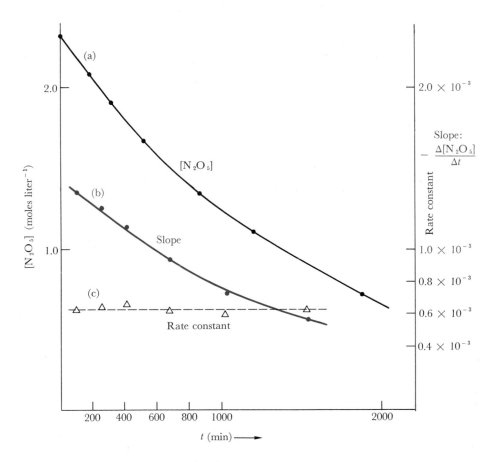

Figure 18–3. The kinetics of the reaction $N_2O_5 \rightleftarrows 2NO_2 + \frac{1}{2}O_2$ *from the data in Table 18–1. (a) Plot of concentration of* N_2O_5 *as a function of time. (b) Plot of the negative of the slope of curve (a) as a function of time, or of* $-\Delta[N_2O_5]/\Delta t$ *versus* Δt. *(c) Plot of the slope at any time divided by the concentration at that time, or of* $-(\Delta[N_2O_5]/\Delta t)/[N_2O_5] = k$, *the first-order rate constant. The constancy of this number proves that the reaction follows a first-order rate law.*

NO_2, both are soluble in CCl_4; the other product, O_2, is not. The reaction

$$N_2O_5 \rightarrow 2NO_2 + \tfrac{1}{2}O_2(g)$$

can be followed by measuring the total volume of oxygen gas that bubbles out of the solution.

The data for this reaction are given in Table 18–1, after O_2 volume measurements have been converted to concentrations of N_2O_5 left in the solution. These data are plotted in Figure 18–3 as an example of the way in which concentration data are treated. In this figure are the concentration of N_2O_5 at any time, the rate of change in this concentration, and this rate of change divided by the concentration. This last quantity is equal to the rate constant. That the rate of change divided by the concentration is constant (within the limits of experimental error in the data in Table 18–1) demonstrates that the reaction is indeed first order.

Stoichiometry and Rate Expressions

The reaction

$$2NO(g) + O_2(g) \rightarrow 2NO_2(g)$$

has an observed rate equation of the form

$$\frac{d[NO_2]}{dt} = k[NO]^2[O_2]$$

The reaction is second order in NO and first order in O_2, and is third order overall. The rate equation happens to agree with the stoichiometry of the chemical reaction; this agreement suggests (but does not prove) that this may be a simple one-step reaction involving three molecules.

In contrast, ethanol and decaborane react in solution according to the equation

$$30C_2H_5OH + B_{10}H_{14} \rightarrow 10B(OC_2H_5)_3 + 22H_2$$

One might naively expect this to have a thirty-first order rate expression. In fact, the reaction is second order, first order in each of the two reactants. For the rate of disappearance of ethanol,

$$-\frac{d[C_2H_5OH]}{dt} = k[C_2H_5OH][B_{10}H_{14}]$$

Summary: The Goals of Chemical Kinetics

Some chemical processes are simple one-step reactions involving one, two, or occasionally three molecules. Many more processes are the combination of several such simple reactions. One of the goals of chemical kinetics is to find out what the true molecular mechanism of a complex process is. Why do HI, HBr, and HCl have such different experimental rate equations for a reaction that looks superficially the same in all three cases? To a kineticist the question: What is the mechanism of the reaction? means: What is the sequence of simple reactions that produces the observed kinetics and stoichiometry of the overall reaction? To this question organic and structural chemists have added: What is the geometry of the reaction for each simple step in the overall process? The goal of this inquiry is to predict why the simple reactions proceed as they do and to predict the rates at which they occur.

The theories that have been developed to calculate the rate constants for simple unimolecular and bimolecular reactions are our next topic.

Before you go on. If you want help to understand the effect of concentration on reaction kinetics, refer to Review 14 of *Programed Reviews of Chemical Principles* by Lassila *et al.*

18-3 CALCULATING RATE CONSTANTS FROM MOLECULAR INFORMATION

Let us assume a simple bimolecular reaction,

$$A + B \rightarrow C + D \tag{18-8}$$

with a rate expression,

$$-\frac{d[A]}{dt} = k[A][B]$$

How far can we go in calculating k from the molecular properties of A, B, C, and D? One of the earliest observations was that k varies with temperature; the rate constant is larger, and the rate of reaction is faster, at higher temperatures.

Arrhenius' Activation Energy

If we plot the logarithm of the rate constant against the reciprocal of temperature, we usually obtain a straight line. Although Arrhenius was not the first person to do this, he developed the idea and gave it an explanation. Therefore, such a plot is called an Arrhenius plot. What does it mean in terms of reaction mechanisms?

Van't Hoff and others had been working, in the late 1800's, on the variation of free energy change of reaction and of the equilibrium constant with temperature. They discovered that the equilibrium constant, K_{eq}, varies with absolute temperature, T, and with the heat of reaction in the following way:

$$\frac{d \ln K_{eq}}{dT} = \frac{\Delta H^0}{RT^2} \tag{18-9}$$

This expression can be derived from the Gibbs–Helmholtz equation mentioned in connection with Figure 16–3, and ultimately can be derived rigorously from thermodynamics. During the same period, G. M. Guldberg and P. Waage found that they could derive the equilibrium constant from kinetic arguments. If the forward reaction in Equation 18–8 has the rate

$$\text{Rate}_1 = k_1[A][B]$$

and the reverse reaction has the rate

$$\text{Rate}_2 = k_2[\text{C}][\text{D}]$$

then they assumed that *equilibrium* is the state in which forward and reverse rates are equal, so no net change in the reacting system is occurring with time:

$$[\text{A}][\text{B}]k_1 = k_2[\text{C}][\text{D}]$$

$$K_{\text{eq}} = \frac{k_1}{k_2} = \frac{[\text{C}][\text{D}]}{[\text{A}][\text{B}]}$$

The equilibrium constant, in this argument, is the ratio of the rate constants for the forward and reverse reactions.

This is an erroneous derivation. It is valid only when the reaction is a simple one-step process in which the stoichiometry of the reaction is reflected in the coefficients of the concentration terms in the rate equation. Nevertheless, it is valid for the kind of reactions we are considering here: simple bimolecular reactions. If the equilibrium constant is the ratio of forward and reverse rate constants, Equation 18–9 suggests that the enthalpy of reaction might be the difference between two energies, E_1 and E_2:

$$K_{\text{eq}} = \frac{k_1}{k_2} \qquad \Delta H^0 = E_1 - E_2$$

$$\frac{d \ln (k_1/k_2)}{dT} = \frac{E_1 - E_2}{RT^2}$$

$$\frac{d \ln k_1}{dT} = \frac{E_1}{RT^2}$$

$$\frac{d \ln k_2}{dT} = \frac{E_2}{RT^2}$$

Or, in general,

$$\frac{d \ln k}{dT} = \frac{E_a}{RT^2} \tag{18–10a}$$

The quantity E_a is called the Arrhenius energy of activation. Equation 18–10a can be rearranged to yield

$$\frac{d \ln k}{d(1/T)} = -\frac{E_a}{R} \tag{18–10b}$$

If the Arrhenius energy of activation is not a function of temperature, Equation 18–10b predicts that a plot of $\ln k$ against the reciprocal of the absolute temperature will generate a straight line. This is true for many reactions, and the activation energy is one of the standard experimental

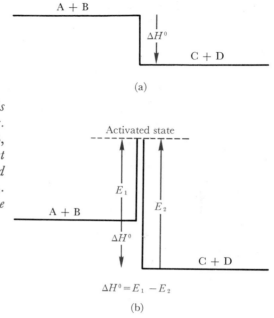

Figure 18–4. (a) The enthalpy or heat of a reaction is the change in enthalpy when reactants become products. (b) The activation energy for the reaction of A and B, E_1, is the energy necessary before A and B will react instead of rebound. The reverse reaction, in which C and D restore A and B, also has an activation energy, E_2. The difference between these two activation energies is the enthalpy of the reaction.

parameters by which a chemical reaction is described. If E_a is not a function of temperature, Equation 18–10b can be integrated to yield

$$k = Ze^{-(E_a/RT)} \tag{18–11}$$

in which Z would be the rate constant if there were no activation energy required.

The activation energy is a barrier that the colliding molecules must surmount if they are to react rather than recoil from one another. We already have used this idea in the thioacetamide reaction of Section 18–1. We postulated that if the thioacetamide and water molecules do not collide head-on with sufficient energy the redistribution of bonds in Figure 18–1(b) and (c) will never occur. Water will recoil from the thioacetamide molecule and no reaction will take place. Now we have experimental evidence, in the form of the temperature dependence of k, that some such threshold energy *is* involved in chemical reactions. Arrhenius' explanation of activation energies assumes that every pair of molecules with energy less than E_a will not react, and every pair with energy greater than E_a will react. The theory is certainly too simple, but it is a beginning.

Nothing is changed in the thermodynamics of the overall reaction, as Figure 18–4 shows. The activation barriers to forward and reverse reactions, E_1 and E_2, are such that their difference, $\Delta H^0 = E_1 - E_2$, is the thermodynamic heat of reaction. The higher the barrier, E_1, the slower the forward reaction will be. However, since E_2 must rise by the same amount as E_1 if their difference is fixed, the reverse reaction is slowed by the same amount. The point of equilibrium is not affected by the individual numerical values of the activation energies for forward and reverse reactions, but only by the difference between them, which is ΔH^0.

Challenge. Enzyme-catalyzed reactions, pressure cookers, and high-altitude cuisine are the topics of Problems 18–1 and 18–9 through 18–11 in the Butler and Grosser book.

Collision Theory of Bimolecular Gas Reactions

The next logical step is to construct a collision theory for gas reactions. A reaction between two molecules occurs, in this theory, when the molecules collide with energy in excess of E_a. A theory could hardly be simpler. There are two questions to be answered before the rate constant can be calculated:

1) How often do two molecules collide per cubic centimeter of gas mixture?
2) In what fraction of these collisions does the combined energy of the two molecules exceed E_a?

The collision frequency can be calculated from the simple kinetic theory of gases with the methods that were introduced in Chapter 2. The frequency depends on the concentrations of the two reacting gases, and also on their molecular weights, the distance between the molecular centers on collision, and on the square root of the temperature. Since the molecules move more rapidly at higher temperatures, they collide more often. The fraction of pairs of molecules having energy equal to or greater than E_a upon collision (if we assume the type of Boltzmann distribution of molecular velocities that we saw in Figure 2–11) is

$$e^{-(E_a/RT)}$$

According to simple collision theory, the rate of reaction then is

$$\text{Rate} \quad = (\text{collision frequency}) \times (\text{probability that } E \geq E_a)$$

$$-\frac{d[A]}{dt} = (Z[A][B]) \times e^{-(E_a/RT)}$$

$$-\frac{d[A]}{dt} = Ze^{-(E_a/RT)}[A][B]$$

The rate of a reaction is greater at higher temperatures because collisions are more frequent and because the probability that a colliding pair will have an energy greater than E_a is also higher. The constant, Z, can be calculated from the molecular weights and the diameters of the reacting molecules by approximating them with spheres. The bimolecular rate constant, k, then is

$$k = Ze^{-(E_a/RT)} \tag{18-12}$$

	A + B $\rightarrow$ Products Rate = $Ze^{-(E_a/RT)}[A][B]$				
			$\log_{10} Z$		
Reaction	E_a (kcal)[a]	Observed	Collision theory	Absolute rate theory	ΔH^0 (kcal)[a]
$NO + O_3 \rightarrow NO_2 + O_2$	2.5	11.9	13.7	11.6	−47.8
$NO + Cl_2 \rightarrow NOCl + Cl$	20.3	12.6	14.0	12.1	+20.0
$NO_2 + CO \rightarrow NO + CO_2$	31.6	13.1	13.6	12.8	−54.1
$2NO_2 \rightarrow 2NO + O_2$	26.6	12.3	13.6	12.7	+27.0
$2NOCl \rightarrow 2NO + Cl_2$	24.5	13.0	13.8	11.6	+18.1
$2ClO \rightarrow Cl_2 + O_2$	0.0	10.8	13.4	10.0	−33.0

[a] Per mole of reactants.

This theory is tested in the data in Table 18–2. The Arrhenius activation energy is tabulated for six bimolecular gas reactions, along with the observed preexponential factor, Z, and its theoretical value as calculated from the collision theory and the absolute rate theory that we will discuss in the next sections. Keep in mind that these are the *logarithms* of Z that are tabulated, so a disagreement between theory and experiment of 1.0 means an error by a factor of 10 in the rate constant. The agreement is generally encouraging for so simple a theory that has no assumptions other than those of the kinetic theory of gases. There are discrepancies; for example, the ClO reaction rate is incorrect by a factor of 400. When discrepancies occur, the absolute rate theory usually does a better job of predicting Z than the collision theory does.

In the right column of Table 18–2 are the standard enthalpies or heats of reaction. The relative enthalpy of reactants and products, and the activation barrier between them, are plotted for these reactions in Figure 18–5. Some reactions, such as $NO_2 + CO$, must surmount a considerable activation barrier. For other reactions, the barrier is nonexistent, as with $2ClO$. For others such as $2NO_2$, the barrier to reaction is only the heat of reaction itself, and the reverse reaction has a zero activation energy. The most general case is diagramed at the bottom of Figure 18–5.

Activated Complexes

Before we consider the absolute rate theory, we must look more closely at the state of the reactant molecules as they cross the activation barrier. In the reaction

$$2ClO \rightarrow Cl_2 + O_2$$

the Cl and O atoms are bonded at the start, and the two ClO molecules are too far apart to exert any influence on one another. At the end of the reaction, the Cl atoms are 1.99 Å apart in a Cl_2 molecule, the O atoms are 1.21 Å

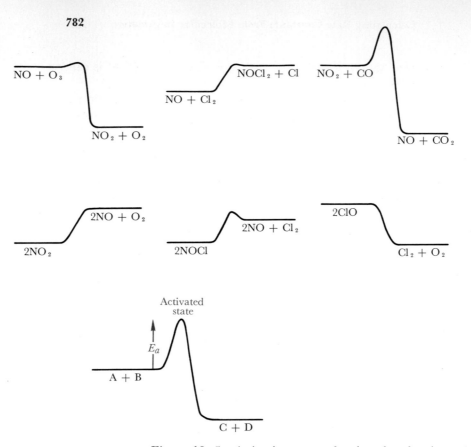

Figure 18–5. Activation-energy barriers for the six reactions tabulated in Table 18–2. Some of these reactions have appreciable barriers; others such as the 2ClO reaction have none at all. The general activation-energy diagram is shown at the bottom.

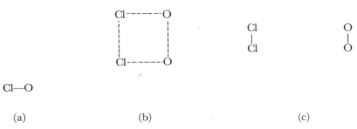

Figure 18–6. (a) The initial state in the reaction $2ClO \rightarrow Cl_2 + O_2$, with two ClO molecules infinitely far apart. (b) A possible transition state. (c) The final state with Cl_2 and O_2 molecules infinitely far apart.

apart in an O_2 molecule, and these two molecules are far apart. What is the intermediate, activated state?

The activated complex is diagramed in Figure 18–6(b). All four atoms are an unspecified distance apart, somewhat farther away than if they were bonding in the stable sense of the term. We can be sure that the activated complex is *not* one in which all four atoms are so far apart that they exert no influence on one another; some sort of a loose complex must exist. The basis for this assertion is a knowledge of the bond energies of the three molecules, and that the activation energy for the 2ClO reaction is zero. The bond energies, or the energies required to separate completely the atoms in a diatomic molecule, are

Molecule	Bond energies
ClO	64.5 kcal mole^{-1}
O_2	118.0 kcal mole^{-1}
Cl_2	57.2 kcal mole^{-1}

If during the reaction the two ClO molecules first were pulled apart, and then the isolated atoms were combined into Cl_2 and O_2, the activation energy for this reaction would be twice the bond energy of ClO, or 129 kcal per 2 moles of ClO. Instead, the activation energy is zero. The activated complex must be a combination of the four atoms such that whatever instability created as Cl and O separate is immediately compensated by the stabilizing influence of associations between Cl and Cl and between O and O.

We can think of the activated complex as an unstable "molecule," with many of the properties of a molecule, except that it decomposes spontaneously either to reactants or to products. The thioacetamide and water molecules in Figure 18–1(b) are in an activated complex, and the energy of this complex is greater than either that of thioacetamide and water or that of acetamide and hydrogen sulfide.

Potential Energy Surfaces

The 2ClO reaction suggests that, in principle, we should be able to calculate the total potential energy of a collection of atoms as a function of their positions in space. This calculation would produce a potential energy surface with hills and plateaus of high energy, and valleys of low energy. Any region of a minimum of potential energy in this plot will represent a stable molecule. Even with four atoms as in the 2ClO reaction we would need, unfortunately, six variables to describe the arrangement of atoms: the bond lengths from each atom to the other three, for example. Our potential energy plot would have to be in seven-dimensional space. This is difficult to visualize and impossible to construct.

We need an example with only two variables so the map can be plotted in three-dimensional space. One of the first maps to be calculated, by Henry

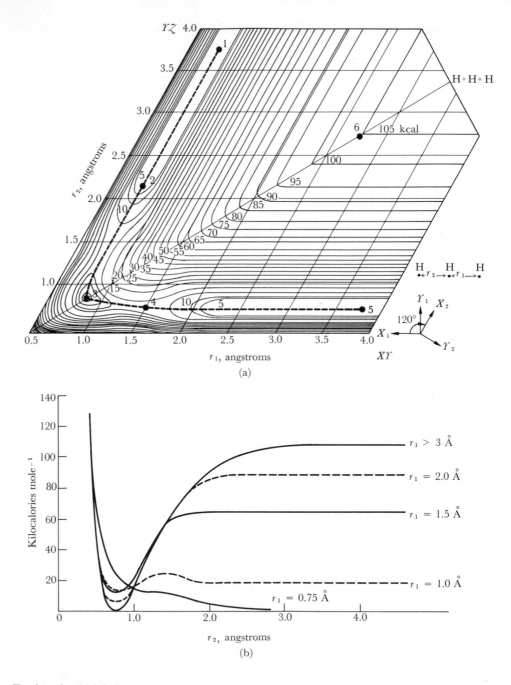

(a)

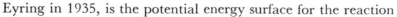

(b)

Eyring in 1935, is the potential energy surface for the reaction

$$H + H_2 \rightleftarrows H_2 + H$$

in which all three atoms are constrained to lie on a straight line. The only variables are the distances from the central hydrogen atom to the other two,

◀ *Figure 18–7. The potential energy of three hydrogen atoms in a straight line, plotted in kilocalories as a function of the separation of the two outside hydrogen atoms from the central one, r_1 and r_2. Contours of equal potential energy in (a) are numbered in kilocalories. The shape of the potential surface is that of two deep valleys parallel to the r_1 and r_2 axes, with sheer walls rising to these axes and with less steep walls rising to a plateau at the upper right corner of (a). The two valleys are connected by a path over a pass or saddle, with the crest of the pass at $r_1 = 0.8 \text{ Å} = r_2$. The calculations for the three-atom system were carried out by semiempirical methods, in 1935, by Henry Eyring and his co-workers. Recently, more exact quantum mechanical calculations have indicated that the depression at the summit of the pass may not be real. Several sections through this potential surface are shown in (b) for different values of r_1. At sufficiently large values of r_1 (over 3 Å), the potential energy curve is that of a diatomic H_2 molecule, virtually unaffected by the third H atom. Compare the potential energy curve for $r_1 > 3$ Å with Figure 10–2. The axes in this plot have been skewed so a marble rolling down a model of this surface would represent accurately the vibrations of the three-atom system. Points 1 through 6, marked in color, correspond to the atomic arrangements shown in Figure 18–8.*

Figure 18–8. Relative positions of the three hydrogen atoms for the six points indicated in color on Figure 18–7(a). Points 1 through 5 represent stages in the reaction of a hydrogen molecule with a hydrogen atom. Point 6 represents the three atoms 2.5 Å apart along a straight line. This state is about 90 kcal less stable than any stage along the reaction pathway. The potential energies in kilocalories are given at the right.

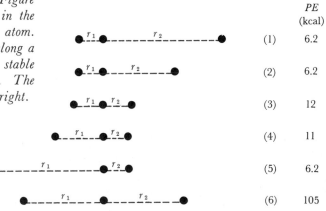

	PE (kcal)
(1)	6.2
(2)	6.2
(3)	12
(4)	11
(5)	6.2
(6)	105

r_1 and r_2:

H—H—H
r_1 r_2

The potential energy of the three-atom system as a function of r_1 and r_2 is shown in Figure 18–7(a). The actual arrangements of the three atoms at the six numbered points marked in color are drawn in Figure 18–8. Sections through this potential energy surface at fixed values of r_1 are shown in Figure 18–7(b).

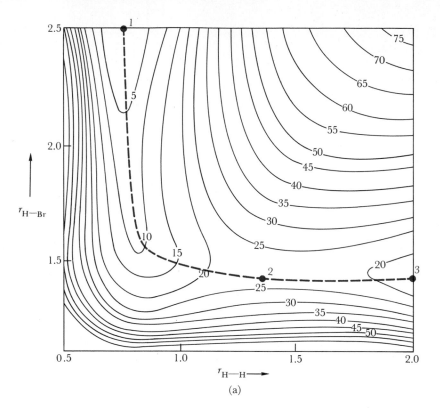

(a)

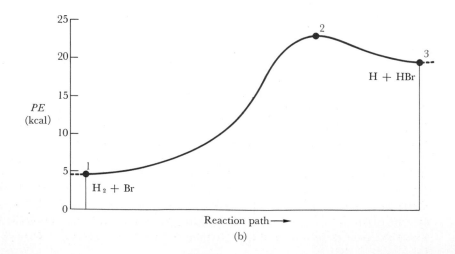

(b)

◀ *Figure 18–9. (a) Potential energy surface for the linear array of H– – – –H– – – –Br. Contours of equal potential energy are marked in kcal per mole of three-atom sets. The reaction pathway for the reaction $H_2 + Br \rightleftharpoons H + HBr$ is marked by a dashed colored line. (b) A potential energy profile of the reaction pathway. Points 1, 2, and 3 correspond to the pathways on the potential energy surface. Point 2 is the activated complex for the reaction. You can verify from the data in Appendix 2 on standard heats of formation of atoms and molecules that the difference in potential energy at Points 1 and 3 is approximately correct.*

If either r_1 or r_2 is large, the three hydrogen atoms exist as a H_2 molecule and an isolated H atom. The potential energy section at constant r_1, for r_1 greater than 3 Å in Figure 18–7(b), is the same as that for an isolated H_2 molecule in Figure 10–2. As an atom approaches a H_2 molecule from the right (Points 1 and 2 of Figures 18–7 and 18–8), the first noticeable effect is an increase in the potential energy of the system of three atoms. The incoming atom is repelled by the molecule, and a more stable situation results if the atom rebounds and moves away again. If the atom has enough kinetic energy to keep approaching the H_2 molecule, it will begin to weaken the H—H bond in the H_2 molecule. At Point 3, both outer atoms are slightly farther from the central one than a normal H—H bond length, but the potential energy of the system of atoms is 6 kcal mole^{-1} higher than that of isolated H_2 and H. Point 3 is the activated complex for the reaction.

The activated complex can decompose either to products or to reactants. There is no reason why the three atoms in the state of Point 3 cannot return to Point 1 as well as proceed to Point 5. What is certain is that the activated complex is unstable and must decompose. Points 1 through 5 in Figure 18–7 are connected by a colored dashed line called a *reaction pathway*. If we plot potential energy along this pathway, an activation energy barrier curve such as those in Figure 18–5 results. Notice that the reaction pathway at all times is a path along a valley between steep walls. It may take 6 kcal of energy to build the activated complex of Point 3, but it takes nearly 100 kcal to separate the atoms as at Point 6.

Similar potential energy surfaces have been calculated for other systems of atoms; the one for $H_2 + Br \rightleftharpoons H + HBr$ is shown in Figure 18–9. Now the shape of the surface is altered because H_2 is more stable than HBr. As the Br atom approaches H_2, it pushes the H atoms apart. The activated complex (Point 2) has the two H atoms twice as far apart as in the H_2 molecule, but has H and Br at nearly their final bond distance in HBr. The activated complex is almost the same as an HBr molecule, and indeed it is only 3 or 4 kcal mole^{-1} less stable than HBr. Calculated potential energy surfaces such as Figure 18–9 are the basis for the common drawings of reaction barriers such as Figure 18–5. A profile of potential energy such as

Figure 18–9(b) is still useful even with reactions of molecules so complicated that we cannot calculate or even plot their complete multidimensional potential energy surface.

Absolute Rate Theory

In the absolute rate theory, reaction takes place when an activated complex breaks down into products. Therefore, the rate of reaction is the product of three factors:

1) The concentration of activated complexes per cubic centimeter.
2) The rate of breakdown of individual complexes or their rate of passage over the activation energy barrier.
3) The probability that a breakdown will form products and not reactants again.

Since the activated complex represents an unstable state of transition between reactants and products, it often is called a *transition state*. The absolute rate theory also is called the transition-state theory. We shall use these terms interchangeably.

The transition-state theory assumes an equilibrium between reactants and the activated complex, usually represented by a superscript double dagger:

$$A + B \rightleftarrows AB^{\ddagger}$$

$$K^{\ddagger} = \frac{[AB^{\ddagger}]}{[A][B]}$$

Hence, the concentration of the activated complex is given by

$$[AB^{\ddagger}] = K^{\ddagger}[A][B]$$

The rate of decomposition is more complicated to calculate, but it turns out to be a universal constant for all bimolecular reactions at a given temperature:

$$\text{Rate of decomposition} = \frac{kT}{h}$$

in which k = Boltzmann's constant, and h = Planck's constant. The probability that a breakdown will be to products and not to reactants is the transmission coefficient, κ. It can be estimated only as having a value between 0.5 and 1.0 in most reactions. Therefore, the overall rate of reaction is

$$-\frac{d[A]}{dt} = \kappa \frac{kT}{h} K^{\ddagger}[A][B]$$

and the rate constant, k_2, is

$$k_2 = \kappa \frac{kT}{h} K^{\ddagger} \tag{18–13}$$

(The symbol k_2 is used here instead of k to represent a bimolecular rate constant to avoid confusion with Boltzmann's constant, k.)

It is possible to calculate the equilibrium constant, $K^{\ddagger}$, between reactants and the activated complex from molecular properties by using statistical mechanics. We shall not even attempt this calculation here; instead we shall look at the thermodynamic interpretation of this rate-constant expression. The equilibrium constant is related to the standard free energy of formation of the activated complex from reactants, and this in turn is related to the standard enthalpy and entropy of the formation of the activated complex:

$$-RT \ln K^{\ddagger} = \Delta G^{0\ddagger} = \Delta H^{0\ddagger} - T \Delta S^{0\ddagger} \qquad (18\text{--}14)$$

Thus, the bimolecular rate constant, k_2, can be written

$$k_2 = \kappa \frac{kT}{h} e^{-(\Delta G^{0\ddagger}/RT)} = \kappa \frac{kT}{h} e^{+(\Delta S^{0\ddagger}/R)} e^{-(\Delta H^{0\ddagger}/RT)} \qquad (18\text{--}15)$$

The enthalpy of activation, $\Delta H^{0\ddagger}$, is nearly the same quantity as the activation energy, E_a. The difference is irrelevant in this discussion. Equation 18–15 indicates that the rate of reaction is slower if the activation energy is large. This result already was obtained in the collision theory; because if the activation energy is large, only a small fraction of the molecules will have enough energy to surmount the barrier and to react rather than rebound upon collision. Equation 18–15 also suggests that the reaction rate is faster if the *entropy* of activation is large. If the activated complex is much more disordered than the reactants, the reaction is enhanced because the equilibrium constant for formation of the complex is large, and more complex is present. In contrast, if the reactants are severely constrained when they combine to make the activated complex, then the reaction is inhibited. We might guess that the entropy of activation in the thioacetamide plus water reaction is negative since the two molecules combine to form one unit in the complex. Both molecules are limited severely in their initial orientations if they are to build the activated complex in Figure 18–1(b) successfully. The entropy of activation in bimolecular reactions is almost always large and negative because the two reactants lose entropy when they combine in the complex. Often the most useful application of absolute rate theory is not to calculate the rate constant directly, but to use the observed rate constant and Equation 18–15 to calculate the entropy of activation. The entropy of activation provides information about the structure of the activated complex. For example, if the calculated entropy of activation is positive, then any mechanism that leads through a tightly organized activated complex must be rejected.

As an example, in the next section we will consider two reactions of the type

$$R_3C\text{---}Br + OH^- \rightarrow R_3C\text{---}OH + Br^-$$

which proceed by different mechanisms, depending on the nature of the R groups. In one mechanism, the Br^- is driven away as OH^- approaches, in the same fashion as the thioacetamide reaction. The activated complex is then a combination of R_3C—Br and OH^-:

$$R_3C\text{---}Br + OH^- \rightarrow \left[\begin{array}{c} R \qquad R \\ \diagdown \quad \diagup \\ Br\cdots C\cdots OH \\ | \\ R \end{array}\right]^- \rightarrow Br^- + R_3C\text{---}OH \qquad (18\text{--}16)$$

This is called a S_N2 reaction, meaning that it is a *substitution* of one group for another, that the groups are *nucleophilic* (donating electrons and attracting nuclei: Lewis bases, in fact), and that *two* molecules are involved in the reaction. The other mechanism is for the R_3C—Br molecule to dissociate spontaneously into Br^- and what is known as a *carbonium ion*, R_3C^+, and for the OH^- ions to react rapidly in a separate step with any free carbonium ions. The activated complex or transition state in this mechanism will be the reactant R_3C—Br just before dissociation:

$$R_3C\text{---}Br \rightarrow [R_3C\text{---}Br]^{\ddagger} \rightarrow R_3C^+ + Br^-$$
$$R_3C^+ + OH^- \rightarrow R_3C\text{---}OH \qquad\qquad (18\text{--}17)$$

This is called a S_N1 mechanism since it is a nucleophilic substitution in which the slowest step is unimolecular. One should be able to distinguish between these two mechanisms by their entropies of activation, calculated from Equation 18–15, and the measured rate constants. The S_N2 mechanism will have a large negative entropy of activation since the activated complex is formed by combining two molecules. In contrast, the S_N1 mechanism will have virtually a zero entropy of activation because the activated complex differs only slightly from the reactant molecule.

Summary: Comparison of Theories

Both the collision theory and the absolute rate theory build upon the idea of an energy of activation that acts as a barrier to reaction. Both, to this extent, are based on the older Arrhenius explanation of the variation of rate constants with temperature. The collision theory focuses on the collision of two reactant molecules, whereas the absolute rate theory deals more with the complex formed after collision and assumes an equilibrium between this complex and the reactants. The collision theory uses the activation energy concept by stating that all molecular pairs that do not have this energy on collision will rebound instead of react. The absolute rate theory postulates instead that a high enthalpy of activation to the complex means that the equilibrium constant and hence the concentration of complexes will be small. If you think of the complex as that which is formed when two molecules have the energy demanded by the collision theory, then the two theories are seen for what they really are: different viewpoints of the same phenomenon.

As was mentioned earlier, the equilibrium constant for activated complex formation, $K^{\ddagger}$, can be calculated from the properties of the reactant molecules and the presumed properties of the complex. This means that the rate constant, k (or k_2), can be calculated from first principles just as it can be in the collision theory. The calculated values from these two theories are compared with observed values in Table 18–2. As you can see, the absolute rate theory usually is a little better in predicting rate constants than is the collision theory. The collision theory is not wrong; it is just too simple a picture of what happens when chemical reactions occur.

18–4 COMPLEX REACTIONS

Most chemical reactions are not simple unimolecular or bimolecular reactions, but are combinations of these. This is why such complicated rate equations as Equations 18–3 or 18–4 arise. Even the hydrogen–iodine reaction, which has been used for over half a century as the classic example of a simple bimolecular reaction (Equation 18–2), is complex.

The Hydrogen–Iodine Reaction

For the reaction

$$H_2 + I_2 \rightarrow 2HI \tag{18–18}$$

the observed rate equation is

$$-\frac{d[H_2]}{dt} = k[H_2][I_2] \tag{18–18a}$$

Above 800°K, side reactions with different mechanisms occur, yet these can be neglected at moderate temperatures. N. N. Semenov and Henry Eyring both have suggested that the true mechanism might not be that of Equation 18–18, but might be a two-step mechanism involving the reversible dissociation of I_2 to $2I$, followed by the trimolecular reaction of I and H_2:

$$I_2 \rightleftarrows 2I \tag{18–19}$$
$$H_2 + 2I \rightarrow 2HI$$

The rate expression for the reaction of one H_2 molecule with two I atoms is

$$-\frac{d[H_2]}{dt} = k'[H_2][I]^2 \tag{18–20}$$

If the dissociation of I_2 is reversible and at equilibrium, we can write an equilibrium-constant expression:

$$K = \frac{[I]^2}{[I_2]} \quad [I]^2 = K[I_2] \tag{18–21}$$

However, substituting for the concentration of I atoms in Equation 18–20 by using the equilibrium of Equation 18–21,

$$-\frac{d[H_2]}{dt} = k'K[H_2][I_2] \qquad (18\text{–}22)$$

produces the *same* rate expression as if the mechanism were really one of bimolecular collision. We thus have two different mechanisms with the same rate expression. How can we choose between them?

The two mechanisms have the same rate equation *so long as* the dissociation of I_2 is at thermal equilibrium, and the amount of I atoms present is given by the thermal equilibrium constant of Equation 18–21. At higher temperatures, more I_2 dissociates, thereby producing the same effect that would have resulted from the greater bimolecular rate constant in the bimolecular mechanism. J. H. Sullivan decided to test the two theories by making the concentration of iodine atoms different from what it normally is in the thermal dissociation of I_2. He did this by dissociating I_2 with 5780-Å light from a mercury vapor lamp. This light should have little effect if the reaction is bimolecular, aside from a slight decrease in I_2 concentration. Conversely, if the trimolecular reaction with I atoms is correct, the rate of reaction should increase with the intensity of irradiating light since more I atoms are being produced.

Sullivan calculated the concentration of I atoms present at several intensities of irradiating light and found that the rate of appearance of HI is proportional to the *square* of the I atom concentration. Therefore, the mechanism of Equation 18–19 is the correct one. The classical $H_2 + I_2$ reaction is a trimolecular reaction masquerading as a simpler bimolecular reaction because of the thermal equilibrium that normally exists between I_2 and $2I$. (At least, until someone even more ingenious designs an experiment that proves that it is a more complicated reaction masquerading as trimolecular. As Sullivan points out [*J. Chem. Phys.* **46**, 73 (1967)], the trimolecular reaction

$$H_2 + 2I \rightarrow 2HI$$

can be replaced by two bimolecular steps:

$$H_2 + I \rightleftarrows H_2I$$
$$H_2I + I \rightarrow 2HI$$

If the first of these is fast and reversible, so that reactants and products are in equilibrium, then the rate expression is the same as for the trimolecular process, and the two mechanisms cannot be distinguished by reaction rates.)

This example makes a point that must be kept in mind at all times. We can never prove that a proposed mechanism is right, only that it has not yet been shown to be wrong. There is always the chance that a more subtle experiment, such as Sullivan's upsetting of thermal equilibrium with light, may uncover the weaknesses in an accepted mechanism. When two theories are presented, the temptation and usually the wiser choice is to opt for the simpler one, at least until the data compel you to do otherwise. But you should always be prepared to change your mind when new data demand it.

Rates and Mechanisms of Substitution Reactions

The reaction of *tert*-butyl bromide with OH^-,

$$(CH_3)_3CBr + OH^- \rightleftarrows (CH_3)_3COH + Br^- \tag{18–23}$$

has the experimental rate expression

$$-\frac{d[(CH_3)_3CBr]}{dt} = k[(CH_3)_3CBr] \tag{18–24}$$

The rate does not appear to depend on the OH^- concentration at all. In contrast, the similar reaction with a less highly substituted carbon atom in ethyl bromide,

$$CH_3CH_2Br + OH^- \rightleftarrows CH_3CH_2OH + Br^- \tag{18–25}$$

has the rate expression that we might expect from the chemical equation:

$$-\frac{d[CH_3CH_2Br]}{dt} = k[CH_3CH_2Br][OH^-] \tag{18–26}$$

Why should these two similar reactions proceed by different mechanisms and have different rate equations? And how is it that the rate in Equation 18–24 can be independent of concentration of one of the reactants?

Reaction 18–23 takes place by the S_N1 mechanism of Equation 18–17. The *tert*-butyl bromide first dissociates in a slow reaction, and the carbonium ion that is formed reacts immediately with OH^-. Whenever a process takes place by a series of rapid steps with one quite slow step, the overall rate of reaction will be controlled by the slow step. The rate in this S_N1 reaction depends entirely on how fast the molecules of $(CH_3)_3CBr$ decompose. The capacity of reacting with carbonium ions by OH^- presumably far exceeds the amount of carbonium ions supplied by the dissociation. The total amount of OH^- present is unimportant.

In contrast, the ethyl bromide reaction takes place by a S_N2 mechanism (Equation 18–16). Here the reaction is between an ethyl bromide molecule and an OH^- ion, and both concentrations affect the reaction rate.

Why do the reactions go with different mechanisms? The S_N2 scheme is possible for ethyl bromide because there is room for three substituents of the C atom (CH_3 and two H), plus OH^- and Br^-. The activated complex

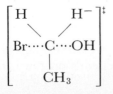

is sterically possible. In contrast, in *tert*-butyl bromide the groups attached to the carbon atom (three CH_3) are large enough that OH^- and Br^- both cannot bind at the same time. The activated complex of the S_N2 reaction is impossible. No reaction can occur until a molecule of *tert*-butyl bromide spontaneously dissociates. The dissociated carbonium ion is then subject to attack, either by Br^- to form the reactants again, or by OH^- to form the product. If Br^- is present only as a result of previous reaction of *tert*-butyl bromide, its concentration is probably much smaller than that of OH^-, and most of the carbonium ions will be converted to *tert*-butyl alcohol, $(CH_3)_3COH$.

In general, a rate expression that disagrees with the stoichiometry of the overall reaction is an indication that the reaction is proceeding by a series of steps. The problem, then, often is to find a set of steps including a slow step that accounts for the observed rate law.

The difference in reaction mechanisms encountered in *tert*-butyl bromide and ethyl bromide also is found in octahedral and square planar complexes of transition metals. Square planar complexes of Pt(II) and other metals can react with new ligands by S_N2 mechanisms because the metal atom is accessible from either side of the plane. The S_N2 mechanism of the reaction

$$Pt(NH_3)_3Cl^+ + Br^- \rightarrow Pt(NH_3)_3Br^+ + Cl^-$$

can be written

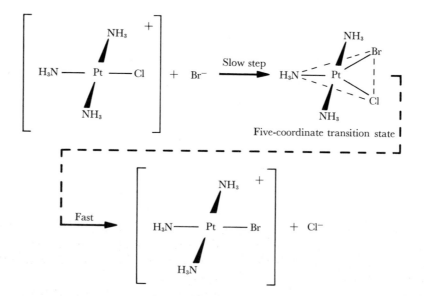

The activated complex is five-coordinated platinum, which breaks down rapidly to products. The rate of the overall reaction depends on the rate of formation of the activated complex. This rate is influenced strongly by the nature of the entering group (Br^- in this example). Ligands capable of forming strong bonds with the central atom are the best entering groups; that is, they displace the leaving group (Cl^- in this example) most rapidly. The ions CN^- and I^- are good entering groups for Pt(II) complexes, whereas NH_3 and H_2O are relatively poor.

It is much more difficult for the six-coordinated octahedral complexes to react by a S_N2 mechanism because six ligands around a central metal, such as Co(III), leave little or no room for the attachment of an entering group in a transition state. Studies of substitution reactions of octahedral Co(III) complexes have established that the important or rate-determining step involves the dissociation of the bond between the Co(III) and the leaving group. The entering group is not involved in this initial dissociation step. For example, in aqueous solution, H_2O displaces Cl^- in the complex $Co(NH_3)_5Cl^{2+}$, thereby producing $Co(NH_3)_5H_2O^{3+}$. The mechanism most consistent with rate studies of this and similar reactions is the S_N1 mechanism, which can be written

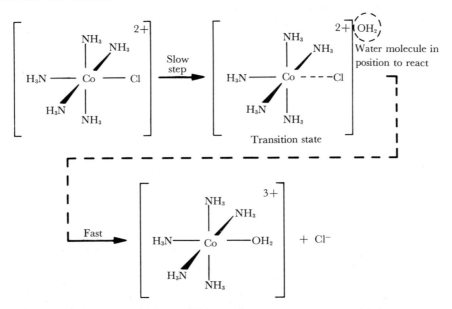

For such a mechanism the entering group plays no significant role in the creation of the transition state and hence in the rate of the overall reaction. A characteristic of most octahedral substitution reactions is the *lack* of influence of entering groups on the rate of reaction.

The rate of exchange of ligands in octahedral complexes depends on the rate of dissociation of a ligand from the complex. The rate constants for exchange of one water molecule for another in aquo complexes,

$$M(H_2O)_6{}^{n+} + H_2O^* \xrightarrow{k} M(H_2O)_5(H_2O^*)^{n+} + H_2O$$

Table 18–3. Reactivities of Complexes at 298°K

Metal ion	Valence d electronic configuration	Aquo complex	Rate constant[a] (sec^{-1})	Class
Cu^{2+}	d^9	$Cu(H_2O)_6^{2+}$	2×10^8	I
Zn^{2+}	d^{10}	$Zn(H_2O)_6^{2+}$	3×10^7	II
Fe^{2+}	d^6	$Fe(H_2O)_6^{2+}$	3×10^6	II
Co^{2+}	d^7	$Co(H_2O)_6^{2+}$	2×10^5	II
Ni^{2+}	d^8	$Ni(H_2O)_6^{2+}$	2.5×10^4	II
Fe^{3+}	d^5	$Fe(H_2O)_6^{3+}$	2.5×10^2	III
Cr^{3+}	d^3	$Cr(H_2O)_6^{3+}$	2×10^{-5}	IV
Co^{3+}	d^6	$Co(NH_3)_5H_2O^{3+}$	6×10^{-6}	IV

[a] Rate of the ligand-exchange reaction

$$M(H_2O)_x + H_2O^* \rightarrow M(H_2O)_{x-1}(H_2O^*) + H_2O$$

are given for several transition-metal ions in Table 18–3. (How can you measure the rate of exchange of one water molecule for another? How do you tell the difference between them? If you cannot think of a way, look back at Chapter 13.)

We can define four categories of ions.

Class I. Exchange of water bound to the metal ion is fast and is controlled by how rapidly the water molecules can diffuse toward and away from the complex ion. First-order rate constants (Table 18–3) are of the order of 10^8 sec^{-1} or larger. Included in this class are the alkali metal ions and the alkaline earth metal ions except Mg^{2+} and Be^{2+}.

Class II. First-order rate constants for water exchange from 10^4 to 10^3 sec^{-1}. The dipositive transition-metal ions and the tripositive lanthanide ions are in this class, along with Mg^{2+}.

Class III. First-order rate constants from 1 to 10^4 sec^{-1}. Tripositive transition-metal ion Fe^{3+}, plus Be^{2+} and Al^{3+}.

Class IV. Relatively inert complexes, with first-order rate constants commonly 10^{-3} to 10^{-6} sec^{-1}, and occasionally as low as 10^{-9} sec^{-1}. Includes Cr^{3+}, Co^{3+}, and Pt^{2+}.

The location of members of these classes in the periodic table is given in Figure 18–10. Why do such large differences in reaction rates exist? The relative rates of reaction of $Cu(H_2O)_6^{2+}$ and $Cr(H_2O)_6^{3+}$ differ by 10^{13}. An event that occurs every 4 hours on the time scale of Cu will have occurred only once on the Cr time scale since the earth came into existence 4.5 billion years ago!

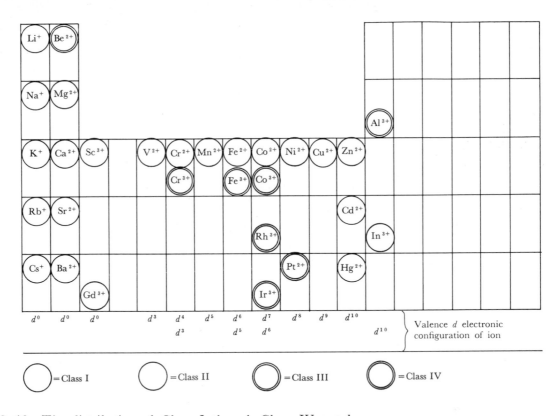

Figure 18–10. The distribution of Class I through Class IV metal ions across the periodic table. The most reactive ions, in Class I, are large and of low charge. The less reactive Class III complexes are small and carry a high charge. Class IV ions have unusually low reactivities that cannot be explained solely by charge and size arguments.

Two important factors are the size and charge of the metal ion. The larger the positive charge on the ion, the more tightly held the ligands will be, and the slower will be the reaction that depends on dissociation of a ligand as a rate-determining step. Similarly, the smaller the ion, the closer the ligands can approach. Electrostatic attraction will be greater, and the ligands will be held more tightly.

The ratio of positive charge on the metal ion, q, to the ionic radius, r, is plotted in Figure 18–11. The trend with increasing charge and decreasing size is apparent. (Look again at Figure 6–9 if you do not remember the trends in ionic radii across the periodic table.) Class I contains all the monopositive ions, and many of the larger dipositive (M^{2+}) ions. Class II contains the smaller dipositive and larger tripositive ions. Class III includes the quite

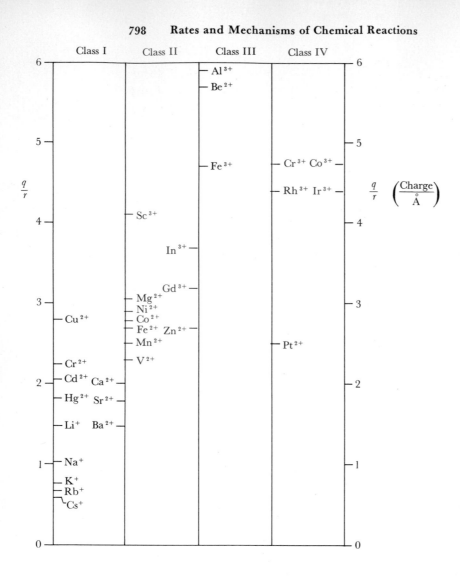

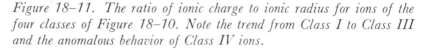

Figure 18–11. The ratio of ionic charge to ionic radius for ions of the four classes of Figure 18–10. Note the trend from Class I to Class III and the anomalous behavior of Class IV ions.

small Be^{2+} ion and the two small ions Fe^{3+} and Al^{3+}. The ions in Class IV appear to be anomalies. We cannot explain their lack of reactivity by simple size and charge arguments.

Class IV ions have d electronic configurations with particularly stable ground states in either octahedral or square planar coordination. For low-spin octahedral complexes, the gap between t_{2g} and e_g energy levels is large (Figure 11–17). States with the three t_{2g} orbitals half filled or fully filled (d^3 and d^6) in such low-spin situations are especially stable and unreactive. The two metal d orbitals that point directly at the octahedral ligands ($d_{x^2-y^2}$ and d_{z^2}) are empty, and the orbitals that are filled or half filled (d_{xy}, d_{yz}, and d_{xz}) point to one side of each ligand. The Cr^{3+} ion is d^3; Co^{3+}, Rh^{3+}, and Ir^{3+} are d^6, and their aquo complexes are low spin. In contrast, the hexa-aquo complex of Fe^{2+} (Table 18–3) is high-spin d^6. Electrons occupy the two orbitals that point at the octahedral ligands, thereby making it easier for the ligands to dissociate and making the S_N1 reaction faster. Similarly, Cu^{2+} and Zn^{2+}, with d^9 and d^{10} configurations, also have electrons in their upper e_g energy level and are especially reactive. In the square planar coordination, the d^8 low-spin electronic configuration of Pt^{2+} exactly fills the four low-lying energy levels (Figure 11–17), and Pt^{2+} is consequently relatively unreactive.

Carbon is relatively unreactive for the same reasons that low-spin complexes of Co(III) are unreactive. In carbon, the four sp^3 bonding orbitals are filled with electrons; thus, there are no empty orbitals of similar energy available for electrons from ligands. The four ligands bonded to a carbon atom have little tendency either to dissociate or to be displaced; hence, reactions of carbon are slow. This behavior accounts for the existence of many compounds of carbon that are not thermodynamically stable. Given enough time, they will decompose to more stable compounds, but the *rate* of decomposition is so slow that they can exist in appreciable quantities in nature. For example, sugars are all thermodynamically unstable relative to water and carbon dioxide, and they liberate free energy when they are oxidized to these products. If carbon were not slow to react, and if the products of photosynthesis immediately decomposed to their thermodynamically more stable constituents, life would be impossible because there would be no way of trapping and saving free energy. Carbon compounds illustrate again the basic difference between *stability* in the thermodynamic sense and *inertness* in the dynamic sense of the term.

Chain Reactions

The reaction

$$H_2 + Br_2 \rightleftarrows 2HBr$$

has the strange rate equation that we already have seen

$$\frac{d[HBr]}{dt} = \frac{k[H_2][Br_2]^{1/2}}{1 + k'([HBr]/[Br_2])} \tag{18-3}$$

For 13 years after this rate law was discovered, no one could account for it.

Then, three groups did so almost simultaneously, those of Henry Eyring, K. F. Herzfeld, and Michael Polanyi. They proposed that the reaction proceeds by a chain mechanism involving two *chain-propagating steps*:

1) $H_2 + Br \xrightarrow{k_1} H + HBr$

2) $H + Br_2 \xrightarrow{k_2} HBr + Br$

When a molecule breaks apart into uncharged fragments having unpaired electrons, the fragments are called radicals. The unpaired electrons (e.g., in H and Br) make the fragments chemically reactive. The atomic product of each of these steps is one reactant for the other step, and they both produce HBr. Thus, HBr results, not from a bimolecular collision, but from an endless chain of Reactions (1) and (2). The first of these two steps is the reaction of Figure 18–9. But where do these atoms of Br and H come from? The Br atoms are postulated to come initially from a *chain-initiating step:*

3) $Br_2 \xrightarrow{k_3} Br + Br$

Why is the dissociation of H_2 not included also? The real reason is that the explanation of Equation 18–3 does not require it, and that if we add it, we obtain the wrong rate expression. We can justify this omission in another way: The dissociation energy of H_2 is 103 kcal mole^{-1}, whereas that of Br_2 is only 45 kcal mole^{-1}.

A large concentration of HBr inhibits the reaction, as we can see from the HBr term in the denominator of Equation 18–3. Moreover, a large concentration of Br_2 counteracts this inhibition. So HBr and Br_2 evidently are competing for the same chemical substance. What might that substance be?

The most likely candidate is hydrogen atoms; the inhibiting reaction would be

4) $H + HBr \xrightarrow{k_4} H_2 + Br$

This is a *chain-inhibiting reaction* and is counteracted if an excess of Br_2 makes Reaction (2) go rapidly, as the rate Equation 18–3 predicts. Finally, the chain is *terminated* by the recombination of Br:

5) $Br + Br \xrightarrow{k_5} Br_2$

How do we obtain Equation 18–3 from these five reactions? If we can do so, this will be a strong argument for the correctness of the chain mechanism, although *not* an absolute proof, as we have seen with HI.

The rate of appearance of HBr is given by

$$\frac{d[HBr]}{dt} = +k_1[H_2][Br] + k_2[H][Br_2] - k_4[H][HBr] \qquad (18\text{–}27)$$

since HBr appears as a result of Reactions (1) and (2), and disappears in Reaction (4). The rates of production of H and Br atoms are given by

$$\frac{d[\text{H}]}{dt} = k_1[\text{H}_2][\text{Br}] - k_2[\text{H}][\text{Br}_2] - k_4[\text{H}][\text{HBr}] \qquad (18\text{–}28)$$

$$\frac{d[\text{Br}]}{dt} = -k_1[\text{H}_2][\text{Br}] + k_2[\text{H}][\text{Br}_2] + k_4[\text{H}][\text{HBr}]$$
$$+ 2k_3[\text{Br}_2] - 2k_5[\text{Br}]^2 \qquad (18\text{–}29)$$

The coefficients of 2 in front of k_3 and k_5 arise because each unit of Reaction (3) produces two Br atoms, and each unit of Reaction (5) removes two Br atoms.

At this point an essential simplification must be made. The actual amount of H and Br atoms present at any time must be small because they are consumed almost at the same rate that they are produced. Soon after the reaction begins, the concentrations of H and Br will reach a *steady state* and will remain constant so long as the reaction continues with a plentiful supply of reactants. Then each of the rate equations in 18–28 and 18–29 can be set equal to zero:

$$0 = k_1[\text{H}_2][\text{Br}] - k_2[\text{H}][\text{Br}_2] - k_4[\text{H}][\text{HBr}] \qquad (18\text{–}30)$$
$$0 = -k_1[\text{H}_2][\text{Br}] + k_2[\text{H}][\text{Br}_2] + k_4[\text{H}][\text{HBr}] + 2k_3[\text{Br}_2] - 2k_5[\text{Br}]^2$$
$$(18\text{–}31)$$

Adding these two equations yields

$$2k_5[\text{Br}]^2 = 2k_3[\text{Br}_2]$$
$$[\text{Br}] = \left(\frac{k_3}{k_5}\right)^{1/2} [\text{Br}_2]^{1/2} \qquad (18\text{–}32)$$

This calculation gives us a steady-state concentration for Br atoms in terms of the concentration of Br_2 molecules.

The HBr rate equation can be rewritten as

$$\frac{d[\text{HBr}]}{dt} = k_1[\text{H}_2][\text{Br}] + \{k_2[\text{Br}_2] - k_4[\text{HBr}]\}[\text{H}] \qquad (18\text{–}33)$$

We can eliminate the H concentration by expressing it in terms of Br concentration from Equation 18–30

$$[\text{H}] = \left(\frac{k_1[\text{H}_2]}{k_2[\text{Br}_2] + k_4[\text{HBr}]}\right)[\text{Br}] \qquad (18\text{–}34)$$

Substituting Equation 18–34 into 18–33, placing everything over a common denominator, and canceling terms yields

$$\frac{d[\text{HBr}]}{dt} = \frac{2k_1k_2[\text{H}_2][\text{Br}_2][\text{Br}]}{k_2[\text{Br}_2] + k_4[\text{HBr}]} \tag{18-35}$$

Dividing top and bottom by $[\text{Br}_2]$, and then eliminating $[\text{Br}]$ with Equation 18–32, yields

$$\frac{d[\text{HBr}]}{dt} = \frac{2k_1(k_3/k_5)^{1/2}[\text{H}_2][\text{Br}_2]^{1/2}}{1 + (k_4/k_2)\dfrac{[\text{HBr}]}{[\text{Br}_2]}} \tag{18-36}$$

This is exactly the experimental rate law, in which the experimental rate constants are related to those for the individual reactions in the chain by

$$k = 2k_1\left(\frac{k_3}{k_5}\right)^{1/2}$$

$$k' = \frac{k_4}{k_2}$$

Now that we know what these two experimental constants mean in terms of the individual reactions, we can give a much fuller interpretation to the rate law, Equation 18–36. Suppose that we could vary the individual rate constants, k_1 to k_5, at will. What effects would these changes have on the overall rate? The overall rate of production of HBr is accelerated if rate constants k_1, k_2, and k_3 are large, or if Reactions (1), (2), and (3) are fast. The first two of these reactions produce HBr; the third prepares the way by making more Br atoms. The production of HBr is slowed if k_4 and k_5 are large, or if the chain-inhibiting and chain-terminating reactions are fast. So long as k_3 and k_5 change together, there is no change in the overall rate of reaction. Reactions (3) and (5) are the opposing initiating and terminating steps. Similarly, so long as k_2 and k_4 change together, the rate is unaffected. This, too, is sensible; for Reactions (2) and (4) are similar in that they both consume an H and produce a Br, but differ in producing HBr in Reaction (2) and removing it in Reaction (4). Inhibition by HBr occurs because Reaction (4) is enhanced, and inhibition is lessened by Br_2 because Reaction (2) is enhanced.

18-5 CATALYSIS

A mixture of hydrogen and oxygen gas can be kept for years without appreciable reaction to produce water. But if a small amount of platinum black is introduced, the mixture explodes. The Pt is a catalyst for the reaction.

As we already have seen in Chapter 12, a *catalyst* is a substance that accelerates the attainment of thermodynamic equilibrium without itself being consumed in the process. It does this by providing an alternative mechanism or pathway for the reaction, with a lower activation energy. If the activation energy for the forward reaction (E_1 in Figure 18–4) is lowered, that of the reverse reaction (E_2) must be lowered by the same amount if the heat of reaction is to be unchanged. A catalyst accelerates both the forward and reverse reaction. It does not change the conditions of equilibrium in a reaction, only the speed of getting there. The Pt catalyst dissociates H_2 gas into hydrogen atoms on the metal surface. These H atoms then react far more rapidly with O_2 molecules that meet them at the metal surface than H_2 molecules do with O_2 in the gas phase.

This is an example of *heterogeneous catalysis*, involving a gas or liquid phase and a surface. Even more common is *homogeneous catalysis*, in which both the reactants and the catalyst are in solution.

Homogeneous Catalysis: Ce^{4+} and Tl^+

The reaction

$$2Ce^{4+} + Tl^+ \rightarrow 2Ce^{3+} + Tl^{3+} \qquad (18–37)$$

is extremely slow even though the free energy change favors it. It is accelerated immensely by small amounts of manganous ion, Mn^{2+}, even though manganous ion is not consumed by the reaction. The reaction in the absence of manganous ion is slow because it requires a three-body collision, which is extremely unlikely. The mechanism with the Mn^{2+} catalyst involves a succession of three two-body collisions instead:

$$\begin{aligned}
Mn^{2+} + Ce^{4+} &\rightarrow Mn^{3+} + Ce^{3+} \\
Mn^{3+} + Ce^{4+} &\rightarrow Mn^{4+} + Ce^{3+} \qquad (18–38) \\
Mn^{4+} + Tl^+ &\rightarrow Mn^{2+} + Tl^{3+}
\end{aligned}$$

The Mn^{2+} ion is a carrier of electrons and a means of converting a difficult one-step process into a series of easy processes. It has the same catalytic function in this reaction that cytochrome *c* does in the oxidation of metabolites in the terminal oxidation chain (Section 11–7). In the language of absolute rate theory, the activated complex in Reaction 18–37 is an arrangement of three ions, and the entropy of activation is extremely low and negative. The rate constant is therefore small. In contrast, the reactions of Equations 18–38 each have an activated complex made from only two ions and considerably *less* negative entropy of activation. Therefore, each rate constant is larger, and the set of three reactions proceeds faster than does Reaction 18–37.

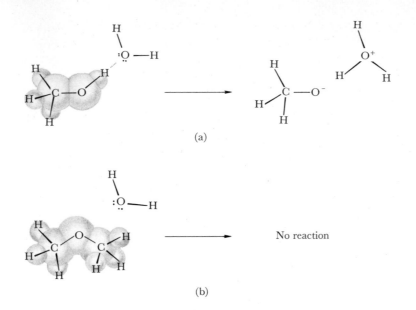

(a)

(b)

Figure 18–12. (a) When methanol is dissolved in water, the proton on the methanol hydroxyl group is small enough that a water molecule can approach it and subject it to a nucleophilic attack. The result is that the O—H bond in methanol is broken and the proton is bound to the solvent. (b) In dimethyl ether, the proton is replaced by a CH$_3$ group. This group is so large and bulky that the water molecule cannot approach close enough to attack it as with methanol. The O—C bond in dimethyl ether does not break, and no CH$_3$O$^-$ is formed. The colored outlines mark the approximate relative sizes of the atoms as measured by van der Waals contact distances between nonbonded atoms.

Acid Catalysis

The ionization of methanol in aqueous solution,

$$CH_3OH + H_2O \rightarrow CH_3O^- + H_3O^+ \tag{18–39}$$

proceeds rapidly (but to a small extent), and in fact too rapidly to measure by bulk sampling methods. In contrast, the analogous reaction with dimethyl ether,

$$CH_3-O-CH_3 + H_2O \rightarrow CH_3O^- + CH_3OH_2^+ \tag{18–40}$$

does not occur to a measurable extent. This is because the hydroxyl hydrogen atom on the alcohol is so small and so exposed that it can be attacked by a nucleophilic reagent such as water (Figure 18–12), whereas the analogous

Table 18–4. *Increase in Rate Constant by Acid Catalysis for the Reaction*
$$CH_3COOCH_3 + H_2O \rightleftarrows CH_3COOH + CH_3OH$$

$-\dfrac{d[CH_3COOCH_3]}{dt} = k[CH_3COOCH_3]$		
Acid	$k/k_{HOAc}^{\,a}$	K_a (Table 5–4)
Acetic, CH_3COOH (formed in reaction)	1.0	1.76×10^{-5}
Formic, $HCOOH$	3.8	1.77×10^{-4}
Dichloroacetic, $CHCl_2COOH$	66.7	3.32×10^{-2}
Trichloroacetic, CCl_3COOH	198.0	0.20
Sulfuric, H_2SO_4	214.0	—
Nitric, HNO_3	267.0	—
Hydrobromic, HBr	284.0	—
Hydrochloric, HCl	290.0	—

a Ratio of rate constant, k, for given acid to that for acetic acid at the same concentration.

—CH_3 group in the ether blocks the approach of H_2O. *Steric hindrance*—the bumping together and mutual repulsion of nonbonded atoms—is the greatest single factor in determining activation energies of reacting molecules. The H^+ ion, being a lone proton, is free from such steric hindrance. It is so small that such barriers to reaction can be avoided, and proton-transfer reactions are usually quite fast. If an alternative mechanism to a given reaction involves proton-transfer steps, it is likely that the reaction will be catalyzed by acids, which provide a source of protons. This is why acid catalysis is so important to chemistry.

The reaction

$$CH_3COOCH_3 + H_2O \rightleftarrows CH_3COOH + CH_3OH$$

has the rate equation

$$-\frac{d[CH_3COOCH_3]}{dt} = k'[H_2O][CH_3COOCH_3]$$
$$= k[CH_3COOCH_3]$$

Although this rate equation is really second order, it is effectively first order since the water concentration in aqueous solution is effectively constant during the reaction. This reaction is catalyzed by acids, and was studied by Ostwald before the turn of the century. Ostwald determined that the rate constant was increased by a factor of 300 by HCl, and that other acids gave the increases shown in Table 18–4. You can see a correlation between the rate constant and the dissociation constant for the acid. We know now that the effectiveness of the acid as a catalyst (as reflected in the rate constant) arises from the concentration of H^+ ion produced by the acid (as reflected in the dissociation constant). All of the completely dissociated strong acids produce approximately the same rate constant for the reaction. As we mentioned in Section 3–5, Arrhenius used Ostwald's data on the catalytic effect of weak acids to support his theory of ionization.

Heterogeneous Catalysis: Carboxypeptidase A

As mentioned previously, catalysis in which the molecules acted upon, or the *substrates*, are bound to the surface of a solid phase is called heterogeneous catalysis. The binding of H_2 molecules to the surface of Pt in the catalyzed synthesis of water is one example. Another class of heterogeneous catalysis is enzymatic reactions. In the carboxypeptidase reaction of Section 12–11, the reaction catalyzed is

$$R'—CO—NH—CHR—COOH + H_2O \rightarrow$$
$$R'—CO—OH + H_2N—CHR—COOH$$

in which R is an amino acid side chain, and R′ represents the remainder of the protein polypeptide chain. The purpose of the enzyme, with its Zn atom at the active site, is to provide an alternative mechanism to a simple solution reaction, a mechanism in which the activation energy is lower. The drawing in Figure 12–23(a) shows an activated complex, with both Zn and Glu helping to weaken the bond between C and N. The water molecule in Figure 12–23(b) can attack the bond much more easily after the bond has been weakened by withdrawal of electrons by the Zn. The attack by water also is made easier by Tyr, which draws a proton from the water molecule. The activated complex in this reaction is a combination of the polypeptide chain being cut, the water molecule, the Zn, and the Glu and Tyr side groups from the enzyme. In addition to providing the necessary chemical groups, the enzyme must put them in the correct position and orientation. Both energy and entropy of activation are changed by enzymatic catalysis.

Challenge. Try Problems 18–21 through 18–24 in the Butler and Grosser book for applications of the Michaelis–Menten mechanism for enzyme catalysis and the biochemically important Krebs cycle.

18–6 SUMMARY

This chapter, like Chapter 12, has been a short introduction to a very large field. It is impossible in this small space to give you more than a brief outline of the problems involved in finding out how molecules react and some of the methods for solving them. Chemical dynamics is presently less well developed than structural chemistry because the problems are basically more difficult. Dynamics needs every shred of information about structure that can be found, but this information is only the starting point for proposing mechanisms of reaction and designing experiments to test them.

The essential question to be answered is, "How does one molecule react with another molecule?" However, because we cannot study individual molecules, we are forced to work with large numbers of molecules, with a distribution of energies, and with unknown relative orientations of the molecules. The energy distribution can be avoided for certain special reactions by molecular-beam methods, but the orientation problem remains.

The rate law or rate expression for the appearance of products does not necessarily agree with what would have been expected from the number of moles of each reactant in the overall equation. When it does agree, this is suggestive evidence that the reaction proceeds in a one-step process as written (although there are pitfalls, as in the HI reaction). When the rate law and overall equation do not agree, as with HBr, this suggests that the overall reaction really proceeds in a series of simpler steps. If one of these steps is much slower than the others, the kinetics of the overall reaction is controlled by this rate-determining step.

The order and the molecularity of a reaction are two quite different quantities that reflect the difference between overall stoichiometry and mechanism. The order of a reaction is simply the sum of the exponents of all the concentration terms in an expression that is a product of such terms. The molecularity of a simple reaction is the number of molecules or ions that collide in that step. An overall, multistep reaction has no molecularity, although its order may be well defined. But the rate expression for the HBr reaction is so complicated that even the concept of reaction order has no meaning, except at low HBr concentrations.

Two theories of simple bimolecular reactions exist: the collision theory and the absolute rate theory. Both are based on Arrhenius' interpretation of the dependence of rate constant on temperature in terms of an energy of activation. The collision theory focuses on the collision that precedes reaction; the absolute rate theory concentrates on the assembly of atoms just after collision but before separation into products. Both give a reasonable explanation of observed rate constants; the absolute rate theory is somewhat better.

Both the enthalpy of activation and the entropy of activation are important in determining the size of the barrier to reaction. Reaction is favored if the enthalpy barrier is low and the entropy of activation is large and positive (or at least not negative). If the activated complex is much more ordered than the reactants, the entropy of activation is large and negative, and reaction is slowed.

A catalyst accelerates a reaction by providing an alternative reaction pathway or mechanism with a lower free energy of activation. It can do this by providing energy for a dissociation or by assisting in the ordering of reactants in the complex. In the first pathway, the enthalpy of activation is lowered (as with H_2 on a Pt surface, or the withdrawal of electrons in a covalent bond by Zn in carboxypeptidase). The second pathway increases the probability that the reactants will be ordered more properly than would be the case by chance in solution. In either case, reaction is more rapid because $\Delta G^{0\ddagger}$ is lower and $K^{\ddagger}$ is larger.

Thermodynamics indicates nothing about the time involved in attaining equilibrium, as we have emphasized several times before. Thermodynamics is concerned only with comparing the initial and final states of a reacting system for quantities such as T, P, V, E, H, S, and G, which are state functions. The change in these quantities is the same whether the reaction takes place in a nanosecond (10^{-9} sec) or an eon (10^9 years), and whether the reaction takes place in one step or in a thousand, so long as the starting conditions and the final conditions are the same. In contrast, kinetics deals with how fast reactions occur. A rock rolling down a hillside will come to a halt and remain stationary forever if it meets a barrier that is even a small fraction of the height of the hill. If the rock is given random disturbances by passers-by, the probability that it will be knocked over the obstacle and will continue down the hill within a specified time depends on the height of the barrier (among other factors). The task of chemical kineticists is to investigate these barriers to chemical reaction, to see what effect they have in slowing reactions, and to find ways to avoid them, either by surmounting them by proper chemical conditions or by circumventing them by catalysis.

SUGGESTED READING

S. W. Benson, *Foundations of Chemical Kinetics*, McGraw-Hill, New York, 1960.

J. O. Edwards, *Inorganic Reaction Mechanisms: An Introduction*, W. A. Benjamin, Menlo Park, Calif., 1965.

H. Eyring and E. M. Eyring, *Modern Chemical Kinetics*, Reinhold, New York, 1963. Slightly above the level of King, but a good reference for the material in this chapter.

A. A. Frost and R. G. Pearson, *Kinetics and Mechanisms*, Wiley, New York, 1961, 2nd ed.

S. Glasstone, K. J. Laidler, and H. Eyring, *The Theory of Rate Processes*, McGraw-Hill, New York, 1941.

E. L. King, *How Chemical Reactions Occur*, W. A. Benjamin, Menlo Park, Calif., 1963. An elementary introduction at the level of this chapter.

QUESTIONS

1 Why does the probability of reaction of thioacetamide and water in Figure 18–1 depend on the relative orientations of the two molecules as they approach each other?

2 Why does the probability of Question 1 depend on the relative velocity of approach of the two molecules?

3 Why does the approach of the water molecule to thioacetamide weaken the C—S bond? Why does it weaken the O—H bonds?

4 Why are the molecules of H_2O [Figure 18–1(a)] and H_2S [Figure 18–1(c)] not drawn the same way?

5 Which of the two factors important in determining the rate of reaction, energy and entropy, is controlled in molecular-beam experiments? How is this done?

6 Why are molecular-beam experiments unsuitable for studying acid–base reactions discussed in Chapter 5? What kinds of reactions can be studied by molecular-beam methods?

7 How is the rate constant for a reaction defined? In the HI rate equation (Equation 18–2), what is the meaning of the d's in numerator and denominator on the left side? What do the chemical symbols in brackets mean?

8 What is the difference between order and molecularity in a chemical reaction? What is the overall order of reactions of Equations 18–23 and 18–25? What is their molecularity? Under what conditions are order and molecularity different? For what kinds of reactions is the concept of molecularity meaningless? For what kinds of reactions is the concept of order meaningless?

9 What physical methods can be used to follow the course of a reaction? What are the relative advantages of physical and chemical techniques for following reaction kinetics?

10 How can you tell when a chemical reaction is first order?

11 Can you suggest an explanation of why the reaction between ethanol and decaborane in Section 18–2 is not thirty-first order? Can you think of a possible mechanism that would account for the observed rate equation?

12 What physical evidence leads to the concept of an energy of activation for a reaction? What is the interpretation of this energy of activation in terms of the reaction mechanism?

13 What is questionable about the derivation of the equilibrium constant in terms of forward and reverse reactions occurring at the same rates? When is this derivation valid?

14 At constant temperature, how does the enthalpy of a reaction change as the activation energy of the forward reaction

changes? How can the activation energy be altered?

15 How is the activation energy used in the collision theory of reaction? In this theory, what factors affect the rate of reaction? In what two ways does temperature influence the rate of reaction in the collision theory?

16 Why, from the evidence in Table 18–2 and Figure 18–5, can we say that the activated complex for the reaction

$$2ClO \rightarrow Cl_2 + O_2$$

is not a complex with four atoms at a great distance from each other?

17 If three hydrogen atoms were spaced at intervals of 2.0 Å along a straight line, what would be the potential energy (expressed as kcal mole^{-1} of such triplets of atoms) as given by Figure 18–7? How much less stable is this state than the highest point in the reaction path from $H + H_2$ to $H_2 + H$?

18 What is an activated complex? What is the activated complex in Figure 18–1? In the absolute rate theory, what assumption is made about the amount of activated complex present?

19 Does a large positive enthalpy of activation favor rapid reaction? Does a large positive entropy of activation favor rapid reaction?

20 Why is it useful to calculate the entropy of activation from measured rate constants and the absolute rate theory? What information does such data give us about reaction mechanisms?

21 What do the symbols mean in S_N1 and S_N2 mechanisms? What is the difference between these mechanisms? What is the activated complex in each mechanism? In which mechanism is the nature of the entering group more important? What factors determine whether a reaction will proceed by S_N1 or S_N2 mechanism?

22 Does the reaction of thioacetamide with water use S_N1 or S_N2 mechanism?

23 Which of the two mechanisms of the preceding two questions will have a larger entropy of activation? What effect will this have on the rate of reaction?

24 Which of the two mechanisms in Questions 21 and 22 will have a larger enthalpy of activation, other factors being the same? What effect will this have on the rate of reaction?

25 In view of the answers to the preceding two questions, why cannot one make dogmatic statements about relative rates of reactions that proceed by S_N1 and S_N2 mechanisms?

26 What is the evidence for the assertion that the reaction of H_2 with I_2 to produce HI is not a simple bimolecular reaction, as had been believed for such a long time?

27 Which mechanism, S_N1 or S_N2, is more likely to be encountered with octahedral complex ions? With square planar ions? Why?

28 How do ionic size and charge affect the rates of reaction of transition-metal complexes? What is the basis for the classification of complex ions in Section 18–4?

29 Why are complex ions of Cr(III) and Co(III) so unreactive? Why are aquo complexes of Na^+ and K^+ more reactive than those of Fe^{3+}?

30 If both Fe(II) and Co(III) have the d^6 electronic configuration, why is the aquo complex of Fe(II) so much more reactive than that of Co(III)? (The greater size of Fe(II) and the smaller

charge are part of the answer, but there is a much more important factor.)

31 With such a small charge-to-radius ratio (Figure 18–11), V(II) should be the most rapidly reacting member of Class II. Instead, its rate constant for ligand exchange in aquo complexes is so small that it could almost be placed in Class III. Both V(II) and Cr(III) are more inert than expected from size and charge. How do you explain this behavior from their electronic configuration? Are compounds of these ions likely to be high spin or low spin? Does such a question have any meaning for these ions? Why, or why not?

32 Why is the rate equation for the reaction of H_2 with Br_2 so much more complicated than that for H_2 and I_2?

33 What is the distinguishing feature of a chain reaction? What is a chain-initiating step? What is a chain-inhibiting step? What chain-inhibiting step is used in nuclear power reactors? (See Section 13–4.)

34 What is the steady-state assumption in solving rate expressions?

35 Why is acid catalysis so common and so effective?

36 Why is the hydrolysis reaction for methyl acetate in aqueous solution a first-order reaction, as commonly measured?

37 In terms of the theories of reaction rates of this chapter, how does a catalyst work? What is the activated complex in a catalytic reaction? How can the catalyst affect the enthalpy of activation? The entropy of activation?

PROBLEMS

1 For the hypothetical reaction

$$2A + 3B \rightarrow 3C + 2D$$

the following rate data were obtained in three experiments at the same temperature:

Initial [A] (mole liter^{-1})	Initial [B] (mole liter^{-1})
0.10	0.10
0.20	0.10
0.20	0.20

Initial rate
(moles of A consumed
liter^{-1} sec^{-1})

0.10
0.40
0.40

(a) Determine the experimental rate equation for the reaction. (b) Calculate the specific rate constant, k. (c) What is the rate of this reaction when $[A] = 0.30$ molar and $[B] = 0.30$ molar?

2 For the hypothetical reaction

$$2A + B \rightarrow 2C$$

the following data were collected in three experiments at 25°C:

Initial $[A]$ (mole liter^{-1})	Initial $[B]$ (mole liter^{-1})
0.10	0.20
0.30	0.40
0.30	0.80

Initial rate
(moles of A consumed
liter^{-1} sec^{-1})

300
3600
14400

(a) What is the experimental rate equation for this reaction? (b) Calculate the specific rate constant for this reaction.

3 In the reaction

$$2NO + Cl_2 \rightarrow 2NOCl$$

the reactants and products are gases at the temperature of the reaction. The following rate data were measured for three experiments:

Initial p_{NO} (torr)	Initial p_{Cl_2} (torr)
380	380
760	760
380	760

Initial rate (atm sec^{-1})

5.1 × 10^{-3}
4.0 × 10^{-2}
1.0 × 10^{-2}

(a) From these data, write the rate equation for this gas reaction. What order is the reaction in NO, Cl$_2$, and overall? (b) Calculate the specific rate constant for this reaction.

4 The reaction $2NO + O_2 \rightarrow 2NO_2$ is first order in oxygen pressure and second order in the pressure of nitric oxide. Write the rate expression.

5 The indicated rate expressions have been obtained for each of the reactions listed below. If these rate expressions hold, even at equilibrium, what are the rate expressions for the reverse reactions at equilibrium?

a) $C_2H_2 + H_2 \rightarrow C_2H_4$
    Rate $= k[H_2]/[C_2H_2]$

b) $C_2H_4 + H_2 \rightarrow C_2H_6$
    Rate $= k[H_2]$

c) $2H_2 + O_2 \rightarrow 2H_2O$
    Rate $= k[H_2][O_2]^{4/3}$

d) $N_2O_5 \rightarrow 2NO_2 + \frac{1}{2}O_2$
    Rate $= k[N_2O_5]$

6 Given the following data for the reaction

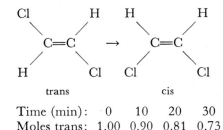

Time (min):	0	10	20	30
Moles trans:	1.00	0.90	0.81	0.73

What is the order of the reaction? How long will it take for half of the trans compound to decompose?

7 The reaction

$$A + B + C \rightarrow D + F$$

was found to be zero order with respect to A. A solution of Reactants A, B, and C was prepared with the following initial concentrations: 0.2M of A, 0.4M of B, and 0.6M of C. The concentration of A in this solution dropped to essentially zero in five minutes. A second solution was prepared with the following initial concentrations: 0.03M of A, 0.4M of B, and 0.6M of C. How long will it take for A to disappear?

8 For the reaction

$$2NO + H_2 \rightarrow N_2O + H_2O$$

the following experimental rate data are collected in three successive experiments at the same temperature:

Initial [NO] (molar)	Initial [H$_2$] (molar)
0.60	0.37
1.20	0.37
1.20	0.74

Initial rate (mole liter^{-1} min^{-1})
0.18
0.72
1.44

Using these experimental data, write the rate expression for the reaction.

9 The reaction

$$2HCrO_4{}^- + 3HSO_3{}^- + 5H^+ \rightarrow 2Cr^{3+} + 3SO_4{}^{2-} + 5H_2O$$

follows the rate equation

$$\text{Rate} = k[HCrO_4{}^-][HSO_3{}^-]^2[H^+]$$

Why isn't the rate proportional to the numbers of ions of each kind that are shown by the equation?

10 The reaction

$$I^- + OCl^- \rightarrow Cl^- + OI^-$$

has the experimental rate equation

$$\text{Rate of disappearance of OCl}^- = k[I^-][OCl^-]$$

How would you describe the order of this reaction?

11 The rate constant in Problem 10 depends on hydroxide ion concentration. For hydroxide ion concentrations of 1.00, 0.50, and 0.25 molar, k is 60, 120,

and 240 liters mole^{-1} sec^{-1}, respectively. What is the order of the reaction with respect to hydroxide ion concentration?

12 The reaction $SO_2Cl_2 \rightarrow SO_2 + Cl_2$ is a first-order reaction with the rate constant $k = 2.2 \times 10^{-5}$ sec^{-1} at 320°C. What fraction of SO_2Cl_2 is decomposed on heating at 320°C for 90 min?

13 For the first-order reaction A $\rightarrow$ C, $k = 5$ min^{-1}. When the first-order reaction D $\rightarrow$ B occurs, only 10% of D decomposes in the same length of time that it takes for 50% of A to decompose in the first reaction. Calculate k for the second reaction.

14 For the decomposition of ammonia, the following data were measured. The top line indicates time in seconds, and the bottom line indicates the corresponding concentration of ammonia, in moles liter^{-1}.

0	1	2	$t_{1/2}$
2.000	1.993	1.987	1.000

For this first-order reaction, write an expression for the rate of reaction and the rate constant, k, and calculate the half-life time, $t_{1/2}$.

15 For the decomposition of N_2O_5 in CCl_4, a plot of log [N_2O_5] against time is a straight line. The rate constant for the reaction is 6.2 $\times$ 10^{-4} sec^{-1} at 45°C. If one begins with 1 mole of N_2O_5 in a 1-liter flask, how long will it take for 20% of the N_2O_5 to decompose? What is the half-life time for the reaction?

16 It often is said that, near room temperature, a reaction rate doubles if the temperature is increased by 10°C. Calculate the activation energy of a reaction whose rate exactly doubles between 27°C and 37°C.

17 What is the activation energy for a reaction for which an increase in temperature from 20°C to 30°C exactly triples the rate constant?

18 The following data give the temperature dependence of the rate constant for the reaction $N_2O_5 \rightarrow 2NO_2 + \frac{1}{2}O_2$. Plot the data and calculate the activation energy of the reaction.

T (°K)	k (sec^{-1})
273	7.87×10^{-7}
298	3.46×10^{-5}
308	1.35×10^{-4}
318	4.98×10^{-4}
328	1.50×10^{-3}
338	4.87×10^{-3}

19 For the decomposition of CH_3I at 285°K, the energy of activation is 43 kcal mole^{-1}. Assuming that the energy of activation is constant, calculate the percent increase in the fraction of molecules with energy greater than E_a when the temperature is increased to 300°K.

20 The rate constant for the decomposition of N_2O_5 in carbon tetrachloride is 6.2×10^{-4} sec^{-1} at 45°C. Calculate the rate constant at 100°C if the activation energy is 24.7 kcal mole^{-1}.

21 Why does it take longer to boil an egg on the top of Mt. San Jacinto (11,000 ft) than in Pasadena (750 ft)? (Smog is not the answer.)

22 For the reaction $Ni + \frac{1}{2}O_2 \rightarrow NiO$, $\Delta H^0 = -59.3$ kcal at 298°K, and the value does not change drastically with temperature. Suppose that the reaction takes place on the surface of Ni so rapidly that all the heat is used to heat the remaining nickel. If at 25°C, one oxygen atom reacts with each 10 Å^2 of surface, what will be the final temperature when a cube of nickel 1.00 cm on an edge reacts with oxygen? The density of nickel is about 9 g cm^{-3}. If the 1-cm cube is ground to form 10^{15} equal cubes and reaction occurs with these, what will be the final temperature? (Ignore the change in heat capacity when nickel oxide is formed. Assume that the law of Dulong and Petit in Chapter 1 is valid.)

23 If a cube of NaCl 1.00 cm on an edge is dissolved in an enormous quantity of water stirred in a tank, it takes six hours before solution is complete. If the cube is ground to a fine powder containing 10^{15} equal spheres, what will be the time required for solution if this time is inversely proportional to the initial area of contact between the NaCl and the water?

24 The reaction

$$H_2 + Cl_2 \rightarrow 2HCl$$

occurs explosively in the presence of light. Assume that this explosion takes place by a chain reaction and is initiated by the formation of (a) hydrogen atoms or (b) chlorine atoms. Using the bond energies of H_2 and Cl_2, calculate the wavelength of light needed in (a) and (b).

25 Calculate the standard free energy change of the reaction

$$2C_6H_6(g) \rightarrow 3CH_4(g) + 9C(gr)$$

at 298°K, and calculate the equilibrium constant, K_{eq}. Other typical reactions that you may have encountered are the precipitation of AgCl from solution,

$$Ag^+ + Cl^- \rightarrow AgCl(s)$$
$$K_{eq} \simeq 10^{10}$$

and the formation of the diammine complex of silver in aqueous solution,

$$Ag^+ + 2NH_3 \rightarrow Ag(NH_3)_2^+$$
$$K_{eq} \simeq 10^8$$

By the criteria of Chapter 15, all three reactions are highly spontaneous. Yet in the laboratory the preceding two reactions proceed essentially instantaneously, whereas the decomposition of benzene to methane and carbon apparently does not proceed at all. Account for this fantastic difference in reaction rates.

26 The decomposition of gaseous N_2O_5 occurs according to the reaction

$$N_2O_5 \rightarrow 2NO_2 + \frac{1}{2}O_2$$

The experimental rate equation is

$$\frac{-d[N_2O_5]}{dt} = k[N_2O_5]$$

and a proposed reaction mechanism is

1) Equilibrium: $N_2O_5 \overset{K}{\rightleftarrows} NO_2 + NO_3$
2) Slow reaction:

$$NO_2 + NO_3 \overset{k_2}{\rightarrow} NO_2 + O_2 + NO$$

3) Fast reaction: $NO + NO_3 \overset{k_3}{\rightarrow} 2NO_2$

a) Show that this mechanism is consistent with the observed rate equation.
b) If $k = 5 \times 10^{-4}$ sec^{-1}, how long does it take for the concentration of N_2O_5 to fall to one tenth its original value?

27 Consider the reaction

$$CH_4 + Cl_2 \overset{light}{\longrightarrow} CH_3\dot{C}l + HCl$$

The mechanism is a chain reaction involving Cl atoms and CH_3 radicals. Which of the following steps does not terminate this chain reaction:

a) $CH_3 + Cl \quad \rightarrow CH_3Cl$
b) $CH_3 + HCl \rightarrow CH_4 + Cl$
c) $CH_3 + CH_3 \rightarrow C_2H_6$
d) $Cl \quad + Cl \quad \rightarrow Cl_2$

28 Assume that the reaction

$$5Br^- + BrO_3^- + 6H^+ \rightarrow$$
$$3Br_2 + 3H_2O$$

proceeds by the mechanism

1) Fast reaction:

$$BrO_3^- + 2H^+ \overset{k_1}{\rightarrow} H_2BrO_3^+$$

2) Fast reaction:

$$H_2BrO_3^+ \overset{k_{-1}}{\longrightarrow} BrO_3^- + 2H^+$$

3) Slow reaction:

$$Br^- + H_2BrO_3^+ \overset{k_2}{\rightarrow}$$
$$Br{-}BrO_2 + H_2O$$

4) Fast reaction:

$$Br{-}BrO_2 + 4H^+ + 4Br^- \overset{k_3}{\rightarrow}$$
$$3Br_2 + 2H_2O$$

Deduce the rate equation that agrees with this mechanism; express the rate constant for the overall reaction in terms of rate constants for the individual steps. The rate equation depends on the concentrations of H^+, Br^-, and BrO_3^-.

29 Consider the reaction

$$5H^+ + [Co(NH_3)_5Cl]^{2+} + [Cr(H_2O)_6]^{2+} \overset{H_2O}{\longrightarrow}$$
$$[Co(H_2O)_6]^{2+} + [Cr(H_2O)_5Cl]^{2+} + 5NH_4^+$$

When this reaction is carried out in the presence of radioisotopically labeled chloride ions, it is found that the radioactive ions do not appear in the product. Keeping in mind that the rate of exchange of ligands bound to Cr^{2+} and Co^{2+} is quite rapid, whereas for Cr^{3+} and Co^{3+} it is quite slow, postulate a mechanism for the reaction [see *J. Am. Chem. Soc.* **75**, 4118 (1953)].

The final impression our mind receives on contemplating these fundamental relations is that of a wonderful mechanism of nature, the functions of which are performed with never-failing certainty, though the mind can only follow them with difficulty, and with a humiliating sense of the incompleteness of its perception.
J. J. Balmer

THE LONG VIEW

A new kind of chemistry is slowly arising. In the past, the end result of chemical effort usually has been a product: plastics, oils, solvents, dyes, pharmaceuticals, fertilizers, insecticides, a better fuel reactor or engine. This period, which should not be denigrated, could be called the era of *product chemistry*. Now, partly because of the growth of the human population and partly because of problems created by product chemistry, we are faced with a new kind of chemical challenge. Future chemists must think about entire *systems* of chemical processes. Environmental chemists who study the pollution problems of the Los Angeles basin are one example of *systems chemists*. Agricultural scientists who study the ecology and the competing uses of an entire geographic area are another example. The line between chemical engineering and chemistry probably will become blurred as chemical engineers improve their methods of studying large chemical systems, and as chemists become more concerned about the ultimate effects of their work.

The naturally occurring chemical systems available for us to study and to use as models are the living systems of this planet.

If Pericles could refer to Athens during its golden age as "the education of Hellas," so we may consider nature as "the education of Chemistry." What we learn about the balance of numerous interlocking chemical reactions may be useful in the design of systems for our own purposes. The ingenuity of a cell as an energy-converting device surpasses anything that we have been able to produce in a flask or an engine. Moreover, the chlorophyll-based photosynthesis system for harvesting energy from the sun is at hand for us to study. If we really understood the *principles* of such interwoven and coupled chemical reactions, we might be able to design nonbiological systems for energy collection that would free us from messy methods such as burning fossil fuels or degrading radioactive elements. The problem is not one of finding energy; there is more of it than we can possibly use. The problem is one of using it *without* degrading our environment in the process.

As another example of systems chemistry, the problem of control of undesirable insects (once we are sure that they are really undesirable) without eliminating all insects, and other small animal life, is one of *selectivity*. The reliance upon insecticides is a blind, shotgun approach. We are beginning to learn about insect sex attractants and chemical lures that are effective in fantastically minute amounts and are directed against only one species of insect. Some day soon it may be possible to control mosquitoes and crop-eating insects by specific chemical methods without otherwise disturbing the balance of nature, which is really a delicate balance of interlocking chemical systems. The screwworm fly, a cattle parasite, has been virtually eradicated in South Texas by a simple application of such methods. Millions of male screwworm flies were grown, sterilized by radiation, and released in the countryside. Every mating of any of these millions of flies with a normal female fly was unproductive. The result was a drastic reduction in the fly population in the next generation. After a few years of such methods, the insects were well under control. The method is hardly more expensive than massive spraying with unselective killing agents, which in this case already had been tried without success. It is also much less of a disturbing factor in the environment.

It is always easier to work *with* an ongoing chemical system than to work in opposition to it, if you can find out how to do it. Finding out how is the coming challenge of chemistry. There are better ways of stopping a locomotive than sticking a steel bar through the drive wheels. We, as chemists, have learned to be productive. Now we must learn to be efficient, and to be clever.

APPENDIXES

APPENDIX 1 USEFUL PHYSICAL CONSTANTS AND CONVERSION FACTORS

PHYSICAL CONSTANTS

Atomic mass unit	$1 \text{ amu} = 1.66053 \times 10^{-24} \text{ g}$
Avogadro's number	$N = 6.022169 \times 10^{23} \text{ mole}^{-1}$
	(^{12}C = exactly 12)
Bohr radius	$a_0 = 0.52918 \text{ Å}$
Boltzmann's constant	$k = 1.38062 \times 10^{-16} \text{ erg deg}^{-1}$ molecule^{-1}
Electron rest mass	$m_e = 0.0004859 \text{ amu} = 9.1095 \times 10^{-28} \text{ g}$
Electronic charge	$e = 4.80325 \times 10^{-10} \text{ esu}$ $(\text{cm}^{3/2}\text{g}^{1/2}\text{sec}^{-1})$
	$e = 1.6021 \times 10^{-19} \text{ coulomb}$
Faraday's constant	$\mathscr{F} = Ne = 96{,}487 \text{ coulombs equivalent}^{-1}$
Gas constant	$R = Nk = 8.3143 \times 10^7 \text{ ergs deg}^{-1}$ mole^{-1}
	$R = 0.082054 \text{ liter atm deg}^{-1} \text{ mole}^{-1}$
	$R = 1.98726 \text{ cal deg}^{-1} \text{ mole}^{-1}$
Neutron rest mass	$m_n = 1.008665 \text{ amu} = 1.67492 \times 10^{-24} \text{ g}$
Planck's constant	$h = 6.6262 \times 10^{-27} \text{ erg sec}$
Proton plus electron	$m_p + m_e = 1.007763 \text{ amu}$
Proton rest mass	$m_p = 1.007277 \text{ amu} = 1.67261 \times 10^{-24} \text{ g}$
Rydberg constant	$R = 109{,}677.581 \text{ cm}^{-1}$
Velocity of light	$c = 2.9979 \times 10^{10} \text{ cm sec}^{-1}$

CONVERSION FACTORS

1 electron volt (eV) = 1.6022×10^{-12} erg
1 erg = 6.2420×10^{11} eV = 2.3901×10^{-11} kcal = 1 g cm^2 sec^{-2}
1 kcal = 4.1840×10^{10} ergs = 2.612×10^{22} eV
1 volt coulomb = 1 joule = 10^7 ergs = 0.23901 cal
1 eV molecule^{-1} = 23.056 kcal mole^{-1} = 8065 cm^{-1}

100 kcal mole^{-1} = 34,982 cm^{-1}

1 atomic unit (au) of energy = 27.21 eV molecule^{-1}

$$= 4.3592 \times 10^{-11} \text{ erg molecule}^{-1}$$

$$= 219,470 \text{ cm}^{-1} = 627.71 \text{ kcal mole}^{-1}$$

1 amu of mass = 931.481 $\times 10^6$ eV of energy = 931.481 MeV

2.303 RT = 1.364 kcal mole^{-1} at 298°K

INTERNATIONAL SYSTEM OF UNITS (SI)

In 1960, the International Bureau of Weights and Measures established the International System of Units (SI) to simplify communication among world scientists. In this text we have not been rigorous about using only SI units, because the traditional units (e.g., angstroms and calories) are still common and you should be familiar with them. However, you should be aware that the trend among scientists is toward the use of strict SI units, and some scientific and engineering journals and textbooks are using them exclusively.

The International System has seven base units: metre (m), kilogram (kg), second (s), ampere (A), kelvin (K), mole (mol), and candela (cd). Supplementary units are radian (rad) for plane angle and steradian (sr) for solid angle. All other SI units are derived from these base and supplementary units. The following table lists examples. For more on the subject, see "The International System of Units (SI)," *National Bureau of Standards Special Publication 330*, 1972, U.S. Government Printing Office, Washington, D. C., and Martin A. Paul, "International System of Units (SI)," *Chemistry* **45,** 14 (1972).

Physical quality	SI unit (symbol)	Conversion factors
Length	metre (m)	1 inch (in) = 0.0254 m (exactly)
		1 angstrom (Å) = 10^{-10} m
Volume	cubic metre (m^3)	1 in^3 = 16.4 cm^3 (approx.)
		1 litre = 10^{-3} m^3
Mass	kilogram (kg)	1 pound (lb) = 0.45359237 kg (exactly)
		1 amu = 1.66053 $\times 10^{-27}$ kg (approx.)
		1 gram (g) = 10^{-3} kg
Time	second (s)	1 day (d) = 86,400 s
		1 hour (h) = 3,600 s
		1 minute (min) = 60 s
Frequency	hertz (Hz = s^{-1})	
Force	newton (N = m kg s^{-2})	1 dyne (dyn) = 10^{-5} N (exactly)

Pressure	pascal $(Pa = N\,m^{-2})$	1 atm = 101,325 Pa (exactly)
		1 torr = 101,325/760 Pa (exactly)
Energy	joule $(J = N\,m)$	1 erg = 10^{-7} J
		1 cal (thermochemical) = 4.184 J (exactly)
		1 electron volt (eV) = 1.60219×10^{-14} J (approx.)
Electric current	ampere (A)	
Quantity of electricity	coulomb $(C = A\,s)$	$e = 1.60219 \times 10^{-19}$ C (approx.)
Thermo-dynamic temperature (T)	kelvin (K)	replaces °K; Celsius temperature $(t) = T - 273.15$ K in °C
Amount of substance	mole (mol)	
Concentration	mole per cubic metre $(mol\,m^{-3})$	1 mole liter^{-1} = 10^3 mol m^{-3} (exactly)

APPENDIX 2 STANDARD ENTHALPIES AND FREE ENERGIES OF FORMATION, AND STANDARD THIRD-LAW ENTROPIES, AT 298°K

This table gives the standard heat (ΔH^0) and free energies (ΔG^0) of formation of compounds from elements in their standard states and the thermodynamic or third-law entropies (S^0) of compounds, all at 298°K. The state of the compound is specified by: (g) = gas; (l) = liquid; (s) = solid; (aq) = aqueous solution. Occasionally the crystal form of the solid is also specified. Compounds are arranged by the group number of a principal element, with metals taking precedence over nonmetals and O and H being considered least important.

This table is an abbreviated version of a more complete one in R. E. Dickerson, *Molecular Thermodynamics*, W. A. Benjamin, Menlo Park, Calif., 1969. Other convenient tabulations are found in the *Chemical Rubber Company Handbook of Chemistry and Physics*, and in *Lange's Handbook of Chemistry*.

	Substance	ΔH^0_{298} (kcal mole^{-1})	ΔG^0_{298} (kcal mole^{-1})	S^0_{298} (cal deg^{-1} mole^{-1})
	H(g)	52.089	48.575	27.393
	H$^+$(aq)	0.0	0.0	0.0
	H$_3$O$^+$(aq)	−68.317	−56.690	16.716
	H$_2$(g)	0.0	0.0	31.211
IA	Li(g)	37.07	29.19	33.143
	Li(s)	0.0	0.0	6.70
	Li$^+$(aq)	−66.554	−70.22	3.4
	LiF(s)	−146.3	−139.6	8.57
	LiCl(s)	−97.70	−91.7	(13.2)
	LiBr(s)	−83.72	−81.2	(16.5)

Substance	ΔH^0_{298}	ΔG^0_{298}	S^0_{298}
LiI(s)	−64.79	−64	−
Na(g)	25.98	18.67	36.715
Na(s)	0.0	0.0	12.2
Na⁺(aq)	−57.279	−62.589	14.4
Na₂(g)	33.97	24.85	55.02
NaO₂(s)	−61.9	−46.5	−
Na₂O(s)	−99.4	−90.0	17.4
Na₂O₂(s)	−120.6	−102.8	(16.0)
NaF(s)	−136.0	−129.3	14.0
NaCl(s)	−98.232	−91.785	17.3
NaBr(s)	−86.030	−83.1	−
NaI(s)	−68.84	−56.7	−
Na₂CO₃(s)	−270.3	−250.4	32.5
K(g)	21.51	14.62	38.296
K(s)	0.0	0.0	15.2
K⁺(aq)	−60.04	−67.46	24.5
KCl(s)	−104.175	−97.592	19.76
KCl(g)	−51.6	−56.2	57.24
Rb(g)	20.51	13.35	40.628
Rb(s)	0.0	0.0	16.6
Rb⁺(aq)	−58.9	−67.45	29.7
RbF(s)	−131.9	−124.3	27.2
RbCl(s)	−102.91	−96.8	−
RbBr(s)	−93.03	−90.38	25.88
RbI(s)	−78.5	−77.8	28.21
Cs(g)	18.83	12.24	41.944
Cs(s)	0.0	0.0	19.8
Cs⁺(aq)	−59.2	−67.41	31.8
CsF(s)	−126.9	−119.5	−
CsCl(s)	−103.5	−96.6	−
CsBr(s)	−94.3	−91.6	29
CsI(s)	−80.5	−79.7	31

	Substance	ΔH^0_{298}	ΔG^0_{298}	S^0_{298}
IIA	Be(g)	76.63	67.60	32.545
	Be(s)	0.0	0.0	2.28
	Be²⁺(aq)	−93	−85.2	−
	Mg(g)	35.9	27.6	35.504
	Mg(s)	0.0	0.0	7.77
	Mg²⁺(aq)	−110.41	−108.99	−28.2
	MgCl₂(s)	−153.40	−141.57	21.4
	MgCl₂·6H₂O(s)	−597.42	−505.65	87.5
	Ca(g)	46.04	37.98	36.99
	Ca(s)	0.0	0.0	9.95
	Ca²⁺(aq)	−129.77	−132.18	−13.2
	CaCO₃(s, calcite)	−288.45	−269.78	22.2
	CaCO₃(s, aragonite)	−288.49	−269.53	21.2
	Sr(g)	39.2	26.3	39.325
	Sr(s)	0.0	0.0	13.0

	Substance	ΔH_{298}^0	ΔG_{298}^0	S_{298}^0
	$Sr^{2+}(aq)$	−130.38	−133.2	−9.4
	$Ba(g)$	41.96	34.60	40.699
	$Ba(s)$	0.0	0.0	16
	$Ba^{2+}(aq)$	−128.67	−134.0	3
	$BaCl_2(s)$	−205.56	−193.8	30
	$BaCl_2 \cdot H_2O(s)$	−278.4	−253.1	40
	$BaCl_2 \cdot 2H_2O(s)$	−349.35	−309.8	48.5
IVB	$Ti(g)$	112	101	43.069
	$Ti(s)$	0.0	0.0	7.24
	$TiO_2(s, \text{rutile III})$	−218.0	−203.8	12.01
	$TiO^{2+}(aq)$		−138	−
	$Ti_2O_3(s)$	−367	−346	18.83
	$Ti_3O_5(s)$	−584	−550	30.92
VIB	$W(g)$	201.6	191.6	41.552
	$W(s)$	0.0	0.0	8.0
VIII	$Fe(g)$	96.68	85.76	43.11
	$Fe(s)$	0.0	0.0	6.49
	$Fe^{2+}(aq)$	−21.0	−20.30	−27.1
	$Fe^{3+}(aq)$	−11.4	−2.53	−70.1
	$Fe_2O_3(s, \text{hematite})$	−196.5	−177.1	21.5
	$Fe_3O_4(s, \text{magnetite})$	−267.9	−242.4	35.0
IB	$Cu(g)$	81.52	72.04	39.744
	$Cu(s)$	0.0	0.0	7.96
	$Cu^+(aq)$	(12.4)	12.0	(−6.3)
	$Cu^{2+}(aq)$	15.39	15.53	−23.6
	$CuSO_4(s)$	−184.00	−158.2	27.1
	$Ag(g)$	69.12	59.84	41.3221
	$Ag(s)$	0.0	0.0	10.206
	$AgCl(s)$	−30.362	−26.224	22.97
	$AgNO_2(s)$	−10.605	4.744	30.62
	$AgNO_3(s)$	−29.43	−7.69	33.68
IIB	$Hg(g)$	14.54	7.59	41.80
	$Hg(l)$	0.0	0.0	18.5
	$HgCl_2(s)$	−55.0	−44.4	(34.5)
	$Hg_2Cl_2(s)$	−63.32	−50.35	46.8
IIIA	$B(g)$	97.2	86.7	36.649
	$B(s)$	0.0	0.0	1.56
	$B_2O_3(s)$	−302.0	−283.0	12.91
	$B_2H_6(g)$	7.5	19.8	55.66
	$B_5H_9(g)$	15.0	39.6	65.88
	$BF_3(g)$	−265.4	−261.3	60.70
	$BF_4^-(aq)$	−365	−343	40
	$BCl_3(g)$	−94.5	−90.9	69.29
	$BCl_3(l)$	−100.0	−90.6	50.0
	$BBr_3(g)$	−44.6	−51.0	77.49

Substance	ΔH^0_{298}	ΔG^0_{298}	S^0_{298}
$BBr_3(l)$	−52.8	−52.4	54.7
$Al(g)$	75.0	65.3	39.303
$Al(s)$	0.0	0.0	6.769
$Al^{3+}(aq)$	−125.4	−115.0	−74.9
$Al_2O_3(s)$	−399.09	−376.77	12.186
$AlCl_3(s)$	−166.2	−152.2	40
$TlI(g)$	8	−3	65.6
$TlI(s)$	−29.7	−29.7	29.4

	Substance	ΔH^0_{298}	ΔG^0_{298}	S^0_{298}
IVA	$C(g)$	171.698	160.845	37.761
	$C(s, \text{diamond})$	0.4532	0.6850	0.5829
	$C(s, \text{graphite})$	0.0	0.0	1.3609
	$CO(g)$	−26.4157	−32.8079	47.301
	$CO_2(g)$	−94.0518	−94.2598	51.061
	$CO_2(aq)$	−98.69	−92.31	29.0
	$CH_4(g)$	−17.889	−12.140	44.50
	$C_2H_2(g)$	54.194	50.0	47.997
	$C_2H_4(g)$	12.496	16.282	52.45
	$C_2H_6(g)$	−20.236	−7.860	54.85
	$C_3H_8(g)$	−24.82	−5.61	64.51
	$n\text{-}C_4H_{10}(g)$	−29.81	−3.75	74.10
	$i\text{-}C_4H_{10}(g)$	−31.45	−4.30	70.42
	$C_6H_6(g)$	19.820	30.989	64.34
	$C_6H_6(l)$	11.718	29.756	41.30
	$HCOOH(g)$	−86.67	−80.24	60.0
	$HCOOH(l)$	−97.8	−82.7	30.82
	$HCOOH(aq)$	−98.0	−85.1	39.1
	$HCOO^-(aq)$	−98.0	−80.0	21.9
	$H_2CO_3(aq)$	−167.0	−149.00	45.7
	$HCO_3^-(aq)$	−165.18	−140.31	22.7
	$CO_3^{2-}(aq)$	−161.63	−126.22	−12.7
	$CH_3COOH(l)$	−116.4	−93.8	38.2
	$CH_3COOH(aq)$	−116.743	−95.51	−
	$CH_3COO^-(aq)$	−116.843	−89.02	−
	$(COOH)_2(s)$	−197.6	−166.8	28.7
	$(COOH)_2(aq)$	−195.57	−166.8	−
	$HC_2O_4^-(aq)$	−195.7	−167.1	36.7
	$C_2O_4^{2-}(aq)$	−197.0	−161.3	12.2
	$HCHO(g)$	−27.7	−26.2	52.26
	$HCHO(aq)$	−	−31.0	−
	$CH_3OH(g)$	−48.10	−38.70	56.8
	$CH_3OH(l)$	−57.036	−39.75	30.3
	$CH_3OH(aq)$	−58.77	−41.88	31.63
	$C_2H_5OH(g)$	−56.27	−40.30	67.4
	$C_2H_5OH(l)$	−66.356	−41.77	38.4
	$CH_3CHO(g)$	−39.76	−31.96	63.5
	$CH_3CHO(aq)$	−49.88	−	−
	$CH_3NH_2(g)$	−6.7	6.6	57.73
	$CH_3SH(g)$	−2.97	0.21	60.90

	Substance	ΔH^0_{298}	ΔG^0_{298}	S^0_{298}
	Si(g)	88.04	77.41	40.120
	Si(s)	0.0	0.0	4.47
	SiO(g)	−26.72	−32.77	49.26
	SiO$_2$(s, quartz)	−205.4	−192.4	10.00
	Ge(g)	78.44	69.50	40.106
	Ge(s)	0.0	0.0	10.14
	Sn(g)	72	64	40.245
	Sn(s, gray)	0.6	1.1	10.7
	Sn(s, white)	0.0	0.0	12.3
	Pb(g)	46.34	38.47	41.890
	Pb(s)	0.0	0.0	15.51
	Pb^{2+}(aq)	0.39	−5.81	5.1
	PbO(s, red)	−52.40	−45.25	16.2
	PbO(s, yellow)	−52.07	−45.05	16.6
VA	N(g)	112.965	108.870	36.6145
	N$_2$(g)	0.0	0.0	45.767
	N$_3^-$(aq)	60.3	77.7	(32)
	NO(g)	21.600	20.719	50.339
	NO$_2$(g)	8.091	12.390	57.47
	NO$_2^-$(aq)	−25.4	−8.25	29.9
	NO$_3^-$(aq)	−49.372	−26.43	35.0
	N$_2$O(g)	19.49	24.76	52.58
	N$_2$O$_2^{2-}$(aq)	−2.59	33.0	6.6
	N$_2$O$_4$(g)	2.309	23.491	72.73
	N$_2$O$_5$(s)	−10.0	32	27.1
	NH$_3$(g)	−11.04	−3.976	46.01
	NH$_3$(aq)	−19.32	−6.36	26.3
	NH$_4^+$(aq)	−31.74	−19.00	26.97
	NH$_4$Cl(s)	−75.38	−48.73	22.6
	(NH$_4$)$_2$SO$_4$(s)	−281.86	−215.19	52.65
	P(g)	75.18	66.71	38.98
	P(s, white)	0.0	0.0	10.6
	P(s, red)	−4.4	−3.3	(7.0)
	P$_4$(g)	13.12	5.82	66.90
	PCl$_3$(g)	−73.22	−68.42	74.49
	PCl$_5$(g)	−95.35	−77.57	84.3
	As(g)	60.64	50.74	41.62
	As(s, gray metal)	0.0	0.0	8.4
	As$_4$(g)	35.7	25.2	69
VIA	O(g)	59.159	54.994	38.469
	O$_2$(g)	0.0	0.0	49.003
	O$_3$(g)	34.0	39.06	56.8
	OH(g)	10.06	8.93	43.888
	OH$^-$(aq)	−54.957	−37.595	−2.52
	H$_2$O(g)	−57.798	−54.635	45.106
	H$_2$O(l)	−68.317	−56.690	16.716
	H$_2$O$_2$(l)	−44.84	−27.240	(22)

	Substance	ΔH^0_{298}	ΔG^0_{298}	S^0_{298}
	$H_2O_2(aq)$	−45.68	−31.470	−
	$S(g)$	53.25	43.57	40.085
	$S(s,\text{ rhombic})$	0.0	0.0	7.62
	$S(s,\text{ monoclinic})$	0.071	0.023	7.78
	$S^{2-}(aq)$	10.0	20.0	−
	$SO(g)$	19.02	12.78	53.04
	$SO_2(g)$	−70.76	−71.79	59.40
	$SO_3(g)$	−94.45	−88.52	61.24
	$H_2S(g)$	−4.815	−7.892	49.15
	$H_2S(aq)$	−9.4	−6.54	29.2
VIIB	$F(g)$	18.3	14.2	37.917
	$F^-(aq)$	−78.66	−66.08	−2.3
	$F_2(g)$	0.0	0.0	48.6
	$HF(g)$	−64.2	−64.7	41.47
	$Cl(g)$	29.012	25.192	39.457
	$Cl^-(aq)$	−40.023	−31.350	13.2
	$Cl_2(g)$	0.0	0.0	53.286
	$ClO^-(aq)$	−	−8.9	10.3
	$ClO_2(g)$	24.7	29.5	59.6
	$ClO_2^-(aq)$	−16.5	−2.56	24.1
	$ClO_3^-(aq)$	−23.50	−0.62	39
	$ClO_4^-(aq)$	−31.41	−2	43.5
	$Cl_2O(g)$	18.20	22.40	63.70
	$HCl(g)$	−22.063	−22.769	44.617
	$HCl(aq)$	−40.023	−31.350	13.2
	$HClO(aq)$	−27.83	−19.110	31.0
	$ClF_3(g)$	−37.0	−27.2	66.61
	$Br(g)$	26.71	19.69	41.8052
	$Br^-(aq)$	−28.90	−24.574	19.29
	$Br_2(g)$	7.34	0.751	58.639
	$Br_2(l)$	0.0	0.0	36.4
	$HBr(g)$	−8.66	−12.72	47.437
	$I(g)$	25.482	16.766	43.184
	$I^-(aq)$	−13.37	−12.35	26.14
	$I_2(g)$	14.876	4.63	62.280
	$I_2(s)$	0.0	0.0	27.9
	$I_2(aq)$	5.0	3.926	−
	$I_3^-(aq)$	−12.4	−12.31	41.5
	$HI(g)$	6.2	0.31	49.314
	$ICl(g)$	4.2	−1.32	59.12
	$ICl_3(s)$	−21.1	−5.40	41.1
	$IBr(g)$	9.75	0.91	61.8
0	$He(g)$	0.0	0.0	30.13
	$Ne(g)$	0.0	0.0	34.45
	$Ar(g)$	0.0	0.0	36.98
	$Kr(g)$	0.0	0.0	39.19
	$Xe(g)$	0.0	0.0	40.53
	$Rn(g)$	0.0	0.0	42.10

APPENDIX 3 A MORE EXACT TREATMENT OF ACID–BASE EQUILIBRIA

In Chapter 5, we introduced some simple acid–base equilibria, with equilibrium expressions uncomplicated enough to solve with the quadratic formula or with approximation methods. In many cases, these expressions were only valid for dilute solutions, or situations in which one component was present in small amounts. These methods ordinarily are good enough, except in unusual circumstances. In this Appendix we shall derive the exact expressions that must be used when the approximations of Chapter 5 fail. We shall see how to handle equilibrium problems when the contribution of protons or hydroxide ions from the dissociation of water cannot be neglected. We shall carry out an exact derivation of the equilibrium equations for weak acids and their salts, and see that the weak-acid equilibria of Section 5–8 and the hydrolysis equilibria of Section 5–9 really are only special cases of the same general process. The exact equations then will permit us to calculate the titration behavior of a weak acid titrated by a strong base, and to compare these results with the strong-acid titration of Section 5–6. Finally, we will see how to handle dissociation equilibria when a dissociating molecule produces more than one proton.

A–1 STRONG AND WEAK ACIDS: THE CONTRIBUTION FROM DISSOCIATION OF WATER

In Section 5–6, we stated that when a strong acid is added to water, the effect is that of adding the same amount of hydrogen ions, since the acid is totally dissociated. But the acid is not the only source of hydrogen ions; water itself dissociates:

$$H_2O \rightleftharpoons H^+ + OH^-$$

with a dissociation constant or ion product of

$$[H^+][OH^-] = K_w = 10^{-14}$$

Is this source of protons important?

For a 0.01-molar solution of nitric acid, the answer is no. The hydrogen ion concentration from the acid is 10^{-2} mole liter^{-1}, and $[H^+]$ from water dissociation even in pure water is only 10^{-7} mole liter^{-1}, one hundred-thousandth as much. Since the added H^+ from the acid will repress water dissociation, the actual contribution of water to the $[H^+]$ will be smaller yet. We can find the concentration of hydroxide ion (which comes only from water dissociation) from the equilibrium expression:

$$[OH^-] = \frac{K_w}{[H^+]_t} = \frac{10^{-14}}{10^{-2}} = 10^{-12}$$

Here $[H^+]_t$ is the total supply of hydrogen ions from all sources.

Protons have no labels. For every hydroxide ion produced by dissociation of water, one hydrogen ion is produced. Thus the concentration of hydrogen ions *coming from the dissociation of water alone* is

$$[H^+]_w = [OH^-] = 10^{-12} \text{ mole liter}^{-1}$$

and this is entirely negligible in comparison with 10^{-2} mole liter^{-1} of hydrogen ions from the nitric acid.

Example. What is the pH of 10^{-6} molar HCl?

Approximate Solution

$$[H^+] = 10^{-6} \text{ mole liter}^{-1} \quad \text{and} \quad pH = 6.0$$

As a check on the dissociation of water:

$$[H^+]_w = [OH^-] = \frac{10^{-14}}{10^{-6}} = 10^{-8} \text{ mole liter}^{-1}$$

The water contribution to the hydrogen ion concentration still is only one percent as large as that from the added acid. Neglecting $[H^+]_w$ only leads to a 1% error in total hydrogen ion concentration.

Exact Solution. An exact solution involves two simultaneous equations, each relating the total hydrogen ion concentration, $[H^+]_t$, to that provided by water dissociation alone, $[H^+]_w$.

Mass balance: $[H^+]_t = 10^{-6} + [H^+]_w$

Water equilibrium: $[H^+]_w = [OH^-] = \dfrac{10^{-14}}{[H^+]_t}$

The mass-balance equation simply states that the total mass of protons is the sum of that provided from each of two sources—acid and water. To solve these two equations, let y represent the total hydrogen ion concentration from all sources, and eliminate $[H^+]_w$:

$$y = 10^{-6} + \frac{10^{-14}}{y}$$

$$y^2 - 10^{-6}y - 10^{-14} = 0$$

$$y = 1.01 \times 10^{-6} \text{ mole liter}^{-1}$$

The true hydrogen ion concentration is 1% higher, because of the contribution from dissociating water, than our approximate calculation predicted. This correction is trivial for strong acids, but can become important when we deal with weak acids.

Weak Acids and Water Dissociation

In Section 5–7, we used the simple dissociation-constant expression

$$K_a = 1.76 \times 10^{-5} = \frac{y^2}{0.0010 - y}$$

to calculate the hydrogen ion concentration of 0.0010-molar acetic acid:

$$[H^+] = y = 1.24 \times 10^{-4} \text{ mole liter}^{-1}$$

Nothing was said about any contribution from the other equilibrium present:

$$K_w = [H^+][OH^-]$$

Are we in trouble by neglecting this component? We can make a quick test to see. Since every dissociation of a water molecule produces one proton and one hydroxide ion, we can follow the concentration of protons from water dissociation by calculating the hydroxide ion concentration. The protons coming from dissociation of acid will repress the water equilibrium, and the hydroxide ion concentration will be only:

$$[OH^-] = \frac{10^{-14}}{1.24 \times 10^{-4}} = 8.1 \times 10^{-11} \text{ mole liter}^{-1}$$

Because this is also the amount of H^+ that comes from the dissociation of water, we are clearly not wrong in omitting this in comparison with H^+ from the acid, 1.24×10^{-4} mole liter^{-1}. For a weaker acid such as HCN the story can be different.

Exercise. Calculate the pH and percent dissociation of HCN in solutions that are 10^{-2}, 10^{-3}, and 10^{-7} molar.

(*Answer:* HCN is an extremely weak acid, with a dissociation constant of $K_a = 4.93 \times 10^{-10}$. The answers to the exercise are:

Concentration (moles liter^{-1}):	10^{-2}	10^{-3}	10^{-7}
pH:	5.7	6.2	8.2
Percent dissociation:	0.02%	0.07%	6.8%)

The preceding exercise yields the surprising result that by making an acid solution sufficiently dilute (10^{-7} molar), we apparently can make its solution basic! This cannot be true. The flaw is that we finally have reduced the hydrogen ion concentration from the acid to the point where it is close to that coming from dissociation of water. The simple equilibrium expression that we have been using is no longer good enough. In the proper treatment for the general acid HA, there are four unknown concentrations, $[H^+]$, $[HA]$, $[A^-]$, and $[OH^-]$, and four equations that relate these unknowns:

Acid dissociation:
$$K_a = \frac{[H^+][A^-]}{[HA]}$$

Water dissociation: $K_w = [H^+][OH^-]$

Mass balance on the acid anion: $c_0 = [HA] + [A^-]$

Charge balance: $[H^+] = [A^-] + [OH^-]$

The key to solving what might seem to be a complicated set of equations is to understand what the charge-balance equation means physically. Since $[OH^-]$ also equals the amount of hydrogen ion produced from dissociation of water, the charge-balance equation shows that the acid anion concentration, $[A^-]$, is less than the total hydrogen ion concentration, $[H^+]$, by just that amount of H^+ that came from water, rather than from HA:

Charge balance: $[A^-] = [H^+] - [OH^-]$

If we represent the desired hydrogen ion concentration by y as before, and use the water-dissociation equation to eliminate $[OH^-]$, then

Charge balance: $[A^-] = y - \dfrac{K_w}{y}$

The rest of the derivation is as before, with the undissociated acid concentration being the initial overall concentration less that which has dissociated:

Mass balance: $[HA] = c_0 - [A^-] = c_0 - y + \dfrac{K_w}{y}$

These two concentrations, $[HA]$ and $[A^-]$, now can be replaced in the acid-dissociation equilibrium expression:

$$K_a = \frac{y(y - K_w/y)}{c_0 - y + K_w/y}$$

As a check on this entire derivation, note that by eliminating the K_w/y terms, we return to the expression that we have been using. If we use the complete equation for the 10^{-7}-molar HCN problem, for which the simple treatment failed, we find $[H^+] = 1.0025 \times 10^{-7}$ mole liter^{-1} and pH = 6.999, which within the accuracy of K_a should be rounded off to pH 7.00. HCN is such a weak acid and is dissociated so little that its contribution to the pool of hydrogen ions at this concentration can be neglected in comparison with that from water molecules!

We now know how to handle the dissociation of water, and know that except for very dilute solutions of very weak acids, water as a proton source can be ignored in comparison with the protons supplied by other acids present. The mathematics that we have used here, involving mass and charge balances, will be of immediate help in solving the equilibrium equations for a solution of a weak acid and its salt.

A-2 WEAK ACID AND SALT: FULL TREATMENT

In Section 5–7, we calculated the pH of a solution of a weak acid, and in Section 5–9, the pH of a solution of the salt of such an acid with a strong base. These and the buffers of Section 5–8 are only different extremes of the same problem: that of finding the pH of a solution of c_a moles liter^{-1} of a weak acid and c_s moles liter^{-1} of its salt with a strong base, where both c_a and c_s can vary from appreciable values to zero. In dealing with buffers, we assumed that c_s was either smaller than c_a or of the same magnitude. In the hydrolysis problem, c_s was appreciable but c_a was zero. In the region between these two extremes, the mathematics (not the chemistry) becomes more difficult. It is worth setting up the general expressions to show how the simpler expressions fall out from them and how the buffer mixture and hydrolysis are two examples of the same phenomena.

The heart of the problem still is the equilibrium expression for a weak acid:

$$K_a = \frac{[H^+][A^-]}{[HA]}$$

The water-dissociation expression is always valid:

$$K_w = [H^+][OH^-]$$

even though it is sometimes unimportant. One mass-balance statement for the acid anion is that the total supply of material, either as anion or as undissociated acid, must equal the sum of the starting concentrations of weak acid and salt:

$$c_a + c_s = [HA] + [A^-]$$

Another mass-balance statement is that the total counter-ion from the salt (let us assume that it is sodium ion) equals the starting salt concentration, since the counter-ion takes no part in the reactions:

$$c_s = [Na^+]$$

The next equation is a charge balance, which states that the overall solution is electrically neutral:

$$[H^+] + [Na^+] = [OH^-] + [A^-]$$

As before, for simplicity let us represent the hydrogen ion concentration by y, and the hydroxide ion concentration by z. The goal in solving these equations is to eliminate $[A^-]$ and $[HA]$ from the acid-equilibrium expression. The charge-balance equation can be rearranged to

$$[A^-] = [Na^+] + y - z = c_s + y - z$$

The first mass-balance equation then can be converted to

$$[HA] = c_a + c_s - [A^-] = c_a + c_s - y + z - c_s = c_a - y + z$$

Substitution into the equilibrium expression completes the derivation:

$$K_a = \frac{y(c_s + y - z)}{(c_a - y + z)}$$

How does this general expression reduce to what we have seen previously? Under acid conditions, the hydroxide ion concentration will be unimportant, and z can be neglected in comparison with y, the hydrogen ion concentration. The general expression becomes

$$K_a = \frac{y(c_s + y)}{c_a - y}$$

which we already have used in the weak acid–salt and buffer problems of Section 5–8. Under basic conditions, for which y can be disregarded in comparison with z, the expression becomes

$$K_a = \frac{y(c_s - z)}{c_a + z} = \frac{K_w(c_s - z)}{z(c_a + z)}$$

or

$$\frac{z(c_a + z)}{(c_s - z)} = \frac{K_w}{K_a} = K_b$$

(Note that we can neglect y only when it is compared with a larger quantity, not when it stands alone as a multiplier.) If the acid concentration, c_a, is zero, the preceding equation becomes the hydrolysis equilibrium expression that we used in Section 5–9.

A–3 TITRATION OF A WEAK ACID BY A STRONG BASE

In Figure 5–5, we plotted the results of titrating 50 ml of a 0.10-molar solution of a strong acid, HNO_3, with a 0.10-molar solution of a strong base.

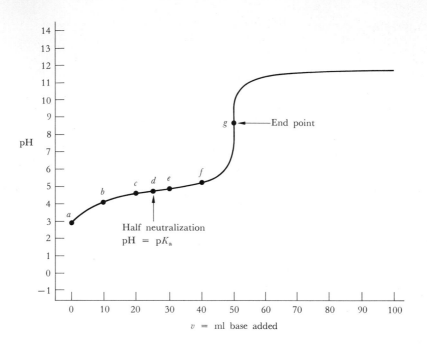

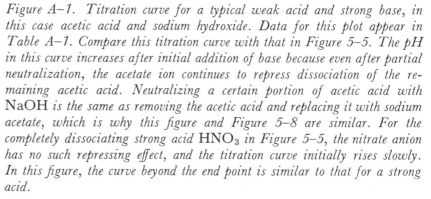

Figure A–1. Titration curve for a typical weak acid and strong base, in this case acetic acid and sodium hydroxide. Data for this plot appear in Table A–1. Compare this titration curve with that in Figure 5–5. The pH in this curve increases after initial addition of base because even after partial neutralization, the acetate ion continues to repress dissociation of the remaining acetic acid. Neutralizing a certain portion of acetic acid with NaOH *is the same as removing the acetic acid and replacing it with sodium acetate, which is why this figure and Figure 5–8 are similar. For the completely dissociating strong acid* HNO_3 *in Figure 5–5, the nitrate anion has no such repressing effect, and the titration curve initially rises slowly. In this figure, the curve beyond the end point is similar to that for a strong acid.*

Now let us repeat this experiment, but with a weak acid, HAc, instead. The general expression that we have just derived is suitable for this purpose, if we consider that the effect of adding sodium hydroxide is to turn acetic acid into sodium acetate by the neutralization reaction:

$$HAc + NaOH \rightarrow H_2O + NaAc$$

or, more accurately:

$$HAc + OH^- \rightarrow H_2O + Ac^-$$

Table A–1.   Titration of 50 ml of 0.10M Acetic Acid by 0.10M Sodium Hydroxide

Point:	a	b	c	d	e	f	g	h
v = ml base added:	0	10	20	25	30	40	45	50
V = total volume:	50	60	70	75	80	90	95	100
[HAc] as added:	0.1000	0.0833	0.0714	0.0667	0.0625	0.0555	0.0526	0.0500
[NaOH] as added:	0.0000	0.0167	0.0286	0.0333	0.0375	0.0444	0.0474	0.0500
$[HAc]_{net} = c_a$:	0.1000	0.0667	0.0428	0.0333	0.0250	0.0111	0.0052	0.0000
$[NaAc]_{net} = c_s$:	0.0000	0.0167	0.0286	0.0333	0.0375	0.0444	0.0474	0.0500
$[H^+]$:	1.33×10^{-3}	7.04×10^{-5}	2.64×10^{-5}	1.76×10^{-5}	1.17×10^{-5}	4.4×10^{-5}	1.9×10^{-5}	1.89×10^{-9}
pH:	2.88	4.15	4.58	4.76	4.93	5.36	6.72	8.68

The calculations for various amounts of added NaOH solution are listed in Table A–1, and the results are plotted in Figure A–1. Point *a* on the plot was obtained from the simple weak-acid equilibrium expression of Section 5–7:

$$K_a = \frac{y^2}{c_a - y}$$

Points *b* through *f* came from the weak acid–salt equilibrium expression of Section 5–8:

$$K_a = \frac{y(c_s + y)}{c_a - y}$$

The end point, *g*, at equal concentrations of acid and base, came from the hydrolysis expression of Section 5–9:

$$K_b = \frac{K_w}{K_a} = \frac{z^2}{c_s - z}$$

All of these equations are simplifications for special conditions of the general expression derived in Section A–2. Beyond the end point, where one in effect is adding more base to a sodium acetate solution, the titration curve is little different from that of a strong acid and base. Only in the region between Points *f* and *g* would one have to struggle with the complete general expression.

The half-neutralization point is equivalent to having identical concentrations of acetic acid and sodium acetate (half of the initial acetic acid has been neutralized by sodium hydroxide). Therefore, the pH at this point is the pK_a of acetic acid. The end point, *g*, does not occur within such a wide range of pH values as in Figure 5–5, and it becomes more important to choose an indicator with a pK_a near 8 or 9. For example, methyl orange could not be used to locate the end point in this acetic acid titration (see Figure 5–7), although phenolphthalein or thymol blue would be ideal.

A–4 POLYPROTIC ACIDS: ACIDS THAT LIBERATE MORE THAN ONE HYDROGEN ION

Sulfuric acid, H_2SO_4, loses one proton as a strong acid with an immeasurably large dissociation constant (in water as the solvent):

$$H_2SO_4 \rightarrow H^+ + HSO_4^-$$

It also can lose a second proton as a weak acid with a measurable dissociation constant:

$$HSO_4^- \rightleftharpoons H^+ + SO_4^{2-} \qquad K_{a_2} = 1.20 \times 10^{-2} \qquad pK_{a_2} = 1.92$$

For carbonic acid, both dissociations are weak:

$$H_2CO_3 \rightleftharpoons H^+ + HCO_3^- \qquad K_{a_1} = 4.3 \times 10^{-7} \qquad pK_{a_1} = 6.37$$
$$HCO_3^- \rightleftharpoons H^+ + CO_3^{2-} \qquad K_{a_2} = 5.61 \times 10^{-11} \qquad pK_{a_2} = 10.25$$

The relative values of K_{a_1} and K_{a_2} are intuitively reasonable. One would expect that HCO_3^-, which already has a negative charge, would be less ready to lose another proton than neutral H_2CO_3.

Phosphoric acid has three dissociations:

$$H_3PO_4 \rightleftharpoons H^+ + H_2PO_4^- \qquad pK_{a_1} = 2.12$$
$$H_2PO_4^- \rightleftharpoons H^+ + HPO_4^{2-} \qquad pK_{a_2} = 7.21$$
$$HPO_4^{2-} \rightleftharpoons H^+ + PO_4^{3-} \qquad pK_{a_3} = 12.67$$

Thus, in a phosphoric acid solution there will be seven ionic and molecular species present: H_3PO_4, $H_2PO_4^-$, HPO_4^{2-}, PO_4^{3-}, H_2O, H^+, and OH^-. Life might appear impossibly complicated, were we not able to make some approximations.

At a pH equal to the pK_a for a dissociation, the two forms of the dissociating species are present in equal concentrations. For the second dissociation of phosphoric acid, for which $pK_{a_2} = 7.21$:

$$K_{a_2} = \frac{[H^+][HPO_4^{2-}]}{[H_2PO_4^-]}$$

$$\log \frac{[HPO_4^{2-}]}{[H_2PO_4^-]} = pH - pK_{a_2}$$

When $pH = pK_{a_2}$, we have the ratio

$$\frac{[HPO_4^{2-}]}{[H_2PO_4^-]} = 1$$

Hence, in a neutral solution, HPO_4^{2-} and $H_2PO_4^-$ are present in about the same concentrations. Very little undissociated H_3PO_4 will be found, since from the first dissociation constant:

$$K_{a_1} = \frac{[H^+][H_2PO_4^-]}{[H_3PO_4]}$$

$$\log \frac{[H_2PO_4^-]}{[H_3PO_4]} = pH - pK_{a_1} = 7.0 - 2.2 = 4.8$$

$$\frac{[H_2PO_4^-]}{[H_3PO_4]} = 6.3 \times 10^4 = 63,000$$

Similarly, little PO_4^{3-} will exist:

$$\log \frac{[PO_4^{3-}]}{[HPO_4^{2-}]} = pH - pK_{a_3} = 7.0 - 12.7 = -5.7$$

$$\frac{[PO_4^{3-}]}{[HPO_4^{2-}]} = 2 \times 10^{-6} = \frac{1}{500,000}$$

The only phosphate species that we have to consider near $pH = 7$ are $H_2PO_4^-$ and HPO_4^{2-}. Similarly, in strong acid solutions near $pH = 3$, only H_3PO_4 and $H_2PO_4^-$ are important. As long as the pK_a's of successive dissociations are separated by three or four units (as they almost always are), matters are simplified.

There is still another simplification. When a polyprotic acid such as carbonic acid, H_2CO_3, dissociates, most of the protons present come from the first dissociation:

$$H_2CO_3 \rightleftarrows H^+ + HCO_3^- \qquad pK_{a_1} = 6.37$$

Since the second dissociation constant is smaller by four orders of magnitude (and the pK_{a_2} larger by four units), the contribution of hydrogen ions from the second dissociation will be only one ten-thousandth as large. Correspondingly, the second dissociation has a negligible effect on the concentration of the product of the first dissociation, HCO_3^-.

Example. At room temperature and 1 atm CO_2 pressure, water saturated in CO_2 has a carbonic acid concentration of approximately 0.040 mole liter^{-1}. Calculate the pH and the concentrations of all carbonate species for a 0.040-molar H_2CO_3 solution.

Solution. Considering initially only the first dissociation:

$$K_{a_1} = 4.3 \times 10^{-7} = \frac{y^2}{0.040 - y} \qquad \text{in which } y = [H^+]$$

From our experience with acetic acid, which has an even larger K_a, we should expect to be able to neglect y in the denominator. The extent of dissociation of an acid with such a small K_a will be very small:

$$y^2 = 4.3 \times 10^{-7} \times 0.040 = 1.72 \times 10^{-8}$$
$$y = 1.31 \times 10^{-4} \text{ mole liter}^{-1}$$

This is the concentration of both hydrogen ion and bicarbonate ion, HCO_3^-:

$$[H^+] = 1.31 \times 10^{-4} \text{ mole liter}^{-1}$$
$$[HCO_3^-] = 1.31 \times 10^{-4} \text{ mole liter}^{-1}$$
$$[H_2CO_3] = 0.040 - 0.00013 = 0.040 \text{ mole liter}^{-1}$$
$$\text{pH} = 4 - 0.12 = 3.88$$

Consequently, carbonated beverages have an acidity somewhere between wine and tomato juice (see Table 5–3). For the second dissociation:

$$HCO_3^- \rightleftarrows H^+ + CO_3^{2-}$$
$$K_{a_2} = 5.6 \times 10^{-11} = \frac{[H^+][CO_3^{2-}]}{[HCO_3^-]}$$

Since this second dissociation has only a minor effect on the first one, we can assume that the hydrogen ion and bicarbonate ion concentrations are effectively the same:

$$[CO_3^{2-}] = \frac{[HCO_3^-]}{[H^+]} \times K_{a_2} = K_{a_2} = 5.6 \times 10^{-11} \text{ mole liter}^{-1}$$

Note the rather surprising result that the concentration of the second dissociation product is equal to the second dissociation constant!

Example.   Calculate the sulfide ion concentration in a solution saturated in H_2S (0.10 mole liter^{-1}), first if the solution is made from distilled water, and second if the solution is made pH $= 3$ with HCl. Use K_a values in Table 5–4.

Solution.   In distilled water, the first dissociation is

$$K_{a_1} = 9.1 \times 10^{-8} = \frac{y^2}{0.10}$$

The dissociation constant is so small that y in the denominator can be neglected immediately. Dissociation will be extremely slight:

$$y = [H^+] = [HS^-] = 9.5 \times 10^{-5} \text{ mole liter}^{-1}$$
$$\text{pH} = 5 - 0.98 = 4.02$$

From the second dissociation:

$$[S^{2-}] = \frac{[HS^-]}{[H^+]} \times K_{a_2} = K_{a_2} = 1.1 \times 10^{-12} \text{ mole liter}^{-1}$$

As with the H_2CO_3 example, the anion produced by the second dissociation has a concentration equal to the second dissociation constant.

In contrast, in HCl solution at pH = 3.0:

$$K_{a_1} = \frac{[H^+][HS^-]}{[H_2S]} = \frac{10^{-3}[HS^-]}{0.10} = 9.1 \times 10^{-8}$$

$$[HS^-] = 9.1 \times 10^{-6} \text{ mole liter}^{-1}$$

$$K_{a_2} = \frac{[H^+][S^{2-}]}{[HS^-]} = \frac{10^{-3}[S^{2-}]}{9.1 \times 10^{-6}} = 1.1 \times 10^{-12}$$

$$[S^{2-}] = \frac{9.1 \times 10^{-6} \times 1.1 \times 10^{-12}}{10^{-3}} = 1.0 \times 10^{-14}$$

The acid has repressed the dissociation of H_2S, making the sulfide ion concentration only one hundredth of what it is in pure water. As we saw in Section 5–10, one can use acids to exert a fine control on sulfide concentration in analytical methods by controlling the pH.

APPENDIX 4
ANSWERS TO EVEN-NUMBERED PROBLEMS

CHAPTER 1

2   0.0949 g
4   1.54×10^{22} molecules
6   24.31
8   141 g; 1.93×10^{24} molecules
10   89.1 g mole^{-1}
12   2.2×10^{19} smaller now.
14   0.960 g; 0.0600 g-atom
16   26.58% K; 35.35% Cr; 38.07% O
18   Ca_3SiO_5
20   51.9 g mole^{-1} (Cr)
22   P_4O_{10}
24   0.283 g S remains
26   7.48 g $CuBr_2$; 4.95 g Br_2 remain
28   6.08×10^{-6} mole of Pb
30   10.0 liters $H_2O(g)$; 5.00 liters O_2 remain
32   Empirical formula, CH_2O; molecular weight = 60 g; molecular formula, $C_2H_4O_2$ (acetic acid, CH_3COOH)
34   H_3C_3O; 109 g mole^{-1}; $H_6C_6O_3$; 1.8×10^{22} molecules
36   Combining weight = 15.3; atomic weight = 45.9
38   Atomic weight = 240; combining capacity = 3

CHAPTER 2

2   4.67
4   619 in.^{3}
6   5.01×10^{13} molecules
8   1.23 atm
10   11.0 g liter^{-1}
12   1.42 g liter^{-1}
14   15.99; CH_4
16   $C_2H_2F_4$
18   $p_{Ar} = 76$ torr; $p_{He} = 764$ torr

20 15.2 torr

22 (a) 1.00 atm; (b) 0.593 atm; (c) 1.593 atm; (d) 0.021 mole; (e) 0.63 ml remains

24 1.59 liters

26 2.79×10^3 cm sec^{-1}; 1.28×10^{-7} °K

28 3101 mph; hydrogen

30 5.65×10^{-14} erg; 7.73×10^{-14} erg; 6.71×10^5 cm sec^{-1}

32 Box 1/Box 2: (a) 0.420; (b) 0.690; (c) 0.290; (d) 0.690; (e) 0.476

CHAPTER 3

2 0.0246 faraday; 2370 coulombs

4 176 coulombs

6 40 hours

8 1.92 g

10 1.61×10^6 kg Al; 1.86 dams

12 -0.56°C

14 (a) -0.930°C; (b) -0.279°C; (c) -0.744°C; (d) -0.558°C

16 The ions are $Pt(NH_3)_4Cl_2^{2+}$ and $2Cl^-$.

18 Radius ratio is not the only factor that determines structure; bonding also is important.

20 1.51 Å

CHAPTER 4

2 a) $Fe_2O_3 + 2Al \rightarrow 2Fe + Al_2O_3$
  b) $Na_2SO_3 + 2HCl \rightarrow 2NaCl + SO_2 + H_2O$
  c) $Mg_3N_2 + 6H_2O \rightarrow 3Mg(OH)_2 + 2NH_3$
  d) $Pb + PbO_2 + 2H_2SO_4 \rightarrow 2PbSO_4 + 2H_2O$

4 a) $2Al + 6HCl \rightarrow 2AlCl_3 + 3H_2$
  b) $4NH_3 + 5O_2 \rightarrow 4NO + 6H_2O$
  c) $3Zn + 2P \rightarrow Zn_3P_2$
  d) $2HNO_3 + Zn(OH)_2 \rightarrow Zn(NO_3)_2 + 2H_2O$

6 $NH_4NO_3 \rightarrow 2H_2O + N_2O$; 2.35 liters

8 (a) 0.128 mole; (b) 0.128 mole; (c) 4.36 g; (d) 2.87 liters

10 $2VO + 3Fe_2O_3 \rightarrow V_2O_5 + 6FeO$
  8.84 g V_2O_5; 2.18 g V_2O_5

12 H_2O is the acid (proton donor); NH_3 is the base.

14 0.054 molar

16 1.13 molal; 1.09 molar; 2.18 normal; 1.07 g ml^{-1}

18 300 ml

20 (a) 2.38 ml; (b) 11 ml; (c) 1200 ml

22 $3.71 \times 10^{-4} M$; $7.42 \times 10^{-4} N$

24 $0.025 M$ Ba^{2+}; $0.020 M$ Cl^-; $[H^+] \simeq 0$; $0.030 M$ OH^-

26 11.97 liters

28 2; $0.509 N$

30 94.5 ml

32 0.23 g   **34** Pyruvic acid

36 2.96 mg   **38** 104

40 135; $CH_3—C_6H_4—COOH$

42 (a) 34.8%; (b) $x = 5$

44 $2NCl_3 \rightarrow N_2 + 3Cl_2 + 110$ kcal; 4.6 kcal

46 Endothermic; $\Delta H^0 = 70$ kcal per mole Ti_3O_5; $3TiO_2 \rightarrow Ti_3O_5 + \frac{1}{2}O_2 - 70$ kcal

48 -76.8 kcal mole^{-1} of Na_2CO_3

50 $\Delta H^0 = -108.1$ kcal   **52** $\Delta H^0 = -91.1$ kcal

CHAPTER 5

2 20

4 $[HI] = 1.88$ moles liter^{-1}; $[H_2] = 0.11$ mole liter^{-1}; $[I_2] = 0.61$ mole liter^{-1}

6 12

8 2.89; 1.71×10^{-5}

10 1.79×10^{-5}; No; $NH_3 + H_2O \rightarrow NH_4^+ + OH^-$; 10.63

12 5.11; 11.11

14 2.22×10^{-5} mole liter^{-1}; 9.35

16 1.1×10^{-2} mole liter^{-1}; 1.96; 4.4%

18 5

20 $OBr^- + H_2O \rightleftharpoons HOBr + OH^-$; $K_b = \dfrac{[HOBr][OH^-]}{[OBr^-]}$;
$K_b = 5.01 \times 10^{-6}$; $K_a = 2.00 \times 10^{-9}$

22 $K_b = 1.99 \times 10^{-11}$; $K_a = 5.03 \times 10^{-4}$

24 9.31   **26** 4.75   **28** 2.75

30 9.08; 9.01; The buffer concentration is insufficient to prevent a sizable pH change.

32 (a) 11.48; (b) 5.48; (c) Yes

34 (a) 0.77; (b) 0.0499; (c) 173; (d) 180; $NH_4^+ < HSO_3^- < CH_3COOH < HF < HNO_2$

36 1.6×10^{-5}

38 2.23×10^{-11} mole liter^{-1}; 5×10^{-21} mole liter^{-1}

40 2.1×10^{-4} mole liter^{-1}

42 11 g

44 -0.20

46 3.1×10^{-16} mole liter^{-1}; 0.031 mole liter^{-1}; By adjusting pH, it is possible to precipitate FeS selectively.

48 12.7; Yes

CHAPTER 6

2 The two lines arise from two different frequencies in each atom's spectrum. Several lines could have been plotted, since each atom has many energy levels.

4 2260 kcal mole^{-1}

6 Nonmetallic properties *decrease* from Si to Pb.

8 (d) is correct. **10** (a) is correct.

12 $CaH_2 + 2H_2O \rightarrow Ca(OH)_2\downarrow + 2H_2\uparrow$

14 CaH_2: ionic; H^- anionic; most basic.

H_2Te: mostly covalent; H^+ cations in aqueous solution; acidic.

GeH_4: covalent; insoluble in water.

H_2S: mostly covalent; H^+ cations in aqueous solution; less acidic than H_2Te.

WH_6: covalent; anionic hydrogens.

CHAPTER 7

2 No. The geometry of a complex is defined as the polygon generated by the substituents bound to the central atom; six substituents generate an octahedron (eight surfaces).

4 6(Si), 3(B), 4(N), 4(Ni), 6(Co);

$+4$(Si), $+3$(B), -3(N), $+2$(Ni), $+2$(Co)

6 XeF_7^-: CN = 7, ON = $+6$;

XeF_8^-: CN = 8, ON = $+7$

8 5(Co), 6(Pt), 3(C), 4(S), 6(Mn)

10 CN = 4, ON = $+2$

12 $+5$(V), $+5$(P), -3(P), $+3$(N), -1(O), -1(H), -3(N),

$+3$(N), $+5$(I), $+1$(Ag)

14 (a) MnO_4^-, SO_3^{2-}, SO_3^{2-}, MnO_4^-; (c) Cl_2 for all; (b) and (d) are not redox reactions.

16 $2MnO_4^- + 6H^+ + 5H_2S \rightarrow 2Mn^{2+} + 8H_2O + 5S$

$+7$; S in H_2S; Mn in MnO_4^-; reduced

18 a) $2MnO_2 + 4KOH + O_2 \rightarrow 2K_2MnO_4 + 2H_2O$

b) $CuCl_4^{2-} + Cu \rightarrow 2CuCl_2^-$

c) $NO_3^- + 4Zn + 10H^+ \rightarrow NH_4^+ + 4Zn^{2+} + 3H_2O$

d) $2ClO_2 + 2OH^- \rightarrow ClO_2^- + ClO_3^- + H_2O$

e) $6Fe^{2+} + Cr_2O_7^{2-} + 14H^+ \rightarrow 6Fe^{3+} + 2Cr^{3+} + 7H_2O$

f) $3Cu + 2NO_3^- + 8H^+ \rightarrow 3Cu^{2+} + 2NO + 4H_2O$

20 $0.0759M$; $0.228N$, 52.68; $0.0759N$, 158.04; $0.304N$, 39.51;

$0.380N$, 31.61

22 52.3

24 (a) 0.50 liter; (b) 0.50 liter; (c) 0.17 liter; (d) 0.25 liter

26 (a) 0.02 normal; (b) 0.02 normal

28 3.13×10^{-4} mole

30 0.0125 molar

32 (a) Mn $+7$ in MnO_4^-, $+2$ in Mn^{2+}; C $+3$ in $(COOH)_2$, $+4$ in CO_2; (b) $2MnO_4^- + 5(COOH)_2 + 6H^+ \rightarrow$ $2Mn^{2+} + 10CO_2 + 8H_2O$; (c) $KMnO_4$: 158.04 g $mole^{-1}$, 31.61 g $equiv^{-1}$; $(COOH)_2$: 90.04 g $mole^{-1}$, 45.02 g $equiv^{-1}$; (d) 0.004; (e) 0.04; (f) 0.500 normal; (g) 0.05 molar; (h) 1.34 g

CHAPTER 8

2 $\lambda = 2.5 \times 10^3$ Å; 7.9×10^{-12} erg; 114 kcal; ultraviolet light

4 6.63×10^{-21} erg; 9.55×10^{-11} kcal; $\lambda = 3.00 \times 10^4$ cm; The energy of radio waves is much less than that of a C—C bond; therefore, such waves would not produce a chemical reaction.

6 $E = h\nu = -\dfrac{k}{n^2}$ and $k = \dfrac{2\pi m_e e^4 Z^2}{h^2}$. Since $h\nu \propto Z^2$, a plot of Z against $\sqrt{\nu}$ should give a straight line.

8 (a)   **10** (e)

12 (a)   **14** (c)

16 -0.85 eV   **18** (e)

20 0, 1, 2, 3   **22** 0, ± 1, ± 2, ± 3; a d electron

24 $\lambda = 7.29 \times 10^{-9}$ cm; The uncertainty in momentum is about five times greater than the momentum itself.

CHAPTER 9

2 a) As: $1s^2 2s^2 2p^6 3s^2 3p^6 4s^2 3d^{10} 4p^3$;
or [Ar] $4s^2 3d^{10} 4p^3$

b) Co^{2+}: [Ar] $4s^0 3d^7$. (Why not [Ar] $4s^2 3d^5$?)

c) Cu: [Ar] $4s^1 3d^{10}$. (Why not [Ar] $4s^2 3d^9$?)

d) S^{2-}: [Ne] $3s^2 3p^6$ or [Ar]

e) Kr: $1s^2 2s^2 2p^6 3s^2 3p^6 3d^{10} 4s^2 4p^6$, or [Ar]

f) C: $1s^2 2s^2 2p^2$

g) W: [Xe] $6s^2 4f^{14} 5d^4$

h) H^+: 0

i) H^-: $1s^2$

j) Cl^-: $1s^2 2s^2 2p^6 3s^2 3p^6$, or [Ar]

4 a) neutral, excited   b) anion, ground

c) neutral, ground   d) neutral, impossible

e) neutral, impossible   f) cation, excited

g) anion, excited   h) cation, impossible

i) neutral, ground   j) cation, ground

6 3.1

8 Although the $3s$ electron of Na is shielded partially by the other 10 electrons, it experiences an effective nuclear charge greater than +1. Therefore, more energy is required to remove it.

10 The energy required to remove the second electron in Mg is greater than that required to remove the first, because the second electron is less shielded from the positively charged nucleus than is the first. The second ionization energy of Mg is smaller than the second ionization energy of Na, because in the case of Na the electron involved is being removed from a noble-gas configuration. Na enters into chemical reactions in the +1 state because the first ionization energy is small, but the second is too large for compounds of Na^{2+} to be formed. The low first ionization energy of Na also accounts for its great reactivity. Mg appears as Mg^{2+} in compounds because the second ionization energy is small enough to permit easy formation of Mg^{2+}. However, the third ionization energy is too great to allow Mg^{3+} to form.

12 The melting points increase in the order given because the compounds in the series become increasingly ionic and, therefore, require an increasing amount of thermal energy to break bonds and cause melting.

14 Cl has the highest electron affinity. It is greater than that of Na and of O because Cl has a greater nuclear charge to attract an electron. It is greater than that of I because the nucleus of Cl is less shielded than the nucleus of I and, therefore, exerts a greater attraction on an electron.

16 P

18 Hydride, MH_2
Oxide, MO
Oxidation state, $+2$
Electronegativity, ≤ 0.9
mp of chloride, $MCl_2 \simeq 1062°C$
Solid MCl_2 will not conduct electricity because the ions are not free to move. Liquid MCl_2 will conduct because the ions are free to move.

CHAPTER 10

2 $Na\cdot$, $\cdot\overset{\cdot}{C}:$, $\cdot\overset{\cdot}{Si}:$, $:\overset{\cdot}{Cl}\cdot$, $:\overset{\cdot\cdot}{\underset{\cdot\cdot}{Kr}}:$

4 $:\overset{\cdot\cdot}{O}::\overset{\cdot\cdot}{O}:$
(The unpaired electrons that actually are present in O_2 would not be predicted from Lewis dot convention.)
$:C:::O:$, $Li\cdot Li^+$, $:C:::N:^-$

6 $:\overset{\cdot\cdot}{\underset{\cdot\cdot}{Cl}}:^- Ba^{2+}:\overset{\cdot\cdot}{\underset{\cdot\cdot}{Cl}}:^-$, $H:\overset{H}{\underset{H}{P}}:H$, $H:\overset{H}{\underset{H}{N}}:H^+$ $:\overset{\cdot\cdot}{\underset{\cdot\cdot}{Cl}}:^-$

$H:\overset{\cdot\cdot}{O}:$
$:\overset{\cdot\cdot}{\underset{\cdot\cdot}{Cl}}:$, $H:\overset{\cdot\cdot}{\underset{H}{O}}:$, $:\overset{\cdot\cdot}{\underset{\cdot\cdot}{O}}:\overset{\cdot\cdot}{\underset{H}{O}}:$

$\overset{\cdot\cdot}{N}$ $\overset{\cdot\cdot}{N}$
$\cdot\overset{\cdot}{\underset{\cdot\cdot}{O}}\cdot$ $\cdot\overset{\cdot}{\underset{\cdot}{O}}:^- \leftrightarrow ^-:\overset{\cdot}{\underset{\cdot}{O}}\cdot$ $\cdot\overset{\cdot}{\underset{\cdot}{O}}\cdot$

8 a)

$$\overset{\overset{-}{\underset{\underset{}{}}{}}\overset{+}{}}{}$$

H : C : N : : : N :

H

(Atoms not marked $+$ or $-$ have zero formal charge.)

b)

$$\overset{+ \quad -}{}$$

H : C : : N : : N :

H

10

$$\left[\begin{array}{c} :\ddot{O}: \\ :\ddot{O}:Br:\ddot{O}: \\ :\ddot{O}: \end{array}\right]^{-} \quad \begin{array}{c} H \\ H:Si:H \\ H \end{array} \quad \left[\begin{array}{c} :\ddot{C}l: \\ :\ddot{C}l:P:\ddot{C}l: \\ :\ddot{C}l: \end{array}\right]^{+} \quad \begin{array}{c} H \\ H:C:\ddot{C}l: \\ :\ddot{C}l: \end{array} \quad \left[\begin{array}{c} :\ddot{F}: \\ :\ddot{F}:B:\ddot{F}: \\ :\ddot{F}: \end{array}\right]^{-}$$

12 $O_2: KK(\sigma_s{}^b)^2(\sigma_s{}^*)^2(\sigma_z{}^b)^2(\pi_{x,y}{}^b)^4(\pi_{x,y}{}^*)^2$

The molecule is paramagnetic. Lewis structure (see Problem 4) would not lead to a prediction of paramagnetism. The heat of dissociation of O_2 should be less than that of NO, since the bond order of O_2 is 2 and that of NO is 2.5 (see Problem 11).

14 N_2

16 Dipole moments: $D = 4.4$ debyes for HF, 7.7 debyes for HI.

18 CO_2 molecule is linear; H_2O is bent.

20 HCl: 17% ionic; CsCl: 75%; TlCl: 29%

22 $Li_2: KK(\sigma_s{}^b)^2$, bond order $= 1$

$Be_2: KK(\sigma_s{}^b)^2(\sigma_s{}^*)^2$, bond order $= 0$; should not exist

24

Molecule	Hybridization	Electronic geometry	Molecular geometry
CH_4	sp^3	tetrahedral	tetrahedral
BF_3	sp^2	trigonal planar	trigonal planar
NF_3	sp^3	tetrahedral	trigonal pyramidal
$ICl_4{}^-$	sp^3d^2	octahedral	square planar
H_2O	sp^3	tetrahedral	angular

26 (a) $CO_3{}^{2-}$; (b) BF_3; (c) NH_3; (d) $ClO_3{}^-$; (e) $ClO_4{}^-$; (f) $SO_4{}^{2-}$; (g) CO_2; (h) SO_2

28

$$CH_3-C\overset{\displaystyle O}{\underset{\displaystyle O^-}{\Big\langle}} \quad \text{and} \quad CH_3-C\overset{\displaystyle O^-}{\underset{\displaystyle O}{\Big\langle}}$$

No methods; the ion does *not* resonate.

Figure 7–9(e): 2.50 debyes. Figure 7–9(f): 1.72 debyes.
Figure 7–9(g): 0.00 debye.

CHAPTER 11

2 The compound is $[Co(NH_3)_5Cl^{2+}]2Cl^-$. It will precipitate two
moles of AgCl per mole of complex. Because the geometry of
the Co(III) complex is octahedral, the five neutral NH_3 groups
and one Cl^- are coordinated to Co. Therefore, two of the three
chlorides must exist outside of the coordination sphere as ions.

4 $PtCl_4 \cdot 3NH_3$ is $[Pt(NH_3)_3Cl_3{}^+]Cl^-$; two ions per mole; cation
is octahedral.
$PtCl_2 \cdot 3NH_3$ is $[Pt(NH_3)_3Cl^+]Cl^-$; two ions per mole; cation
is square planar.

6 dichlorotetraamminecobalt(III) bromide; potassium hexa-
cyanochromate(III); sodium tetrachlorocobaltate(II)

8 a) $[Al(H_2O)_5OH]Cl_2$
b) $Na_3[Co(CO_3)_3]$
c) $Na_4[Fe(CN)_6]$
d) $(NH_4)_3[Co(NO_2)_6]$

10 There are five geometrical isomers, one of which has an optical
isomer.

12

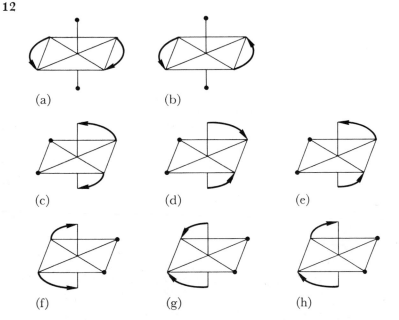

(a)   (b)

(c)   (d)   (e)

(f)   (g)   (h)

where $\cdot$ = Cl^- and $\rightarrow$ = $H_2N—CH_2—CH(CH_3)—NH_2$. All
other possibilities can be converted to these eight by turning
the ion around in the suitable way. Structures (f), (g), and (h)
differ from (c), (d), and (e), respectively, by a mirror reflection.
Structure (a) has a plane of symmetry.

14 (d) follows; the others do not.

16 a) 3, 2, 1, 0, 0, 1
   b) 3, 4, 5, 4, 4, 3

18 *cis*-$Pt(NH_3)_2Cl_4$ uses d^2sp^3 hybrid orbitals and the other two complexes use dsp^2. All electrons are paired in each case, and all are inner complexes.

20 Valence bond:

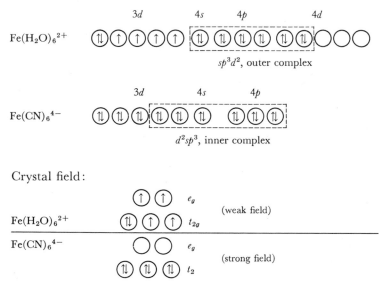

Crystal field:

$Fe(H_2O)_6{}^{2+}$ e_g (weak field) t_{2g}

$Fe(CN)_6{}^{4-}$ e_g (strong field) t_2

The crystal field theory assumes electrostatic repulsion of ligand electron pairs and metal d orbitals, whereas the valence bond theory assumes the metal–ligand bond is covalent and is formed from the overlap of hybridized metal orbitals and ligand orbitals. The valence bond theory and the crystal field theory can be used to rationalize the number of unpaired electrons in a complex. Valence bond theory offers no explanation of spectra, whereas crystal field theory does so quite adequately.

22 a) Octahedral; d^2sp^3 hybridization, using $3d_{z^2}$, $3d_{x^2-y^2}$, $4s$, $4p_x$, $4p_y$, and $4p_z$ atomic orbitals.
   b) Hexaamminecobalt(III) chloride
   c)

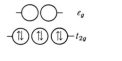 e_g
t_{2g}

 e_g
t_{2g}

low-spin
diamagnetic
(correct)

high-spin
paramagnetic

d) $Co(NH_3)_6{}^{2+}$

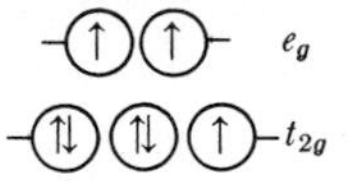

The high-spin form is preferred because the crystal-field splitting energy for Co^{2+} is less than that for Co^{3+}.

24

$[NH_3]$	$[Cu^{2+}(aq)]$
0.10	2.5×10^{-10}
0.50	4.0×10^{-13}
1.00	2.5×10^{-14}
3.00	3.1×10^{-16}

To keep $[Cu^{2+}]$ less than $10^{-15}M$, $[NH_3]$ must exceed $2.24M$.

26 Mn^{2+}, $pH = 5.26$
Fe^{2+}, $pH = 4.00$
Ag^+, $pH = 6.00$
$Cr^{3+} > Fe^{2+} > Mn^{2+} > Ag^+$
The greater the ionic charge, the greater the acidity.

28 9.2×10^{-6} mole liter^{-1}; 4.3×10^{-10} mole liter^{-1}; 2.5×10^{-14} mole liter^{-1}

30 1.3×10^{-5} mole liter^{-1}; 2×10^{-13} mole liter^{-1}, which is small compared to $[Ag^+]$ and can be neglected.

32 0.1 mole liter^{-1}

CHAPTER 12

2 Possible structures:

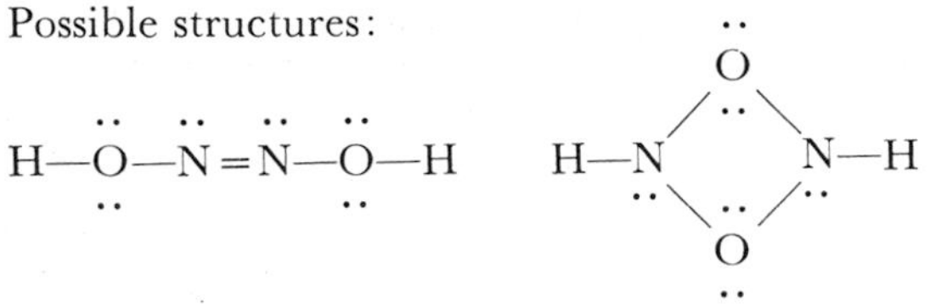

4 2,3-dimethylbutane
6 1,3-butadiene
8 1-bromo-1-chloro-2-methylpropene
10 formic acid or methanoic acid
12 acetaldehyde or ethanal
14 acetone, propanone, or dimethyl ketone
16

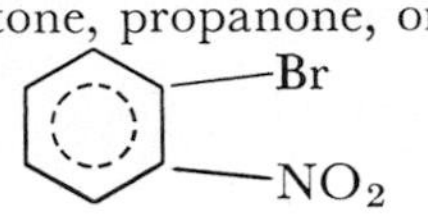

18 $CH_3-CH_2-CH_2-CH_2-NH_2$

20

22 CH_3—SH

24 Na^+ ^-OCC—CH_2—CH_3

26 H_2N—CH_2—CH_2—NH_2

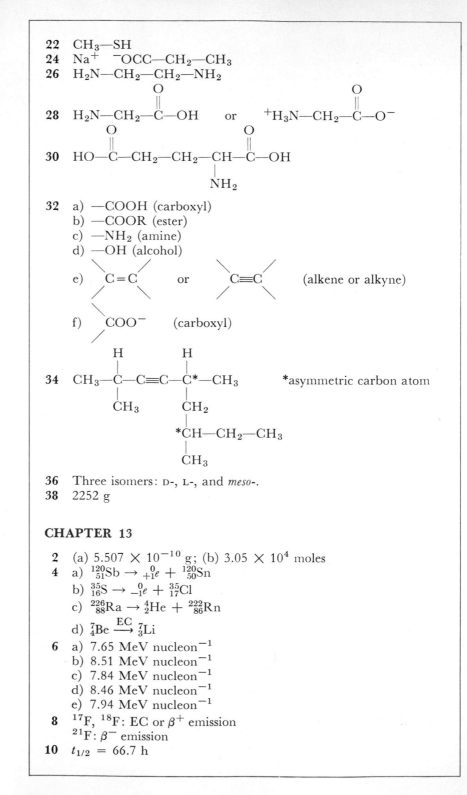

28 H_2N—CH_2—$\overset{\overset{\displaystyle O}{\|}}{C}$—OH   or ^+H_3N—CH_2—$\overset{\overset{\displaystyle O}{\|}}{C}$—$O^-$

30 HO—$\overset{\overset{\displaystyle O}{\|}}{C}$—$CH_2$—$CH_2$—$\underset{\underset{\displaystyle NH_2}{|}}{CH}$—$\overset{\overset{\displaystyle O}{\|}}{C}$—OH

32 a) —COOH (carboxyl)
  b) —COOR (ester)
  c) —NH_2 (amine)
  d) —OH (alcohol)

  e) C$=$C or C$\equiv$C (alkene or alkyne)

  f) COO$^-$ (carboxyl)

34 CH_3—$\overset{\overset{\displaystyle H}{|}}{C}$—C$\equiv$C—$\overset{\overset{\displaystyle H}{|}}{C^*}$—$CH_3$ *asymmetric carbon atom

$\underset{\underset{\displaystyle CH_3}{|}}{}\underset{}{}$

with CH_3 and CH_2—$^*CH$—CH_2—CH_3 with CH_3

36 Three isomers: D-, L-, and *meso*-.

38 2252 g

CHAPTER 13

2 (a) 5.507×10^{-10} g; (b) 3.05×10^4 moles

4 a) $^{120}_{51}Sb \rightarrow \,^{0}_{+1}e + \,^{120}_{50}Sn$
  b) $^{35}_{16}S \rightarrow \,^{0}_{-1}e + \,^{35}_{17}Cl$
  c) $^{226}_{88}Ra \rightarrow \,^{4}_{2}He + \,^{222}_{86}Rn$
  d) $^{7}_{4}Be \xrightarrow{\text{EC}} \,^{7}_{3}Li$

6 a) 7.65 MeV nucleon^{-1}
  b) 8.51 MeV nucleon^{-1}
  c) 7.84 MeV nucleon^{-1}
  d) 8.46 MeV nucleon^{-1}
  e) 7.94 MeV nucleon^{-1}

8 ^{17}F, ^{18}F: EC or β^+ emission
  ^{21}F: β^- emission

10 $t_{1/2} = 66.7$ h

12 $\lambda = 0.0677$ Å

14 $^{230}_{90}$Th, $^{226}_{88}$Ra, $^{222}_{86}$Rn, $^{218}_{84}$Po, $^{214}_{82}$Pb, $^{214}_{83}$Bi, $^{210}_{81}$Tl, $^{210}_{82}$Pb, $^{210}_{83}$Bi, $^{210}_{84}$Po, $^{206}_{82}$Pb

16   2.05 days

18   0.537 g

20   3940 years

22 1.50×10^9 years

CHAPTER 14

2 1.3×10^{15} sec^{-1}; 2.3×10^{-5} cm; 4.3×10^4 cm^{-1}; ultraviolet

4   338 cal mole^{-1}; 4.0 Å

CHAPTER 15

2 $\Delta H^0 = 17.889$ kcal
      $\Delta E = \Delta H - \Delta PV = \Delta H - \Delta nRT$
          $= -123$ kcal $- 3.5$ kcal $= -127$ kcal

4 $\Delta H^0 = -56.6$ kcal mole^{-1} compared to measured value of -56.27 kcal mole^{-1}

6 $\Delta H^0 = -57.8$ kcal mole^{-1} compared to measured value of -57.798 kcal mole^{-1}

8   a) vapor
     b) vapor
     c) 3 heads and 2 tails
     d) mixture

10   Hydrogen bonding in liquid phase lowers entropy of liquid HF and hence increases ΔS_{vap}.

12   Nonspontaneous;
      $\Delta G^0 = +78.7$ kcal; $\Delta H^0 = +117.2$ kcal; $\Delta S^0 = +129.3$ e.u.;
      $T\Delta S^0 = +38.5$ kcal; $\Delta H^0 - T\Delta S^0 = +78.8$ kcal
      Enthalpy opposes reaction; entropy favors reaction; enthalpy dominates.

14 $\Delta G^0 = -2.64$ kcal for equation as written. Spontaneous.
      $K_{eq} = 87$.

16 $\Delta G^0 = -0.6850$ kcal mole^{-1}; The conversion is kinetically unfavorable.

18 $\Delta G^0 = -21.98$ kcal; $\Delta H^0 = -42.28$ kcal;
      $\Delta S^0 = (\Delta H^0 - \Delta G^0)/T = -68.1$ e.u.;
      ΔS^0 (tables) $= -68.0$ e.u.
      Order increases and entropy decreases when a solid is formed from two gases.

20 $\Delta H^0_{vap} = 8.10$ kcal mole^{-1} at 25°C, and 7.30 kcal mole^{-1} at 80°C. Liquid molecules at 80°C have more energy, thus less energy is needed for evaporation. In terms of Le Chatelier's principle, an increase in temperature causes the equilibrium position to shift toward more gas formation.

22 $\Delta G^0 = -39.8$ kcal mole^{-1} of CH_3OH; The reaction would become less spontaneous.

24 51 torr; 0.80 for benzene, 0.20 for toluene

26 $\Delta G^0 = -7.952$ kcal; $K_{eq} = \dfrac{a_{NH_3}^2}{a_{N_2} a_{H_2}^3} = 6.8 \times 10^5$

28 20 g NaCl

30 152 mm Hg

32 1720 g; 102.8°C

34 $C_7H_6O_2$; Benzoic acid exists as a dimer in benzene.

CHAPTER 16

2 $4H_2O(g) + 3Fe(s) \rightleftarrows 4H_2(g) + Fe_3O_4(s)$

$$K_{eq} = \frac{a_{H_2}^4}{a_{H_2O}^4}$$

4 $K_{eq} = 5 \times 10^{59}$; Yes; $\Delta G = -80$ kcal

6 An increase in pressure will lower the freezing point.

8 Less; more; decrease; no effect

10 $K_{eq} = 8.81$; less conversion; more; less; less conversion but K_{eq} is not affected

12 $\Delta G^0 = +6.49$ kcal; $K_{eq} = 1.74 \times 10^{-5}$ compared to 1.76×10^{-5} in Table 5–4.

14 $K_{eq} = \dfrac{[H_3O^+][Cre^-]}{[HCre]} = 1.0 \times 10^{-10}$

$[Cre^-] = 1.0 \times 10^{-5}$ mole liter^{-1}

pH = 5

$\Delta G^0 = 13.64$ kcal

16 At 20°C, $K_{sp} = 21.8$ and $\Delta G^0 = -1.78$ kcal.
At 100°C, $K_{sp} = 116$ and $\Delta G^0 = -3.52$ kcal.
From Gibbs–Helmholtz plot, $\Delta H^0 \simeq +4.59$ kcal. The process is endothermic.

18 $\alpha = \left(\dfrac{K_{eq}}{P + K_{eq}}\right)^{1/2}$; 1.95×10^{-7}

CHAPTER 17

2 0.118 V; 5.44 kcal

4 (a) +1.10 V; (b) +0.85 V; (c) −0.72 V

6 a) $2Ag + Cu^{2+} \rightarrow 2Ag^+ + Cu$
This reaction is spontaneous in the *reverse* direction.
b) $\mathscr{E}^0 = -0.46$ V
c) Cu to Ag

8 a) Spontaneous: $Ni + Sn^{2+} \rightarrow Ni^{2+} + Sn$
b) Sn is cathode, Ni is anode
c) $\mathscr{E}^0 = +0.09$ V
d) $\mathscr{E} = +0.15$ V
(*Note:* Ni data must be obtained from a handbook.)

10 $2H_2O_2 \rightarrow 2H_2O + O_2$; $\mathscr{E}^0 = 1.77 - 0.68 = 1.10$ V

Since $\mathscr{E}^0$ is positive, the disproportionation is spontaneous.

12 $+1.36$ V; $+0.99$ V; $+0.96$ V

a) MnO_4^- is best oxidizing agent; NO is best reducing agent.

b) $\mathscr{E}^0 = +1.49 - 1.42 = +0.07$; Yes, barely.

c) $\mathscr{E}^0 = +0.96 - 1.42 = -0.46$; No.

d) $\mathscr{E}^0 = +0.96 - 1.99 = -0.03$; No.

e) $\mathscr{E}^0 = +1.36 - 1.42 = -0.06$; No.

f) $\mathscr{E}^0 = +1.36 - 0.99 = +0.37$; Yes.

g) $\mathscr{E}^0 = +1.49 - 1.36 = +0.13$; Yes.

14 $\mathscr{E}$ will decrease as Cu^{2+} is removed by complexing with ethylenediamine.

16 a) $2Ag^+ + H_2SO_3 + H_2O \rightarrow 2Ag + SO_4^{2-} + 4H^+$

b) $H_2SO_3|SO_4^{2-}|\;|Ag^+|Ag$

c) $\mathscr{E}^0 = +0.80 - 0.20 = +0.60$ V

d) $K_{eq} = 2.0 \times 10^{20}$ e) $Q = 1.1 \times 10^3$

18 $Fe + Cu^{2+} \rightarrow Fe^{2+} + Cu$

$\mathscr{E}^0 = +0.34 + 0.41 = +0.75$ V

$\mathscr{E} = +0.81$ V

20 Voltage will decrease.

$Q = [Zn^{2+}]/[Cu^{2+}] = 2.15 \times 10^3$

22 a) $\mathscr{E} = +0.059$ V

b) Cathode: $1.0M$ Ag^+; Anode: $0.1M$ Ag^+

c) $\Delta G = -1.36$ kcal

24 $Zn + Cu^{2+} \rightleftharpoons Zn^{2+} + Cu$

a) $\mathscr{E}$ decreases. $[Cu(NH_3)_4{}^{2+}$ formed]

b) $\mathscr{E}$ increases. $[Zn(NH_3)_4{}^{2+}$ formed]

c) $\mathscr{E}$ decreases. (Cu complex more stable than Zn complex)

d) $\mathscr{E}$ increases. (Zn^{2+} removed as ZnS)

e) Opposing influences; Difficult to predict, but voltage probably increases.

26 48.9; 2.40×10^3; 5.8×10^6

28 5×10^{-13}

CHAPTER 18

2 a) Rate $= \dfrac{-d[A]}{dt} = k[A][B]^2$

b) $k = 75,000$ liter^2 mole^{-2} sec^{-1}

4 $\dfrac{dp_{NO_2}}{dt} = kp^2{}_{NO}p_{O_2}$

6 First order; $k = 0.0105$ min^{-1}; $t_{1/2} = 66$ min

8 Rate $= k[NO]^2[H_2]$

10 First order in I^- and in OCl^-; second order overall

12 0.11

14 $\dfrac{-d[NH_3]}{dt} = k[NH_3]$

$k = 0.00338$ sec^{-1}; $t_{1/2} = 205$ sec

16 2800 cal mole^{-1}

18 $E_a = 4.63$ kcal mole^{-1}

20 0.20 sec^{-1}

22 32°C; 95°C

24 a) 2.78×10^{-5} cm

b) 5.01×10^{-5} cm

26 a) Rate$_2 = k_2[NO_2][NO_3] = \dfrac{k_2 k_1}{k_{-1}} [N_2O_5] = k[N_2O_5]$

b) 5×10^3 sec

28 Rate$_2 = k[BrO_3^-][H^+]^2[Br^-]$

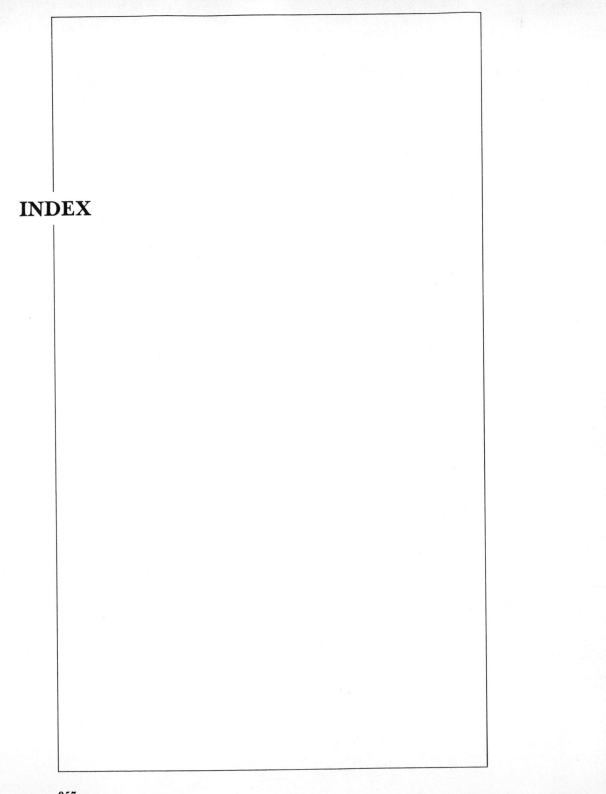

INDEX

INDEX

ATOMIC MASSES OF ELEMENTS

REFERRED TO $^{12}C = 12$ (EXACTLY)

NAME	SYMBOL	ATOMIC NUMBER	ATOMIC WEIGHT	NAME	SYMBOL	ATOMIC NUMBER	ATOMIC WEIGHT
Actinium	Ac	89	(227)	Francium	Fr	87	(223)
Aluminum	Al	13	26.9815[a]	Gadolinium	Gd	64	157.25[b]
Americium	Am	95	(243)	Gallium	Ga	31	69.72
Antimony	Sb	51	121.75[b]	Germanium	Ge	32	72.59[b]
Argon	Ar	18	39.948[b]	Gold	Au	79	196.9665[a]
Arsenic	As	33	74.9216[a]	Hafnium	Hf	72	178.49[b]
Astatine	At	85	~210	Helium	He	2	4.00260[a]
Barium	Ba	56	137.34[b]	Holmium	Ho	67	164.9303[a]
Berkelium	Bk	97	(247)	Hydrogen	H	1	1.0080[a]
Beryllium	Be	4	9.01218[a]	Indium	In	49	114.82
Bismuth	Bi	83	208.9806[a]	Iodine	I	53	126.9045[a]
Boron	B	5	10.81[a]	Iridium	Ir	77	192.22[b]
Bromine	Br	35	79.904[a]	Iron	Fe	26	55.847[b]
Cadmium	Cd	48	112.40	Krypton	Kr	36	83.80
Calcium	Ca	20	40.08	Lanthanum	La	57	138.9055[b]
Californium	Cf	98	(251)	Lawrencium	Lr	103	(257)
Carbon	C	6	12.011[a]	Lead	Pb	82	207.2[a]
Cerium	Ce	58	140.12	Lithium	Li	3	6.941[a]
Cesium	Cs	55	132.9055[a]	Lutetium	Lu	71	174.97
Chlorine	Cl	17	35.453[a]	Magnesium	Mg	12	24.305[a]
Chromium	Cr	24	51.996[a]	Manganese	Mn	25	54.9380[a]
Cobalt	Co	27	58.9332[a]	Mendelevium	Md	101	(256)
Copper	Cu	29	63.546[a]	Mercury	Hg	80	200.59[b]
Curium	Cm	96	(247)	Molybdenum	Mo	42	95.94[b]
Dysprosium	Dy	66	162.50	Neodymium	Nd	60	144.24[b]
Einsteinium	Es	99	(254)	Neon	Ne	10	20.179[b]
Erbium	Er	68	167.26	Neptunium	Np	93	237.0482[a]
Europium	Eu	63	151.96	Nickel	Ni	28	58.71[b]
Fermium	Fm	100	(257)	Niobium	Nb	41	92.9064[a]
Fluorine	F	9	18.9984[a]	Nitrogen	N	7	14.0067[a]